Statistical Methods for Reliability Data

WILEY SERIES IN PROBABILITY AND STATISTICS
APPLIED PROBABILITY AND STATISTICS SECTION

Established by WALTER A. SHEWHART and SAMUEL S. WILKS

Editors: *Vic Barnett, Ralph A. Bradley, Noel A. C. Cressie, Nicholas I. Fisher, Iain M. Johnstone, J. B. Kadane, David G. Kendall, David W. Scott, Bernard W. Silverman, Adrian F. M. Smith, Jozef L. Teugels; J. Stuart Hunter, Emeritus*

A complete list of the titles in this series appears at the end of this volume.

Statistical Methods for Reliability Data

WILLIAM Q. MEEKER
Department of Statistics
Iowa State University

LUIS A. ESCOBAR
Department of Experimental Statistics
Louisiana State University

A Wiley-Interscience Publication
JOHN WILEY & SONS, INC.
New York • Chichester • Weinheim • Brisbane • Singapore • Toronto

This text is printed on acid-free paper. ∞

Copyright © 1998 by John Wiley & Sons, Inc. All rights reserved.

Published simultaneously in Canada.

No part of this publication may be reproduced, stored in a retrieval system or transmitted in any form or by any means, electronic, mechanical, photocopying, recording, scanning or otherwise, except as permitted under Sections 107 or 108 of the 1976 United States Copyright Act, without either the prior written permission of the Publisher, or authorization through payment of the appropriate per-copy fee to the Copyright Clearance Center, 222 Rosewood Drive, Danvers, MA 01923, (978) 750-8400, fax (978) 750-4744. Requests to the Publisher for permission should be addressed to the Permissions Department, John Wiley & Sons, Inc., 605 Third Avenue, New York, NY 10158-0012, (212) 850-6011, fax (212) 850-6008, E-Mail: PERMREQ@WILEY.COM.

Library of Congress Cataloging-in-Publication Data:
Meeker, William Q.
    Statistical methods for reliability data / William Q. Meeker, Luis
A. Escobar.
      p. cm. — (Wiley series in probability and statistics.
Applied probability and statistics)
    "A Wiley-Interscience publication."
    Includes bibliographical references and index.
    ISBN 0-471-14328-6 (cloth : alk. paper)
    1. Reliability (Engineering)—Statistical methods. I. Escobar,
Luis A. II. Title. III. Series.
TS173.M44   1998
$620'.00452$—dc21   97-39270

Printed in the United States of America

10 9 8 7 6 5 4 3 2

To Karen, Katherine, and my parents
W.Q.M.

To Lida, Juan, Catalina, Daniela,
and my mother Inés
L.A.E.

Contents

Preface   xv

Acknowledgments   xxi

1. Reliability Concepts and Reliability Data   1

    1.1. Introduction, 2
    1.2. Examples of Reliability Data, 4
    1.3. General Models for Reliability Data, 15
    1.4. Repairable Systems and Nonrepairable Units, 19
    1.5. Strategy for Data Collection, Modeling, and Analysis, 20

2. Models, Censoring, and Likelihood for Failure-Time Data   26

    2.1. Models for Continuous Failure-Time Processes, 27
    2.2. Models for Discrete Data from a Continuous Process, 32
    2.3. Censoring, 34
    2.4. Likelihood, 36

3. Nonparametric Estimation   46

    3.1. Introduction, 47
    3.2. Estimation from Singly Censored Interval Data, 47
    3.3. Basic Ideas of Statistical Inference, 48
    3.4. Confidence Intervals from Complete or Singly Censored Data, 50
    3.5. Estimation from Multiply Censored Data, 52
    3.6. Pointwise Confidence Intervals from Multiply Censored Data, 54
    3.7. Estimation from Multiply Censored Data with Exact Failures, 57

3.8. Simultaneous Confidence Bands, 60
  3.9. Uncertain Censoring Times, 64
  3.10. Arbitrary Censoring, 65

4. **Location-Scale-Based Parametric Distributions** 75

  4.1. Introduction, 76
  4.2. Quantities of Interest in Reliability Applications, 76
  4.3. Location-Scale and Log-Location-Scale Distributions, 78
  4.4. Exponential Distribution, 79
  4.5. Normal Distribution, 80
  4.6. Lognormal Distribution, 82
  4.7. Smallest Extreme Value Distribution, 83
  4.8. Weibull Distribution, 85
  4.9. Largest Extreme Value Distribution, 86
  4.10. Logistic Distribution, 88
  4.11. Loglogistic Distribution, 89
  4.12. Parameters and Parameterization, 90
  4.13. Generating Pseudorandom Observations from a Specified Distribution, 91

5. **Other Parametric Distributions** 97

  5.1. Introduction, 97
  5.2. Gamma Distribution, 98
  5.3. Generalized Gamma Distribution, 99
  5.4. Extended Generalized Gamma Distribution, 101
  5.5. Generalized F Distribution, 102
  5.6. Inverse Gaussian Distribution, 103
  5.7. Birnbaum–Saunders Distribution, 105
  5.8. Gompertz–Makeham Distribution, 108
  5.9. Comparison of Spread and Skewness Parameters, 110
  5.10. Distributions with a Threshold Parameter, 111
  5.11. Generalized Threshold-Scale Distribution, 113
  5.12. Other Methods of Deriving Failure-Time Distributions, 115

6. **Probability Plotting** 122

  6.1. Introduction, 122
  6.2. Linearizing Location-Scale-Based Distributions, 123
  6.3. Graphical Goodness of Fit, 127

    6.4. Probability Plotting Positions, 128
    6.5. Probability Plots with Specified Shape Parameters, 136
    6.6. Notes on the Application of Probability Plotting, 141

7. Parametric Likelihood Fitting Concepts: Exponential Distribution   153

    7.1. Introduction, 153
    7.2. Parametric Likelihood, 155
    7.3. Confidence Intervals for θ, 159
    7.4. Confidence Intervals for Functions of θ, 163
    7.5. Comparison of Confidence Interval Procedures, 164
    7.6. Likelihood for Exact Failure Times, 165
    7.7. Data Analysis with No Failures, 167

8. Maximum Likelihood for Log-Location-Scale Distributions   173

    8.1. Introduction, 173
    8.2. Likelihood, 174
    8.3. Likelihood Confidence Regions and Intervals, 177
    8.4. Normal-Approximation Confidence Intervals, 186
    8.5. Estimation with Given σ, 192

9. Bootstrap Confidence Intervals   204

    9.1. Introduction, 204
    9.2. Bootstrap Sampling, 205
    9.3. Exponential Distribution Confidence Intervals, 208
    9.4. Weibull, Lognormal, and Loglogistic Distribution Confidence Intervals, 212
    9.5. Nonparametric Bootstrap Confidence Intervals, 217
    9.6. Percentile Bootstrap Method, 226

10. Planning Life Tests   231

    10.1. Introduction, 232
    10.2. Approximate Variance of ML Estimators, 236
    10.3. Sample Size for Unrestricted Functions, 238
    10.4. Sample Size for Positive Functions, 239
    10.5. Sample Sizes for Log-Location-Scale Distributions with Censoring, 240

11. Parametric Maximum Likelihood: Other Models   254

11.1. Introduction, 255
11.2. Fitting the Gamma Distribution, 256
11.3. Fitting the Extended Generalized Gamma Distribution, 257
11.4. Fitting the BISA and IGAU Distributions, 260
11.5. Fitting a Limited Failure Population Model, 262
11.6. Truncated Data and Truncated Distributions, 266
11.7. Fitting Distributions that Have a Threshold Parameter, 273

12. Prediction of Future Random Quantities   289

12.1. Introduction, 290
12.2. Probability Prediction Intervals (θ Given), 292
12.3. Statistical Prediction Interval (θ Estimated), 293
12.4. The (Approximate) Pivotal Method for Prediction Intervals, 296
12.5. Prediction in Simple Cases, 298
12.6. Calibrating Naive Statistical Prediction Bounds, 300
12.7. Prediction of Future Failures from a Single Group of Units in the Field, 304
12.8. Prediction of Future Failures from Multiple Groups of Units with Staggered Entry into the Field, 308

13. Degradation Data, Models, and Data Analysis   316

13.1. Introduction, 317
13.2. Models for Degradation Data, 317
13.3. Estimation of Degradation Model Parameters, 326
13.4. Models Relating Degradation and Failure, 327
13.5. Evaluation of $F(t)$, 328
13.6. Estimation of $F(t)$, 331
13.7. Bootstrap Confidence Intervals, 332
13.8. Comparison with Traditional Failure-Time Analyses, 333
13.9. Approximate Degradation Analysis, 336

14. Introduction to the Use of Bayesian Methods for Reliability Data   343

14.1. Introduction, 344
14.2. Using Bayes's Rule to Update Prior Information, 344

10.6. Test Plans to Demonstrate Conformance with a Reliability Standard, 247
10.7. Some Extensions, 250

CONTENTS

14.3. Prior Information and Distributions, 345
14.4. Numerical Methods for Combining Prior Information with a Likelihood, 350
14.5. Using the Posterior Distribution for Estimation, 356
14.6. Bayesian Prediction, 358
14.7. Practical Issues in the Application of Bayesian Methods, 362

15. System Reliability Concepts and Methods 369

15.1. Introduction, 369
15.2. System Structures and System Failure Probability, 370
15.3. Estimating System Reliability from Component Data, 380
15.4. Estimating Reliability with Two or More Causes of Failure, 382
15.5. Other Topics in System Reliability, 386

16. Analysis of Repairable System and Other Recurrence Data 393

16.1. Introduction, 394
16.2. Nonparametric Estimation of the MCF, 396
16.3. Nonparametric Comparison of Two Samples of Recurrence Data, 404
16.4. Parametric Models for Recurrence Data, 406
16.5. Tools for Checking Point-Process Assumptions, 409
16.6. Maximum Likelihood Fitting of Poisson Process, 412
16.7. Generating Pseudorandom Realizations from an NHPP Process, 417
16.8. Software Reliability, 419

17. Failure-Time Regression Analysis 427

17.1. Introduction, 428
17.2. Failure-Time Regression Models, 429
17.3. Simple Linear Regression Models, 432
17.4. Standard Errors and Confidence Intervals for Regression Models, 435
17.5. Regression Model with Quadratic μ and Nonconstant σ, 439
17.6. Checking Model Assumptions, 443
17.7. Models with Two or More Explanatory Variables, 447
17.8. Product Comparison: An Indicator-Variable Regression Model, 450
17.9. The Proportional Hazards Failure-Time Model, 455
17.10. General Time Transformation Functions, 459

18. Accelerated Test Models — 466

- 18.1. Introduction, 466
- 18.2. Use-Rate Acceleration, 470
- 18.3. Temperature Acceleration, 471
- 18.4. Voltage and Voltage-Stress Acceleration, 479
- 18.5. Acceleration Models with More than One Accelerating Variable, 484
- 18.6. Guidelines for the Use of Acceleration Models, 487

19. Accelerated Life Tests — 493

- 19.1. Introduction, 493
- 19.2. Analysis of Single-Variable ALT Data, 495
- 19.3. Further Examples, 504
- 19.4. Some Practical Suggestions for Drawing Conclusions from ALT Data, 515
- 19.5. Other Kinds of Accelerated Tests, 517
- 19.6. Potential Pitfalls of Accelerated Life Testing, 522

20. Planning Accelerated Life Tests — 534

- 20.1. Introduction, 535
- 20.2. Evaluation of Test Plans, 538
- 20.3. Planning Single-Variable ALT Experiments, 540
- 20.4. Planning Two-Variable ALT Experiments, 547
- 20.5. Planning ALT Experiments with More than Two Experimental Variables, 558

21. Accelerated Degradation Tests — 563

- 21.1. Introduction, 564
- 21.2. Models for Accelerated Degradation Test Data, 565
- 21.3. Estimating Accelerated Degradation Test Model Parameters, 566
- 21.4. Estimation of Failure Probabilities, Distribution Quantiles, and Other Functions of Model Parameters, 567
- 21.5. Confidence Intervals Based on Bootstrap Samples, 568
- 21.6. Comparison with Traditional Accelerated Life Test Methods, 569
- 21.7. Approximate Accelerated Degradation Analysis, 574

22. Case Studies and Further Applications — 582

- 22.1. Dangers of Censoring in a Mixed Population, 583
- 22.2. Using Prior Information in Accelerated Testing, 586

22.3.	An LFP/Competing Risk Model, 590	
22.4.	Fatigue-Limit Regression Model, 593	
22.5.	Planning Accelerated Degradation Tests, 597	

Epilogue — 602

Appendix A. Notation and Acronyms — 609

Appendix B. Some Results from Statistical Theory — 617

- B.1. cdfs and pdfs of Functions of Random Variables, 617
- B.2. Statistical Error Propagation—The Delta Method, 619
- B.3. Likelihood and Fisher Information Matrices, 621
- B.4. Regularity Conditions, 621
- B.5. Convergence in Distribution, 623
- B.6. Outline of General ML Theory, 625

Appendix C. Tables — 629

References — 645

Author Index — 665

Subject Index — 671

Preface

Over the past 10 years there has been a heightened interest in improving quality, productivity, and reliability of manufactured products. Global competition and higher customer expectations for safe, reliable products are driving this interest. To meet this need, many companies have trained their design engineers and manufacturing engineers in the appropriate use of designed experiments and statistical process monitoring/control. Now reliability is being viewed as the product feature that has the potential to provide an important competitive edge. A current industry concern is in developing better processes to move rapidly from product conceptualization to a cost-effective highly reliable product. A reputation for unreliability can doom a product, if not the manufacturing company.

Data collection, data analysis, and data interpretation methods are important tools for those who are responsible for product reliability and product design decisions. This book describes and illustrates the use of proven traditional techniques for reliability data analysis and test planning, enhanced and brought up to date with modern computer-based graphical, analytical, and simulation-based methods. The material in this book is based on our interactions with engineers and statisticians in industry as well as on courses in applied reliability data analysis that we have taught to MS-level statistics and engineering students at both Iowa State University and Louisiana State University.

Audience and Assumed Knowledge

We have designed this book to be useful to statisticians and engineers working in industry as well as to students in university engineering and statistics programs. The book will be useful for on-the-job training courses in reliability data analysis. There is challenge in addressing such a wide-ranging audience. Communications among engineers and statisticians, however, is not only necessary but essential in the industrial research and development environment. We hope that this book will aid such communication. To produce a book that will appeal to both engineers and statisticians, we have placed primary focus on applications, data, concepts, methods, and interpretation. We use simple computational examples to illustrate ideas and concepts but, as in practical applications, rely on computers to do most of the computations. We have also included a collection of exercise problems at the end of each chapter.

These exercises will give readers a chance to test their knowledge of basic material, to explore conceptual ideas of reliability testing, data analysis, and interpretation, and to see possible extensions of the material in the chapters.

It will be helpful for readers to have had a previous course in intermediate statistical methods covering basic ideas of statistical modeling and inference, graphical methods, estimation, confidence intervals, and regression analysis. Only the simplest concepts of calculus are used in the main body of the text (e.g., probability for a continuous random variable is computed as area under a density curve; a first derivative is a slope or a rate of change; a second derivative is a measure of curvature). Appendix B and some advanced exercises use calculus, linear algebra, basic optimization ideas, and basic statistical theory. Concepts, however, are presented in a relaxed and intuitive manner that we hope will also appeal to interested nonstatisticians. Throughout the book we have attempted to avoid the heavy language of mathematical statistics.

A detailed understanding of underlying statistical theory is *not* necessary to apply the methods in this book. Such details are, however, often important to understanding how to extend methods to new situations or developing new methods. Appendix B, at the end of the book, outlines the general theory and provides references to more detailed information. Also, many derivations and interesting extensions are covered in advanced guided exercises at the end of each chapter.

Particularly challenging exercises (i.e., exercises requiring knowledge of calculus or statistical theory) are marked with a triangle (▲). Exercises requiring computer programming (beyond the use of standard statistical packages) are marked with a diamond (◆).

Special Features of the Book

Special features of this book include the following:

1. We emphasize general methods that can be applied to the wide range of problems found in industrial reliability data analysis—specifically, nonparametric estimation of a failure-time distribution function, probability plotting, and maximum likelihood estimation of important reliability characteristics (failure probabilities, distribution quantiles, and hazard functions), and associated statistical intervals. In the basic chapters (3, 6, 7, 8, 17, and 19), we apply these methods to the most frequently encountered models in reliability data analysis. In special chapters (which can be skipped without loss of continuity or understanding), we apply the general methods to important but less frequently occurring situations (e.g., problems involving truncation and prediction).

2. Throughout the book we use computer graphics for displaying data, for displaying the results of analyses, and for explaining technical concepts.

3. We use simulation methods to complement large-sample asymptotic theory (practical sample sizes are often small to moderate in size). We explain and illustrate modern, more accurate (but computationally demanding) methods of inference: likelihood and bootstrap methods for constructing statistical intervals.

PREFACE   xvii

4. For both nonparametric and parametric analyses, we illustrate the use of general likelihood-based methods of handling *arbitrarily* censored data (including left, right, and interval censoring with overlapping intervals) and truncated data that frequently arise in statistical reliability studies.
5. We provide methods for planning reliability studies (length of test, number of specimens, and levels of experimental factors).
6. We cover methods for analyzing *degradation* data. Such data are becoming increasingly important where there are requirements for extremely high reliability.
7. Almost all of our examples and exercises use real data, including many data sets that have not previously appeared in any book. In order to protect proprietary information, some data have been changed by a scale factor and, in some cases, generic product names have been used (e.g., Device-A, Component-B, Alloy-A).
8. Numerical examples in this book were done using the S-PLUS system for graphics and data analysis (a product of MathSoft, Inc., Seattle, WA). A suite of special S-PLUS functions was developed in parallel with this book. Although we have not included explicit information about software use in the chapters, the suite of special S-PLUS functions and a listing of the S-PLUS commands used to do the examples in the book are available from the authors via anonymous ftp at the Wiley ftp site. Instructions about how to access the software are given below.

How to Download the Software Examples
The Wiley public ftp site includes special S-PLUS function examples created for the applications discussed in this book. The files can be accessed through either a standard ftp program or the ftp client of a Web browser using the http protocol. You can access the files from a Web browser through the following address:

http://www.wiley.com/products/subject/mathematics

On the Mathematics and Statistics home page you will see a link to the ftp Software Archive, which includes a link to information about the book and access to the software.
To gain ftp access, type the following at your Web browser's URL address input box:

ftp://ftp.wiley.com

You can set an ID of anonymous; no password is required.
The files are located in the **public/sci_tech_med/reliability** directory. Be sure to also download and read the README.TXT file, which includes directions on how to install and use the program.
If you need further information about downloading the files, you can reach Wiley's tech support line at 212-850-6753.

Other Software to Use with the Book

Today there are many commercial statistical software packages. Unfortunately, only a few of these packages have adequate capabilities for doing reliability data analysis (e.g., the ability to do nonparametric and parametric estimation with censored data). Nelson (1990a, pages 237–240) outlines the capabilities of a number of commercial and noncommercial packages that were available at that time. As software vendors become more aware of their customers' needs, capabilities in commercial packages are improving. Here we describe briefly the capabilities of a few packages that we and our colleagues have found to be useful.

MINITAB (1997), SAS PROC RELIABILITY (1997), SAS JMP (1995), S-PLUS (1996), and a specialized program called WinSMITH (Abernethy 1996) can do nonparametric and parametric product censored data analysis to estimate a single distribution (Chapters 3, 6, 7, and 8). SAS JMP can also analyze data with more than one failure mode (Chapter 15). MINITAB, SAS PROC RELIABILITY, SAS JMP, and S-PLUS can do parametric regression and accelerated life test analyses (Chapters 17 and 19), as well as semiparametric Cox proportional hazards regression analysis. SAS PROC RELIABILITY can, in addition, do the nonparametric repairable systems analyses (Chapter 16).

Overview and Paths Through the Book

There are many possible paths that readers and instructors might take through this book. Chapters 1–16 cover single distribution models without any explanatory variables. Chapters 17–21 describe failure-time regression models. Chapter 22 presents case studies that illustrate, in the context of real problems, the integration of ideas presented throughout the book. This chapter also usefully illustrates how some of the general methods presented in the earlier chapters can be extended and adapted to deal with new problems.

Chapters 1–3 and 6–8 provide basic material that will be of interest to almost all readers and should be read in sequence. Chapter 4 discusses parametric failure-time models based on location-scale distributions and Chapter 5 covers more advanced distributional models. It is possible to use only a light reading of Chapter 4 and to skip Chapter 5 altogether before proceeding on to the important methods in Chapters 6–8. Chapter 9 explains and illustrates the use of bootstrap (simulation-based) methods for obtaining confidence intervals. Chapter 10 focuses on test planning: evaluating the effects of choosing sample size and length of observation. Chapters 11–16 cover a variety of special more advanced topics for single distribution models. Some of the material in Chapter 5 is prerequisite for the material in Chapter 11, but it is possible simply to work in Chapter 11, referring back to Chapter 5 only as needed. Otherwise, each of Chapters 10 through 14 has only material up to Chapter 8 as prerequisite. Chapter 15 introduces some important system reliability concepts and shows how the material in the first part of the book can be used to make statistical statements about the reliability of a system or a population of systems. Chapter 16 explains and illustrates the fundamental ideas behind analyzing system-repair and other recurrence data (as opposed to data on components and other replaceable units).

There are several groups of chapters on special topics that can be read in sequence.

- **Accelerated testing.** Chapter 17 introduces models and methods for regression analysis (assessing the effects of explanatory variables) for failure-time data. Chapter 18 introduces physically based reliability models used in accelerated testing. Chapter 19 shows how to analyze data from accelerated life tests. Chapter 20 describes test planning for regression and accelerated test applications with censored data.
- **Degradation analysis.** Chapter 13 provides methods for analyzing degradation reliability data. Use of degradation data in accelerated tests is covered in Chapter 21. Chapter 22 contains a case study that describes a method for planning accelerated degradation tests.
- **Bayesian methods.** Chapter 14 introduces concepts and applications of Bayesian methods for failure-time data. A case study in Chapter 22 extends these ideas to regression with an application to accelerated life test data analysis.

Appendix A provides a summary and index of notation used in the book. Appendix B outlines the general maximum likelihood and other statistical theory on which the methods in the book are based. Appendix C gives tables for some of the larger data sets used in our examples.

Use as a Textbook

A two-semester course would be required to cover thoroughly all of the material in the book. For a one-semester course, aimed at engineers and/or statisticians, an instructor could cover Chapters 1–4, 6–8, and 17–19, along with selected material from the appendices (according to the background of the students), and a few other chapters according to interests and tastes.

This book could be used as the basis for workshops or short courses aimed at engineers or statisticians working in industry. For an audience with a working knowledge of basic statistical tools, Chapters 1–3, key sections in Chapter 4, and Chapters 6–8 could be covered in one day. If the purpose of the short course is to introduce the basic ideas and illustrate with examples, then some material from Chapters 17–20 could also be covered. For a less experienced audience or for a more relaxed presentation, allowing time for exercises and discussion, two days would be needed to cover this material. Extending the course to three or four days would allow covering selected material in Chapters 9–22.

WILLIAM Q. MEEKER
LUIS A. ESCOBAR

Ames, Iowa
Baton Rouge, Louisiana
April 1998

Acknowledgments

A number of individuals provided helpful comments on all or part of draft versions of this book. In particular, we would like to acknowledge Chuck Annis, Hwei-Chun Chou, William Christensen, Necip Doganaksoy, Tom Dubinin, Michael Eraas, Shuen-Lin Jeng, Gerry Hahn, Joseph Lu, Michael LuValle, Enid Martinets, Silvia Morales, Peter Morse, Dan Nordman, Steve Redman, Ernest Scheuer, Ananda Sen, David Steinberg, Mark Vandeven, Kim Wentzlaff, and a number of anonymous reviewers. Over the years, we have benefited from numerous technical discussions with Vijay Nair. Vijay provided a number of suggestions that substantially improved several parts of this book. We would also like to acknowledge our students (too numerous to mention by name) for their penetrating questions, high level of interest, and useful suggestions for improving our courses.

We would like to make special acknowledgment to Wayne Nelson, who gave us detailed feedback on earlier versions of most of the chapters of this book. Additionally, much of our knowledge in this area has it roots in our interactions with Wayne and Wayne's outstanding books and other publications in the area of reliability and reliability data analysis.

Parts of this book were written while the first author was visiting the Department of Statistics and Actuarial Science at the University of Waterloo and the Department of Experimental Statistics at Louisiana State University. Support and use of facilities during these visits are gratefully acknowledged.

We greatly benefited from facilities, traveling support, and encouragement from Jeff Hooper and Michèle Boulanger at AT&T Bell Laboratories, Gerald Hahn at General Electric Corporate Research and Development Center, Enrique Villa and Victor Pérez Abreu at CIMAT in Guanajuato, México, Guido E. Del Pino at Universidad Católica in Santiago, Chile, Héctor Allende at Universidad Técnica Federico Santa María in Valparaiso, Chile, Yves L. Grize in Basel, Switzerland, and Stephan Zayac, Ford Motor Company.

Dean L. Isaacson, Head, Department of Statistics, Iowa State University, Lynn R. LaMotte, former Head, and E. Barry Moser, Interim Head, Department of Experimental Statistics, Louisiana State University, provided helpful encouragement and support to both authors. We would also like to thank our secretaries Denise Riker and Elaine Miller for excellent support and assistance while writing this book.

Finally, we would like to thank our wives and children for their love, patience, and understanding during the recent years in which we worked most weekends and many evenings to complete this project.

W. Q. M.
L. A. E.

Statistical Methods for
Reliability Data

CHAPTER 1

Reliability Concepts and Reliability Data

Objectives

This chapter explains:

- Basic ideas behind product reliability.
- Reasons for collecting reliability data.
- Distinguishing features of reliability data.
- General models for reliability data.
- Examples of reliability data and the motivation for the collection of the data.
- A general strategy that can be used for data analysis, modeling, and inference from reliability data.

Overview

This chapter introduces some of the basic concepts of product reliability. Section 1.1 explains the relationship between quality and reliability and outlines how statistical studies are used to obtain information that can be used to assess and improve product reliability. Section 1.2 presents examples to illustrate studies that resulted in different kinds of reliability data. These examples are used in data analysis and exercises in subsequent chapters. Section 1.3 explains, in general terms, important qualitative aspects of statistical models that are used to describe populations and processes in reliability applications. Section 1.4 emphasizes the important distinction between studies focusing on data from repairable systems and nonrepairable units. Section 1.5 describes a general strategy for exploring, analyzing, and drawing conclusions from reliability data. This strategy is illustrated in examples throughout the book and in the case studies in Chapter 22.

1.1 INTRODUCTION

1.1.1 Quality and Reliability

Rapid advances in technology, development of highly sophisticated products, intense global competition, and increasing customer expectations have put new pressures on manufacturers to produce high-quality products. Customers expect purchased products to be reliable and safe. Systems, vehicles, machines, devices, and so on should, with high probability, be able to perform their intended function under usual operating conditions, for some specified period of time.

Technically, reliability is often defined as the probability that a system, vehicle, machine, device, and so on will perform its intended function under operating conditions, for a specified period of time. Improving reliability is an important part of the larger overall picture of improving product quality. There are many definitions of quality, but general agreement that an unreliable product is *not* a high-quality product. Condra (1993) emphasizes that "reliability is quality over time."

Modern programs for improving reliability of existing products and for assuring continued high reliability for the next generation of products require quantitative methods for predicting and assessing various aspects of product reliability. In most cases this will involve the collection of reliability data from studies such as laboratory tests (or designed experiments) of materials, devices, and components, tests on early prototype units, careful monitoring of early-production units in the field, analysis of warranty data, and systematic longer-term tracking of products in the field.

1.1.2 Reasons for Collecting Reliability Data

There are many possible reasons for collecting reliability data. Examples include the following:

- Assessing characteristics of materials over a warranty period or over the product's design life.
- Predicting product reliability.
- Predicting product warranty costs.
- Providing needed inputs for system-failure risk assessment.
- Assessing the effect of a proposed design change.
- Assessing whether customer requirements and government regulations have been met.
- Tracking the product in the field to provide information on causes of failure and methods of improving product reliability.
- Supporting programs to improve reliability through the use of laboratory experiments, including accelerated life tests.
- Comparing components from two or more different manufacturers, materials, production periods, operating environments, and so on.
- Checking the veracity of an advertising claim.

1.1.3 Distinguishing Features of Reliability Data

Reliability data can have a number of special features requiring the use of special statistical methods. For example:

- Reliability data are typically censored (exact failure times are not known). The most common reason for censoring is the frequent need to analyze life test data before all units have failed. More generally, censoring arises when actual response values (e.g., failure times) cannot be observed for some or all units under study. Thus censored observations provide a bound or bounds on the actual failure times.
- Most reliability data are modeled using distributions for positive random variables like the exponential, Weibull, gamma, and lognormal. Relatively few applications use the normal distribution as a model for product life.
- Inferences and predictions involving extrapolation are often required. For example, we might want to estimate the proportion of the population that will fail after 900 hours, based on a test that runs only 400 hours (extrapolation in time). Also we might want to estimate the time at which 1% of a product population will fail at 50°C based on tests at 85°C (extrapolation in operating conditions).
- It is often necessary to use past experience or other scientific or engineering judgment to provide information as input to the analysis of data or to a decision-making process. This information may take the form of a physically based model and/or the specification of one or more parameters (e.g., physical constants or materials properties) of such a model. This is also a form of extrapolation from the past to the present or future behavior of a process or product.
- Typically, the traditional parameters of a statistical model (e.g., mean and standard deviation) are *not* of primary interest. Instead, design engineers, reliability engineers, managers, and customers are interested in specific measures of product reliability or particular characteristics of a failure-time distribution (e.g., failure probabilities, quantiles of the life distribution, failure rates).
- Especially with censored data, model fitting requires computer implementation of numerical methods, and often there is no *exact* theory for statistical inferences.
- Integrated software to do all of the needed analyses is not available yet. There are useful, but limited, capabilities in commercial packages like BMDP, MINITAB, SAS, S-PLUS, SYSTAT, and WinSMITH. The examples in this book were done with extensions of the S-PLUS system, as described in the preface.

This book emphasizes the analysis of data from studies conducted to assess or improve product reliability. Data from reliability studies, however, closely resemble data from time-to-event studies in other areas of science and industry including biology, ecology, medicine, economics, and sociology. The methods of analysis in these other areas are the same or similar to those used in reliability data analysis. Some synonyms for reliability data are failure-time data, life data, survival data (used in medicine and biological sciences), and event-time data (used in the social sciences).

1.2 EXAMPLES OF RELIABILITY DATA

This section describes examples and data sets that illustrate the wide range of applications and characteristics of reliability data. These and other examples are used in subsequent chapters to illustrate the application of statistical methods for analyzing and drawing conclusions from such data.

1.2.1 Failure-Time Data with no Explanatory Variables

In many applications reliability data will be collected on a sample of units that are assumed to have come from a particular process or population and to have been tested or operated under nominally identical conditions. More realistically, there are physical differences among units (e.g., strength or hardness) and operating conditions (e.g., temperature, humidity, or stress) and these contribute to the variability in the data. The assumption used in drawing inferences from such *single distribution* data is that these differences accurately reflect the variability in life caused by the actual differences in the population or process of interest.

Example 1.1 Ball Bearing Fatigue Data. Lieblein and Zelen (1956) describe and give data from fatigue endurance tests for deep-groove ball bearings. The ball bearings came from four different major bearing companies. There was disagreement in the industry on the appropriate parameter values to use to describe the relationship between fatigue life and stress loading. The main objective of the study was to estimate values of the parameters in the equation relating bearing life to load.

The data shown in Table 1.1 are a subset of $n = 23$ bearing failure times for units tested at one level of stress, reported and analyzed by Lawless (1982). Figure 1.1 shows that the data are skewed to the right. Because of the lower bound on cycles (or time) to failure at zero, this distribution shape is typical of reliability data. Figure 1.2 illustrates the failure pattern over time. □

Modern electronic systems may contain anywhere from hundreds to hundreds of thousands of integrated circuits (ICs). In order for such a system to have high reliability, it is necessary for the individual ICs and other components to have extremely high reliability, as in the following example.

Example 1.2 Integrated Circuit Life Test Data. Meeker (1987) reports the results of a life test of $n = 4156$ integrated circuits tested for 1370 hours at accelerated conditions of 80°C and 80% relative humidity. The accelerated conditions were used

Table 1.1. Ball Bearing Failure Times in Millions of Revolutions

17.88	28.92	33.00	41.52	42.12	45.60
48.40	51.84	51.96	54.12	55.56	67.80
68.64	68.64	68.88	84.12	93.12	98.64
105.12	105.84	127.92	128.04	173.40	

Data from Lawless (1982).

EXAMPLES OF RELIABILITY DATA

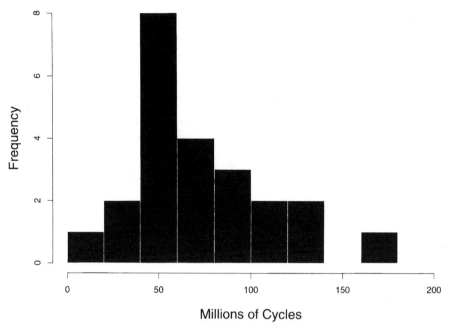

Figure 1.1. Histogram of the ball bearing failure data.

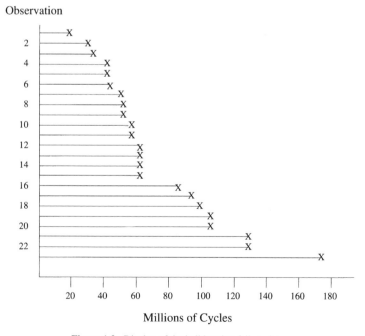

Figure 1.2. Display of the ball bearing failure data.

Table 1.2. Integrated Circuit Failure Times in Hours

.10	.10	.15	.60	.80	.80
1.20	2.50	3.00	4.00	4.00	6.00
10.00	10.00	12.50	20.00	20.00	43.00
43.00	48.00	48.00	54.00	74.00	84.00
94.00	168.00	263.00	593.00		

When the test ended at 1370 hours, there were 4128 unfailed units. Data from Meeker (1987).

to shorten the test by causing defective units to fail more rapidly. The primary purpose of the experiment was to estimate the proportion of defective units being manufactured in the current production process and to estimate the amount of "burn-in" time that would be required to remove most of the defective units from the product population. The reliability engineers were also interested in whether it might be possible to get the needed information about the state of the production process, in the future, using much shorter tests (say, 200 or 300 hours). The data are reproduced in Table 1.2. There were 25 failures in the first 100 hours, three more between 100 and 600 hours, and no more failures out to 1370 hours, when the test was terminated. Ties in the data indicate that failures were detected at inspection times. A subset of the data is depicted in Figure 1.3. □

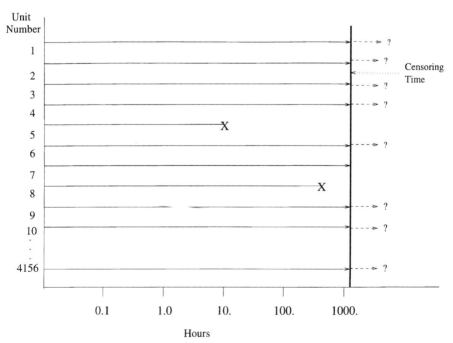

Figure 1.3. General failure pattern of the integrated circuit life test, showing a subset of the data where 28 out of 4156 units failed in the 1370-hour test.

EXAMPLES OF RELIABILITY DATA

Table 1.3. Failure Data from a Circuit Pack Field Tracking Study

Operating Hours		Number Failing	
Interval Endpoint			
Lower	Upper	Vendor 1	Vendor 2
0	1	10	unknown
1	2	1	unknown
2	5	3	unknown
5	10	1	unknown
10	20	2	unknown
20	50	6	unknown
50	100	3	unknown
100	200	2	unknown
200	500	8	unknown
500	1,000	4	unknown
1,000	2,000	5	2
2,000	5,000	6	5
5,000	6,000	3	6
6,000	7,000	9	11
7,000	8,000	10	7
8,000	9,000	16	14
9,000	10,000	7	10
10,000	11,000	unknown	14

After 10,000 hours of operation, there were 4897 unfailed packs for Vendor 1 and after 11,000 hours of operation there were 4924 unfailed packs for Vendor 2.

Example 1.3 Circuit Pack Reliability Field Trial. Table 1.3 gives information on the number of failures observed during periodic inspections in a field trial of early-production circuit packs employing new technology devices. The circuit packs were manufactured under the same design, but by two different vendors. The trial ran for 10,000 hours. The 4993 circuit packs from Vendor 1 came straight from production. The 4993 circuit packs from Vendor 2 had already seen 1000 hours of burn-in testing at the manufacturing plant under operating conditions similar to those in the field trial. Such circuit packs were sold at a higher price because field reliability was supposed to have been improved by the burn-in screening of circuit packs containing defective components. Failures during the first 1000 hours of burn-in were not recorded. This is the reason for the unknown entries in the table and for having information out to 11,000 hours for Vendor 2. The data in Table 1.3 is for the first failure in a position; information on circuit packs replaced after initial failure in a position was not part of the study.

Inspections were costly and were spaced more closely at the beginning of the study because more failures were expected there. The early "infant mortality" failures were caused by component defects in a small proportion of the circuit packs. Such failures are typical for an immature product. For such products, burn-in of circuit packs can be used to weed out most of the packs with weak components. Such burn-in, however, is expensive, and one of the manufacturer's goals was to develop robust design and manufacturing processes that would eliminate or reduce, as quickly as possible, the occurrence of such defects in future generations of similar products.

There were several goals for this study:

- Determine if there was an important difference in the reliability of the products from the two different vendors.
- Determine the specific causes of failures so that the information could be used to improve product design or manufacturing methods.
- Estimate the circuit pack "hazard function" (a measure of failure propensity defined in Chapter 2) out to 10,000 hours.
- Estimate the point at which the hazard function levels off. After this point in time, burn-in would not be useful for improving the reliability of the circuit packs.
- Judge if and when the burn-in period can be used effectively to improve early-life reliability.
- Estimate the failure-time distribution for the first 10,000 hours of life (the warranty period for the product).
- Estimate the proportion of units that will fail in the first 50,000 hours of life (expected technological life of the units). □

Example 1.4 Diesel Generator Fan Failure Data. Nelson (1982, page 133) gives data on diesel generator fan failures. Failures in 12 of 70 generator fans were reported at times ranging between 450 hours and 8750 hours. Of the 58 units that did not fail, the reported running times (i.e., censoring times) ranged between 460 and 11,500 hours. Different fans had different running times because units were introduced into service at different times and because their use-rates differed. The data are reproduced in Appendix Table C.1. Figure 1.4 provides an initial graphical representation of the data. Figure 1.5 shows the censoring data. The data were collected to answer questions like:

- What percentage of the units will fail under warranty?
- Would the fan failure problem get better or worse in the future? In reliability terminology, does hazard function (sometimes called failure rate) for fans increase or decrease with fan age? □

Example 1.5 Heat Exchanger Tube Crack Data. Nuclear power plants use heat exchangers to transfer energy from the reactor to steam turbines. A typical heat exchanger contains thousands of tubes through which steam flows continuously when the heat exchanger is in service. With age, heat exchanger tubes develop cracks, usually due to some combination of stress-corrosion and fatigue. A heat exchanger can continue to operate safely when the cracks are small. If cracks get large enough, however, leaks can develop, and these could lead to serious safety problems and expensive, unplanned plant shut-down time. To protect against having

EXAMPLES OF RELIABILITY DATA

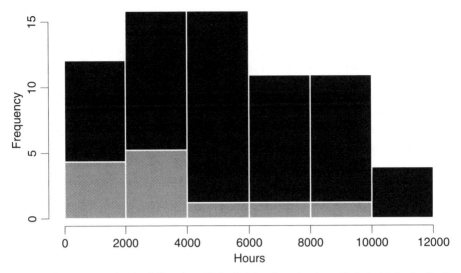

Figure 1.4. Histogram showing failure times (light shade) and running times (dark shade) for the diesel generator fan data.

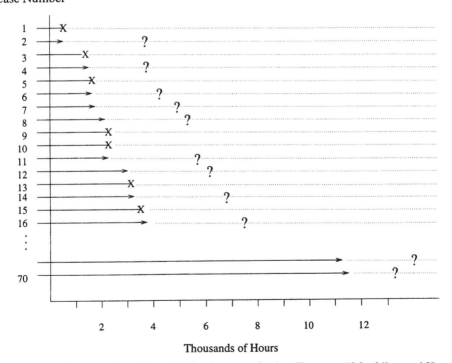

Figure 1.5. Failure pattern in a subset of the diesel generator fan data. There were 12 fan failures and 58 right-censored observations.

leaks, heat exchangers are taken out of service periodically so that its tubes (and other components) can be inspected with nondestructive evaluation techniques. At the end of each inspection period, tubes with detected cracks are plugged so that water will no longer pass through them. This reduces plant efficiency but extends the life of the expensive heat exchangers. With this in mind, heat exchangers are built with extra capacity and can remain in operation up until the point where a certain percentage (e.g., 5%) of the tubes have been plugged.

Figure 1.6 illustrates the inspection data, available at the end of 1983, from three different power plants. At this point in time, Plant 1 had been in operation for 3 years, Plant 2 for 2 years, and Plant 3 for only 1 year. Because all of the heat exchangers were manufactured according to the same design and specifications and because the heat exchangers were operated in generating plants run under similar tightly controlled conditions, it seemed that it should be reasonable to combine the data from the different plants for the sake of making inferences and predictions about the time-to-crack distribution of the heat exchanger tubes. Figure 1.7 illustrates the same data displayed in terms of amount of operating time instead of calendar time.

The engineers were interested in predicting tube life of a larger population of tubes in similar heat exchangers in other plants, for purposes of proper accounting and depreciation and so that the company could develop efficient inspection and replacement strategies. They also wanted to know if the tube failure rate was constant

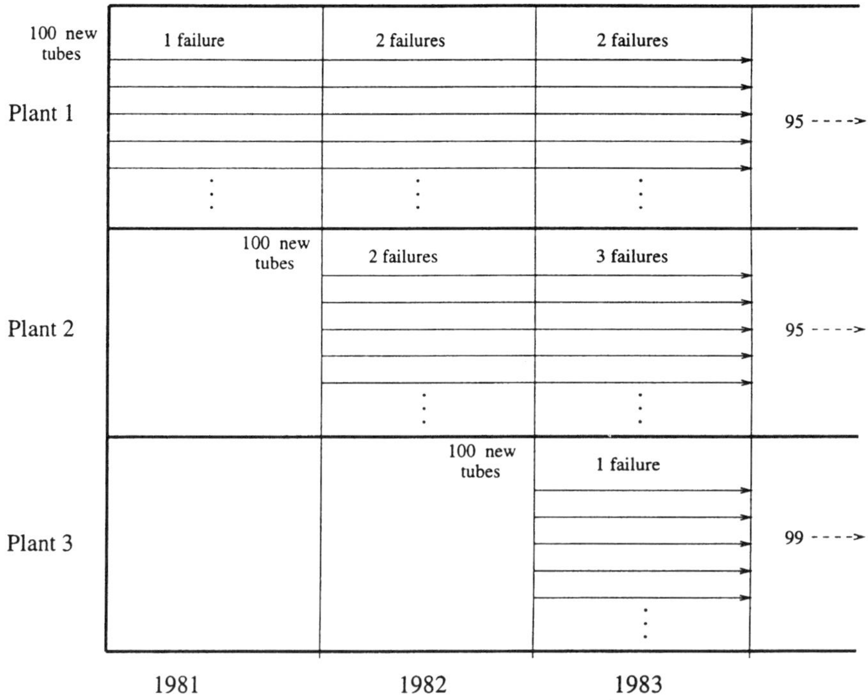

Figure 1.6. Heat exchanger tube crack inspection data in calendar time.

EXAMPLES OF RELIABILITY DATA   11

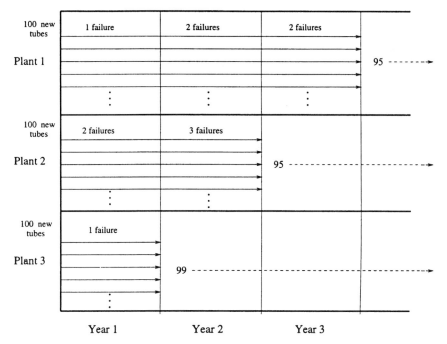

Figure 1.7. Heat exchanger tube crack inspection data in operating time.

over time or if suspected wearout mechanisms (corrosion and fatigue) would, as suspected, begin to cause failures to occur with higher frequency as the heat exchanger ages. □

Example 1.6   Transmitter Vacuum Tube Data. Table 1.4 gives life data for a certain kind of transmitter vacuum tube (designated as "V7" within a particular transmitter design). Although solid-state electronics has made vacuum tubes obsolete for most applications, such tubes are still widely used in the output stage of high-

Table 1.4.   Failure Times for the V7 Transmitter Tube

Days		Number Failing
Interval Endpoint		
Lower	Upper	
0	25	109
25	50	42
50	75	17
75	100	7
100	∞	13

Data from Davis (1952).

power transmitters. These data were originally analyzed in Davis (1952). As seen in many practical situations, the exact failure times were not reported. Instead, we have only the number of failures in each interval or bin. Such data are known as grouped data, interval data, binned data, or read-out data. □

Example 1.7 Turbine Wheel Crack Initiation Data. Nelson (1982) describes a study to estimate the distribution of time to crack initiation for turbine wheels. Each of 432 wheels was inspected once to determine if it had started to crack or not. At the time of the inspections, the wheels had different amounts of service time (age). A unit found to be *cracked* at its inspection was *left*-censored at its age (because the crack had initiated at some unknown point before its inspection age). A unit found to be *uncracked* at its inspection was *right*-censored at its age (because a crack would be initiated at some unknown point after that age). The data in Table 1.5, taken from Nelson (1982), show the number of cracked and uncracked wheels in different age categories, showing the midpoint of the time intervals given by Nelson. The data were put into intervals to facilitate simpler analyses.

In some applications components with an initiated crack could continue in service for rather long periods of time with the expectation that in-service inspections, scheduled frequently enough, could detect cracks before they grow to a size that could cause a safety hazard.

The important objectives of the study were to obtain information that could be used to:

- Estimate the distribution of the time to crack initiation.
- Schedule in-service inspections.
- Assess whether the wheel's crack initiation rate is increasing as the wheels age. An increasing rate would suggest preventive replacement of the wheels by some age when the risk of cracking gets too high.

Table 1.5. Turbine Wheel Inspection Data Summary at Time of Study

100-hours of Exposure Interval	Interval Midpoint	# Cracked	# Not Cracked
0–8	4	0	39
8–12	10	4	49
12–16	14	2	31
16–20	18	7	66
20–24	22	5	25
24–28	26	9	30
28–32	30	9	33
32–36	34	6	7
36–40	38	22	12
40–44	42	21	19
44+	46	21	15

Data from Nelson (1982), page 409.

EXAMPLES OF RELIABILITY DATA   13

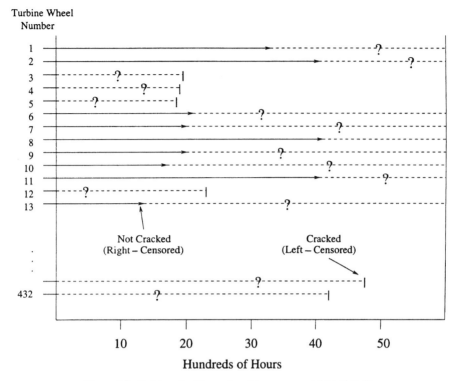

Figure 1.8. Turbine wheel inspection data summary at time of study.

The failure/censoring pattern of these data is quite different from the previous examples and is illustrated in Figure 1.8. The analysts did not know the initiation time for any of the wheels. Instead, all they knew about each wheel was its age and whether a crack had initiated or not. □

1.2.2 Failure-Time Data with Explanatory Variables

Example 1.8 Printed Circuit Board Accelerated Life Test Data. Meeker and LuValle (1995) give data from an accelerated life test on failure of printed circuit boards. The purpose of the experiment was to study the effect of the stresses on the failure-time distribution and to predict reliability under normal operating conditions. More specifically, the experiment was designed to study a particular failure mode— the formation and growth of conductive anodic filaments between copper-plated through-holes in the printed circuit boards. Actual growth of the filaments could not be monitored. Only failure time (defined as a short circuit) could be observed directly. Special test boards were constructed for the experiment. The data described here are part of the results of a much larger experiment aimed at determining the effects of temperature, relative humidity, and electric field on the reliability of printed circuit boards.

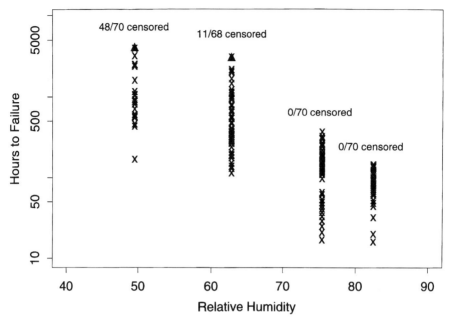

Figure 1.9. Scatter plot of printed circuit board accelerated life test data.

Spacing between the holes in the test boards was chosen to simulate the spacing in actual printed circuit boards. Each test vehicle contained three identical 8×18 matrices of holes with alternate columns charged positively and negatively. These matrices, or "boards," were the observational units in the experiment. Data analysis indicated that any clustering effect of boards within test boards was small enough to ignore in the study.

Meeker and LuValle (1995) give the number of failures that was observed in each of a series of 4-hour and 12-hour long intervals over the life test period. This experiment resulted in interval-censored data because only the interval in which each failure occurred was known. In this example all test units had the same inspection times. A graph of the data in Figure 1.9 plots the midpoint of the intervals containing failures versus relative humidity. The graph shows that failures occur earlier at higher levels of humidity. □

Example 1.9 Accelerated Test of Spacecraft Nickel–Cadmium Battery Cells. Brown and Mains (1979) present the results of an extensive experiment to evaluate the long-term performance of rechargable nickel–cadmium battery cells that were to be used in spacecraft. The study used eight experimental factors. The first five factors shown in the table were environmental or accelerating factors (set to higher than usual levels to obtain failure information more quickly). The other three factors were product-design factors that could be adjusted in the product design to optimize performance and reliability of the batteries to be manufactured. The experiment ran 82

GENERAL MODELS FOR RELIABILITY DATA

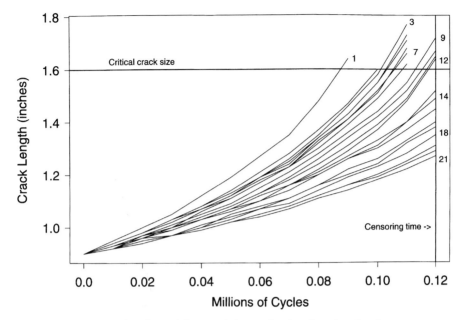

Figure 1.10. Alloy-A fatigue crack size as a function of number of cycles.

batteries, each containing 5 individual cells. Each battery was tested at a combination of factor levels determined according to a central composite experimental plan (see page 487 of Box and Draper, 1987, for information on central composite experimental designs). □

1.2.3 Degradation Data with no Explanatory Variables

Example 1.10   Fatigue Crack-Size Data.   Figure 1.10 and Appendix Table C.14 give the size of fatigue cracks as a function of number of cycles of applied stress for 21 test specimens. This is an example of degradation data. The data were reported originally in Hudak, Saxena, Bucci, and Malcolm (1978). The data were collected to obtain information on crack growth rates for the alloy. The data in Appendix Table C.14 were obtained visually from Figure 4.5.2 of Bogdanoff and Kozin (1985, page 242). For our analysis in the examples in Chapter 13, we will refer to these data as Alloy-A and assume that a crack of size 1.6 inches is considered to be a failure. □

1.3   GENERAL MODELS FOR RELIABILITY DATA

1.3.1   Definition of the Target Population or Process

Unless there is a clear definition of the target process or population, conclusions from a statistical study will appear fuzzy. Clear definition of the target population or process also allows precise statements about assumptions needed for the validity of conclusions to be drawn from a study.

As suggested by Deming (1975), statistical studies can be divided, broadly, into two different categories:

- **Enumerative studies** answer questions about *populations* that consist of a finite set of identifiable units. In the product reliability context, these units may be in service in the field or they may be stored in boxes in a warehouse. Typically, the statistical study is conducted by selecting a random sample from the population, carefully evaluating the units in the sample, and then making an inference or inferences about the larger population from which the sample was taken.
- **Analytic studies** answer questions about *processes* that generate units or other output over time. Again, in the reliability context, interest might center on the life distribution of electric motors that will be produced, in the future, by a particular production process.

Although the statistical data presentation and analysis methods may appear to be the same or very similar for these two different types of studies, the underlying assumptions required to make inferences (and thus statements of conclusions from a study) are quite different. In an enumerative study, the key assumption is that the sampling frame (list of population units from which the sample will be randomly selected) accurately represents the actual units in the population. In an analytic study, there is no population. Instead, the key assumption needed for inferences about characteristics of the process is that the process will behave in the future as it has in the past. Most reliability studies are analytic studies. For a more detailed description of these ideas, other examples, and references, see Deming (1975) and Chapter 1 of Hahn and Meeker (1991).

1.3.2 Causes of Failure and Degradation Leading to Failure

Many failure modes can be traced to some underlying degradation process. For example, fatigue cracks will initiate and grow in a steel frame if there are sufficiently high stresses. Tread on automobile tires and friction material on automobile brake pads and clutches wear with use. Corrosion causes thinning of walls of pipes in a chemical reactor. Filament material in operating incandescent light bulbs evaporates over time.

Traditionally, most statistical studies of product reliability have been based on failure-time data. For some reliability tests, however, it is possible to record the actual level of degradation on units as a function of time. Such data, particularly in applications where few or no failures are expected, can provide considerably more reliability information than would be available from traditional failure-time data. For most products it is difficult, expensive, or impossible to obtain degradation measurements and only (censored) failure-time data will be available. Thus because of its continuing importance in reliability analysis, most of the material in this book focuses on failure-time data. Chapters 13 and 21, however, describe methods for

GENERAL MODELS FOR RELIABILITY DATA   17

using degradation data for making inferences on reliability. Examples of degradation data are given there and in Section 1.2.3.

Not all failures can be traced to degradation; some product failures are caused by sudden accidents. For example, a tire may be punctured by a nail in the road, or a computer modem may fail from a lightning-induced charge on an unprotected telephone line.

Especially when the goal of a reliability study is to develop a highly reliable product or to improve the reliability of an existing product, it is important to consider the cause or causes of product failure (sometimes known as "modes" of failure). Understanding the physical and chemical mechanisms (including sources of variability in these mechanisms) and random risk factors leading to failure can suggest methods for eliminating failure modes or reducing the probability of a failure mode, thereby improving reliability.

1.3.3 Environmental Effects on Reliability

Environmental factors play an important part in product reliability. Automobiles corrode more rapidly in geographic areas with heavy use of salt on icy roads. An automobile battery would be expected to last longer in the warm climate of Florida than in the stressfully cold climate of Alaska. Due to increased heat and ultraviolet ray exposure, paints and other coating materials degrade more rapidly in the sunny southern parts of the United States. Driving automobiles on poorly maintained roads will cause fatigue failures of certain components to occur more rapidly than on smooth roads. Electronic components installed in the engine compartment of an automobile are subjected to much higher failure-causing heat, humidity, and vibration than are similar components installed in an air-conditioned office. Closely related is the effect of harsher-than-usual handling or operation of a product. For example, some household sump pumps are designed for a 50% duty cycle. If such a pump, in an emergency situation, has to run continuously, the temperature of the electric motor's components will become exceedingly high and the motor's life will be much shorter than expected. Excessive acceleration and braking of an automobile will lead to excessive wear on brake pads, relative to the number of miles driven. Attaching a trailer to an automobile can put additional strain on the engine and transmission as well as on parts of the electrical system.

A large proportion of product reliability problems result from unanticipated failure modes caused by environmental effects that were not part of the initial reliability-evaluation program. When making an assessment of reliability it is important to consider environmental effects. Data from designed experiments or field-tracking studies can be used to assess the effect that anticipated environmental factors and operational variables will have on reliability.

In some applications it is possible to protect products from harsh environments. Alternatively, products can be designed to be robust enough to withstand the harshest expected environments. Such products may have increased cost but could be expected to have exceedingly high reliability in more benign environments. One of the challenges of product design is to discover and develop economical means of building in robustness to environmental and other factors that manufacturers and users are unable to control. See the Epilogue of this book for further discussion and Hamada (1995a,b) for a description of some particular examples.

1.3.4 Definition of Time Scale

The life of many products can be measured on more than one scale. For example, the lifetimes of many automobile components are measured in terms of distance driven; others are measured in terms of calendar age. Light bulb life is typically measured in terms of the number of hours of use, but the number of on–off cycles could also affect life length. For factory life tests of products like washing machines and toasters, life would be measured in number of use-cycles. For data from the field, information on the number of use-cycles may not be available for individual units; time in service and average-use profiles are more commonly available.

As suggested by these examples, the choice of a time for measuring product life is often suggested by an underlying process leading to failure, even if the degradation process cannot be observed directly. For example, in a population of washing machines with different use-rates, wear on washing machine components is more directly related to use-cycles than to months in service. There would be more relative variability in the months-in-service data than in the number-of-use-cycles data.

In some cases there may be more than one measure of product life. For example, the ability of an automobile battery to hold a charge depends on the battery's age *and* on the number of charge/discharge cycles it has seen. There are a number of possible methods that could be used for handling such data. For example, it may be possible to estimate (directly or indirectly) the effect that both use-cycles and charge/discharge cycles have on degradation of battery-cell components (as described in Chapter 13) and use this information to develop a suitable measure of the amount of battery life as a function of these variables. Alternatively one could develop a statistical model that uses the observed number of charge/discharge cycles to help explain the variability in the time to failure measured in real time (see Chapter 17).

1.3.5 Definitions of Time Origin and Failure Time

When conducting a study of lifetimes of a product or material it is important to define clearly the begin-point and the endpoint of life. The definition may be arbitrary but should be purposeful. For example, in a constant-burn life test of incandescent light bulbs, the definitions would be clear and unambiguous. The beginning of life of a refrigerator installed in the field may, however, be more difficult to define or determine. Possibilities, with varying degrees of accessibility and relevance, include the date manufactured, date of sale, or reported date of installation. Similarly, end of life of customers' automobile tires often depends on subjective judgment on the amount of remaining tread, made at a convenient time (e.g., when the automobile is being serviced). In life tests of fluorescent light bulbs manufacturers define failure as the time when a bulb reaches 60% of its initial lumens output.

1.4 REPAIRABLE SYSTEMS AND NONREPAIRABLE UNITS

It is important to distinguish between data from and models for the following two situations:

- The time of failure (or other clearly specified event) for nonrepairable units or components (including data in *nonrepairable* components within a *repairable* system), or time to *first* failure of a system (whether it is repaired or not).
- A sequence of reported system-failure times (or the times of other events) for a repairable system.

Both of these applications can be important in reliability analyses. The models and data analysis methods appropriate for these two different areas are, however, generally quite different.

1.4.1 Reliability Data from Components and Other Nonrepairable Units

Data from *nonrepairable* units arise from many different kinds of reliability studies. Examples include:

- Laboratory tests to study durability, wear, or other lifetime properties of particular materials or components.
- Operational life tests on complete systems or subsystems, conducted before a product is released to customers when information is obtained on components and subsystems that are replaced upon failure.
- Data from customer field operation of larger integrated systems or subsystems, especially, when information is obtained on components and subsystems that are replaced upon failure.

When reliability tests are conducted on larger systems and subsystems (even those that may be repaired), it is essential that component-level information on cause of failure be obtained if the purpose of the data collection is improvement of system reliability, as opposed to mere assessment of overall system reliability.

In some simple situations it might be possible to assume that failure-time data from a sample of nonrepairable units or components can be modeled as a sample from a particular population or manufacturing process having a single failure-time distribution. In other situations, failure time depends on explanatory variables (which we will denote collectively by a vector x) such as environmental variables, operating conditions, manufacturer, and date manufactured. Starting in Chapter 17, the focus of this book will turn to models and methods that use explanatory variables in the modeling and analysis of reliability data. In more complicated situations, the failure-time distribution may depend on the age (or more accurately on the physical condition) of the system in which it is installed.

1.4.2 Repairable Systems Reliability Data

The purpose of some reliability studies is to describe the failure trends and patterns of an *overall system* or population of systems. System failures are followed by system repairs and data consist of a sequence of failure times for one or more copies of the system. When a single component or subsystem in a larger system is repaired or replaced after a failure, the distribution of the time to the next system failure will depend on the overall state of the system at the time just before the current failure and the nature of the repair. Thus repairable system data, in many situations, could be described with models that allow for changes in the state of the system over time or for dependencies between failures over time. There are also simpler models that describe a system's failure intensity (rate of occurrence of failures) as a function of system age and, perhaps, other explanatory variables. Such models are also useful for describing the failure-time distribution of repairable systems. Basic models and methods of analysis of repairable system data are covered in Chapter 16.

1.5 STRATEGY FOR DATA COLLECTION, MODELING, AND ANALYSIS

Reliability studies involving, for example, laboratory experimentation or field tracking require careful planning and execution. Mistakes can be extremely costly in terms of material and time, not to mention the possibility of drawing erroneous conclusions. Even if a mistake is recognized, rarely will there be enough time or money to comfortably repeat a flawed reliability study.

The rest of this book develops and illustrates the use of statistical methods for reliability data analysis. Chapters 2–8 describe basic models and reliability data analysis. Chapters 10 and 20 and Section 22.5 describe methods of planning reliability studies that will provide the desired degree of precision for estimating or predicting reliability. The other chapters describe more advanced methods and models for analyzing reliability data.

1.5.1 Planning a Reliability Study

Discussion of technical details for planning reliability studies is delayed until data analysis methods have been covered. This is necessary because proper planning depends on knowledge and, in many cases, simulated use of analysis methods *before* the final study plan is specified.

The initial stages of a reliability study should include:

- Careful definition of the problem to be solved (including a precise specification of the target population or process) and the questions to be answered by the study, in particular, the estimates to be obtained.
- Consideration of the resources available for the study (time, money, materials, equipment, personnel, etc.).

STRATEGY FOR DATA COLLECTION, MODELING, AND ANALYSIS   21

- Design of the experiment or study, including a careful assessment of precision of estimates as a function of the size of the study (i.e., sample size and expected number of failures). Because estimation precision generally depends on unknown model parameters, making such an assessment requires planning values of unknown population and process characteristics.
- In some situations, when little is known about the target population or process, it is often useful to conduct a pilot study to obtain the information needed for success of the main study.
- In new or unfamiliar situations, it is useful, before the test, to conduct a trial analysis of data simulated from a proposed model suggested from available information, engineering judgment, or previous experience. These ideas are illustrated in Chapters 10 and 20 and in Section 22.5.

1.5.2 Strategy for Data Analysis and Modeling

After the data collection has been completed, or at various points in time during the study, available data will be analyzed. The data analyses described in this book illustrate the steps in the following general strategy.

- Begin the analysis by looking at the data without making any distributional or other strong model assumptions. This will allow information to pass to the analyst without distortion that could be caused by making inappropriate model assumptions. The primary tool for these initial steps is graphical analysis, as illustrated in Section 1.2 and in examples throughout this book. Chapters 3 and 6 introduce other graphical analysis methods.
- For many applications it will be useful to fit one or more parametric models to the data, for purposes of description, estimation, or prediction. Generally this process progresses from simple to more elaborate models, depending on the purpose of the study, the amount of data, and other information that is available. In some cases it might be desirable or necessary to combine current data with previous data or other prior information. Chapter 4 describes some simple commonly used distributions for reliability data. Chapter 5 describes more advanced distributional models for reliability data. Methods of fitting such distributions to data are described starting in Chapter 7.
- Before using a fitted model for estimation or prediction, one should examine appropriate diagnostics and use other tools for assessing the adequacy of model assumptions. Graphical tools are especially useful for this purpose. It is important to remember that, especially in situations where there is little data, it will be difficult to detect small-to-moderate departures from model assumptions and that just because we have no strong evidence against model assumptions, does not mean that those assumptions can be trusted.
- If there are no obvious departures from the assumed model, one will generally proceed, with caution, to estimating parameters or predicting future outcomes (e.g., number of failures). For most reliability applications, such estimates and predictions include statistical intervals to reflect uncertainty and variability.

- In addition to using graphical methods for initial analyses and for diagnostics, it is helpful to display *results* of the analysis graphically, including estimates or predictions and uncertainty bounds (e.g., confidence or prediction intervals).
- Finally, it may be possible to use the results of the study to draw conclusions about product reliability, perhaps contingent on particular model assumptions. For some model assumptions it is possible to use the available data to assess the adequacy of the assumption (e.g., adequacy of model fit within the range of the data). For other assumptions, the data may provide no information about model adequacy. In situations where there is no information to assess the adequacy of assumptions, it is useful (even important) to vary assumptions and assess the *impact* that such perturbations have on final answers. The additional uncertainty uncovered by such sensitivity analyses should be reported along with conclusions.

BIBLIOGRAPHIC NOTES

Nelson (1982, 1990a) and Lawless (1982) provide other interesting reliability data sets. Hahn and Meeker (1982a,b) describe basic concepts and outline potential pitfalls of life data analysis. Chapter 1 of Hahn and Meeker (1991) provides a detailed discussion of implicit assumptions that are required to draw valid inferences from statistical studies. These assumptions parallel those needed to make inferences from reliability data. Ansell and Phillips (1989) describe some of the practical problems that arise in the analysis of reliability data. Nelson (1990a, page 237) outlines the capabilities of a variety of software packages that have procedures for analyzing censored reliability data. Kalbfleisch and Lawless (1988), Lawless and Kalbfleisch (1992), and Baxter and Tortorella (1994) describe some technical methods for dealing with field reliability data.

Kordonsky and Gertsbakh (1993) discuss the important problem of choosing appropriate time scales when analyzing reliability data. They formalize and discuss the concepts of optimal and good time scales and they show how to find a best linear combination of observable time scales when the criterion is minimizing the coefficient of variation of the time scale. Kordonsky and Gertsbakh (1995a) discuss theoretical aspects of finding a best time scale for monitoring systems whose failure has serious consequences (airplanes, nuclear power plants, etc.). They present a method for calculating this scale when the observed data are complete (noncensored) on two observable time scales like operational time and number of cycles. Kordonsky and Gertsbakh (1995b) generalize these results to the case in which the data are censored. For another view on combining multiple time scales into a single time scale for life testing with complete data, see Farewell and Cox (1979).

EXERCISES

1.1. Discuss the assumptions that would be needed to take the heat exchanger data for calendar time in Figure 1.6 and convert them to operating-time data shown in Figure 1.7 and then use these data for purposes of analysis and inference on the life distribution of heat exchanger tubes of the type in these exchangers.

1.2. It has been argued both that quality is a part of reliability and that reliability is a part of quality. Discuss the relationship between these two disciplines. To help make your discussion concrete, use your knowledge of a particular product to help express your ideas.

1.3. In the development and presentation of traditional statistical methods, description and inference are often presented in terms of means and variances (or standard deviations) of distributions.
   (a) Use some of the examples in this chapter to explain why, in many applications, reliability or design engineers would be more interested in the time at which 1% (or some smaller percentage) of a particular component will fail instead of the time at which 50% would fail.
   (b) Explain why means and variances of time to failure may not be of such high interest in reliability studies.
   (c) Give at least one example of a product for which mean time to failure would be of interest. Explain why.

1.4. Consider the following situations. For each, discuss the reasons why the study might be considered to be either analytic or enumerative. For each example outline the assumptions needed so that the sample data will be useful for making inferences about the population or process of interest.
   (a) Ten light bulbs were selected at random points in time from a production process. One of these bulbs was then selected at random and put aside for future use. The other nine bulbs were tested until failure. The data from the nine failures were to be used to construct a prediction interval for the last bulb.
   (b) A company has entered into a contract to buy a large lot of light bulbs. The price will be determined as a function of the average failure time of a random sample of 100 bulbs.
   (c) Example 1.5.
   (d) Example 1.9.

1.5. An important part of quantifying product reliability is specification of an appropriate time scale (or time scales) on which life should be measured (e.g., hours of operation, cycles of operation). For each of the following products, suggest and give reasons for an appropriate scale (or scales) on which one might measure life for the following products. Also, discuss possible environmental factors that might affect the lifetime of individual units.
   (a) Painted surface of an automobile.
   (b) Automobile lead–acid battery.
   (c) Automobile windshield wipers.

(d) Automobile tires.

(e) Incandescent light bulb.

1.6. For each of the products listed in Exercise 1.5, explain your best understanding of the underlying failure mechanism. Also, describe possible ways in which an analyst could define failure.

1.7. For each of the following, discuss whether field failures of the unit or product should be considered to be a failure of a repairable system, a failure of replaceable unit within a system, or both. Explain why.

(a) Automobile alternator.

(b) Video cassette recorder.

(c) Microwave oven.

(d) Home air conditioner.

(e) Hand-held calculator.

(f) Clothes dryer.

1.8. For each of the products listed in Exercise 1.7, describe the range of environments that the product might encounter in use and the effect that environment could have on the product's reliability.

1.9. Consider the turbine wheel data in Table 1.5.

(a) Compute the proportion cracked at each level of exposure.

(b) How do you explain the fact that the proportion of cracked wheels decreases as age increases at 14, 30, and 42 hundred hours of exposure?

(c) Discuss the assumptions that one would have to make in order to answer the questions raised by the objectives listed in Example 1.7.

1.10. Construct a histogram for the V7 tube data in Table 1.4. Discuss alternative methods of handling the last open-ended time interval. What information does this plot provide?

1.11. Figure 1.4 shows both the number of failures and the number of censored observations in each of six time intervals. Explain why such a graphical display needs to be interpreted differently than a histogram of uncensored failure times like Figure 1.1.

1.12. A telephone electronic switching system contains a large number of nominally identical circuit packs. When a circuit pack fails, only a small part of the system's functionality is lost. Failed packs are replaced, as soon as possible, with new circuit packs. All of the circuit packs have serial numbers and detailed records are kept so that the failure times are known for all packs that fail and so that the running times are known for all of the packs that do not fail. In practice, to assess circuit pack reliability, it would be common to treat the circuit packs in the system as a sample from a larger population of circuit packs and use the data to make inferences about the larger population.

(a) List three distinct different *precise* definitions for the larger population or process that could be of interest to reliability engineers, design engineers, or financial managers.

(b) For each of the definitions in part (a), state the assumptions that must be satisfied to make the desired inferences about the circuit pack life distribution. Comment on the reasonableness of these assumptions and how departures from the assumptions could result in misleading conclusions.

(c) Assume that you have been given the above description of a switching system and have been asked to attend a meeting where a study is to be planned to monitor, continuously, the early-life reliability (defined as the first 1000 hours) of circuit packs. The purpose of the study is to determine the effect of recent design and manufacturing process changes on circuit pack reliability. Data will be obtained from three particular systems that are physically close to the design and manufacturing facilities of the circuit pack manufacturer. Before offering advice on the plan you will need further information. Prepare a list of questions that you would ask of design engineers, reliability engineers, and manufacturing engineers who will be attending the meeting.

1.13. Explain how graphical methods can be used to complement analytical methods of data analysis.

1.14. There was a considerable amount of censoring at low levels of humidity in the printed circuit accelerated life test data shown in Figure 1.9. Explain how such censoring can obscure important information about the relationship between humidity and time to failure.

CHAPTER 2

Models, Censoring, and Likelihood for Failure-Time Data

Objectives

This chapter explains:

- Models for continuous failure-time processes.
- Models for the discrete data from these continuous failure-time processes.
- Common censoring mechanisms that restrict our ability to observe all of the failure times that might occur in a reliability study.
- Principles of likelihood and how likelihood is related to the probability of the observed data.
- How likelihood ideas can be used to make inferences from reliability data.

Overview

This chapter introduces basic concepts of modeling failure-time processes. Section 2.1 explains the basic relationships among cumulative distributions, densities, survival, hazard, and quantile functions for modeling of continuous failure-time processes. These relationships are used extensively in subsequent chapters and they are essential background to read the rest of the book. Section 2.2 describes the modeling of discrete data that arise from our limited ability to observe continuous processes. This section also explains briefly the importance of censoring, censoring mechanisms, and important assumptions about censoring mechanisms, needed for proper application of the methodology in the book. Section 2.4 introduces likelihood-based statistical methods. This section provides general rules for writing the likelihood for reliability data with several kinds of censoring. This section is an integral part of the methodology used in the book and it should be read by most readers.

2.1 MODELS FOR CONTINUOUS FAILURE-TIME PROCESSES

As explained in Chapter 1, the most widely used metric for reliability of a product is its failure-time distribution and the most commonly collected reliability data contain information on the failure times of samples of materials, components, or of complete systems. This chapter presents basic models for such data.

Most failure-time processes are modeled on a continuous scale. This section describes some common models for describing such processes. The symbol T will be used to denote a nonnegative, continuous random variable describing the failure time of a unit or system.

2.1.1 Failure-Time Distribution Functions

The probability distribution for failure time T can be characterized by a cumulative distribution function, a probability density function, a survival function, or a hazard function. These functions are described below and illustrated, for a typical failure-time distribution, in Figure 2.1. The choice of which function or functions to use depends on convenience of model specification, interpretation, or technical development. All of these functions are important for one purpose or another.

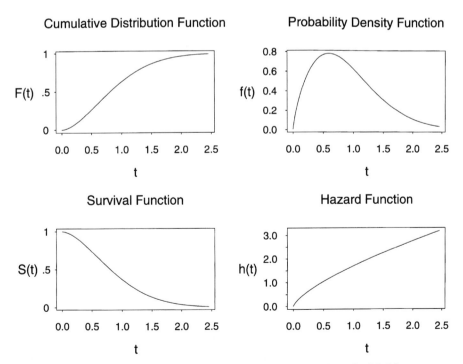

Figure 2.1. Failure-time cdf, pdf, sf, and hf corresponding to Examples 2.1–2.8.

Cumulative Distribution Function

The cumulative distribution function (cdf) of T, $F(t) = \Pr(T \leq t)$, gives the probability that a unit will fail before time t. Alternatively, $F(t)$ can be interpreted as the proportion of units in the population (or taken from some stationary process) that will fail before time t. [Here a stationary process is defined as one that generates units that have a $F(t)$ that does not change over time.]

Example 2.1 cdf. The NW corner of Figure 2.1 shows the particular cdf $F(t) = 1 - \exp(-t^{1.7})$ for t between 0 and 2.5. □

Probability Density Function

The probability density function (pdf) for a continuous random variable T is defined as the derivative of $F(t)$ with respect to t: $f(t) = dF(t)/dt$. The pdf can be used to represent relative frequency of failure times as a function of time. Although the pdf is less important than the other functions for applications in reliability, it is used extensively in the development of technical results. As illustrated in Figure 2.3, the cdf at t is computed as the area under the pdf from 0 to t, giving the probability of failing before t. That is, $F(t) = \int_0^t f(x)\,dx$.

Example 2.2 pdf. Corresponding to Example 2.1, the NE corner of Figure 2.1 shows the particular pdf $f(t) = dF(t)/dt = 1.7t^{.7}\exp(-t^{1.7})$ for t between 0 and 2.5. □

Survival Function

The survival function (sf), also known as the reliability function, is the complement of the cdf, $S(t) = \Pr(T > t) = 1 - F(t) = \int_t^\infty f(x)\,dx$, and gives the probability of surviving beyond time t.

Example 2.3 sf. Corresponding to Example 2.1, the SW corner of Figure 2.1 shows the particular sf $S(t) = 1 - F(t) = \exp(-t^{1.7})$ for t between 0 and 2.5. □

Hazard Function

The hazard function (hf), also known as the hazard rate, the instantaneous failure rate function, and by other names, is defined by

$$h(t) = \lim_{\Delta t \to 0} \frac{\Pr(t < T \leq t + \Delta t \mid T > t)}{\Delta t} = \frac{f(t)}{1 - F(t)}.$$

The hazard function expresses the propensity to fail in the next small interval of time, given survival to time t. That is, for small Δt,

$$h(t) \times \Delta t \approx \Pr(t < T \leq t + \Delta t \mid T > t). \tag{2.1}$$

Example 2.4 hf. Corresponding to Example 2.1, the SE corner of Figure 2.1 shows the particular hf $h(t) = f(t)/[1 - F(t)] = 1.7 \times t^{.7}$ for t between 0 and 2.5. □

MODELS FOR CONTINUOUS FAILURE-TIME PROCESSES

The hazard function can be interpreted as a failure rate in the following sense. If there is a large number of items [say, $n(t)$] in operation at time t, then $n(t) \times h(t)$ is approximately equal to the number of failures per unit time [or $h(t)$ is approximately equal to the number of failures per unit time per unit at risk]. The hazard function has units of fraction failed per unit time. Because of its close relationship with failure processes and maintenance strategies, some reliability engineers think of modeling failure time in terms of $h(t)$. The "bathtub curve" shown in Figure 2.2 provides a useful conceptual model for the hazard of some product populations. There may be early failures of units with quality-related defects (infant mortality). During much of the useful life of a product, the hazard may be approximately constant because failures are caused by external shocks that occur at random. Late-life failures are due to wearout. Many reliability studies focus on one side or the other of this curve.

Cumulative Hazard Function

For some purposes it is useful to define the function

$$H(t) = \int_0^t h(x)\,dx,$$

commonly known as the cumulative hazard function. The cdf or survival function for T can be obtained from the hazard function. For example, for any continuous distribution

$$F(t) = 1 - \exp[-H(t)] = 1 - \exp\left[-\int_0^t h(x)\,dx\right]. \tag{2.2}$$

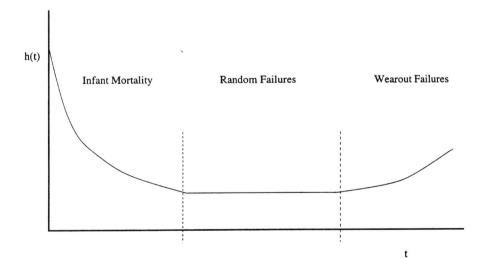

Figure 2.2. Bathtub curve hazard function.

Average Hazard Rate
The average hazard rate between times t_1 and t_2 is

$$\text{AHR}(t_1, t_2) = \frac{\int_{t_1}^{t_2} h(u)\, du}{t_2 - t_1} = \frac{H(t_2) - H(t_1)}{t_2 - t_1}$$

and can be viewed as a typical hazard rate value over the interval. Also, if $F(t_2) - F(t_1)$ is small (say, less than .1), then

$$\text{AHR}(t_1, t_2) \approx \frac{F(t_2) - F(t_1)}{t_2 - t_1}. \quad (2.3)$$

An important special case arises when $t_1 = 0$, giving

$$\text{AHR}(t) = \frac{\int_0^t h(u)\, du}{t} = \frac{H(t)}{t} \approx \frac{F(t)}{t} \quad (2.4)$$

and the approximation is good for small $F(t)$, say, $F(t) < .10$. The right-hand sides of (2.3) and (2.4) provide simple interpretation for the AHR expressions. In either case, AHR can be interpreted as the approximate fraction failing per unit time over the specified interval. Of course, if one is really interested in computing the fraction failing, this is easy to do directly without any approximation.

Hazard Rate in FITs
Especially in high-reliability electronics applications, it is common to express hazard rates in units of FITs. A FIT rate is defined as the hazard function in units of 1/hours, multiplied by 10^9. FITs (failures in time) were originally used to describe hazard rates corresponding to components for which $h(t)$ is *constant* over time (a model that will be described in Section 4.4). In such applications, and when the number of components at risk is large relative to the number that will fail, FITs can be interpreted, for example, as a prediction for the number of failures per billion hours of operation or the number of failures per 1000 hours of operation per one million units at risk.

The use of FIT rates has carried over to the more modern and realistic failure models with nonconstant $h(t)$. In this case it is important to distinguish between a FIT rate for a particular point in time [$h(t) \times 10^9$] or an average from beginning of life to a particular point in time [$\text{AHR}(t) \times 10^9$]; both uses are common. Because these FIT rates can be vastly different, the distinction is important.

Example 2.5 Constant-Hazard FIT Rate. In a large computing network there are 165,000 copies of a particular component that are at risk to fail. The manufacturer of the components claims that the component hazard is constant over time at 15 FITs. Thus $h(t) = 15 \times 10^{-9}$ failures per unit per hour for all time t measured in units of hours. A prediction for the number of failures from this component in 1 year (8760 hours) of operation is $15 \times 10^{-9} \times 165,000 \times 8760 \approx 217$. □

MODELS FOR CONTINUOUS FAILURE-TIME PROCESSES

Example 2.6 Nonconstant-Hazard FIT Rate. The manufacturer of a particular integrated circuit device claims that the device's hazard function is $h(t) = 1.8 \times 10^{-7} \times t^{-.8}$ and time t is measured in units of hours. From this, the FIT rate for this population of components at 1 hour is $1.8 \times 10^{-7} \times (1)^{-.8} \times 10^9 = 180$ FITs. The FIT rate for this population of components at 10,000 hours is $1.8 \times 10^{-7} \times (10000)^{-.8} \times 10^9 = .1136$ FITs. By simple integration, the average hazard rate to 10,000 hours is $\text{AHR}(10000) = (1.8/.2) \times 10^{-7} \times (10000)^{-.8} = 5.68 \times 10^{-10}$ or .568 FITs. □

2.1.2 The Quantile Function and Distribution Quantiles

The quantile t_p is the inverse of the cdf; it is the time at which a specified proportion p of the population fails. For example, $t_{.20}$ is the time by which 20% of the population will fail. This is illustrated in Figure 2.3. By definition, the cdf $F(t)$ is nondecreasing. This leaves two possibilities.

- When $F(t)$ is strictly increasing there is a unique value t_p that satisfies $F(t_p) = p$, and we write $t_p = F^{-1}(p)$.
- When $F(t)$ is constant (i.e., flat) over some interval or intervals, there can be more than one solution t to the equation $F(t) = p$. Taking t_p equal to the smallest value of t satisfying $F(t) = p$ is the standard convention.

In general, for $0 < p < 1$, we define the p quantile of $F(t)$ as the *smallest* time t such that $\Pr(T \leq t) = F(t) \geq p$.

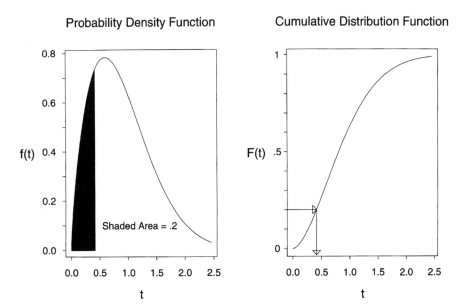

Figure 2.3. Plots showing that the quantile function is the inverse of the cdf.

Example 2.7 Quantile Function. Using $p = F(t) = 1 - \exp(-t^{1.7})$ from Example 2.1, solving for t gives $t_p = [-\log(1-p)]^{1/1.7}$ and $t_{.2} = [-\log(1-.2)]^{1/1.7} = .414$, as shown in Figure 2.3. □

2.2 MODELS FOR DISCRETE DATA FROM A CONTINUOUS PROCESS

Most failure-time processes are modeled on a continuous scale. Because of inherent limitations in measurement precision, however, failure-time data are *always* discrete. Limitations in ability to observe or in a measuring instrument's ability to record can cause data to be censored or truncated, as illustrated in the examples in Section 1.2. The rest of this chapter develops a general structure for modeling such data. Subsequent sections describe different kinds of observations that arise in reliability data analysis and show how to compute the likelihood (or "probability of the data").

2.2.1 Multinomial Failure-Time Model

Because all data are discrete, it is convenient to partition the time line $(0, \infty)$ into $m + 1$ observation intervals. The partitioning depends on inspection times, measurement precision, and/or roundoff; it can be expressed as follows:

$$(t_0, t_1], (t_1, t_2], \ldots, (t_{m-1}, t_m], (t_m, t_{m+1}), \tag{2.5}$$

where $t_0 = 0$ and $t_{m+1} = \infty$. This partition is illustrated in Figure 2.4. For example, if failure times are recorded to the nearest hour, then each interval would be 1 hour long, up until t_m, the last recording. In general, these intervals need not be of equal length. Observe that the last interval is of infinite length. Define

$$\pi_i = \Pr(t_{i-1} < T \le t_i) = F(t_i) - F(t_{i-1}) \tag{2.6}$$

as the multinomial probability that a unit will fail in interval i. Note that $\pi_i \ge 0$ and $\sum_{j=1}^{m+1} \pi_j = 1$. The survival function evaluated at t_i is $S(t_i) = \Pr(T > t_i) = 1 - F(t_i) = \sum_{j=i+1}^{m+1} \pi_j$. Then

$$p_i = \Pr(t_{i-1} < T \le t_i \mid T > t_{i-1}) = \frac{F(t_i) - F(t_{i-1})}{1 - F(t_{i-1})} = \frac{\pi_i}{S(t_{i-1})} \tag{2.7}$$

is the conditional probability that a unit will fail in interval i, given that the unit was still operating at the beginning of interval i. Thus $p_{m+1} = 1$ but the only restriction on $p_1, \ldots, p_m$ is $0 \le p_i \le 1$.

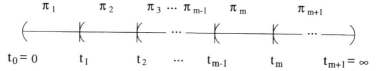

Figure 2.4. Partitioning of time into nonoverlapping intervals.

MODELS FOR DISCRETE DATA FROM A CONTINUOUS PROCESS   33

2.2.2 Multinomial cdf

Using (2.7), it is easy to show that

$$S(t_i) = \prod_{j=1}^{i} [1 - p_j], \quad i = 1, \ldots, m + 1. \tag{2.8}$$

This result and those following are important for data analysis methods developed in Chapter 3. Then the cdf of T, evaluated at t_i, can be expressed as

$$F(t_i) = 1 - \prod_{j=1}^{i} [1 - p_j], \quad i = 1, \ldots, m + 1$$

or as

$$F(t_i) = \sum_{j=1}^{i} \pi_j, \quad i = 1, \ldots, m + 1. \tag{2.9}$$

Thus $\pi = (\pi_1, \ldots, \pi_{m+1})$ or $p = (p_1, \ldots, p_m)$ are alternative sets of *basic parameters* to model discrete failure-time data.

Example 2.8 *Computation of* $F(t_i)$, $S(t_i)$, π_i, *and* p_i. Table 2.1 shows values of $F(t_i)$, $S(t_i)$, π_i, and p_i based on cdf $F(t) = 1 - \exp(-t^{1.7})$ used in Examples 2.1–2.7. The quantities in the table illustrate the use of (2.6), (2.7), (2.8), and (2.9) for inspections at .5, 1, 1.5, 2, and 2.5 (note that some arithmetic using values in the table may be off a little in the last digit due to the limited precision in the three digits shown in the table). Figure 2.5 shows, graphically, the relationship between the π values and $F(t)$ for this example. □

Table 2.1. Illustration of Probabilities for the Multinomial Failure-Time Model Computed from $F(t) = 1 - \exp(-t^{1.7})$

i	t_i	$F(t_i)$	$S(t_i)$	π_i	p_i	$1 - p_i$
0	.0	.000	1.000			
1	.5	.265	.735	.265	.265	.735
2	1.0	.632	.368	.367	.500	.500
3	1.5	.864	.136	.231	.629	.371
4	2.0	.961	.0388	.0976	.715	.285
5	∞	1.000	.0000	.0388	1.000	.000
				1.000		

34 MODELS, CENSORING, AND LIKELIHOOD FOR FAILURE-TIME DATA

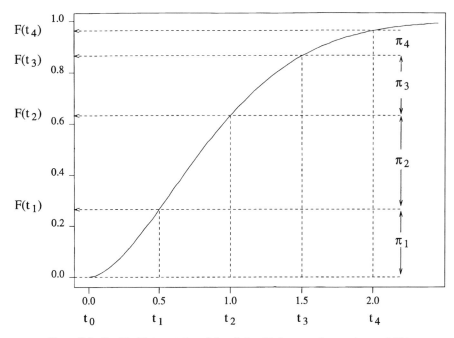

Figure 2.5. Graphical interpretation of the relationship between the π_i values and $F(t)$.

2.3 CENSORING

2.3.1 Censoring Mechanisms

Censoring restricts the ability to observe failure times exactly. As illustrated in the examples in Chapter 1, censoring is common in reliability data analysis and arises for a number of different reasons.

- Generally there are constraints on the length of life tests or other reliability studies and, as a result, data have to be analyzed before all units have failed. Removing unfailed units from test at a prespecified time is known as "time censoring" or "Type I censoring." Units may be tested simultaneously or in sequence (e.g., because of a limited number of test positions). Examples 1.2 and 1.3 illustrate time censoring.
- A life test that is terminated after a specified number of failures results in "failure censoring," also known as "Type II censoring." Although the statistical properties of estimates from failure-censored data are simpler than the corresponding properties from time-censored data, failure-censored tests are less common in practice.
- In many life tests, failures are discovered only at times of inspection. Interval-censored observations consist of upper and lower bounds on a failure time. Such

data are also known as inspection data, grouped data, or read-out data. If a unit has failed at its first inspection, it is the same as a left-censored observation. If a unit has not failed by the time of the last inspection, it is right-censored, the upper endpoint of the interval being ∞. See Examples 1.5 and 1.6. If each unit has only one inspection time (perhaps differing from unit to unit), and where the observation is on whether the unit failed or not, the data are known as quantal-response data, as in Example 1.7.

- Some products have more than one cause of failure. If primary interest is focused on one particular cause of failure, failure from other causes (sometimes known as competing risks) can, in some situations, be viewed as a form of *random right censoring*. This kind of random censoring can lead to multiple right censoring where some failure times and censoring times are intermixed as in Example 3.8.
- In some situations units are introduced into the field or put on test at different times. This is known as staggered entry. If the data are to be analyzed at a point in time when not all units have failed, the data will, usually, be multiply right-censored with some failure times again exceeding some of the running times as in Example 1.4. Censoring due to staggered entry of units is a type of *systematic multiple censoring*.

If it is a reasonable approximation that units manufactured over the period of time came from the same process, the data could be pooled together and analyzed to make inferences about that process. Often, however, a process or product design will change over time and pooling such data could lead to misleading conclusions. Caution is advised and it is good practice to look for time trends in data. The case study in Section 22.1 illustrates such a situation.

2.3.2 Important Assumptions on Censoring Mechanisms

Use of most models and methods to analyze censored data implies important assumptions about the nature of the censoring and its relationship to the failure process. Simply stated, a censoring time (i.e., the time at which we stop observing a unit that has not failed) can be either random or predetermined. In order for standard censored data analysis methods to be valid, it is necessary that the censoring time of a unit depend only on the history of the observed failure-time process. Using future events (or indicators of future events) to stop observing a unit could introduce bias. This cause-of-censoring assumption would be violated, for example, if units were taken off test before actual failure, but in response to some precursor to a future failure (e.g., increase in vibration for an electrical motor). For the standard censoring mechanisms described in Section 2.3.1, the stopping times depend only on the history of the observed failure-time process. Relatedly, standard methods of analyzing censored data require the assumption that censoring is noninformative. This implies that the censoring times of units provide no information about the failure-time distribution.

2.4 LIKELIHOOD

2.4.1 Likelihood-Based Statistical Methods

The general idea of likelihood inference is to fit models to data by entertaining model–parameter combinations for which the probability of the data is large. Model–parameter combinations with relatively high probabilities are more plausible than combinations with low probability. Likelihood methods provide general and versatile tools for fitting models to data. The methods can be applied with a wide variety of parametric and nonparametric models with censored, interval, and truncated data. It is also possible to fit models with explanatory variables (i.e., regression analysis).

There is a well-developed large-sample likelihood theory for regular models that provides straightforward methods for fitting models to data. The theory guarantees that these methods are, in large samples, statistically efficient (i.e., yield the most accurate estimates). These properties are approximate in moderate and small sample sizes, and various studies have shown that likelihood methods generally perform as well as other available methods. With censored data, "large sample" really means "large number of failures" and a typical guideline for large is 20 or more, but this really depends on the problem and the questions to be answered.

Likelihood theory can be extended to more complicated *nonregular* models and the basic concepts are similar. Also, much current statistical research is focused on the development of more refined, but computationally intensive, methods that will work better for smaller sample sizes.

2.4.2 Specifying the Likelihood Function

The likelihood function is either equal to or approximately proportional to the probability of the data. This section describes a general method of computing the probability of a given data set. Then, for a given set of data and specified model, the likelihood is viewed as a function of the unknown model parameters (where we can use either the π_i values or the p_i values in the multinomial model introduced in Section 2.2). The form of the likelihood function will depend on factors like:

- The assumed probability model.
- The form of available data (censored, interval censored, etc.).
- The question or focus of the study. This includes issues relating to identifiability of parameters (i.e., the data's ability or inability to estimate certain features of a statistical model).

The total likelihood can be written as the joint probability of the data. Assuming n independent observations, the sample likelihood is

$$L(\boldsymbol{p}) = L(\boldsymbol{p}; \text{DATA}) = C \prod_{i=1}^{n} L_i(\boldsymbol{p}; \text{data}_i), \qquad (2.10)$$

where $L_i(\boldsymbol{p}; \text{data}_i)$ is the probability of the observation i, data_i is the data for observation i, and $\boldsymbol{p}$ is the vector of parameters to be estimated. To estimate $\boldsymbol{p}$ from

LIKELIHOOD

the available DATA, we find the values of p that maximize $L(p)$. In the usual situations where the constant term C in (2.10) does not depend on p, one can simply take $C = 1$ for purposes of estimating p (see Section 2.4.4 for more information on C). The likelihood in (2.10) can also be written as a function of the multinomial cell probabilities π. Similarly, if there is a specified parametric form for $F(t; \theta)$ the likelihood can be written as a function of the parameters θ. We use p here because Chapter 3 illustrates the direct estimation of p.

2.4.3 Contributions to the Likelihood Function

Figure 2.6 illustrates the intervals of uncertainty for examples of left-censored, interval-censored, and right-censored observations. The likelihood contributions for each of these cases, shown in Table 2.2, is simply the probability of failing in the corresponding interval of uncertainty.

Interval-Censored Observations

If a unit's failure time is known to have occurred between times t_{i-1} and t_i, the probability of this event is

$$L_i(p) = \int_{t_{i-1}}^{t_i} f(t)\,dt = F(t_i) - F(t_{i-1}). \tag{2.11}$$

The three middle rows in Table 1.4 are examples of interval-censored observations.

Example 2.9 Likelihood for an Interval Censored Observation. Refer to Figure 2.6 and Table 2.1. If a unit is still operating at the $t = 1.0$ inspection but a

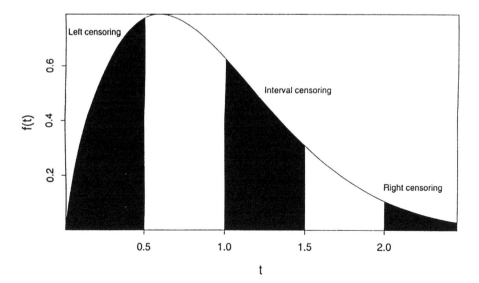

Figure 2.6. Likelihood contributions for different kinds of censoring.

Table 2.2. Contributions to Likelihood for Life Table Data

Censoring Type	Range	Likelihood
d_i observations interval-censored in t_{i-1} and t_i	$t_{i-1} < T \leq t_i$	$[F(t_i) - F(t_{i-1})]^{d_i}$
ℓ_i observations left-censored at t_i	$T \leq t_i$	$[F(t_i)]^{\ell_i}$
r_i observations right-censored at t_i	$T > t_i$	$[1 - F(t_i)]^{r_i}$

failure is found at the $t = 1.5$ inspection, then the likelihood (probability) for the interval-censored observation is $\pi_i = F(1.5) - F(1.0) = .231$. □

Although most data arising from observation of a continuous-time process can be thought of as having occurred in intervals similar to (t_{i-1}, t_i), the following important special cases warrant separate consideration.

Left-Censored Observations
Left-censored observations occur in life test applications when a unit has failed at the time of its first inspection; all that is known is that the unit failed before the inspection time (e.g., the first row of Table 1.4). In other situations, left-censored observations arise when the exact value of a response has not been observed and we have, instead, an upper bound on that response. Consider, for example, a measuring instrument that lacks the sensitivity needed to measure observations below a known threshold (e.g., a noise floor in an ultrasonic measuring system). When the measurement is taken, if the signal is below the instrument threshold, all that is known is that the measurement is less than the threshold. If there is an upper bound t_i for observation i, causing it to be left-censored, the probability and likelihood contribution of the observation is

$$L_i(\boldsymbol{p}) = \int_0^{t_i} f(t)\,dt = F(t_i) - F(0) = F(t_i). \tag{2.12}$$

Equation (2.7) shows how L_i can be written as a function of $\boldsymbol{p}$. Alternatively, (2.6) shows how L_i can be written as a function of $\boldsymbol{\pi}$. Note that a left-censored observation can also be considered to be an interval-censored observation between 0 and t_i.

Example 2.10 *Likelihood of a Left-Censored Observation.* Refer to Figure 2.6 and Table 2.1. If a failure is found at the first inspection time $t = .5$, then the likelihood (probability) for the left-censored observation is $F(.5) = .265$.
□

Right-Censored Observations
Right censoring is common in reliability data analysis. For example, the last bin in Table 1.4 contains all lifetimes greater than 100 days. The observations in this bin

LIKELIHOOD

are right-censored because all that is known about the failure times in this bin is that they were greater than 100 days.

If there is a lower bound t_i for the ith failure time, the failure time is somewhere in the interval (t_i, ∞). Then the probability and likelihood contribution for this right-censored observation is

$$L_i(p) = \int_{t_i}^{\infty} f(t)\, dt = F(\infty) - F(t_i) = 1 - F(t_i). \tag{2.13}$$

Example 2.11 Likelihood of a Right-Censored Observation. Refer to Figure 2.6 and Table 2.1. If a unit has not failed by the last inspection at $t = 2$, then the likelihood (probability) for the right-censored observation is $1 - F(2) = .0388$. □

Total Likelihood

The total likelihood, or joint probability of the DATA, for n independent observations is

$$L(p; \text{DATA}) = C \prod_{i=1}^{n} L_i(p; \text{data}_i) \tag{2.14}$$

$$= C \prod_{i=1}^{m+1} [F(t_i)]^{\ell_i} [F(t_i) - F(t_{i-1})]^{d_i} [1 - F(t_i)]^{r_i},$$

where $n = \sum_{j=1}^{m+1} (d_j + r_j + \ell_j)$ and C is a constant depending on the sampling inspection scheme but not on the parameters p. So we can take $C = 1$. We want to find p so that $L(p)$ is large. The p that maximizes $L(p)$ provides a maximum likelihood estimate of $F(t)$. For some problems, it will be more convenient to write the likelihood and do the optimization in terms of π. As described in Section 2.2.1, either set of basic parameters can be used.

2.4.4 Form of the Constant Term C

The form of constant term C in (2.10) and (2.14) depends on the underlying sampling and censoring mechanisms and is difficult to characterize in general. For our multinomial model, assuming inspection data and *no losses* (i.e., no right-censored observations before the last interval),

$$C = \frac{n!}{d_1! \cdots d_{m+1}!},$$

which is the usual multinomial coefficient. Another important special case arises when we increase the number of intervals, approaching continuous inspection. Then with an underlying continuous failure-time process (so there will be no ties), all d_i values will be either 0 or 1 depending on whether there is a failure or not in interval i. In this case C reduces to $n!$, corresponding to the number of permutations of the n order statistics. With Type I single-time censoring at t_m and no more than one failure in any of the intervals before t_m, $C = n!/r_{m+1}!$, where $r_{m+1} = d_{m+1}$ is the number of right-censored observations, all of which are beyond t_m.

Because, for most models, C is a constant that does not depend on the model parameters, it is common practice to take $C = 1$ and suppress C from likelihood expressions and computations.

2.4.5 Likelihood Terms for General Reliability Data

Although some reliability data sets are reported in life table form (e.g., Table 1.4), other data sets report only the times or the intervals in which failures actually occurred or observations were censored. For such data sets there is an alternative, more general form for writing the likelihood. This form of the likelihood is commonly used as input for computer software for analyzing failure-time data. In general, observation i consists of an interval $(t_i^L, t_i]$, $i = 1, \ldots, n$, that contains the failure time T for unit i in the sample. The intervals $(t_i^L, t_i]$ may overlap and their union may not cover the entire time line $(0, \infty)$. In general, $t_i^L \neq t_{i-1}$. Assuming that the censoring is at t_i the likelihood for individual observations can be computed as shown in Table 2.3; the joint likelihood for the DATA with n independent observations is

$$L(\boldsymbol{p}; \text{DATA}) = \prod_{i=1}^{n} L_i(\boldsymbol{p}; \text{data}_i).$$

Some of the failure times or intervals may appear more than once in a data set. Then w_j is used to denote the frequency (weight or multiplicity) of such identical observations and

$$L(\boldsymbol{p}; \text{DATA}) = \prod_{j=1}^{k} \left[L_j(\boldsymbol{p}; \text{data}_j) \right]^{w_j}. \tag{2.15}$$

Chapter 3 shows how to compute the maximum likelihood estimate of $F(t)$ without having to make any assumption about the underlying distribution of T. Starting in Chapter 7 we show how to estimate a small number of unknown parameters from a more highly structured parametric model for $F(t)$.

Table 2.3. Contributions to the Likelihood for General Failure Time Data

Type of Observation	Characteristic	Likelihood of a Single Response $L_i(\boldsymbol{p}; \text{data}_i)$
Interval-censored	$t_i^L < T \leq t_i$	$F(t_i) - F(t_i^L)$
Left-censored at t_i	$T \leq t_i$	$F(t_i)$
Right-censored at t_i	$T > t_i$	$1 - F(t_i)$

2.4.6 Other Likelihood Terms

The likelihood contributions used in (2.14) and (2.15) will cover the vast majority of reliability data analysis problems that arise in practice. There are, however, other kinds of observations and corresponding likelihood contributions that can arise and these can be handled with only a slight extension of this framework.

Random Censoring in the Intervals

Until now, it has been assumed that right censoring occurs at the end of the inspection intervals. If C is a random censoring time, an observation is censored *in* the interval $(t_{i-1}, t_i]$ if $t_{i-1} < C \leq t_i$ and $C \leq T$. Similarly, an observation is a failure in that interval if $t_{i-1} < T \leq t_i$ and $T \leq C$. To account for right-censored observations that occur at unknown random points in the intervals, one usually assumes that the censoring is determined by a random variable C with pdf $f_C(t)$ and cdf $F_C(t)$ and that the failure time T and censoring time C are statistically independent. (But it is important to recognize that making such an assumption does not make it so!) Then for continuous T, the joint probability (likelihood) for r_i right-censored observations in $(t_{i-1}, t_i]$ and d_i failures in $(t_{i-1}, t_i]$ is

$$L_i(p; \text{data}_i) = \{\Pr[(T \leq C) \cap (t_{i-1} < T \leq t_i)]\}^{d_i} \{\Pr[(C \leq T) \cap (t_{i-1} < C \leq t_i)]\}^{r_i}$$

$$= \left\{ \int_{t_{i-1}}^{t_i} f_T(t)\,[1 - F_C(t)]\,dt \right\}^{d_i} \times \left\{ \int_{t_{i-1}}^{t_i} f_C(t)\,[1 - F_T(t)]\,dt \right\}^{r_i}. \quad (2.16)$$

Example 2.12 Battery Failure Data with Multiple Failure Modes. Morgan (1980) presents data from a study conducted on 68 battery cells. The purpose of the test was to determine early causes of failure, to determine which causes reduce product life the most, and to estimate failure-time distributions. Each test cell was subjected to automatic cycling (charging and discharging) at normal operating conditions. Some survived until the end of the test and others were removed before failure for physical examination. The original data giving precise times of failure or removal were not available. Instead, the data in Appendix Table C.6 provide a useful summary. By the nature of this summary, however, the removals (censoring times) do not occur at the ends of the intervals (as in the examples in Chapter 1). □

Truncated Data

In some reliability studies, observations may be *truncated*. Truncation, which is similar to but different from censoring, arises when observations are actually observed only when they take on values in a particular range. For observations that fall outside the certain range, the *existence* is not known (and this is what distinguishes truncation from censoring). Equivalently, sampling from a truncated distribution leads to truncated data. Examples and appropriate likelihood-based methods for handling truncated data, based on conditional probabilities, will be given in Section 11.6.

BIBLIOGRAPHIC NOTES

Theory for likelihood inference based on grouped and multinomial data has been given, for example, by Kulldorff (1961), Rao (1973), and Elandt-Johnson and Johnson (1980). Aalen and Husebye (1991), in a biomedical context, describe a general structure for observation stopping times that can be viewed as the cause of censoring. They explain the conditions under which the likelihood methods in this chapter are appropriate and give examples of stopping rules that could lead to biased inferences. Lagakos (1979), Kalbfleisch and Prentice (1980), and Lawless (1982, Chapter 1) also discuss these issues.

EXERCISES

2.1. Although the diesel generator fan failure times in Appendix Table C.1 were reported as exact failures, the ties suggest that the data are really discrete due to rounding or because failures were found on inspection. Suggest appropriate partitioning of the time line to reflect the true discrete nature of the data. Explain how you arrived at this partitioning. Use this partitioning to develop an expression for the discrete-data likelihood.

2.2. It is possible for a continuous cdf to be constant over some intervals of time.
   (a) Give an example of a physical situation that would result in a cdf $F(t)$ that is constant over some values of t.
   (b) Sketch such a cdf and its corresponding pdf.
   (c) For your example, explain why the convention for defining quantiles given in Section 2.1.2 is sensible. Are there alternative definitions that would also be suitable?

2.3. Consider a random variable with cdf $F(t) = t/2, 0 < t \leq 2$. Do the following:
   (a) Derive expressions for the corresponding pdf and hazard functions.
   (b) Use the results of part (a) to verify the relationship given in (2.2).
   (c) Sketch (or use the computer to draw) the cdf and pdf functions.
   (d) Sketch (or use the computer to draw) the hazard function. Give a clear intuitive reason for the behavior of $h(t)$ as $t \to 2$. *Hint:* By the time $t = 2$, all units in the population must have failed.
   (e) Derive an expression for t_p, the p quantile of $F(t)$, and use this expression to compute $t_{.4}$. Illustrate this on your plots of the cdf and pdf functions.
   (f) Compute $\Pr(.1 < T \leq .2)$ and $\Pr(.8 < T \leq .9)$. Illustrate or indicate this probabilities on your graphs.
   (g) Compute $\Pr(.1 < T \leq .2 \mid T > .1)$ and $\Pr(.8 < T \leq .9 \mid T > .8)$. Compare your answers with the approximation in (2.1).
   (h) Explain the results in part (g) and give a general result on the relationship between $\Pr(t < T \leq t + \Delta t \mid T > t)$ and the approximation in (2.1).

EXERCISES   43

2.4. Consider a cdf $F(t) = 1 - \exp[-(t/\eta)^\beta]$, $t > 0$, $\eta > 0$, $\beta > 0$. (This is the cdf of the Weibull distribution, which will be discussed in detail in Chapter 4.)
   (a) Derive an expression for the pdf $f(t)$.
   (b) Derive an expression for the hazard function $h(t)$.
   (c) Sketch (or use the computer to draw) the cdf, pdf, and hazard functions for $\eta = 1$ and $\beta = .5, 1,$ and 2.

2.5. Consider a cdf $F(t) = 1 - \exp(-t)$, $t > 0$. Do the following:
   (a) Derive expressions for the corresponding pdf and hazard functions.
   (b) Sketch (or use the computer to draw) the cdf, pdf, and hazard functions.
   (c) Derive an expression for t_p, the p quantile of $F(t)$, and use this expression to compute $t_{.1}$. Illustrate this on your plots of the cdf and pdf functions.
   (d) Compute $\Pr(.1 < T \leq .2)$. Illustrate this probability on your graphs. Also compute $\Pr(.1 < T \leq .2 \mid T > .1)$. Compare your answer with the approximation in (2.1).

▲2.6. Consider a continuous random variable with cdf $F(t) = 1 - \exp(-t)$ and the partitioning time points $t_0 = 0$, $t_1 = .1$, $t_2 = .2$, $t_3 = .5$, $t_4 = 1$, $t_5 = 2$, $t_6 = \infty$ to do the following:
   (a) Sketch (or use the computer to make) a graph of $F(t)$ over the range $0 < t \leq 10$.
   (b) Compute and make a table of values of $F(t_i)$, $S(t_i)$, and π_i, p_i at t_i for $i = 1, \ldots, 6$.

2.7. An electronic system contains 20 copies of a particular integrated circuit that is at risk to failure during operation. The manufacturer of the integrated circuit claims that its average hazard rate over the first 2 years of operation is 75 FITs. For the 1500 systems just put into operation, compute a prediction for the total number of these integrated circuits that will fail over the next 2 years of operation.

2.8. Write an expression for the likelihood of the turbine wheel data in Table 1.5. Also give an explicit expression for the constant term $\mathcal{C}$. How would the likelihood differ if we knew the actual service time of each of the inspected turbine wheels?

2.9. Consider the V7 vacuum tube data in Table 1.4.
   (a) Explain why the failures in the interval 0–25 days could be considered to be left-censored observations.
   (b) Explain why the failures in the interval 100–∞ days are right-censored observations.

2.10. Write down an expression for the likelihood of the V7 tube data in Table 1.4. Also give an explicit expression for the constant term $\mathcal{C}$.

2.11. A test facility with 20 test positions is being used to conduct a life test of a newly designed battery. Each battery will be tested until failure or until it has accumulated 100 charge/recharge cycles or until it is taken off test for some other reason. When a battery fails, it will be replaced with a new unit, keeping the 20 test positions busy. At several randomly occurring times during the test it will be necessary to remove one or more unfailed units from test. The removed units will be used for other experiments and demonstrations but will not be returned to the life test. Removed units will be treated as censored observations (all that is known is that the unit did not fail by the time the unit was removed from test). The following list suggests some methods that might be used for choosing which battery to remove from test. For each method of choosing, explain whether the censoring mechanism is "fair" or not (i.e., a censoring method that will not lead to undue bias for making inferences about the distribution of battery life). If it is a "fair" censoring method, explain why the method might be better than the other suggested methods (acknowledging that this would depend on the purpose of the test). If the selection method will result in bias, explain the direction of the bias.

- A battery selected at random.
- The battery with the most running time.
- The battery with the least running time.
- The battery with the lowest measured capacity (as measured at the end of each cycle).

2.12. Consider the life test described in Exercise 2.11. Generally, experimenters would want to assume that there would be no differences among the 20 test positions. Describe the consequences of incorrectly making such an assumption and how one could detect such differences and/or protect against such consequences.

▲2.13. Show that the pdf, cdf, survival, hazard, and cumulative hazard are mathematically equivalent descriptions of a continuous distribution in the sense that given any of these functions the other four are completely determined.

▲2.14. Consider the setting given in Section 2.2.
(a) Prove that equation (2.7) is true.
(b) Show that
$$\pi_1 = p_1,$$
$$\pi_i = p_i \prod_{j=1}^{i-1} [1 - p_j], \quad i = 2, \ldots, m,$$
$$\pi_{m+1} = \prod_{j=1}^{m} [1 - p_j].$$

EXERCISES

(c) Provide an argument to show that if $\pi_1 > 0, \ldots, \pi_{m+1} > 0$, then $0 < p_i < 1$ is the only restriction on the p_i values for $i = 1, \ldots, m$.

(d) Prove that equation (2.8) is true.

▲2.15. Consider the special case of (2.16), where $f_C(t)$ is a probability mass function assigning all of its probability to points t_i ($i = 1, \ldots, m$).

(a) Show, in this case, that (2.16) reduces to

$$L_i(p; \text{data}_i) = \{[F(t_i) - F(t_{i-1})][1 - F_C(t_{i-1})]\}^{d_i} \times \{f_C(t_i)[1 - F(t_i)]\}^{r_i}$$
$$= [f_C(t_i)]^{r_i} [1 - F_C(t_{i-1})]^{d_i} \times [F(t_i) - F(t_{i-1})]^{d_i} [1 - F(t_i)]^{r_i}.$$

(b) Give conditions under which parts of this likelihood term $L_i(p; \text{data}_i)$ can be absorbed into the likelihood constant C so that this likelihood term will correspond to the $L_i(p; \text{data}_i)$ in (2.15).

▲2.16. If a continuous random variable T has a cdf $F(t) = \Pr(T \leq t)$, then it is easy to show that the transformed random variable $F(T)$ follows a 0–1 uniform distribution. A similar property for random variables is that the cumulative hazard transformation $H(T)$ follows an exponential distribution. Show this.

CHAPTER 3

Nonparametric Estimation

Objectives

This chapter explains:

- Simple statistical methods, based on the binomial distribution, to estimate a cdf $F(t)$ from interval and singly right-censored data, without having to assume an underlying parametric distribution. This is called "nonparametric" estimation.
- Standard errors of the nonparametric estimator and approximate confidence intervals for $F(t)$.
- Life table methods to extend nonparametric estimation to allow for combinations of interval-censored and multiply right-censored data.
- The Kaplan–Meier nonparametric estimator for data with observations reported as exact failure times.
- A generalized nonparametric estimator of $F(t)$ for arbitrary censoring (including combinations and mixtures of exact failure times with left, right, and interval censoring).

Overview

The nonparametric (model-free) estimates described in this chapter are used throughout this book as a tool for reliability data analysis. Section 3.2 starts with a simple method that applies to problems with complete data or single censoring. Section 3.3 explains the basic ideas of statistical inference and introduces the ideas behind the use of confidence intervals, another statistical tool used throughout this book. Confidence intervals for complete data or single censoring are described in Section 3.4. The methods are generalized to the commonly encountered multiple censoring in Sections 3.5, 3.6, and 3.7. Simultaneous confidence bands (used for helping to choose a model in Chapter 6) are presented in Section 3.8. Sections 3.9 and 3.10 provide other general methods for handling more complicated kinds of censoring.

3.1 INTRODUCTION

As explained in Section 1.5.2, data analysis should begin with analytical and graphical tools that do not require strong model assumptions. Such methods allow the data to be interpreted without distortion that might be caused by using inadequate model assumptions. This chapter describes methods for computing nonparametric estimates and confidence intervals for $F(t)$. In some cases, such estimates are all that will be needed for an analysis. In other situations a nonparametric analysis provides an intermediate step toward a more highly structured model allowing more precise or more extensive inferences, provided that the additional assumptions of such a model are valid.

3.2 ESTIMATION FROM SINGLY CENSORED INTERVAL DATA

This section shows how to compute a nonparametric estimate of a cdf from interval-censored data when either all units fail or all of the right censoring is at one point at the end of the study (known as single right censoring).

Example 3.1 Plant 1 Heat Exchanger Data. Figure 3.1 shows the Plant 1 data from Example 1.5. As a first step in our analysis of the heat exchanger data we want to estimate $F(t)$ for just Plant 1. □

Most studies involving inspection start at time zero with an initial sample of units. Information is available on the status of the units at the end of each time interval. Let n be the initial number of units (sample size) and let d_i denote the number of units that *died* or failed in the ith interval $(t_{i-1}, t_i]$. The nonparametric estimator $\widehat{F}(t_i)$, based on the simple binomial distribution, is

$$\widehat{F}(t_i) = \frac{\text{\# of failures up to time } t_i}{n} = \frac{\sum_{j=1}^{i} d_j}{n}. \tag{3.1}$$

Example 3.2 Nonparametric Estimator of $F(t)$. The data from Plant 1 ($n = 100, d_1 = 1, d_2 = 2, d_3 = 2$) give

$$\widehat{F}(1) = 1/100, \quad \widehat{F}(2) = 3/100, \quad \widehat{F}(3) = 5/100.$$

This estimate is shown graphically with the • symbols in Figure 3.2. □

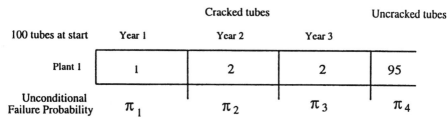

Figure 3.1. Plant 1 heat exchanger data.

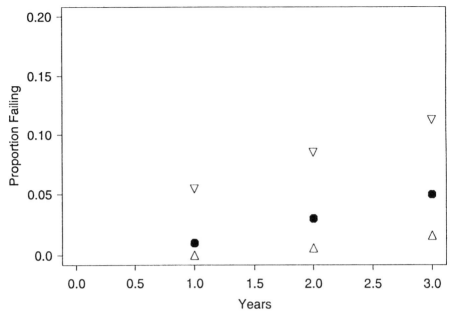

Figure 3.2. Plot of the nonparametric estimate for the Plant 1 heat exchanger data with pointwise approximate 95% confidence intervals based on binomial theory.

In general, this nonparametric estimator $\widehat{F}(t_i)$ is defined at all values of t_i (upper endpoints of all intervals). Additionally, if interval i is known to have no failures, then $\widehat{F}(t) = \widehat{F}(t_{i-1})$ for $t_{i-1} \leq t \leq t_i$. If interval i is known to contain one or more failures, $\widehat{F}(t)$ increases from $\widehat{F}(t_{i-1})$ to $\widehat{F}(t_i)$ in the interval $(t_{i-1}, t_i]$. In this case $\widehat{F}(t)$ is undefined for $t_{i-1} < t < t_i$. Intuitively, this is because we do not know the exact location of the failure(s) within the interval $(t_{i-1}, t_i]$ and thus we have no information on how $\widehat{F}(t)$ is increasing in the interval. By using the binomial distribution, it is easy to show that $\widehat{F}(t_i)$ is the maximum likelihood estimator of $F(t)$.

3.3 BASIC IDEAS OF STATISTICAL INFERENCE

3.3.1 The Sampling Distribution of $\widehat{F}(t_i)$

Estimates like $\widehat{F}(t_i), i = 1, \ldots, m$, computed from a set of sample data can be interpreted in at least two different ways.

- Estimates can be viewed as descriptive of the *particular data set* used to compute the estimate. This is known as "descriptive statistics."
- More commonly, there is interest *beyond* the particular sample units and estimates are used to make inferences about the process or larger existing population of units from which the sample units were chosen at random. This is an example of "inferential statistics."

BASIC IDEAS OF STATISTICAL INFERENCE

In inferential applications, an estimate [say, $\widehat{F}(t_i)$ at a particular time t_i] will deviate from $F(t_i)$, the actual population or process cdf at t_i. The standard (non-Bayesian[1]) approach to quantifying the possible size of the difference between $\widehat{F}(t_i)$ and $F(t_i)$ is to consider what would happen if the inferential *procedure* (sampling and estimation) were repeated a large number of times, each time getting different data and thus a different $\widehat{F}(t_i)$. The distribution of $\widehat{F}(t_i)$ values is called a sampling distribution, and this distribution provides insight into the probable deviation between $\widehat{F}(t_i)$ and $F(t_i)$.

3.3.2 Confidence Intervals

A point estimate, by itself, can be misleading, as it may or may not be close to the quantity being estimated. Confidence intervals are one of the most useful ways of quantifying uncertainty due to "sampling error" arising from limited sample sizes. Confidence intervals, however, generally do not quantify possible errors and biases arising from an inadequate model or other invalid model assumptions.

Confidence intervals have a specified "level of confidence," typically 90% or 95%, expressing one's confidence (*not probability*) that a specific interval contains the quantity of interest. A specific interval either contains the quantity of interest or not; the truth is unknown. It is important to recognize that the confidence level pertains to a probability statement about the performance of the confidence interval *procedure* rather than a statement about any particular interval. See Chapter 2 of Hahn and Meeker (1991) for further discussion of the interpretation of statistical intervals.

"Coverage probability" is the probability that a confidence interval *procedure* will result in an interval containing the quantity of interest. When the specified level of confidence [generically $100(1 - \alpha)\%$] is not equal to the coverage probability, the procedure and resulting intervals are said to be *approximate*. In some simple problems coverage probability for a given procedure can be computed analytically and, correspondingly, *exact* confidence interval methods can be developed. "Conservative" procedures have a coverage probability that is *at least* as large as the specified confidence level. In most practical problems involving censored data, there are no "exact" confidence interval procedures. There are, however, a number of useful, and often simple, approximate methods. Better approximations generally require more computations. The adequacy of the approximations can be checked with repeated simulation. In turn, often simulation can be used to obtain better approximations, as described in Chapter 9. Section 3.6 describes simple methods for computing approximate confidence intervals. Chapter 9 presents a more accurate simulation-based approach.

---

[1] Bayesian methods of statistical inference, described in Chapter 14, are based on the specification of a prior distribution to describe prior knowledge or opinion about the model parameters. As explained in Chapter 14, this alternative approach to inference leads to inference statements with a somewhat different interpretation.

3.4 CONFIDENCE INTERVALS FROM COMPLETE OR SINGLY CENSORED DATA

3.4.1 Pointwise Binomial-Based Confidence Interval for $F(t_i)$

A conservative $100(1 - \alpha)\%$ confidence interval $[\underline{F}(t_i), \widetilde{F}(t_i)]$ for $F(t_i)$ based on binomial sampling is

$$\underline{F}(t_i) = \left\{ 1 + \frac{(n - n\widehat{F} + 1)\mathcal{F}_{(1-\alpha/2;\, 2n-2n\widehat{F}+2,\, 2n\widehat{F})}}{n\widehat{F}} \right\}^{-1}, \qquad (3.2)$$

$$\widetilde{F}(t_i) = \left\{ 1 + \frac{n - n\widehat{F}}{(n\widehat{F} + 1)\mathcal{F}_{(1-\alpha/2;\, 2n\widehat{F}+2,\, 2n-2n\widehat{F})}} \right\}^{-1},$$

where $\widehat{F} = \widehat{F}(t_i)$ and $\mathcal{F}_{(p;\nu_1,\nu_2)}$ is the p quantile of the F distribution with (ν_1, ν_2) degrees of freedom. Elementary statistics textbooks provide tables of F distribution quantiles. The confidence interval in (3.2) is conservative in the sense that the coverage probability is greater than or equal to $1 - \alpha$. Theory for this confidence interval is given, for example, in Brownlee (1960, pages 119–120).

A one-sided approximate $100(1 - \alpha)\%$ confidence bound for $F(t)$ can be obtained by replacing $\mathcal{F}_{(1-\alpha/2)}$ with $\mathcal{F}_{(1-\alpha)}$ in (3.2) and using the appropriate endpoint of the two-sided confidence interval. For example, a conservative 90% confidence interval from (3.2) can also be viewed as two conservative 95% one-sided confidence bounds.

Example 3.3 Binomial Confidence Interval for F(t). To illustrate the computation of the binomial confidence intervals from (3.2), we compute a 95% confidence interval for $F(2)$ at which point 3 of the 100 tubes had failed. Then with $n = 100$ and $n\widehat{F}(2) = 3$, $\mathcal{F}_{(.975;196,6)} = 4.8831$, and $\mathcal{F}_{(.975;8,194)} = 2.2578$, substituting into (3.2) gives

$$\underline{F}(2) = \left\{ 1 + \frac{(100 - 3 + 1)\mathcal{F}_{(.975;200-6+2,6)}}{3} \right\}^{-1} = .0062,$$

$$\widetilde{F}(2) = \left\{ 1 + \frac{100 - 3}{(3 + 1)\mathcal{F}_{(.975;6+2,200-6)}} \right\}^{-1} = .0852.$$

Thus we are (at least) 95% confident that the probability of failing before the end of 2 years is between .0062 and .0852. □

3.4.2 Pointwise Normal-Approximation Confidence Interval for $F(t_i)$

For a specified value of t_i, a simpler approximate $100(1 - \alpha)\%$ confidence interval for $F(t_i)$ is

$$[\underline{F}(t_i), \widetilde{F}(t_i)] = \widehat{F}(t_i) \pm z_{(1-\alpha/2)} \widehat{\text{se}}_{\widehat{F}}, \qquad (3.3)$$

where $z_{(p)}$ is the p quantile of the standard normal distribution and

$$\widehat{\text{se}}_{\widehat{F}} = \sqrt{\widehat{F}(t_i)[1 - \widehat{F}(t_i)]/n} \qquad (3.4)$$

CONFIDENCE INTERVALS FROM COMPLETE OR SINGLY CENSORED DATA

is an estimate of the standard error of $\widehat{F}(t_i)$. This confidence interval is based on the assumption that

$$Z_{\widehat{F}} = \frac{\widehat{F}(t_i) - F(t_i)}{\widehat{se}_{\widehat{F}}}$$

can be adequately approximated by a NOR(0, 1) (standard normal) distribution. For this approximation to be adequate, $n\widehat{F}(t_i)$ should be at least 5 to 10 and no more than $n - 5$ or $n - 10$. Otherwise the approximation will be crude, and it is even possible to get confidence limits that are outside the interval 0 to 1. The computations for (3.3) are, however, simple and can be done easily by hand.

Example 3.4 *Normal-Approximation Confidence Interval for the Plant 1 F(t).*
To illustrate these intervals we compute an approximate 95% confidence interval for $F(3)$ for the Plant 1 heat exchanger tubes. At $t = 3$, 5 of the 100 tubes had failed. Then with $n = 100$, $\widehat{F}(3) = .05$, and $z_{.975} = 1.960$, substituting into (3.3) gives

$$[\underline{F}(3), \widetilde{F}(3)] = .05 \pm 1.960 \times .02179 = [.0073, \quad .0927],$$

where $\widehat{se}_{\widehat{F}} = \sqrt{.05(1 - .05)/100} = .02179$ is an estimate of the standard error of $\widehat{F}(t_3)$.

Table 3.1 also shows confidence intervals for $F(1)$ and $F(2)$ (both the conservative and the normal-approximation). Note that the normal-approximation intervals for

Table 3.1. Nonparametric Estimates and Approximate Confidence Intervals for the Heat Exchanger Tube $F(t)$

					Pointwise Approximate Confidence Intervals	
Year	t_i	d_i	$\widehat{F}(t_i)$	$\widehat{se}_{\widehat{F}}$	$\underline{F}(t_i)$	$\widetilde{F}(t_i)$
(0–1]	1	1	.01	.00995		
95% Confidence Intervals for $F(1)$						
Based on binomial theory					[.0003, .0545]	
Based on $Z_{\widehat{F}} \stackrel{.}{\sim} \text{NOR}(0, 1)$					[−.0095, .0295]	
(1–2]	2	2	.03	.01706		
95% Confidence Intervals for $F(2)$						
Based on binomial theory					[.0062, .0852]	
Based on $Z_{\widehat{F}} \stackrel{.}{\sim} \text{NOR}(0, 1)$					[−.0034, .0634]	
(2–3]	3	2	.05	.02179		
95% Confidence Intervals for $F(3)$						
Based on binomial theory					[.0164, .1128]	
Based on $Z_{\widehat{F}} \stackrel{.}{\sim} \text{NOR}(0, 1)$					[.0073, .0927]	

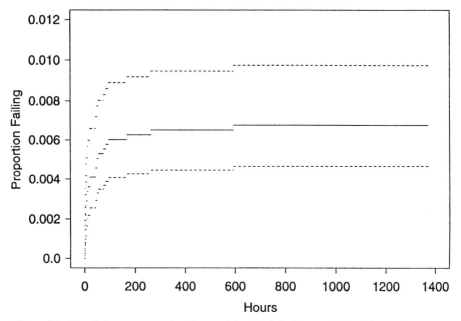

Figure 3.3. Plot of the nonparametric estimate of the cdf of the integrated circuit failure times. Also shown are a set of normal-approximation 95% confidence intervals.

$F(1)$ and $F(2)$ have nonsensical negative lower endpoints. Section 3.6 provides a method of improving the normal-approximation interval for situations in which the conservative binomial interval does not apply. □

Example 3.5 Nonparametric Confidence Intervals for Integrated Circuit Failure Data. The integrated circuit failure data in Table 1.2 are singly censored at 1370 hours. Ties in the data suggest that failures were found at points in times where there was an inspection. For this example, however, as is commonly done in practice when the intervals are small relative to the spread in the data, this discreteness in the data will be ignored. Thus the simple binomial methods can be used to estimate $F(t)$ at any specified value of t. Figure 3.3 shows the nonparametric estimate along with normal-approximation 95% confidence intervals for each estimated point. The normal-approximation intervals are nonsymmetric because they are based on an approximation-improving transformation explained in Section 3.6. □

3.5 ESTIMATION FROM MULTIPLY CENSORED DATA

This section shows how to compute a nonparametric estimate of a cdf from data with multiple right censoring (failures occur after some units have been censored). Suppose that an initial sample of n units start operating at time zero. If a unit does not fail in interval i, it is either censored at the end of interval i or it continues into interval $i + 1$. Information is available on the status of the units at the end of each interval. The intervals may be large or small and need not be of equal length, as long as the intervals for different units do not overlap (Section 3.10 extends the method to data with different, overlapping intervals, which arise, for example, when units are not subject to the same inspection schedule). Let d_i denote the number of units that *died* or failed in the ith interval $(t_{i-1}, t_i]$. Also, let r_i denote the number of units that survive interval i and are *right*-censored at t_i. The units that are alive at the beginning of interval i are called the "risk set" for interval i (i.e., those at risk to failure) and the size of this risk set at the beginning of interval i is

$$n_i = n - \sum_{j=0}^{i-1} d_j - \sum_{j=0}^{i-1} r_j, \quad i = 1, \ldots, m, \quad (3.5)$$

where m is the number of intervals and it is understood that $d_0 = 0$ and $r_0 = 0$. An estimator of the conditional probability of failing in interval i, given that a unit enters this interval, is the sample proportion failing

$$\widehat{p}_i = \frac{d_i}{n_i}, \quad i = 1, \ldots, m.$$

Substituting these into (2.8) provides an estimator of the survival function:

$$\widehat{S}(t_i) = \prod_{j=1}^{i} [1 - \widehat{p}_j], \quad i = 1, \ldots, m. \quad (3.6)$$

Then the corresponding nonparametric estimator of $F(t_i)$ is

$$\widehat{F}(t_i) = 1 - \widehat{S}(t_i), \quad i = 1, \ldots, m. \quad (3.7)$$

Here $\widehat{p}_i$ is the maximum likelihood (ML) estimator of the conditional probability p_i from (2.7). This implies that $\widehat{F}(t_i)$ is the ML estimator of $F(t_i)$ (see Exercise 3.22). The nonparametric estimator $\widehat{F}(t_i)$ is defined at all t_i values (endpoints of all observation intervals). Additionally, if interval i is known to have zero failures, then $\widehat{F}(t_i) = \widehat{F}(t_{i-1})$ for $t_{i-1} \leq t \leq t_i$. If interval i is known to contain one or more failures, $\widehat{F}(t)$ increases from $\widehat{F}(t_{i-1})$ to $\widehat{F}(t_i)$ in the interval $(t_{i-1}, t_i]$ but, as before, $\widehat{F}(t)$ is not defined over the interval. Note that when there are no censored observations before the last failure, (3.7) is numerically equivalent to (3.1).

Example 3.6 *Nonparametric Estimate of $F(t)$ for the Pooled Heat Exchanger Tube Data.* Returning to the heat exchanger data from Example 1.5, Figure 3.4 displays the pooled data from Figure 1.7 across the three different plants. For each year of operation, the bottom of Figure 3.4 shows the number in the risk set (in the upper left-hand corner), the number that cracked in each interval, and the number censored at the end of each interval. Table 3.2 illustrates numerical computations. Figures 3.5 and 3.8 show the estimate with different sets of approximate confidence intervals and bands that are explained in Sections 3.6.3 and 3.8, respectively. □

54 NONPARAMETRIC ESTIMATION

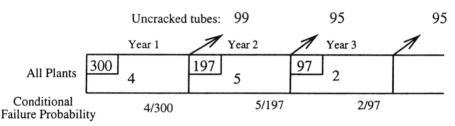

Figure 3.4. Pooling of the heat exchanger data in preparation for computing the nonparametric estimate of $F(t)$. The number of units at risk in each cell is shown in the small rectangles.

Table 3.2. Calculations for the Nonparametric Estimate of $F(t_i)$ for the Pooled Heat Exchanger Tube Data

Year	t_i	Failed d_i	Censored r_i	Entered n_i	$\widehat{p}_i$	$1 - \widehat{p}_i$	$\widehat{S}(t_i)$	$\widehat{F}(t_i)$
(0–1]	1	4	99	300	4/300	296/300	.9867	.0133
(1–2]	2	5	95	197	5/197	192/197	.9616	.0384
(2–3]	3	2	95	97	2/97	95/97	.9418	.0582

3.6 POINTWISE CONFIDENCE INTERVALS FROM MULTIPLY CENSORED DATA

3.6.1 Approximate Variance of $\widehat{F}(t_i)$

Because $\widehat{F}(t_i) = 1 - \widehat{S}(t_i)$, $\text{Var}[\widehat{F}(t_i)] = \text{Var}[\widehat{S}(t_i)]$. Using the delta method approach in Appendix Section B.2, a first-order Taylor series approximation for $\widehat{S}(t_i)$ is

$$\widehat{S}(t_i) \approx S(t_i) + \sum_{j=1}^{i} \left.\frac{\partial S}{\partial q_j}\right|_{q_j} (\widehat{q}_j - q_j),$$

where $q_j = 1 - p_j$. Because the $\widehat{q}_j$ values are approximately uncorrelated binomial proportions (the $\widehat{q}_j$ values are asymptotically, as $n \to \infty$, uncorrelated), it follows that

$$\text{Var}[\widehat{F}(t_i)] = \text{Var}[\widehat{S}(t_i)] \approx [S(t_i)]^2 \sum_{j=1}^{i} \frac{p_j}{n_j(1-p_j)}. \tag{3.8}$$

The right-hand side of (3.8) is also an asymptotic (large-sample approximate) variance denoted by $\text{Avar}[\widehat{F}(t_i)]$. This can be shown by using the large-sample approximation in Appendix Sections B.6.1 and B.6.3.

3.6.2 Greenwood's Formula

Substituting $\widehat{p}_j$ for p_j and $\widehat{S}(t_i)$ for $S(t_i)$ in (3.8) gives the following variance estimator:

$$\widehat{\text{Var}}[\widehat{F}(t_i)] = \widehat{\text{Var}}[\widehat{S}(t_i)] = [\widehat{S}(t_i)]^2 \sum_{j=1}^{i} \frac{\widehat{p}_j}{n_j(1-\widehat{p}_j)}. \tag{3.9}$$

This is known as "Greenwood's formula." An estimator of the standard error of $\widehat{F}(t_i)$ is

$$\widehat{\text{se}}_{\widehat{F}} = \sqrt{\widehat{\text{Var}}[\widehat{F}(t_i)]}. \tag{3.10}$$

Note that when there are no censored observations before t_i, (3.10) is numerically equivalent to (3.4).

3.6.3 Pointwise Normal-Approximation Confidence Interval for $F(t_i)$

Because $\widehat{F}(t)$ is defined only at the upper endpoint of intervals that contain failures, $\widehat{F}(t)$ is generally estimated only at such points [if there are no failures in an interval, $\widehat{F}(t)$ remains constant over that interval]. For a specified upper endpoint t_i at which an estimate of $F(t)$ is desired, a normal-approximation $100(1-\alpha)\%$ confidence interval for $F(t_i)$ is

$$[\underaccent{\tilde}{F}(t_i), \widetilde{F}(t_i)] = \widehat{F}(t_i) \pm z_{(1-\alpha/2)} \widehat{\text{se}}_{\widehat{F}}, \tag{3.11}$$

where $z_{(p)}$ is the p quantile of the standard normal distribution. In general, a one-sided approximate $100(1-\alpha)\%$ confidence bound can be obtained by replacing $z_{(1-\alpha/2)}$ with $z_{(1-\alpha)}$ and using the appropriate endpoint of the two-sided confidence interval.

The approximate confidence intervals from (3.3) or (3.11) are based on the assumption that the distribution of

$$Z_{\widehat{F}} = \frac{\widehat{F}(t_i) - F(t_i)}{\widehat{\text{se}}_{\widehat{F}}} \tag{3.12}$$

can be approximated adequately by a NOR(0, 1) distribution. Then

$$\Pr[z_{(\alpha/2)} < Z_{\widehat{F}} \le z_{(1-\alpha/2)}] \approx 1 - \alpha \qquad (3.13)$$

implies that

$$\Pr[\widehat{F}(t_i) - z_{(1-\alpha/2)}\widehat{\operatorname{se}}_{\widehat{F}} < F(t_i) \le \widehat{F}(t_i) - z_{(\alpha/2)}\widehat{\operatorname{se}}_{\widehat{F}}] \approx 1 - \alpha. \qquad (3.14)$$

This gives the approximate coverage probability for intervals computed with the procedure in (3.11). Note that in (3.14), $F(t_i)$ is fixed while $\widehat{F}(t_i)$ and $\widehat{\operatorname{se}}_{\widehat{F}}$ are random. The approximation in (3.14) is a large-sample approximation and improves with increasing sample size. Appendix Section B.5 provides more information on such large-sample approximations.

When the sample size is not large, however, the distribution of $Z_{\widehat{F}}$ may be badly skewed and the normal distribution may not provide an adequate approximation, particularly in the tails of the distribution [where $\widehat{F}(t)$ is close to 0 or 1]. For example, it is possible that (3.11) gives $\utilde{F}(t) < 0$ or $\widetilde{F}(t) > 1$, a result that is outside the possible range for $F(t)$. Generally a better approximation might be obtained by using the logit transformation ($\operatorname{logit}(p) = \log[p/(1 - p)]$) and basing the confidence intervals on the distribution of

$$Z_{\operatorname{logit}(\widehat{F})} = \frac{\operatorname{logit}[\widehat{F}(t_i)] - \operatorname{logit}[F(t_i)]}{\widehat{\operatorname{se}}_{\operatorname{logit}(\widehat{F})}}. \qquad (3.15)$$

Because $\operatorname{logit}[\widehat{F}(t_i)]$, like a standard normal random variable, is unrestricted (i.e., ranges between $-\infty$ and ∞), (3.15) can be expected to be closer to NOR(0, 1) than (3.12). This leads (the needed steps are left as an exercise) to the two-sided approximate $100(1 - \alpha)\%$ confidence interval

$$[\utilde{F}(t_i), \widetilde{F}(t_i)] = \left[\frac{\widehat{F}}{\widehat{F} + (1 - \widehat{F}) \times w}, \frac{\widehat{F}}{\widehat{F} + (1 - \widehat{F})/w}\right], \qquad (3.16)$$

where $w = \exp\{z_{(1-\alpha/2)}\widehat{\operatorname{se}}_{\widehat{F}}/[\widehat{F}(1 - \widehat{F})]\}$. The endpoints of this interval will always lie between 0 and 1. A one-sided approximate $100(1 - \alpha)\%$ confidence bound can be obtained by replacing $z_{(1-\alpha/2)}$ with $z_{(1-\alpha)}$ and using the appropriate endpoint of the two-sided confidence interval.

Example 3.7 *Normal-Approximation Confidence Intervals for the Heat Exchanger Data.* This example illustrates the computation of standard errors and nonparametric approximate confidence intervals for the heat exchanger data, using both large-sample approximations in this section. For the failure probability at $t_i = t_1 = 1$, we have $\widehat{F}(1) = .0133$ and

$$\widehat{\operatorname{Var}}[\widehat{F}(1)] = (.9867)^2 \left[\frac{.0133}{300(.9867)}\right] = .0000438.$$

Then $\widehat{\text{se}}_{\widehat{F}} = \sqrt{.0000438} = .00662$ and the approximate 95% confidence interval for $F(1)$ from (3.11) is

$$[\underline{F}(1), \widetilde{F}(1)] = .0133 \pm 1.960(.00662) = [.0003, .0263].$$

The corresponding interval from (3.16), based on the logit transformation, is

$$[\underline{F}(1), \widetilde{F}(1)] = \left[\frac{.0133}{.0133 + (1 - .0133) \times w}, \frac{.0133}{.0133 + (1 - .0133)/w} \right]$$
$$= [.0050, .0350],$$

where $w = \exp\{1.960(.00662)/[.0133(1 - .0133)]\} = 2.687816$. Differences between the two methods are large enough to be of practical importance. The intervals based on the logit transformation are expected to provide a better approximation to the nominal 95% confidence level.

For the failure probability at $t_i = t_2 = 2$, we have

$$\widehat{\text{Var}}[\widehat{F}(2)] = (.9616)^2 \left[\frac{.0133}{300(.9867)} + \frac{.0254}{197(.9746)} \right] = .0001639$$

so that $\widehat{\text{se}}_{\widehat{F}} = \sqrt{.0001639} = .0128$. The approximate 95% confidence intervals for $F(2)$ are

$$[\underline{F}(2), \widetilde{F}(2)] = .0384 \pm 1.960(.0128) = [.0133, .0635]$$

and

$$[\underline{F}(2), \widetilde{F}(2)] = \left[\frac{.0384}{.0384 + (1 - .0384) \times w}, \frac{.0384}{.0384 + (1 - .0384)/w} \right]$$
$$= [.0198, .0730],$$

where $w = \exp\{1.960(.0128)/[.0384(1 - .0384)]\} = 1.972739$.

Table 3.3 gives and Figure 3.5 shows the nonparametric estimates for $F(t_i)$ and pointwise approximate 95% confidence intervals. The intervals are not symmetric around the estimates because of the logit transformation. The intervals are wide because of the heavy censoring and the small number of failures. □

3.7 ESTIMATION FROM MULTIPLY CENSORED DATA WITH EXACT FAILURES

Failures are often reported at exact times. In such cases, the reported times are denoted by t_i. This section shows how to apply the methods of Sections 3.5 and 3.6 to estimate $F(t)$ for such exact failures.

Table 3.3. Summary of Calculations for Nonparametric Approximate Confidence Intervals for $F(t)$ for the Pooled Heat Exchanger Tube Data

Year	t_i	$\widehat{F}(t_i)$	$\widehat{\text{se}}_{\widehat{F}}$	Pointwise Confidence Intervals
(0–1]	1	.0133	.00662	
95% Confidence Intervals for F(1)				
Based on $Z_{\text{logit}(\widehat{F})} \stackrel{.}{\sim} \text{NOR}(0,1)$				[.0050, .0350]
Based on $Z_{\widehat{F}} \stackrel{.}{\sim} \text{NOR}(0,1)$				[.0004, .0133]
(1–2]	2	.0384	.0128	
95% Confidence Intervals for F(2)				
Based on $Z_{\text{logit}(\widehat{F})} \stackrel{.}{\sim} \text{NOR}(0,1)$				[.0198, .0730]
Based on $Z_{\widehat{F}} \stackrel{.}{\sim} \text{NOR}(0,1)$				[.0133, .0635]
(2–3]	3	.0582	.0187	
95% Confidence Intervals for F(3)				
Based on $Z_{\text{logit}(\widehat{F})} \stackrel{.}{\sim} \text{NOR}(0,1)$				[.0307, .1076]
Based on $Z_{\widehat{F}} \stackrel{.}{\sim} \text{NOR}(0,1)$				[.0216, .0949]

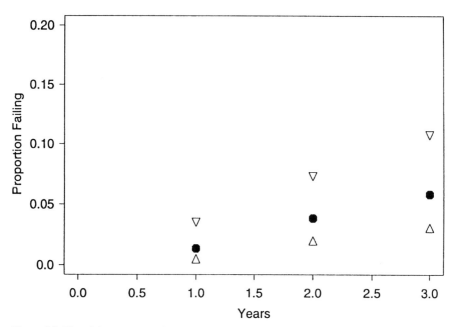

Figure 3.5. Plot of the nonparametric estimate for the heat exchanger data along with a set of pointwise logistic-transform normal-approximation 95% confidence intervals for $F(t)$.

Example 3.8 Shock Absorber Failure Data. Appendix Table C.2 gives the failure times (in number of kilometers of use) of vehicle shock absorbers, first reported in O'Connor (1985). The table shows two different failure modes, denoted by M1 and M2. Engineers responsible for shock absorber manufacturing and reliability would be interested in the distribution of time to failure for the individual failure modes. Engineers responsible for higher-level automobile system reliability and choosing among alternative vendors would be more interested in the overall failure-time distribution for the part. □

Exact failure times arise from a continuous inspection process (or, perhaps, from having used a very large number of closely-spaced inspections). In the limit, as the number of inspections increases and the width of the inspection intervals approaches zero, failures are concentrated in a relatively small number of intervals. Most intervals will not contain any failures. $\widehat{F}(t)$ is *constant* over all intervals that have no failures. Thus with small intervals, $\widehat{F}$ will become a step function with gaps over the intervals where there were failures and with jumps at the upper endpoint of these intervals. In the limit, as the width of the intervals approaches 0, the size of the gaps approaches 0 and the step function increases at the reported failure times. This limiting case of the interval-based nonparametric estimator is generally known as the product-limit or Kaplan–Meier estimator.

Example 3.9 Nonparametric Estimator and Normal-Approximation Confidence Intervals for the Shock Absorber Data. For the data from Example 3.8 and Appendix Table C.2 we do not differentiate between the two different failure modes. Instead, we estimate the time to failure when both mode M1 and M2 are acting. Table 3.4 illustrates the computations for the product-limit estimator up to 12,200

Table 3.4. Nonparametric Estimates for the Shock Absorber Data up to 12,200 km

t_i (km)	Failed d_j	Censored r_j	Entered n_j	$\widehat{p}_j$	$1 - \widehat{p}_j$	$\widehat{S}(t_i)$	$\widehat{F}(t_i)$
6,700	1	0	38	1/38	37/38	.9737	.0263
6,950	0	1	37				
7,820	0	1	36				
8,790	0	1	35				
9,120	1	0	34	1/34	33/34	.9451	.0549
9,660	0	1	33				
9,820	0	1	32				
11,310	0	1	31				
11,690	0	1	30				
11,850	0	1	29				
11,880	0	1	28				
12,140	0	1	27				
12,200	1	0	26	1/26	25/26	.9087	.0913

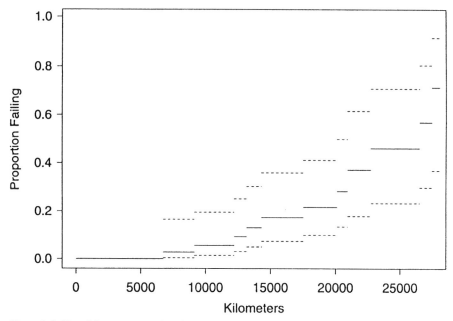

Figure 3.6. Plot of the nonparametric estimate for the shock absorber data along with a set of pointwise logistic-transform normal-approximation 95% confidence intervals for $F(t)$.

km. Figure 3.6 shows the nonparametric estimator and a set of pointwise approximate 95% confidence intervals for $F(t)$. Estimated standard errors were computed using (3.10) and the confidence intervals were computed using the logistic-transformation method in Section 3.6.3. The wide confidence intervals indicate a high degree of uncertainty. □

3.8 SIMULTANEOUS CONFIDENCE BANDS

3.8.1 Motivation

The *pointwise* confidence intervals defined in Section 3.6.3 are useful for making a statement about $F(t_i)$ at *one* particular specified value of t_i (even though it is common practice to plot a set of such intervals). In many applications, however, it is necessary to quantify the sampling uncertainty, simultaneously, over a range of values of t. To do this, we can use simultaneous confidence bands for $F(t)$. As explained in Chapter 6, simultaneous confidence bands are particularly useful for judging the magnitude of observed departures from fitted parametric models.

The overall coverage probability for the collection of pointwise intervals (e.g., Figure 3.6) is generally less than that for any individual interval. Plotting an estimate of $F(t)$ showing imultaneous confidence bands more accurately reflects the uncertainty over the range of times displayed on the plot.

3.8.2 Large-Sample Simultaneous Confidence Bands for $F(t)$

Approximate $100(1 - \alpha)\%$ simultaneous confidence bands for $F(t)$ can be obtained from

$$\left[\tilde{F}(t), \tilde{\tilde{F}}(t)\right] = \widehat{F}(t) \pm e_{(a,b,1-\alpha/2)} \widehat{\text{se}}_{\widehat{F}(t)} \quad \text{for all} \quad t \in [t_L(a), t_U(b)], \quad (3.17)$$

where the range $[t_L(a), t_U(b)]$ is a complicated function of the censoring pattern in the data, as described in Section 3.8.3. With no censoring the range of t is given by the values of t for which $a \leq \widehat{F}(t) \leq b$. The approximate factors $e_{(a,b,1-\alpha/2)}$ given in Table 3.5 were computed from a large-sample approximation given in Nair (1984). Because the factor $e_{(a,b,1-\alpha/2)}$ is the same for all values of t, this family of bands is known as the "equal precision" or "EP" simultaneous confidence bands. The factors $e_{(a,b,1-\alpha/2)}$ are larger than the corresponding pointwise normal-approximation $z_{(1-\alpha/2)}$ values. Thus the width of the simultaneous bands, at any given point t, is wider than the corresponding pointwise confidence interval at that point. This is as expected (and necessary) to account for the simultaneous nature of the bands afforded by (3.17).

Simultaneous approximate confidence bands like those defined in (3.17) are based on the approximate distribution of

$$Z_{\max \widehat{F}} = \max_{t \in [t_L(a), t_U(b)]} \left[\frac{\widehat{F}(t) - F(t)}{\widehat{\text{se}}_{\widehat{F}(t)}}\right]. \quad (3.18)$$

As explained in Section 3.6.3 for the pointwise confidence intervals, it is generally better to compute the simultaneous confidence bands based on the logit transformation

Table 3.5. Factors $e_{(a,b,1-\alpha/2)}$ for the EP Simultaneous Approximate Confidence Bands

Limits		Confidence Level			
a	b	.80	.90	.95	.99
.005	.995	2.86	3.12	3.36	3.85
.01	.995	2.84	3.10	3.34	3.83
.05	.995	2.76	3.03	3.28	3.77
.1	.995	2.72	3.00	3.25	3.75
.005	.99	2.84	3.10	3.34	3.83
.01	.99	2.81	3.07	3.31	3.81
.05	.99	2.73	3.00	3.25	3.75
.1	.99	2.68	2.96	3.21	3.72
.005	.95	2.76	3.03	3.28	3.77
.01	.95	2.73	3.00	3.25	3.75
.05	.95	2.62	2.91	3.16	3.68
.1	.95	2.56	2.85	3.11	3.64
.005	.9	2.72	3.00	3.25	3.75
.01	.9	2.68	2.96	3.21	3.72
.05	.9	2.56	2.85	3.11	3.64
.1	.9	2.48	2.79	3.06	3.59

of $\widehat{F}$. These can be computed from

$$\left[\underset{\approx}{F}(t), \widetilde{F}(t)\right] = \left[\frac{\widehat{F}(t)}{\widehat{F}(t) + [1-\widehat{F}(t)] \times w}, \frac{\widehat{F}(t)}{\widehat{F}(t) + [1-\widehat{F}(t)]/w}\right], \quad (3.19)$$

where $w = \exp\{e_{(a,b,1-\alpha/2)}\widehat{\mathrm{se}}_{\widehat{F}}/[\widehat{F}(1-\widehat{F})]\}$. The endpoints of these bands will always lie between 0 and 1. The bands computed from (3.19) are based on the approximate distribution of the random function

$$Z_{\max \mathrm{logit}(\widehat{F})} = \max_{t \in [t_L(a), t_U(b)]} \left[\frac{\mathrm{logit}[\widehat{F}(t)] - \mathrm{logit}[F(t)]}{\widehat{\mathrm{se}}_{\mathrm{logit}[\widehat{F}(t)]}}\right]. \quad (3.20)$$

The bands that we have computed for our examples use this approximation.

3.8.3 Determining the Time Range for Simultaneous Confidence Bands for $F(t)$

Specifying the quantities a and b determines the range $[t_L(a), t_U(b)]$ over which simultaneous confidence bands for $F(t)$ are defined. Let

$$\widehat{\sigma}(t) = n \sum_{j: t_j \le t} \frac{d_j}{n_j(n_j - d_j)} \quad \text{and} \quad \widehat{K}(t) = \frac{\widehat{\sigma}(t)}{1 + \widehat{\sigma}(t)},$$

where the summation is over j such that $t_j \le t$. Then the simultaneous confidence bands have a range covering all values of t such that $a \le \widehat{K}(t) \le b$. The function $\widehat{K}(t)$ behaves like a nonparametric estimate $\widehat{F}(t)$: it is nondecreasing, $0 \le \widehat{K}(t) \le 1$, and $\widehat{K}(t) = \widehat{F}(t)$ when there is no censoring.

Example 3.10 Simultaneous Confidence Bands for the Shock Absorber Life cdf. Figure 3.7 is similar to Figure 3.6, but it displays approximate 95% simultaneous confidence bands for $F(t)$ instead of a set of pointwise confidence intervals. Note that the upper limit of the simultaneous confidence bands is constant from 6700 km to 12,200 km, even though there is a failure and a corresponding jump in $\widehat{F}(t)$ at 9120 km. This is due to an adjustment made to (3.19) so that the simultaneous confidence bands for $F(t)$ do not decrease, thus agreeing with the nondecreasing characteristic of cdfs. If the upper band is decreasing on the left, it is made flat from $t_L(a)$ to the point of the minimum. If the lower band is decreasing on the right, it is made flat from the point of maximum to $t_U(b)$. These adjustments, if needed, give tighter, more sensible bands and have no effect on the actual coverage probability of the simultaneous bands. □

Example 3.11 Simultaneous Confidence Bands for the Pooled Heat Exchanger Tube Data. Figure 3.8 for the heat exchanger data shows simultaneous confidence bands. As expected, the simultaneous bands are much wider than the set of pointwise confidence intervals in Figure 3.5. □

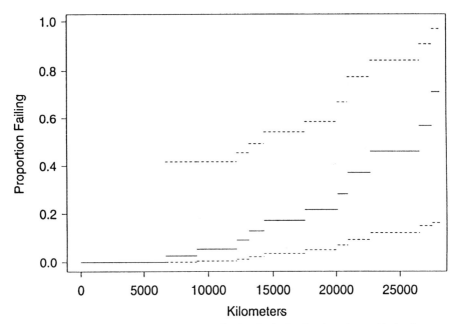

Figure 3.7. Plot of the nonparametric estimate for the shock absorber data along with simultaneous logistic-transform normal-approximation 95% confidence bands for $F(t)$ computed from (3.19).

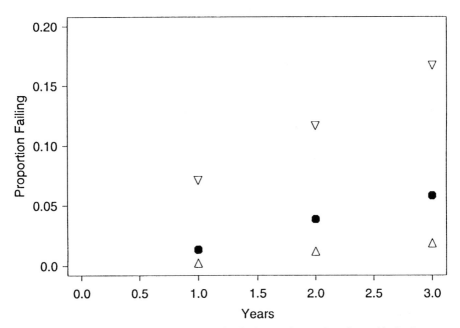

Figure 3.8. Plot of the nonparametric estimate for the heat exchanger data along with simultaneous logistic-transform normal-approximation 95% confidence intervals for $F(t)$ computed from (3.19).

3.9 UNCERTAIN CENSORING TIMES

The methods in earlier sections of this chapter assume that all left- and right-censored observations occur at upper endpoints of the intervals. If all of the censoring times are *known*, this is not a serious restriction because the data intervals can be defined to accommodate all of the data. If, however, censoring times are known only to be *within* specified intervals, the risk set is decreasing over the intervals in a manner that cannot be specified precisely. If the censoring times are random and the form of the distribution is known, a likelihood estimation method could be based on the likelihood in (2.16). Without such knowledge, another approach is needed.

Two extreme methods of handling the censored observations in the intervals are:

- Assume that all censored observations are removed at t_i, the upper endpoint of the interval. This gives $\widehat{p}_i = d_i/n_i$ as used in Section 3.5. This estimate of p_i is biased upward.
- Assume that all censored observations are removed at t_{i-1}, the lower endpoint of the interval. This gives $\widehat{p}_i = d_i(n_i - r_i)$. This estimate of p_i is biased downward.

A commonly used compromise is $\widehat{p}_i = d_i/(n_i - r_i/2)$, the harmonic mean of the two more extreme estimates. These compromise estimates $\widehat{p}_i$ can be substituted into (3.6) and (3.9) leading through (3.7) and (3.10) to the "actuarial" or "life table" nonparametric estimate $\widehat{F}(t)$ and the corresponding standard error $\widehat{se}_{\widehat{F}}$.

Example 3.12 *Nonparametric Estimate for the Prototype Battery Data with Uncertain Censoring Times.* The prototype battery failure data in Example 2.12 and Appendix Table C.6 have both failure times and censored observations within

Table 3.6. Calculations for the Nonparametric Life Table Estimate of $F(t_i)$ for the Prototype Battery Data

Interval in Hours	Failed in	Censored in	Entered	Adjusted at Risk				
$(t_{i-1}, t_i]$	$(t_{i-1}, t_i]$	$(t_{i-1}, t_i]$	$(t_{i-1}, t_i]$					
$(t_{i-1}, t_i]$	d_i	r_i	n_i	$n_i - r_i/2$	$\widehat{p}_i$	$1 - \widehat{p}_i$	$\widehat{S}(t_i)$	$\widehat{F}(t_i)$
(0 – 50]	1	5	68	65.5	1/65.5	64.5/65.5	.985	.015
(50 – 100]	0	6	62	59	0/59	59/59	.985	.015
(100 – 150]	1	1	56	55.5	1/55.5	54.5/55.5	.967	.033
(150 – 200]	4	6	54	51	4/51	47/51	.891	.109
(200 – 250]	1	2	44	43	1/43	42/43	.870	.130
(250 – 300]	1	1	41	40.5	1/40.5	39.5/40.5	.849	.151
(300 – 350]	1	2	39	38	1/38	37/38	.827	.173
(350 – 400]	4	2	36	35	4/35	31/35	.732	.268
(450 – 500]	4	3	30	28.5	4/28.5	24.5/28.5	.629	.371
(500 – 550]	2	1	23	22.5	2/22.5	20.5/22.5	.573	.427
(550 – 600]	2	0	20	20	2/20	18/20	.516	.484
⋮	⋮	⋮	⋮	⋮	⋮	⋮	⋮	⋮

the given intervals. Here we consider the life distribution of the batteries, without distinguishing among the different failure modes. Table 3.6 illustrates the computations for the nonparametric life table estimate of $F(t)$ up to 600 hours. □

3.10 ARBITRARY CENSORING

The nonparametric estimate in (3.6) works only for some kinds of censoring patterns (e.g., multiple right censoring and interval censoring with intervals that do not overlap). When censoring is more complicated, an alternative is needed. This need arises because, with complicated censoring, we do not know the n_i values in (3.5).

The Peto–Turnbull estimator provides the needed generalization of the nonparametric ML estimator that can be used for:

- Arbitrary censoring (e.g., combinations of left and right censoring and interval censoring with overlapping intervals).
- Truncated data (as described in Section 11.6).

The basic idea is to write the likelihood as in (2.10) and to maximize this likelihood to estimate the vector p or π from which one can compute an estimate of $F(t)$. We illustrate the basic idea in the following example. With the simple right-censoring patterns used previously, the Peto–Turnbull estimator is equivalent to the nonparametric estimator defined in Sections 3.5 and 3.7.

Example 3.13   Nonparametric Estimate of the Turbine Wheel Distribution of Time to Crack Initiation Based on Inspection Data.   As explained in Example 1.7, the turbine wheel inspection data on time-to-crack-initiation can be viewed as a collection of overlapping right- and left-censored observations. Figure 3.9 plots the raw observed proportion failing as a function of hours of exposure. Due to random variability, this crude estimate of $F(t)$ is decreasing in several places. The true cdf is, of course, a nondecreasing function of time. Although it is *not* possible to use the product-limit estimator to compute the nonparametric maximum likelihood estimate of $F(t)$, the general maximum likelihood approach introduced in Section 2.4 still works. Figure 3.10 illustrates the basic parameters π_i used in computing the nonparametric estimate of $F(t_i)$ for this example. Using terms like those in (2.12) and (2.13) and the data summarized in Table 1.5 leads to

$$L(\pi) = L(\pi; \text{DATA}) = C \times [\pi_1]^0 \times [\pi_2 + \cdots + \pi_{12}]^{39}$$
$$\times [\pi_1 + \pi_2]^4 \times [\pi_3 + \cdots + \pi_{12}]^{49}$$
$$\times [\pi_1 + \pi_2 + \pi_3]^2 \times [\pi_4 + \cdots + \pi_{12}]^{31}$$
$$\vdots$$
$$\times [\pi_1 + \cdots + \pi_{11}]^{21} \times [\pi_{12}]^{15},$$

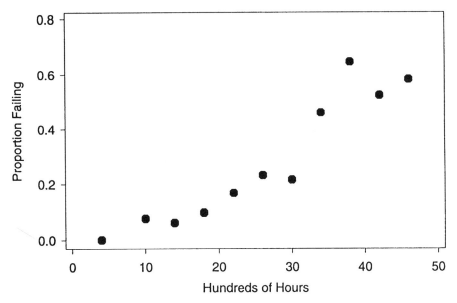

Figure 3.9. Plot of proportions failing versus hours of exposure for the turbine wheel inspection data.

where $\pi_{12} = 1 - \sum_{i=1}^{11} \pi_i$. The elements of $\boldsymbol{\pi} = (\pi_1, \ldots, \pi_{11})$ that maximize $L(\boldsymbol{\pi})$ give $\widehat{\boldsymbol{\pi}}$, the ML estimator of $\boldsymbol{\pi}$. Substituting the elements of $\widehat{\boldsymbol{\pi}}$ into (2.9) provides the nonparametric estimator of $F(t)$. The estimate, which is nondecreasing, is plotted in Figure 3.11. The pointwise confidence intervals in Figure 3.11 were computed using (3.19). The needed $\widehat{\text{se}}_{\widehat{F}}$ values were computed based on general methods given in Appendix Section B.6.4. □

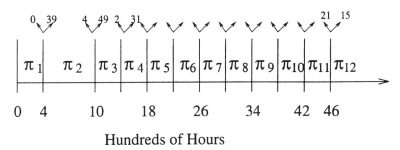

Figure 3.10. Basic parameters used in computing the nonparametric ML estimate of $F(t_i)$ for the turbine wheel inspection data.

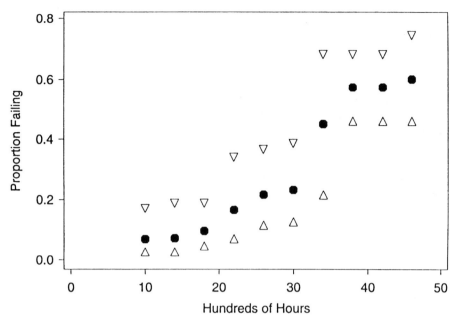

Figure 3.11. Plot of the nonparametric ML estimate for the turbine wheel inspection data along with a set of pointwise logistic-transform normal-approximation 95% confidence intervals for $F(t)$.

BIBLIOGRAPHIC NOTES

For the estimator given in (3.6), Kaplan and Meier (1958) allowed the width of the intervals in (2.5) to approach zero and the number of intervals to approach ∞. This is the origin of the alternative name "product-limit estimator." This estimator is also widely known as the "Kaplan–Meier estimator." Kaplan and Meier (1958), Elandt-Johnson and Johnson (1980, page 172), and Lawless (1982, page 74) provide more detailed justification for (3.6) as an ML estimator.

Nelson (1969, 1972, 1982) defined and illustrated the use of a nonparametric estimator for the cumulative hazard function. Corresponding asymptotic theory is given by Aalen (1976). This provides an alternative estimator for $F(t)$, sometimes referred to as the "Nelson–Aalen estimator." The Nelson–Aalen nonparametric estimator is asymptotically equivalent, as the sample size increases, to the product-limit nonparametric estimator. Some properties of this estimator are explored in Exercise 3.23.

Fleming and Harrington (1992), and Anderson, Borgan, Gill, and Keiding (1993) give detailed treatment of the general theory for both the product-limit estimator and the Nelson–Aalen estimator and outline the related literature with biomedical applications.

Nair (1981) gives asymptotic theory for the simultaneous confidence bands described in Section 3.8. Nair (1984) used simulation to compare these bands with alternative bands suggested in the literature. Weston and Meeker (1990) suggest and use simulation to evaluate the modification to Nair's bands, based on the logit transformation. Their results showed that the logit transformation provides a better approximation to the nominal coverage probability.

Peto (1973) defined the nonparametric maximum likelihood estimator for arbitrary censoring, including complicated overlapping interval-censored data. Turnbull (1976) further generalized the estimator to cover "truncated data" (to be discussed in Section 11.6) and suggested an EM (expectation-maximization) algorithm to compute the estimate. Gentleman and Geyer (1994) describe asymptotic theory and computational issues for this estimator.

Thomas and Grunkemeier (1975) compare several different nonparametric confidence intervals for $F(t)$. They conclude that confidence intervals based on inverting a likelihood ratio test for $F(t)$ are more accurate than intervals based on the normal approximation. Owen (1990) describes "empirical likelihood," providing a theoretical basis for nonparametric likelihood ratio confidence regions. Li (1995a), using the Thomas and Grunkemeier (1975) problem as a starting point, provides theory and an algorithm for computing nonparametric confidence intervals based on inverting a likelihood ratio test for $F(t)$. The work is extended to truncated data in Li (1995b).

EXERCISES

3.1. Use the ball bearing life test data in Table 1.1 to do the following:
   (a) Compute a nonparametric estimate of the population fraction failing by 75 million cycles.
   (b) Use the conservative interval in (3.2) to compute an approximate 90% confidence interval for the population fraction failing by 75 million cycles.
   (c) Use the normal-approximation method in (3.3) to compute an approximate 90% confidence interval for the population fraction failing by 75 million cycles.
   (d) Comparing the intervals from parts b and c, what do you conclude about the adequacy of the normal-approximation method for these data?

3.2. Repeat Exercise 3.1, using the population fraction failing by 25 million cycles. Why does the normal-approximation method not work so well in this case?

3.3. Show how (3.11) follows from (3.14) and how (3.14) follows from (3.12).

3.4. Parida (1991) gives data from a load-controlled high-cycle fatigue test conducted on 130 chain links. The 130 links were selected randomly from a population of different heats used to manufacture the links. Each link was tested until failure or until it had run for 80 thousand cycles, whichever came first. There were 10 failures—one each reported at 33, 46, 50, 59, 62, 71, 74, and 75 thousand cycles and 2 reported at 78 thousand cycles. The other 120 links had not failed by 80 thousand cycles.
   (a) Use (3.1) to compute the nonparametric estimate of $F(t)$ and corresponding standard errors.

(b) Compute a set of pointwise approximate 90% confidence intervals for $F(t)$. Explain the proper interpretation of these intervals.

(c) For the first three failures, compare the numerical estimates from (3.7) with the numerical estimates from (3.1).

(d) The original paper reported the number of cycles to failure, as given above. Suggest reasons why the numbers of cycles to failures were not given with more precision and the effect that this has on the results of the analysis.

(e) The original paper reported that the tested units had been selected from a random sample of heats. What might have happened in the experiment if all of the sample links had been selected from just one or two heats?

(f) The original paper did not report the order in which the tests were run. Typically, fatigue tests require the use of one or a few expensive test stands and tests are done in sequence. The order in which the failures occurred was not described in the original paper. Is it possible that there was some useful information in knowing the order in which the 130 units had been tested? Discuss.

3.5. The supplier of an electromechanical control for a household appliance ran an accelerated life test on sample controls. In the test, 25 controls were put on test and run until failure or until 30 thousand cycles had been accumulated. Failures occurred at 5, 21, and 28 thousand cycles. The other 22 controls did not fail by the end of the test.

(a) Compute and plot a nonparametric estimate for $F(t)$.

(b) Compute an approximate 95% confidence interval for the probability that an electromechanical device from the same production process, tested in the same way, would fail before 30 thousand cycles. Use the conservative binomial distribution approach.

(c) Compute an approximate 95% confidence interval for the probability that an electromechanical device from the same production process, tested in the same way, would fail before 30 thousand cycles. Use the normal-approximation method based on $Z_{\widehat{F}(30)} \overset{\cdot}{\sim} \text{NOR}(0, 1)$.

(d) Explain why, in this situation, the approach in part (b) would be preferred to the approach in part (c).

(e) The appliance manufacturer is really interested in the probability of the number of days to failure for its product. Use-rate differs from household to household, but the average rate is 2.3 cycles per day. What can the manufacturer say about the proportion of devices that would fail in 10 years of operation (the expected technological life of the product)?

(f) Refer to part (e). Describe an appropriate model to use when use-rate varies in the population of units. To simplify, start by assuming that there are only two different use-rates. Discuss, using appropriate expressions.

3.6. Over the past 18 months, ten separate copies of an electronic system have been deployed in earth orbit, where repair is impossible. Continuous remote monitoring, however, provides information on the state of the system and each of its main subsystems. Each system contains three nominally identical devices and it was learned, after deployment, that these devices are, in the system's environment, failing unexpectedly. The failures cause degradation to the overall system operation. For future systems that are to be deployed, the problem will be fixed, but owners of the systems have asked for information on the amount of degradation that can be expected in future years of operation among these currently deployed. To date, 5 of the 30 devices have failed. Due to the staggered entry of the systems into service, the available data are multiply censored. The following table summarizes the available information with times given in hours. Times of unfailed units are marked with a "+."

System	Device 1	Device 2	Device 3
1	564 +	564 +	564 +
2	1321 +	1104	1321 +
3	1933 +	1933 +	1933 +
4	1965 +	1965 +	1965 +
5	2578 +	2345	2578 +
6	3122 +	3122 +	3122 +
7	5918 +	5918 +	4467
8	7912 +	7912 +	6623
9	8156 +	8156 +	8156 +
10	7885	12229 +	12229 +

(a) Compute a nonparametric estimate of $F(t)$, the life distribution of the devices, assuming that the devices are operating and failing independently.

(b) Plot the nonparametric estimate of $F(t)$.

(c) Compute pointwise approximate 95% confidence intervals for $F(t)$ and add these to your plot.

(d) Explain why it might be that the 30 devices are not operating and failing independently and how this would affect conclusions drawn from the data.

(e) Describe possible reasons why this failure-causing problem was not discovered earlier and what might be done to minimize the chance of such problems occurring in the future. What questions would you ask of a client who wants you to help interpret the information in the failure data?

3.7. Consider the Plant 1 heat exchanger data in Figure 3.4.
  (a) Write the likelihood for these data in terms of π_1, π_2, π_3, and π_4.
  (b) Write the likelihood for these data in terms of p_1, p_2, p_3, and p_4.

EXERCISES

▲3.8. For a given t_i, $i = 1, \ldots, n$, show that the expression in (3.1) is the maximum likelihood estimator for $\widehat{F}(t_i)$.

▲3.9. The expression in (3.4) was obtained by evaluating the square root of $\text{Var}[\widehat{F}(t_i)]$ at $\widehat{F}(t_i)$. Show this by deriving the expression for $\text{Var}[\widehat{F}(t_i)]$.

◆3.10. Some computer programs (e.g., statistical packages and spreadsheets) can be used to generate pseudorandom samples from a uniform distribution. Let $U_1, \ldots, U_n$ denote such a sample. Then $T_1 = -\log(1 - U_1), \ldots, T_n = -\log(1 - U_n)$ is a pseudorandom sample from an exponential distribution (to be described in more detail in Chapter 4). Simulate a sequence of 50 such samples each of size $n = 200$. For each sample:
   (a) Compute and plot $\widehat{F}$.
   (b) Make a histogram of the 50 values of $\widehat{F}(1)$.
   (c) Make a histogram of the 50 values of $Z_{\widehat{F}(1)}$.
   (d) Make a histogram of the 50 values of $Z_{\text{logit}(\widehat{F})}$.
   (e) Compare the histograms in parts (c) and (d). Which statistic seems to be better approximated by a NOR(0, 1) distribution?

◆3.11. Repeat Exercise 3.10 using samples of size $n = 20$ (and perhaps other values of n). Compare the plots with those from Exercise 3.10 and describe how sample size affects the distribution of statistics like $\widehat{F}$, $Z_{\widehat{F}}$, and $Z_{\text{logit}(\widehat{F})}$.

3.12. Weis, Caldararu, Snyder, and Croitoru (1986) report on the results of a life test on silicon photodiode detectors in which 28 detectors were tested at 85°C and 40 volts reverse bias. These conditions, which were more stressful than normal use conditions, were used in order to get failures quickly. Specified electrical tests were made at 0, 10, 25, 75, 100, 500, 750, 1000, 1500, 2000, 2500, 3000, 3600, 3700, and 3800 hours to determine if the detectors were still performing properly. Failures were found after the inspections at 2500 (1 failure), 3000 (1 failure), 3500 (2 failures), 3600 (1 failure), 3700 (1 failure), and 3800 (1 failure). The other 21 detectors had not failed after 3800 hours of operation. Use these data to estimate the life distribution of such photodiode detectors running at the test conditions.
   (a) From the description given above, the data would be useful for making inferences about what particular populations or process? Explain your reasoning.
   (b) Compute and plot a nonparametric estimate of the cdf for time to failure at the test conditions.
   (c) Compute standard errors for the nonparametric estimate in part (b).
   (d) Compute a set of pointwise approximate 95% confidence intervals for $F(t)$ and add these to your plot.
   (e) Compute simultaneous approximate 95% confidence bands for $F(t)$ over the complete range of observation.

(f) Provide a careful explanation of the differences in interpretation and application of the pointwise confidence intervals and the simultaneous confidence bands.

▲**3.13.** Use the delta method (Appendix Section B.2) and the assumptions given in Section 3.6.1 to derive (3.8).

▲**3.14.** Show how (3.16) follows from (3.15). Begin by using the delta method (e.g., Appendix Section B.2) to obtain an expression for $\widehat{se}_{\text{logit}(\widehat{F})}$ as a function of $\widehat{se}_{\widehat{F}}$.

3.15. Example 3.9 illustrates the computations for the nonparametric estimation of the cdf for the shock absorber data up to 12,220 km. Complete the computations for the rest of the data (i.e., out to 28,100 km). Use $S(12200) = .9086984$ to continue the cumulative product in (3.6).
  (a) Plot the nonparametric estimate out to 28,100 km.
  (b) Compute $\widehat{se}_{\widehat{F}}$ out to 28,100 km.
  (c) Compute a set of pointwise approximate 90% confidence intervals for $F(t)$ out to 28,100 km and add these to the plot in part (a).
  (d) Explain why, with right-censored data, for the nonparametric estimation method, there is only a limited range of time over which we can estimate $F(t)$.

▲**3.16.** Show that with single censoring (i.e., all failures precede the first censoring time) that (3.7) simplifies to (3.1) and that (3.9) simplifies to (3.4).

3.17. Example 3.6 illustrated the computations for the nonparametric and approximate confidence intervals for $F(1)$ and $F(2)$ for the pooled heat exchanger tube data. Complete similar computations for $F(3)$.

3.18. Explain why the nonparametric estimate of $F(t)$ is a set of points for the heat exchanger data in Example 3.6 but a step function for the shock absorber data in Example 3.9.

3.19. Use the data in Table 1.4 to do the following for the V7 transmitter tube:
  (a) Compute an estimate of the conditional probability of failing for each cell.
  (b) Compute a nonparametric estimate for $F(t)$ for each cell.
  (c) Plot the estimate of $F(t)$ along with a set of pointwise approximate 95% confidence intervals.
  (d) For each interval, compute and plot an estimate of the probability of failing in that interval given a unit enters the interval.
  (e) Explain how these estimates could be used to plan for preventive maintenance for a group of radio transmitters, each with one such tube.

EXERCISES

3.20. Use the diesel engine fan data in Appendix Table C.1 to compute the product-limit nonparametric estimate of $F(t)$ using (3.6).

3.21. Example 3.12 illustrated the computations for the nonparametric estimate of the cdf for prototype batteries with both failure times and censored observations within the given intervals. Complete the computations for the rest of the data in Table C.6 (i.e., out to 1700 hours). Use $\widehat{S}(600) = .51608$ to continue the cumulative product in (3.6).

(a) Plot the nonparametric estimate out to 1700 hours.

(b) Compute $\widehat{\text{se}}_{\widehat{F}}$ out to 1700 hours.

(c) Compute a set of approximate pointwise 90% confidence intervals for $F(t)$ out to 1700 hours and add these to the plot in part (a).

▲3.22. Consider the model in Section 2.2.1 and the data collection method described in Section 3.5.

(a) Show that the likelihood of the data, as a function of the parameters, is

$$L(\underline{\pi}) = \pi_1^{d_1} \times \pi_2^{d_2} \times \cdots \times \pi_m^{d_m} \times [S(t_1)]^{r_1} \times [S(t_2)]^{r_2} \times \cdots \times [S(t_m)]^{r_m}.$$

(b) Show that in terms of the parameters $\boldsymbol{p} = (p_1, \ldots, p_m)$,

$$L(\boldsymbol{p}) = \prod_{j=1}^{m} p_j^{d_j}(1 - p_j)^{n_j - d_j},$$

where $n_j = n - \sum_{i=0}^{j-1} d_i - \sum_{i=0}^{j-1} r_i$, with the understanding that $d_0 = 0$ and $r_0 = 0$.

(c) Show that the maximum likelihood estimators of the parameters are

$$\widehat{p}_j = \frac{d_j}{n_j}, \quad j = 1, \ldots, m.$$

(d) Show that the observed information matrix for the parameters $\boldsymbol{p}$ is diagonal and that the jth diagonal element of the matrix is equal to

$$-\frac{\partial^2 \log[L(\boldsymbol{p})]}{\partial p_j^2} = \frac{n_j}{p_j(1 - p_j)}$$

evaluated at $\widehat{\boldsymbol{p}}$. This shows that, asymptotically (in large samples), the components of $\widehat{\boldsymbol{p}}$ are uncorrelated and $\text{Var}(\widehat{p}_j) = p_j(1 - p_j)/n_j$. Use these results and the delta method to derive Greenwood's formula as given in (3.9).

▲3.23. Consider the relationship $S(t_i) = \exp[-H(t_i)]$, where $H(t)$ is the cumulative hazard function. Note that a nonparametric ML estimator (based on the product-limit estimator) of $H(t)$ without assuming a distributional form is

$$\widehat{H}(t_i) = -\sum_{j=1}^{i} \log(1-\widehat{p}_j) \approx \sum_{j=1}^{i} \widehat{p}_j = \sum_{j=1}^{i} \frac{d_j}{n_j} = \widehat{\widehat{H}}(t_i)$$

$\widehat{\widehat{H}}(t_i)$ is known as the Nelson–Aalen estimator of $H(t_i)$. Thus $\widehat{\widehat{F}}(t_i) = 1 - \exp[-\widehat{\widehat{H}}(t_i)]$ is another nonparametric estimator for $F(t_i)$.

(a) Give conditions to assure a good agreement between $\widehat{H}(t_i)$ and $\widehat{\widehat{H}}(t_i)$ and thus between $\widehat{F}(t_i)$ and $\widehat{\widehat{F}}(t_i)$.
(b) Use the delta method to compute approximate expressions for $\text{Var}[\widehat{H}(t_i)]$ and $\text{Var}[\widehat{\widehat{H}}(t_i)]$. Comment on the expression(s) you get.
(c) Compute Nelson–Aalen estimate of $F(t)$ and compare with the estimate computed in Exercise 3.20. Describe similarities and differences.
(d) Show that $\widehat{\widehat{H}}(t_i) < \widehat{H}(t_i)$ and that $\widehat{\widehat{F}}(t_i) < \widehat{F}(t_i)$.
(e) Describe suitable modifications of the estimator that can be used when failure and censoring times are grouped into common intervals.

CHAPTER 4

Location-Scale-Based Parametric Distributions

Objectives

This chapter explains:

- Important ideas behind parametric models in the analysis of reliability data.
- Motivation for important functions of model parameters that are of interest in reliability studies.
- The location-scale family of probability distributions.
- Properties and the importance of the exponential distribution.
- Properties and the importance of log-location-scale distributions such as the Weibull, lognormal, and loglogistic distributions.
- How to generate pseudorandom data from a specified distribution (such random data are used in simulation evaluations in subsequent chapters).

Overview

This chapter introduces some basic ideas of parametric modeling and the most important parametric distributions. Parametric distributions are used extensively in subsequent chapters. Section 4.1 explains some of the basic concepts and motivation for using parametric models. Section 4.2 describes important functions of parameters like failure probabilities and distribution quantiles. Section 4.3 introduces the important location-scale family of distributions. Sections 4.4–4.11 give detailed information on these and the important log-location-scale distributions. Subsequent chapters require at least a basic understanding of the characteristics and notation for the exponential, Weibull, and lognormal distributions. Applications for the other distributions follow without difficulty. Physical motivation for these and the other distributions is helpful in practical modeling applications. Section 4.12 describes alternative choices for parameters. Section 4.13 describes methods for generating simulated values from a specified distribution. In various parts of this book we will use simulation to develop and explore data analysis methods.

4.1 INTRODUCTION

As we saw in Chapter 3, it is possible to make certain kinds of inferences without having to assume a particular parametric form for a failure-time distribution. There are, however, many problems in reliability data analysis where it is either useful or essential to use a parametric distribution form. This chapter describes a number of simple probability distributions that are commonly used to model failure-time processes. Chapter 5 does the same for other important and useful, but more complicated, distributions. The discussion in these chapters concerns underlying continuous-time models, although much of the material also holds for discrete-time models.

As explained in Chapter 2, a natural model for a continuous random variable, say, T, is the cumulative distribution function (cdf). Specific examples given in this chapter and in Chapter 5 are of the form $\Pr(T \le t) = F(t; \boldsymbol{\theta})$, where $\boldsymbol{\theta}$ is a vector of parameters. In this book, we use T to denote positive random variables like failure time, so that $T > 0$; correspondingly, we will use Y to denote unrestricted random variables so that $-\infty < Y = \log(T) < \infty$. Unlike the "basic parameters" in $\boldsymbol{\pi}$ and $\boldsymbol{p}$ used in the "nonparametric" formulation in Chapters 2 and 3, the parametric models described in this chapter will have a $\boldsymbol{\theta}$ containing a small fixed number of parameters. The most commonly used parametric probability distributions have between one and four parameters, although there are some distributions with more than four parameters. More complicated models could contain many more parameters involving mixtures, competing failure modes, or other combinations of distributions or models that include explanatory variables. One simple example that we will use later in this chapter is the exponential distribution for which

$$\Pr(T \le t) = F(t; \theta) = 1 - \exp\left(-\frac{t}{\theta}\right), \quad t > 0 \qquad (4.1)$$

where θ is the single scalar parameter of the distribution (equal to the mean or first moment, in this example).

Use of parametric distributions complements nonparametric techniques and provides the following advantages:

- Parametric models can be described concisely with just a few parameters, instead of having to report an entire curve.
- It is possible to use a parametric model to extrapolate (in time) to the lower or upper tail of a distribution.
- Parametric models provide smooth estimates of failure-time distributions.

In practice it is often useful to do various parametric and nonparametric analyses of a data set.

4.2 QUANTITIES OF INTEREST IN RELIABILITY APPLICATIONS

Starting in Chapter 7, we will focus on the problem of *estimating* the parameters $\boldsymbol{\theta}$ and important functions of $\boldsymbol{\theta}$. In this section we describe ideas behind parameterization of

QUANTITIES OF INTEREST IN RELIABILITY APPLICATIONS 77

a probability distribution and describe a number of particular functions of parameters that are of interest for reliability analysis.

In most practical problems, interest centers on quantities that are functions of $\boldsymbol{\theta}$ and the ML estimates of these functions will *not* depend on the particular parameterization that is used to specify the parametric model. The quantities of interest discussed here extend the list introduced in Chapter 2, and now these quantities will be expressed as functions of the small set of parameters $\boldsymbol{\theta}$. Specifically, for distributions of positive and continuous random variables (there are similar definitions for discrete and/or nonpositive random variables):

- The "probability of failure" $p = \Pr(T \leq t) = F(t; \boldsymbol{\theta})$ by a specified t. For example, if T is the time of failure of a unit, then p is the probability that the unit will fail before t.
- The "p quantile" of the distribution of T is the smallest value t such that $F(t; \boldsymbol{\theta}) \geq p$. We will express the p quantile as $t_p = F^{-1}(p; \boldsymbol{\theta})$. For the failure-time example, t_p is the time at which $100p\%$ of the units in the product population will have failed. The median is equal to $t_{.5}$.
- The "hazard function" (hf) is defined as

$$h(t) = \frac{f(t; \boldsymbol{\theta})}{1 - F(t; \boldsymbol{\theta})}. \tag{4.2}$$

As described in Section 2.1.1, the hazard function is of particular interest in reliability applications because it indicates, for surviving units, the propensity to fail in the following small interval of time, as a function of age.

- The mean life (also known as the "average," "expectation," or "first moment") of T

$$\mathrm{E}(T) = \int_0^\infty t f(t; \boldsymbol{\theta}) \, dt = \int_0^\infty [1 - F(t)] \, dt \tag{4.3}$$

is a measure of the center of $f(t; \boldsymbol{\theta})$. When $f(t; \boldsymbol{\theta})$ is highly skewed, the mean may differ appreciably from other measures of central tendency like the median. The mean is sometimes, but not always, one of the distribution parameters. For some pdfs, the value of the integral will be infinite. Then it is said that the mean of T "does not exist." When T is time to failure, the mean is sometimes referred to as the MTTF, for mean time to failure.

- The variance (also known as the "second central moment") of T

$$\mathrm{Var}(T) = \int_0^\infty [t - \mathrm{E}(T)]^2 f(t; \boldsymbol{\theta}) \, dt$$

is a measure of spread of the distribution of T. $\mathrm{Var}(T)$ is the average squared deviation of T from its mean. Again, if the value of the integral is infinite, it is said that the variance of T "does not exist." The quantity $\mathrm{SD}(T) = \sqrt{\mathrm{Var}(T)}$, known as the "standard deviation" of T, is easier to interpret because it has the same units as T.

- The unitless quantity $\gamma_2 = \mathrm{SD}(T)/\mathrm{E}(T)$, known as the "coefficient of variation" of T, is useful for comparing the relative amount of variability in different distributions. The quantity $1/\gamma_2 = \mathrm{E}(T)/\mathrm{SD}(T)$ is sometimes known as the "signal-to-noise ratio."
- The unitless quantity

$$\gamma_3 = \frac{\int_0^\infty [t - \mathrm{E}(T)]^3 f(t; \boldsymbol{\theta})\, dt}{[\mathrm{Var}(T)]^{3/2}},$$

known as the "standardized third central moment" or "coefficient of skewness" of T, is a measure of the skewness in the distribution of T. When a distribution is symmetric, $\gamma_3 = 0$. It is, however, possible to have $\gamma_3 = 0$ for a distribution that is not perfectly symmetric (e.g., the Weibull distribution, discussed in in Section 4.8, has $\gamma_3 = 0$ when $\beta = 3.602$, but the distribution is only approximately symmetric). Usually, however, when γ_3 is positive (negative), the distribution of T is skewed to the right (left).

For reliability applications, quantiles, failure probabilities, and the hazard function are typically of higher interest than distribution moments. In subsequent chapters we will describe *point estimation* and, at the same time, emphasize methods of obtaining *confidence intervals* (for scalars) and *confidence regions* (for simultaneous inference on a vector of two or more quantities) for parameters and important functions of parameters. Confidence intervals and regions quantify the uncertainty in parameter estimates arising from the fact that inferences are generally based on only a finite number of observations from the process or population of interest.

4.3 LOCATION-SCALE AND LOG-LOCATION-SCALE DISTRIBUTIONS

A random variable Y belongs to the location-scale family of distributions if its cdf can be expressed as

$$\Pr(Y \le y) = F(y; \mu, \sigma) = \Phi\left(\frac{y - \mu}{\sigma}\right),$$

where Φ does not depend on any unknown parameters. In this case we say that $-\infty < \mu < \infty$ is a location parameter and that $\sigma > 0$ is a scale parameter. Substitution shows that Φ is the cdf of Y when $\mu = 0$ and $\sigma = 1$. Also, Φ is the cdf of $(Y - \mu)/\sigma$. Location-scale distributions are important for a number of reasons including:

- Many of the widely used statistical distributions are either location-scale distributions or closely related. These distributions include the exponential, normal, Weibull, lognormal, loglogistic, logistic, and extreme value distributions.
- Methods of data analysis and inference, statistical theory, and computer software developed for the location-scale family can be applied to any of the members of the family.
- Theory for location-scale distributions is relatively simple.

In cases where Φ does depend on one or more unknown parameters (as with a number of the distributions described in Chapter 5), Y is not a member of the location-scale family, but the location-scale structure and notation will still be useful for us.

A random variable T belongs to the log-location-scale family distribution if $Y = \log(T)$ is a member of the location-scale family. The Weibull, lognormal, and loglogistic distributions are the most important members of this family.

4.4 EXPONENTIAL DISTRIBUTION

When T has an exponential distribution, we indicate this by $T \sim \text{EXP}(\theta, \gamma)$. The two-parameter exponential distribution (to distinguish it from the more commonly used one-parameter exponential distribution) has cdf, pdf, and hf

$$F(t; \theta, \gamma) = 1 - \exp\left(-\frac{t - \gamma}{\theta}\right),$$

$$f(t; \theta, \gamma) = \frac{1}{\theta} \exp\left(-\frac{t - \gamma}{\theta}\right),$$

$$h(t; \theta, \gamma) = \frac{1}{\theta}, \quad t > \gamma,$$

where $\theta > 0$ is a scale parameter and γ is both a location and a threshold parameter. For $\gamma = 0$ this is the well-known one-parameter exponential distribution (and often known simply as the exponential distribution). When T has this simpler distribution, we indicate it by $T \sim \text{EXP}(\theta)$. The cdf, pdf, and hf are graphed in Figure 4.1 for $\theta = .5, 1$, and 2 and $\gamma = 0$.

For integer $m > 0$, $\text{E}[(T - \gamma)^m] = m! \, \theta^m$. Thus the mean and variance of the exponential distribution are, respectively, $\text{E}(T) = \gamma + \theta$ and $\text{Var}(T) = \theta^2$. The p quantile of the exponential distribution is $t_p = \gamma - \log(1 - p) \, \theta$.

The one-parameter exponential distribution, where $\gamma = 0$, is the simplest distribution that is commonly used in the analysis of reliability data. The exponential distribution has the important characteristic that its hf is constant (does not depend on time t). A constant hf implies that, for an unfailed unit, the probability of failing in the next small interval of time is independent of the unit's age. Physically, a constant hf suggests that the population of units under consideration is not wearing out or otherwise aging. The exponential distribution is a popular distribution for some kinds of electronic components (e.g., capacitors or robust, high-quality integrated circuits). This exponential distribution would *not* be appropriate for a population of electronic components having failure-causing quality defects (such defects are difficult to rule out completely and are a leading cause of electronic system reliability problems). On the other hand, the exponential distribution might be useful to describe failure times for components that exhibit physical wearout if the wearout does not show up until long after the expected technological life of the system in which the compo-

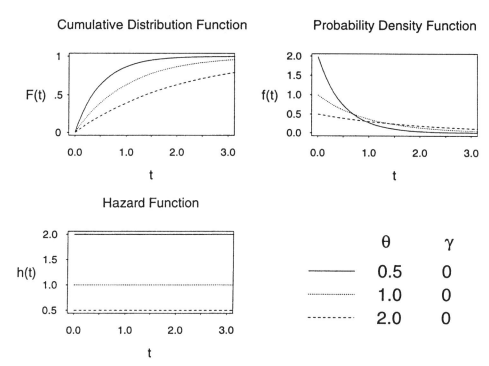

Figure 4.1. Exponential cdf, pdf, and hf for $\theta = .5$, 1, and 2 and $\gamma = 0$.

nent would be installed (e.g., electronic components in computing equipment having failures caused by random external events).

Under very special circumstances, the exponential distribution may be appropriate for the times between system failures, arrivals in a queue, and other interarrival time distributions. Specifically, the exponential distribution is the distribution of interval times of a homogeneous Poisson process. See Chapter 3 of Thompson (1988) and Chapter 16 for more information on homogeneous Poisson processes.

The exponential distribution is usually *inappropriate* for modeling the life of mechanical components (e.g., bearings) subject to some combination of fatigue, corrosion, or wear. It is also usually inappropriate for electronic components that exhibit wearout properties over their technological life (e.g., lasers and filament devices). A distribution with an increasing hf is, in such applications, usually more appropriate. Similarly, for populations containing mixtures of good and bad units the population hf may decrease with life because, as the bad units fail and leave the population, only the stronger units are left.

4.5 NORMAL DISTRIBUTION

When Y has a normal distribution, we indicate this by $Y \sim \text{NOR}(\mu, \sigma)$. The normal distribution is a location-scale distribution with cdf and pdf

NORMAL DISTRIBUTION

$$F(y; \mu, \sigma) = \Phi_{\text{nor}}\left(\frac{y - \mu}{\sigma}\right),$$

$$f(y; \mu, \sigma) = \frac{1}{\sigma}\phi_{\text{nor}}\left(\frac{y - \mu}{\sigma}\right), \quad -\infty < y < \infty,$$

where $\phi_{\text{nor}}(z) = (1/\sqrt{2\pi})\exp(-z^2/2)$ and $\Phi_{\text{nor}}(z) = \int_{-\infty}^{z} \phi_{\text{nor}}(w)\,dw$ are, respectively, the pdf and cdf for the standardized NOR($\mu = 0, \sigma = 1$) distribution. Here $-\infty < \mu < \infty$ is a location parameter and $\sigma > 0$ is a scale parameter. When there is no useful simplification of the hf definition in (4.2), as with the normal distribution, the definition will not be repeated. The normal distribution pdf, cdf, and hf are graphed in Figure 4.2 for $\mu = 5$ and $\sigma = .3, .5, .8$.

For integer $m > 0$, $E[(Y - \mu)^m] = 0$ if m is odd and $E[(Y - \mu)^m] = m!\sigma^m/[2^{m/2}(m/2)!]$ if m is even. From this, the mean and variance of the normal distribution are, respectively, $E(Y) = \mu$ and $\text{Var}(Y) = \sigma^2$. The p quantile of the normal distribution is $y_p = \mu + \Phi_{\text{nor}}^{-1}(p)\sigma$, where $\Phi_{\text{nor}}^{-1}(p) = z_p$ is the p quantile of the standard normal distribution.

As a model for variability, the normal distribution has a long history of use in many areas of application. This is due to the simplicity of normal distribution theory and the central limit theorem. The central limit theorem states that the distribution of the sum of a large number of independent identically distributed random quantities

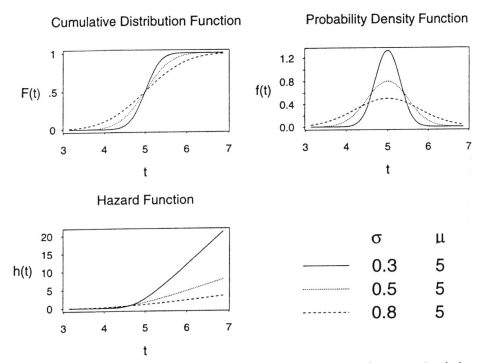

Figure 4.2. Normal cdf, pdf, and hf with location parameter (mean) $\mu = 5$ and scale parameter (standard deviation) $\sigma = .3, .5,$ and $.8$.

has, approximately, a normal distribution. In reliability data analysis, the use of the normal distribution is, however, less common. As seen from Figure 4.2, the normal distribution has an increasing hf that begins to increase rapidly near, but before, the point of median life. The normal distribution has proved to be a useful distribution for certain life data when $\mu > 0$ and the coefficient of variation (σ/μ) is small. Examples include electric filament devices (e.g., incandescent light bulbs and toaster heating elements) and strength of wire bonds in integrated circuits (component strength is often used as an easy-to-obtain surrogate measure or indicator of eventual reliability). Also, as described in Section 4.6, the normal distribution is often a useful model for the logarithms of failure times (see the next section).

4.6 LOGNORMAL DISTRIBUTION

When T has a lognormal distribution, we indicate this by $T \sim \text{LOGNOR}(\mu, \sigma)$. If $T \sim \text{LOGNOR}(\mu, \sigma)$ then $Y = \log(T) \sim \text{NOR}(\mu, \sigma)$. The lognormal cdf and pdf are

$$F(t; \mu, \sigma) = \Phi_{\text{nor}}\left[\frac{\log(t) - \mu}{\sigma}\right], \quad (4.4)$$

$$f(t; \mu, \sigma) = \frac{1}{\sigma t}\phi_{\text{nor}}\left[\frac{\log(t) - \mu}{\sigma}\right], \quad t > 0, \quad (4.5)$$

where ϕ_{nor} and Φ_{nor} are pdf and cdf for the standardized normal. The median $t_{.5} = \exp(\mu)$ is a scale parameter and $\sigma > 0$ is a shape parameter. The lognormal cdf, pdf, and hf are graphed in Figure 4.3 for $\sigma = .3, .5,$ and $.8$ and $\mu = 0$, corresponding to the median $t_{.5} = \exp(\mu) = 1$.

The most common definition of the lognormal distribution uses base e (natural) logarithms. Base 10 (common) logarithms are also used in some areas of application. Bottom-line answers for important reliability metrics (e.g., estimates of failure probabilities, failure rates, and quantiles) will not depend on the base that is used. The definition of the parameters μ (mean of the *logarithm* of T) and σ (standard deviation of the *logarithm* of T) will, however, depend on the base that is used. For this reason it is important to make consistent use of one particular base. In this book we will generally use base e (natural) logarithms for the lognormal distribution definition.

For integer $m > 0, \text{E}(T^m) = \exp(m\mu + m^2\sigma^2/2)$. From this it follows that the mean and variance of the lognormal distribution are, respectively, $\text{E}(T) = \exp(\mu + .5\sigma^2)$ and $\text{Var}(T) = \exp(2\mu + \sigma^2)[\exp(\sigma^2) - 1]$. The quantile function of the lognormal distribution is $t_p = \exp[\mu + \Phi_{\text{nor}}^{-1}(p)\sigma]$.

The lognormal distribution is a common model for failure times. Following from the central limit theorem (mentioned in Section 4.5), application of the lognormal distribution could be justified for a random variable that arises from the product of a number of identically distributed independent positive random quantities. It has been suggested that the lognormal is an appropriate model for time to failure caused by a degradation process with combinations of random rate constants that combine

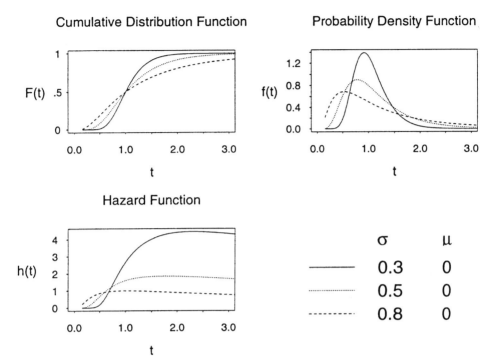

Figure 4.3. Lognormal cdf, pdf, and hf for scale parameter $t_{.5} = \exp(\mu) = 1$ and for shape parameter $\sigma = .3, .5,$ and $.8$.

multiplicatively (e.g., see the models in Chapter 13). The lognormal distribution is widely used to describe time to fracture from fatigue crack growth in metals. As shown in Figure 4.3 (also see Exercise 4.19), the lognormal $h(t)$ starts at 0, increases to a point in time, and then decreases eventually to zero. For large σ, $h(t)$ reaches a maximum early in life and then decreases. For this reason, the lognormal distribution is often used as a model for a population of electronic components that exhibits a decreasing hf. It has been suggested that early-life "hardening" of certain kinds of materials or components might lead to such an hf. The lognormal distribution also arises as the time to failure distribution of certain degradation processes, as described in Chapter 13. The lognormal distribution described in this section is sometimes referred to as the "two-parameter lognormal distribution" to distinguish it from the three-parameter lognormal distribution described in Section 5.10.2.

4.7 SMALLEST EXTREME VALUE DISTRIBUTION

When the random variable Y has a smallest extreme value distribution, we indicate this by $Y \sim \text{SEV}(\mu, \sigma)$. The SEV cdf, pdf, and hf are

$$F(y; \mu, \sigma) = \Phi_{\text{sev}}\left(\frac{y - \mu}{\sigma}\right),$$

$$f(y; \mu, \sigma) = \frac{1}{\sigma} \phi_{sev}\left(\frac{y-\mu}{\sigma}\right),$$

$$h(y; \mu, \sigma) = \frac{1}{\sigma} \exp\left(\frac{y-\mu}{\sigma}\right), \quad -\infty < y < \infty,$$

where $\Phi_{sev}(z) = 1 - \exp[-\exp(z)]$ and $\phi_{sev}(z) = \exp[z - \exp(z)]$ are the cdf and pdf, respectively, for standardized SEV ($\mu = 0, \sigma = 1$). Here $-\infty < \mu < \infty$ is the location parameter and $\sigma > 0$ is the scale parameter. The SEV cdf, pdf, and hf are graphed in Figure 4.4 for $\mu = 50$ and $\sigma = 5, 6,$ and 7.

The mean, variance, and quantile functions of the smallest extreme value distribution are $E(Y) = \mu - \sigma\gamma$, $\text{Var}(Y) = \sigma^2\pi^2/6$, and $y_p = \mu + \Phi_{sev}^{-1}(p)\sigma$, where $\Phi_{sev}^{-1}(p) = \log[-\log(1-p)]$ and $\gamma \approx .5772$ is Euler's constant.

Figure 4.4 shows that the smallest extreme value distribution pdf is skewed to the left. Although most failure-time distributions are skewed to the right, distributions of strength will sometimes be skewed to the left (because of a few weak units in the lower tail of the distribution, but a sharper upper bound for the majority of units in the upper tail of the strength population). The SEV distribution may have physical justification arising from an extreme value theorem. Namely, it is the limiting standardized distribution of the minimum of a large number of random variables from a certain class of distributions (this class includes the normal distribution as a special case). If σ is small relative to μ the SEV distribution can be used as a life

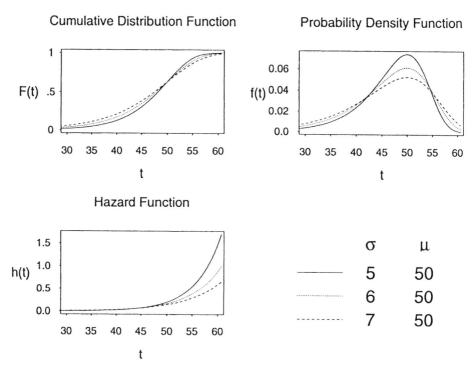

Figure 4.4. Smallest extreme value cdf, pdf, and hf with $\mu = 50$ and $\sigma = 5, 6,$ and 7.

distribution. The exponentially increasing hf suggests that the SEV would be suitable for modeling the life of a product that experiences very rapid wearout after a certain age. The distributions of logarithms of failure times can often be modeled with the SEV distribution; see Section 4.8. Also see the closely related Gompertz–Makeham distribution in Section 5.8.

4.8 WEIBULL DISTRIBUTION

The Weibull distribution cdf is often written as

$$\Pr(T \leq t; \eta, \beta) = 1 - \exp\left[-\left(\frac{t}{\eta}\right)^{\beta}\right], \quad t > 0. \tag{4.6}$$

For this parameterization, $\beta > 0$ is a shape parameter and $\eta > 0$ is a scale parameter as well as the .632 quantile. The practical value of the Weibull distribution stems from its ability to describe failure distributions with many different commonly occurring shapes. As illustrated in Figure 4.5, for $0 < \beta < 1$, the Weibull has a decreasing hf. With $\beta > 1$, the Weibull has an increasing hf.

For integer $m > 0$, $E(T^m) = \eta^m \Gamma(1 + m/\beta)$, where $\Gamma(\kappa) = \int_0^\infty z^{\kappa-1} \exp(-z)\, dz$ is the gamma function. From this it follows that the mean and variance of the Weibull

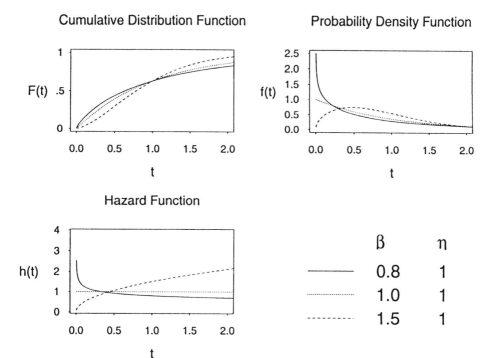

Figure 4.5. Weibull cdf, pdf, and hf for $t_{.632} = \eta = \exp(\mu) = 1$ and $\beta = 1/\sigma = .8, 1,$ and 1.5.

distribution are, respectively, $E(T) = \eta\Gamma(1 + 1/\beta)$ and $Var(T) = \eta^2[\Gamma(1 + 2/\beta) - \Gamma^2(1 + 1/\beta)]$. The Weibull p quantile is $t_p = \eta[-\log(1 - p)]^{1/\beta}$. Note that when $\beta = 1$, the cdf in (4.6) reduces to an exponential distribution with scale parameter $\theta = \eta$.

It is convenient to use a simple alternative parameterization for the Weibull distribution. This alternative parameterization is based on the relationship between the Weibull distribution and the smallest extreme value distribution described in Section 4.7. In particular, if T has a Weibull distribution, then $Y = \log(T) \sim \text{SEV}(\mu, \sigma)$, where $\sigma = 1/\beta$ is the scale parameter and $\mu = \log(\eta)$ is the location parameter. Thus when T has a Weibull distribution, we indicate this by $T \sim \text{WEIB}(\mu, \sigma)$. In this form, the Weibull cdf, pdf, and hf can be written as

$$F(t; \mu, \sigma) = \Phi_{\text{sev}}\left[\frac{\log(t) - \mu}{\sigma}\right], \tag{4.7}$$

$$f(t; \mu, \sigma) = \frac{1}{\sigma t} \phi_{\text{sev}}\left[\frac{\log(t) - \mu}{\sigma}\right] = \frac{\beta}{\eta}\left(\frac{t}{\eta}\right)^{\beta-1} \exp\left[-\left(\frac{t}{\eta}\right)^{\beta}\right],$$

$$h(t; \mu, \sigma) = \frac{1}{\sigma \exp(\mu)}\left[\frac{t}{\exp(\mu)}\right]^{1/\sigma - 1} = \frac{\beta}{\eta}\left(\frac{t}{\eta}\right)^{\beta-1}, \quad t > 0.$$

Then the Weibull p quantile is $t_p = \exp[\mu + \Phi_{\text{sev}}^{-1}(p)\,\sigma]$. The Weibull/SEV relationship parallels the lognormal/normal relationship. The SEV parameterization is useful because location-scale distributions are easier to work with in general. As mentioned in Section 4.3, transforming the Weibull distribution into an SEV distribution allows the use of general results for location-scale distributions, which apply directly to all such distributions, including the Weibull, lognormal, and some other distributions.

The theory of extreme values shows that the Weibull distribution can be used to model the minimum of a large number of independent positive random variables from a certain class of distributions. Thus extreme value theory also suggests that the Weibull distribution may be suitable. The more common justification for its use is empirical: the Weibull distribution can be used to model failure-time data with decreasing or increasing hf. The Weibull distribution described in this section is sometimes referred to as the "two-parameter Weibull distribution" to distinguish it from the three-parameter Weibull distribution described in Section 5.10.2.

4.9 LARGEST EXTREME VALUE DISTRIBUTION

When Y has a largest extreme value distribution, we indicate this by $Y \sim \text{LEV}(\mu, \sigma)$. The largest extreme value distribution cdf, pdf, and hf are

$$F(y; \mu, \sigma) = \Phi_{\text{lev}}\left(\frac{y - \mu}{\sigma}\right),$$

$$f(y; \mu, \sigma) = \frac{1}{\sigma}\phi_{\text{lev}}\left(\frac{y - \mu}{\sigma}\right),$$

LARGEST EXTREME VALUE DISTRIBUTION

$$h(y; \mu, \sigma) = \frac{\exp\left(-\frac{y-\mu}{\sigma}\right)}{\sigma \left\{\exp\left[\exp\left(-\frac{y-\mu}{\sigma}\right)\right] - 1\right\}}, \quad -\infty < y < \infty,$$

where $\Phi_{\text{lev}}(z) = \exp[-\exp(-z)]$ and $\phi_{\text{lev}}(z) = \exp[-z - \exp(-z)]$ are cdf and pdf for the standardized LEV($\mu = 0, \sigma = 1$) distribution. Here $-\infty < \mu < \infty$ is a location parameter and $\sigma > 0$ is a scale parameter. The LEV cdf, pdf, and hf are graphed in Figure 4.6 for $\mu = 10$ and $\sigma = 5, 6,$ and 7.

The mean, variance, and quantile functions of the largest extreme value distribution are $E(Y) = \mu + \sigma\gamma$, $\text{Var}(Y) = \sigma^2 \pi^2/6$, and $y_p = \mu + \Phi_{\text{lev}}^{-1}(p)\sigma$, where $\Phi_{\text{lev}}^{-1}(p) = -\log[-\log(p)]$. Note the close relationship between LEV and SEV: if $Y \sim \text{LEV}(\mu, \sigma)$ then $-Y \sim \text{SEV}(-\mu, \sigma)$ and $\Phi_{\text{lev}}^{-1}(p) = -\Phi_{\text{sev}}^{-1}(1-p)$.

The theory of extreme values shows that the LEV distribution can be used to model the maximum of a large number of random variables from a certain class of distributions (which includes the normal distribution). As shown in Figure 4.6, the largest extreme value pdf is skewed to the right. The LEV hf always increases but is bounded in the sense that $\lim_{t \to \infty} h(t; \mu, \sigma) = 1/\sigma$. Although most failure-time distributions are skewed to the right, the LEV distribution is not commonly used as a model for failure times. This is because the LEV distribution (like the SEV and normal

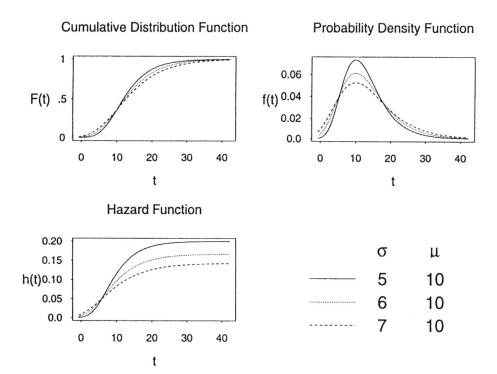

Figure 4.6. Largest extreme value cdf, pdf, and hf with $\mu = 10$ and $\sigma = 5, 6,$ and 7.

distributions) has positive probability of negative observations and there are a number of other right-skewed distributions that do not have this property. Nevertheless, the LEV distribution could be used as a model for life if σ is small relative to $\mu > 0$.

4.10 LOGISTIC DISTRIBUTION

When Y has a logistic distribution, we indicate this by $Y \sim \text{LOGIS}(\mu, \sigma)$. The logistic distribution is a location-scale distribution with cdf, pdf, and hf

$$F(y; \mu, \sigma) = \Phi_{\text{logis}}\left(\frac{y-\mu}{\sigma}\right),$$

$$f(y; \mu, \sigma) = \frac{1}{\sigma}\phi_{\text{logis}}\left(\frac{y-\mu}{\sigma}\right),$$

$$h(y; \mu, \sigma) = \frac{1}{\sigma}\Phi_{\text{logis}}\left(\frac{y-\mu}{\sigma}\right), \quad -\infty < y < \infty,$$

where $\Phi_{\text{logis}}(z) = \exp(z)/[1+\exp(z)]$ and $\phi_{\text{logis}}(z) = \exp(z)/[1+\exp(z)]^2$ are the cdf and pdf, respectively, for a standardized $\text{LOGIS}(\mu = 0, \sigma = 1)$. Here $-\infty < \mu < \infty$ is a location parameter and $\sigma > 0$ is a scale parameter. The logistic cdf, pdf, and hf are graphed in Figure 4.7 for location parameter $\mu = 15$ and scale parameter $\sigma = 1$, 2, and 3.

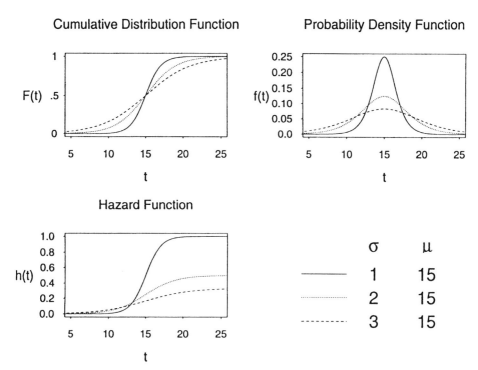

Figure 4.7. Logistic cdf, pdf, and hf with $\mu = 15$ and $\sigma = 1, 2$, and 3.

For integer $m > 0$, $E[(Y-\mu)^m] = 0$ if m is odd, and $E[(Y-\mu)^m] = 2\sigma^m (m!)[1 - (1/2)^{m-1}]\sum_{i=1}^{\infty}(1/i)^m$ if m is even. From this $E(Y) = \mu$ and $Var(Y) = \sigma^2\pi^2/3$. The p quantile is $y_p = \mu + \Phi_{\text{logis}}^{-1}(p)\sigma$, where $\Phi_{\text{logis}}^{-1}(p) = \log[p/(1-p)]$ is the p quantile of the standard logistic distribution.

The shape of the logistic distribution is very similar to that of the normal distribution; the logistic distribution has slightly "longer tails." In fact, it would require an extremely large number of observations to assess whether data come from a normal or logistic distribution. The main difference between the distributions is in the behavior of the hf in the upper tail of the distribution, where the logistic hf levels off, approaching $1/\sigma$ for large y. For some purposes, the logistic distribution has been preferred to the normal distribution because its cdf can be written in a simple closed form. With modern software, however, it is not any more difficult to compute probabilities from a normal cdf.

4.11 LOGLOGISTIC DISTRIBUTION

When T has a loglogistic distribution, we indicate this by $T \sim \text{LOGLOGIS}(\mu, \sigma)$. If $T \sim \text{LOGLOGIS}(\mu, \sigma)$ then $Y = \log(T) \sim \text{LOGIS}(\mu, \sigma)$. The loglogistic cdf, pdf, and hf are

$$F(t; \mu, \sigma) = \Phi_{\text{logis}}\left[\frac{\log(t) - \mu}{\sigma}\right],$$

$$f(t; \mu, \sigma) = \frac{1}{\sigma t}\phi_{\text{logis}}\left[\frac{\log(t) - \mu}{\sigma}\right],$$

$$h(t; \mu, \sigma) = \frac{1}{\sigma t}\Phi_{\text{logis}}\left[\frac{\log(t) - \mu}{\sigma}\right], \quad t > 0,$$

where ϕ_{logis} and Φ_{logis} are the pdf and cdf, respectively, for a standardized LOGIS, defined in Section 4.10. The median $t_{.5} = \exp(\mu)$ is a scale parameter and $\sigma > 0$ is a shape parameter. The LOGLOGIS cdf, pdf and hf are graphed in Figure 4.8 for scale parameter $\exp(\mu) = 1$ and $\sigma = .2, .4$, and $.6$.

For integer $m > 0$, $E(T^m) = \exp(m\mu)\Gamma(1 + m\sigma)\Gamma(1 - m\sigma)$, where $\Gamma(x)$ is the gamma function. From this $E(T) = \exp(\mu)\Gamma(1 + \sigma)\Gamma(1 - \sigma)$ and $Var(T) = \exp(2\mu)[\Gamma(1+2\sigma)\Gamma(1-2\sigma) - \Gamma^2(1+\sigma)\Gamma^2(1-\sigma)]$. Note that for values of $\sigma \geq 1$, the mean of T does not exist and for $\sigma \geq 1/2$, the variance of T does not exist. The p quantile function is $t_p = \exp[\mu + \Phi_{\text{logis}}^{-1}(p)\sigma]$, where $\Phi_{\text{logis}}^{-1}(p)$ is defined in Section 4.10.

Corresponding to the similarity between the logistic and normal distributions, the shape of the loglogistic distribution is similar to that of the lognormal distribution.

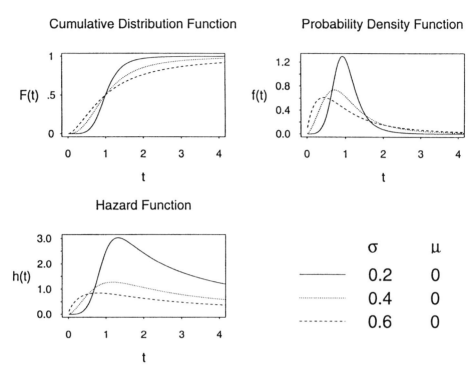

Figure 4.8. Loglogistic cdf, pdf, and hf for $t_{.5} = \exp(\mu) = 1$ and $\sigma = .2, .4,$ and $.6$.

4.12 PARAMETERS AND PARAMETERIZATION

The choice of $\boldsymbol{\theta}$, a set of parameters (the values of which are usually unknown) to describe a particular model, is somewhat arbitrary and may depend on tradition, on physical interpretation, or on having a model parameterization with desirable computational properties for estimating parameters. For example, the exponential distribution can be written in terms of its mean θ, as in (4.1), or its constant hazard $\lambda = 1/\theta$. The μ, σ notation for the Weibull distribution allows us to see connections with other location-scale-based distributions. The traditional parameters of a normal distribution are $\theta_1 = \mu$ and $\theta_2 = \sigma > 0$, the mean and standard deviation, respectively. An alternative with no restrictions on the range of the parameters would be $\theta_1 = \mu$ and $\theta_2 = \log(\sigma)$. Another parameterization, which may have better numerical properties for estimation with heavily censored data sets, is $\theta_1 = \mu + z_p\sigma$ and $\theta_2 = \log(\sigma)$, where z_p is the p quantile of the standard normal distribution. The best value of p to use depends on the amount of censoring. In particular, if the sample contains failure times with no censoring, choose $p = .5$ with $z_p = 0$ because then $\widehat{\theta}_1$ (the maximum likelihood estimate of the mean) and $\widehat{\theta}_2$ (the maximum likelihood estimate of the log standard deviation) would be statistically independent (this is a well-known result from statistical theory). Exercise 8.20 explores this issue more thoroughly.

4.13 GENERATING PSEUDORANDOM OBSERVATIONS FROM A SPECIFIED DISTRIBUTION

Simulation (or Monte Carlo simulation) methods are becoming increasingly important for many applications of statistics and, indeed, quantitative analysis in general. In particular, it is possible to determine, through simulation, numerical quantities that are difficult or impossible to compute by purely analytical means. This book uses a simulation approach in a number of methods, examples, and exercises. A pseudorandom number generator is the basic building block of any simulation application. This section will show some simple methods for generating pseudorandom numbers from specified probability distributions. The bibliographic notes at the end of this chapter give references for more technical details and more advanced methods of generating pseudorandom numbers from specified distributions.

4.13.1 Uniform Pseudorandom Number Generator

Most computers, data analysis software, and spreadsheets provide a pseudorandom number generator for the uniform distribution on $(0, 1)$ [denoted by UNIF$(0, 1)$]. This distribution has its probability distributed uniformly from $(0, 1)$. The cdf and pdf of the UNIF$(0, 1)$ distribution are $F_U(u) = u$ and $f_U(u) = 1$, $0 < u < 1$. Pseudorandom numbers from the UNIF$(0, 1)$ distribution can be used easily to generate random numbers from other distributions, both discrete and continuous.

4.13.2 Pseudorandom Observations from Continuous Distributions

Suppose $U_1, \ldots, U_n$ is a pseudorandom sample from a UNIF$(0, 1)$. Then if $t_p = F_T^{-1}(p)$ is the quantile function for the distribution of the random variable T from which a sample of pseudorandom numbers is desired, $T_1 = F_T^{-1}(U_1), \ldots, T_n = F_T^{-1}(U_n)$ is a pseudorandom sample from F_T. For example, to generate a pseudorandom sample from the Weibull distribution for specified parameters η and β, first obtain the UNIF$(0, 1)$ pseudorandom sample $U_1, \ldots, U_n$ and then compute $T_1 = \eta[-\log(1 - U_1)]^{1/\beta}, \ldots, T_n = \eta[-\log(1 - U_n)]^{1/\beta}$. Similarly, for the lognormal distribution the pseudorandom sample can be obtained from $T_1 = \exp[\mu + \Phi_{\text{nor}}^{-1}(U_1)\sigma], \ldots, T_n = \exp[\mu + \Phi_{\text{nor}}^{-1}(U_n)\sigma]$.

4.13.3 Efficient Generation of Censored Pseudorandom Samples

This section shows how to generate pseudorandom *censored* samples from a specified cdf $F(t; \boldsymbol{\theta})$. Such samples are useful for implementing simulations like those used throughout the book and for bootstrap methods like those described in Chapter 9.

General Approach
Let $U_{(i)}$ denote the ith order statistic from a random sample of size n from a UNIF$(0, 1)$ distribution. Using the properties of order statistics, the conditional distribution of

$U_{(i)}$ given $U_{(i-1)}$ is

$$\Pr\left[U_{(i)} \leq u | U_{(i-1)} = u_{(i-1)}\right] = 1 - \left[\frac{1-u}{1-u_{(i-1)}}\right]^{(n-i+1)}, \quad u \geq u_{(i-1)}.$$

Let U be a pseudorandom UNIF(0, 1) variable. Then using the method described in Section 4.13.2, given $U_{(i-1)}$ (where $U_{(0)} = 0$), a pseudorandom observation $U_{(i)}$ is

$$U_{(i)} = 1 - [1 - U_{(i-1)}] \times (1-U)^{1/(n-i+1)}, \quad i = 1, \ldots, n.$$

Pseudorandom uniform order statistics generated in this way can be used to generate failure- and time-censored samples.

Failure-Censored Samples
The algorithm to generate a pseudorandom failure-censored sample (Type II censored) with n units and r failures is as follows:

1. Generate $U_1, \ldots, U_r$ pseudorandom observations from a UNIF(0, 1).
2. Compute the uniform pseudorandom order statistics

$$U_{(1)} = 1 - [1 - U_{(0)}] \times (1-U_1)^{1/n}$$
$$U_{(2)} = 1 - [1 - U_{(1)}] \times (1-U_2)^{1/(n-1)}$$
$$\vdots$$
$$U_{(r)} = 1 - [1 - U_{(r-1)}] \times (1-U_r)^{1/(n-r+1)}$$

3. The pseudorandom sample from $F(t; \boldsymbol{\theta})$ is

$$T_{(i)} = F^{-1}[U_{(i)}; \boldsymbol{\theta}], \quad i = 1, \ldots, r.$$

For example, for a log-location-scale based cdf with $F(t; \boldsymbol{\theta}) = \Phi[(\log(t) - \mu)/\sigma]$,

$$T_{(i)} = \exp\left\{\mu + \sigma \Phi^{-1}[U_{(i)}]\right\}, \quad i = 1, \ldots, r.$$

Time-Censored Samples
The algorithm to generate a pseudorandom time-censored sample (Type I censored) with n units and censoring time t_c uses the formulas in Section 4.13.3 to generate $T_{(1)}, T_{(2)}, \ldots$ sequentially. The process continues until a failure-time observation, say, $T_{(i)}$, exceeds t_c. This yields a censored sample consisting of $T_{(1)}, \ldots, T_{(i-1)}$ failure times and $(n - i + 1)$ censored observations. Specifically, define $U_{(0)} = 0$, start with $i = 1$, and generate the sequence as follows:

1. Generate a new pseudorandom observation U_i from a UNIF(0, 1). Compute $U_{(i)} = 1 - [1 - U_{(i-1)}] \times (1 - U_i)^{1/(n-i+1)}$ and $T_{(i)} = F^{-1}[U_{(i)}; \boldsymbol{\theta}]$.

2. If $T_{(i)} > t_c$, stop; the sample consists of the failure times $T_{(1)}, \ldots, T_{(i-1)}$ and $(n - i + 1)$ censored observations.
   3. If $T_{(i)} \leq t_c$, increment i and return to step 1.

Note that if $T_{(1)} > t_c$, there are no failures before t_c.

4.13.4 Pseudorandom Observations for Discrete Distributions

The same general idea used in Section 4.13.2 can be used to generate data from a discrete distribution. The process, however, can be a little more complicated if the discrete distribution quantiles cannot be computed directly. The multinomial distribution described in Section 2.2.1 is a good example. Starting with the values $\pi_1, \ldots, \pi_m$, compute $F(t_i)$ using (2.9). Then for each $U_i, i = 1, \ldots, n$, T_i is the smallest value of t such that $F(t) \geq U_i$. If m is not too large, it is possible to use a look-up table to determine T_i as a function of U_i. Otherwise, it is necessary to search through the possible values of $F(t)$ to find the first one exceeding U_i.

BIBLIOGRAPHIC NOTES

Johnson, Kotz, and Balakrishnan (1994, 1995) provide detailed information on a wide range of different continuous probability distribution functions. Evans, Hastings, and Peacock (1993) provide a brief description and summary of properties of a large number of parametric distributions including most, but not all, of the distributions outlined in this chapter. Crow and Shimizu (1988) provide detailed information on the lognormal distribution. Galambos (1978) is an important reference for the asymptotic theory of extreme value distributions, providing extensive theory and background. Balakrishnan (1991) gives detailed information on the logistic distribution. Kennedy and Gentle (1980) provide detailed information on generation of pseudorandom numbers from the uniform distribution and a variety of other special distributions. Morgan (1984) and Ripley (1987) do the same and also provide useful material on how to do stochastic simulations. Kennedy and Gentle (1980, pages 225–227) and Castillo (1988, pages 58–63) describe methods for generating pseudorandom order statistics.

EXERCISES

4.1. Show that for a continuous $F(t)$, $h(t) = 1/\theta, t > 0$ is a constant (i.e., not depending on t) if and only if $T \sim \text{EXP}(\theta)$.

4.2. Derive expressions for the mean, variance, and quantile functions of the exponential distribution.

4.3. Derive the expression for $\Phi_{\text{sev}}^{-1}(p)$ based on the expression for $\Phi_{\text{sev}}(z)$ in Section 4.7.

4.4. Show that if Y has a SEV(μ, σ) distribution then $-Y$ has a LEV$(-\mu, \sigma)$ distribution.

▲**4.5.** Let $T \sim$ WEIB(μ, σ), $\eta = \exp(\mu)$, and $\beta = 1/\sigma$.
(a) For $m > 0$, show that $E(T^m) = \eta^m \Gamma(1 + m/\beta)$, where $\Gamma(x)$ is the gamma function.
(b) Use the result in (a) to show that $E(T) = \eta \Gamma(1 + 1/\beta)$ and Var$(T) = \eta^2[\Gamma(1 + 2/\beta) - \Gamma^2(1 + 1/\beta)]$.

4.6. Consider the Weibull distributions with parameters $\eta = 10$ years and $\beta = .5$, 1, 2, and 4.
(a) Compute (using a computer if available) and graph the Weibull hf for t ranging between 0 and 10.
(b) Explain the practical interpretation of the hf at $t = 1$ and $t = 10$ years.
(c) Compute and plot the Weibull cdfs over the same range of t. For which shape parameter value is the probability of failing the largest at 1 year? At 10 years? Explain.

4.7. Consider the Weibull $h(t)$. Note that when $\beta = 1$, $h(t)$ is constant and that when $\beta = 2$, $h(t)$ increases linearly. Show that if:
(a) $0 < \beta < 1$, then $h(t)$ is decreasing in t.
(b) $1 < \beta < 2$, then $h(t)$ is concave increasing.
(c) $\beta > 2$, then $h(t)$ is convex increasing.

4.8. Starting with equation (4.6), show that the distribution of $Y = \log(T)$ is SEV$[\log(\eta), 1/\beta]$.

4.9. Derive the expression for $\Phi_{\text{logis}}^{-1}(p)$ based on the expressions for $\Phi_{\text{logis}}(z)$ given in Section 4.10.

4.10. Even though, theoretically, the SEV, LEV, LOGIS, and NOR distributions can take on negative values, the probability of nonpositive outcomes is negligible for certain combinations of parameters.
(a) For the combinations of parameter values for the SEV distribution shown in Figure 4.4, compute $\Pr(Y \leq 0)$.
(b) For the SEV, LEV, LOGIS, and NOR distributions, derive a general expression relating μ and σ, and guaranteeing that $\Pr(Y \leq 0) \leq \epsilon$.

4.11. Show that if T is LOGNOR(μ, σ) then $1/T$ is LOGNOR$(-\mu, \sigma)$.

4.12. The exponential distribution is said to possess a "memoryless" property. This memoryless property implies that a used unit is just as reliable as one that is new—that there is no wearout. Probabilistically this memoryless property can be stated as $\Pr(T \leq \delta) = \Pr(T \leq t_0 + \delta \mid T > t_0)$ for any $t_0 > 0$. Show

EXERCISES

that for a continuous random variable, this memoryless property holds if and only if $T \sim \text{EXP}(\theta)$.

▲4.13. Show that if Y is $\text{SEV}(\mu, \sigma)$ then $E(Y) = \mu - \sigma\gamma$ and $\text{Var}(Y) = \sigma^2\pi^2/6$, where $\gamma \approx .5772$, in this context, is known as Euler's constant. Observe that from integral tables one gets $\int_0^\infty \log(x)\exp(-x)\,dx = -\gamma$ and

$$\int_0^\infty [\log(x)]^2 \exp(-x)\,dx = \frac{\pi^2}{6} + \gamma^2.$$

▲4.14. Assume that T is $\text{LOGNOR}(\mu, \sigma)$ and m is an arbitrary real number.
   (a) Show that $E(T^m) = \exp(\mu m + .5\sigma^2 m^2)$.
   (b) Use the result in (a) to show that $E(T) = \exp(\mu + .5\sigma^2)$ and $\text{Var}(T) = \exp[2\mu + \sigma^2][\exp(\sigma^2) - 1]$.

4.15. The coefficient of variation, γ_2, is a useful scale-free measure of relative variability for a random variable.
   (a) Derive an expression for the coefficient of variation for the Weibull distribution.
   (b) Compute γ_2 for all combinations of $\beta = .5, 1, 3, 5$ and $\eta = 50, 100$. Also, draw (or use the computer to draw) a graph of the Weibull pdfs for the same combinations of parameters.
   (c) Explain the effect that changes in η and β have on the shape of the Weibull density and the effect that they have on γ_2.

4.16. The coefficient of skewness, γ_3, is a useful scale-free measure of skewness in the distribution of a random variable. Do Exercise 4.15 for the coefficient of skewness.

◆4.17. Generate 500 pseudorandom observations from a "parent" lognormal distribution with $\mu = 5$ and $\sigma = .8$.
   (a) Compare the histogram of the observations with a plot of the parent lognormal density function.
   (b) Compute the sample median of the 500 observations and compare it with the median of the parent lognormal distribution.
   (c) Compute the sample mean of the 500 observations and compare it with the mean of the parent lognormal distribution.
   (d) Compute the sample standard deviation of the 500 observations and compare with the standard deviation of the parent lognormal distribution.

4.18. Repeat Exercise 4.17 using a Weibull distribution with $\eta = 100$ and $\beta = .5$. Comment on how the results differ from those with the lognormal distribution.

▲4.19. Consider a lognormal distribution with cdf $F(t; \mu, \sigma)$.

(a) Show that for any values of μ and σ, $h(t)$ always has the following characteristics: $\lim_{t\to\infty} h(t) = 0$, $\lim_{t\to 0} h(t) = 0$, and $h(t)$ has a unique maximum at a point t_{max}, with $0 < t_{max} < \infty$.

(b) Show that t_{max} satisfies the relationship

$$h(t_{max}) = \frac{1}{t_{max}}\left[1 + \frac{\log(t_{max}) - \mu}{\sigma^2}\right].$$

(c) Use the result in (b) to show that

$$\exp(\mu - \sigma^2) \leq t_{max} \leq \exp(\mu - \sigma^2 + 1)$$

and

$$\Phi_{nor}(-\sigma) \leq F(t_{max}; \mu, \sigma) \leq \Phi_{nor}\left(-\sigma + \frac{1}{\sigma}\right).$$

(d) Comment on the effect that σ has on how "early" or "late" in time the lognormal hf reaches its maximum. In particular, show that (i) for large values of σ, $h(t)$ is increasing only on an interval of negligible probability; and (ii) for small values of σ, $h(t)$ is increasing in an interval that has at least a probability of about 50%.

(e) Plot the hf when the parameter (μ, σ) values are $(0, 5)$; $(0, 1/5)$; $(1, 5)$; $(1, 1/5)$. Comment on the adequacy of the probability bounds given in (c).

▲4.20. Show that for any value of the parameters (μ, σ), the normal $h(y)$ is always increasing.

CHAPTER 5

Other Parametric Distributions

Objectives

This chapter explains:

- The properties and importance of various parametric distributions that cannot be transformed into a location-scale distribution.
- Threshold-parameter distributions.
- How some statistical distributions can be determined by applying basic ideas of probability theory to physical properties of a failure process, system, or population of units.

Overview

This chapter is a continuation of Chapter 4, describing more advanced parametric distributions. This chapter is a prerequisite only for Chapter 11 and may otherwise be omitted. Section 5.2 describes the gamma distribution while Section 5.3 describes the generalized gamma and the extended generalized gamma distributions. The generalized gamma distribution contains the lognormal and Weibull distributions as special cases and is thus useful for statistical assessment of the best fitting distribution. Sections 5.5–5.11 describe and compare a variety of other potentially useful parametric distributions. Some of these were developed on the basis of physical theory. Section 5.12 describes other methods of deriving useful probability distributions from physical or other considerations.

5.1 INTRODUCTION

Chapter 4 introduced a number of important probability distributions, all of which belong to or could be transformed into a distribution that belongs to the location-scale family of distributions. This chapter describes a number of additional probability distributions that have also been useful in reliability data analysis.

It is not necessary to be familiar with all of the formulas and other details in this chapter in order to use these distributions in reliability applications. They are included to show some of the connections among various probability distributions that are commonly used in reliability modeling and to provide background for the application of these distributions in Chapters 6 and 11.

5.2 GAMMA DISTRIBUTION

5.2.1 Gamma cdf, pdf, and Hazard Function

When T has a gamma distribution, we indicate this by $T \sim \text{GAM}(\theta, \kappa)$. The gamma distribution cdf and pdf are

$$F(t; \theta, \kappa) = \Gamma_I\left(\frac{t}{\theta}; \kappa\right), \tag{5.1}$$

$$f(t; \theta, \kappa) = \frac{1}{\Gamma(\kappa)\theta}\left(\frac{t}{\theta}\right)^{\kappa-1} \exp\left(-\frac{t}{\theta}\right), \quad t > 0, \tag{5.2}$$

where $\theta > 0$ is a scale parameter and $\kappa > 0$ is a shape parameter. Here $\Gamma_I(v; \kappa)$ is the incomplete gamma function defined by

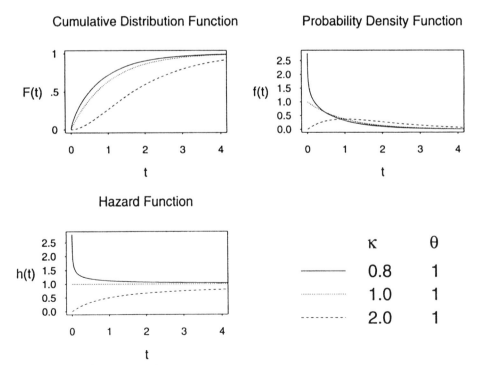

Figure 5.1. Gamma cdf, pdf, and hf for $\theta = 1$ and $\kappa = .8, 1,$ and 2.

… GENERALIZED GAMMA DISTRIBUTION … 99

$$\Gamma_I(v;\kappa) = \frac{\int_0^v x^{\kappa-1}\exp(-x)\,dx}{\Gamma(\kappa)}, \quad v > 0. \tag{5.3}$$

The gamma pdf, cdf, and hazard functions are graphed in Figure 5.1. The gamma distribution can be useful for modeling certain life distributions. Letting $\kappa = 1$ gives the exponential distribution as a special case. As shown in Figure 5.1, the gamma distribution hazard function can be either decreasing (when $\kappa < 1$) or increasing (when $\kappa > 1$), in either case approaching a constant level late in life [i.e., $\lim_{t\to\infty} h(t;\theta,\kappa) = 1/\theta$]. Also, the sum of κ independent exponential random variables with mean θ has a gamma distribution with parameters θ and κ. This property can be used to motivate the use of the gamma distribution in some applications.

5.2.2 Gamma Moments and Quantiles

For any $m \geq 0$, $\mathrm{E}(T^m) = \theta^m \Gamma(m+\kappa)/\Gamma(\kappa)$. From this it follows that $\mathrm{E}(T) = \theta\kappa$ and $\mathrm{Var}(T) = \theta^2\kappa$. The gamma p quantile is $t_p = \theta\Gamma_I^{-1}(p;\kappa)$, where $\Gamma_I^{-1}(p;\kappa)$ is the inverse of the incomplete gamma function defined in (5.3). That is, $\Gamma_I\left[\Gamma_I^{-1}(p;\kappa);\kappa\right] = p$.

5.2.3 Gamma Standardized Parameterization

The gamma cdf and pdf can also be written as follows:

$$F(t;\theta,\kappa) = \Phi_{\lg}[\log(t) - \mu;\kappa].$$

$$f(t;\theta,\kappa) = \frac{1}{t}\phi_{\lg}[\log(t) - \mu;\kappa], \quad t > 0,$$

where $\mu = \log(\theta)$ and

$$\Phi_{\lg}(z;\kappa) = \Gamma_I[\exp(z);\kappa],$$

$$\phi_{\lg}(z;\kappa) = \frac{1}{\Gamma(\kappa)}\exp[\kappa z - \exp(z)]$$

are, respectively, the cdf and pdf for the standardized loggamma variable $Z = \log(T/\theta) = \log(T) - \mu$. Unlike the standardized distributions used in Chapter 4, these standardized distributions depend on the shape parameter κ.

5.3 GENERALIZED GAMMA DISTRIBUTION

5.3.1 Generalized Gamma cdf and pdf

The generalized gamma distribution contains the exponential, gamma, Weibull, and lognormal distributions as special cases. When T has a generalized gamma distribution we indicate this by $T \sim \mathrm{GENG}(\theta,\beta,\kappa)$. As will be demonstrated in Chapter 11, the GENG distribution is useful for helping to choose among these special-case distributions. The cdf and pdf for the generalized gamma distribution are

$$F(t; \theta, \beta, \kappa) = \Gamma_I\left[\left(\frac{t}{\theta}\right)^\beta; \kappa\right], \tag{5.4}$$

$$f(t; \theta, \beta, \kappa) = \frac{\beta}{\Gamma(\kappa)\theta}\left(\frac{t}{\theta}\right)^{\kappa\beta-1}\exp\left[-\left(\frac{t}{\theta}\right)^\beta\right], \quad t > 0,$$

where $\theta > 0$ is a scale parameter, $\beta > 0$ and $\kappa > 0$ are shape parameters, and $\Gamma_I(v, \kappa)$ is the incomplete gamma function given in (5.3).

5.3.2 Generalized Gamma Moments and Quantiles

For $m \geq 0$, $E(T^m) = \theta^m\,\Gamma\left(m/\beta + \kappa\right)/\Gamma(\kappa)$. From this,

$$E(T) = \frac{\theta\,\Gamma(1/\beta + \kappa)}{\Gamma(\kappa)},$$

$$\text{Var}(T) = \theta^2\left[\frac{\Gamma\left(2/\beta + \kappa\right)}{\Gamma(\kappa)} - \frac{\Gamma^2\left(1/\beta + \kappa\right)}{\Gamma^2(\kappa)}\right].$$

The p quantile is $t_p = \theta\left[\Gamma_I^{-1}(p; \kappa)\right]^{1/\beta}$.

5.3.3 Special Cases of the Generalized Gamma Distribution

In this section we show the relationship between the GENG(θ, β, κ) and the well-known distributions that are special cases.

- When $\beta = 1$, $T \sim \text{GAM}(\theta, \kappa)$.
- When $\kappa = 1$, $T \sim \text{WEIB}(\theta, \beta)$.
- When $(\beta, \kappa) = (1, 1)$, $T \sim \text{EXP}(\theta)$.
- As $\kappa \to \infty$, $T \sim \text{LOGNOR}[\log(\theta) + \log(\kappa)/\beta, 1/(\beta\sqrt{\kappa})]$.

5.3.4 Generalized Gamma Reparameterization for Numerical Stability

The parameterization in terms of (θ, β, κ) is generally numerically unstable for fitting the distribution to data. Farewell and Prentice (1977) recommend the alternative parameterization

$$\mu = \log(\theta) + \frac{1}{\beta}\log(\lambda^{-2}), \quad \sigma = \frac{1}{\beta\sqrt{\kappa}}, \quad \lambda = \frac{1}{\sqrt{\kappa}}.$$

This parameterization is numerically stable if there is little or no censoring. Using the cdf in (5.4) and defining $\omega = [\log(t) - \mu]/\sigma$ gives

$$F(t; \theta, \beta, \kappa) = \Gamma_I\left[\lambda^{-2}\exp(\lambda w); \lambda^{-2}\right] = \Phi_{lg}\left[\lambda\omega + \log(\lambda^{-2}); \lambda^{-2}\right],$$

$$f(t; \theta, \beta, \kappa) = \frac{\lambda}{\sigma t}\phi_{lg}\left[\lambda\omega + \log(\lambda^{-2}); \lambda^{-2}\right], \quad t > 0,$$

where $-\infty < \mu < \infty$, $\sigma > 0$, and $\lambda > 0$. In this parameterization the quantile function is

$$t_p = \exp\left\{\mu + \frac{\sigma}{\lambda}\log\left[\lambda^2\Gamma_I^{-1}\left(p; \frac{1}{\lambda^2}\right)\right]\right\}.$$

5.4 EXTENDED GENERALIZED GAMMA DISTRIBUTION

Using the alternative stable parameterization and allowing λ to become negative generalizes the GENG to what we will call the extended generalized gamma distribution, enlarging the family to include other distributions as special cases. Then T has an EGENG(μ, σ, λ) distribution with pdf and cdf given by

$$F(t; \mu, \sigma, \lambda) = \begin{cases} \Phi_{lg}\left[\lambda\omega + \log(\lambda^{-2}); \lambda^{-2}\right] & \text{if } \lambda > 0 \\ \Phi_{nor}(\omega) & \text{if } \lambda = 0 \\ 1 - \Phi_{lg}\left[\lambda\omega + \log(\lambda^{-2}); \lambda^{-2}\right] & \text{if } \lambda < 0, \end{cases} \quad (5.5)$$

$$f(t; \mu, \sigma, \lambda) = \begin{cases} \dfrac{|\lambda|}{\sigma t}\phi_{lg}\left[\lambda\omega + \log(\lambda^{-2}); \lambda^{-2}\right] & \text{if } \lambda \neq 0 \\ \dfrac{1}{\sigma t}\phi_{nor}(\omega) & \text{if } \lambda = 0, \end{cases} \quad (5.6)$$

where $t > 0$, $\omega = [\log(t) - \mu]/\sigma$, $-\infty < \mu < \infty$, $-\infty < \lambda < \infty$, and $\sigma > 0$. Note that if $T \sim \text{EGENG}(\mu, \sigma, \lambda)$ and $c > 0$ then $cT \sim \text{EGENG}[\mu + \log(c), \sigma, \lambda]$. Thus $\exp(\mu)$ is a scale parameter and σ and λ are shape parameters. For any given fixed value of λ, the EGENG distribution is a log-location-scale distribution.

5.4.1 Extended Generalized Gamma Moments and Quantiles

The moments for the EGENG can be obtained using

$$E(T^m) = \begin{cases} \dfrac{\exp(m\mu)(\lambda^2)^{m\sigma/\lambda}\Gamma\left[\lambda^{-1}(m\sigma + \lambda^{-1})\right]}{\Gamma(\lambda^{-2})} & \text{if } m\lambda\sigma + 1 > 0, \lambda \neq 0 \\ +\infty & \text{if } m\lambda\sigma + 1 \leq 0, \lambda \neq 0. \end{cases}$$

When $\lambda = 0$, $E(T^m) = \exp[m\mu + (1/2)(m\sigma)^2]$. Using these, it is easy to compute the mean, variance, and other central moments for the EGENG distribution.

Inverting (5.5) gives the EGENG p quantile

$$t_p = \exp[\mu + \sigma \times \omega(p; \lambda)],$$

where $\omega(p; \lambda)$ is the p quantile of $[\log(T) - \mu]/\sigma$ given by

$$\omega(p; \lambda) = \begin{cases} (1/\lambda) \log\left[\lambda^2 \Gamma_I^{-1}(p; \lambda^{-2})\right] & \text{if } \lambda > 0 \\ \Phi_{\text{nor}}^{-1}(p) & \text{if } \lambda = 0 \\ (1/\lambda) \log\left[\lambda^2 \Gamma_I^{-1}(1 - p; \lambda^{-2})\right] & \text{if } \lambda < 0. \end{cases}$$

5.4.2 Special Cases of the Extended Generalized Gamma Distribution

The EGENG(μ, σ, λ) distribution has the following important special cases:

- If $\lambda > 0$, then EGENG(μ, σ, λ) = GENG(μ, σ, λ).
- If $\lambda = 1$, $T \sim$ WEIB(μ, σ).
- If $\lambda = 0$, $T \sim$ LOGNOR(μ, σ).
- If $\lambda = -1$, $1/T \sim$ WEIB$(-\mu, \sigma)$ (i.e., T has a reciprocal Weibull, also known as the Fréchet distribution of *maxima*).
- When $\lambda = \sigma$, $T \sim$ GAM(θ, κ), where $\theta = \lambda^2 \exp(\mu)$ and $\kappa = \lambda^{-2}$.
- When $\lambda = \sigma = 1$, $T \sim$ EXP(θ), where $\theta = \exp(\mu)$.

5.5 GENERALIZED F DISTRIBUTION

5.5.1 Background

The generalized F distribution is a four-parameter distribution that includes the GENG family and the loglogistic, among other distributions, as special cases. The distribution is useful for choosing among the special case distributions.

5.5.2 Generalized F cdf and pdf Functions

When T has a generalized F distribution we write $T \sim$ GENF(μ, σ, κ, r). The GENF cdf and pdf are

$$F(t; \mu, \sigma, \kappa, r) = \Phi_{\text{lf}}\left[\frac{\log(t) - \mu}{\sigma}; \kappa, r\right],$$

$$f(t; \mu, \sigma, \kappa, r) = \frac{1}{\sigma t} \phi_{\text{lf}}\left[\frac{\log(t) - \mu}{\sigma}; \kappa, r\right], \quad t > 0,$$

where

$$\phi_{\text{lf}}(z; \kappa, r) = \frac{\Gamma(\kappa + r)}{\Gamma(\kappa)\Gamma(r)} \frac{(\kappa/r)^\kappa \exp(\kappa z)}{[1 + (\kappa/r)\exp(z)]^{\kappa+r}}$$

is the pdf of the central log F distribution with 2κ numerator and $2r$ denominator degrees of freedom and Φ_{lf} is the corresponding cdf (for which there is, in general, no closed-form expression). Also, $\phi_{\text{lf}}(z; \kappa, r)$ and $\Phi_{\text{lf}}(z; \kappa, r)$ are the pdf and cdf of $Z = [\log(T) - \mu]/\sigma$. Note that $\exp(\mu)$ is a scale parameter while $\sigma > 0$, $\kappa > 0$, and $r > 0$ are shape parameters. When $\kappa = r$, $\phi_{\text{lf}}(z; \kappa, r)$ is symmetric about $z = 0$.

5.5.3 Generalized F Moments and Quantiles

For $m \geq 0$,

$$E(T^m) = \begin{cases} \dfrac{\exp(m\mu)\,\Gamma(\kappa + m\sigma)\,\Gamma(r - m\sigma)}{\Gamma(\kappa)\,\Gamma(r)} \left(\dfrac{r}{\kappa}\right)^{m\sigma} & \text{if } r > m\sigma \\ \infty & \text{otherwise.} \end{cases}$$

$E(T)$ and $\text{Var}(T)$ can be computed directly from this expression. The p quantile of the generalized F is $t_p = \exp(\mu)[\mathcal{F}_{(p;2\kappa,2r)}]^\sigma$, where $\mathcal{F}_{(p;2\kappa,2r)}$ is the p quantile of an F distribution with $(2\kappa, 2r)$ degrees of freedom. The expression for the quantile follows directly from the fact that if $T \sim \text{GENF}(\mu, \sigma, \kappa, r)$, then $T = \exp(\mu) V^\sigma$ or equivalently $\log(T) = \mu + \sigma \log(V)$, where V has an F distribution with $(2\kappa, 2r)$ degrees of freedom. Finally, observe that for fixed (κ, r), the variable T has a log-location-scale distribution where the "standardized" distribution is the log of an F random variable with $(2\kappa, 2r)$ degrees of freedom.

5.5.4 Special Cases of the Generalized F Distribution

The $\text{GENF}(\mu, \sigma, \kappa, r)$ has a number of important special cases:

- $1/T \sim \text{GENF}(-\mu, \sigma, r, \kappa)$ (i.e., the reciprocal of T is also GENF).
- When $(\mu, \sigma) = (0, 1)$ then T has an F distribution (sometimes known as "Snedecor's F distribution") with 2κ numerator and $2r$ denominator degrees of freedom.
- When $(\kappa, r) = (1, 1)$, $\text{GENF}(\mu, \sigma, \kappa, r) = \text{LOGLOGIS}(\mu, \sigma)$.
- As $r \to \infty$, in the limit, $T \sim \text{GENG}[\exp(\mu)/\kappa^\sigma, 1/\sigma, \kappa]$.
- For $\kappa = 1$ and as $r \to \infty$, in the limit, $T \sim \text{WEIB}[\exp(\mu), 1/\sigma]$.
- When $\kappa = 1$, T has a Burr type XII distribution with cdf

$$F(t; \mu, \sigma, r) = 1 - \dfrac{1}{\left[1 + \dfrac{1}{r}\left[\dfrac{t}{\exp(\mu)}\right]^{1/\sigma}\right]^r}, \quad t > 0,$$

where $r > 0$ and $\sigma > 0$ are shape parameters and $\exp(\mu)$ is a scale parameter.
- As $\kappa \to \infty$ and $r \to \infty$, $T \overset{\cdot}{\sim} \text{LOGNOR}\left(\mu, \sigma\sqrt{(\kappa + r)/(\kappa r)}\right)$.

5.6 INVERSE GAUSSIAN DISTRIBUTION

5.6.1 Background

A common parameterization for the cdf of the inverse Gaussian (IGAU) distribution is (e.g., Chhikara and Folks, 1989)

$$\Pr(T \le t; \theta, \lambda) = \Phi_{\text{nor}}\left[\frac{(t-\theta)\sqrt{\lambda}}{\theta\sqrt{t}}\right] + \exp\left(\frac{2\lambda}{\theta}\right)\Phi_{\text{nor}}\left[-\frac{(t+\theta)\sqrt{\lambda}}{\theta\sqrt{t}}\right], \quad t > 0, \tag{5.7}$$

where $\theta > 0$ and $\lambda > 0$ are parameters having the same units as T.

The inverse Gaussian distribution was originally given by Schrödinger (1915) as the distribution of the first passage time in Brownian motion. The parameters θ and λ relate to the Brownian motion parameters as follows. Consider a Brownian process $B(t) = ct + dW(t), t > 0$, where c and d are constants and $W(t)$ is a Wiener process. Let T be the first passage time of a specified level b_0 [i.e., the first time that $B(t) \ge b_0$]. This leads directly to (5.7), where $\theta = b_0/c$ and $\sqrt{\lambda} = b_0/d$. Tweedie (1945) gives more details on this approach to deriving the IGAU distribution. Wald (1947) derived the inverse Gaussian as a limiting form for the distribution of the sample size for a sequential probability ratio test.

5.6.2 Inverse Gaussian cdf and pdf

For life data analysis (and other modeling applications) it is often more convenient to reparameterize the distribution so that it has a scale parameter and a shape parameter (instead of having two shape parameters that depend on the units in which time is measured). If T has an inverse Gaussian distribution, we denote this by $T \sim \text{IGAU}(\theta, \beta)$. The IGAU cdf and pdf are

$$F(t; \theta, \beta) = \Phi_{\text{ligau}}[\log(t/\theta); \beta], \tag{5.8}$$

$$f(t; \theta, \beta) = \frac{1}{t}\phi_{\text{ligau}}[\log(t/\theta); \beta], \quad t > 0,$$

where $\theta > 0$ is a scale parameter and $\beta = \lambda/\theta > 0$ is a unitless shape parameter.
Here

$$\Phi_{\text{ligau}}(z; \beta) = \Phi_{\text{nor}}\left\{\sqrt{\beta}\left[\frac{\exp(z)-1}{\exp(z/2)}\right]\right\} + \exp(2\beta)\Phi_{\text{nor}}\left\{-\sqrt{\beta}\left[\frac{\exp(z)+1}{\exp(z/2)}\right]\right\},$$

$$\phi_{\text{ligau}}(z; \beta) = \frac{\sqrt{\beta}}{\exp(z/2)}\phi_{\text{nor}}\left\{\sqrt{\beta}\left[\frac{\exp(z)-1}{\exp(z/2)}\right]\right\}, \quad -\infty < z < \infty$$

are the cdf and pdf, respectively, of $\log(T/\theta)$, the log standardized inverse Gaussian distribution. Figure 5.2 shows the IGAU cdf, pdf, and hazard functions.

5.6.3 Inverse Gaussian Moments and Quantiles

For integer $m > 0$,

$$\mathrm{E}(T^m) = \theta^m \sum_{i=0}^{m-1} \frac{(m-1+i)!}{i!(m-1-i)!}\left(\frac{1}{2\beta}\right)^i.$$

BIRNBAUM–SAUNDERS DISTRIBUTION

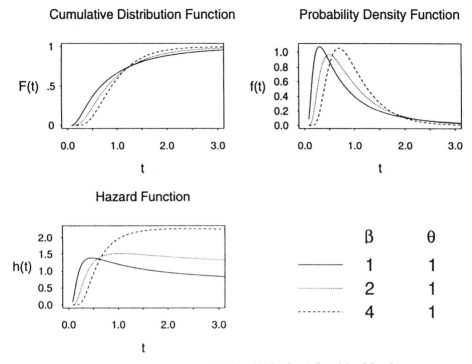

Figure 5.2. Inverse Gaussian cdf, pdf, and hf for $\beta = 1, 2,$ and 4 and $\theta = 1$.

From this $E(T) = \theta$ and $Var(T) = \theta^2/\beta$. The p quantile is $t_p = \theta \exp[\Phi^{-1}_{\text{ligau}}(p; \beta)]$. There is no simple closed-form equation for $\Phi^{-1}_{\text{ligau}}(p; \beta)$, so it must be computed by inverting $p = \Phi_{\text{ligau}}(z; \beta)$ numerically.

5.6.4 Inverse Gaussian Distribution Properties

The inverse Gaussian distribution has the following properties:

- If $T \sim \text{IGAU}(\theta, \beta)$ and $c > 0$ then $cT \sim \text{IGAU}(c\theta, \beta)$.
- The IGAU hazard function $h(t; \theta, \beta)$ is unimodal, $h(0; \theta, \beta) = 0$, and

$$\lim_{t \to \infty} h(t; \theta, \beta) = \beta/(2\theta).$$

- For large values of β, the IGAU distribution is very similar to a $\text{NOR}(\theta, \theta/\sqrt{\beta})$.

5.7 BIRNBAUM–SAUNDERS DISTRIBUTION

5.7.1 Background

The Birnbaum–Saunders distribution was derived by Birnbaum and Saunders (1969) based on a model for the number of cycles necessary to force a fatigue crack to grow to a critical size that would cause fracture.

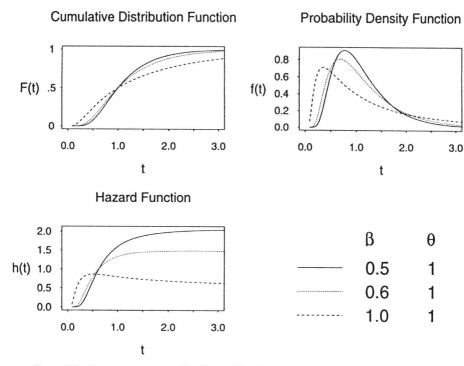

Figure 5.3. Birnbaum–Saunders cdf, pdf, and hf for shape parameter $\beta = .5, .6,$ and 1 and $\theta = 1$.

5.7.2 Birnbaum–Saunders cdf and pdf

If T has a Birnbaum–Saunders distribution, we denote this by $T \sim \text{BISA}(\theta, \beta)$. The BISA cdf and pdf are

$$F(t; \theta, \beta) = \Phi_{\text{nor}}(z),$$

$$f(t; \theta, \beta) = \frac{\sqrt{\frac{t}{\theta}} + \sqrt{\frac{\theta}{t}}}{2\beta t} \phi_{\text{nor}}(z), \quad t > 0,$$

where $\theta > 0$ is a scale parameter, $\beta > 0$ is a shape parameter, and

$$z = \frac{1}{\beta}\left(\sqrt{\frac{t}{\theta}} - \sqrt{\frac{\theta}{t}}\right).$$

Figure 5.3 shows the BISA cdf, pdf, and hazard function.

5.7.3 Compact Parameterization for Birnbaum–Saunders

Sometimes it is useful to write the BISA cdf and pdf as follows:

BIRNBAUM–SAUNDERS DISTRIBUTION

$$F(t; \theta, \beta) = \Phi_{\text{lbisa}}[\log(t/\theta); \beta],$$

$$f(t; \theta, \beta) = \frac{1}{t} \phi_{\text{lbisa}}[\log(t/\theta); \beta], \quad t > 0,$$

where

$$\Phi_{\text{lbisa}}(z; \beta) = \Phi_{\text{nor}}(v),$$

$$\phi_{\text{lbisa}}(z; \beta) = \left[\frac{\exp(z/2) + \exp(-z/2)}{2\beta} \right] \phi_{\text{nor}}(v),$$

and

$$v = \frac{1}{\beta}[\exp(z/2) - \exp(-z/2)], \quad -\infty < z < \infty.$$

5.7.4 Birnbaum–Saunders Moments and Quantiles

The mth moment of the Birnbaum–Saunders distribution is

$$E(T^m) = \theta^m \sum_{i=0}^{m} \beta^{2(m-i)} \frac{[2(m-i)]!}{[2^{3(m-i)}](m-i)!} \sum_{k=0}^{m-i} \binom{2m}{2k} \binom{m-k}{i}.$$

It follows that

$$E(T) = \theta\left(1 + \frac{\beta^2}{2}\right) \quad \text{and} \quad \text{Var}(T) = (\theta\beta)^2 \left(1 + \frac{5\beta^2}{4}\right).$$

The p quantile can be expressed as

$$t_p = \frac{\theta}{4} \left\{ \beta \, \Phi_{\text{nor}}^{-1}(p) + \sqrt{4 + \left[\beta \, \Phi_{\text{nor}}^{-1}(p)\right]^2} \right\}^2. \tag{5.9}$$

5.7.5 Properties of the Birnbaum–Saunders Distribution

The Birnbaum–Saunders distribution has the following properties:

- If $T \sim \text{BISA}(\theta, \beta)$ and $c > 0$ then $cT \sim \text{BISA}(c\theta, \beta)$.
- If $T \sim \text{BISA}(\theta, \beta)$ then $1/T \sim \text{BISA}(\theta^{-1}, \beta)$.
- The BISA hazard function $h(t; \theta, \beta)$ is not always increasing. Also, it is easy to show that $h(0; \theta, \beta) = 0$ and $\lim_{t \to \infty} h(t; \theta, \beta) = 1/(2\theta\beta^2)$. Through numerical experiments with a wide range of parameters it appears that $h(t; \theta, \beta)$ is always unimodal.

There is a close relationship between the BISA and the IGAU distributions. Simply stated, the BISA distribution was from a discrete-time degradation process. The

IGAU is based on an underlying continuous-time stochastic process. Nevertheless, comparison of Figures 5.2 and 5.3 shows that the shapes of the two distributions are very similar. Desmond (1986) describes the relationship in more detail. The BISA and the IGAU distributions are similar to the lognormal distribution (Section 4.6) in shape and behavior. This can be seen by comparing Figures 4.3, 5.2, and 5.3. A direct comparison fitting the models to data is illustrated in Figure 11.6.

5.8 GOMPERTZ–MAKEHAM DISTRIBUTION

The Gompertz–Makeham (or GOMA) distribution has an increasing hazard function and is used to model human life in middle age and beyond. A common parameterization for this distribution is

$$\Pr(T \leq t; \gamma, \kappa, \lambda) = 1 - \exp\left[-\frac{\lambda \kappa t + \gamma \exp(\kappa t) - \gamma}{\kappa}\right], \quad t > 0,$$

where $\gamma > 0, \kappa > 0, \lambda \geq 0$, and all the parameters have units that are the reciprocal of the units of t.

An alternative representation for the cdf is

$$\Pr(T \leq t; \gamma, \kappa, \lambda) = 1 - \left[\frac{1 - \Phi_{\text{sev}}\left(\frac{t - \mu}{\sigma}\right)}{1 - \Phi_{\text{sev}}\left(\frac{-\mu}{\sigma}\right)}\right] \exp(-\lambda t), \quad t > 0, \qquad (5.10)$$

where $\mu = -(1/\kappa)\log(\gamma/\kappa)$ and $\sigma = 1/\kappa$. When $\lambda = 0$, (5.10) reduces to the Gompertz distribution, corresponding to an SEV distribution truncated at $t = 0$ (i.e., an SEV random variable, conditional on being positive). The GOMA distribution satisfies a requirement for a positive random variable and has a hazard function similar to that of the SEV. In fact, as indicated in the following section, except for an additive constant, the hazard function of the Gompertz–Makeham distribution agrees with the hazard function of an SEV truncated at the origin. See Exercise 5.10 for another interpretation.

5.8.1 Gompertz–Makeham Scale/Shape Parameterization and cdf, pdf, and Hazard Functions

In order to separate out the scale parameter from the shape parameters, we parameterize in terms of $[\theta, \zeta, \eta] = [1/\kappa, \log(\kappa/\gamma), \lambda/\kappa]$ and say that $T \sim \text{GOMA}(\theta, \zeta, \eta)$ if

$$F(t; \theta, \zeta, \eta) = \Phi_{\text{lgoma}}\left[\log\left(\frac{t}{\theta}\right); \zeta, \eta\right],$$

$$f(t; \theta, \zeta, \eta) = \frac{1}{t} \phi_{\text{lgoma}}\left[\log\left(\frac{t}{\theta}\right); \zeta, \eta\right],$$

$$h(t; \theta, \zeta, \eta) = \frac{\eta}{\theta} + \frac{\exp(-\zeta)}{\theta} \exp\left(\frac{t}{\theta}\right), \quad t > 0.$$

Here θ is a scale parameter, ζ and η are unitless shape parameters (not depending on the time scale), and

$$\Phi_{\text{lgoma}}(z; \zeta, \eta) = 1 - \exp\{\exp(-\zeta) - \exp[\exp(z) - \zeta] - \eta \exp(z)\},$$

$$\phi_{\text{lgoma}}(z; \zeta, \eta) = \exp(z)\{\eta + \exp[\exp(z) - \zeta]\}[1 - \Phi_{\text{lgoma}}(z; \zeta, \eta)]$$

are, respectively, the standardized cdf and pdf of $Z = \log(T/\theta)$. These functions are graphed in Figure 5.4. The p quantile is $t_p = \theta \exp[\Phi_{\text{lgoma}}^{-1}(p; \zeta, \eta)]$. There is no simple closed-form equation for $\Phi_{\text{lgoma}}^{-1}(p; \zeta, \eta)$, so it must be computed by inverting $p = \Phi_{\text{lgoma}}(z; \zeta, \eta)$ numerically.

5.8.2 Gompertz–Makeham Distribution Properties

Some properties of the Gompertz–Makeham distribution are:

- $h(0; \theta, \zeta, \eta) = (1/\theta)[\eta + \exp(-\zeta)]$ and $h(t; \theta, \zeta, \eta)$ increases with t at an exponential rate.
- If $T \sim \text{GOMA}(\theta, \zeta, \eta)$ and $c > 0$ then $cT \sim \text{GOMA}(c\theta, \zeta, \eta)$.

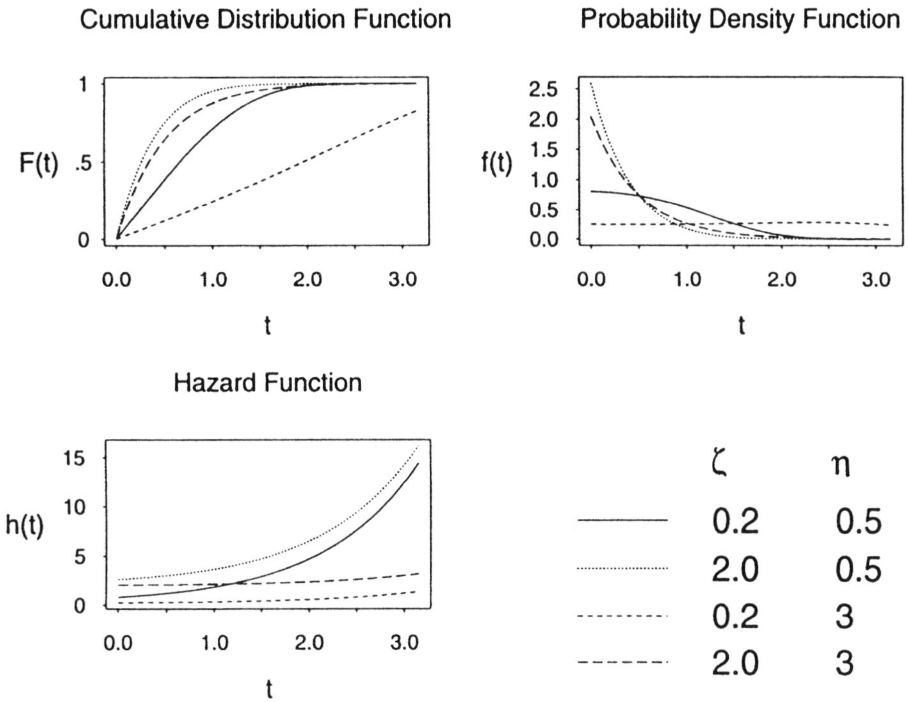

	ζ	η
——	0.2	0.5
··········	2.0	0.5
- - - -	0.2	3
– – – –	2.0	3

Figure 5.4. Gompertz–Makeham cdf, pdf, and hf for $\theta = 1$, $\zeta = .2$ and 2, and $\eta = .5$ and 3.

5.9 COMPARISON OF SPREAD AND SKEWNESS PARAMETERS

Figure 5.5 is similar to the figure on page 27 of Cox and Oakes (1984). It plots γ_3, the standardized third central moment (a unitless indication of skewness, defined in Section 4.2) against γ_2, the coefficient of variation (the standardized second central moment, a unitless indication of spread, also defined in Section 4.2). The curves in this graph indicate the wide range of shapes that the corresponding distribution can take across values of its shape parameter. Any distribution lying above the $\gamma_3 = 0$ line will tend to be skewed to the right. Distributions lying below this line tend to be skewed to the left (e.g., the Weibull distribution with large β). We can also see the similarities and differences among the different distributions that we have described. For example, the lines for the Birnbaum–Saunders and the inverse Gaussian distributions are not too far apart and the Weibull and gamma distributions cross at the exponential distribution point. Note that the generalized F (GENF) distribution spans the shapes of the specific distributions (but not all of the specific distributions are special cases of the GENF).

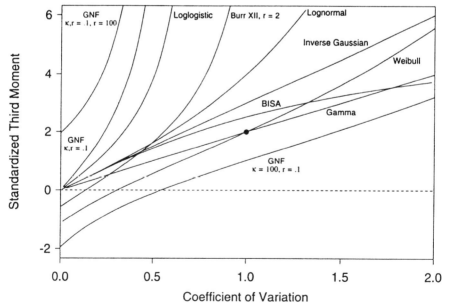

Figure 5.5. Standardized third moment versus coefficient of variation for different values of the shape parameters for failure-time distributions. The Burr type XII distribution with $r = 2$ is equivalent to the GENF$(\mu, \sigma, 1, 2)$. The • marks the exponential distribution point.

5.10 DISTRIBUTIONS WITH A THRESHOLD PARAMETER

5.10.1 Background

The exponential, lognormal, Weibull, gamma, and generalized gamma distributions, as defined previously, all are defined on the positive real line $(0, \infty)$. That is, the pdf $f(t) > 0$ for all $t > 0$ and $f(t) = 0$ for $t < 0$. Correspondingly, the cdf begins increasing at $t = 0$. All of these and other similar distributions can be generalized by the addition of a threshold parameter, which we denote by γ, to shift the beginning of the distribution away from 0. These distributions are particularly useful for fitting skewed distributions that are shifted far to the right of 0.

5.10.2 The cdf and pdf for Distributions with a Threshold

The cdf and pdf of a log-location-scale distribution with a threshold can be expressed as

$$F(t; \mu, \sigma, \gamma) = \Phi\left[\frac{\log(t - \gamma) - \mu}{\sigma}\right], \tag{5.11}$$

$$f(t; \mu, \sigma, \gamma) = \frac{1}{\sigma(t - \gamma)} \phi\left[\frac{\log(t - \gamma) - \mu}{\sigma}\right], \quad t > \gamma, \tag{5.12}$$

where Φ is a completely specified cdf and ϕ is the pdf corresponding to Φ. For a particular log-location-scale distribution, substitute the appropriate Φ and ϕ. For example, to obtain the three-parameter lognormal distribution, substitute Φ_{nor} and ϕ_{nor}. Figure 5.6 shows the three-parameter lognormal pdfs for $\mu = 0$ and $\sigma = .5$ with $\gamma = 1, 2,$ and 3. Similarly, for the three-parameter Weibull distribution, substitute Φ_{sev} and ϕ_{sev} giving

$$\Pr(T \leq t) = F(t; \alpha, \beta, \gamma) = \Phi_{sev}\left[\frac{\log(t - \gamma) - \mu}{\sigma}\right]$$

$$= 1 - \exp\left[-\left(\frac{t - \gamma}{\eta}\right)^\beta\right], \quad t > \gamma,$$

where $\sigma = 1/\beta$ and $\mu = \log(\eta)$. Sometimes γ is called a "guarantee parameter" because with $\gamma > 0$, failure is impossible before time γ. More generally, however, there is no mathematical reason to restrict γ to be positive.

The properties of the distributions with nonzero γ are closely related to the properties of the distributions with $\gamma = 0$ given in the earlier section in this chapter. In general, γ is added to the expectation and quantiles of the distribution with threshold equal to 0 to obtain the corresponding expectation and quantiles of the distribution with a given γ. Because changing γ simply shifts the distribution on the time axis, there is no effect on the distribution's spread or shape. Thus $\text{Var}(T)$ does not change with changes in γ and it can be obtained directly from the distribution with $\gamma = 0$. Section 11.7 will describe methods of fitting threshold distributions to data.

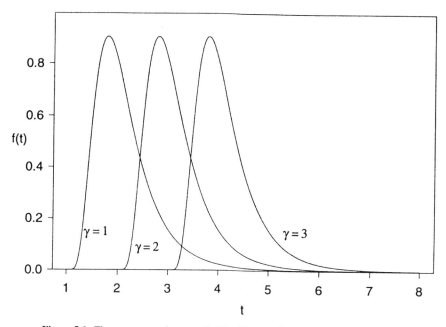

Figure 5.6. Three-parameter lognormal pdfs with $\sigma = .5$, $\mu = 0$, and $\gamma = 1, 2$, and 3.

5.10.3 Embedded Distributions

For some values of the parameters (μ, σ, γ), the threshold distribution is very similar to a two-parameter location-scale distribution, as described below. To facilitate the description of these embedded distributions, we use the reparameterization

$$\alpha = \gamma + \eta \quad \text{and} \quad \varsigma = \sigma\eta,$$

where $\eta = \exp(\mu)$. Then the cdf for the threshold log-location-scale distribution becomes

$$F(t; \alpha, \sigma, \varsigma) = \Phi\left[\log(1 + \sigma z)^{1/\sigma}\right] \quad \text{for } z > -1/\sigma,$$

where $z = (t - \alpha)/\varsigma$. As $\sigma \to 0$ from above, then $(1 + \sigma z)^{1/\sigma} \to \exp(z)$, and the *limiting* distribution or embedded distribution is

$$F(t; \alpha, 0, \varsigma) = \Phi(z) \quad \text{for } -\infty < t < \infty.$$

For example, when $\Phi = \Phi_{\text{sev}}$ the embedded distribution is the SEV distribution and when $\Phi = \Phi_{\text{nor}}$ the embedded distribution is the normal distribution. It is important to observe that the embedded distributions are not members of the original family (i.e., the embedded SEV distribution is not a Weibull distribution and the embedded normal distribution is not a lognormal distribution, except in the limit as $\sigma \to 0$ from above). In terms of the original parameters, embedded distributions arise when σ^{-1}, $\exp(\mu)$, and $-\gamma$ are all approaching $+\infty$ at rates such that $\sigma \times \exp(\mu)$ and $\gamma + \exp(\mu)$ approach finite values.

5.11 GENERALIZED THRESHOLD-SCALE DISTRIBUTION

5.11.1 Background

After reparameterization of a threshold-scale distribution to $(\alpha, \sigma, \varsigma)$, the parameter space can be enlarged to include the embedded distributions. This is accomplished by including the limiting case $\sigma = 0$. It is also useful to enlarge the parameter space such that the limiting distributions are interior points of the parameter space. This is achieved by allowing σ to take any value in $(-\infty, \infty)$. We call such a distribution a generalized threshold-scale (or GETS) distribution.

5.11.2 Generalized Threshold-Scale cdf and pdf

The cdf of the GETS distribution is

$$F(t; \alpha, \sigma, \varsigma) = \begin{cases} \Phi\left[\log(1 + \sigma z)^{1/\sigma}\right] & \text{for } \sigma > 0, z > -1/\sigma \\ \Phi(z) & \text{for } \sigma = 0, -\infty < z < \infty \\ 1 - \Phi\left[\log(1 + \sigma z)^{1/|\sigma|}\right] & \text{for } \sigma < 0, z < -1/\sigma, \end{cases} \quad (5.13)$$

where $z = (t - \alpha)/\varsigma$, $-\infty < \sigma < \infty$, $-\infty < \alpha < \infty$, and $\varsigma > 0$. The corresponding pdf is

$$f(t; \alpha, \sigma, \varsigma) = \begin{cases} \phi\left[\log(1 + \sigma z)^{1/|\sigma|}\right] \times \dfrac{1}{\varsigma(1 + \sigma z)}, & \text{for } \sigma \neq 0 \\ \phi(z) \times \dfrac{1}{\varsigma} & \text{for } \sigma = 0, \end{cases}$$

Note that the restrictions on z in (5.13) define the range of values of t having positive density, as a function of α, σ, and ς. In particular, for $\sigma > 0$, $t > \alpha - \varsigma/\sigma$ and for $\sigma < 0$, $t < \alpha - \varsigma/\sigma$.

5.11.3 Generalized Threshold-Scale Quantiles

Inverting the cdf in (5.13) gives the GETS p quantile

$$t_p = \alpha + \varsigma \times w(\sigma, p),$$

where

$$w(\sigma, p) = \begin{cases} \dfrac{\exp[\sigma\Phi^{-1}(p)] - 1}{\sigma} & \text{for } \sigma > 0 \\ \Phi^{-1}(p) & \text{for } \sigma = 0 \\ \dfrac{\exp\{|\sigma|\Phi^{-1}(1-p)\} - 1}{\sigma} & \text{for } \sigma < 0. \end{cases}$$

5.11.4 Special Cases of the Generalized Threshold-Scale Distribution

In this section we show the relationship between the GETS($\alpha, \sigma, \varsigma$) and some well-known distributions that are special cases.

- The location-scale distributions, including the normal, logistic, SEV, and LEV, are obtained by using the appropriate definition of Φ along with $\sigma = 0$, giving

$$F(t; \alpha, 0, \varsigma) = \Phi[(t - \alpha)/\varsigma].$$

- The threshold log-location-scale distributions are obtained with $\sigma > 0$, giving

$$F(t; \alpha, \sigma, \varsigma) = \Phi\{[\log(t - \gamma) - \mu]/\sigma\}, \quad t > \gamma,$$

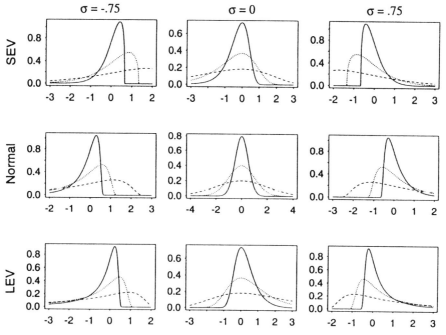

Figure 5.7. SEV-GETS, NOR-GETS, and LEV-GETS pdfs with $\alpha = 0$; $\sigma = -.75, 0, .75$; and $\varsigma = .5$ (least disperse), 1, and 2 (most disperse).

where $\gamma = \alpha - \varsigma/\sigma$ and $\mu = \log(\varsigma/\sigma)$. Using $\Phi = \Phi_{\text{nor}}$ gives the three-parameter lognormal distribution. Using $\Phi = \Phi_{\text{sev}}$ gives the three-parameter Weibull distribution (also known as Weibull-type distribution for *minima*). With $\Phi = \Phi_{\text{lev}}$ one obtains the Fréchet for *maxima* with a threshold parameter.

- The reflection (negative) of the threshold log-location-scale distributions are obtained with $\sigma < 0$, giving

$$F(t; \alpha, \sigma, \varsigma) = 1 - \Phi\{[\log(-t - \gamma) - \mu]/|\sigma|\}, \quad t < -\gamma$$

where $\gamma = -(\alpha - \varsigma/\sigma)$ and $\mu = \log(-\varsigma/\sigma)$. With $\Phi = \Phi_{\text{nor}}$ this gives the negative of a lognormal with a threshold parameter. With $\Phi = \Phi_{\text{sev}}$ this gives the negative of a Weibull (also known as a Weibull-type distribution for *maxima*) with a threshold.

Figure 5.7 shows pdfs for the SEV-GETS, NOR-GETS, and LEV-GETS distributions with $\alpha = 0$; $\sigma = -.75, 0, .75$; and $\varsigma = .5$ (least disperse), 1, and 2 (most disperse).

5.12 OTHER METHODS OF DERIVING FAILURE-TIME DISTRIBUTIONS

There are a number of general methods for deriving other failure-time distributions. The basic idea behind these methods is to model the physical system or the physical/chemical processes leading to failure.

5.12.1 Finite (Discrete) Mixture Distributions

Mixtures of distributions often arise in practice. For example, components may be manufactured over a period of time, using two different machines. Physical characteristics and thus reliability of the components from the two different machines may be different, but it may be impossible otherwise to distinguish between the components made with the different machines. For example, 40% of all units in a population were manufactured at plant A and have a lognormal life distribution with $\exp(\mu_A) = 60$ thousand hours and $\sigma_A = .7$. The other 60% of the units are manufactured at plant B and have a lognormal distribution with $\exp(\mu_B) = 70$ and $\sigma_B = .7$. Then the population cdf is

$$F(t) = .4 \times \Phi_{\text{nor}}\left[\frac{\log(t) - \log(60)}{.7}\right] + .6 \times \Phi_{\text{nor}}\left[\frac{\log(t) - \log(70)}{.7}\right]$$

Product mixtures also result from different environments. For some products, such as toasters, the environment may be assumed to be homogeneous. Other products might be subject to widely different operating environments (e.g., automobile batteries used in Florida versus those used in Alaska).

An extreme product mixture situation arises when a failure type can occur only in a subset of the population, for example, on those units for which an operator skips a step in an assembly operation or units that are made from a particular batch of raw materials or that include a particular optional accessory, such as an air-conditioning unit for an automobile. This situation is closely related to the concept of immunity from a failure mode discussed in Section 11.5.

More generally, the cdf and pdf of units in a population consisting of a mixture of units from k different populations can be expressed as $F(t; \boldsymbol{\theta}) = \sum_i \xi_i F_i(t; \boldsymbol{\theta}_i)$ and $f(t; \boldsymbol{\theta}) = \sum_i \xi_i f_i(t; \boldsymbol{\theta}_i)$, respectively, where $\boldsymbol{\theta} = (\boldsymbol{\theta}_1, \boldsymbol{\theta}_2, \ldots, \xi_1, \xi_2, \ldots), 0 \leq \xi_i \leq 1$, and $\sum_i \xi_i = 1$. There may be some components of the $\boldsymbol{\theta}_i$ that are common across the components (or "subpopulations") of the mixture. Others will differ from component to component. In general, however, such mixtures tend to have a large number of parameters and this usually makes estimation difficult.

In some situations it may be possible to estimate the parameters of the individual components of a mixture distribution. This task is facilitated when it is possible to identify the individual population from which sample units originated. When identification is not possible it may be difficult to separate out the different components from the available data unless there is considerable "separation" in the components and/or enormous amounts of data. In other situations, to answer certain questions of interest, it is sufficient to fit a simple single distribution to describe the overall mixture. Indeed, it can be said that all populations and processes are mixtures. However, because individual components may not be identifiable and/or may not be of interest, we use a single distribution to describe the mixture. There are, however, some potential pitfalls of fitting a simple distribution to what is really data from a mixture of different subpopulations. If data collection methods or censoring tends to overrepresent certain subpopulations of the mixture, seriously misleading conclusions are possible. Hahn and Meeker (1982b) describe and illustrate these pitfalls with an example. Also see the case study in Section 22.1.

5.12.2 Compound (Continuous Mixture) Distributions

An important class of probability models arises from distributions in which one or more of the parameters is modeled by a continuous random variable. These distributions are called compound distributions and can also be viewed as a continuous mixture of a family of distributions.

Suppose that T, for a fixed value of a scalar parameter θ_1, has a distribution with density $f_{T|\theta_1}(t; \boldsymbol{\theta})$ but that θ_1, an element from the vector $\boldsymbol{\theta} = (\theta_1, \boldsymbol{\theta}_2)$, is itself random from unit to unit, according to a distribution with density $f_{\theta_1}(\vartheta; \boldsymbol{\theta}_3)$, where $\boldsymbol{\theta}_3$ is a parameter vector that has no elements in common with $\boldsymbol{\theta}$. Then the cdf of the unconditional distribution of T, a compound distribution, is

$$F(t; \boldsymbol{\theta}_2, \boldsymbol{\theta}_3) = \Pr(T \leq t) = \int_{-\infty}^{\infty} \Pr(T \leq t \mid \theta_1 = \vartheta) f_{\theta_1}(\vartheta; \boldsymbol{\theta}_3) d\vartheta$$

$$= \int_{-\infty}^{\infty} F_{T|\theta_1 = \vartheta}(t; \boldsymbol{\theta}) f_{\theta_1}(\vartheta; \boldsymbol{\theta}_3) d\vartheta$$

and the corresponding pdf is

$$f(t; \boldsymbol{\theta}_2, \boldsymbol{\theta}_3) = \int_{-\infty}^{\infty} f_{T \mid \theta_1 = \vartheta}(t; \boldsymbol{\theta}) f_{\theta_1}(\vartheta; \boldsymbol{\theta}_3) \, d\vartheta.$$

Extension to a vector of parameters $\boldsymbol{\theta}_1$ is straightforward. Many important applications, however, use only a scalar θ_1.

Example 5.1 Pareto Distribution. If the units in a population have an exponential distribution, but with a failure rate that varies from unit to unit according to a GAM(θ, κ) distribution, then the unconditional time to failure of a unit selected at random from the population has a Pareto distribution of the form

$$F(t; \theta, \kappa) = 1 - \frac{1}{(1 + \theta t)^{\kappa}}, \quad t > 0. \tag{5.14}$$

The proof of this result is left as an exercise. □

Of course, it is possible to define and use compound distributions that do not have simple closed-form expressions. We will illustrate the use of such distributions in Section 22.4. For other examples of compound distributions, see page 163 of Johnson, Kotz, and Balakrishnan (1994).

5.12.3 Power Distributions

Distributions of minima and maxima of iid random variables provide a useful method for generating distributions of random variables that have a number of important applications.

Minimum-Type Distributions
If κ is a positive constant and W is a random variable with cdf F_W, we say that Y is a minimum-type distribution generated from W if for all y

$$\Pr(Y \leq y) = F_Y(y) = 1 - [1 - F_W(y)]^{\kappa}$$

or, equivalently, $S_Y(y) = [S_W(y)]^{\kappa}$, where $S_W = 1 - F_W$ is the survival or reliability function. An important special case is when κ is an integer. In this situation F_Y is the cdf of the minimum of κ independent observations from F_W.

It can be shown that Y is a minimum-type distribution generated from W if and only if the hazard functions of Y and W are proportional, that is, $h_Y(y) = \kappa \times h_W(y)$. Also if $W \sim$ WEIB(μ, σ) then $Y \sim$ WEIB$[\mu - \sigma \log(\kappa), \sigma]$. But in general (i.e., when W is not Weibull distributed) Y is not a member of the same family as W.

Maximum-Type Distributions
For κ and W defined as in minimum-type distributions, we say that Y is a maximum-type distribution generated from W if for all y,

$$\Pr(Y \leq y) = F_Y(y) = [F_W(y)]^{\kappa}.$$

An important case is when κ is an integer. Then F_Y is the cdf of the maximum of κ independent observations from F_W.

5.12.4 Distributions Based on Stochastic Components of Physical/Chemical Degradation Models

We saw in Sections 5.6.1 and 5.7 how distributions of time to failure can be derived from details of randomness in particular physical phenomena. Chapters 13 and 18 provide other examples.

BIBLIOGRAPHIC NOTES

Johnson, Kotz, and Balakrishnan (1994, 1995) provide detailed information on a wide range of different continuous probability distribution functions. Evans, Hastings, and Peacock (1993) provide a brief description and summary of properties of a large number of parametric distributions including most, but not all, of the distributions outlined in this chapter. Chhikara and Folks (1989) give detailed information on the inverse Gaussian distribution.

Although the generalized gamma distribution had appeared in the literature earlier, Prentice (1974) was the first to provide a parameterization and operational method for estimation of the distribution parameters that works well with moderate sample sizes, as long as censoring is not too heavy. Liu, Meeker, and Escobar (1998) suggest and illustrate the use of a parameterization that is stable even for heavy censoring. Farewell and Prentice (1977) showed how to use the generalized gamma distribution in problems of parametric distribution discrimination. Prentice (1975) extends the generalized gamma distribution to the generalized F distribution.

Cheng and Iles (1990) describe some statistical problems created by embedded distributions that arise as limiting cases of the threshold (three-parameter) gamma, inverse Gaussian, loglogistic, lognormal, and Weibull distributions, providing motivation for the GETS family of distributions. See also Nakamura (1991) for a different view of embedded distributions within threshold families.

Titterington, Smith, and Makov (1985) and Everitt and Hand (1981) provide detailed information on finite mixture distributions. Nelson and Doganaksoy (1995) describe the power lognormal distribution, including methods for estimation. Barlow and Proschan (1975) provide a detailed and extensive discussion of classes of distributions that are based on different hazard function behaviors like IHR and DHR.

EXERCISES

5.1. If the times to failure in a population are adequately described by a distribution with a decreasing hazard function, one might think that the surviving units in the population are getting better with time. In fact, decreasing hazard functions are common for certain solid-state electronic components and elec-

EXERCISES 119

tronic systems. Weaker units fail early, after which the hazard decreases. For a mixture of two exponential distributions with $\gamma = 0$ but different values of θ (say, $\theta_1 = 1$ and $\theta_2 = 5$), and equal proportions from the two populations, do the following:

(a) Obtain an expression for the cdf of the mixture.
(b) Obtain an expression for the pdf of the mixture.
(c) Use the previous two parts to derive an expression for the hazard function of the mixture.
(d) Graph the mixture hazard from $t = 0$ to $t = 10$.
(e) What is the shape of the mixture hazard? What is the intuition for this result?
(f) In what sense is the mixed exponential population "improving" with time (as suggested by the decreasing hazard function)?

5.2. In some applications a sample of failure times comes from a mixture of subpopulations.

(a) Write down the expression for the cdf $F(t)$ for a mixture of two exponential distributions with means $\theta_1 = 1$ and $\theta_2 = 10$ (subpopulations 1 and 2, respectively) with ξ being the proportion from subpopulation 1.
(b) For $\xi = 0, .1, .5, .9,$ and 1, compute the mixture $F(t)$ for a number of values of t ranging between 0 and 30. Plot these distributions on one graph.
(c) Plot $\log(t)$ versus $\log\{-\log[1 - F(t)]\}$ for each $F(t)$ computed in part (b). Comment on the shapes of the mixtures of exponential distributions, relative to a pure exponential distribution or a Weibull distribution.
(d) Plot the hazard function $h(t)$ of the mixture distributions in part (b).
(e) Qualitatively, what do the Weibull plots in part (c) suggest about the hazard function of a mixture of two exponential distributions?

5.3. Show that the exponential distribution is a special case of the gamma distribution given in Section 5.2.

▲5.4. Refer to Exercise 5.1. Show that a mixture of two exponential distributions with different θ values will always have a decreasing hazard function.

◆5.5. Conduct a numerical/graphical comparison of the shapes of the hazard function for a population consisting of the mixture of two Weibull distributions. Investigate all combinations of the parameters $\alpha_1 > \alpha_2 = 1, 2; \beta_1, \beta_2 = .5, 2; \xi_1 = .01, .4$. What do you conclude?

▲5.6. Let $T_i, i = 1, \ldots, \kappa$, be κ independent random variables from the EXP(θ) distribution. Show that the random variable $\sum_{i=1}^{\kappa} T_i$ has a GAM(θ, κ) distribution.

▲5.7. Section 5.3.3 gives important special cases of the generalized gamma distribution. Use direct substitution to show the relationships to the WEIB, EXP, and GAM distributions. Use a limiting agreement to show the relationship to the LOGNOR distribution.

▲5.8. As described in Example 5.1, the Pareto distribution can be derived as a gamma mixture of exponential distributions.
  (a) Show this by deriving (5.14).
  (b) Take the first derivative of (5.14) with respect to t to obtain the Pareto pdf.
  (c) Plot the Pareto hazard function for several different combinations of θ and κ.

▲5.9. Derive the expression for the Birnbaum–Saunders p quantile given in (5.9).

▲5.10. As an interpretation of the Gompertz–Makeham distribution, suppose that the failure time of a device is determined by which of the following two events happens first:

  - Wearout at time W, which can be modeled by an SEV(μ, σ), left truncated at time zero.
  - An accident at time R, which can be modeled by an EXP(λ).

  In other words, the device failure time is $T = \min\{W, R\}$. For this exercise, also suppose that W and R are independent.
  (a) Show that the cdf for T is the same as the Gompertz–Makeham cdf, having the form

  $$\Pr(T \le t; \mu, \sigma, \lambda) = 1 - \left[\frac{1 - \Phi_{\text{sev}}\left(\frac{t - \mu}{\sigma}\right)}{1 - \Phi_{\text{sev}}\left(\frac{-\mu}{\sigma}\right)}\right] \exp\left(-\frac{t}{\lambda}\right), \quad t > 0$$

  $$= \Phi_{\text{lgoma}}\left[\log\left(\frac{t}{\theta}\right); \zeta, \eta\right],$$

  where $\theta = \sigma$, $\zeta = \mu/\sigma$, and $\eta = \sigma/\lambda$. Hint: One can write $F_T(t) = 1 - [1 - F_W(t)][1 - F_R(t)]$. Explain why.
  (b) Show that the hazard function for T is $h_T(t; \mu, \sigma, \lambda) = h_R(t; \lambda) + h_W(t; \mu, \sigma)$. Also show that this coincides with the hazard function of a GOMA(θ, ζ, η).

▲5.11. Show that the hazard relationship in part (b) of Exercise 5.10 holds in general when $T = \min\{W, R\}$, where W and R are any two independent continuous random variables.

EXERCISES

5.12. Let $T_{(1)}$ denote the minimum of m independent Weibull random variables with parameters $\mu_i, i = 1, \ldots, m$, and constant σ. Show that $T_{(1)}$ has a Weibull distribution.

▲5.13. Use the lognormal base (i.e., use $\Phi = \Phi_{nor}$) GETS(η, σ, ζ) distribution given in Section 5.10.1 to do the following:
  (a) Plot the cdfs and pdfs for all possible combinations of the parameters $\eta = -.5, .5; \sigma = -1, 1;$ and $\zeta = 1$.
  (b) For fixed $\eta = 0$ and $\zeta = 1$, draw the cdfs for several small values of σ, say, $\sigma = \pm.1, \pm.01, \pm.001$, and compare with a NOR(0, 1) cdf. How well do the GETS cdfs approach the normal cdf as $\sigma \to 0$?

▲5.14. Consider the GETS(η, σ, ζ) distribution as given in Section 5.10.1.
  (a) Show that when $\sigma \to 0$ the GETS cdf approaches $\Phi(z)$. *Hint:* When $\sigma \to 0, (1/\sigma) \log(1 + \sigma z) \to z$.
  (b) Show that, in terms of the parameters (γ, α, η) of Section 5.10.1, when σ approaches 0 from above (below) then the γ is approaching $+\infty$ ($-\infty$) and $(\gamma + \alpha) \to \eta$.

CHAPTER 6

Probability Plotting

Objectives

This chapter explains:

- Applications of probability plots.
- Basic probability plotting concepts for both complete and censored data.
- How to analytically linearize a cdf on special plotting scales.
- How to plot a modified nonparametric estimate of $F(t)$ on probability paper and how to use such a plot to judge the adequacy of a particular parametric distribution.
- Analytical and simulation methods of separating useful information from "noise" when using a probability plot to assess the reasonableness of a particular distributional model.
- Graphical estimates of important reliability characteristics like failure probabilities and distribution quantiles.

Overview

This chapter presents the important topic of probability plotting. Probability plots are used throughout this book to present data, guide modeling, and present the results of analyses. Sections 6.1–6.4 explain the basic concepts of probability plotting. Section 6.5 describes some useful extensions to the standard probability plots, while Section 6.6 explains additional aspects of the practical application of probability plots, including the use of simulation to help interpret such plots.

6.1 INTRODUCTION

Probability plots are an important tool for analyzing data and have been particularly popular in the analysis of life data.

6.1.1 Purposes of Probability Plots

In practical applications, probability plots are used to:

- Assess the adequacy of a particular distributional model.
- Provide *nonparametric* graphical estimates of probabilities and distribution quantiles.
- Obtain graphical estimates of *parametric* model parameters (e.g., by fitting a straight line through the points on a probability plot).
- Display the *results* of a parametric maximum likelihood fit along with the data.

In addition, probability plots often reveal information about a population, a process, or data that might otherwise escape detection.

6.1.2 Probability Plotting Scales: Linearizing a cdf

The figures in Chapter 4 show that cdfs from different distributions have similar shapes. Thus distinguishing, by eye, among cdfs from different distributions is not easy. Probability plots use special scales on which a cdf of a particular distribution plots as a straight line.

The plot of $\{t$ versus $F(t)\}$ can be linearized by finding transformations of $F(t)$ and t such that the relationship between the transformed variables is linear. Then the transformed axes can be relabeled in terms of the original probability and time variables. The resulting probability scale is generally nonlinear and is called the "probability scale." The data scale is usually a log scale or a linear scale, depending on the particular distribution and type of probability plot.

Probability paper was developed initially to allow data analysts to plot data, obtain estimates, and assess fit of a particular model by comparing with a straight line. Normal and Weibull probability papers have been widely used in practice. However, because there are many different combinations of possible probability and data axes that might be needed, it is useful to have a computer implementation of probability plotting methods like those described in this chapter.

6.2 LINEARIZING LOCATION-SCALE-BASED DISTRIBUTIONS

The quantile function for $F(t)$ provides a convenient starting point for finding the transformation needed for linearizing a cdf. We illustrate the ideas in this section with a subset of the location-scale-based distributions from Chapter 4. The approach, however, is similar for other location-scale-based distributions. In Section 6.5, we extend the method to non-location-scale distributions.

6.2.1 Linearizing the Exponential cdf

The quantile function for the two-parameter exponential distribution (see Section 4.4) is

$$t_p = \gamma - \log(1 - p)\theta.$$

This implies that $\{t_p \text{ versus } -\log(1 - p)\}$ plots as a straight line, as illustrated in Figure 6.1. Because we plot time on the horizontal axis and p on the vertical axis (corresponding to the traditional cdf plots used in Chapters 1–5), γ is the intercept on the time scale [because $-\log(1 - p) = 0$ when $p = 0$] and the slope on the time versus quantile scales is equal to $1/\theta$. In this case, θ determines the slope of the line and γ determines the horizontal position of the line.

The linear scale on the right-hand side of the plot, corresponding to $-\log(1 - p)$, is useful for graphically estimating the slope. Because this scale is unnecessary for the other applications listed in Section 6.1.1 and because it is bad style to encumber graphs with unnecessary scales, such scales will be displayed only in selected examples in this book.

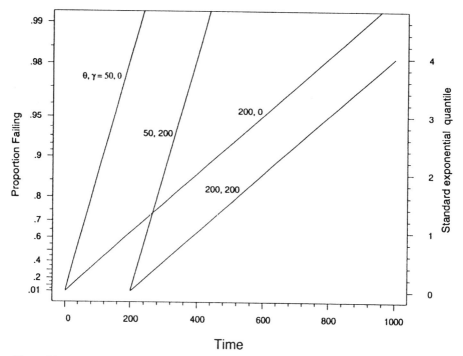

Figure 6.1. Exponential probability plot (exponential distribution probability scale) showing exponential cdfs as straight lines for $\theta = 50, 200$ and $\gamma = 0, 200$.

LINEARIZING LOCATION-SCALE-BASED DISTRIBUTIONS

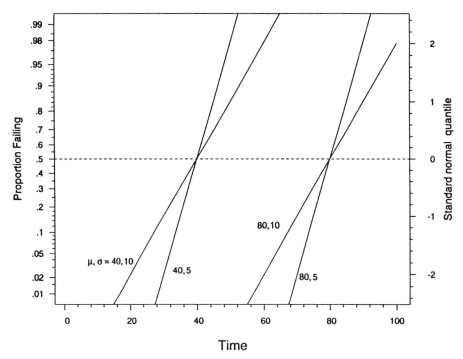

Figure 6.2. Normal probability plot (normal distribution probability scale) showing normal cdfs as straight lines for $\mu = 40, 80$ and $\sigma = 5, 10$.

6.2.2 Linearizing the Normal cdf

As shown in Section 4.5, the quantile function for the normal distribution is

$$y_p = \mu + \Phi_{nor}^{-1}(p)\sigma,$$

where $\Phi_{nor}^{-1}(p)$ is the p quantile of the standard normal distribution. As illustrated in Figure 6.2, this implies that $\{y_p \text{ versus } \Phi_{nor}^{-1}(p)\}$ plots as a straight line. The normal mean μ (location parameter) can be read from the time scale at the point where the cdf intersects the $\Phi_{nor}^{-1}(p) = 0$ line [the right-hand scale is $\Phi_{nor}^{-1}(p)$ and the horizontal dashed line shows that $\Phi_{nor}^{-1}(p) = 0$ at $p = .5$]. The slope of the line on the time versus quantile scales is $1/\sigma$. Any normal cdf plots as a straight line with positive slope. Correspondingly, any straight line with positive slope corresponds to a normal distribution. Note the symmetry of the probability scale above and below .5, following from the symmetry of the normal distribution pdf (shown in Figure 4.2).

6.2.3 Linearizing the Lognormal cdf

Lognormal probability plots are closely related to normal probability plots. As shown in Section 4.6, the quantile function for the lognormal distribution is $t_p = \exp[\mu + \Phi_{nor}^{-1}(p)\sigma]$, where $\Phi_{nor}^{-1}(p)$ is the p quantile of the standard normal distribution. This leads to

$$\log(t_p) = \mu + \Phi_{\text{nor}}^{-1}(p)\sigma.$$

As illustrated in Figure 6.3, this implies that $\{\log(t_p) \text{ versus } \Phi_{\text{nor}}^{-1}(p)\}$ plots as a straight line. The lognormal scale parameter (median) $t_{.5} = \exp(\mu)$ can be read from the time scale at the point where the cdf intersects the $\Phi_{\text{nor}}^{-1}(p) = 0$ line [the horizontal dashed line shows that $\Phi_{\text{nor}}^{-1}(p) = 0$ at $p = .5$]. The slope of the line on the time versus quantile scales is $1/\sigma$. Any lognormal cdf plots as a straight line with positive slope. Correspondingly, any straight line with positive slope corresponds to a lognormal distribution.

The lognormal data scale is a logarithmic scale. The lognormal probability scale is the same as that on normal probability plots (Figure 6.2). The base-10 log-time scale on the top of the graph is preferred by some engineers. This scale, along with the linear $\Phi_{\text{nor}}^{-1}(p)$ scale on the right-hand side of the plot, facilitates the computation of σ for the base-10 lognormal distribution (see Section 4.6).

Example 6.1 Reading Parameter Values from a Probability Plot. To illustrate the process of reading parameter values from a probability plot, refer to

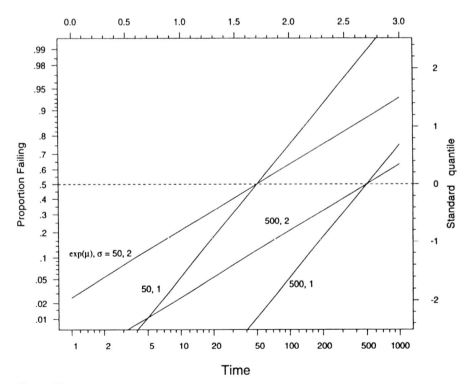

Figure 6.3. Lognormal probability plot (normal distribution probability scales) showing lognormal cdfs as straight lines for $\exp(\mu) = 50, 500$ and $\sigma = 1, 2$.

the exp(μ), σ = 50, 2 line on Figure 6.3. The value of the median exp(μ) of the lognormal distribution corresponding to this line can be read from the time axis where the line crosses the dotted line. Then exp(μ) = 50, which corresponds to μ = $\log_{10}(50)$ = 1.69897 for the base-10 lognormal distribution and μ = $\log(50)$ = 3.912023 for the base-e lognormal distribution. To find sigma, one needs to find the reciprocal of the slope of the line. Start with the base-10 log-time scale and for best resolution, use the extreme endpoints (in this case $\log_{10}(1)$ = 0 and $\log_{10}(1000)$ = 3). The corresponding standardized quantile values can be read from the right-hand axis as approximately -1.97 and 1.5. Then for the base-10 lognormal distribution

$$\sigma \approx \left(\frac{3 - 0}{1.5 - (-1.97)}\right) = .865$$

and for the base-e lognormal distribution $\sigma \approx \log(10) \times .865 = 1.99$, where the factor $\log(10)$ = 2.302585 converts logarithms from base-10 to logarithms in base-e.

□

6.2.4 Linearizing the Weibull cdf

As shown in Section 4.8, the quantile function for the Weibull distribution can be expressed (showing both the location-scale and the common parameterization) as $t_p = \exp[\mu + \Phi_{\text{sev}}^{-1}(p)\sigma] = \eta[-\log(1-p)]^{1/\beta}$, where $\Phi_{\text{sev}}^{-1}(p) = \log[-\log(1-p)]$. This leads to

$$\log(t_p) = \mu + \log[-\log(1 - p)]\sigma = \log(\eta) + \log[-\log(1 - p)]\frac{1}{\beta}.$$

As illustrated in Figure 6.4, this implies that {$\log(t_p)$ versus $\log[-\log(1-p)]$} plots as a straight line. The Weibull scale parameter $\eta = \exp(\mu)$ can be read from the time scale at the point where the cdf intersects the $\log[-\log(1-p)] = 0$ line (indicated by the horizontal dashed line at $p \approx .632$). The slope of the line on the log time versus quantile scales is $\beta = 1/\sigma$. Any Weibull cdf plots as a straight line with positive slope. Correspondingly, any straight line with positive slope corresponds to a Weibull distribution. Exponential cdfs plot as straight lines with slopes equal to 1. Note the log scale for time (with a linear time scale, this would be a smallest extreme value distribution probability plot).

6.3 GRAPHICAL GOODNESS OF FIT

Assessment of distributional adequacy is an important application of probability plots. As shown in Section 6.4, this is done by plotting the nonparametric estimate $\widehat{F}(t)$ on the linearizing probability scales and assessing departures from a straight line. Such probability plots can be made even more useful by plotting, in addition, simultaneous

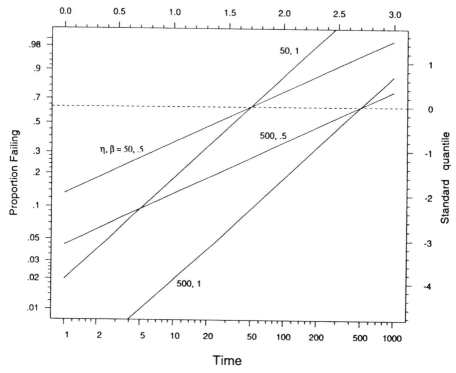

Figure 6.4. Weibull probability plot (smallest extreme value distribution probability scale) showing Weibull cdfs as straight lines for $\eta = 50, 500$ and $\beta = .5, 1$.

confidence bands like those presented in Section 3.8. Based on the available data, any possible $F(t)$ within these bands is, statistically, consistent with the data. On probability paper for a particular distribution, if it is possible to draw a straight line all the way between the bands, then the distribution is consistent with the data. We will use simultaneous confidence bands on all probability plots in this chapter.

6.4 PROBABILITY PLOTTING POSITIONS

To construct a probability plot, one must decide how to plot the nonparametric estimate of $F(t)$ on the probability scales described in Section 6.2. With exact times, it has been traditional to plot each failure time against an estimate of the probability of failing at that time. To follow this tradition, we plot an estimate of $F(t)$ at some specified points in time—typically the failure times when they are reported and the upper endpoints of inspection intervals for inspection data. Then we need to define "plotting positions," consisting of a corresponding estimate of $F(t)$, at these points in time.

6.4.1 Criteria for Choosing Plotting Positions

Criteria for choosing plotting positions should depend on the application or purpose for constructing the probability plot. The following are some possible applications that will suggest criteria.

Checking Distributional Assumptions. Probability plotting is used to check if the observed data are well approximated by the postulated parametric distribution $F(t; \boldsymbol{\theta})$. For this purpose, some bias in the slope and location of the fitted line is not a serious problem. For this reason, it is generally suggested that, for assessing distributional assumptions, the choice of plotting positions, in moderate-to-large samples, is not so important.

Estimation of Parameters. If the purpose of the probability plot is to use a fitted line to estimate parameters of a particular distribution (by using the slope and intercept of a line drawn through the data points), the "best" plotting positions will depend on the assumed underlying model and the functions to be estimated (e.g., which quantile or moment is of interest). For complete data, letting i index the ordered observations, there is some general agreement that the plotting positions

$$p_i = \frac{i - .5}{n}$$

provide a good choice for general-purpose use in probability plotting.

Display of Maximum Likelihood Fits with Data. As shown in Chapter 7, maximum likelihood (ML) fitting of parametric models is a convenient and general method for obtaining estimates and predictions from censored data. One important application of a probability plot is to display the ML fit graphically and to compare with the corresponding nonparametric estimate. In this case an important criterion is that the line "fit" the points well when the assumed model being fit with ML agrees with the data. With a poor choice of plotting positions, the ML line may not fit the plotted points. Then the probability plot can give the false impression that the parametric model and the data disagree, even though any difference between the points and the line will generally be small relative to sampling error variability that would be observed by repeating the sampling process. Plotting simultaneous confidence bands on the probability plot will indicate the amount of sampling variability one might expect to see.

6.4.2 Choice of Plotting Positions

There are three cases to consider: (1) continuous inspection (or small inspection intervals resulting in exact failures), (2) interval-censored data with relatively large intervals, and (3) arbitrarily censored data, which could include combinations of left censoring, right censoring, and overlapping failure intervals.

Continuous Inspection Data and Single Censoring

With continuous inspection and single right censoring (or complete data), the nonparametric estimate $\widehat{F}(t_{(i)}) = i/n$ is a step function increasing by an amount $1/n$

at each reported failure time. Let $t_{(1)}, t_{(2)}, \ldots$ be the ordered failure times. From a plot of the step function (e.g., Figure 6.5), we see that $\widehat{F}(t)$ steps up at each reported failure time. Plotting at the bottom (top) of the step would lead to bias in the plotted points and the ML line would tend to be above (below) the plotted points. Also, in situations where the last reported time is a failure, it is not possible to plot a point at $\widehat{F}(t) = 1$ (the value of the nonparametric estimate at the last failure). A reasonable compromise plotting position is the midpoint of the jump

$$\frac{i - .5}{n} = \frac{1}{2}\left[\widehat{F}(t_{(i)}) + \widehat{F}(t_{(i-1)})\right]. \tag{6.1}$$

Another justification for this definition of plotting positions [estimator of $F(t)$ at t] is that the median of the ith order statistic (i.e., the ith largest observation) in a sample of size n is approximately $F^{-1}[(i - .5)/n]$. For complete or singly censored data, this plotting position has been useful for a variety of different distributions and purposes (e.g., pages 293–294 of Hahn and Shapiro, 1967 for discussion of alternative plotting positions).

Example 6.2 Probability Plots of Fatigue Life Data for Alloy T7987. Table 6.1 gives the fatigue life (rounded to the nearest thousand cycles) for 67 specimens of Alloy T7987 that failed before having accumulated 300 thousand cycles of testing. There were, in addition, 5 "runout" specimens that survived until 300 thousand cycles without failure. Figure 6.5 shows a plot of $\widehat{F}(t)$, the nonparametric estimate of the

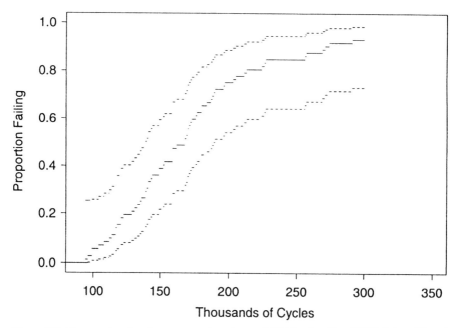

Figure 6.5. Linear-scales plot of nonparametric estimate of $F(t)$ for the Alloy T7987 fatigue life and simultaneous approximate 95% confidence bands for $F(t)$.

Table 6.1. Number of Cycles (in Thousands) of Fatigue Life for 67 of 72 Alloy T7987 Specimens that Failed Before 300 Thousand Cycles

94	96	99	99	104	108	112	114	117	117
118	121	121	123	129	131	133	135	136	139
139	140	141	141	143	144	149	149	152	153
159	159	159	159	162	168	168	169	170	170
171	172	173	176	177	180	180	184	187	188
189	190	196	197	203	205	211	213	224	226
227	256	257	269	271	274	291			

fatigue life cdf. Some of the step increases are integer multiples of $1/n$ because of the ties resulting from rounding. The points in the Weibull probability plot in Figure 6.6 are, for each reported failure point, plotted at a probability corresponding to half the jump-height of each step in Figure 6.5. Figure 6.6 indicates that the Weibull distribution does not provide a good fit to the data. Figure 6.7 indicates that the lognormal distribution provides a much better fit than the Weibull distribution. Both distributions, however, show concave behavior in the lower tail, an indication that a threshold parameter for either the Weibull or lognormal distribution would improve the fit to the data. This is investigated further in Section 6.5. □

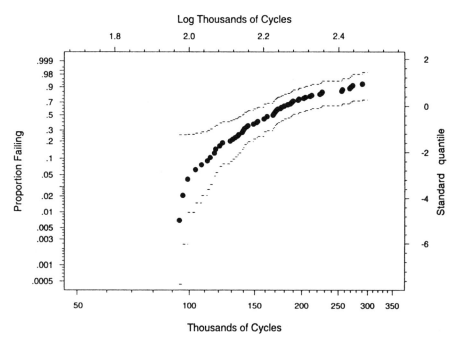

Figure 6.6. Weibull probability plot of the Alloy T7987 fatigue life data and simultaneous approximate 95% confidence bands for $F(t)$.

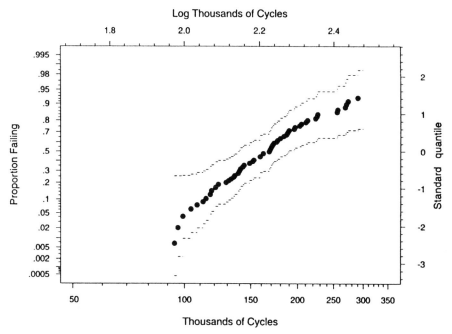

Figure 6.7. Lognormal probability plot for the Alloy T7987 fatigue life and simultaneous approximate 95% confidence bands for $F(t)$.

Continuous Inspection Data and Multiple Censoring

With continuous inspection and multiple right censoring the usual nonparametric estimate $\widehat{F}(t)$ is again a step function with steps at each reported failure time (but, due to censoring between failures, the step increases may be different from $1/n$). Corresponding to the definition for single censoring in (6.1), we modify $\widehat{F}(t_{(i)})$ to get plotting positions for multiple censoring as $\{t_{(i)}$ versus $p_i\}$ with

$$p_i = \tfrac{1}{2}[\widehat{F}(t_{(i)}) + \widehat{F}(t_{(i-1)})]. \tag{6.2}$$

Example 6.3   Comparison of Weibull and Lognormal Probability Plots for the Shock Absorber Data.   For the shock absorber data in Example 3.8 and Appendix Table C.2, the nonparametric estimate of $F(t)$ is given in Figure 3.6. Figures 6.8 and 6.9 show, respectively, Weibull and lognormal probability plots of these data, along with approximate 95% nonparametric simultaneous confidence bands. The Weibull distribution appears to provide a better description of these data. With the large amount of uncertainty expressed by the simultaneous confidence bands, however, we certainly could not rule out the lognormal distribution as an adequate distribution.

□

Interval-Censored Inspection Data

With interval-censored data, if there are failures in each interval, $\widehat{F}(t)$ is defined at the upper endpoint of each interval (see Section 3.7). Let $(t_0, t_1], \ldots, (t_{m-1}, t_m]$ be the

PROBABILITY PLOTTING POSITIONS

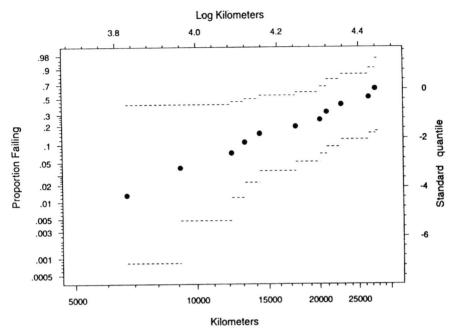

Figure 6.8. Weibull probability plot of the shock absorber data with simultaneous approximate 95% confidence bands for $F(t)$.

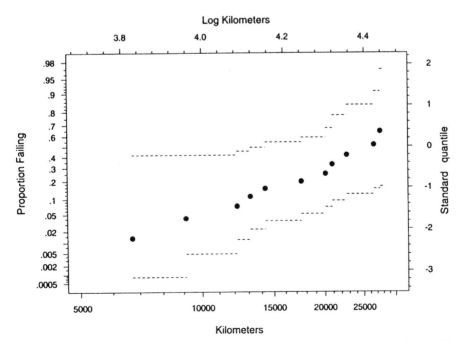

Figure 6.9. Lognormal probability plot of the shock absorber data with simultaneous approximate 95% confidence bands for $F(t)$.

intervals preceding the m inspection times. The upper endpoints of the inspection intervals $t_i, i = 1, 2, \ldots$, are convenient plotting times. For corresponding plotting positions here use $p_i = \widehat{F}(t_i)$. The justification for this choice is that, with no censoring, from standard binomial theory,

$$\mathrm{E}[\widehat{F}(t_i)] = F(t_i). \tag{6.3}$$

With losses (multiple censoring), (6.3) will be approximately true. When there are no censored observations beyond t_m, $F(t_m) = 1$ and this point cannot be plotted on probability paper.

Example 6.4 Exponential Probability Plot for the Heat Exchanger Tube Data.
Figure 6.10 is an exponential probability plot of the heat exchanger data showing the nonparametric estimate of $F(t)$ computed in Example 3.6. Also shown are 95% nonparametric simultaneous confidence bands for $F(t)$. These bands are very wide due to the small number of observed cracks in the combined heat exchanger data. The bands are not symmetric because we used the logit transformation (described in Section 3.6.3) to improve the large-sample approximation. The points on this plot fall roughly along a straight line, indicating that there is no evidence here to contradict an exponential distribution assumption. Of course, the width of the simultaneous confidence bands for $F(t)$ also indicates that it is certainly possible that the heat exchanger tube life has a distribution far from the exponential distribution. □

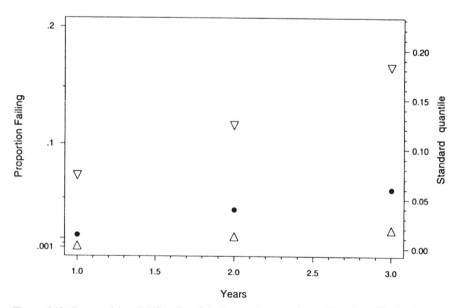

Figure 6.10. Exponential probability plot of the heat exchanger tube cracking data with simultaneous approximate 95% confidence bands for $F(t)$.

Arbitrarily Censored Data

With mixtures of left censoring, right censoring, and observations reported as exact failures, $\widehat{F}(t)$ can consist of a mixture of sets of points and horizontal lines of increasing height. Such estimates require a compromise between the other two cases.

Example 6.5 Turbine Wheel Data. The turbine wheel data from Example 1.7 and given in Table 1.5 consist of a set of overlapping left- and right-censored observations. Figure 6.11 is an exponential probability plot of the turbine wheel data. Figure 6.12 is a lognormal probability plot of the same data. Both plots contain 95% simultaneous nonparametric confidence bands.

It is clear that the lognormal distribution fits these data better than the exponential distribution. For these data, the Weibull probability plot (not shown here) was very similar to the lognormal probability plot. The great width of the 95% simultaneous confidence bands indicates, however, that none of these distributions could be ruled out. As a practical matter, however, it is generally more conservative to use the more general Weibull or lognormal distributions. Unless the assumption could be based on physical experience with related data or other information apart from the data, use of the exponential distribution with such sparse data would generally give an unrealistically small indication of sampling uncertainty. □

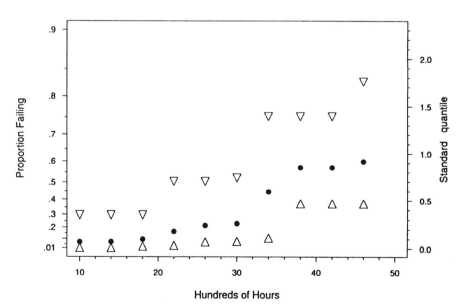

Figure 6.11. Exponential probability plot of the turbine wheel inspection data with simultaneous approximate 95% confidence bands for $F(t)$.

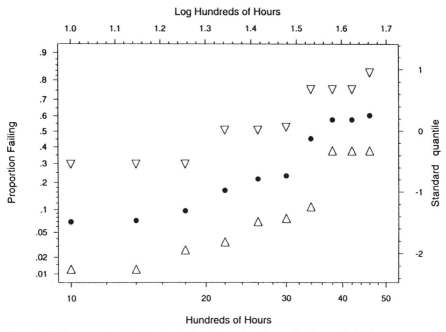

Figure 6.12. Lognormal probability plot of the turbine wheel inspection data with simultaneous approximate 95% confidence bands for $F(t)$.

6.5 PROBABILITY PLOTS WITH SPECIFIED SHAPE PARAMETERS

The methods in Section 6.2 can extend to constructing probability plots for distributions that are not members of the log-location-scale family. In addition to the other applications described in Section 6.4.1, such plots help graphically identify the possibility of improving fit by using a nonzero threshold parameter (Sections 5.10.1 and 11.7).

Some distributions are not in the location-scale or log-location-scale families and cannot be transformed into such a distribution (e.g, the gamma and generalized gamma distributions and other distributions covered in Chapter 5). Such distributions have one or more unknown shape parameters (if a distribution has a single shape parameter whose value is assumed known, the distribution can be considered to be a location-scale distribution). It is still possible to construct a probability plot for distributions with an unknown shape parameter, but the plotting scales depend on the given value or estimate for the shape parameter. There are two approaches to specifying an unknown shape parameter for a probability plot:

- Plot the data with different given values of the shape parameter in an attempt to find a value that will give a probability plot that is nearly linear.
- Use parametric maximum likelihood methods to estimate the shape parameter and use the estimated value to construct probability plotting scales. This is discussed starting in Chapter 7 and continuing in Chapters 8 and 11.

6.5.1 Linearizing the Gamma cdf

As shown in Section 5.2.2, the quantile function for the gamma distribution is $t_p = \Gamma_I^{-1}(p; \kappa)\theta$. The quantile function for the three-parameter gamma distribution, allowing the distribution to start at γ instead of 0 (Section 5.10.1), is

$$t_p = \gamma + \Gamma_I^{-1}(p; \kappa)\theta.$$

This implies that $\{t_p$ versus $\Gamma_I^{-1}(p; \kappa)\}$ plots as a straight line. In contrast to the exponential probability plot, the probability scale for the gamma probability plot depends on specification of the shape parameter κ. As with the exponential probability plots (described in Section 6.2.1), γ is the intercept on the time scale [because $\Gamma_I^{-1}(p; \kappa) = 0$ when $p = 0$]. When plotting time is on the horizontal axis, the slope of the cdf line equals $1/\theta$. Thus changing θ changes the slope of the line, and changing γ changes the horizontal position of the line.

Example 6.6 Gamma Probability Plots for Fatigue Data for Alloy T7987. Here we return to the Alloy T7987 fatigue life data introduced in Example 6.2. Figure 6.13 shows gamma probability plots with shape parameter $\kappa = .8, 1.2, 2,$ and 5. These plots show the effect that choosing different gamma shape parameters will have on the curvature in the probability plot. Among these shape parameters, $\kappa = 2$ seems to give the best fit to the data. Each of these plots also indicates the need for a threshold parameter γ that is approximately 90 (Table 6.1 shows that the smallest observation in the data set was 94). There is, for these data, some physical justification for a threshold parameter. For some alloys, the amount of time that it takes for a fatigue crack to initiate and grow to failure may be on the order of hundreds of thousands of cycles, particularly if deformation caused by loading is primarily elastic. □

6.5.2 Linearizing the Weibull cdf Using a Linear Time Scale and Specified Shape Parameter

The quantile function for the three-parameter Weibull distribution

$$t_p = \gamma + \eta[-\log(1-p)]^{1/\beta}$$

can be obtained by inverting the cdf given in Section 5.10.2. This expression shows that $\{t_p$ versus $[-\log(1-p)]^{1/\beta}\}$ plots as a straight line. Unlike the standard log-time-scale Weibull probability plot described in Section 6.2.4, the probability scale for the linear-time-scale Weibull probability plot requires a given value of the shape parameter β. However, the plots provide instead a graphical estimate of the threshold parameter γ (which was previously constrained to be 0). As with the gamma and

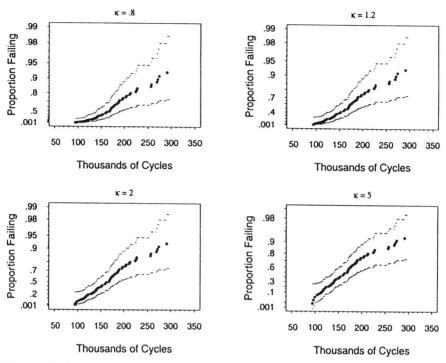

Figure 6.13. Gamma probability plots with κ = .8, 1.2, 2, and 5 for the Alloy T7987 fatigue life data with simultaneous approximate 95% confidence bands for $F(t)$.

exponential probability plots, γ is the intercept on the time scale (because $[-\log(1 - p)]^{1/\beta} = 0$ when $p = 0$). When time is on the horizontal axis, the slope of the cdf line is equal to $1/\eta$. As with the gamma and exponential probability plots, changing η changes the slope of the line, and changing γ changes the horizontal position of the line.

Example 6.7 Comparison of Log- and Linear-Time-Scale Weibull Probability Plots for Fatigue Life Data for Alloy T7987. Although the standard log-data-scale Weibull probability plot in Figure 6.6 (with threshold parameter $\gamma = 0$, implicitly) indicated a poor fit, the linear-data-scale Weibull probability plot with specified $\beta = 1.4$ (this value was determined by trial to provide the best fit visually) in Figure 6.14 indicates that a Weibull distribution with a shape parameter $\beta = 1.4$ and threshold parameter of approximately $\gamma = 90$ will provide a good fit to the data.

□

6.5.3 Linearizing the Generalized Gamma cdf

As shown in Section 5.3.2, the quantile function for the generalized gamma distribution (GENG) is $t_p = \theta[\Gamma_I^{-1}(p;\kappa)]^{1/\beta}$. This leads to

PROBABILITY PLOTS WITH SPECIFIED SHAPE PARAMETERS

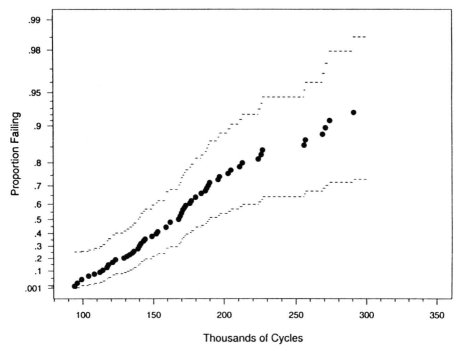

Figure 6.14. Linear-scale Weibull plot with $\beta = 1.4$ for the Alloy T7987 fatigue life with simultaneous approximate 95% confidence bands for $F(t)$.

$$\log(t_p) = \log(\theta) + \log[\Gamma_I^{-1}(p; \kappa)]\frac{1}{\beta},$$

implying that $\{\log(t_p)$ versus $\log[\Gamma_I^{-1}(p; \kappa)]\}$ plots as a straight line. Unlike the standard Weibull probability plot in Section 6.2.4, the probability scale for the GENG probability plot requires a given value of the shape parameter κ. The scale parameter θ is the intercept on the time scale, corresponding to the time where the cdf crosses the horizontal line at $\log[\Gamma_I^{-1}(p; \kappa)] = 0$. The slope of the line on the graph with time on the horizontal axis is β.

Example 6.8 GENG Probability Plots for the Ball Bearing Fatigue Data.
Example 1.1 introduced data on the number of revolutions to failure for 23 ball bearings. Figure 6.15 shows GENG probability plots with specified values of $\kappa = .1, 1, 4,$ and 20. The linear right-hand axis shows the gamma standard quantiles corresponding to $\log[\Gamma_I^{-1}(p; \kappa)]$. As explained in Section 5.3.3, the value of $\kappa = 1$ corresponds to a Weibull distribution and $\kappa \to \infty$ corresponds to the lognormal distribution. The value $\kappa = 20$ was chosen to correspond, roughly, to the lognormal distribution. None of the values of κ could be ruled out, but values greater than 1 seem to fit better. The value of $\kappa = 4$ was chosen by trial and error as a compromise between the Weibull and lognormal distributions. These examples show that the ranges of the standard

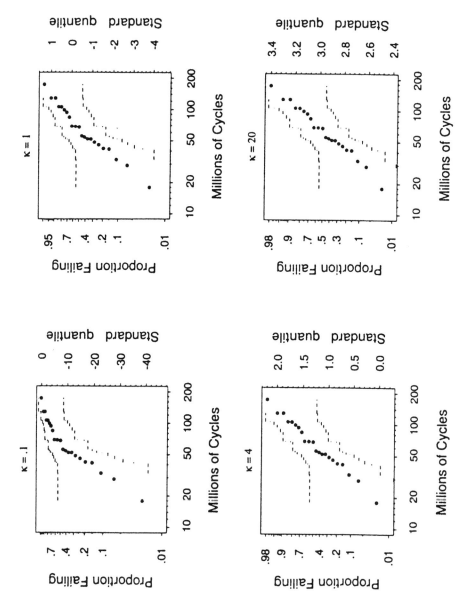

Figure 6.15. GENG probability plots of the ball bearing fatigue data with specified $\kappa = .1, 1, 4,$ and 20.

quantile scales depend strongly on the specified value of κ. Relatedly, the value of p corresponding to $\log[\Gamma_I^{-1}(p;\kappa)] = 0$ depends strongly on κ (and for $\kappa = 20$ it is off the scale). This is an indication of the potential problems, alluded to in Section 5.3.4, associated with statistically estimating the three traditional GENG parameters. □

6.5.4 Summary of Probability Plotting Methods

Table 6.2 summarizes the linearizing transformations given in Sections 6.2 and 6.5. This table also indicates which parameters need to be specified and which can be estimated from the slope and time-scale intercept of a fitted line.

6.6 NOTES ON THE APPLICATION OF PROBABILITY PLOTTING

6.6.1 Using Simulation to Help Interpret Probability Plots

- When the points on the probability plot follow a curved pattern (as in Figure 6.6), a smooth curve drawn through the points will still provide a useful graphical nonparametric estimate of the cdf. The curve provides quantile or failure probability estimates. For some, this kind of plot is easier to interpret than the cdf plot with linear probability axes (compare Figure 6.5 with Figure 6.6).
- As we have illustrated in our examples, analysts should try probability plotting with different assumed distributions and compare the results. Of course, finding a probability plot that indicates a good fit to the data does not guarantee that the model will be adequate for the desired purpose. This is a judgment that must be made in the context of the particular application.
- When assessing linearity, one must generally allow for the fact that, for most distributions, there will be more variability in the extreme observations. Judgment about the departure from linearity to expect comes with experience. Even experienced data analysts find, however, that it helpful to either (1) Plot simultaneous nonparametric confidence bands (e.g., the methods described in Section 3.8 and Section 6.3) to help assess the sampling uncertainty in the nonparametric estimate of $F(t)$ or (2) Use simulation methods to assess sampling variability directly. For example, one could generate simulated censored data from a particular distribution and plot the nonparametric estimates from a series of such data sets to get a sense of the deviations from linearity that one would expect under specific assumed distributions.

To illustrate the use of simulation, Figure 6.16 shows probability plots of simulated normal distribution samples of size $n = 10, 20,$ and 40. There are five probability plots from each sample size to allow an assessment of the repeatability or consistency of such plots. What we see is that there is very little consistency in the $n = 10$ plots. Even though the data were normally distributed, the pattern in the plots can deviate importantly from a straight line. With the larger sample sizes, however, there is more consistency across the repeated plots, except for the variability in the tails of the

Table 6.2. Summary of Probability Plot Scales to Linearize cdfs

Family	cdf	Linearizing Transformation		Specified Parameter	Identified Parameters	
		Time (Data) Scale	Probability Scale		Time-Scale Intercept	Slope
Exponential	$1 - \exp\left(-\dfrac{t-\gamma}{\theta}\right)$	t_p	$-\log(1-p)$		γ	$\dfrac{1}{\theta} \approx \dfrac{1}{t_{.63}}$
Smallest extreme value	$\Phi_{\text{sev}}\left(\dfrac{y-\mu}{\sigma}\right)$	y_p	$\Phi_{\text{sev}}^{-1}(p)$		$\mu \approx y_{.63}$	$\dfrac{1}{\sigma}$
Weibull (two-parameter)	$\Phi_{\text{sev}}\left[\dfrac{\log(t)-\mu}{\sigma}\right]$	$\log(t_p)$	$\Phi_{\text{sev}}^{-1}(p)$		$\eta = e^{\mu} \approx t_{.63}$	$\beta = \dfrac{1}{\sigma}$
Weibull (three-parameter)	$\Phi_{\text{sev}}\left[\dfrac{\log(t-\gamma)-\mu}{\sigma}\right]$	t_p	$\exp[\Phi_{\text{sev}}^{-1}(p)\sigma]$	$\beta = \dfrac{1}{\sigma}$	γ	$\dfrac{1}{\eta} \approx \dfrac{1}{t_{.63}}$
Normal	$\Phi_{\text{nor}}\left(\dfrac{y-\mu}{\sigma}\right)$	y_p	$\Phi_{\text{nor}}^{-1}(p)$		$\mu = y_{.5}$	$\dfrac{1}{\sigma}$
Lognormal (two-parameter)	$\Phi_{\text{nor}}\left[\dfrac{\log(t)-\mu}{\sigma}\right]$	$\log(t_p)$	$\Phi_{\text{nor}}^{-1}(p)$		$e^{\mu} = t_{.5}$	$\dfrac{1}{\sigma}$
Lognormal (three-parameter)	$\Phi_{\text{nor}}\left[\dfrac{\log(t-\gamma)-\mu}{\sigma}\right]$	t_p	$\exp[\Phi_{\text{nor}}^{-1}(p)\sigma]$	σ	γ	$e^{-\mu} = \dfrac{1}{t_{.5}}$
Gamma (three-parameter)	$\Gamma_{\text{I}}\left(\dfrac{t-\gamma}{\theta}; \kappa\right)$	t_p	$\Gamma_{\text{I}}^{-1}(p; \kappa)$	κ	γ	$\dfrac{1}{\theta}$
Generalized gamma	$\Gamma_{\text{I}}\left[\left(\dfrac{t}{\theta}\right)^{\beta}; \kappa\right]$	$\log(t_p)$	$\log[\Gamma_{\text{I}}^{-1}(p; \kappa)]$	κ	θ	β

The functions defined under "cdf" and "Probability Scale" are defined in Chapters 4 and 5.

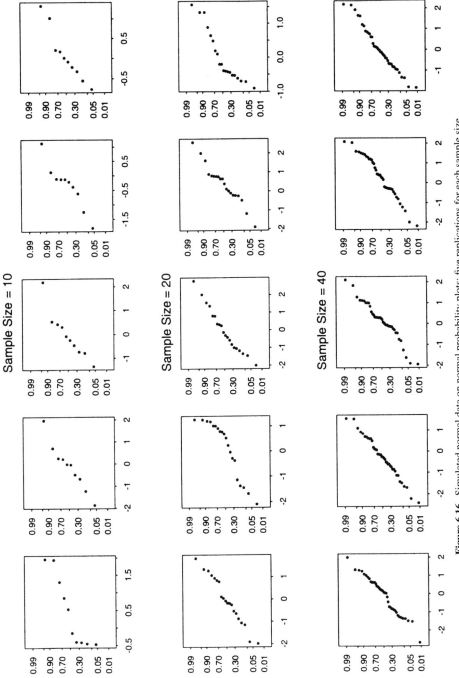

Figure 6.16. Simulated normal data on normal probability plots: five replications for each sample size.

distribution. Figure 6.17 has similar normal probability plots, but in that case the simulated data were from an exponential distribution. In this case we see that some of the plots with $n = 10$ do not deviate too much from a straight line and, to some extent, are similar to the probability plots of the normal data in Figure 6.16. For the larger sample sizes, however, there is enough consistency to indicate that samples of size 20 to 40 are sufficiently large to distinguish between data from exponential and normal distributions.

Inexperienced analysts tend to expect plots to be straighter than they are. Simulations of this kind can and should be used to help analysts "calibrate" their interpretation of probability plots, particularly in unfamiliar situations.

6.6.2 Possible Reason for a Bend in a Probability Plot

Probability plots with a sharp bend or change in slope generally indicate an abrupt change in a failure process. Causes for such behavior could include two or more failure modes or a mixture of different subpopulations. Such causes should be investigated and will often suggest how to improve product reliability.

Example 6.9 Bleed System Failure. Appendix Table C.7 gives failure and running time for 2256 bleed systems. The data were abstracted from Abernethy, Breneman, Medlin, and Reinman (1983) who present an analysis similar to the one done here.

The top row of Figure 6.18 shows a plot of the nonparametric estimate and a corresponding Weibull probability plot of the data. The different slopes on this probability plot before and after 600 hours suggest some kind of change. Closer examination of the data showed that 9 of the 19 failures had occurred at Base D. In the bottom row of Figure 6.18, separate analyses of the Base D data and the data from the other bases indicated different life distributions. The large slope ($\beta \approx 5$) for Base D indicated strong wearout behavior. The relatively small slope for the other bases ($\beta \approx .85$) suggested infant mortality or accidental failures. After investigation it was determined that the early-failure problem at Base D was caused by salt air (Base D was near the ocean). A change in maintenance procedures there solved the dominant bleed system reliability problem. □

6.6.3 Use of Grid Lines and Special Scales on Probability Plots

In the past, most data were plotted by hand on pre-prepared probability paper that contained grid lines. The grid lines make it easier to plot points by hand and allowed one to more precisely read numbers from the plot. While grid lines are useful, some analysts feel that grid lines can get in the way of interpreting a plot. Computer programs should provide an option to include grid lines or not. Because our interest is primarily in graphical perception of the information on a plot, we will generally not use grid lines on our computer-generated data analysis plots. If we are interested in particular numbers that would be read from the graph, the numbers are available from tabular computer output. For purposes of illustration, however, the following

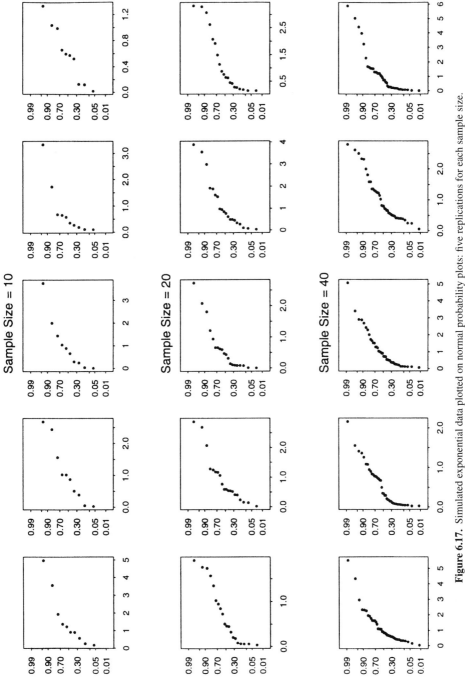

Figure 6.17. Simulated exponential data plotted on normal probability plots: five replications for each sample size.

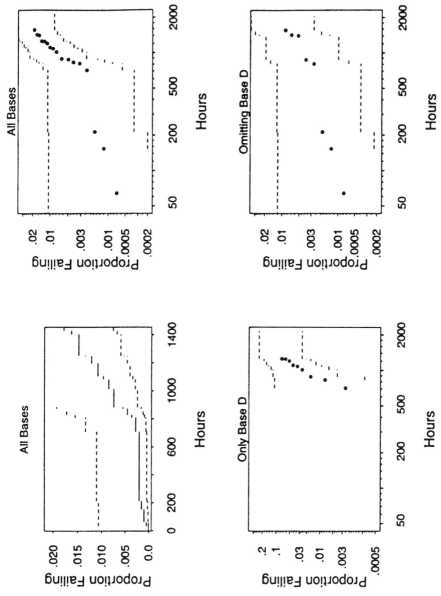

Figure 6.18. Bleed system data: (NW) linear plot of the cdf, (NE) Weibull probability plot for all bases, (SW) Weibull probability plots for Base D alone, and (SE) Weibull plot for all bases except Base D.

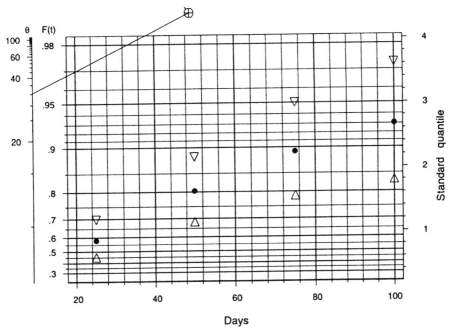

Figure 6.19. Exponential probability plot of the V7 transmitter tube failure data with simultaneous approximate 95% confidence bands for $F(t)$.

example uses probability plots with grid lines. It is also possible to put special scales on probability paper to facilitate graphical estimation of the parameter related to the slope of a line on the plot. Such scales are on some commercial probability papers but can also be put on computer-generated plots.

Example 6.10   V7 Transmitter Tube Failure Data.   Figure 6.19 is an exponential probability plot of the V7 transmitter tube data from Example 1.6. We see some departure from a straight line in the plot, but the width of the confidence bands makes it clear that this could be the result of random variability.

Figures 6.20 and 6.21 are, respectively, Weibull and lognormal probability plots of the V7 transmitter tube failure data. Comparing Figures 6.19, 6.20, and 6.21 suggests that none of these distributions can be ruled out but that the lognormal distribution provides the best fit among these distributions. Vacuum tubes have parts (filaments and cathode coatings) that will deteriorate with use, suggesting that the exponential distribution would not be an appropriate model.

Figures 6.19, 6.20, and 6.21 also contain special scales that allow one to graphically estimate θ, β, and σ, respectively, without doing any computations. To do this, draw a line, as shown, parallel to a line through the data points, going through the mark "⊕" and read, respectively, the estimates $\theta = 32$, $\beta = .75$, and $\sigma = 1.06$ from the scales on the left-hand side of the graphs.   □

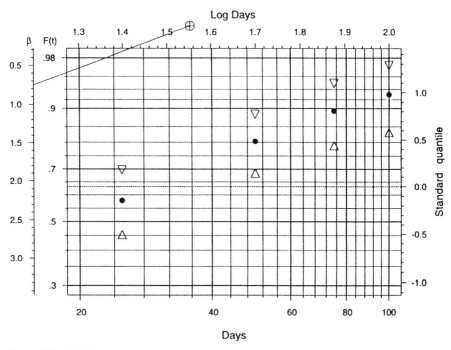

Figure 6.20. Weibull probability plot of the V7 transmitter tube failure data with simultaneous approximate 95% confidence bands for $F(t)$.

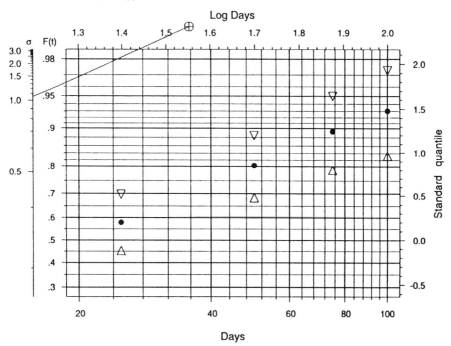

Figure 6.21. Lognormal probability plot of the V7 transmitter tube failure data with simultaneous approximate 95% confidence bands for $F(t)$.

148

BIBLIOGRAPHIC NOTES

Most of the literature on methods for probability plotting is concerned with complete (uncensored) data. Chapter 8 of Hahn and Shapiro (1967) provides a nice summary of basic theory and methods and illustrates, with simulated data, the variability that one expects to see in the points on a probability plot. Harter (1984) reviews some history concerning the choice of plotting positions. Chernoff and Lieberman (1954), Blom (1958), and Barnett (1976) used good estimation of model parameters as a criterion for choosing plotting positions. David (1981, page 208) gives an excellent review of the results on these last three papers. For multiply censored data Lawless (1982, page 88) suggests the use of the half-step correction to the nonparametric estimate of $F(t)$ defined in (6.2). Nelson and Thompson (1971) provide Weibull probability papers that can be copied and used for making probability plots "by hand." Such papers are also available commercially.

For multiply censored data Nelson (1972, 1982, Chapter 4) proposed the use of a hazard plot. A hazard plot can be viewed as a type of probability plot with special plotting positions corresponding to the Nelson–Aalen nonparametric estimate of the cdf (see Exercise 3.23). Nelson (1982, page 135) suggests modified hazard plotting positions obtained by averaging the hazard step function at the jumps. These are similar to the modified plotting positions in (6.2). Nelson comments: "The modified positions agree better with a distribution fitted by maximum likelihood." An alternative that would serve the same purpose would average the estimates in the probability scale instead, similar to (6.2). Wilk, Gnanadesikan, and Huyett (1962a) show how to construct probability plots for the gamma distribution. Nair (1981) describes, in more detail, the theory and applications of simultaneous confidence bands as a tool for assessing distributional goodness of fit.

EXERCISES

6.1. For the LOGLOGIS(μ, σ) distribution with cdf

$$F(t) = \Phi_{\text{logis}}\left[\frac{\log(t) - \mu}{\sigma}\right], \quad t > 0; \quad -\infty < \mu < \infty, \quad \sigma > 0,$$

(a) Find the probability scales that will linearize all the cdfs in the logistic family.

(b) Use the scales to generate a properly labeled graph, and display the LOGLOGIS(1, 1) and the LOGLOGIS(1, 2) cdfs.

(c) What quantile of this distribution corresponds to the scale parameter $\exp(\mu)$?

6.2. Starting with an ordinary piece of graph paper with linear divisions, perform the following steps to create Weibull probability paper with time ranging between 10 and 1000 and probability ranging between .001 and .999. Refer to

Figure 6.4 for an example. Alternatively, program a spreadsheet or statistical package to do the same thing with computer graphics.
- **(a)** Find values of $\log[-\log(1-p)]$ for $p = .001$ and $p = .999$. Use these to develop a linear axis on the right-hand side of the graph.
- **(b)** For selected values of p between .001 and .999 (e.g., .001, .01, .1, .3, .5, .7, .9, .99, .999) compute $\log[-\log(1-p)]$. Find this value on the right-hand side axis to determine the location of the p label on the left-hand side axis.
- **(c)** Find values of $\log(t)$ for $t = 10$ and $t = 1000$. Use these to develop a linear axis for $\log(t)$ on the top of the page.
- **(d)** For selected values of t between 10 and 1000 (e.g., 10, 20, 50, 100, 200, 500, 1000), compute $\log(t)$ and use the location on the top axis to determine the corresponding locations for the time labels on the bottom axis.

6.3. Consider the scale parameter η for the Weibull distribution.
- **(a)** Show that $\eta = \exp(\mu)$ for the Weibull distribution is approximately equal to the .63 quantile.
- **(b)** Discuss the practical importance of estimating η for a population of integrated circuits to be installed in new personal computers.
- **(c)** Is it possible to get a good graphical estimate of η from a probability plot based on a life test for which only 3.5% of the integrated circuits failed by the end of the test?
- **(d)** For what Weibull "parameters" (i.e., functions of η and β) can one get good graphical estimates from such data?

6.4. Use the following 10 simulated observations from a Weibull distribution with $\eta = 1$ and $\beta = 2$ (so that $\mu = 0$ and $\sigma = .5$) to make a Weibull probability plot and use it to obtain graphical estimates of the parameters η and β. $t_i = .74, 1.21, .22, .37, 1.28, .73, .99, .67, .71, .33$. How do the estimates compare with the "true parameter values"?

6.5. Consider the ball bearing fatigue data given in Example 1.1 and Table 1.1.
- **(a)** Compute a nonparametric estimate of $F(t)$, the proportion of units failing as function of time. Plot your estimate on paper with linear scales.
- **(b)** Make a lognormal probability plot of the data. This is accomplished by ordering the failure times in increasing order, $t_{(1)} \leq \cdots \leq t_{(23)}$. Then plot $t_{(i)}$ versus $(i - .5)/n$ on lognormal probability paper.
- **(c)** Do the same as in part (b) but on Weibull probability paper.
- **(d)** Comment on the adequacy of the lognormal and Weibull models to describe these data.

6.6. Use the answers to Exercise 3.6 to do the following:

EXERCISES   151

   (a) Make a Weibull probability plot displaying the device failure data.
   (b) Use the plotted points to estimate the proportion of devices that will fail before 10,000 hours of operation.
   (c) Comment on whether the Weibull distribution fits the data well.
   (d) Use the slope and location of this line to estimate the Weibull distribution parameters.
   (e) Use the plotted points to estimate the proportion of devices that will fail before 100,000 hours. Comment on the usefulness of this estimate.

6.7. A sample of 100 specimens of a titanium alloy were subjected to a fatigue test to determine time to crack initiation. The test was run up to a limit of 100,000 cycles. The observed times of crack initiation (in units of 1000 of cycles) were 18, 32, 39, 53, 59, 68, 77, 78, 93. No crack had initiated in any of the other 91 specimens.
   (a) Compute a nonparametric estimate, $\widehat{F}(t)$, of the cdf $F(t)$ using both the simple binomial method and the Kaplan–Meier method (in this case these two methods provide the same answer).
   (b) Plot $\widehat{F}(t)$ on linear axes.
   (c) Use $\widehat{F}(t)$ to compute plotting positions and plot the data on Weibull paper. Use the plot to obtain an estimate of the Weibull shape parameter β.
   (d) Comment on the adequacy of the Weibull distribution.
   (e) Comment on the adequacy of the available data if the purpose of the experiment was to estimate $t_{.1}$.

6.8. For the high-cycle fatigue life data in Exercise 3.4, construct probability plots for the exponential, lognormal, Weibull, and gamma distributions (trying several values of κ for the gamma distribution). Which distributions appear suitable for describing the shape of the distribution in the lower tail?

6.9. Using the life test data on silicon photodiode detectors from Exercise 3.12, construct probability plots for the exponential, Weibull, and lognormal distributions. Which distributions look like they might provide an adequate model for photodiode detector life?

6.10. Figures 6.3 and 6.4 have horizontal lines at the standardized quantile value of 0.
   (a) For Figure 6.3, explain why the dotted line crosses the $F(t)$ scale at $F(t) = .5$.
   (b) For Figure 6.4, compute the value of $F(t)$ where $F(t)$ crosses the dotted line corresponding to 0 on the standardized quantile scale.

6.11. Using the linear scales on the top and right of Figure 6.7, we can use a straight line drawn through the data points to obtain graphical estimates of the lognormal distribution fit to the Alloy T7987 data. Use these estimates to compute a parametric estimate of $F(200)$ by substituting them into (4.4).

6.12. Use Figures 6.5, 6.6, and 6.7 to obtain nonparametric graphical estimates of $F(200)$ for the Alloy T7987 data. Are the answers similar? Explain why or why not. Compare your answers with those obtained in Exercise 6.11. Explain the reason for observed differences.

6.13. Use the linear scales on the top and right of Figure 6.8 to compute graphical estimates of the Weibull distribution parameters for the shock absorber data. Note that the linear axis on top of the plot gives base 10 logarithms.

▲6.14. Consider the family of gamma distributions with scale parameter θ and shape parameter κ, as in equation (5.1). Show that for a fixed value of κ, the probability plotting scales $\{t_p, \Gamma_1^{-1}(p;\kappa)\}$ provide linearizing scales of the distribution for all values of the parameter θ.

▲6.15. Consider an uncensored sample (i.e., all observations reported as exact failures) $t_{(1)} \leq \cdots \leq t_{(n)}$ of failure times used to make a Weibull probability plot. Let $(t_{(i)}, p_i)$, $i = 1, \ldots, n$, be the points on the probability plot, where p_i is defined in (6.2). A simple method for estimating the Weibull (η, β) parameters is the following. Use least squares to fit a straight line through the points using $\log(t_{(i)})$ as the response (y) and $\Phi_{\text{sev}}^{-1}(p_i)$ as the explanatory variable (x). Then use the intercept and the slope of the line, respectively, to estimate the parameters μ and σ. Denote these estimates by $\widehat{\sigma}_{\text{ols}}$ and $\widehat{\mu}_{\text{ols}}$. Then estimates for (η, β) are $(\exp(\widehat{\mu}_{\text{ols}}), 1/\widehat{\sigma}_{\text{ols}})$.
(a) Derive the equations for the estimates $\widehat{\sigma}_{\text{ols}}$ and $\widehat{\mu}_{\text{ols}}$.
(b) Do the assumptions that assure optimality of the ordinary least squares estimators hold in this case? Give details for your answer.
(c) Is the standard R-squared statistic used in regression a useful measure of goodness of fit for this problem? Why or why not?

CHAPTER 7

Parametric Likelihood Fitting Concepts: Exponential Distribution

Objectives

This chapter explains:

- Likelihood for a parametric model using discrete data.
- Likelihood for samples containing right- and left-censored observations.
- Use of parametric likelihood as a tool for data analysis and inference about a single population or process.
- The use of likelihood and normal-approximation confidence intervals for model parameters and other quantities of interest.
- The density approximation to the likelihood for observations reported as exact failures.

Overview

This chapter introduces some basic ideas of parametric maximum likelihood (ML) methods. The ideas presented here are used throughout the rest of the book. Section 7.2 shows how to construct the likelihood (probability of the data) function and describes the basic ideas behind using this function to estimate a parameter. Sections 7.3–7.5 describe methods for computing confidence intervals for parameters and functions of parameters. Section 7.6 describes the commonly used probability density in the construction of a likelihood function. Section 7.7 shows how to get a confidence bound on the exponential distribution parameter even if there are no failures.

7.1 INTRODUCTION

As explained in Chapter 4, parametric distributions, when used appropriately, can provide a simple, parsimonious, versatile, visually appealing failure-time model. ML is perhaps the most versatile method for fitting statistical models to data. The appeal of ML stems from the fact that it can be applied to a wide variety of statistical models and kinds of data (e.g., continuous, discrete, categorical, censored, truncated), where other popular methods, like least squares, are not, in general, satisfactory. In typical applications, the goal is to use a parametric statistical model to describe a set of data or a process or population that generated a set of data. Modern computing hardware and software have tremendously expanded the feasible areas of application for ML methods.

Statistical theory (see the bibliographic notes at the end of this chapter and Appendix Section B.6.1 for some references) shows that, under standard regularity conditions, ML estimators are "optimal" in large samples. More specifically, this means that ML estimators are consistent and asymptotically (as the sample size increases) efficient. That is, among consistent competitors to ML estimators, none has a smaller asymptotic variance.

Chapter 14 describes and illustrates the use of the closely related Bayesian methods that allow one to incorporate prior information into the model fitting and estimation process. Besides the Bayesian methods (which require specification of a prior distribution for the unknown parameters), there is no general theory that suggests alternatives to ML that will be optimal in finite samples. Comparisons in specific cases have shown that, for practical purposes, and without incorporating prior information, it is difficult to improve on ML methods.

This chapter emphasizes methods, concepts, examples, and interpretation of data. Appendix Section B.6 outlines the general theory.

Example 7.1   Time Between α-Particle Emissions of Americium-241.   Although not from the area of reliability, this example is analogous to certain special reliability applications in which the distribution of time between events can be described with an exponential distribution.

Berkson (1966) investigates the randomness of α-particle emissions of americium-241 (which has a half-life of about 458 years). Physical theory suggests that, over a short period of time, the interarrival times of observed particles would be independent and come from an exponential distribution

$$F(t; \theta) = 1 - \exp\left(-\frac{t}{\theta}\right), \qquad (7.1)$$

where θ is the mean time between arrivals. The corresponding homogeneous Poisson process that counts the number of emissions on the real-time line (see Section 4.4 for more information on the exponential distribution and Chapter 16 for more information on the Poisson process) has arrival rate or intensity $\lambda = 1/\theta$. For the interarrival times of α particles, λ is proportional to the americium-241 decay rate, size of the sample, the counter size and efficiency, and so on.

The data consisted of 10,220 observed interarrival times of α particles (time unit equal to 1/5000 second). The observed interarrival times were put into intervals (or bins) running from 0 to 4000 time units with interval lengths ranging from 25 to 100 time units, with one additional interval for observed times exceeding 4000 time units. To save space in our analysis, this example uses a smaller number of larger bins;

PARAMETRIC LIKELIHOOD

Table 7.1. Binned α-Particle Interarrival Time Data in 1/5000 Second

Time Interval Endpoint		Interarrival Times Frequency of Occurrence			
		All Times	Random Samples of Times		
Lower	Upper	$n = 10220$	$n = 2000$	$n = 200$	$n = 20$
0	100	1609	292	41	3
100	300	2424	494	44	7
300	500	1770	332	24	4
500	700	1306	236	32	1
700	1000	1213	261	29	3
1000	2000	1528	308	21	2
2000	4000	354	73	9	0
4000	∞	16	4	0	0

reducing the number of bins in this way will not seriously affect the precision of ML estimates. These data are shown in Table 7.1.

To illustrate the effects of sample size on the inferences, simple random samples (i.e., each interval-censored interarrival time having equal probability) of sizes $n = 2000$, 200, and 20 were drawn with replacement from these interarrival times. The following examples compare the results that one obtains with these different sample sizes. When focusing on just one sample, the sample of size $n = 200$ interarrival times is used. □

Statistical modeling, in practice, is an iterative procedure of fitting proposed models in search of a model that provides an adequate description of the population or process of interest, without being unnecessarily complicated. Application of ML methods generally starts with a set of data and a tentative statistical model for the data. The tentative model is often suggested by the initial graphical analysis (Chapter 6), physical theory, previous experience with similar data, or other expert knowledge.

Example 7.2 Probability Plot for the α-Particle Data. Figure 7.1 shows an exponential probability plot of the sample with $n = 200$. The approximate linearity of the plot indicates that the exponential distribution provides a good fit to these data. This is reinforced by the simultaneous nonparametric confidence bands. □

7.2 PARAMETRIC LIKELIHOOD

7.2.1 Probability of the Data

Proceeding from the ideas introduced in Section 2.4.2, but now using a parametric model (as described in Chapter 4), the likelihood function can be viewed as the *probability of the observed data*, written as a function of the model's parameters. For a parametric model, the number of parameters is usually small relative to the

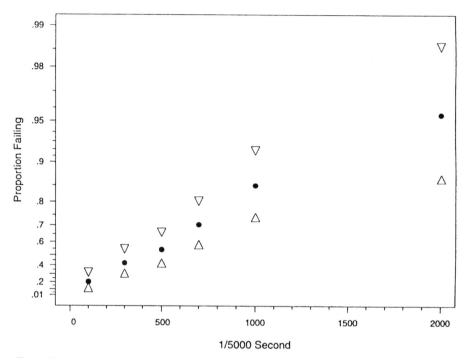

Figure 7.1. Exponential probability plot of the $n = 200$ sample of α-particle interarrival time data. The plot also shows simultaneous nonparametric approximate 95% confidence bands.

nonparametric models described and used in Chapters 2 and 3. The exponential distribution in (7.1) has only one parameter.

For a set of n independent observations, the likelihood function can be written as the following joint probability

$$L(\theta) = L(\theta; \text{DATA}) = C \prod_{i=1}^{n} L_i(\theta; \text{data}_i). \tag{7.2}$$

As described in Section 2.4.4, the quantity C in (7.2) is a constant term that does not depend on the data or on θ (in general θ can be a vector, but in this chapter it is a scalar). As in Chapter 3, for computational purposes, let $C = 1$. The likelihood contribution terms $L_i(\theta; \text{data}_i)$ were explained in detail in Section 2.4.3. For example, if a failure time is known to have occurred between times t_{i-1} and t_i, the probability of this event is

$$L_i(\theta; \text{data}_i) = L_i(\theta) = \int_{t_{i-1}}^{t_i} f(t; \theta)\, dt = F(t_i; \theta) - F(t_{i-1}; \theta). \tag{7.3}$$

For a given set of data, $L(\theta)$ can be viewed as a function of θ. The dependence of $L(\theta)$ on the data will be understood and is usually suppressed in notation. The values

PARAMETRIC LIKELIHOOD   **157**

of θ for which $L(\theta)$ is relatively large are more plausible than values of θ for which the probability of the data is relatively small. There may or may not be a unique value of θ that maximizes $L(\theta)$. Regions in the space of θ with relatively large $L(\theta)$ can be used to define confidence regions for θ. One can also use ML to estimate *functions* of θ. The rest of this chapter shows how to make these concepts operational for the single-parameter exponential distribution, using simple examples for illustration. Subsequent chapters treat models with two or more parameters.

7.2.2 Likelihood Function and Its Maximum

Given a sample of n independent observations, denoted generically by data$_i$, $i = 1, \ldots, n$, and a specified model, the total likelihood $L(\theta)$ for the sample is given by equation (7.2). For some purposes, it is convenient to use the log likelihood $\mathcal{L}_i(\theta) = \log[L_i(\theta)]$. For all practical problems $\mathcal{L}(\theta)$ will be representable in computer memory without special scaling [which is not so for $L(\theta)$ because of possible extreme exponent values], and some theory for ML is developed more naturally in terms of sums like

$$\mathcal{L}(\theta) = \log[L(\theta)] = \sum_{i=1}^{n} \mathcal{L}_i(\theta)$$

rather than in terms of the products in equation (7.2). Note that the maximum of $\mathcal{L}(\theta)$, if one exists, occurs at the same value of θ as the maximum of $L(\theta)$.

Example 7.3 Likelihood for the α-Particle Data. In this example, the unknown parameter θ is a scalar and this makes the analysis particularly simple and provides a useful first example to illustrate basic concepts. Substituting equation (7.1) into (7.3) and (7.3) into (7.2) gives the exponential distribution likelihood function (joint probability) for interval data (e.g., Table 7.1) as

$$L(\theta) = \prod_{i=1}^{n} L_i(\theta) = \prod_{i=1}^{n} [F(t_i; \theta) - F(t_{i-1}; \theta)] \qquad (7.4)$$

$$= \prod_{j=1}^{8} [F(t_j; \theta) - F(t_{j-1}; \theta)]^{d_j} = \prod_{j=1}^{8} \left[\exp\left(-\frac{t_{j-1}}{\theta}\right) - \exp\left(-\frac{t_j}{\theta}\right)\right]^{d_j},$$

where d_j is the number of interarrival times in interval j. Note that in the first line of (7.4), the product is over the n observed times and in the second line, it is over the 8 bins into which the data have been grouped. □

The ML estimate of θ is found by maximizing $L(\theta)$. When there is a unique global maximum, $\widehat{\theta}$ denotes the value of θ that maximizes $L(\theta)$. In general, however, the maximum may not be unique. The function $L(\theta)$ may have multiple local maxima or can have flat spots along which $L(\theta)$ changes slowly, if at all. Such flat spots may or may not be at the maximum value of $L(\theta)$. The shape and magnitude of $L(\theta)$ relative to $L(\widehat{\theta})$ over all possible values of θ describe the information on θ that is contained in data$_i$, $i = 1, \ldots, n$.

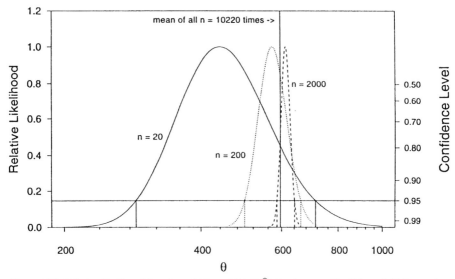

Figure 7.2. Relative likelihood functions $R(\theta) = L(\theta)/L(\widehat{\theta})$ for the $n = 20$, 200, and 2000 samples and ML estimate for the $n = 10{,}220$ sample of the α-particle data. Vertical lines give corresponding approximate 95% likelihood confidence intervals.

Example 7.4 Relative Likelihood for the α-Particle Data. Figure 7.2 shows the *relative likelihood functions*

$$R(\theta) = \frac{L(\theta)}{L(\widehat{\theta})}$$

for the samples of size $n = 2000$, 200, and 20 and a vertical line at the mean of all $n = 10{,}220$ times. These functions allow one to judge the probability of the data for values of θ, *relative* to the probability at the ML estimate. For example, $R(\theta_0) = .1$ implies that the probability of the data is 10 times larger at $\widehat{\theta}$ than at θ_0. The next section explains how to use $R(\theta)$ to compute confidence intervals for θ.

Figure 7.2 indicates that the spread of the likelihood function tends to decrease as the sample size increases. The relative likelihood functions for the larger samples are much tighter than those for the smaller samples, indicating that the larger samples contain more information about θ. The θ value at which the different likelihood functions are maximized is random and depends, in this comparison, on the results of the sampling described in Example 7.1. The four $\widehat{\theta}$ values differ, but they are consistent with the variability that one would expect from random sampling using the corresponding sample sizes. □

7.2.3 Comparison of α-Particle Data Analyses

Figure 7.3 shows another exponential probability plot of the $n = 200$ sample. The solid line on this graph is the ML estimate of the exponential cdf $F(t; \theta)$. The dotted lines are drawn through a set of pointwise parametric normal-approximation

CONFIDENCE INTERVALS FOR θ

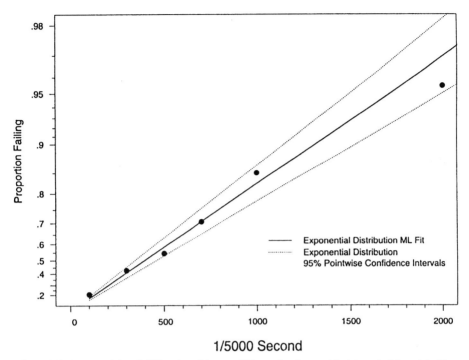

Figure 7.3. Exponential probability plot of the $n = 200$ sample of α-particle interarrival time data. The plot also shows the parametric exponential ML estimate and approximate 95% confidence intervals for $F(t; \theta)$.

95% confidence intervals for $F(t; \theta)$; these parametric intervals will be explained in Section 7.3. Table 7.2 summarizes the results of fitting exponential distributions to the four different samples in Table 7.1; it includes ML estimates, standard errors, and confidence intervals. Section 7.3 provides results specifically for θ, the mean (which is also the .632 quantile, $t_{.632}$) of the exponential distribution. Section 7.4.1 shows how to obtain similar results for $\lambda = 1/\theta$, the arrival intensity rate (per unit of time).

7.3 CONFIDENCE INTERVALS FOR θ

7.3.1 Likelihood Confidence Intervals for θ

The likelihood function provides a versatile method for assessing the information that the data contains on parameters, or functions of parameters. Specifically, the likelihood function provides a generally useful method for finding approximate confidence intervals for parameters and functions of parameters.

An approximate $100(1 - \alpha)\%$ likelihood-based confidence interval for θ is the set of all values of θ such that

$$-2\log[R(\theta)] \leq \chi^2_{(1-\alpha;1)}$$

Table 7.2. Comparison of α-Particle ML Results

	All Times $n = 10{,}220$	Sample of Times		
		$n = 2000$	$n = 200$	$n = 20$
ML estimate $\widehat{\theta}$	596.3	612.8	572.3	440.2
Standard error $\widehat{\text{se}}_{\widehat{\theta}}$	6.084	14.13	41.72	101.0
Approximate 95% Confidence Intervals for θ				
Based on the likelihood	[584, 608]	[586, 641]	[498, 662]	[289, 713]
Based on $Z_{\log(\widehat{\theta})} \stackrel{.}{\sim} \text{NOR}(0, 1)$	[584, 608]	[586, 641]	[496, 660]	[281, 690]
Based on $Z_{\widehat{\theta}} \stackrel{.}{\sim} \text{NOR}(0, 1)$	[584, 608]	[585, 640]	[490, 653]	[242, 638]
ML estimate $\widehat{\lambda} \times 10^5$	168	163	175	227
Standard error $\widehat{\text{se}}_{\widehat{\lambda} \times 10^5}$	1.7	3.8	13	52
Approximate 95% Confidence Intervals for $\lambda \times 10^5$				
Based on the likelihood	[164, 171]	[156, 171]	[151, 201]	[140, 346]
Based on $Z_{\log(\widehat{\lambda})} \stackrel{.}{\sim} \text{NOR}(0, 1)$	[164, 171]	[156, 171]	[152, 202]	[145, 356]
Based on $Z_{\widehat{\lambda}} \stackrel{.}{\sim} \text{NOR}(0, 1)$	[164, 171]	[156, 171]	[149, 200]	[125, 329]

or, equivalently, the set defined by

$$R(\theta) \geq \exp\left[-\chi^2_{(1-\alpha;1)}/2\right].$$

The theoretical justification for this interval is given in Appendix Section B.6.5.

Example 7.5 Likelihood Confidence Intervals for the Mean Time Between Arrivals of α Particles. Figure 7.2 illustrates likelihood confidence intervals. The horizontal line at $\exp[-\chi^2_{(.95;1)}/2] = .147$ corresponds to approximate 95% confidence intervals. The vertical lines dropping from the respective curves give the endpoints of the confidence intervals for the different samples. Table 7.2 gives numerical values of likelihood-based approximate 95% confidence intervals (as well as intervals based on other methods to be explained subsequently). Figure 7.2 shows that increasing sample size tends to reduce confidence interval length. Approximate (large-sample) theory shows that confidence interval length under standard regularity conditions is approximately proportional to $1/\sqrt{n}$ (see Appendix Section B.6.1 for an outline of this theory and Chapter 10 for methods of choosing the sample size to control the width of a confidence interval). □

A one-sided approximate $100(1 - \alpha)\%$ confidence bound can be obtained by drawing the horizontal line at $\exp[-\chi^2_{(1-2\alpha;1)}/2]$ and using the appropriate endpoint of the resulting two-sided confidence interval.

Example 7.6   One-Sided Likelihood-Based Confidence Bounds for the Mean Time Between Arrivals of α Particles.   Referring to Figure 7.2, the horizontal line at $\exp[-\chi^2_{(.95;1)}/2] = .147$ would provide one-sided approximate 97.5% confidence bounds for θ. For one-sided approximate 95% confidence bounds the line would be drawn at $\exp[-\chi^2_{(.90;1)}/2] = .259$ (corresponding to .90 on the right-hand scale on Figure 7.2). □

7.3.2   Relationship Between Confidence Intervals and Significance Tests

Significance testing (sometimes called hypothesis testing) is a statistical technique widely used in many areas of science. The basic idea is to assess the reasonableness of a claim or hypothesis about a model or parameter value, relative to observed data. One can test a hypothesis by first constructing a $100(1 - \alpha\%)$ confidence interval for the quantity of interest and then checking to see if the interval encloses the hypothesized value or not. If not, then the hypothesis is rejected "at the α level of significance." If the interval encloses the hypothesized value, then the appropriate conclusion is that the data are consistent with the hypothesis (it is important, however, to note that failing to reject a hypothesis is not the same as saying that the hypothesis is true—see the following examples). Most practitioners find confidence intervals much more informative than the yes/no result of an significance test. See pages 39–40 of Hahn and Meeker (1991) and other references given there for further discussion of this subject.

To be more formal, a likelihood ratio test for a single-parameter model can be done by comparing the maximum of the likelihood under the "null hypothesis" to the maximum of the likelihood over all possible values for the parameter. A likelihood much smaller under the null hypothesis provides evidence to refute the hypothesis. Specifically, for the exponential distribution, the single-point null hypothesis $\theta = \theta_0$ should be rejected if

$$-2\log[L(\theta_0)/L(\widehat{\theta})] > \chi^2_{(1-\alpha;1)}, \qquad (7.5)$$

where $\widehat{\theta}$ is the ML estimate of θ. Rejection implies that the data are not consistent with the null hypothesis. Using the definition given in Section 7.3.1, it is easy to see that a likelihood-based confidence interval is the set of all values of θ that would not be rejected under the likelihood ratio test defined in (7.5).

Example 7.7 Likelihood-Ratio Test for the Mean Time Between Arrivals of α Particles. Suppose that investigators conducted the α-particle experiment to test the hypothesis that the mean time between arrivals of α particles is $\theta = 650$. Based on the confidence intervals for $n = 200$ in Table 7.2, we would have to conclude that there is not enough evidence to reject this hypothesis. Correspondingly,

$$-2\log\left[L(650)/L(572.3)\right] = 2.94 < \chi^2_{(.95;1)} = 3.84$$

again showing that there is not sufficient evidence in the $n = 200$ sample to reject the hypothesis. Using the $n = 2000$ sample, however, does provide sufficient evidence to reject the hypothesis that $\theta = 650$ at the 5% level of significance, as 650 is not in the 95% confidence interval. □

7.3.3 Normal-Approximation Confidence Intervals for θ

A $100(1-\alpha)\%$ normal-approximation confidence interval for θ is

$$[\underline{\theta},\ \widetilde{\theta}] = \widehat{\theta} \pm z_{(1-\alpha/2)}\widehat{se}_{\widehat{\theta}}. \tag{7.6}$$

A one-sided approximate $100(1-\alpha)\%$ confidence bound can be obtained by replacing $z_{(1-\alpha/2)}$ with $z_{(1-\alpha)}$ in (7.6) and using the appropriate endpoint of the resulting two-sided confidence interval.

An estimate of the standard error of $\widehat{\theta}$ is typically computed from the "observed information" as

$$\widehat{se}_{\widehat{\theta}} = \sqrt{\left[-\frac{d^2\mathcal{L}(\theta)}{d\theta^2}\right]^{-1}}, \tag{7.7}$$

where the second derivative is evaluated at $\widehat{\theta}$. This computation is a special case of (B.10) in Appendix Section B.6.4. The second derivative measures curvature of $\mathcal{L}(\theta)$ at $\widehat{\theta}$. If $\mathcal{L}(\theta)$ is approximately quadratic, large curvature implies a narrow likelihood and thus a small estimate of the standard error of $\widehat{\theta}$.

The approximate confidence interval in (7.6) is based on the assumption that the distribution of

$$Z_{\widehat{\theta}} = \frac{\widehat{\theta} - \theta}{\widehat{se}_{\widehat{\theta}}} \tag{7.8}$$

can be approximated by a NOR(0, 1) distribution. Then

$$\Pr[z_{(\alpha/2)} < Z_{\widehat{\theta}} \le z_{(1-\alpha/2)}] \approx 1 - \alpha, \tag{7.9}$$

which implies

$$\Pr[\widehat{\theta} - z_{(1-\alpha/2)}\widehat{se}_{\widehat{\theta}} < \theta \le \widehat{\theta} + z_{(1-\alpha/2)}\widehat{se}_{\widehat{\theta}}] \approx 1 - \alpha \tag{7.10}$$

CONFIDENCE INTERVALS FOR FUNCTIONS OF θ

because $z_{(1-\alpha/2)} = -z_{(\alpha/2)}$. The approximation is usually better for large samples but may be poor for small samples. See Appendix Section B.5 for more information on such large-sample approximations.

Example 7.8 Normal-Approximation Confidence Intervals for the Mean Time Between Arrivals of α Particles. For the $n = 200$ α-particle data $\widehat{se}_{\hat{\theta}} = 41.72$ and an approximate 95% confidence interval for θ based on the assumption that $Z_{\hat{\theta}} \sim \text{NOR}(0, 1)$ is

$$[\underset{\sim}{\theta},\ \widetilde{\theta}] = 572.3 \pm 1.960(41.72) = [490,\ 653].$$

Thus we are 95% confident that θ is in this interval. □

An alternative approximate confidence interval for positive quantities like θ is

$$[\underset{\sim}{\theta},\ \widetilde{\theta}] = [\hat{\theta}/w,\ \hat{\theta} \times w], \tag{7.11}$$

where $w = \exp(z_{(1-\alpha/2)}\widehat{se}_{\hat{\theta}}/\hat{\theta})$. This interval is based on the assumption that the distribution of

$$Z_{\log(\hat{\theta})} = \frac{\log(\hat{\theta}) - \log(\theta)}{\widehat{se}_{\log(\hat{\theta})}} \tag{7.12}$$

can be approximated by a NOR(0, 1) distribution, where $\widehat{se}_{\log(\hat{\theta})} = \widehat{se}_{\hat{\theta}}/\hat{\theta}$ is obtained by using the delta method in Appendix Section B.2. The confidence interval in (7.11) follows because an approximate $100(1 - \alpha)\%$ confidence interval for $\log(\theta)$ is

$$[\underset{\sim}{\log(\theta)},\ \widetilde{\log(\theta)}] = \log(\hat{\theta}) \pm z_{(1-\alpha/2)}\widehat{se}_{\log(\hat{\theta})}.$$

For a parameter, like θ, that must be positive, (7.11) is often suggested as providing positive interval endpoints and probably a more accurate approximate interval than (7.6). Although there is no guarantee that (7.11) will be more accurate than (7.6) in a particular setting, the sampling distribution of $Z_{\log(\hat{\theta})}$ is usually more symmetric than that of $Z_{\hat{\theta}}$ and the log transformation ensures that the lower endpoint of the confidence interval will be positive [which is not always so for confidence intervals based on (7.6)].

Example 7.9 Normal-Approximation Confidence Intervals for the Mean Time Between Arrivals of α Particles. For the α-particle data, an approximate 95% confidence interval for θ based on the assumption that $Z_{\log(\hat{\theta})} \sim \text{NOR}(0, 1)$ is

$$[\underset{\sim}{\theta},\ \widetilde{\theta}] = [572.3/1.1536,\ 572.3 \times 1.1536] = [496,\ 660],$$

where $w = \exp\{1.960 \times 41.72/572.3\} = 1.1536$. □

7.4 CONFIDENCE INTERVALS FOR FUNCTIONS OF θ

For one-parameter distributions like the exponential, confidence intervals for θ can be translated directly into confidence intervals for monotone functions of θ.

7.4.1 Confidence Intervals for the Arrival Rate

The arrival rate $\lambda = 1/\theta$ is a *decreasing* function of θ. Thus the upper limit $\widetilde{\theta}$ is substituted for θ to get a lower limit for λ and vice versa. The confidence interval for λ obtained in this manner will contain λ if and only if the corresponding interval for θ contains θ. Thus the confidence interval for λ has the same confidence level as the interval for θ.

Example 7.10 Likelihood-Based Confidence Intervals for the Arrival Rate of α Particles. Using the likelihood-based confidence interval for the $n = 200$ sample in Table 7.2,

$$[\underset{\sim}{\lambda}, \ \widetilde{\lambda}] = [1/\widetilde{\theta}, \ 1/\underset{\sim}{\theta}] = [.00151, \ .00201].$$

Also, the ML estimate of λ is obtained as $\widehat{\lambda} = 1/\widehat{\theta} = .00175$. □

7.4.2 Confidence Intervals for $F(t; \theta)$

Because $F(t; \theta)$ is a decreasing function of θ, the confidence interval for the exponential distribution $F(t_e; \theta)$ for a particular t_e is

$$[\underset{\sim}{F}(t_e), \ \widetilde{F}(t_e)] = [F(t_e; \widetilde{\theta}), \ F(t_e; \underset{\sim}{\theta})].$$

One can compute a set of pointwise confidence intervals for a range of values of t. In this case (unlike in Section 3.8), the set can also be interpreted as simultaneous confidence bands for the entire exponential cdf $F(t; \theta)$. This is because θ is the only unknown parameter for this model and the bands will contain the unknown exponential cdf $F(t; \theta)$ if and only if the corresponding confidence interval for θ contains the unknown true θ.

Example 7.11 Confidence Intervals for the α-Particle Time Between Arrivals cdf. The dotted lines in Figure 7.3 are drawn through a set of pointwise normal-approximation 95% confidence intervals for the exponential $F(t; \theta)$. □

In subsequent chapters where models have more than one parameter (as with the nonparametric simultaneous confidence bands in Section 3.8), a collection of intervals must be handled differently because the confidence level applies only to the process of constructing an interval for a single point in time t_e. Generally, making a simultaneous statement would require either a wider set of bands or a lower level of confidence.

7.5 COMPARISON OF CONFIDENCE INTERVAL PROCEDURES

For the particle arrival data, Table 7.2 compares approximate 95% confidence intervals for θ based on the likelihood, the normal approximation $Z_{\log(\widehat{\theta})} \overset{\cdot}{\sim} \text{NOR}(0, 1)$, and the normal approximation $Z_{\widehat{\theta}} \overset{\cdot}{\sim} \text{NOR}(0, 1)$. Statistical theory suggests that, in large samples, the log likelihood will be approximately quadratic with the approximation improving as the sample size increases and that all of the different procedures for computing confidence intervals will give similar answers. This is consistent with the results in Table 7.2. For the sample with $n = 20$, however, there are some rather large differences among the procedures.

Simulation studies have shown that the computationally demanding likelihood procedure can be expected to provide better intervals (i.e., an actual coverage probability closer to the nominal confidence level). Also, between the other two simple-to-compute procedures, the normal-approximation procedure based on $\log(\hat{\theta})$ provides a better approximation.

It is possible to improve slightly the normal-approximation procedure in (7.8) by using the p quantile of the Student's t distribution, $t_{(p;\nu)}$, in place of the standard normal quantile, $z_{(p)}$ in (7.6) or (7.11). For complete data, $n - 1$ is an obvious choice for the degrees of freedom ν. This is also a reasonable choice for censored data (a correction for censoring might be contemplated, but no generally useful rule is known). The improvement afforded by using $t_{(p;\nu)}$ instead of $z_{(p)}$ is negligible in samples larger than 30 or so because $t_{(p;\nu)}$ approaches $z_{(p)}$ for large ν. In samples with fewer than 30 or even 50 observations, any normal-approximation procedure can be rather crude (especially for the single-sided coverage probabilities, which are important in the common situation where the cost of being outside the interval differs from one side to the other). Usually, however, the normal-approximation procedures are quick, useful, and adequate for exploratory work. When more accurate confidence interval approximations are required (e.g., for reporting final results), one should use likelihood procedures or other procedures based on simulation (to be described in Chapter 9).

7.6 LIKELIHOOD FOR EXACT FAILURE TIMES

7.6.1 Correct Likelihood for Observations Reported as Exact Failures

Consider the diesel generator fan data in Appendix Table C.1. Although time is a continuous variable and the failure times were initially reported as exact times, these data (as with most data) are actually discrete. In this case, the reported failure times were rounded to the nearest 10 hours. Thus the "correct likelihood" is one for interval-censored data (7.3). For example, with the exponential distribution, the likelihood contribution (probability) of the failure recorded at 450 hours is

$$L_1(\theta) = F(455; \theta) - F(445; \theta).$$

7.6.2 Using the Density Approximation for Observations Reported as Exact Values

The traditional and commonly used form of the likelihood for an observation, say, the ith, reported as an "exact" failure at time t_i, is

$$L_i(\theta) = f(t_i; \theta), \qquad (7.13)$$

where $f(t_i; \theta) = dF(t_i; \theta)/dt$ is the assumed pdf for the random variable T. The density approximation in (7.13) is convenient, easy-to-use, and, in some simple special cases, yields closed-form equations for ML estimates.

The use of the density approximation (7.13) instead of the correct discrete likelihood can be justified as follows. For most statistical models, the contribution to the likelihood (i.e., probability of the data) of observations reported as exact values can, for small $\Delta_i > 0$, be approximated by

$$[F(t_i; \theta) - F(t_i - \Delta_i; \theta)] \approx f(t_i; \theta)\Delta_i, \tag{7.14}$$

where Δ_i does not depend on θ. Because the right-hand sides of (7.14) and (7.13) differ by a factor of Δ_i, when the density approximation is used, the approximate likelihood in (7.13) differs from the probability in (7.14) by a constant scale factor. As long as the approximation in (7.14) is adequate and because Δ_i does not depend on θ, however, the general character (i.e., the shape and the location of the maximum) of the likelihood is not affected.

7.6.3 ML Estimates for the Exponential θ Based on the Density Approximation

The density approximation to the likelihood generally provides an adequate approximation for the exponential distribution. For a sample consisting of only right-censored observations and observations reported as exact failure times (and no left-censored or interval-censored observations), it is easy to show that the ML estimate of θ is computed as

$$\widehat{\theta} = \frac{TTT}{r}, \tag{7.15}$$

where $TTT = \sum_{i=1}^{n} t_i$ is known as the "total time on test" and where $t_i, i = 1, \ldots, n$, are the reported failure times for units that failed and the running (or censoring) time for the right-censored observations. Note that the sum runs over *all* of the failures *and* each of the censoring times. In this case, an estimate of standard error of $\widehat{\theta}$ is a special case of (7.7) and is computed as

$$\widehat{\operatorname{se}}_{\widehat{\theta}} = \sqrt{\left[-\frac{d^2 \mathcal{L}(\theta)}{d\theta^2}\right]^{-1}} = \sqrt{\frac{\widehat{\theta}^2}{r}} = \frac{\widehat{\theta}}{\sqrt{r}}, \tag{7.16}$$

where r is the number of failures.

Example 7.12 Comparison of ML Estimates for the Fan Data Based on the Correct Likelihood and the Density Approximation. Fitting the exponential distribution to the diesel generator fan data in Appendix Table C.1 using the "correct" likelihood contribution in (7.3) and the density approximation in (7.13) gives, to seven decimal places, the same answers [$\log(\widehat{\theta}) = 10.26476$ and $\widehat{\operatorname{se}}_{\log(\widehat{\theta})} = .2886757$].

Although the agreement is nearly exact in this case, with other distributions, the degree of agreement will not be so good unless the Δ_i are small. □

7.6.4 Confidence Intervals for the Exponential with Complete Data or Failure Censoring

Suppose that a life test starts at time 0 and that all failures are reported as exact failures. If the test continues until all units have failed or if the test is terminated after a prespecified number r failures (Type II censoring), then the ML estimate of θ is $\hat{\theta} = TTT/r$ and an exact $100(1 - \alpha)\%$ confidence interval for θ can be computed from

$$[\underset{\sim}{\theta}, \tilde{\theta}] = \left[\frac{2(TTT)}{\chi^2_{(1-\alpha/2;2r)}}, \frac{2(TTT)}{\chi^2_{(\alpha/2;2r)}}\right].$$

This interval is based on the fact that $2(TTT/\theta) \sim \chi^2_{(2r)}$. This confidence interval procedure has exactly the specified confidence level $100(1 - \alpha)\%$. Lawless (1982, page 127) and Bain and Engelhardt (1991, page 122) provide technical details and justification for this method. With time (Type I) censoring, the procedure still provides a useful approximation.

Example 7.13 Confidence Interval for the Mean Life of a New Insulating Material. A life test for a new insulating material used 25 specimens. The specimens were tested simultaneously at 30 kV (considerably higher than the rated voltage of 20 kV). The test was run until 15 of the specimens failed (failure or Type II censoring). The failure times were recorded as 1.08, 12.20, 17.80, 19.10, 26.00, 27.90, 28.20, 32.20, 35.90, 43.50, 44.00, 45.20, 45.70, 46.30, and 47.80 hours. The total time on test for these data is $1.08 + 12.20 + \cdots + 47.80 + 10 \times 47.80 = 950.88$ hours and thus the ML estimate of θ is $950.88/15 = 63.392$ hours. A 95% confidence interval for θ is

$$[\underset{\sim}{\theta}, \tilde{\theta}] = \left[\frac{2(950.88)}{\chi^2_{(.975;30)}}, \frac{2(950.88)}{\chi^2_{(.025;30)}}\right] = \left[\frac{1901.76}{46.98}, \frac{1901.76}{16.79}\right] = [40.48, 113.26].$$

An estimate of the standard error of $\hat{\theta}$ using (7.16) is $\widehat{se}_{\hat{\theta}} = \sqrt{(63.392)^2/15} = 16.37$. □

7.7 DATA ANALYSIS WITH NO FAILURES

For data on high-reliability components, it is possible that there will be no failures. The ML estimate of the exponential distribution mean is then $\hat{\theta} = \infty$ (and sometimes it is said that the ML estimate "does not exist"). This is not a useful answer because, generally, it is known that there would have been failures if the period of testing had been extended. With zero failures and an assumed exponential distribution it is, however, possible to obtain a *lower* confidence bound for θ. In particular, if there are no failures in a life test with total time on test TTT (defined in Section 7.6.3), a conservative $100(1-\alpha)\%$ lower confidence bound on θ is

$$\utilde{\theta} = \frac{2(TTT)}{\chi^2_{(1-\alpha;2)}} = \frac{TTT}{-\log(\alpha)} \tag{7.17}$$

because $\chi^2_{(1-\alpha;2)} = -2\log(\alpha)$. This bound is based on the fact that under the exponential failure-time distribution, with immediate replacement of failed units, the number of failures observed in a life test with a fixed total time on test has a Poisson distribution.

As in Section 7.4, this confidence bound can be translated into a lower confidence bound for functions of θ like t_p for specified p or an upper confidence bound for $F(t_e)$ for a specified t_e. Unless TTT is large, however, the resulting bound may not be very informative. See Nelson (1985) for justification and further discussion of this method.

Example 7.14 *Analysis of the Diesel Generator Fan Data Assuming Removal After 200 Hours of Service.* For this example, suppose that each of the diesel generator fans described in Example 1.4 had been removed unfailed after 200 hours of service. There was a total of 70 fans in the study. None failed before 200 hours of service. Thus $TTT = 14{,}000$ hours. A conservative 95% lower confidence bound on θ is

$$\utilde{\theta} = \frac{2(TTT)}{\chi^2_{(.95;2)}} = \frac{28000}{5.991} = 4674.$$

Using the entire data set, the point estimate of θ was 28,701 with a likelihood-based approximate 95% lower confidence bound of $\utilde{\theta} = 18{,}485$ hours. This shows how little information comes from a short test with zero or few failures.

A conservative 95% upper confidence bound on $F(10000; \theta)$, the probability of failing before 10,000 hours, is $\widetilde{F}(10000) = F(10000; \utilde{\theta}) = 1 - \exp(-10000/4674) = .882$. Again, due to the limited amount of information from the short test, this upper bound is not very useful for making a statement on the fan's 10,000-hour reliability [of course, the only lower bound that can be computed for $F(10000; \theta)$ is 0, which also is not very useful]. □

BIBLIOGRAPHIC NOTES

The ML method dates back to early work by Fisher (1925) and has been used as a method for constructing estimators ever since. Statistical theory for ML is briefly covered in most textbooks on mathematical statistics. Some particularly useful references for the approach used here are Kulldorff (1961), Kempthorne and Folks (1971), Rao (1973), and Cox and Hinkley (1974), Lehmann (1983), Nelson (1982), Lawless (1982), Casella and Berger (1990), and Lindsey (1996). Epstein and Sobel (1953) described the fundamental ideas of using the exponential distribution as a model for life data.

Anscombe (1964) and Sprott (1973) suggest using the transformed parameter $\nu = \theta^{-1/3}$ for the exponential parameter because it will make the third derivative of exponential log likelihood function nearly 0. The important implication is that this transformation makes the exponential log likelihood function nearly symmetric and approximately quadratic in ν. In this scale, the normal-approximation inferences provide an excellent approximation to the likelihood-based inferences. Meeker and Escobar (1995) show that the normal-approximation intervals can be viewed as approximations to likelihood intervals obtained by using a quadratic approximation to the log likelihood function.

Meeker (1986) uses asymptotic variances to show that binning data will not seriously affect the precision of ML estimates as long as there are a reasonable number of bins (say, more than 5 or 6) and that the bins are chosen so that the number of observations in the different bins is not too uneven.

For examples of failures of the "density approximation" to the likelihood, see Le Cam (1990), Friedman and Gertsbakh (1980), and Meeker and Escobar (1994). Each of these references describes a model or models for which the "density approximation" likelihood has a path or paths in the parameter space along which $L(\theta)$ approaches ∞. In these cases, using the "correct" likelihood based on discrete data eliminates the singularity in $L(\theta)$.

EXERCISES

7.1. Consider the case of n observations reported as "exact" failure times from an EXP(θ) distribution.

  (a) Show that the ML estimate, $\widehat{\theta}$, of θ is the sample mean, say, $\bar{t} = \sum_{i=1}^{n} t_i / n$.

  (b) Show that the relative likelihood has the simple expression

  $$R(\theta) = \exp(n) \left(\frac{\bar{t}}{\theta} \right)^n \exp\left(-\frac{n\bar{t}}{\theta} \right).$$

  (c) Explain how to use $R(\theta)$ to obtain an approximate confidence interval for θ based on inverting a likelihood ratio test (i.e., assume that when evaluated at the true value of θ, $-2\log[R(\theta)] \overset{.}{\sim} \chi_1^2$).

  (d) Suppose that $n = 4$ and $\bar{t} = .87$. Obtain the ML estimate and an approximate 90% confidence interval for θ using the method in part (c). Plot $R(\theta)$ to facilitate your work.

7.2. Let $t_1, \ldots, t_r$ be the failure times in a singly time-censored sample of size n from an EXP(θ) distribution. Suppose that the prespecified censoring time is t_c. Then the total time on test is $TTT = \sum_{i=1}^{r} t_i + (n-r)t_c$.

(a) Write an expression for the log likelihood using the density approximation for the observations reported as exact failures.
(b) Show that the ML estimate is $\widehat{\theta} = TTT/r$.
(c) Use the result of part (a) to show that the ML estimate of θ is equal to ∞ (or "does not exist") when all the observations are censored.
(d) Derive an expression for the relative likelihood similar to the one given in part (b) of Problem 7.1 (note that in this case TTT/r plays the role of $\bar{t}$ and r the role of n).

7.3. Refer to Exercise 7.2. Show that if there are no failures by time t_c, then $L(\theta)$ always increases as a function of θ and thus $L(\theta)$ does not have a maximum.

7.4. A large electronic system contains thousands of copies of a particular component, which we will refer to as Component-A (each system is custom-made to order and the actual number of components varies with the size of the particular system). The failure rate for Component-A is small, but because there are so many of the components in operation, failures are reported from time to time. A field-tracking study was conducted to estimate the failure rate of a new design for Component-A (which was intended to provide better reliability at lower cost), to see if there was any real improvement. A number of systems were put into service simultaneously and monitored for 1000 hours. The total number of copies of Component-A in all of the systems was 9432. Failures of Component-A were observed at 345, 556, 712, and 976 hours. Failure mode analysis suggested that the failures were due to random shocks rather than any kind of quality defects or wearout. This, along with past experience with simpler components, suggested that an exponential distribution might be an appropriate model for the lifetime of Component-A.
(a) Compute the ML estimate for the exponential mean θ for Component-A.
(b) Compute an approximate 95% confidence interval for θ.
(c) Explain the interpretation of the confidence interval obtained in part (b).
(d) Explain the interpretation of the hazard function $\lambda = 1/\theta$. In what way can this quantity be used to compute failure rates for Component-A or for the overall system?
(e) Use the results of part (b) to obtain an approximate 95% confidence interval for λ. Express this in units of FITs (failures per 10^9 hours of operation).
(f) The hazard for the old part was $\lambda_{old} = 8.5 \times 10^{-7}$ (or 850 FITs). How strong is the evidence that reliability has improved with the new design? Describe how you would phrase a statement about the relative improvement.
(g) Compute and plot the exponential $\widehat{F}(t)$ from 100 to 10,000 hours on Weibull paper. Comment on the usefulness of this plot.

EXERCISES

7.5. The manufacturer of computer hard disks reports in its promotional literature an "MTBF" figure. This figure is obtained by taking a sample of units from each day's production, putting the units on test for a period t_c (perhaps 1 week), pooling the available data over several months, and computing

$$\widehat{\text{MTBF}} = \frac{TTT}{r},$$

where TTT is the total time on test of all units that were tested and r is the observed number of failures (typically a very small number of drives). Reported figures are typically numbers like MTBF = 30 years. Comment on the usefulness and validity of the use of this figure for characterizing the reliability of disk drives.

7.6. Nelson (1982, page 105) provides data on minutes to breakdown for an insulating fluid. There were 11 tests at 30 kV. After 100 minutes, there were 7 breakdowns (failures) at the following times (in minutes): 7.74, 17.05, 20.46, 21.02, 22.66, 43.40, 47.30. The other 4 units had not failed.

(a) Make a Weibull probability plot and an exponential probability plot for these data.

(b) What do you think about the suggestion of using an exponential distribution, EXP(θ), to model the data?

(c) Assuming an exponential distribution, obtain the ML estimate of θ and give an estimate of $\text{se}_{\widehat{\theta}}$.

(d) Compute an approximate 95% confidence interval for θ.

(e) Compute the ML estimate of $t_{.1}$, the .1 quantile of the time to breakdown distribution.

(f) Compute an approximate 95% confidence interval for $t_{.1}$.

7.7. A life test for a new insulating material used 50 specimens. The specimens were tested simultaneously at 40 kV (considerably higher than the rated voltage of 20 kV). The test was run until 10 of the specimens failed (this is known as "failure" or Type II censoring). The failure times were recorded as 8, 11, 12, 13, 19, 21, 28, 34, 36, and 44 hours. The engineers responsible for the reliability believe, based on previous experience with similar materials tested under similar conditions, that the failure-time distribution at 40 kV can be described by an EXP(θ) distribution.

(a) Construct an exponential probability plot of the data. Does the plot provide any evidence that the exponential distribution is inadequate?

(b) Compute TTT, the total time on test, and $\widehat{\theta}$, the ML estimate for θ.

(c) Compute an estimate of the standard error of $\widehat{\theta}$.

(d) Compute 95% confidence intervals for θ based on $Z_{\widehat{\theta}} \overset{\cdot}{\sim}$ NOR(0, 1), $Z_{\log(\widehat{\theta})} \overset{\cdot}{\sim}$ NOR(0, 1), and the exact distribution of $2(TTT/\theta)$. Which of these intervals would you feel comfortable using? Why?

(e) For this problem, is there any extrapolation involved in estimating θ? Explain.

(f) Compute 95% confidence intervals for $t_{.1}$, $h(50; \theta)$, and $F(50; \theta)$ based on the exact distribution of $2(TTT/\theta)$. Is there any extrapolation involved in these intervals? Explain.

7.8. Use the results from Exercise 7.7 to compute and plot (use an exponential plot) the ML estimate of $F(t; \theta)$ and 95% simultaneous parametric confidence bands for $F(t; \theta)$, based on the exact distribution of $2(TTT/\theta)$.

7.9. A life test was conducted for the same insulating material described in Exercise 7.7. Again, 50 specimens were tested, but at 25 kV. The test ran for 20 hours without any failures. The test had to be terminated at this time so that the test equipment could be used for other experiments.

(a) Compute the TTT and show why the ML estimate for θ is equal to ∞.

(b) Compute a conservative 95% lower confidence bound for θ.

(c) For this problem, is there any extrapolation involved in computing the lower confidence bound for θ? Explain.

(d) Use the result in part (b) to compute a conservative 95% lower confidence bound for $t_{.1}$, and conservative 95% upper confidence bounds for $h(50; \theta)$ and $F(50; \theta)$.

▲**7.10.** Refer to Exercise 7.1 to show that in this case $2n\widehat{\theta}/\theta$ has a χ^2 distribution with $2n$ degrees of freedom and use this to obtain an expression for a $100(1-\alpha)\%$ confidence interval for θ.

▲**7.11.** Refer to Exercise 7.2. Derive an expression for the bias of $\widehat{\theta}$ [i.e., $E(\widehat{\theta}) - \theta$], conditional on the existence of the estimator (i.e., conditional on $r > 0$) and show that the bias is negligible when t_c is large.

▲**7.12.** Refer to Exercise 7.2. Derive (7.16).

▲**7.13.** Consider n observations with inspection data from an EXP(θ) distribution with $0 < \theta < \infty$. Show the following:

(a) The ML estimate of θ does not exist (i.e., there is no unique maximum of the likelihood function) when $d_1 = n$ or $d_{m+1} = n$.

(b) When $d_i = n$ ($i \neq 1, m+1$), the ML estimator is

$$\widehat{\theta} = \frac{t_i - t_{i-1}}{\log(t_i) - \log(t_{i-1})}.$$

CHAPTER 8

Maximum Likelihood for Log-Location-Scale Distributions

Objectives

This chapter explains:

- Likelihood methods for fitting log-location-scale distributions (especially the Weibull and lognormal distributions).
- Likelihood confidence intervals/regions for model parameters and for *functions* of model parameters.
- Normal-approximation confidence intervals/regions.
- Estimation and confidence intervals for log-location-scale distributions with a given shape parameter.

Overview

This chapter extends the methods in Chapter 7 to two-parameter distributions that are based on location-scale distributions. These distributions, including the popular lognormal and Weibull distributions, are the workhorses of parametric reliability modeling. Section 8.2 shows how to compute the likelihood for these distributions and censored data. The distributions and estimation methods used in this chapter are also used in the regression and accelerated testing chapters (Chapters 17 and 19). Sections 8.3 and 8.4 show how to compute confidence regions and intervals for parameters and functions of the parameters. Section 8.5 shows the effect of using a given value of the scale parameter σ instead of estimating it as if it were unknown.

8.1 INTRODUCTION

This chapter describes fitting of and making inferences from the two-parameter Weibull and lognormal life distributions (Section 11.7 discusses estimation of the parameters of the three-parameter versions of these distributions). The methods also apply to other location-scale distributions or distributions that can be transformed into a location-scale form including the normal, smallest extreme value, largest extreme value, logistic, and loglogistic distributions.

For these distributions and exact failure times (i.e., when the amount of roundoff is small relative to the variability in the data), the density approximation in (7.13) adequately approximates the correct discrete likelihood on the left-hand side of (7.14). Thus we will use (7.13) to represent the likelihood for exact observations in this chapter.

8.2 LIKELIHOOD

8.2.1 Likelihood for Location-Scale Distributions

The likelihood for a sample $y_1, \ldots, y_n$ from a location-scale distribution for a random variable $-\infty < Y < \infty$, consisting of exact (i.e., not censored) and right-censored observations can be written as

$$L(\mu, \sigma) = \prod_{i=1}^{n} L_i(\mu, \sigma; \text{data}_i) = \prod_{i=1}^{n} [f(y_i; \mu, \sigma)]^{\delta_i} [1 - F(y_i; \mu, \sigma)]^{1-\delta_i}$$

$$= \prod_{i=1}^{n} \left[\frac{1}{\sigma} \phi\left(\frac{y_i - \mu}{\sigma}\right) \right]^{\delta_i} \times \left[1 - \Phi\left(\frac{y_i - \mu}{\sigma}\right) \right]^{1-\delta_i},$$

where

$$\delta_i = \begin{cases} 1 & \text{if } y_i \text{ is an exact observation} \\ 0 & \text{if } y_i \text{ is a right-censored observation.} \end{cases}$$

As defined in Section 4.3, for a particular location-scale distribution, substitute the appropriate Φ and ϕ. For the smallest extreme value distribution substitute Φ_{sev} and ϕ_{sev}. Similarly, for the normal distribution, substitute Φ_{nor} and ϕ_{nor}, and for the logistic distribution substitute Φ_{logis} and ϕ_{logis}. Left-censored and interval-censored observations could be factored in, using these same functions, as described in Section 2.4.3.

When there is no censoring, the normal distribution likelihood simplifies and it is possible to solve explicitly to obtain the values of μ and σ that maximize this likelihood. This is a standard exercise in most textbooks on mathematical statistics.

8.2.2 Likelihood for the Lognormal, Weibull, and Other Log-Location-Scale Distributions

Because the logarithm of lognormal, Weibull, and loglogistic random variables follow corresponding location-scale distributions, the likelihoods for these distributions can also be written in terms of the standardized location-scale distributions. In particular, for a sample consisting of exact failure times and right-censored observations, the

LIKELIHOOD

likelihood can be expressed as

$$L(\mu, \sigma) = \prod_{i=1}^{n} \left\{ \frac{1}{\sigma t_i} \phi \left[\frac{\log(t_i) - \mu}{\sigma} \right] \right\}^{\delta_i} \times \left\{ 1 - \Phi \left[\frac{\log(t_i) - \mu}{\sigma} \right] \right\}^{1-\delta_i}, \quad (8.1)$$

where δ_i again indicates whether observation i is a failure or a right-censored observation. Left-censored and interval-censored observations could be factored in as described in Section 2.4.3. It is important to note that some computer programs omit the $1/t_i$ term in the density part of the likelihood. Because this term does not depend on the unknown parameters, this has no effect on the location of the ML estimates. It does, however, affect the reported value of the likelihood (or more commonly the log likelihood) at the maximum, and therefore caution should be used when comparing values of log likelihoods computed with different software.

Again, the particular Φ and ϕ functions determine the distribution to be used. For the Weibull distribution, use Φ_{sev} and ϕ_{sev}; for the lognormal, use Φ_{nor} and ϕ_{nor}; and for the loglogistic, use Φ_{logis} and ϕ_{logis}.

Example 8.1 Shock Absorber Data Likelihood and ML Estimates. This example returns to the shock absorber data in Examples 3.8 and 3.9. The Weibull and lognormal probability plots with nonparametric simultaneous confidence bands in Figures 6.8 and 6.9 indicated no strong preference for either distribution. Figure 8.1 is a contour plot of the relative likelihood function $R(\mu, \sigma) = L(\mu, \sigma)/L(\hat{\mu}, \hat{\sigma})$ for

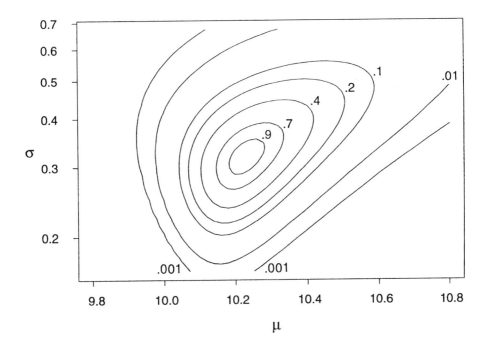

Figure 8.1. Weibull relative likelihood for the shock absorber data.

the Weibull distribution model. The surface is well behaved with a unique maximum (i.e., $\hat{\mu} = 10.23$ and $\hat{\sigma} = .3164$) defining the ML estimates. The orientation of the contours indicates some positive correlation between $\hat{\mu}$ and $\hat{\sigma}$. □

Fitting a Weibull (lognormal) distribution is equivalent to fitting a straight line through the data on a Weibull (lognormal) probability plot, using the ML criterion to choose the line. Generally when fitting a distribution with ML, it is convenient to use a probability plot that also shows the fitted distribution.

Example 8.2   Comparison of Weibull and Lognormal Distribution Fits to the Shock Absorber Data.   Figures 8.2 and 8.3 give Weibull and lognormal probability plots, respectively, with the corresponding ML estimates of $F(t)$ represented by the straight lines in the plots. The dotted lines are drawn through a set of 95% pointwise normal-approximation confidence intervals for $F(t)$ (computed as described in Section 8.4.3). The curved line going through the points on the lognormal probability plot is the corresponding ML estimate of the Weibull $F(t)$.

Table 8.1 gives ML estimates, standard errors, and confidence intervals for both the Weibull and lognormal distributions. The following sections describe methods for computing these standard errors and confidence intervals. The parameters μ and σ are not directly comparable because these parameters have different interpretations in the Weibull and lognormal distributions. Comparing Figures 8.2 and 8.3 and the

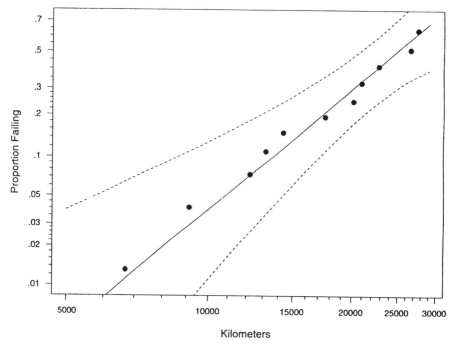

Figure 8.2. Weibull probability plot of shock absorber failure times (both failure modes) with maximum likelihood estimates and pointwise approximate 95% confidence intervals for $F(t)$.

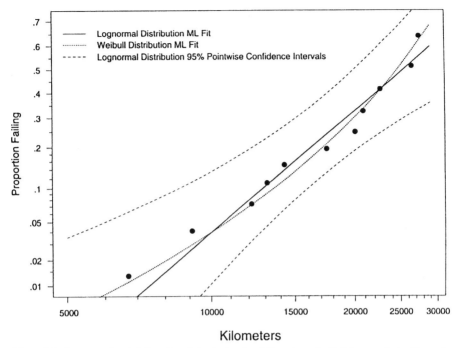

Figure 8.3. Lognormal probability plot of shock absorber failure times (both failure modes) with maximum likelihood estimate and pointwise approximate 95% confidence intervals for $F(t)$. The curved line going through the points is the Weibull maximum likelihood estimate.

estimates of $t_{.1}$ and $F(10000)$ indicates good agreement for inferences from these two distributions *within the range of the data*. Although Figures 8.2 and 8.3 suggest a slightly better fit for the Weibull distribution, as methods described in Section 11.3 show, however, these data could reasonably have arisen from either of these two distributions.

In situations like this, one should be cautious about making inferences in the tails of the distributions and outside the range of the data. The estimates there can be importantly different *and* the data do not strongly suggest one model over the other. In general, it is useful and important to fit different distributions to compare results on questions of interest. □

8.3 LIKELIHOOD CONFIDENCE REGIONS AND INTERVALS

8.3.1 Joint Confidence Regions for μ and σ

As described in Appendix Section B.6.6, any of the constant-likelihood contour lines on Figure 8.1 define an approximate joint confidence region for μ and σ that can be calibrated accurately, even in moderately small samples (e.g., 15–20 failures), by using the large-sample χ^2 approximation for the distribution of the likelihood-ratio statistic. For a two-dimensional relative likelihood (or two-dimensional profile

Table 8.1. Comparison of Shock Absorber Estimates and Confidence Intervals

	Distribution	
	Weibull	Lognormal
ML estimate $\hat{\mu}$	10.23	10.14
Standard error $\widehat{\text{se}}_{\hat{\mu}}$	.1099	.1442
Approximate 95% Confidence Intervals for μ		
Based on the likelihood	[10.06, 10.54]	[9.91, 10.53]
Based on $Z_{\hat{\mu}} \stackrel{.}{\sim} \text{NOR}(0, 1)$	[10.01, 10.45]	[9.86, 10.43]
ML estimate $\hat{\sigma}$	.3164	.5301
Standard error $\widehat{\text{se}}_{\hat{\sigma}}$	.07317	.1127
Approximate 95% Confidence Intervals for σ		
Based on the likelihood	[.210, .527]	[.367, .858]
Based on $Z_{\log(\hat{\sigma})} \stackrel{.}{\sim} \text{NOR}(0, 1)$	[.201, .498]	[.349, .804]
Based on $Z_{\hat{\sigma}} \stackrel{.}{\sim} \text{NOR}(0, 1)$	[.173, .460]	[.309, .751]
ML estimate $\hat{t}_{.1}$	13602	12910
Standard error $\widehat{\text{se}}_{\hat{t}_{.1}}$	1982	1667
Approximate 95% Confidence Intervals for $t_{.1}$		
Based on the likelihood	[9400, 17300]	[9400, 16300]
Based on $Z_{\log(\hat{t}_{.1})} \stackrel{.}{\sim} \text{NOR}(0, 1)$	[10200, 18100]	[10000, 16600]
Based on $Z_{\hat{t}_{.1}} \stackrel{.}{\sim} \text{NOR}(0, 1)$	[9700, 17500]	[9600, 16200]
ML estimate $\hat{F}(10000)$	.03908	.03896
Standard error $\widehat{\text{se}}_{\hat{F}(10000)}$	.02480	.02561
Approximate 95% Confidence Intervals for $F(10000)$		
Based on the likelihood	[.0092, .1136]	[.0085, .1159]
Based on $Z_{\text{logit}(\hat{F})} \stackrel{.}{\sim} \text{NOR}(0, 1)$	[.0110, .1292]	[.0105, .1342]
Based on $Z_{\hat{F}} \stackrel{.}{\sim} \text{NOR}(0, 1)$	[−.0095, .0877]	[−.0112, .0892]

LIKELIHOOD CONFIDENCE REGIONS AND INTERVALS

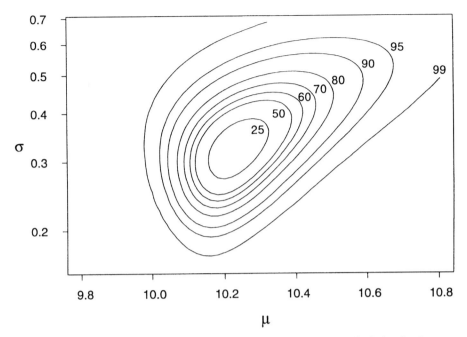

Figure 8.4. Weibull likelihood joint confidence regions for μ and σ for the shock absorber data.

likelihood), the region $R(\theta_i, \theta_j) > \exp[-\chi^2_{(1-\alpha;2)}/2] = \alpha$ provides an approximate $100(1-\alpha)\%$ joint confidence region for θ_i and θ_j.

Example 8.3 Joint Confidence Region for Shock Absorber (μ, σ). In Figure 8.1, the region $R(\mu, \sigma) > \exp(-\chi^2_{(.90;2)}/2) = .1$ provides an approximate 90% joint likelihood-based confidence region for μ and σ. Figure 8.4, similar to Figure 8.1, plots contours of constant values of $100\Pr\{\chi^2_2 \leq -2\log[R(\mu, \sigma)]\}$ giving approximate confidence levels corresponding to joint likelihood-based confidence regions for μ and σ. □

8.3.2 Individual Confidence Intervals for μ and σ

The profile likelihood for a single parameter summarizes the sample information for that parameter and provides likelihood confidence intervals.

Confidence Interval for μ
The profile likelihood for μ is

$$R(\mu) = \max_{\sigma} \left[\frac{L(\mu, \sigma)}{L(\hat{\mu}, \hat{\sigma})} \right]. \tag{8.2}$$

The interval over which $R(\mu) > \exp[-\chi^2_{(1-\alpha;1)}/2]$ is an approximate $100(1-\alpha)\%$ confidence interval for μ. The general theory for likelihood confidence intervals is

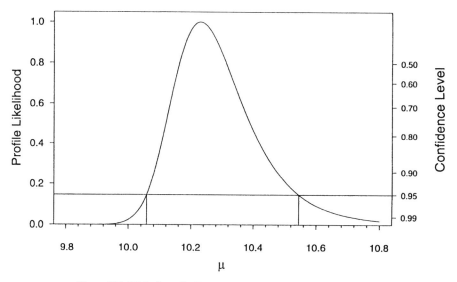

Figure 8.5. Weibull profile likelihood $R(\mu)$ for the shock absorber data.

given in Appendix Section B.6.6. Intuitively, we want to include all values of μ that have high likelihood. Using (8.2), for every fixed value of μ, find the point of *highest* relative likelihood by maximizing the relative likelihood with respect to σ. This gives the profile likelihood level for that value of μ. Values of μ with high profile likelihood are more plausible than those with lower values of profile likelihood.

Example 8.4 Profile Likelihood $R(\mu)$ for the Shock Absorber Data. Figure 8.5 shows $R(\mu)$ and indicates how to obtain the likelihood-based approximate 95% confidence interval; the right-hand scale indicates the appropriate position for drawing the horizontal line to obtain intervals with other levels of confidence. Numerical values of the confidence interval endpoints are given in Table 8.1, for comparison with other intervals that are described later in this chapter. □

Example 7.6 shows how to use the profile likelihood to obtain one-sided confidence bounds.

Confidence Interval for σ
The profile likelihood for σ is

$$R(\sigma) = \max_{\mu} \left[\frac{L(\mu, \sigma)}{L(\widehat{\mu}, \widehat{\sigma})} \right].$$

The interval over which $R(\sigma) > \exp[-\chi^2_{(1-\alpha;1)}/2]$ is an approximate $100(1-\alpha)\%$ confidence interval for σ.

Example 8.5 Profile Likelihood $R(\sigma)$ for the Shock Absorber Data. Figure 8.6 shows $R(\sigma)$ and indicates how to obtain the likelihood-based approximate 95% confidence interval. Table 8.1 gives numerical values. A corresponding interval

LIKELIHOOD CONFIDENCE REGIONS AND INTERVALS 181

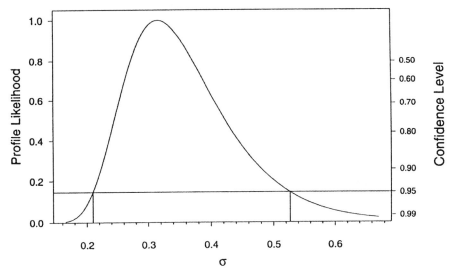

Figure 8.6. Weibull profile likelihood $R(\sigma)$ for the shock absorber data.

for $\beta = 1/\sigma$ can be obtained by taking the reciprocal of the endpoints of the interval for σ. This interval provides strong evidence that $\sigma < 1$ (or $\beta > 1$), indicating that the shock absorber population has a hazard function that increases with age. This is consistent with the suggestion that shock absorbers tend to wear out. □

8.3.3 Likelihood Confidence Intervals for Functions of μ and σ

The parameters for a statistical model are often chosen for convenience, by tradition, so that the parameters have scientific meaning, or for numerical reasons. This chapter, for convenience and consistency, uses the location and scale parameters μ and σ as the basic distribution parameters as these are most commonly used to describe location-scale distributions. Interest, however, often centers on functions of these parameters like probabilities $p = F(t) = \Phi[(\log(t) - \mu)/\sigma]$ or distribution quantiles like $t_p = F^{-1}(p) = \exp[\mu + \Phi^{-1}(p)\sigma]$. In general, the ML estimator of a function $g(\mu, \sigma)$ is $\widehat{g} = g(\widehat{\mu}, \widehat{\sigma})$. Due to this *invariance* property of ML estimators, likelihood-based methods can, in principle, be applied, as described above, to make inferences about such functions. For any function of interest, this can be done by defining a one-to-one transformation (or reparameterization), $g(\mu, \sigma) = [g_1(\mu, \sigma), g_2(\mu, \sigma)]$, that contains the function of interest among its elements. Either of the new parameters may be identical to the old ones. Using this method to compute confidence intervals for the elements of $g(\mu, \sigma)$ requires that the first partial derivatives of $g(\mu, \sigma)$ be continuous. Then ML fitting can be carried out and profile plots can be made for this new parameterization in a manner that is the same as that described above for (μ, σ). This provides a procedure for obtaining ML estimates and likelihood confidence intervals for any scalar or vector function of (μ, σ). If one can readily compute $g(\mu, \sigma)$ and its inverse, this method is simple to implement. Otherwise, iterative numerical methods for obtaining the inverse are needed, requiring more computing time.

Confidence Interval for t_p

The profile likelihood for the p quantile $t_p = \exp[\mu + \Phi^{-1}(p)\sigma]$ is

$$R(t_p) = \max_{\sigma} \left[\frac{L(t_p, \sigma)}{L(\widehat{\mu}, \widehat{\sigma})} \right],$$

where the likelihood under the reparameterized model, $L(t_p, \sigma)$, is obtained by substituting $\log(t_p) - \Phi^{-1}(p)\sigma$ for μ in the expression (8.1) for $L(\mu, \sigma)$.

Example 8.6 Profile Likelihood for Weibull Quantiles from the Shock Absorber Data. Figure 8.7 shows a contour plot for the relative likelihood $R(t_{.1}, \sigma)$. The plot gives us an immediate sense of the plausible ranges of values for these two quantities. In contrast to Figure 8.1, the orientation of the contours indicates a negative correlation between $\widehat{t}_{.1}$ and $\widehat{\sigma}$. Figure 8.8 shows the profile likelihood for $t_{.1}$ with an approximate 95% likelihood-based confidence interval indicated. Table 8.1 gives numerical values. □

Confidence Intervals for $F(t_e)$

Likelihood-based confidence intervals for $F(t_e)$, the population failure probability at a specified time t_e, can be found in a similar manner. In particular, the profile

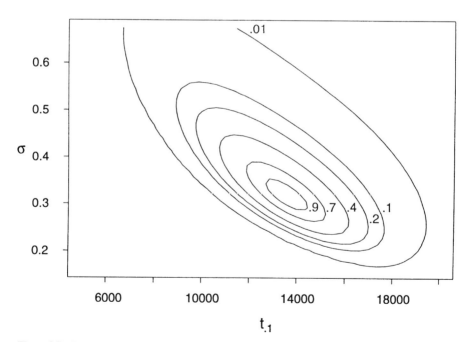

Figure 8.7. Contour plot of Weibull relative likelihood $R(t_{.1}, \sigma)$ for the shock absorber data (parameterized with $t_{.1}$ and σ).

LIKELIHOOD CONFIDENCE REGIONS AND INTERVALS 183

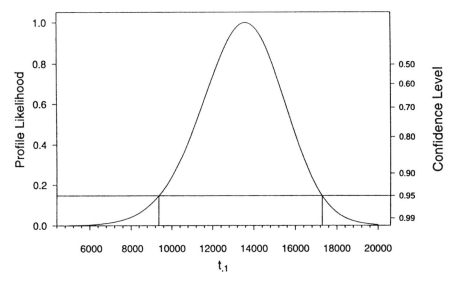

Figure 8.8. Weibull profile likelihood $R(t_{.1})$ for the shock absorber data.

likelihood for $F(t_e)$ is

$$R[F(t_e)] = \max_{\sigma} \left\{ \frac{L[F(t_e), \sigma]}{L(\widehat{\mu}, \widehat{\sigma})} \right\},$$

where $L[F(t_e), \sigma]$ for the reparameterized model is obtained by substituting $\log(t_e) - \Phi^{-1}[F(t_e)]\sigma$ for μ in the expression (8.1) for $L(\mu, \sigma)$.

Example 8.7   Profile Likelihood $R[F(t_e)]$ for the Shock Absorber Data. Figures 8.9 and 8.10 give profile likelihoods for $F(10000)$ and $F(20000)$, the probabilities that a shock absorber will fail by $t_e = 10,000$ and $t_e = 20,000$ kilometers, respectively. Note that the profile likelihood for $F(20000)$ is more symmetric than the profile likelihood for $F(10000)$. One explanation for this is that large-sample approximations (which lead to, among other things, approximate symmetry of the likelihood) tend to be worse in the tails of the distribution.   □

8.3.4 Relationship Between Confidence Intervals and Significance Tests

As described in Section 7.3.2, there is a close relationship between a confidence interval and a hypothesis test for a single parameter or other quantity of interest. With two or more parameters or other quantities of interest, there is a similar close relationship between a confidence region and a joint hypothesis test. This section describes the link between confidence intervals, confidence regions, and hypothesis tests for simple location-scale distributions. Corresponding general theory is outlined in Appendix Sections B.6.5 and B.6.6. Application of these methods to other models is usually straightforward and various applications are described in subsequent chapters.

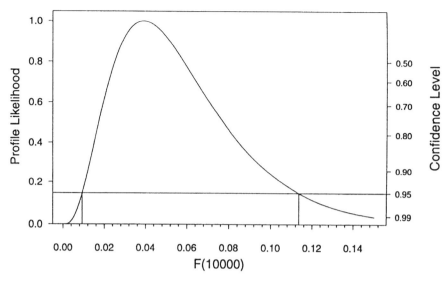

Figure 8.9. Weibull profile likelihood $R[F(10000)]$ for the shock absorber data.

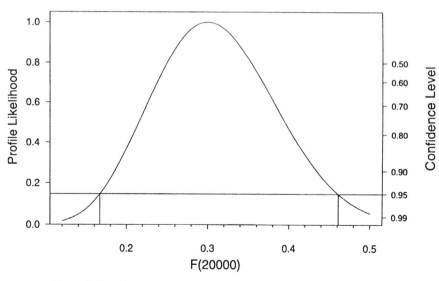

Figure 8.10. Weibull profile likelihood $R[F(20000)]$ for the shock absorber data.

LIKELIHOOD CONFIDENCE REGIONS AND INTERVALS

Although it can still be argued that confidence regions are more informative than the yes/no result of a hypothesis test, a joint confidence region for more than two quantities of interest can be difficult to display and interpret.

Formally, a likelihood-ratio test of a hypothesis, for a two-parameter model, can be done by comparing the likelihood under the null hypothesis with the maximum of the likelihood. The point null hypothesis (μ_0, σ_0) should be rejected if

$$-2\log\left[\frac{L(\mu_0, \sigma_0)}{L(\hat{\mu}, \hat{\sigma})}\right] > \chi^2_{(1-\alpha;2)}, \tag{8.3}$$

where $(\hat{\mu}, \hat{\sigma})$ is the ML estimate of (μ, σ). There are two degrees of freedom for this statistic because there are two free parameters in the full model but zero free parameters with μ_0 and σ_0 fixed; the difference is two. Then, according to the definition given in (8.3), a likelihood-ratio-based confidence region is the set of all values of (μ_0, σ_0) that would not be rejected under the likelihood-ratio test defined in (8.3).

Example 8.8   Likelihood-Ratio Test for the Shock Absorber Weibull Parameters.   Suppose that investigators conducted the shock absorber study to test the hypothesis that the data from a new design are consistent with a historical Weibull distribution ($\mu_0 = 10.1, \sigma_0 = .35$), corresponding to extensive past experience with the old design. Substituting into (8.3) gives

$$-2\log\left[\frac{L(10.1, .35)}{L(10.23, .3164)}\right] = 2.73 < \chi^2_{(.95;2)} = 5.99,$$

showing that the data are consistent with the hypothesized values at the 5% level of significance. Note also that $(\mu_0 = 10.1, \sigma_0 = .35)$ lies *inside* the 95% confidence region shown in Figure 8.4. If, however, the hypothesized value had been ($\mu_0 = 10.0, \sigma_0 = .5$), the appropriate conclusion would have been that the data do provide sufficient evidence to reject the hypothesis at the 1% (or smaller) level of significance, because ($\mu_0 = 10.0, \sigma_0 = .5$) does not lie in the 99% confidence region in Figure 8.4. □

For testing just one parameter (or a single function of the two parameters), there is also a correspondence between a subset likelihood-ratio test and the profile likelihood functions used in Sections 8.3.2 and 8.3.3. For example, we would reject, at the $100\alpha\%$ level of significance, the hypothesis that $\sigma = \sigma_0$ if

$$-2\log\left\{\max_{\mu}\left[\frac{L(\mu, \sigma_0)}{L(\hat{\mu}, \hat{\sigma})}\right]\right\} > \chi^2_{(1-\alpha;1)}. \tag{8.4}$$

In this case there is one degree of freedom because there are two free parameters in the full model minus the one free parameter μ with $\sigma = \sigma_0$ constrained, leaving one degree of freedom in the optimization.

Example 8.9 Likelihood-Ratio Test for σ. Suppose that someone has asked whether the shock absorber data are consistent with the hypothesis of an exponential distribution (i.e., $\sigma = 1$). This hypothesis should be rejected at the 1% level of significance, because (8.4) yields

$$-2\log\left\{\max_\mu\left[\frac{L(\mu, 1)}{L(10.23, .3164)}\right]\right\} = 14.86 > \chi^2_{(.99;1)} = 6.63.$$

Also note that $\sigma_0 = 1$ is outside the profile likelihood region in Figure 8.6. Thus the shock absorber data are not consistent with the exponential distribution. □

8.4 NORMAL-APPROXIMATION CONFIDENCE INTERVALS

Normal-approximation confidence intervals are easy to compute and, at present, are used in most commercial statistical packages. Using ML estimates of the model parameters and of the variance–covariance matrix of the ML estimates, it is possible to compute confidence intervals for parameters and functions of parameters with a hand-calculator. The main shortcomings of normal-approximation confidence intervals are: (1) they have actual coverage probabilities that can be importantly different from the nominal specification, unless the number of failures is large; and (2) unlike the likelihood-based intervals, they depend on the transformation used for the parameter, as illustrated in the following examples. With moderate-to-large samples they are useful for preliminary confidence intervals, where rapid interactive analysis is important.

Normal-approximation confidence regions are based on a *quadratic approximation* to the log likelihood and are adequate when the log likelihood is approximately quadratic over the confidence region. With large samples, under the usual regularity conditions (Appendix Section B.4), the log likelihood is approximately quadratic, and thus the normal-approximation and the likelihood-based intervals will be in close agreement. The sample size required to have an adequate approximation is not easy to characterize because it depends on the model, on the amount of censoring and truncation (Section 11.6), and on the particular quantity of interest. In some extreme examples, with heavy censoring, a sample size n on the order of thousands is not sufficient for a good approximation. With censoring, it is usually better to describe the adequacy of large-sample approximations in terms of the number of failures instead of the sample size. When the quadratic approximation to the log likelihood is poor, likelihood-based intervals (Section 8.3) or bootstrap-based intervals (Chapter 9) should be used instead, especially when reporting final results.

8.4.1 Parameter Variance–Covariance Matrix

Appendix Section B.6.7 describes the general theory for computing confidence intervals based on the large-sample approximate normal distribution of the ML estimators. These intervals require an estimate of the variance–covariance matrix for the ML estimates of the model parameters. For the location-scale distribution, one computes the

NORMAL-APPROXIMATION CONFIDENCE INTERVALS

local estimate $\widehat{\Sigma}_{\widehat{\theta}}$ of $\Sigma_{\widehat{\theta}}$ as the inverse of the *observed* information matrix

$$\widehat{\Sigma}_{\widehat{\mu},\widehat{\sigma}} = \begin{bmatrix} \widehat{\text{Var}}(\widehat{\mu}) & \widehat{\text{Cov}}(\widehat{\mu},\widehat{\sigma}) \\ \widehat{\text{Cov}}(\widehat{\mu},\widehat{\sigma}) & \widehat{\text{Var}}(\widehat{\sigma}) \end{bmatrix} = \begin{bmatrix} -\dfrac{\partial^2 \mathcal{L}(\mu,\sigma)}{\partial \mu^2} & -\dfrac{\partial^2 \mathcal{L}(\mu,\sigma)}{\partial \mu \partial \sigma} \\ -\dfrac{\partial^2 \mathcal{L}(\mu,\sigma)}{\partial \sigma \partial \mu} & -\dfrac{\partial^2 \mathcal{L}(\mu,\sigma)}{\partial \sigma^2} \end{bmatrix}^{-1}, \quad (8.5)$$

where the partial derivatives are evaluated at $\mu = \widehat{\mu}$ and $\sigma = \widehat{\sigma}$.

The intuitive motivation for this estimator is a generalization of the likelihood curvature ideas described in Section 7.3.3. The partial second derivatives describe the curvature of the log likelihood at the ML estimate. More curvature in the log likelihood surface implies a more concentrated likelihood near $\widehat{\mu}, \widehat{\sigma}$, and this implies more precision.

Example 8.10 Estimate of Variance–Covariance Matrix for the Shock Absorber Data Weibull ML Estimates. For the shock absorber data and the Weibull distribution model,

$$\widehat{\Sigma}_{\widehat{\mu},\widehat{\sigma}} = \begin{bmatrix} .01208 & .00399 \\ .00399 & .00535 \end{bmatrix}. \quad (8.6)$$

An estimate of the correlation between $\widehat{\mu}$ and $\widehat{\sigma}$ is $\widehat{\rho}_{\widehat{\mu},\widehat{\sigma}} = \widehat{\text{Cov}}(\widehat{\mu},\widehat{\sigma})/\sqrt{\widehat{\text{Var}}(\widehat{\mu})\widehat{\text{Var}}(\widehat{\sigma})}$ $= .4963$. This positive correlation is reflected in the orientation of the likelihood contours in Figure 8.1. □

8.4.2 Confidence Intervals for Model Parameters

Approximating the distribution of $Z_{\widehat{\mu}} = (\widehat{\mu} - \mu)/\widehat{\text{se}}_{\widehat{\mu}}$ by a NOR(0, 1) distribution yields an approximate $100(1-\alpha)\%$ confidence interval for μ as

$$[\underset{\sim}{\mu}, \widetilde{\mu}] = \widehat{\mu} \pm z_{(1-\alpha/2)} \widehat{\text{se}}_{\widehat{\mu}}. \quad (8.7)$$

A one-sided approximate $100(1-\alpha)\%$ confidence bound can be obtained by replacing $z_{(1-\alpha/2)}$ with $z_{(1-\alpha)}$ and using the appropriate endpoint of the two-sided confidence interval.

After constructing a confidence interval for a particular parameter (or other quantity), it is simple to transform the endpoints of the interval to get a confidence interval for the desired monotone function of that parameter. For example, an approximate $100(1-\alpha)\%$ confidence interval for $\eta = \exp(\mu)$ [still based on the $Z_{\widehat{\mu}} \overset{\cdot}{\sim}$ NOR(0, 1) approximation] is $[\underset{\sim}{\eta}, \widetilde{\eta}] = [\exp(\underset{\sim}{\mu}), \exp(\widetilde{\mu})]$.

Example 8.11 Normal-Approximation Confidence Interval for the Shock Absorber Weibull Scale Parameter. For the shock absorber data, $\widehat{\text{se}}_{\widehat{\mu}} = \sqrt{.01208} = .1099$ and

$$[\underset{\sim}{\mu}, \widetilde{\mu}] = 10.23 \pm 1.960(.1099) = [10.01, \quad 10.45]$$

is an approximate 95% confidence interval for μ. From this, the corresponding confidence interval for the Weibull scale parameter $\eta = \exp(\mu)$ is

$$[\underset{\sim}{\eta}, \widetilde{\eta}] = [\exp(\underset{\sim}{\mu}), \quad \exp(\widetilde{\mu})] = [22{,}350, \quad 34{,}386].$$

Note that, due to round off, exponentiating the rounded answers 10.01 and 10.45 would give somewhat different answers. Because the Weibull scale parameter η is approximately the .63 quantile of the distribution, this interval tells us that we are (approximately) 95% confident that the interval from 22,350 to 34,386 km encloses the point in time where 63% of the population will fail. □

Because σ is a positive parameter, it is common practice to use the log transformation to obtain a confidence interval. Approximating the sampling distribution of $Z_{\log(\widehat{\sigma})} = [\log(\widehat{\sigma}) - \log(\sigma)]/\widehat{\text{se}}_{\log(\widehat{\sigma})}$ by a NOR(0, 1) distribution, an approximate $100(1-\alpha)\%$ confidence interval for σ is

$$[\underset{\sim}{\sigma}, \widetilde{\sigma}] = [\widehat{\sigma}/w, \quad \widehat{\sigma} \times w], \tag{8.8}$$

where $w = \exp[z_{(1-\alpha/2)}\widehat{\text{se}}_{\widehat{\sigma}}/\widehat{\sigma}]$ and $\widehat{\text{se}}_{\widehat{\sigma}} = \sqrt{\widehat{\text{Var}}(\widehat{\sigma})}$.

Example 8.12 *Normal-Approximation Confidence Intervals for the Shock Absorber Weibull Shape Parameter.* For the shock absorber example, $\widehat{\text{se}}_{\widehat{\sigma}} = \sqrt{.005353} = .07316$ and an approximate 95% confidence interval for σ is

$$[\underset{\sim}{\sigma}, \widetilde{\sigma}] = [.3164/1.5733, \quad .3164 \times 1.5733] = [.201, \quad .498],$$

where $w = \exp[1.960(.07316)/.3164] = 1.5733$. The corresponding approximate 95% confidence interval for the Weibull shape parameter $\beta = 1/\sigma$ is

$$[\underset{\sim}{\beta}, \widetilde{\beta}] = [1/\widetilde{\sigma}, \quad 1/\underset{\sim}{\sigma}] = [1/.498, \quad 1/.201] = [2.01, \quad 4.97].$$

Note that because the reciprocal transformation is *decreasing* the upper endpoint for σ translates into the lower endpoint for β and vice versa. Both sets of intervals use the approximation $Z_{\log(\widehat{\sigma})} \overset{.}{\sim} \text{NOR}(0, 1)$. Note that the interval for β provides strong evidence that $\beta > 1$, implying an increasing hazard function and suggesting wearout behavior for the shock absorbers.

Comparison in Table 8.1 shows that the confidence interval for σ based on the approximation $Z_{\log(\widehat{\sigma})} \overset{.}{\sim} \text{NOR}(0, 1)$ agrees well with the likelihood-based interval but differs considerably from the untransformed normal-approximation interval based on the approximation of $Z_{\widehat{\sigma}} \overset{.}{\sim} \text{NOR}(0, 1)$. This suggests that using the log transformation for computing confidence intervals for positive parameters like σ provides a better procedure. □

8.4.3 Normal-Approximation Confidence Intervals for Functions of μ and σ

Following the general theory in Appendix Section B.6.7, a normal-approximation confidence interval for a function of μ and σ, say, $g_1 = g_1(\mu, \sigma)$, can be based on the large-sample approximate NOR(0, 1) distribution of $Z_{\widehat{g}_1} = (\widehat{g}_1 - g_1)/\widehat{\text{se}}_{\widehat{g}_1}$. Then an approximate $100(1 - \alpha)\%$ confidence interval for g_1 is

$$[\underline{g}_1, \ \widetilde{g}_1] = \widehat{g}_1 \pm z_{(1-\alpha/2)} \widehat{\text{se}}_{\widehat{g}_1}, \tag{8.9}$$

where, using a special case of (B.10),

$$\widehat{\text{se}}_{\widehat{g}_1} = \sqrt{\widehat{\text{Var}}(\widehat{g}_1)}$$

$$= \left[\left(\frac{\partial g_1}{\partial \mu} \right)^2 \widehat{\text{Var}}(\widehat{\mu}) + 2 \left(\frac{\partial g_1}{\partial \mu} \right) \left(\frac{\partial g_1}{\partial \sigma} \right) \widehat{\text{Cov}}(\widehat{\mu}, \widehat{\sigma}) + \left(\frac{\partial g_1}{\partial \sigma} \right)^2 \widehat{\text{Var}}(\widehat{\sigma}) \right]^{1/2}. \tag{8.10}$$

The partial derivatives in (8.10) should be evaluated at $\mu = \widehat{\mu}$ and $\sigma = \widehat{\sigma}$.

Confidence Interval for a Distribution Quantile t_p

An approximate $100(1 - \alpha)\%$ confidence interval for $t_p = \exp[\mu + \Phi^{-1}(p)\sigma]$ based on the large-sample approximate NOR(0, 1) distribution of $Z_{\log(\widehat{t}_p)} = [\log(\widehat{t}_p) - \log(t_p)]/\widehat{\text{se}}_{\log(\widehat{t}_p)}$ is

$$[\underline{t}_p, \ \widetilde{t}_p] = \left[\widehat{t}_p/w, \ \widehat{t}_p \times w \right], \tag{8.11}$$

where $w = \exp[z_{(1-\alpha/2)} \widehat{\text{se}}_{\widehat{t}_p}/\widehat{t}_p]$. Applying (8.10) gives

$$\widehat{\text{se}}_{\widehat{t}_p} = \sqrt{\widehat{\text{Var}}(\widehat{t}_p)} = \sqrt{\widehat{t}_p^2 \widehat{\text{Var}}[\log(\widehat{t}_p)]}$$

$$= \widehat{t}_p \left\{ \widehat{\text{Var}}(\widehat{\mu}) + 2\Phi^{-1}(p)\widehat{\text{Cov}}(\widehat{\mu}, \widehat{\sigma}) + [\Phi^{-1}(p)]^2 \widehat{\text{Var}}(\widehat{\sigma}) \right\}^{1/2}. \tag{8.12}$$

Example 8.13 *Normal-Approximation Confidence Intervals for the Shock Absorber Weibull .1 Quantile.* The ML estimate for the Weibull distribution .1 quantile is

$$\widehat{t}_{.1} = \exp\left[\widehat{\mu} + \Phi^{-1}_{\text{sev}}(.1)\widehat{\sigma}\right] = \exp[10.23 + (-2.2504).3164] = 13602$$

and substituting into (8.12) gives

$$\widehat{\text{se}}_{\widehat{t}_{.1}} = 13602 \left[.01208 + 2(-2.2504)(.00399) + (-2.2504)^2(.005353)\right]^{1/2} = 1982.$$

An approximate 95% confidence interval for $t_{.1}$ based on $Z_{\log(\widehat{t}_{.1})} \stackrel{.}{\sim} \text{NOR}(0, 1)$ is

$$[\underline{t}_{.1}, \ \widetilde{t}_{.1}] = [13602/1.3306, \quad 13602 \times 1.3306 = [10{,}223, \quad 18{,}098],$$

where $w = \exp[1.960(1982)/13602] = 1.3306$.

Comparison in Table 8.1 suggests that both of the normal-approximation intervals deviate in the same direction from the likelihood-based interval. This is related to the left-skewed shape of $R(t_{.1})$, as seen in Figure 8.8. The log transformation on $t_{.1}$ does not improve the symmetry of the profile likelihood for this example. □

Confidence Interval for F(t)
Let t_e be a specified time at which an estimate of $F(t)$ is desired. The ML estimate for $F(t_e)$ is $\widehat{F}(t_e) = F(t_e; \widehat{\mu}, \widehat{\sigma}) = \Phi(\widehat{\zeta}_e)$, where $\widehat{\zeta}_e = [\log(t_e) - \widehat{\mu}]/\widehat{\sigma}$. An approximate confidence interval can be obtained from

$$[\widetilde{F}(t_e), \widetilde{F}(t_e)] = \widehat{F}(t_e) \pm z_{(1-\alpha/2)} \widehat{\text{se}}_{\widehat{F}}, \tag{8.13}$$

where applying (8.10) gives

$$\widehat{\text{se}}_{\widehat{F}} = \frac{\phi(\widehat{\zeta}_e)}{\widehat{\sigma}} \left[\widehat{\text{Var}}(\widehat{\mu}) + 2\widehat{\zeta}_e \widehat{\text{Cov}}(\widehat{\mu}, \widehat{\sigma}) + \widehat{\zeta}_e^2 \widehat{\text{Var}}(\widehat{\sigma}) \right]^{1/2}. \tag{8.14}$$

The interval in (8.13) is based on the NOR(0, 1) approximation for $Z_{\widehat{F}} = [\widehat{F}(t_e) - F(t_e)]/\widehat{\text{se}}_{\widehat{F}}$. With a small to moderate number of failures, however, the approximation could be poor; endpoints of the interval might even fall outside the range $0 \leq F(t_e) \leq 1$. A confidence interval procedure based on a transformation $g_1 = g_1(F)$ would have a confidence level closer to the nominal $100(1 - \alpha)\%$ if $Z_{\widehat{g}_1} = (g_1 - \widehat{g}_1)/\widehat{\text{se}}_{\widehat{g}_1}$ has a distribution that is closer than $Z_{\widehat{F}}$ to NOR(0, 1). Usually g_1 is a function of F chosen such that g_1 ranges from $(-\infty, \infty)$, the same range as the normal distribution. For estimating probabilities, for example, the logit transformation

$$\widehat{g}_1 = \text{logit}\left[F(t_e; \widehat{\mu}, \widehat{\sigma})\right] = \log\left\{ \frac{F(t_e; \widehat{\mu}, \widehat{\sigma})}{1 - F(t_e; \widehat{\mu}, \widehat{\sigma})} \right\}$$

does this. Then we find a confidence interval for logit(F) and transform the endpoints of this interval to obtain a confidence interval based on the approximate normal distribution of $Z_{\text{logit}(\widehat{F})}$. Using the inverse logit transformation gives the $100(1 - \alpha)\%$ confidence interval for $F(t_e)$ as

$$[\widetilde{F}(t_e), \widetilde{F}(t_e)] = \left[\frac{\widehat{F}}{\widehat{F} + (1 - \widehat{F}) \times w}, \frac{\widehat{F}}{\widehat{F} + (1 - \widehat{F})/w} \right], \tag{8.15}$$

where $w = \exp\{(z_{(1-\alpha/2)} \widehat{\text{se}}_{\widehat{F}})/[\widehat{F}(1 - \widehat{F})]\}$. The endpoints of this interval will always be between 0 and 1.

Example 8.14 *Normal-Approximation Confidence Interval for F(t) for the Weibull Distribution Fit to Shock Absorber Data.* Table 8.1 gives normal-approximation confidence intervals for $F(t_e)$ for $t_e = 10{,}000$ km based on $Z_{\text{logit}(\widehat{F})}$ and $Z_{\widehat{F}}$ following, approximately, NOR(0, 1) distributions. Comparison shows that the interval from (8.15) agrees well with the likelihood-based interval, but the interval from (8.13) has a negative lower endpoint, a clear indication of an inadequate approximation. □

Confidence Interval for the Hazard Function h(t)

Let t_e be a specified point in time at which an estimate of the hazard function h is desired. The ML estimate for $h(t_e)$ is

$$\widehat{h}(t_e) = h(t_e; \widehat{\mu}, \widehat{\sigma}) = \frac{\phi(\widehat{\zeta}_e)}{t_e \widehat{\sigma}[1 - \Phi(\widehat{\zeta}_e)]},$$

where $\widehat{\zeta}_e = [\log(t_e) - \widehat{\mu}]/\widehat{\sigma}$. Following Section 8.4.3, an approximate $100(1 - \alpha)\%$ confidence interval based on the NOR(0, 1) large-sample approximation to the distribution of $Z_{\log(\widehat{h})} = \{\log[\widehat{h}(t_e)] - \log[h(t_e)]\}/\widehat{se}_{\log(\widehat{h})}$ is

$$[\underline{h}(t_e), \widetilde{h}(t_e)] = [\widehat{h}(t_e)/w, \widehat{h}(t_e) \times w], \tag{8.16}$$

where $w = \exp[z_{(1-\alpha/2)}\widehat{se}_{\widehat{h}}/\widehat{h}(t_e)]$ and $\widehat{se}_{\widehat{h}}$ can be obtained by using (8.10).

Example 8.15 Weibull Hazard Function Estimate for the Shock Absorber Data. Figure 8.11 shows the ML estimate and a set of pointwise approximate 95% confidence intervals for the Weibull hazard function, computed from the shock absorber data. In this example, $\widehat{\beta} > 1$ ($\widehat{\sigma} < 1$) so $h(t)$ is increasing. Note that the Weibull $h(t)$ is always linear on log–log axes. □

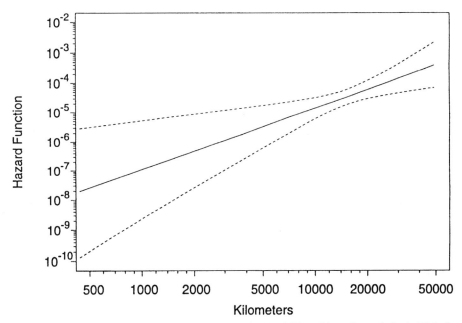

Figure 8.11. ML estimate and pointwise normal-approximation 95% confidence intervals for the Weibull hazard function for the shock absorber data.

8.4.4 Improved Normal-Approximation Confidence Intervals

As mentioned in Section 7.5, the normal-approximation intervals like those in (8.7) can be improved slightly by using $t_{(p;\nu)}$ instead of $z_{(p)}$. Corresponding to the well-known procedure of constructing a confidence interval for the mean of a normal distribution, with complete data (no censoring), confidence intervals with exact coverage probabilities are available for the mean (median) of a normal (lognormal) distribution. Such intervals are obtained by using $t_{(p;\nu)}$ instead of $z_{(p)}$ and substituting $[n/(n-1)]^{1/2}\text{se}_{\hat{\mu}}$ for $\text{se}_{\hat{\mu}}$ in (8.7). When there is no censoring, exact intervals are also available for normal/lognormal distribution quantiles, using the noncentral-t distribution (e.g., see Chapter 4 of Hahn and Meeker, 1991, or Table 3 in Odeh and Owen, 1980). It is, in fact, possible to obtain exact confidence intervals for any location-scale or log-location-scale distribution parameters or quantiles with complete data or Type II censoring. General tables for the factors to replace the normal-approximation $z_{(p)}$ are not generally available (see Robinson, 1983, for references to limited tables that are available). Procedures described in Chapter 9 can, however, be used to compute such factors with simulation.

8.5 ESTIMATION WITH GIVEN σ

When such information is available, one may use a given value for σ (or $\beta = 1/\sigma$ for the Weibull distribution) instead of estimating σ. The effect is to provide considerably more precision from limited data. The danger is the given value of σ may be seriously incorrect, resulting in misleading conclusions.

8.5.1 Lognormal/Normal Distribution with Given σ

For the lognormal/normal distribution with given σ and when there is no censoring, $\mathcal{L}(\mu)$ will be quadratic in μ [which implies that $L(\mu)$ will have the shape of a normal density]. This may be approximately so for other distributions under certain regularity conditions (see Appendix Section B.4) and large samples. When there is censoring, $L(\mu)$ is no longer exactly quadratic. Also, there is no closed-form solution to find the ML estimates for μ and iterative methods are required to compute the ML estimate $\hat{\mu}$.

8.5.2 Weibull/Smallest Extreme Value Distribution with Given σ

With a given value of $\sigma = 1/\beta$ it is possible to transform Weibull random variables to exponential random variables and to use the simpler methods for this distribution. If $T_1,\ldots,T_n$ have a Weibull distribution with shape parameter $\beta = 1/\sigma$ and scale parameter $\eta = \exp(\mu)$, then $T_1^\beta,\ldots,T_n^\beta$ have an exponential distribution with mean $\theta = \eta^\beta$.

In simple situations, the available data consist of a sample of n observations $t_1,\ldots,t_n$ of which r are exact failure times and $n - r$ are the running times of unfailed units. From this it follows, as an extension of (7.15) in Section 7.6.3, that the maximum likelihood estimate of the Weibull scale parameter, for fixed $\sigma = 1/\beta$, is

$$\widehat{\eta} = \left(\frac{\sum_{i=1}^{n} t_i^{\beta}}{r}\right)^{1/\beta}. \tag{8.17}$$

An estimate of the standard error of $\widehat{\eta}$ is

$$\widehat{\text{se}}_{\widehat{\eta}} = \frac{\widehat{\eta}}{\beta}\sqrt{\frac{1}{r}}. \tag{8.18}$$

It is interesting to note here that the ML estimate and its standard error can be computed without knowing which of the t_i correspond to failure times and which correspond to running times. Data of this kind arise commonly, for example, when the service times of individual units that fail are not known, but when there is knowledge, generally, of the total service time of all of the units in the field. Without a given value for β, such data are of limited value for estimating the failure-time distribution.

The simple formulas in (8.17) and (8.18) hold only for combinations of right censoring and observations reported as exact failures. For other kinds of data, there are still useful gains in precision from using a given value of β, but iterative methods are, in general, needed to compute the ML estimates and standard errors. See Nelson (1985) for justification of these methods and another example.

Example 8.16 Bearing-Cage Field Data. Appendix Table C.5 gives bearing-cage fracture times for 6 failed units as well as running times for 1697 units that had accumulated various amounts of service time without failing. The data and an analysis appear in Abernethy, Breneman, Medlin, and Reinman (1983). These data represent a population of units that had been introduced into service over time. There were concerns about the adequacy of the bearing-cage design. Analysts wanted to use these initial data to decide if a redesign would be needed to meet the design-life specification. This requirement was that $t_{.1}$ (referred to as B10 in some references) be at least 8000 hours. Management also wanted to know how many additional failures could be expected in the next year from the population of units currently in service. This second issue will be explored in Chapter 12.

Figure 8.12 shows four Weibull probability plots with different superimposed fitted Weibull distributions (solid lines) and approximate 95% normal-approximation pointwise confidence intervals (dotted lines). In the NW corner is a fitted Weibull distribution in which the shape parameter β was estimated. In the other three plots, the Weibull shape parameter was fixed at a specified value ($\beta = 1.5, 2,$ and 3). The ML estimate of β is 2.035 while the ML estimate of $t_{.1}$ is 3903 hours (considerably below the design life of 8000 hours). A 95% likelihood confidence interval for $t_{.1}$ is [2093, 22,144] hours, indicating that the design life *might* be much more than 8000 hours. The poor precision (wide interval) is due to the small number of failures.

Using a given value of β provides much more precision. There is, however, some risk that the given β is seriously incorrect, which could lead to seriously incorrect conclusions. For example, using $\beta = 1.5$ results in a much more optimistic estimate

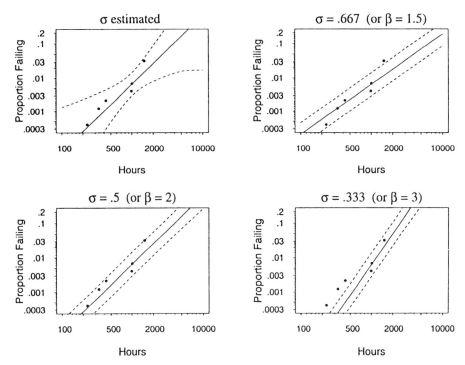

Figure 8.12. Weibull probability plots of the bearing-cage fracture data with Weibull ML estimates and sets of 95% pointwise confidence intervals for $F(t_e)$ with estimated $\hat{\sigma} = 1/\hat{\beta} = .491$, and given values $\beta = 1.5, 2,$ and 3.

of bearing-cage reliability. Using $\beta = 2$ or 3, however, gives a strong indication that the design-life requirement had not been met. □

The most striking conclusion from the previous example, and true in general, is the higher degree of precision obtained by using a given Weibull shape parameter β, especially outside the range of the data. Estimation of $t_{.1}$ required extrapolating from the proportion .055 (the maximum value of the nonparametric estimate) out to .1 (not an unreasonable amount of extrapolation for this kind of application). When there has been much accumulated experience with a product and its particular failure mode (or modes), it may be possible to safely use a particular given value for the Weibull shape parameter for analysis and decision making. Generally it is a good idea to use a plausible range of values, as shown in Figure 8.12. If the range of plausible β values is close to that suggested by the data themselves (as in the example), then the combined uncertainties will be approximately the same as reflected in the confidence intervals of the β-estimated model. If the given value of β is closer to the true β than is $\hat{\beta}$, then it is possible to achieve important increases in precision using the given value of β.

ESTIMATION WITH GIVEN σ 195

Chapter 14 describes Bayesian methods of data analysis that will allow a more formal method of incorporating prior uncertain knowledge about parameters, such as β, into an analysis.

8.5.3 Weibull/Smallest Extreme Value Distribution with Given $\beta = 1/\sigma$ and Zero Failures

ML estimates for the Weibull distribution cannot be computed unless the available data contain one or more failures. Section 7.7, however, showed how to compute a *lower* confidence bound on the exponential distribution mean when there are no failures. The resulting precision may be poor but still useful for some practical purposes. With a given Weibull shape parameter, the same ideas can be used to obtain a lower confidence bound on the Weibull scale parameter η [or, correspondingly, the SEV location parameter $\mu = \log(\eta)$].

For a sample of n units on test with running times $t_1, \ldots, t_n$ and no failures, a conservative $100(1-\alpha)\%$ lower confidence bound for η is

$$\underset{\sim}{\eta} = \left(\frac{2 \sum_{i=1}^{n} t_i^\beta}{\chi^2_{(1-\alpha;2)}} \right)^{1/\beta} = \left(\frac{\sum_{i=1}^{n} t_i^\beta}{-\log(\alpha)} \right)^{1/\beta} \tag{8.19}$$

because $\chi^2_{(1-\alpha;2)} = -2\log(\alpha)$. As in Section 7.7, this bound is based on the fact that, under the exponential distribution, with immediate replacement of failed units, the number of failures observed in a life test with a fixed total time on test has a Poisson distribution.

As described in Sections 7.4 and 7.7, $\underset{\sim}{\eta}$ can be translated into a lower confidence bound for increasing functions of η like t_p for specified p. Similarly, $\underset{\sim}{\eta}$ can be translated into an *upper* confidence bound for decreasing functions of η like $\widetilde{F}(t_e)$ for a specified t_e. Precision is a function of $\sum_{i=1}^{n} t_i^\beta$. Unless there are many large t_i values, the resulting confidence bound may largely be uninformative. See Nelson (1985) for justification and further discussion of this method.

Example 8.17 Determination of Component Safe-Life. A metal component in a ship's propulsion system (which we will refer to as Component-B) is known to fail from fatigue-caused fracture after some period of time in service. To minimize the probability of unscheduled downtime, the component is usually replaced during certain scheduled (for other purposes) preventive maintenance. Because of persistent reliability problems, the component was redesigned to have a longer service life. Previous experience with this and other components using the same alloy and similar designs suggests that the Weibull shape parameter is near $\beta = 2$, and almost certainly between 1.5 and 2.5. A number of the newly designed components were put into service during the past year. No failures have been reported with the new design. The service engineers asked that the data be used to assess whether the replacement age might be increased from 2000 hours to 4000 hours. The running times of the in-service components are given in Table 8.2. Using a fixed value $\beta = 2$, an approximate

Table 8.2. Early Production Failure-Free Running Times for Component-B

Hours	500	1000	1500	2000	2500	3000	3500	4000
Number of Units	10	12	8	9	7	9	6	3

Staggered entry data, with no reported failures.

95% lower confidence bound on η is

$$\underset{\sim}{\eta} = \left(\frac{2\sum_{i=1}^{n} t_i^2}{\chi^2_{(.95;2)}}\right)^{1/2} = \left(\frac{2 \times (10 \times 500^2 + \cdots + 3 \times 4000^2)}{5.99}\right)^{1/2}$$

$$= \left(\frac{2 \times 314750000}{5.99}\right)^{1/2} = 10250.$$

As described in Section 7.4.2, because there is only one unknown parameter in this problem, η and the given β can be substituted into the Weibull cdf to provide simultaneous upper confidence bounds on the entire cdf (or quantile function). This is illustrated in Figure 8.13. In particular, for $\beta = 2$, an upper confidence bound on the probability of failure before 4000 hours is obtained by substituting $\underset{\sim}{\eta}$ into (4.6) giving $\widetilde{F}(4000) = 1 - \exp[-(4000/10250)^2] = .141$. Also shown in this figure are

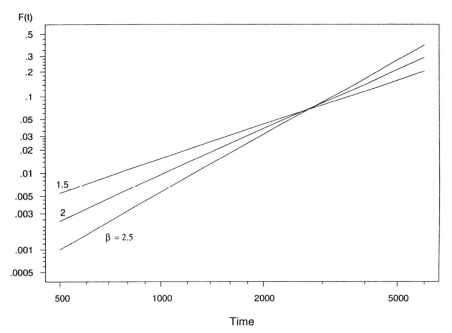

Figure 8.13. Weibull distribution 95% upper confidence bounds on $F(t)$ for Component-B with different fixed values for the Weibull shape parameter.

similar lines drawn for the upper bounds corresponding to given values of $\beta = 1.5$ and 2.5. This provides sensitivity analysis with respect to the uncertainty in β.

The conclusion from the confidence bounds is that the redesigned component may well have improved the life distribution enough to extend the replacement interval to 4000 hours. There is not, however, sufficient evidence to demonstrate this level of improvement. The probability of failure could be as small as 0 or (using the pessimistic $\beta = 2.5$) as large as $\widetilde{F}(4000) = 1 - \exp[-(4000/7925)^{2.5}] = .166$ because $\underset{\sim}{\eta} = 7925$ for $\beta = 2.5$. □

BIBLIOGRAPHIC NOTES

Escobar and Meeker (1992) give the partial derivatives of the log likelihood needed in (8.5) for location-scale distributions for observations reported as exact failure as well as right-, left-, and interval-censored observations. These expressions are useful for computing estimates of variance–covariance matrices like that shown in (8.5).

Ostrouchov and Meeker (1988) compare normal-approximation and likelihood confidence intervals for Weibull distribution parameters and percentiles. They show that likelihood intervals provide actual coverage probabilities that are closer to nominal confidence levels when there is a small to moderate number of failures in the sample. Doganaksoy (1995) and Doganaksoy and Schmee (1993) describe the advantages of likelihood confidence intervals and show that certain corrected likelihood intervals can provide important improvements in confidence level accuracy when the number of failures is small (say, less than 10). Meeker and Escobar (1995) compare likelihood and normal-approximation confidence intervals (also see Exercise 8.17).

Cheng and Iles (1983, 1988) provide simultaneous confidence bands for $F(t)$ for complete data from a Weibull or lognormal distribution. Escobar and Meeker (1998b) provide extensions and related technical results.

EXERCISES

8.1. Use the ball bearing data from Example 1.1 to do the following:

(a) Fit a lognormal distribution to the data. To facilitate the computations, use the following summary of the data:

$$\sum_{i=1}^{23} y_i = 95.46, \quad \sum_{i=1}^{23}(y_i - \bar{y})^2 = 6.26,$$

where $y_i = \log(t_i)$ and $\bar{y} = \sum_{i=1}^{23} y_i/23$. Plot the fitted lognormal distribution along with the nonparametric estimate on lognormal probability plot paper like that used in Exercise 6.5b.

(b) A computer program gives the following ML estimates for a Weibull distribution: $\hat{\mu} = 4.41$, $\hat{\sigma} = .476$. Do the same as in (a) but for the Weibull distribution.

(c) Comment on the adequacy of the lognormal and Weibull distributions to fit these data.

8.2. Use the heat exchanger tube crack data in Figure 1.7 to do the following:
  (a) Fit an exponential distribution to the data. (Note that this could be done by fitting a Weibull distribution with β constrained to be 1.)
  (b) Fit a Weibull distribution to the data.
  (c) Compare the values of the log likelihood from the two different distributions. How do they compare and what does this suggest?

8.3. Refer to Exercise 8.2 and use the fitted Weibull and exponential distributions to do a likelihood-ratio test to see if the data are consistent with a Weibull shape parameter $\beta = 1$. What is the implication of this hypothesis and what is your conclusion?

8.4. Refer to Exercise 8.2 and use the Weibull distribution to:
  (a) Compute an approximate 95% confidence interval for $F(2)$, the proportion of cracked tubes after 2 years of service, based on $Z_{\widehat{F}(2)} \stackrel{.}{\sim} \text{NOR}(0, 1)$.
  (b) Compute an approximate 95% confidence interval for $F(2)$, based on $Z_{\text{logit}[\widehat{F}(2)]} \stackrel{.}{\sim} \text{NOR}(0, 1)$.
  (c) Explain why you might prefer to report one of these intervals over the other.

8.5. Return to the fatigue crack-initiation test in Exercise 6.7. Fit a Weibull distribution with a given shape parameter $\beta = 2$.
  (a) Compute the ML estimate of η. What is the *practical* interpretation of this estimate?
  (b) Obtain the estimate $\widehat{\text{se}}_{\widehat{\eta}}$.
  (c) Compute a conservative 95% confidence interval for η.
  (d) Plot the Weibull estimate of $F(t)$ along with the nonparametric estimate. Comment on the adequacy of the Weibull distribution to describe the data.
  (e) What is an estimate of the .1 quantile of the time-to-initiation distribution?
  (f) Compute a conservative 95% confidence interval for the .1 quantile of the time-to-initiation distribution.

8.6. Suppose that $\widehat{g}$ is the ML estimate of a quantity g, and that $\widehat{\text{se}}_{\widehat{g}}$ is an estimated standard error. Show that a confidence interval for g, based on $Z_{\log(\widehat{g})} = [\log(\widehat{g}) - \log(g)]/\widehat{\text{se}}_{\log(\widehat{g})} \stackrel{.}{\sim} \text{NOR}(0, 1)$, has the form $[\underset{\sim}{g}, \widetilde{g}] = [\widehat{g}/w, \widehat{g} \times w]$, where $w = \exp[z_{(1-\alpha/2)}\widehat{\text{se}}_{\widehat{g}}/\widehat{g}]$.

8.7. A particular type of integrated circuit (IC) is known to have an electromigration-related failure mode. A life test was conducted at a temperature of 120°C in order to learn more about the life distribution and when failures might be expected to occur. A total of 20 ICs were tested and 5 failed before 500 hours, when the test was stopped. Failure times were at 252, 315, 369, 403, and 474 hours. ML estimates of the lognormal parameters are $\widehat{\mu} = 6.56$ and $\widehat{\sigma} = .534$. The variance–covariance matrix estimate for $\widehat{\mu}$ and $\widehat{\sigma}$ is

$$\widehat{\Sigma}_{\widehat{\mu},\widehat{\sigma}} = \begin{bmatrix} .0581 & .0374 \\ .0374 & .0405 \end{bmatrix}.$$

Recall that $\exp(\widehat{\mu}) = \exp(6.56) = 706$ hours is the ML estimate of the lognormal median.

(a) Make a lognormal probability plot of the failure data.

(b) Compute ML estimates of the lognormal $F(200)$ and $F(1000)$ and use these to draw a line representing the lognormal estimate of $F(t)$.

(c) Use $\widehat{\mu}$ and $\widehat{\sigma}$ to compute an estimate of the mean of the lognormal distribution (equation given in Section 4.6). Compare this with the estimate of the lognormal median. Comment on the difference.

(d) Compute $\widehat{t}_{.1}$, the ML estimate of the .1 quantile of the life distribution.

(e) Compute the standard error for $\widehat{t}_{.1}$. Explain the interpretation of this quantity.

(f) Compute an approximate 95% confidence interval for $t_{.1}$. Include this interval in the plot in part (a). Explain the interpretation of this interval and the justification for the approximate method that you use.

8.8. A life test was run on 20 prototype high-power RF transmitting tubes. In order to obtain tube life information more quickly, the tubes were tested at higher than usual levels of voltage. The tubes were tested simultaneously until failure or until 1.50 thousand hours. Failures were observed at .82, .99, 1.06, 1.08, 1.24, 1.39, 1.40 thousand hours. The other 13 tubes ran until 1.5 thousand hours without failure. Based on experience with life tests on similar products, the engineers believe that the Weibull shape parameter is $\beta = 3$. Do the following with β fixed at at this value.

(a) Make a Weibull probability plot of these data. Determine if the data are consistent with the specified value of β.

(b) Compute the ML estimate of the Weibull scale parameter η. Compute and graph the ML estimate of $F(t)$ on the probability plot constructed in part (a).

(c) Compute a 95% confidence interval for η. What is the practical interpretation of this interval?

(d) Compute the ML estimate and a 95% confidence interval for $F(t)$ at 1000 hours.

(e) Explain the interpretation of the confidence interval from part (d).

8.9. Redo Exercise 8.8, but use $\beta = 2$. Comment on the differences in the results and the potential effect of using an incorrect value of $\beta = 3$, if the actual Weibull shape parameter is $\beta = 2$.

8.10. Refer to Exercise 7.4. Suppose now that failure analysis and previous experience suggest that component–A life can be modeled adequately with a Weibull failure-time distribution.
  (a) Compute and plot the ML estimate of $F(t)$ with a Weibull distribution using given $\beta = 3$.
  (b) Compute and plot the ML estimate of $F(t)$ with a Weibull distribution using given $\beta = .33$.
  (c) Compute and plot the Weibull $\widehat{F}(t)$ and pointwise 95% confidence intervals, estimating β from the data.
  (d) Compare the plots constructed above. Use these plots and describe the consequences of using a seriously incorrect value of β.

8.11. Use the diesel locomotive fan failure data in Appendix Table C.1 to do the following:
  (a) Fit an exponential distribution to the fan data.
  (b) Fit a Weibull distribution to the fan data.
  (c) Do a likelihood-ratio test that the Weibull shape parameter is equal to 1. How would the conclusion affect fan replacement policy?
  (d) Compute an approximate 95% confidence interval for $t_{.1}$, the time at which 10% of the fan population will fail, based on $Z_{\widehat{t}_{.1}} \overset{\cdot}{\sim} \text{NOR}(0, 1)$.
  (e) Compute an approximate 95% confidence interval for $t_{.1}$, based on $Z_{\log(\widehat{t}_{.1})} \overset{\cdot}{\sim} \text{NOR}(0, 1)$.

8.12. Nelson (1982, page 529) analyzes failure data to compare two different snubber designs (a snubber is a component in a toaster). The data are in Appendix Table C.4.
  (a) Use probability plots and maximum likelihood fits to assess the adequacy of different parametric distributions for the data from the old design.
  (b) Analyze the new-design data in the same way to assess if there is evidence that the distributions differ. Section 17.8 describes analytical methods for such comparisons.

8.13. Consider a sample of n observations with $t_1, \ldots, t_r$ reported as exact failure times (suppose that $2 \leq r \leq n$) and $n - r$ observations censored at a common time t_c. Provide a simple expression for the Weibull log likelihood in the $\eta = \exp(\mu)$ and $\beta = 1/\sigma$ parameterization.

▲8.14. Refer to Exercise 8.13.
  (a) Derive an expression for the ML estimate of the Weibull distribution parameter η with right censoring at time t_c, for a given value of β.

(b) Explain how Weibull ML methods, when the shape parameter is given, are related to inference for the exponential distribution.

▲8.15. Refer to Exercise 8.13.
(a) Take partial derivatives of the log likelihood with respect to β and η.
(b) Show that the Weibull ML estimates can be obtained by solving the following two equations:

$$\left[\frac{\sum_{i=1}^{r} t_i^\beta \log(t_i) + (n-r)t_c^\beta \log(t_c)}{\sum_{i=1}^{r} t_i^\beta + (n-r)t_c^\beta} - \frac{1}{\beta} \right] - \frac{1}{r}\sum_{i=1}^{r} \log(t_i) = 0,$$

$$\eta^\beta = \frac{1}{r}\left[\sum_{i=1}^{r} t_i^\beta + (n-r)t_c^\beta \right].$$

Note that the first equation does not contain η and is thus easy to solve numerically for β.
(c) Use the equations in part (b) to determine the smallest sample size (n) and the smallest number of failures (r) that are needed for the ML estimates of η and β to be unique. *Hint:* Use of numerical and graphical methods can provide insight and suggest an appropriate analytical solution to this problem.

8.16. Write the Weibull likelihood for a sample containing both exact failure times and right-censored values, using σ and t_p as the distribution parameters (for some specified fixed $0 < p < 1$). Explain how to use this likelihood to compute a profile likelihood for t_p.

▲8.17. Consider a sample of n failure times, $t_1, \ldots, t_n$, that can be fit to a lognormal distribution.
(a) Write an expression for the likelihood and show that it is a function only of $\sum_{i=1}^{n} \log(t_i)$, $\sum_{i=1}^{n} [\log(t_i)]^2$, n, μ, and σ.
(b) Derive simple expressions for ML estimates of the lognormal parameters.
(c) Derive an expression for the relative likelihood $R(\mu, \sigma)$.
(d) Derive a simple expression for the profile likelihood $R(\mu)$.
(e) Derive a simple expression for the profile likelihood $R(\sigma)$.
(f) Challenge: Derive the profile likelihood $R(t_p)$, where t_p is the lognormal p quantile.

▲8.18. Use (8.10) to derive an expression for the standard error estimate $\widehat{se}_{\widehat{h}(t_e)}$ needed in (8.16). Show how this expression simplifies for the Weibull distribution.

▲8.19. For a location-scale distribution, under certain regularity conditions, in large samples, -2 times the logarithm of the relative likelihood function, $R(\mu, \sigma) =$

$L(\mu, \sigma)/L(\widehat{\mu}, \widehat{\sigma})$, when evaluated at the true μ and σ, has approximately a χ_2^2 distribution.

(a) Show how to use this property to test if a specified pair of parameter values, say, μ_0 and σ_0, are consistent with the data.

(b) Explain how to use the function $R(\mu, \sigma)$ to obtain an approximate joint confidence region for μ, σ.

(c) Show why, for a particular α, the $R(\mu, \sigma) = \alpha$ contour of the two-dimensional relative likelihood function (or a two dimensional profile likelihood function) provides an approximate $100(1 - \alpha)\%$ joint confidence region for μ, σ.

▲8.20. For the shock absorber data, the orientation of the contours in Figure 8.1 indicates some positive correlation between $\widehat{\mu}$ and $\widehat{\sigma}$. The orientation of the contours in Figure 8.7 indicates a negative correlation between $\widehat{t}_{.1}$ and $\widehat{\sigma}$.

(a) Use the numerical results in (8.6) to compute an estimate of the correlation between $\widehat{t}_{.1}$ and $\widehat{\sigma}$.

(b) For which value of p will $\widehat{t}_p$ and $\widehat{\sigma}$ be approximately uncorrelated?

8.21. As explained in Section 8.5.1, data analysis with the Weibull distribution is simpler and estimates are more precise if the Weibull shape parameter is given.

(a) Explain the dangers of using a specified Weibull shape parameter value when analyzing life data.

(b) A reliability engineer for a project claims that the Weibull shape parameter is known for a population of components. How would you, as an analyst, approach the problem of working with the engineer to make a reliability prediction based on a limited censored sample?

8.22. A random sample of five new automobile horns was taken from early production. The horns were tested simultaneously in a simulated use environment (with heat, humidity, salt air, and vibration) until each horn had reached a total of 600 thousand cycles. There were no failures in the test. The reliability specification for the horns says that $t_{.01}$ for the horn's life distribution should be at least 200 thousand cycles. The dominant failure mode for similar horns, manufactured in the past, was stress-corrosion-induced fatigue cracking and this failure mode had a Weibull shape parameter of $\beta = 1/\sigma = 2.3$.

(a) Using the given value of β, compute a lower 95% confidence bound for $t_{.01}$ of the horn's life distribution.

(b) Do the results contradict the reliability specification? Do the results demonstrate the reliability specification? What is the difference?

(c) Suggest, based on physical considerations, arguments to convince someone that it is possible that the test might lack validity for testing the reliability specification.

EXERCISES 203

8.23. Analysis of field data has suggested that a particular engine bearing is an important life-limiting component. The responsible engineers believe that improving the bearing's reliability would have an important effect on overall engine reliability. A redesigned bearing was tested extensively in a bench life test with simulated loads. Ten bearings were tested, each for 500 hours, with no reported failures. On the basis of previous field failure data for this class of bearing, a Weibull distribution with a shape parameter of $\beta = 2.3$ had provided an adequate description for the life of this bearing. Find a 95% lower confidence bound for the .01 quantile of the bearing life distribution.

8.24. Consider the problem described in Exercise 8.23 but suppose that the presumed distribution was lognormal with $\sigma = .55$.
  (a) Use the nonparametric method in Chapter 3 to obtain an approximate 95% upper confidence bound for $F(500)$.
  (b) Use the nonparametric upper confidence bound from part (a) and the lognormal distribution assumption to obtain a 95% lower confidence bound for the .01 quantile of the bearing life distribution.
  (c) Explain the relationship between the interval given in part (b) and the corresponding method for the Weibull distribution explained in Section 8.5.3.

CHAPTER 9

Bootstrap Confidence Intervals

Objectives

This chapter explains:

- The use of computer simulation to obtain confidence intervals. Such intervals are known as bootstrap confidence intervals.
- Methods for generating bootstrap samples.
- How to obtain and interpret simulation-based *parametric pointwise* bootstrap confidence intervals.
- How to obtain and interpret simulation-based *nonparametric pointwise* bootstrap confidence intervals.

Overview

This chapter describes and illustrates simulation-based bootstrap methods of finding confidence intervals. Especially with limited data, these intervals generally provide procedures with coverage probabilities that are closer to the nominal confidence level, when compared with the commonly used normal-approximation methods. Section 9.2 provides a general overview of bootstrap sampling methods. Sections 9.3 and 9.4, covering parametric bootstrap methods, build heavily on the confidence interval methods presented in Chapters 7 and 8. An understanding of the basic ideas in these chapters is important. The methods in this chapter can, however, be applied in a straightforward manner to parametric models used in the other chapters of this book. Correspondingly, the nonparametric bootstrap methods in Section 9.5 build on material from Chapter 3.

9.1 INTRODUCTION

The normal-approximation confidence intervals described in earlier chapters are adequate for most casual or informal analyses, particularly when the sample size is large (or, with right-censored data, when the number of failures is large). There are, however, computationally intensive methods that can provide better approximate confidence intervals. We have seen (in Chapters 7 and 8) that likelihood-based methods can be expected to out-perform the normal-approximation intervals. Simulation provides another important method to obtain exact or more accurate approximate confidence intervals. This chapter shows how to use bootstrap methods to obtain bootstrap confidence intervals for the inferential models used in Chapters 3, 7, and 8.

Bootstrap intervals, when used properly, can be expected to be more accurate than the normal-approximation methods and competitive with the likelihood-based methods. In subsequent chapters we will apply bootstrap methods for models where other reasonable alternatives do not exist (e.g., when likelihood-based methods are too demanding computationally).

9.2 BOOTSTRAP SAMPLING

9.2.1 General Idea

As explained in Section 3.3.2, a confidence interval procedure is judged on the basis of how well the procedure would perform if it were repeated over and over again. In particular, a confidence interval should (on the average) not be too wide (for a one-sided bound we can say that the bound should not be too far away from a point estimate) and the coverage probability (probability that the interval contains the quantity of interest) should be equal or close to the nominal coverage probability $1 - \alpha$. The idea of bootstrap sampling is to simulate the repeated sampling process and use the information from the distribution of appropriate statistics in the bootstrap samples to compute the needed confidence interval (or intervals), reducing the reliance on large-sample approximations.

For example, let θ be a parameter of interest. When computing a confidence interval for θ, instead of assuming that $Z_{\log(\widehat{\theta})}$ in (7.12) has a NOR(0, 1) distribution, we can use computer simulation to get an approximation to the *actual* distribution of $Z_{\log(\widehat{\theta})}$ (for estimating a positive parameter, the log transformation can also provide an important advantage when using this simulation-based method). If the approximation is better, the coverage probability will be closer to the nominal $1 - \alpha$. For some simple situations (complete data or Type II censored data from a location-scale or a log-location-scale distribution) the distribution of statistics like $Z_{\log(\widehat{\theta})}$ does not depend on the unknown value of θ. Then the resulting confidence intervals are "exact" in the sense that the actual coverage probability is the same as the nominal $1 - \alpha$; otherwise the coverage probability is approximately equal to $1 - \alpha$, but the approximation, in most cases, is better than assuming that $Z_{\log(\widehat{\theta})}$ has a NOR(0, 1) distribution and the approximation improves as the sample size increases.

We would like to obtain the distribution of appropriate statistics like $Z_{\log(\widehat{\theta})}$ by simulating from the actual population (or generating data from the actual process). Not being able to do this (because we do not know the exact character of the true population or process), we generate data based on information in the sample data. It is necessary to generate a large number (denoted by B) of "bootstrap samples" that can be used to approximate sampling distributions of interest. Due to the use of random (actually pseudorandom) samples, if the bootstrap method is applied twice to the same problem, there will be some differences between the answers obtained. With B chosen large enough, such differences will be negligible. When the goal is to compute confidence intervals, the usual recommendation is to use between $B = 2000$ and $B = 5000$ bootstrap samples (larger values of B are recommended for estimating the more extreme quantiles of the bootstrap distribution that are required for higher confidence levels). To reduce possible effects of simulation variability in our examples we have used $B = 10{,}000$, but such a large number should not be necessary for most practical applications.

9.2.2 Bootstrap Sampling Methods

There are several different methods for generating the needed bootstrap samples $\text{DATA}_j^*, j = 1, \ldots, B$.

- Figure 9.1 illustrates the fully "parametric" bootstrap sampling procedure. This method simulates each sample of size n from the assumed parametric distribution, using the ML estimates computed from the actual data to replace the unknown parameters. That is, sampling is from $F(t; \widehat{\theta})$. Observations are censored according to the specified censoring mechanism. The disadvantage of this

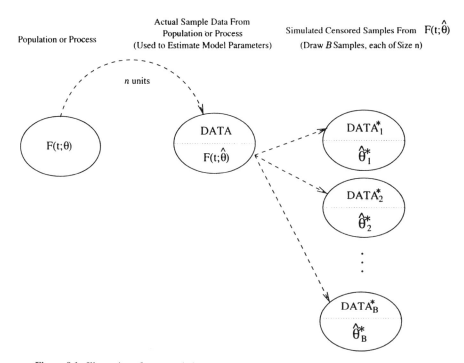

Figure 9.1. Illustration of parametric bootstrap (simulation) sampling for parametric inference.

method is that it requires complete specification of the censoring process. In simple problems (when all units will be run to failure or with Type I or Type II censoring) this presents no difficulties. The specification is, however, more difficult for complicated systematic or random censoring. Often the needed information (like the times that failed units would have been censored or inspected had they not failed) may be unknown.

- Figure 9.2 illustrates the simpler "nonparametric" bootstrap sampling scheme. In this method, each sample of size n is obtained by sampling, with replacement, from the actual data cases in the original data set. Specifically, to obtain DATA^*_j, sample *with* replacement from the data cases in DATA until n (the sample size) cases have been selected. In each draw, each data case in DATA has an equal probability of being chosen. New "bootstrap estimates" are computed for each sample of size n. The entire process is repeated for $j = 1, \ldots, B$ bootstrap samples. This method is simple to use (because the method depends only on the censored data and does not require explicit specification of the censoring mechanism) and generally, with moderate to large samples, provides results that are close to the fully parametric approach. When the number of distinct sample observations in DATA is very small (say, less than 7), however, the distribution of the bootstrap statistics will be noticeably discrete and the fully parametric approach would be preferable.

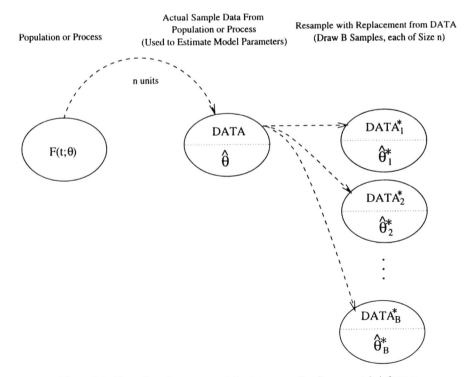

Figure 9.2. Illustration of nonparametric bootstrap sampling for parametric inference.

Unless otherwise noted, the nonparametric bootstrap sampling method illustrated in Figure 9.2 will be used.

It is useful to save the bootstrap estimates $\widehat{\theta}_j^*, j = 1,\ldots,B$, and the corresponding bootstrap standard error estimates (or the variance–covariance matrix when there is more than one model parameter) in a computer file so that they can be used subsequently in different ways without having to recompute the ML estimates (the computationally intensive part of bootstrap methods). In new and unfamiliar situations it can be useful to examine graphically the distribution of bootstrap statistics, as we will do in the following examples.

9.3 EXPONENTIAL DISTRIBUTION CONFIDENCE INTERVALS

This section shows how to apply bootstrap methods to the exponential distribution used in Chapter 7. Table 9.1 provides a comparison of the results for the three different sample sizes for the α-particle data from Example 7.3. Because there is a more interesting contrast among the methods for small sample sizes, the following examples give details of the application of the bootstrap methods for the $n = 20$ sample. As predicted by theory, with large samples, the different methods result in similar intervals.

For the exponential distribution, instead of assuming $Z_{\log(\widehat{\theta})} \stackrel{.}{\sim} \text{NOR}(0, 1)$, Monte Carlo simulation-based methods can be used to obtain a better approximation to the distribution of $Z_{\log(\widehat{\theta})}$. For example, the bootstrap approximation for the distribution

Table 9.1. Comparison of Likelihood and Bootstrap Approximate Confidence Intervals for the α-Particle Interarrival Times

	All Times	Sample of Times		
	$n = 10{,}220$	$n = 2000$	$n = 200$	$n = 20$
Approximate 95% Confidence Intervals for θ				
Based on the likelihood	[584, 608]	[586, 641]	[498, 662]	[289, 713]
Based on $Z_{\log(\widehat{\theta})} \stackrel{.}{\sim} Z_{\log(\widehat{\theta}^*)}$	[584, 608]	[586, 641]	[496, 670]	[309, 694]
Based on $Z_{\widehat{\theta}} \stackrel{.}{\sim} Z_{\widehat{\theta}^*}$	[584, 608]	[586, 641]	[496, 670]	[309, 694]
Approximate 95% Confidence Intervals for $\lambda \times 10^5$				
Based on the likelihood	[164, 171]	[156, 171]	[151, 201]	[140, 346]
Based on $Z_{\widehat{\theta}} \stackrel{.}{\sim} Z_{\widehat{\theta}^*}$	[164, 171]	[156, 171]	[149, 202]	[144, 324]
Based on $Z_{\log(\widehat{\theta})} \stackrel{.}{\sim} Z_{\log(\widehat{\theta}^*)}$	[164, 171]	[156, 171]	[149, 202]	[144, 324]

of $Z_{\log(\widehat{\theta})}$ in (7.12) can be obtained by simulating B bootstrap samples of size n and computing ML estimates for each bootstrap sample. Then, for each bootstrap sample, compute $Z_{\log(\widehat{\theta}^*_j)} = [\log(\widehat{\theta}^*_j) - \log(\widehat{\theta})]/\widehat{\mathrm{se}}_{\log(\widehat{\theta}^*_j)}$, where $\log(\widehat{\theta}^*_j)$ is the jth bootstrap ML estimate of $\log(\widehat{\theta})$ and $\widehat{\mathrm{se}}_{\log(\widehat{\theta}^*_j)}$ is the corresponding standard error estimate. The bootstrap ML estimates $\log(\widehat{\theta}^*)$ and $\widehat{\mathrm{se}}_{\log(\widehat{\theta}^*)}$ are computed as in Section 7.3.3, but from the bootstrap samples DATA$^*_j$, $j = 1, \ldots, B$. Because such intervals are based on the distribution of t-like statistics (i.e., statistics computed in a manner that is similar to t-statistics used with normal distribution models with no censoring), the method is called the "bootstrap-t" method.

Example 9.1 Bootstrap Sample for the α-Particle Mean Time Between Arrivals. Following Example 7.3, Figure 9.3 is an exponential probability plot showing $F(t; \widehat{\theta})$ from the original $n = 20$ sample (the thicker, longer line) and 50 (out of $B = 10{,}000$) of the bootstrap estimates $F(t; \widehat{\theta}^*)$, each computed from bootstrap samples of size $n = 20$, fit to the exponential distribution. As explained in Section 6.2.1, each bootstrap estimate of $\widehat{\theta}^*$ gives a line on the exponential probability plot. Figure 9.3 provides insight into the amount of variability that one would expect to see in estimates of various different quantities of interest (e.g., failure probabilities at given times), based on repeated samples of size $n = 20$. □

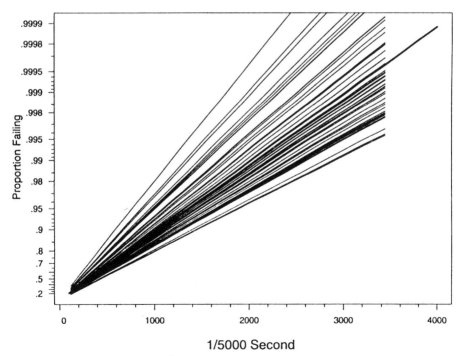

Figure 9.3. Exponential plot of $F(t; \widehat{\theta})$ from the original sample (shown with the thicker, longer line) and 50 (out of $B = 10{,}000$) $F(t; \widehat{\theta}^*)$ computed from exponential bootstrap samples for the α-particle data.

9.3.1 Bootstrap Confidence Intervals for θ

An approximate $100(1-\alpha)\%$ confidence interval for θ based on the assumption that the simulated distribution of $Z_{\hat{\theta}^*}$ provides a good approximation to the distribution of $Z_{\hat{\theta}}$ is

$$[\underset{\sim}{\theta}, \tilde{\theta}] = \left[\hat{\theta} - z_{\hat{\theta}^*_{(1-\alpha/2)}} \widehat{se}_{\hat{\theta}},\ \hat{\theta} - z_{\hat{\theta}^*_{(\alpha/2)}} \widehat{se}_{\hat{\theta}}\right], \tag{9.1}$$

where $z_{\hat{\theta}^*_{(p)}}$ is the p quantile of the distribution of $Z_{\hat{\theta}^*}$ and $\hat{\theta}$ and $\widehat{se}_{\hat{\theta}}$ are the estimates from the original sample [same as those used in (7.6) and (7.11)]. The justification for (9.1) is similar to the justification of (7.6) given in Section 7.3.3 except that, unlike NOR(0, 1), the distribution of $Z_{\hat{\theta}^*}$ is not symmetric.

Because θ is a positive parameter and because the estimated standard error of $\hat{\theta}$ is proportional to $\hat{\theta}$, a better bootstrap procedure can be expected by basing confidence intervals on the assumption that the simulated distribution of $Z_{\log(\hat{\theta}^*)}$ provides a good approximation to the distribution of $Z_{\log(\hat{\theta})}$. In particular, analogous to (7.11),

$$[\underset{\sim}{\theta}, \tilde{\theta}] = [\hat{\theta}/\underset{\sim}{w}, \hat{\theta}/\tilde{w}], \tag{9.2}$$

where $\underset{\sim}{w} = \exp[z_{\log(\hat{\theta}^*)_{(1-\alpha/2)}} \widehat{se}_{\hat{\theta}}/\hat{\theta}]$, $\tilde{w} = \exp[z_{\log(\hat{\theta}^*)_{(\alpha/2)}} \widehat{se}_{\hat{\theta}}/\hat{\theta}]$, and $z_{\log(\hat{\theta}^*)_{(p)}}$ is the p quantile of the distribution of $Z_{\log(\hat{\theta}^*)}$. Again, the justification for (9.2) is similar to the justification of (7.11) given in Section 7.3.3 except that $\underset{\sim}{w} \neq \tilde{w}$ because, unlike the NOR(0, 1), the distribution of $Z_{\log(\hat{\theta}^*)}$ is not symmetric.

Example 9.2 Bootstrap Confidence Interval for α-Particle Mean Time Between Arrivals. Figure 9.4 gives histograms of the $B = 10{,}000$ values of $\hat{\theta}^*$, $Z_{\hat{\theta}^*}$, and $Z_{\log(\hat{\theta}^*)}$. Figure 9.4 also shows the cumulative distribution of $Z_{\log(\hat{\theta}^*)}$ indicating, with the dashed lines, the .025 and .975 quantiles of the distribution of $Z_{\log(\hat{\theta}^*)}$. Numerically these quantiles are $z_{\log(\hat{\theta}^*)_{(.025)}} = -1.9858$ and $z_{\log(\hat{\theta}^*)_{(.975)}} = 1.5483$, respectively. Substituting these quantiles and the values of $\hat{\theta}$ and $\widehat{se}_{\hat{\theta}}$ obtained from the original $n = 20$ sample (see Table 7.2) into (9.2) gives

$$[\underset{\sim}{\theta}, \tilde{\theta}] = [440.2/1.4265,\ 440.2/.6341] = [309,\ 694],$$

where

$$\underset{\sim}{w} = \exp(1.5483(101.0)/440.2) = 1.4265$$

and

$$\tilde{w} = \exp(-1.9858(101.0)/440.2) = .6341.$$

Table 9.1 compares this interval with the interval based on the likelihood-ratio method. □

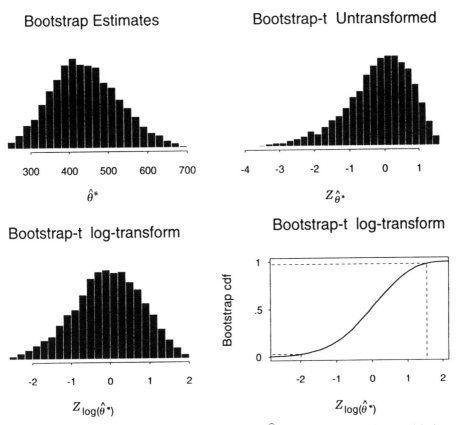

Figure 9.4. Histograms of exponential bootstrap samples of $\hat{\theta}^*$, $Z_{\hat{\theta}^*}$, and $Z_{\log(\hat{\theta}^*)}$ for the α-particle data and the empirical cdf of $Z_{\log(\hat{\theta}^*)}$.

9.3.2 Bootstrap Confidence Intervals for Functions of θ

For the exponential distribution (or other distributions with just one unknown parameter), the simple method given in Section 7.4 can be used to obtain confidence intervals for monotone functions of θ. As with the intervals in Chapter 7, the resulting confidence interval(s) inherit the coverage properties of the particular interval that was used for θ.

Example 9.3 Bootstrap Confidence Intervals for the Arrival Rate of α Particles. Using the bootstrap confidence interval for θ from the $n = 20$ sample in Table 7.2,

$$[\utilde{\lambda}, \tilde{\lambda}] = [1/\tilde{\theta}, \ 1/\utilde{\theta}] = [.00144, \ .00324].$$

Although this interval is for λ, the method is based on the assumption that $Z_{\log(\hat{\theta})} \stackrel{.}{\sim} Z_{\log(\hat{\theta}^*)}$. □

9.3.3 Comparison of Confidence Interval Methods

For the α-particle arrival data, Table 9.1 compares 95% approximate confidence intervals for θ based on:

- The likelihood-ratio method.
- The bootstrap approximation $Z_{\log(\widehat{\theta})} \overset{\cdot}{\sim} Z_{\log(\widehat{\theta}^*)}$.
- The bootstrap approximation $Z_{\widehat{\theta}} \overset{\cdot}{\sim} Z_{\widehat{\theta}^*}$.

For this example, there is almost no difference between the bootstrap methods based on $Z_{\widehat{\theta}} \overset{\cdot}{\sim} Z_{\widehat{\theta}^*}$ and $Z_{\log(\widehat{\theta})} \overset{\cdot}{\sim} Z_{\log(\widehat{\theta}^*)}$ (but this will not be true in general). The comparison shows that, in this case, the likelihood-ratio method and the bootstrap methods are in close agreement for all sample sizes except $n = 20$. For $n = 20$, the likelihood-based intervals are wider and simulation results (details not shown here) indicate that their coverage probability is larger than nominal.

9.4 WEIBULL, LOGNORMAL, AND LOGLOGISTIC DISTRIBUTION CONFIDENCE INTERVALS

Because of its simple specification, we continue to use the bootstrap resampling scheme illustrated in Figure 9.2. For samples with very small numbers failing (say, less than 7), it would be better to use the fully parametric method illustrated in Figure 9.1. Similar to the normal-approximation intervals in Section 8.4, it is important to base the bootstrap-t confidence intervals on bootstrap evaluations of the distributions of t-like statistics employing transformations that transform statistics to an unlimited range: $Z_{\widehat{\mu}}$, $Z_{\log(\widehat{\sigma})}$, $Z_{\log(\widehat{t}_1)}$, $Z_{\text{logit}[\widehat{F}(t)]}$, and $Z_{\log(\widehat{h})}$.[1]

Example 9.4 Bootstrap Sample for the Shock Absorber Example. Following Example 8.1, for each of $B = 10{,}000$ bootstrap samples, bootstrap estimates $\widehat{\mu}^*$, $\widehat{\sigma}^*$ were computed. Figure 9.5 shows the first 1000 $(\widehat{\mu}^*, \widehat{\sigma}^*)$ pairs, as well as the most extreme pair (which happened not to be among the first 1000). The correlation between $\widehat{\mu}^*$ and $\widehat{\sigma}^*$ is evidence of positive correlation between $\widehat{\mu}$ and $\widehat{\sigma}$ (which we also saw in Example 8.10). Estimates $\widehat{\Sigma}_{\widehat{\mu}^*, \widehat{\sigma}^*}$ of the variance–covariance matrix also were obtained so that bootstrap standard errors could be computed. All results were stored in a computer file to allow rapid computation of the various bootstrap confidence intervals to be described in this section.

Figure 9.6 shows the Weibull cdf ML estimate for the original sample (thick longer line) and for the first 50 bootstrap samples. Studying this plot (even with the relatively small number of bootstrap samples shown) provides a clear picture of the precision with which it will be possible to estimate different quantiles and failure probabilities.

□

---

[1] Actually what is really needed is a transformation that will make the studentized bootstrap statistic behave approximately like a pivotal statistic.

WEIBULL, LOGNORMAL, LOGLOGISTIC DISTRIBUTION CONFIDENCE INTERVALS 213

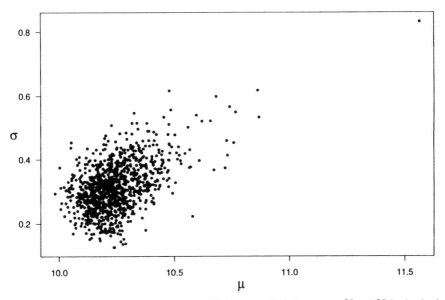

Figure 9.5. Scatter plot of 1000 (out of $B = 10{,}000$) bootstrap Weibull estimates $\widehat{\mu}^*$ and $\widehat{\sigma}^*$ for the shock absorber example.

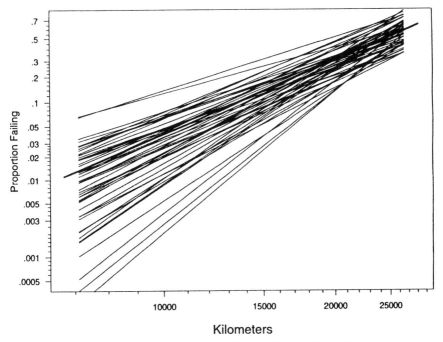

Figure 9.6. Weibull plot of $F(t; \widehat{\mu}, \widehat{\sigma})$ from the original sample (shown with the thicker, longer line) and 50 (out of $B = 10{,}000$) $F(t; \widehat{\mu}^*, \widehat{\sigma}^*)$ bootstrap Weibull estimates computed from bootstrap samples for the shock absorber example.

9.4.1 Construction of Confidence Intervals for Parameters

Following the approach used in Section 9.3, the bootstrap approximation for the distribution of $Z_{\hat{\mu}}$ can be obtained by using the bootstrap samples to compute B values of $Z_{\hat{\mu}_j^*} = (\hat{\mu}_j^* - \hat{\mu})/\widehat{\text{se}}_{\hat{\mu}_j^*}$ where $\hat{\mu}_j^*$ is the jth bootstrap estimate of $\hat{\mu}$ and $\widehat{\text{se}}_{\hat{\mu}_j^*}$ is the corresponding standard error estimate. Generally, B should be between 2000 and 4000 (but we continue to use $B = 10{,}000$ in our examples to further reduce the effects of Monte Carlo variability). The needed bootstrap estimates are computed as in Section 8.4, but from the bootstrap samples DATA_j^*, $j = 1, \ldots, B$. An approximate $100(1 - \alpha)\%$ confidence interval for μ based on the assumption that the simulated distribution of $Z_{\hat{\mu}^*}$ provides a good approximation to the distribution of $Z_{\hat{\mu}}$ is

$$[\underset{\sim}{\mu}, \ \widetilde{\mu}] = \left[\hat{\mu} - z_{\hat{\mu}^*_{(1-\alpha/2)}} \widehat{\text{se}}_{\hat{\mu}}, \ \hat{\mu} - z_{\hat{\mu}^*_{(\alpha/2)}} \widehat{\text{se}}_{\hat{\mu}}\right], \qquad (9.3)$$

where $z_{\hat{\mu}^*_{(p)}}$ is the p quantile of the distribution of $Z_{\hat{\mu}^*}$ and $\hat{\mu}$ and $\widehat{\text{se}}_{\hat{\mu}}$ are the estimates from the original sample [same as those used in (8.7)]. The justification for (9.3) is similar to the justification of (3.11) given in Section 3.6.3.

Example 9.5 Bootstrap Confidence Interval for the Shock Absorber Weibull μ Parameter. Following Example 9.4, Figure 9.7 gives histograms of the $B = 10{,}000$ values of $\hat{\mu}^*$ and $Z_{\hat{\mu}^*}$. Also shown is the cumulative distribution of $Z_{\hat{\mu}^*}$ indicating, with the dashed lines, the .025 and .975 quantiles of the distribution of $Z_{\hat{\mu}^*}$. Numerically these quantiles are $z_{\hat{\mu}^*_{(.025)}} = -2.363$ and $z_{\hat{\mu}^*_{(.975)}} = 1.252$, respectively. From this, using some previous results from Example 8.11 and substitution into (9.3) gives

$$[\underset{\sim}{\mu}, \ \widetilde{\mu}] = [10.23 - 1.252(.1099), \ 10.23 + 2.363(.1099)] = [10.09, \ 10.49].$$

Table 9.2 shows that these values are consistent with the likelihood-based interval. The corresponding confidence interval for the Weibull scale parameter $\eta = \exp(\mu)$ can be computed by exponentiating the endpoints of this interval, as shown in Example 8.11. □

An approximate $100(1 - \alpha)\%$ confidence interval for σ can be computed in a manner that is similar to (8.8). In particular $[\underset{\sim}{\sigma}, \ \widetilde{\sigma}] = [\hat{\sigma}/w, \ \hat{\sigma}/\widetilde{w}]$, where $w = \exp[z_{\log(\hat{\sigma}^*)_{(1-\alpha/2)}} \widehat{\text{se}}_{\hat{\sigma}}/\hat{\sigma}]$ and $\widetilde{w} = \exp[z_{\log(\hat{\sigma}^*)_{(\alpha/2)}} \widehat{\text{se}}_{\hat{\sigma}}/\hat{\sigma}]$ and $z_{\log(\hat{\sigma}^*)_{(p)}}$ is the p quantile of the distribution of $Z_{\log(\hat{\sigma}^*)}$. This interval is based on the assumption that the simulated distribution of $Z_{\log(\hat{\sigma}^*)}$ provides a good approximation to the distribution of $Z_{\log(\hat{\sigma})}$.

Example 9.6 Bootstrap Confidence Interval for the Shock Absorber Weibull σ Parameter. Following Example 9.4, Figure 9.8 gives histograms of the $B = 10{,}000$ values of $\hat{\sigma}^*$, $Z_{\hat{\sigma}^*}$, and $Z_{\log(\hat{\sigma}^*)}$. Figure 9.8 also shows the cumulative distribution of $Z_{\log(\hat{\sigma}^*)}$ indicating, with the dashed lines, the .025 and .975 quantiles of the distribution of $Z_{\log(\hat{\sigma}^*)}$. Numerically these values are $z_{\log(\hat{\sigma}^*)_{(.025)}} = -2.458$ and $z_{\log(\hat{\sigma}^*)_{(.975)}} = 1.589$, respectively. From this, using some previous results from Example 8.12, $[\underset{\sim}{\sigma}, \ \widetilde{\sigma}] = [.3164/w, \ .3164/\widetilde{w}] = [.2191, \ .5585]$, where $w =$

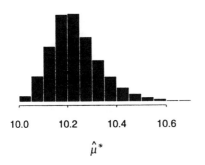
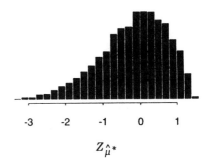
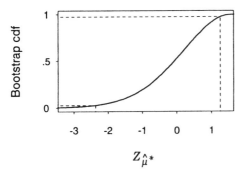

Figure 9.7. Bootstrap distributions of Weibull $\hat{\mu}^*$ and $Z_{\hat{\mu}^*}$ based on $B = 10{,}000$ bootstrap samples for the shock absorber example.

$\exp[1.589(.07316)/.3164] = 1.4440$ and $\widetilde{w} = \exp[-2.458(.07316)/.3164] = .56645$. Table 9.2 also shows the bootstrap interval based on $Z_{\hat{\sigma}^*}$. Both bootstrap methods deviate somewhat from the normal-approximation methods, but agreement is best with the likelihood method. The corresponding confidence interval for the Weibull shape parameter $\beta = 1/\sigma$ can be computed as shown in Example 8.12. □

9.4.2 Confidence Intervals for Functions of Parameters

Bootstrap confidence intervals for functions of parameters can be computed by following the general approach used in Section 8.4.3 for normal-approximation intervals, but using quantiles of the bootstrap estimates of appropriate Z distributions, as illustrated in Section 9.4.1. The following examples describe and illustrate this procedure.

Example 9.7 Bootstrap Confidence Intervals for the Shock Absorber Weibull $F(t)$. Following Example 9.4, Figure 9.9 gives histograms of the $B = 10{,}000$ values

Table 9.2. Comparison of Likelihood and Parametric Bootstrap Approximate Confidence Interval Methods for the Shock Absorbers

	Distribution	
	Weibull	Lognormal
Approximate 95% *Confidence Intervals for μ*		
Based on the likelihood	[10.06, 10.54]	[9.91, 10.53]
Based on $Z_{\hat{\mu}} \stackrel{.}{\sim} Z_{\hat{\mu}^*}$	[10.09, 10.49]	[9.96, 10.47]
Approximate 95% *Confidence Intervals for σ*		
Based on the likelihood	[.210, .527]	[.367, .858]
Based on $Z_{\log(\hat{\sigma})} \stackrel{.}{\sim} Z_{\log(\hat{\sigma}^*)}$	[.219, .559]	[.383, 1.01]
Based on $Z_{\hat{\sigma}} \stackrel{.}{\sim} Z_{\hat{\sigma}^*}$	[.220, .560]	[.387, .999]
Approximate 95% *Confidence Intervals for $t_{.1}$*		
Based on the likelihood	[9400, 17300]	[9400, 16300]
Based on $Z_{\log(\hat{t}_{.1})} \stackrel{.}{\sim} Z_{\log(\hat{t}_{.1}^*)}$	[8700, 17200]	[8100, 16100]
Based on $Z_{\hat{t}_{.1}} \stackrel{.}{\sim} Z_{\hat{t}_{.1}^*}$	[8300, 17300]	[7800, 16200]
Approximate 95% *Confidence Intervals for $F(10000)$*		
Based on the likelihood	[.0092, .1136]	[.0085, .1159]
Based on $Z_{\text{logit}(\hat{F})} \stackrel{.}{\sim} Z_{\text{logit}(\hat{F}^*)}$	[.0104, .1209]	[.0091, .1292]
Based on $Z_{\hat{F}} \stackrel{.}{\sim} Z_{\hat{F}^*}$	[.0044, .2229]	[.0015, .6128]

of $\widehat{F}^*(t_e)$, $Z_{\hat{F}^*}$, and $Z_{\text{logit}(\hat{F}^*)}$ for $t_e = 10{,}000$ km. Figure 9.9 also shows the cumulative distribution of $Z_{\text{logit}(\hat{F}^*)}$ indicating, with the dashed lines, the .025 and .975 quantiles of the distribution of $Z_{\text{logit}(\hat{F}^*)}$. Numerically these values are $z_{\text{logit}(\hat{F}^*)_{(.025)}} = -1.845$ and $z_{\text{logit}(\hat{F}^*)_{(.975)}} = 2.045$, respectively. From this, using $\widehat{F}(t_e) = .03908$ and $\widehat{\text{se}}_{\hat{F}} = .02480$ (from Table 8.1) and substituting the bootstrap distribution quantiles for the NOR(0, 1) quantiles in (8.15) gives the values in Table 9.2.

Table 9.2 shows that the confidence intervals based on the distribution of $Z_{\text{logit}(\hat{F}^*)}$ are generally close to those obtained with the likelihood-based method. Intervals based on the distribution of $Z_{\hat{F}^*}$ are, however, quite different from *any* of the other intervals. The primary reason for this is the extreme lower tail of the $Z_{\hat{F}^*(t_e)}$ distribution, shown in Figure 9.9 (the minimum value of $Z_{\hat{F}^*(t_e)}$ being approximately -20). In particular, $z_{\hat{F}^*(t_e)_{(.025)}} = -7.411$. This indicates the importance, particularly for restricted quantities like $0 \le \widehat{F}(t) \le 1$, of using an appropriate transformation that will cause the distribution of the Z^* bootstrap-t statistics to be less sensitive to the unknown actual parameters of the underlying model. □

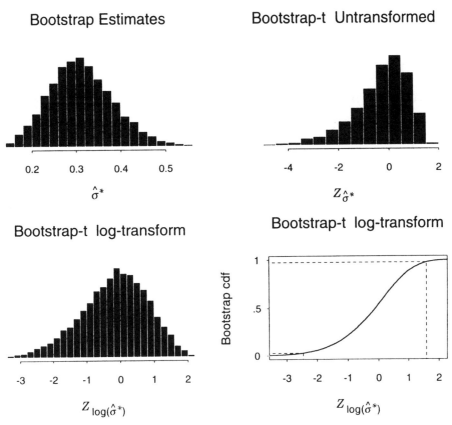

Figure 9.8. Bootstrap distributions of $\hat{\sigma}^*$, $Z_{\hat{\sigma}^*}$, and $Z_{\log(\hat{\sigma}^*)}$ based on $B = 10{,}000$ bootstrap samples for the shock absorber example.

9.4.3 Comparison of Confidence Interval Methods

Table 9.2 contains numerical values of approximate confidence intervals for μ, σ, $F(10000)$, and $t_{.1}$ based on the likelihood and bootstrap methods. Relative to the width of the intervals, the differences among the methods are not large, except that the use of the logit transformation on the interval for $F(10000)$ has a big effect. Comparison with the likelihood-based interval makes it clear that the use of the logit transformation provides an important improvement in this case.

9.5 NONPARAMETRIC BOOTSTRAP CONFIDENCE INTERVALS

9.5.1 Nonparametric Bootstrap Sampling

As explained in Section 9.2, to use the bootstrap method, it is necessary to generate a large number (denoted by B) of bootstrap samples that can be used to approximate the

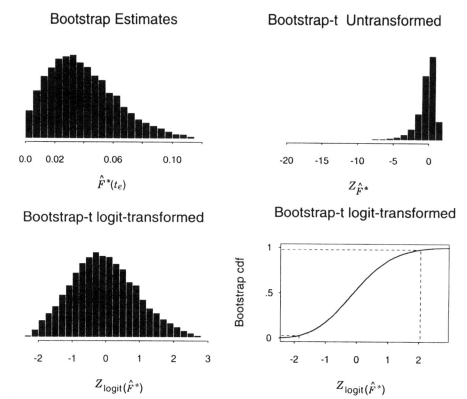

Figure 9.9. Weibull model bootstrap distributions of $\widehat{F}^*(t_e)$, $Z_{\widehat{F}^*}$, and $Z_{\text{logit}(\widehat{F}^*)}$ for $t_e = 10{,}000$ km based on $B = 10{,}000$ bootstrap samples for the shock absorber example.

sampling distributions of interest. Figure 9.10 illustrates the nonparametric bootstrap sampling method with nonparametric estimation.

Example 9.8 Bootstrap Samples for $F(t)$ from the Heat Exchanger Tube Data. For this example $B = 10{,}000$ bootstrap samples were generated. For each of the B bootstrap samples, bootstrap estimates $\widehat{F}^*(t)$ and $\widehat{\text{se}}_{\widehat{F}^*}$ were computed. Figure 9.11 shows $\widehat{F}^*(t)$ values for the first 50 of the $B = 10{,}000$ bootstrap samples. Note the discreteness (limited number of outcomes) in the bootstrap estimates, especially at the end of the first year of operation. □

9.5.2 A Limitation and Warning on the Application of Bootstrap Methods for Nonparametric Estimation

The justification for the bootstrap is based on large-sample theory. Even with large samples, however, there can be difficulties in the tails of the sample. For the nonparametric bootstrap, there will be a separate bootstrap distribution at each t_i for which there was one or more failures in the original sample. These bootstrap distributions

NONPARAMETRIC BOOTSTRAP CONFIDENCE INTERVALS

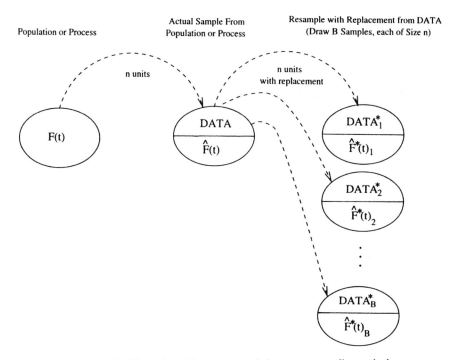

Figure 9.10. Illustration of the nonparametric bootstrap resampling method.

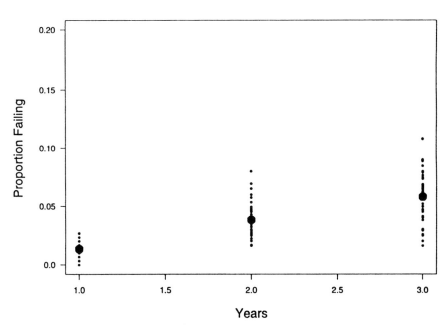

Figure 9.11. Plot of $\widehat{F}(t)$ (large dots) and $\widehat{F}^*(t)$ values for the first 50 of the $B = 10{,}000$ bootstrap samples (small dots) for the pooled-data heat exchanger tube example.

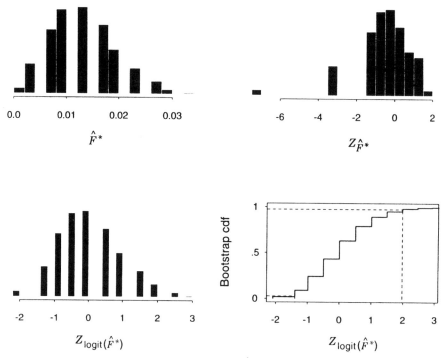

Figure 9.12. Nonparametric bootstrap distributions of $\widehat{F}^*(t_i)$, $Z_{\widehat{F}^*}$, and $Z_{\text{logit}(\widehat{F}^*)}$ at $t_i = 1$ year for $B = 10{,}000$ bootstrap samples from the pooled heat exchanger tube data. There were 167 occurrences of $\widehat{F}^*(1) = 0$ that were adjusted to $.5/n_i$, where $n_1 = 300$ is the size of the risk set at the beginning of year 1.

are approximately continuous outside the tails of the sample data (because the number of different possible outcomes in the bootstrap sampling process is large). In the lower tail of the observed sample (i.e., at smaller t_i values), however, the bootstrap distribution can be far from continuous. As we will see in the examples, the standard bootstrap methods are not useful at such points on the distribution; plotting the observed bootstrap distribution identifies potential problems of this kind.

Example 9.9 Bootstrap Sampling Distributions from the Heat Exchanger Tube Data. After 1 year of operation, there were only 4 failed tubes across the heat exchangers. As shown in Figures 9.11 and 9.12 the number of different outcomes in the bootstrap sampling at 1 year is small. At 2 years, the number of possible outcomes is large enough for the distribution to be approximately continuous. □

9.5.3 Distribution of Bootstrap Statistics

The accuracy of bootstrap approximation confidence intervals, like the normal-approximation intervals, will, in general, depend on the transformation used in the procedure. To allow use of the logit transformation and plotting of the bootstrap estimates one can adjust any $\widehat{F}^*(t_i) = 0$ to $\widehat{F}^*(t_i) = .5/n_i$ and any $\widehat{F}^*(t_i) = 1$ to $\widehat{F}^*(t_i) = (n_i - .5)/n_i$, where n_i is the size of the risk set at the beginning of interval i. These adjustments will have no effect on confidence intervals unless the frequency of such events exceeds $\alpha/2$ (in which case, the nonparametric bootstrap method should not be used anyway because of the extreme discreteness in the distribution).

Example 9.10 Distribution of Bootstrap Statistics for the Heat Exchanger Tube Data. Figure 9.12 gives histograms of the $B = 10{,}000$ values of $\widehat{F}^*(t_i)$, $Z_{\widehat{F}^*} = (\widehat{F}^* - \widehat{F})/\widehat{\text{se}}_{\widehat{F}^*}$, and $Z_{\text{logit}(\widehat{F}^*)} = [\text{logit}(\widehat{F}^*) - \text{logit}(\widehat{F})]/\widehat{\text{se}}_{\text{logit}(\widehat{F}^*)}$ at $t_i = 1$ year of operation. Also shown is the cumulative distribution of $Z_{\text{logit}(\widehat{F}^*)}$ indicating, with the dashed lines, the .025 and .975 quantiles of the distribution of $Z_{\text{logit}(\widehat{F}^*)}$. Numerically these quantiles are $z_{\text{logit}(\widehat{F}^*)(.025)} = -1.394$ and $z_{\text{logit}(\widehat{F}^*)(.975)} = 1.972$, respectively. Figure 9.13 does the same for $t_i = 2$ years of operation. Because of the relatively small number of possible outcomes at $t_i = 1$, the bootstrap distributions there are far from continuous. At $t_i = 2$, however, the distribution has the appearance of being approximately continuous. Corresponding plots for $t_i = 3$ were very similar to the

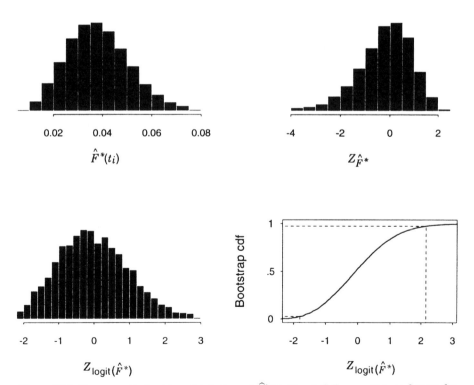

Figure 9.13. Nonparametric bootstrap distributions of $\widehat{F}^*(t_i)$, $Z_{\widehat{F}^*}$, and $Z_{\text{logit}(\widehat{F}^*)}$ at $t_i = 2$ years for $B = 10{,}000$ bootstrap samples from the pooled heat exchanger tube data. There were 3 occurrences of $\widehat{F}^*(2) = 0$ that were adjusted to $.5/n_2$, where $n_2 = 197$ is the size of the risk set at the beginning of year 2.

plots for $t_i = 2$. Comparing across the three time points, the bootstrap distribution of $Z_{\text{logit}(\widehat{F}^*)}$ is much more stable [i.e., has a more consistent shape and spread for different values of $F(t_i)$] than that of $Z_{\widehat{F}^*}$. For this reason, a procedure based on the $Z_{\text{logit}(\widehat{F})} \overset{\cdot}{\sim} Z_{\text{logit}(\widehat{F}^*)}$ can be expected to provide better approximate confidence intervals. □

9.5.4 Pointwise Nonparametric Bootstrap Confidence Intervals for $F(t_i)$

To obtain nonparametric bootstrap confidence intervals for $F = F(t_i)$ at t_i, we modify the approach in Section 3.6.3, substituting bootstrap estimates of the quantiles of the distribution of $Z_{\text{logit}(\widehat{F}^*)}$ in place of the NOR(0, 1) quantiles. Specifically, a two-sided approximate $100(1 - \alpha)\%$ bootstrap confidence interval based on $Z_{\text{logit}(\widehat{F})} \overset{\cdot}{\sim} Z_{\text{logit}(\widehat{F}^*)}$ is

$$[\underline{F}(t_i), \widetilde{F}(t_i)] = \left[\frac{\widehat{F}}{\widehat{F} + (1 - \widehat{F}) \times \underline{w}}, \frac{\widehat{F}}{\widehat{F} + (1 - \widehat{F}) \times \widetilde{w}} \right], \quad (9.4)$$

where

$$\underline{w} = \exp\{z_{\text{logit}(\widehat{F}^*)_{(1-\alpha/2)}} \widehat{\text{se}}_{\widehat{F}} / [\widehat{F}(1 - \widehat{F})]\} \text{ and } \widetilde{w} = \exp\{z_{\text{logit}(\widehat{F}^*)_{(\alpha/2)}} \widehat{\text{se}}_{\widehat{F}} / [\widehat{F}(1 - \widehat{F})]\}.$$

This formula is similar to (3.16), with an important difference. For the bootstrap interval, there are separate values of w for the upper and lower tails of the distribution. This is because the distribution of $Z_{\text{logit}(\widehat{F}^*)}$, unlike the NOR(0, 1) distribution, is not symmetric.

Example 9.11 *Nonparametric Bootstrap Confidence Interval for the Heat Exchanger Tube Time-to-Crack cdf.* Consider estimating the fraction of heat exchanger tubes that crack after 1 year of operation (i.e., $t_i = 1$). Using the quantiles of the distribution of $Z_{\text{logit}(\widehat{F}^*)}$, previous results from Example 3.7, and substituting into (9.4), we find

$$[\underline{F}(t_i), \widetilde{F}(t_i)] = \left[\frac{.0133}{.0133 + (1 - .0133) \times \underline{w}}, \frac{.0133}{.0133 + (1 - .0133) \times \widetilde{w}} \right]$$
$$= [.0050, .0266],$$

where

$$\underline{w} = \exp\{1.972(.00662)/[.0133(1 - .0133)]\} = 2.704,$$
$$\widetilde{w} = \exp\{-1.394(.00662)/[.0133(1 - .0133)]\} = .4950.$$

Nonparametric bootstrap confidence intervals based on $Z_{\widehat{F}} \overset{\cdot}{\sim} Z_{\widehat{F}^*}$ are computed similarly.

Table 9.3 gives the numerical results for bootstrap and normal-approximation methods for $t_e = 1, 2, 3$. Comparison shows that the $Z_{\text{logit}(\widehat{F})} \overset{\cdot}{\sim} \text{NOR}(0, 1)$ and $Z_{\text{logit}(\widehat{F})} \overset{\cdot}{\sim} Z_{\text{logit}(\widehat{F}^*)}$ methods are not too different. Because it is based on a more

Table 9.3. Results of Calculations for Nonparametric Approximate Confidence Intervals for F and Comparison with Normal-Approximation Intervals for the Heat Exchanger Data

Year	t_e	$\widehat{F}(t_e)$	$\widehat{\mathrm{se}}_{\widehat{F}}$	Pointwise Confidence Intervals
(0–1]	1	.0133	.00662	
Approximate 95% Confidence Intervals for $F(1)$				
Based on $Z_{\mathrm{logit}(\widehat{F})} \stackrel{.}{\sim} \mathrm{NOR}(0,1)$				[.0050, .0350]
Based on $Z_{\widehat{F}} \stackrel{.}{\sim} \mathrm{NOR}(0,1)$				[.0003, .0263]
Based on $Z_{\mathrm{logit}(\widehat{F})} \stackrel{.}{\sim} Z_{\mathrm{logit}(\widehat{F}^*)}$				[.0050, .0266]
Based on $Z_{\widehat{F}} \stackrel{.}{\sim} Z_{\widehat{F}^*}$				[.0038, .0332]
(1–2]	2	.0384	.0128	
Approximate 95% Confidence Intervals for $F(2)$				
Based on $Z_{\mathrm{logit}(\widehat{F})} \stackrel{.}{\sim} \mathrm{NOR}(0,1)$				[.0198, .0730]
Based on $Z_{\widehat{F}} \stackrel{.}{\sim} \mathrm{NOR}(0,1)$				[.0133, .0635]
Based on $Z_{\mathrm{logit}(\widehat{F})} \stackrel{.}{\sim} Z_{\mathrm{logit}(\widehat{F}^*)}$				[.0188, .0702]
Based on $Z_{\widehat{F}} \stackrel{.}{\sim} Z_{\widehat{F}^*}$				[.0171, .0770]
(2–3]	3	.0582	.0187	
Approximate 95% Confidence Intervals for $F(3)$				
Based on $Z_{\mathrm{logit}(\widehat{F})} \stackrel{.}{\sim} \mathrm{NOR}(0,1)$				[.0307, .1076]
Based on $Z_{\widehat{F}} \stackrel{.}{\sim} \mathrm{NOR}(0,1)$				[.0216, .0949]
Based on $Z_{\mathrm{logit}(\widehat{F})} \stackrel{.}{\sim} Z_{\mathrm{logit}(\widehat{F}^*)}$				[.0302, .1097]
Based on $Z_{\widehat{F}} \stackrel{.}{\sim} Z_{\widehat{F}^*}$				[.0282, .1168]

precise evaluation, the bootstrap interval procedure based on $Z_{\mathrm{logit}(\widehat{F})} \stackrel{.}{\sim} Z_{\mathrm{logit}(\widehat{F}^*)}$ can be expected to have better coverage properties.

For this example, the observed frequencies of $\widehat{F}^* = 0$ were much less than $\alpha/2 = .025$ for all values of t_i and so the continuity adjustment described in Section 9.5.3 has no effect on the confidence intervals. □

Example 9.12 Bootstrap Confidence Intervals for $F(t)$ from the Shock Absorber Data. The application of the bootstrap method for this example is similar to the approach used in Section 9.5.1. There are, however, important differences in the character of the bootstrap distributions because the shock absorber failure times were reported as exact failures (i.e., not grouped into inspection intervals). Thus we could take as t_e, any particular time of interest out to 28,100 km, the last time of observation.

Again, $B = 10{,}000$ bootstrap samples were generated. For each of the B bootstrap samples, nonparametric bootstrap estimates $\widehat{F}^*$ and $\widehat{\text{se}}_{\widehat{F}^*}$ were computed. Figure 9.14 shows $\widehat{F}(t_e)$ as thick solid lines and $\widehat{F}^*(t_e)$ for the first 50 of the $B = 10{,}000$ bootstrap samples as thin dashed lines. Note the discreteness in the $\widehat{F}^*(t_e)$ bootstrap estimates at the earlier points in time. This figure alone gives a sense of the sampling variability in both $\widehat{F}^*(t_e)$ and $\widehat{F}(t_e)$.

Figures 9.15, 9.16, and 9.17 show more detail at three particular points in time (time of the 1st, 3rd, and 7th failures). In this sequence of figures we see the bootstrap distributions of $\widehat{F}^*(t_e)$ and $Z_{\text{logit}(\widehat{F}^*)}$ become progressively closer to continuous as we move from $t = 6700$ to $t = 20{,}100$. The distribution of the untransformed $Z_{\widehat{F}^*}$ is somewhat more erratic, again indicating the importance of using a transformation to make the distribution of the Z statistic more stable over the possible values of the underlying model.

Figure 9.18 is similar to Figure 3.6 but also shows the set of approximate 95% pointwise bootstrap confidence intervals for $F(t_e)$ as dotted lines, to compare with the normal-approximation intervals shown with dashed lines. □

Because of the highly discrete distribution for the first two segments, the bootstrap confidence intervals there are not credible. Beyond this point, however, the differences

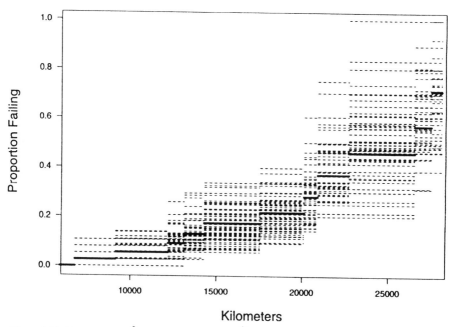

Figure 9.14. Nonparametric $\widehat{F}(t)$ (thick solid line) and $\widehat{F}^*(t)$ for 50 of the $B = 10{,}000$ bootstrap samples from the shock absorber data (thin dashed lines).

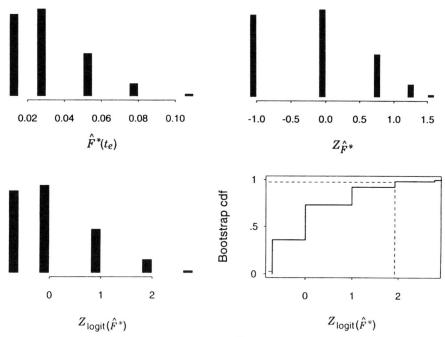

Figure 9.15. Nonparametric bootstrap distributions of $\widehat{F}^*(t_e)$, $Z_{\widehat{F}^*}$, and $Z_{\text{logit}(\widehat{F}^*)}$ at $t_e = 6700$ km (the 1st failure time) for $B = 10{,}000$ bootstrap samples from the shock absorber data. There were 3635 occurrences of $\widehat{F}^*(t_e) = 0$ that were adjusted to $.5/n_1$.

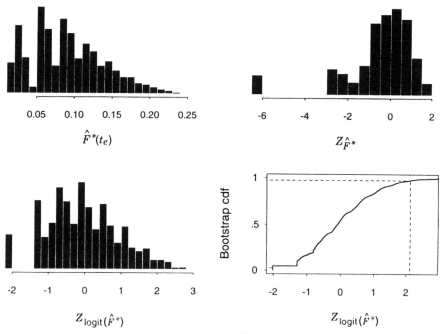

Figure 9.16. Nonparametric bootstrap distributions of $\widehat{F}^*(t_e)$, $Z_{\widehat{F}^*}$, and $Z_{\text{logit}(\widehat{F}^*)}$ at $t_e = 12{,}200$ km (the 3rd failure time) for $B = 10{,}000$ bootstrap samples from the shock absorber data. There were 446 occurrences of $\widehat{F}^*(t_e) = 0$ that were adjusted to $.5/n_3$.

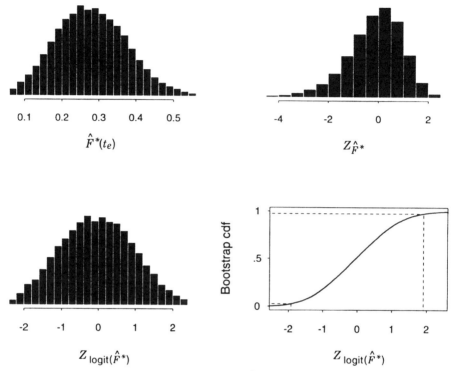

Figure 9.17. Nonparametric bootstrap distributions of $\widehat{F}^*(t_e)$, $Z_{\widehat{F}^*}$, and $Z_{\text{logit}(\widehat{F}^*)}$ at $t_e = 20{,}100$ km (the 7th failure time) for $B = 10{,}000$ bootstrap samples from the shock absorber data. There was one occurrence of $\widehat{F}^*(t_e) = 0$ that was adjusted to $.5/n_7$.

between the two methods are not large. The bootstrap intervals are, for some time points, narrower (this will not be true in general, however) and sometimes shifted up somewhat. The important advantage of bootstrap intervals is that, except for the first few segments, they provide intervals based on approximations that are generally better than the simpler normal-approximation approach.

9.6 PERCENTILE BOOTSTRAP METHOD

Sections 9.3–9.5 describe and illustrate the use of the "bootstrap-t" method. When it is not easy to compute the standard error for an estimate, the percentile bootstrap, as described in Efron and Tibshirani (1993), provides a simple useful alternative. In its simplest form, the percentile method uses appropriate percentiles of the bootstrap distribution to define the confidence interval. Suppose that $\widehat{\theta}_j^*$, $j = 1, \ldots, B$, is the bootstrap sample for a parameter θ. Then the $100(1 - \alpha)\%$ percentile bootstrap interval for θ is

$$\left[\underset{\sim}{\theta},\ \widetilde{\theta}\right] = \left[\widehat{\theta}_{[l]}^*,\ \widehat{\theta}_{[u]}^*\right],$$

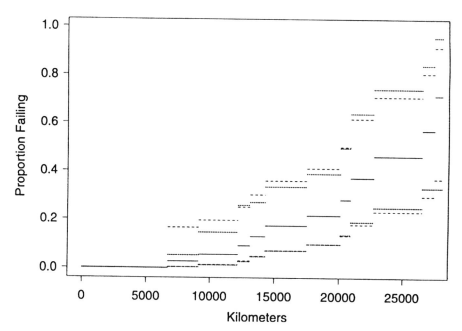

Figure 9.18. Plot of the nonparametric estimate for the shock absorber data comparing the normal approximation $Z_{\text{logit}(\widehat{F})} \overset{\cdot}{\sim} \text{NOR}(0, 1)$ (dashed lines) from Chapter 3 and the bootstrap $Z_{\text{logit}(\widehat{F})} \overset{\cdot}{\sim} Z_{\text{logit}(\widehat{F}^*)}$ (dotted lines) pointwise 95% approximate confidence intervals for $F(t_e)$.

where $\widehat{\theta}^*_{[j]}$, $j = 1, \ldots, B$, is the bootstrap sample ordered from smallest to largest, $l = B \times (\alpha/2)$ and $u = B \times (1 - \alpha/2)$. Generally l and u would be rounded to the next lowest and next highest integer, respectively. Alternatively, B can be chosen so that l and u are integers.

The percentile bootstrap method has the advantage that the interval provided by the procedure does not depend on the transformation scale of the parameter (this property is called "transformation preserving" in Efron and Tibshirani, 1993). In small samples the percentile method is generally not as accurate as the bootstrap-t with an appropriate transformation. Efron and Tibshirani (1993) give methods for adjusting the percentile method for bias and nonconstant variance, generally providing accuracy comparable with the bootstrap-t method. Section 13.7 gives a detailed algorithm for the bias-corrected percentile method and an example of its use. In Section 21.5 the percentile method is used to obtain confidence intervals for accelerated degradation test analysis.

BIBLIOGRAPHIC NOTES

With complete data or Type II censoring (including progressive censoring) and location-scale distributions, the distribution of the t-like statistics, such as $Z_{\widehat{\mu}}$, $Z_{\log(\widehat{\sigma})}$, and $Z_{\log(\widehat{t}_{.1})}$, depend only on the sample size and the fixed number of failures. This implies that the parametric bootstrap approach, using the bootstrap sampling approach illustrated in Figure 9.1, provides confidence intervals that have exactly the nominal level of confidence. This approach to computing confidence intervals is described, for example, in Robinson (1983). For Type I censoring, in general, "exact" methods do not exist.

Bootstrap methods for obtaining confidence intervals have become extremely popular in recent years. The early work of Efron (e.g., Efron, 1982) generated a large amount of research activity. The books by Hall (1992) and Efron and Tibshirani (1993) describe and illustrate most of the important theory and methods. Shao and Tu (1995) is another important reference providing theory on which bootstrap methods are based. Nonparametric methods for censored data are described in Efron (1981) and Akritas (1986).

Jeng and Meeker (1998) describe a large simulation that was used to compare various methods of constructing confidence bounds for Weibull distribution parameters and quantiles based on Type I censored data (with Type II censored data, the parametric bootstrap method has exact coverage probabilities). Their conclusions indicate that simulation-based methods, for most situations, provide important improvements over normal-approximation methods, especially when interest is primarily on one side or the other of an interval, as is generally the case in reliability applications.

An appendix of Efron and Tibshirani (1993) provides S-PLUS functions for computing bootstrap intervals. If S-PLUS is not available or for special problems, the needed computations can be programmed by the analyst. Buckland (1985) provides a Fortran algorithm. Other programming languages or a high-level programming environment can be used, but efficiency may suffer with simulations programmed largely in interpretive languages.

EXERCISES

▲9.1. Consider the bootstrap sampling distribution described in Example 9.12. Consider the bootstrap distribution of events at the point $t_e = 6700$ km (the 1st failure time). Refer to Figure 9.15.

(a) The number of observed failures before $t_e = 6700$ km in a bootstrap sample has a binomial distribution. What are the parameters of this distribution?

(b) List all of the possible outcomes in the sample space of part (a) and compute the probability associated with each event. This will require the use of a computer package for computing binomial probabilities.

(c) Use the results in parts (a) and (b) to compute the true distribution of $\widehat{F}^*(t_e)$, $Z_{\widehat{F}^*}$, and $Z_{\text{logit}(\widehat{F}^*)}$ at $t_e = 6700$. (As we did in these examples, substitute $\widehat{F}^* = .5/38$ for the event of 0 failures.) Compare your answers with the histograms in Figure 9.15.

▲9.2. Refer to Exercise 9.1. Do the same things for $t_e = 9120$ km (the 2nd failure time). Note that for this exercise a generalization of the binomial distribution is

EXERCISES 229

required. This is because it is important to take into consideration the number of censored observations between $t_e = 6700$ and $t_e = 9120$.

◆9.3. Write a computer program to do the following, using the shock absorber data from Example 9.4 to illustrate the program's use:
(a) Sample with replacement to obtain a sample of size $n = 38$ shock absorbers from the 38 rows of Appendix Table C.2.
(b) Compute the Weibull distribution ML bootstrap estimates $(\widehat{\mu}^*, \widehat{\sigma}^*)$ for this sample.
(c) Compute the bootstrap studentized quantile estimate $Z_{\log(\widehat{t}_{.1}^*)} = [\log(\widehat{t}_{.1}^*) - \log(\widehat{t}_{.1})]/\widehat{se}_{\log(\widehat{t}_{.1}^*)}$.
(d) Repeat steps (a) to (c) 2000 times to get 2000 values of $Z_{\log(\widehat{t}_{.1}^*)}$. Make a histogram of these values to see the bootstrap distribution of $Z_{\log(\widehat{t}_{.1}^*)}$. Find the .025 and .975 quantiles of this distribution.
(e) Use the results in part (d) to compute a bootstrap confidence interval for $t_{.1}$. Compare with the interval in Table 9.2. Why are they not exactly the same? What could be done to assure better agreement in such a comparison?

9.4. Explain the possible advantages and disadvantages of using the bootstrap sampling methods illustrated in Figures 9.1 and 9.2 when the goal is to obtain a confidence interval for the quantile of a Weibull distribution under the following conditions with n small (say, 10) and n large (say, 100):
(a) A complete sample of n failure times.
(b) A Type II sample consisting of $r = n/2$ failures out of n units.
(c) A Type I censored sample, starting with n units and censored at a point in time close to the median of the distribution.
(d) A randomly censored sample, starting with n units where units are censored at random points in time such that about 50% of the units will be observed to fail and the others will be censored.

▲9.5. Let $t_{(1)} < \cdots < t_{(r)}$ denote the observed failure times from a life test that was censored at time t_c. Suppose that the underlying failure-time distribution is a log-location-scale distribution so that $T \sim \Phi\{[\log(t) - \mu]/\sigma\}$, where Φ is a standardized location-scale cdf. Suppose that $1 \le r \le n$ (at least one failure) and let $\widehat{\mu}$ and $\widehat{\sigma}$ be the ML estimators of the parameters. Show that the distributions of $\widehat{\sigma}/\sigma$, $(\widehat{\mu} - \mu)/\sigma$, and $(\widehat{\mu} - \mu)/\widehat{\sigma}$ depend only on the cdf Φ, the sample size n, and the proportion of censoring, $q_c = 1 - \Phi\{[\log(t_c) - \mu]/\sigma\}$.

This can be accomplished with the following steps (as illustrated for the distribution of $\widehat{\sigma}/\sigma$).
(a) Denote by R the random number of failures in the life test. Show that conditional on having one or more failures the distribution of R is

$$\Pr(R = r) = \frac{\binom{n}{r}(1 - q_c)^r q_c^{n-r}}{1 - q_c^n}, \quad r = 1, \ldots, n.$$

(b) Show that conditional on having one or more failures the distribution of $\hat{\sigma}/\sigma$ is

$$\Pr\left(\frac{\hat{\sigma}}{\sigma} \leq v\right) = \sum_{r=1}^{n} \Pr\left[\left(\frac{\hat{\sigma}}{\sigma} \leq v\right) \text{ and } (R = r)\right]$$

$$= \sum_{r=1}^{n} \Pr\left[\left(\frac{\hat{\sigma}}{\sigma} \leq v\right) \middle| R = r\right] \Pr(R = r).$$

(c) Write the likelihood for the censored sample of r failures and show that for each fixed r ($r = 1, \ldots, n$) and n, the ML estimator is a function of $t_{(j)}$, $j = 1, \ldots, r$, and t_c, say, $\hat{\sigma} = H_{n:r}[\log(t_{(1)}), \ldots, \log(t_{(r)}), t_c]$.

(d) Use the location-scale property of the distribution of T to show that, for each fixed r and n, the estimator is "equivariant" in the sense that

$$(\hat{\sigma}/\sigma) = H_{n:r}[z_{(1)}, \ldots, z_{(r)}, q_c],$$

where $z_{(j)} = [\log(t_{(j)}) - \mu]/\sigma$ is the jth order statistic from a random variable with cdf, $\Phi(z)/(1 - q_c)$, $z \leq [\log(t_c) - \mu]/\sigma$.

9.6. Refer to Figure 9.6. Using the ideas of the percentile bootstrap discussed in Section 9.6, determine, approximately, a 95% confidence interval for $t_{.2}$ of the shock absorber failure-time distribution.

9.7. Refer to Figure 9.6. Using the ideas of the percentile bootstrap discussed in Section 9.6, determine, approximately, a 95% confidence interval for $F(10000)$ of the shock absorber failure-time distribution.

CHAPTER 10

Planning Life Tests

Objectives

This chapter explains:

- Basic ideas for planning a life test or field-tracking study.
- The use of simulation to get an indication of how the results of a life test or other study might look, to see how the data might be analyzed, and to get a rough idea of the precision that can be expected for a proposed test plan.
- The use of large-sample approximate methods to assess the precision of the results that will be obtained from a future reliability study.
- How to determine an approximate sample size that provides a specified degree of precision.
- How to assess the trade-offs involving sample size and study length.
- The use of simulation to check and "calibrate" the easier-to-use large-sample approximate methods.
- Methods for assessing sensitivity of test planning conclusions to unknown inputs that must be provided.

Overview

This chapter provides tools for evaluating and controlling estimation precision for a life test when censored data are expected. For those interested primarily in data analysis methods, this chapter can be skipped. Section 10.1 introduces the basic ideas of test planning and uses simulation to illustrate and explain the effect that sample size has on sampling variability. Simulation is an extremely important tool for test planning. Section 10.2 shows how to compute approximate sampling variability directly. Sections 10.3 and 10.4 show how to find the sample size needed to control sampling variability (or precision) and illustrate the ideas for the normal and exponential distribution. Section 10.5 applies these methods to problems involving Type I censored data with the Weibull and lognormal distributions. Section 10.6 describes methods for planning a test to demonstrate conformance with a specified reliability standard. Section 10.7 describes some extensions to other types of censoring and related sample size problems.

10.1 INTRODUCTION

10.1.1 Basic Ideas

Because life tests and reliability field-tracking studies are expensive, it is essential to plan them carefully. Frequently asked questions include "How many units do I need to test in order to estimate the .1 quantile of life?" or "How long do I need to run the life test?" Simply put, more test units and more test time will generate more information, which improves the precision of estimates. Precision and other test plan properties depend, however, on the actual model and its parameters. In order to describe the kind of results that might be expected from a particular test plan, it is necessary to have some "planning information" about the life distribution to be estimated. Having such information makes it possible to assess the effect that sample size and test length will have on the outcome of a particular test plan. Such planning information is typically obtained from design specifications, expert opinion, or previous experience with similar products or materials. Here the superscript □ is used to denote a planning value of a population or process quantity.

Example 10.1 Engineering "Planning Values" and Assumed Distribution for Planning an Insulation Life Test. A manufacturer wants to estimate the .1 quantile of the life distribution of a newly developed insulation. Tests are run on small specimens and at higher than usual electrical stress (specified in kilovolts/mm) to cause failures to occur sooner. The amount of time available for the life test is 1000 hours. Engineering has provided the following information in order to help plan the life test.

- They expect that about 12% of the specimens will fail in the first 500 hours of the test and about 20% of the specimens will fail by the end of 1000 hours (i.e., the proportion failing at the censoring time should be in the neighborhood of $p_c^\square = .2$).
- For purposes of test planning, the engineers will use a Weibull distribution to describe failure-time variability, but they also want to make an assessment using the lognormal distribution (they would be concerned if the answers differ too much).

Substituting the above planning information into the Weibull distribution cdf (4.7) and solving for μ and σ provides "planning values" $\mu^\square = 8.774$ [or $\eta^\square = \exp(8.774) = 6464$ hours], and $\sigma^\square = 1.244$ (or $\beta^\square = 1/1.244 = .8037$). A simple way of getting these values graphically is to plot the two planning failure probability points (500, .12) and (1000, .2) on probability paper, draw a straight line through the points, and read off parameter planning values, as discussed in Section 6.6.3. Figure 10.1 illustrates this for the Weibull distribution. Similarly, using the lognormal distribution cdf (4.4) gives $\mu^\square = 8.658$ and $\sigma^\square = 2.079$. □

INTRODUCTION

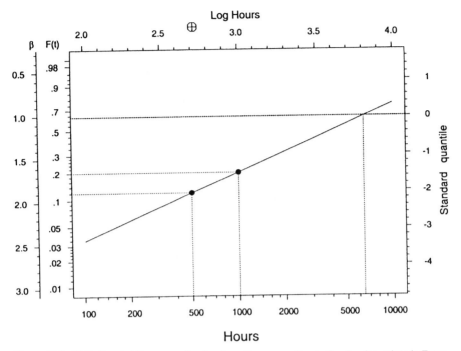

Figure 10.1. Weibull probability paper showing the cdf corresponding to the planning values in Example 10.1.

10.1.2 Simulation of a Proposed Test Plan

Simulation provides a powerful, insightful tool for planning experiments. The following steps outline a useful simulation method for helping to plan a life test.

- Use the chosen model and planning values of the distribution parameters to simulate data from the proposed life test.
- Analyze the data, perhaps fitting more than one distribution.
- Assess precision of estimates. This can be done initially by computing approximate confidence intervals, as would be done for the real data.
- Simulate and fit distributions to many samples to assess the sample-to-sample differences. Such multiple simulations provide an assessment of estimation precision. This assessment does not depend on the usual large-sample approximations.
- Repeat the simulation–evaluation process with different sample sizes to gauge the actual sample size and test length requirements to achieve the desired precision.
- Repeat the simulation–evaluation process with different input "planning values" over the range of their uncertainty.

Example 10.2   Illustration of Simulations of Insulation Life Tests. Figures 10.2, 10.3, and 10.4 show plots of ML estimates obtained from 30 simulated samples of size $n = 5$, $n = 50$, and $n = 500$, respectively, from a Weibull distribution with $\mu = 8.774$ and $\sigma = 1.244$ (shown with the thicker, longer line). The dashed vertical line at $t_c = 1000$ indicates the censoring time (end of the test). The horizontal line at $p = .1$ provides a better visualization of the distribution of estimates of $t_{.1}$. These graphs illustrate a number of interesting and important points about the effect that sample size will have on our ability to make inferences. In particular:

- For the $n = 5$ estimates in Figure 10.2, there is enormous variability in the ML estimates. In fact, for 13 of the simulated data sets, there are no ML estimates because all units were censored [the probability of all units being censored in a sample of $n = 5$ units is $(1 - .2)^5 = .328$]. This is a strong indication that the usual large-sample approximate confidence intervals, in this situation, will be seriously inadequate. The standard deviation of the 17 values of $\log(\widehat{t}_{.1})$ (for those samples that had 1 or more failures) was 1.36.
- The estimates for $n = 50$ in Figure 10.3 indicate much more accurate estimation. The spread in the ML estimates of $t_{.1}$ might be small enough for some applications. The standard deviation of the 30 values of $\log(\widehat{t}_{.1})$ was .408.

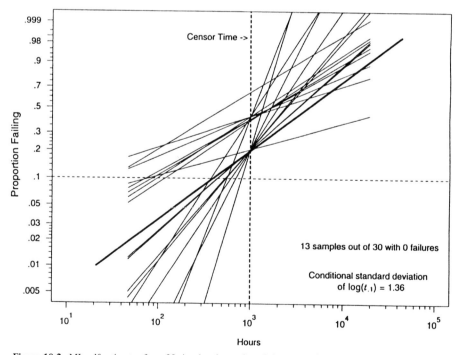

Figure 10.2. ML cdf estimates from 30 simulated samples of size $n = 5$ from a Weibull distribution with $\mu^\square = 8.774$ and $\sigma^\square = 1.244$ (shown with the thicker, longer line).

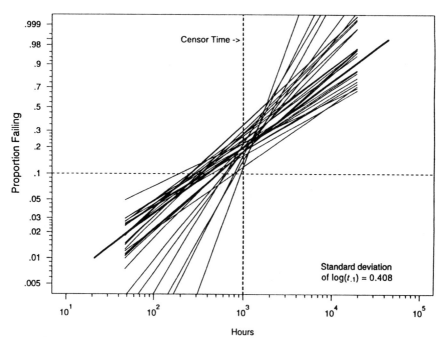

Figure 10.3. ML cdf estimates from 30 simulated samples of size $n = 50$ from a Weibull distribution with $\mu^\square = 8.774$ and $\sigma^\square = 1.244$ (shown with the thicker, longer line).

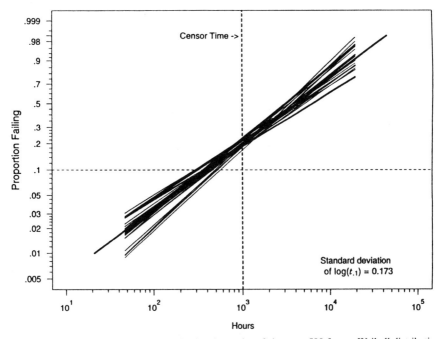

Figure 10.4. ML cdf estimates from 30 simulated samples of size $n = 500$ from a Weibull distribution with $\mu^\square = 8.774$ and $\sigma^\square = 1.244$ (shown with the thicker, longer line).

- As shown in Figure 10.4, increasing the sample size to $n = 500$ provides a substantial reduction in sampling variability. The standard deviation of the 30 values of $\log(\widehat{t}_{.1})$ was .173.

In general, simulation is an easy, useful method of assessing variability. □

To control the standard deviation of an estimator to a specified degree of precision, it is possible to interpolate among simulated values at different sample sizes. Statistical theory (e.g., Section 10.2.2) tells us that the sampling variance of estimators is approximately proportional to $1/n$, where n is the sample size. This suggests a linear relationship between $1/\sqrt{n}$ and the standard deviation. Section 10.2 suggests a more direct approach to controlling precision.

10.1.3 Uncertainty in Planning Values

Life test planning requires specification of a model and "planning values" for the model parameters. Of course, these model parameters are generally unknown and, indeed, this is usually the reason for conducting the life test. Typically, however, planning values can be obtained from some combination of experience with similar products, design specifications, and engineering judgment. As suggested in Section 10.1.2, one could evaluate plans over a range of planning values. An alternative would be to use a Bayesian approach by specifying a prior probability distribution to describe the uncertainty in the unknown model parameters, in effect, averaging over the plans suggested by the prior distribution.

10.2 APPROXIMATE VARIANCE OF ML ESTIMATORS

10.2.1 Motivation for Use of Large-Sample Approximations of Test Plan Properties

In contrast to the use of simulation for assessing properties of proposed test plans, large-sample approximations provide:

- Simple expressions that allow one to compute directly the approximate precision of a specified estimator as a function of sample size.
- Simple approximate expressions for the needed sample size as a function of the specified precision of an estimator.
- Simple tables or graphs of variance factors that provide insight and allow easy assessments of trade-offs in test planning decisions (e.g., sample size and test length).

The remainder of this chapter describes the basic ideas behind the approximate large-sample formulas, illustrates their use in the development of simple-to-use figures for test planning, and shows how to fine-tune test plans by using simulation methods.

10.2.2 Basic Large-Sample Approximations

This section summarizes some important ideas given in more detail in Appendix Section B.6.1. Under standard regularity conditions (see Appendix Section B.4), for a model with parameters $\boldsymbol{\theta} = (\theta_1, \ldots, \theta_k)$, the following results hold approximately for large samples:

- ML estimators $\widehat{\boldsymbol{\theta}}$ follow a multivariate normal distribution with mean vector $\boldsymbol{\theta}$ and covariance matrix $\Sigma_{\widehat{\boldsymbol{\theta}}}$ [abbreviated $\widehat{\boldsymbol{\theta}} \stackrel{.}{\sim} \text{MVN}(\boldsymbol{\theta}, \Sigma_{\widehat{\boldsymbol{\theta}}})$].
- The large-sample approximate covariance matrix can be computed from $\Sigma_{\widehat{\boldsymbol{\theta}}} = I_{\boldsymbol{\theta}}^{-1}$, where

$$I_{\boldsymbol{\theta}} = \text{E}\left[-\frac{\partial^2 \mathcal{L}(\boldsymbol{\theta})}{\partial \boldsymbol{\theta}\, \partial \boldsymbol{\theta}'}\right] = \sum_{i=1}^{n} \text{E}\left[-\frac{\partial^2 \mathcal{L}_i(\boldsymbol{\theta})}{\partial \boldsymbol{\theta}\, \partial \boldsymbol{\theta}'}\right] \qquad (10.1)$$

is the Fisher information matrix. Recall from Section 7.3.3 that more curvature in the log likelihood implies more precision for estimation. The actual amount of curvature at the maximum of a likelihood function is, in general, random, depending on the sample data. The matrix $I_{\boldsymbol{\theta}}$ can be viewed as the expected amount of curvature in the sample log likelihood function at its maximum. The inverse of this curvature matrix provides the approximate large-sample covariance matrix that, given the model and "planning values" for $\boldsymbol{\theta}$, can be used for test planning.

In most practical problems, interest will center on one or more scalar functions of the parameters, say, $g = g(\boldsymbol{\theta})$. Then, in large samples, the distribution of $\widehat{g} = g(\widehat{\boldsymbol{\theta}})$ can be approximated by a normal distribution, $\widehat{g} \stackrel{.}{\sim} \text{NOR}[g(\boldsymbol{\theta}), \text{Ase}(\widehat{g})]$, where, from the delta method (see Appendix Sections B.2 and B.6.3)

$$\text{Avar}(\widehat{g}) = \left[\frac{\partial g(\boldsymbol{\theta})}{\partial \boldsymbol{\theta}}\right]' \Sigma_{\widehat{\boldsymbol{\theta}}} \left[\frac{\partial g(\boldsymbol{\theta})}{\partial \boldsymbol{\theta}}\right]. \qquad (10.2)$$

When the function $g(\boldsymbol{\theta})$ is *positive* for all $\boldsymbol{\theta}$, then it is generally better to use an alternate form of the delta method approximation in which

$$\log[g(\widehat{\boldsymbol{\theta}})] \stackrel{.}{\sim} \text{NOR}\{\log[g(\boldsymbol{\theta})], \text{Ase}[\log(\widehat{g})]\},$$

where

$$\text{Avar}[\log(\widehat{g})] = \left(\frac{1}{g}\right)^2 \text{Avar}(\widehat{g}).$$

The approximate standard errors for $\widehat{g}$ and $\log(\widehat{g})$ are, respectively,

$$\text{Ase}(\widehat{g}) = \frac{1}{\sqrt{n}}\sqrt{V_{\widehat{g}}} \quad \text{and} \quad \text{Ase}[\log(\widehat{g})] = \frac{1}{\sqrt{n}}\sqrt{V_{\log(\widehat{g})}}$$

where the variance factors $V_{\widehat{g}} = n\text{Avar}(\widehat{g})$ and $V_{\log(\widehat{g})} = n\text{Avar}[\log(\widehat{g})]$ may (and usually do) depend on the actual value of $\boldsymbol{\theta}$ but they do *not* depend on n. Thus it is easy to choose n to control Ase($\cdot$). To compute $V_{\widehat{g}}$ and $V_{\log(\widehat{g})}$, one uses planning values $\boldsymbol{\theta}^{\square}$, as described in Example 10.1.

10.3 SAMPLE SIZE FOR UNRESTRICTED FUNCTIONS

When $-\infty < g(\boldsymbol{\theta}) < \infty$, an approximate $100(1 - \alpha)\%$ confidence interval for $g(\boldsymbol{\theta})$, using a reexpression of (8.9), is

$$[\underset{\sim}{g}, \widetilde{g}] = \widehat{g} \pm z_{(1-\alpha/2)}(1/\sqrt{n})\sqrt{\widehat{V}_{\widehat{g}}} = \widehat{g} \pm D, \quad (10.3)$$

where $z_{(p)}$ is the p quantile of the standard normal distribution and $\widehat{V}_{\widehat{g}}$ is $V_{\widehat{g}}$ evaluated at $\widehat{\boldsymbol{\theta}}$. The actual confidence interval *half-width* $D = (\widetilde{g} - \underset{\sim}{g})/2$ is a convenient measure of confidence interval precision. To compute the sample size needed for a specified degree of precision, let D_T denote a specified target value for D, replace $\widehat{V}_{\widehat{g}}$ by $V_{\widehat{g}}^{\square}$ in (10.3), and solve for n, giving

$$n = \frac{z_{(1-\alpha/2)}^2 V_{\widehat{g}}^{\square}}{D_T^2}, \quad (10.4)$$

where $V_{\widehat{g}}^{\square}$ is $V_{\widehat{g}}$ evaluated at $\boldsymbol{\theta}^{\square}$. To obtain n, one needs to specify $1 - \alpha$, D_T, t_c, and the planning values $\boldsymbol{\theta}^{\square}$ needed to compute $V_{\widehat{g}}^{\square}$.

Test plans with this sample size provide confidence intervals for $g(\boldsymbol{\theta})$ with the following characteristics:

- In repeated samples, approximately $100(1 - \alpha)\%$ of the intervals will contain $g(\boldsymbol{\theta})$.
- In repeated samples, $\widehat{V}_{\widehat{g}}$ is random because it depends on $\widehat{\boldsymbol{\theta}}$ (which depends on the sample data). If $\widehat{V}_{\widehat{g}} > V_{\widehat{g}}^{\square}$, then confidence interval width $2D$ is greater than the target $2D_T$.
- The probability that the realized interval width $2D$ is greater than the target width $2D_T$ is near .5.

Example 10.3 *Sample Size Needed to Estimate the Mean of Light Bulb Life.*
The life of some types of incandescent light bulbs can be modeled adequately with a normal distribution. Depending on the particular design, mean life might be on the order of 1000 hours with a standard deviation under 200 hours. To satisfy a request from marketing, it was desired to plan a life test that would estimate mean life of light bulbs so that a 95% confidence interval has a half-width that is approximately 30 hours. The product engineers are willing to assume that life is adequately described by a normal distribution with a standard deviation no larger than $\sigma^{\square} = 200$ hours and there is enough time to let all of the bulbs fail before analyzing the data.

From elementary statistics, $\widehat{\mu} = \bar{x}$ so $V_{\widehat{\mu}} = n\text{Var}(\bar{x}) = \sigma^2$ and $V_{\widehat{\mu}}^{\square} = (\sigma^{\square})^2 = (200)^2$. Substituting this and $D_T = 30$ into (10.4) shows that the number of bulbs needed is

$$n = \frac{z_{(1-\alpha/2)}^2 V_{\widehat{\mu}}^\square}{D_T^2} = \frac{(1.96)^2(200)^2}{30^2} \approx 171. \qquad \square$$

10.4 SAMPLE SIZE FOR POSITIVE FUNCTIONS

When $g(\boldsymbol{\theta}) > 0$ for all $\boldsymbol{\theta}$, using a reexpression of (7.11) and (8.11), an approximate $100(1 - \alpha)\%$ confidence interval for $\log[g(\boldsymbol{\theta})]$ is

$$\left[\log(\underset{\sim}{g}),\ \widetilde{\log}(g)\right] = \log(\widehat{g}) \pm (1/\sqrt{n})z_{(1-\alpha/2)}\sqrt{\widehat{V}_{\log(\widehat{g})}}$$
$$= \log(\widehat{g}) \pm \log(R).$$

Exponentiation yields a confidence interval $[\underset{\sim}{g},\ \widetilde{g}\,] = [\widehat{g}/R,\ \widehat{g}R\,]$ for g. Here

$$R = \exp\left[\frac{1}{\sqrt{n}} z_{(1-\alpha/2)}\sqrt{\widehat{V}_{\log(\widehat{g})}}\right] = \frac{\widehat{g}}{\underset{\sim}{g}} = \frac{\widetilde{g}}{\widehat{g}} = \sqrt{\widetilde{g}/\underset{\sim}{g}} \qquad (10.5)$$

is a convenient measure of confidence interval precision. Let R_T denote a target value for the precision factor R. Typical values for R are numbers like 1.2 and 1.5, indicating approximate expected deviation of 20% or 50%, respectively, between the estimate and the upper (or lower) confidence bound. Replacing $\widehat{V}_{\log(\widehat{g})}$ with $V_{\log(\widehat{g})}^\square$ in (10.5) and solving for sample size n gives

$$n = \frac{z_{(1-\alpha/2)}^2 V_{\log(\widehat{g})}^\square}{[\log(R_T)]^2}. \qquad (10.6)$$

Similar to the sample size formula for unrestricted functions of parameters, this sample size provides confidence intervals for $g(\boldsymbol{\theta})$ with the following characteristics:

- In repeated samples, approximately $100(1 - \alpha)\%$ of the intervals will contain $g(\boldsymbol{\theta})$.
- In repeated samples, $\widehat{V}_{\log(\widehat{g})}$ is random because it depends on $\widehat{\boldsymbol{\theta}}$ (which depends on the sample data). If $\widehat{V}_{\log(\widehat{g})} > V_{\log(\widehat{g})}^\square$ then $R = \sqrt{\widetilde{g}/\underset{\sim}{g}}$ will be greater than R_T.
- The realized precision factor R will be greater than the target R_T with a probability that is near .5.

Example 10.4 Sample Size Needed to Estimate the Mean of an Exponential Distribution for Insulation Life. A newly developed electrical insulation requires a life test to estimate the mean life of specimens at highly accelerated conditions. That is, the test will be run at higher than usual voltage to get failure information quickly. It is possible to use simultaneous testing of all units but the test must be completed in only 500 hours. Insulation engineers have been able to suggest a planning value of $\theta^\square = 1000$ hours. The experimenters need to choose the sample size to be large enough so that a 95% confidence interval will have endpoints that are approximately 50% away from the estimated mean (so $R_T = 1.5$).

From Section 7.6.3, the ML estimate for the exponential mean will be computed as $\widehat{\theta} = TTT/r$, where TTT is the total time on test and r is the number of failures. It follows, as a special case of (10.1) and (10.2), that the scaled asymptotic (large-sample approximate) variance of $\widehat{\theta}$ is

$$V_{\widehat{\theta}} = n\text{Avar}(\widehat{\theta}) = \frac{n}{\mathrm{E}\left[-\dfrac{\partial^2 \mathcal{L}(\theta)}{\partial \theta^2}\right]} = \frac{\theta^2}{1 - \exp\left(-\dfrac{t_c}{\theta}\right)}.$$

Then using the delta method,

$$V^{\square}_{\log(\widehat{\theta})} = \frac{V^{\square}_{\widehat{\theta}}}{(\theta^{\square})^2} = \frac{1}{1 - \exp\left(-\dfrac{500}{1000}\right)} = 2.5415.$$

Thus the number of needed specimens for the test is

$$n = \frac{z^2_{(1-\alpha/2)} V^{\square}_{\log(\widehat{\theta})}}{[\log(R_T)]^2} = \frac{(1.96)^2 (2.5415)}{[\log(1.5)]^2} \approx 60. \qquad \square$$

10.5 SAMPLE SIZES FOR LOG-LOCATION-SCALE DISTRIBUTIONS WITH CENSORING

10.5.1 Large-Sample Approximate Variance–Covariance Matrix for Location-Scale Parameters

This section specializes the computation of sample sizes to situations in which:

- T has a log-location-scale distribution with parameters (μ, σ).
- The life test is to be Type I right-censored at time t_c.

In this case, the large-sample approximate covariance matrix can be computed as

$$\Sigma_{(\widehat{\mu},\widehat{\sigma})} = \begin{bmatrix} \text{Avar}(\widehat{\mu}) & \text{Acov}(\widehat{\mu}, \widehat{\sigma}) \\ \text{Acov}(\widehat{\mu}, \widehat{\sigma}) & \text{Avar}(\widehat{\sigma}) \end{bmatrix} = \frac{1}{n} \begin{bmatrix} V_{\widehat{\mu}} & V_{(\widehat{\mu},\widehat{\sigma})} \\ V_{(\widehat{\mu},\widehat{\sigma})} & V_{\widehat{\sigma}} \end{bmatrix} = I^{-1}_{(\mu,\sigma)},$$

where $I_{(\mu,\sigma)}$ is the Fisher information matrix for (μ, σ). Appendix Table C.20 provides, for the lognormal/normal distributions, the following as a function of the standardized censoring time $\zeta_c = [\log(t_c) - \mu]/\sigma$:

- $100\Phi(\zeta_c)$, the population percentage failing by the standardized censoring time ζ_c.
- The scaled large-sample approximate variance–covariance factors $(1/\sigma^2)V_{\widehat{\mu}}$, $(1/\sigma^2)V_{\widehat{\sigma}}$, and $(1/\sigma^2)V_{(\widehat{\mu},\widehat{\sigma})}$.

- $\rho_{(\widehat{\mu},\widehat{\sigma})} = V_{(\widehat{\mu},\widehat{\sigma})}/\sqrt{V_{\widehat{\mu}}V_{\widehat{\sigma}}}$, the large-sample approximate correlation between the ML estimators $\widehat{\mu}$ and $\widehat{\sigma}$.
- The scaled Fisher information matrix elements f_{11}, f_{22}, and f_{12}. The scaled Fisher information matrix for a single observation from the corresponding location-scale distribution is

$$\frac{\sigma^2}{n} I_{(\mu,\sigma)} = \begin{bmatrix} f_{11} & f_{12} \\ f_{12} & f_{22} \end{bmatrix}.$$

- For σ given, the large-sample approximate variance factor for $\widehat{\mu}$ is $(1/\sigma^2)V_{\widehat{\mu}|\sigma} = (n/\sigma^2)\text{Avar}(\widehat{\mu}|\sigma) = 1/f_{11}$. For μ given, the factor for $\widehat{\sigma}$ is $(1/\sigma^2)V_{\widehat{\sigma}|\mu} = (n/\sigma^2)\text{Avar}(\widehat{\sigma}|\mu) = 1/f_{22}$.

10.5.2 Sample Size to Estimate Parameters when T is Log-Location-Scale

To compute needed sample sizes for estimating μ and σ under Type I censoring for the lognormal distribution, the variance factors in Appendix Table C.20 can be used directly in the sample size formulas (10.4) and (10.6). Algorithm LSINF by Escobar and Meeker (1994) provides the f_{ij} elements for the smallest extreme value (Weibull), normal (lognormal), and logistic (loglogistic) distributions. These elements facilitate computation of quantities like those in Appendix Table C.20 and allow easy programming of the computations in this chapter. Section 10.5.3 shows how to use the variance factors from Appendix Table C.20 or Algorithm LSINF to compute variance factors for ML estimates of functions of μ and σ. Sections 10.7.1 and 10.7.2 show how to use the scaled Fisher information matrix elements to compute variance factors for Type II censoring and multiple censoring.

Example 10.5 *Sample Size Needed to Estimate the Shape Parameter of a Weibull Distribution for Insulation Life.* Recall the test situation described in Example 10.1, where it was expected that about 20% of the insulation specimens would fail in the 1000-hour test and that 12% would fail in the first 500 hours, giving the Weibull parameter planning values $\mu^\square = 8.774$ and $\sigma^\square = 1.244$ [or $\eta^\square = \exp(8.774) = 6464$ and $\beta^\square = 1/1.244 = .8037$]. Suppose that the engineers need a test plan that estimates the Weibull shape parameter $\beta = 1/\sigma$ such that a 95% confidence interval has endpoints that are approximately 50% away from the ML estimate (so $R_T = 1.5$). From Table 1 of Meeker and Nelson (1977) or using Algorithm LSINF from Escobar and Meeker (1994), using as input $\zeta^\square = [\log(1000) - 8.774]/1.244 = -1.5$ or $\Phi_{\text{sev}}(-1.5) = .20$ (the proportion failing by the end of the test), gives $V^\square_{\log(\widehat{\beta})} = V^\square_{\log(\widehat{\sigma})} = [1/(\sigma^\square)^2]V^\square_{\widehat{\sigma}} = 4.74$. Thus

$$n = \frac{z^2_{(1-\alpha/2)} V^\square_{\log(\widehat{\sigma})}}{[\log(R_T)]^2} = \frac{(1.96)^2(4.74)}{[\log(1.5)]^2} \approx 111$$

is the number of specimens that should be tested. □

Example 10.6 Sample Size Needed to Estimate σ of a Lognormal Distribution for Insulation Life. We use the same inputs as in Example 10.5 but assume that the underlying distribution is lognormal. Using the lognormal planning values from Example 10.1 gives $n \approx 83$. Note that this sample size is not directly comparable to that from Example 10.5 because the σ parameters have different meanings. □

10.5.3 Large-Sample Approximate Variance for Estimators of Functions of Location-Scale Parameters

A special case of the Taylor-series approximation in Appendix equation (B.9) can be used to compute variance factors for ML estimates of functions of μ and σ; namely,

$$V_{\widehat{g}} = \left(\frac{\partial g}{\partial \mu}\right)^2 V_{\widehat{\mu}} + \left(\frac{\partial g}{\partial \sigma}\right)^2 V_{\widehat{\sigma}} + 2\left(\frac{\partial g}{\partial \mu}\right)\left(\frac{\partial g}{\partial \sigma}\right) V_{(\widehat{\mu},\widehat{\sigma})}, \quad (10.7)$$

$$V_{\log(\widehat{g})} = \left(\frac{1}{g}\right)^2 V_{\widehat{g}}, \quad \text{if } g > 0,$$

$$V_{\exp(\widehat{g})} = g^2 \, V_{\widehat{g}}.$$

Here the large-sample approximate variance factors $V_{\widehat{\mu}}$ and $V_{\widehat{\sigma}}$ and the large-sample approximate covariance factor $V_{(\widehat{\mu},\widehat{\sigma})}$ depend on the assumed location-scale distribution and the standardized censoring time $\zeta_c = [\log(t_c) - \mu]/\sigma$.

10.5.4 Sample Size to Estimate a Quantile when T is Log-Location-Scale (μ, σ)

To find the sample size needed to estimate $t_p > 0$ with a specified degree of precision, it is convenient to work with $g(\boldsymbol{\theta}) = \log(t_p) = \mu + \Phi^{-1}(p)\sigma$, the logarithm of the p quantile of T. Here $\Phi^{-1}(p)$ is the p quantile of the standardized random variable $Z = [\log(T) - \mu]/\sigma$. The sample size n is obtained as a special case of (10.6) and it is given by

$$n = \frac{z_{(1-\alpha/2)}^2 V^{\square}_{\log(\widehat{t}_p)}}{[\log(R_T)]^2}.$$

$V^{\square}_{\log(\widehat{t}_p)}$ is obtained by evaluating

$$V_{\log(\widehat{t}_p)} = V_{\widehat{\mu}} + \left[\Phi^{-1}(p)\right]^2 V_{\widehat{\sigma}} + 2\Phi^{-1}(p) V_{(\widehat{\mu},\widehat{\sigma})} \quad (10.8)$$

[a special case of (10.7)] at the planning values $\zeta_c^{\square} = [\log(t_c) - \mu^{\square}]/\sigma^{\square}$ and $\sigma^{\square}$. To obtain n one also needs to specify Φ and a target value R_T for $R = \widetilde{g}/\widehat{g} = \widehat{g}/g$.

For the Weibull distribution, Figure 10.5 gives a plot of the large-sample approximate variance factor $(1/\sigma^2)V_{\log(\widehat{t}_p)}$ versus the quantile of interest p. The lines correspond to different values of the expected proportion failing in the life test, $p_c = \Pr(T \leq t_c) = \Pr(Z \leq \zeta_c)$. Figure 10.6 is a similar plot for the lognormal distribution.

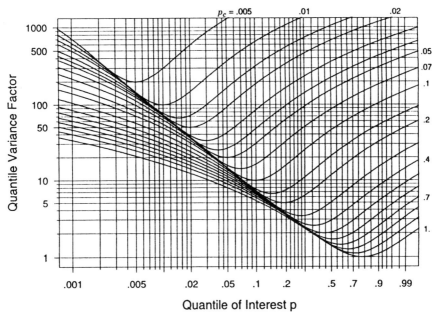

Figure 10.5. Large-sample approximate variance factor $(1/\sigma^2)V_{\log(\hat{t}_p)}$ for ML estimation of Weibull quantiles as a function of p_c (the population proportion failing by censoring time t_c) and p (the quantile of interest).

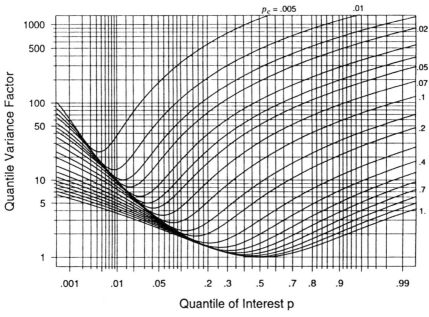

Figure 10.6. Large-sample approximate variance factor $(1/\sigma^2)V_{\log(\hat{t}_p)}$ for ML estimation of lognormal quantiles as a function of p_c (the population proportion failing by censoring time t_c) and p (the quantile of interest).

Close inspection of Figures 10.5 and 10.6 indicates the following:

- Although the factor values differ, the behavior of these plots is similar across distributions.
- Looking vertically, using any particular quantile of interest p, shows that increasing the test length (increasing the expected proportion of failures) always reduces the variance. After a point, however, the returns diminish (indicated by p_c lines that are closer together).
- The point of diminishing returns for a longer test is somewhat beyond the quantile being estimated. For example, if the goal of the test is to estimate the .1 quantile, then important gains in precision can be obtained by running the test until 15% or so of the units fail. There is, however, little additional improvement in precision from running the test much longer. Also, running the test far beyond 10% failing could introduce bias into the estimate of the .1 quantile.

Example 10.7 Sample Size Needed to Estimate $t_{.1}$ of a Weibull Distribution for Insulation Life. Refer to the insulation evaluation problem described in Examples 10.1 and 10.5. Suppose now that the engineers want to obtain a test plan that will estimate the Weibull $t_{.1}$ such that a 95% confidence interval will have endpoints that are approximately 50% away from the ML estimate of $t_{.1}$ (so $R_T = 1.5$). By taking variance factors from Table 1 of Meeker and Nelson (1977) or using Algorithm LSINF from Escobar and Meeker (1994), and using (10.8) or by taking the quantile variance factor directly from Figure 10.5 (entering with $p_c^\square = .2$ and $p = .1$) gives $[1/(\sigma^\square)^2]V^\square_{\log(\hat{t}_p)} = 7.28$ so $V^\square_{\log(\hat{t}_p)} = 7.28 \times (1.244)^2 = 11.266$. Thus

$$n = \frac{z^2_{(1-\alpha/2)} V^\square_{\log(\hat{t}_{.1})}}{[\log(R_T)]^2} = \frac{(1.96)^2(11.266)}{[\log(1.5)]^2} \approx 263$$

is the number of specimens that should be tested. □

Example 10.8 Sample Size Needed to Estimate $t_{.1}$ of a Lognormal Distribution Used to Describe Insulation Life. Using the same inputs as in Example 10.7, but assuming that the underlying distribution is lognormal, using the lognormal planning values from Example 10.1, and using Figure 10.6 to obtain $[1/(\sigma^\square)^2]V^\square_{\log(\hat{t}_p)}$ gives $n \approx 208$. Figure 10.7 is a lognormal probability plot showing ML estimates from 30 replications of this proposed test plan. The range of estimates of $t_{.1}$ indicates that the proposed plan will provide the desired degree of precision. □

10.5.5 Sample Sizes to Estimate the Hazard Function When T has a Log-Location-Scale Distribution

When T has a log-location-scale distribution, the hazard function of T evaluated at t_e can be expressed as

$$g(\boldsymbol{\theta}) = h(t_e; \boldsymbol{\theta}) = \frac{\phi(\zeta_e)}{t_e \sigma [1 - \Phi(\zeta_e)]}.$$

SAMPLE SIZES FOR LOG-LOCATION-SCALE DISTRIBUTIONS WITH CENSORING

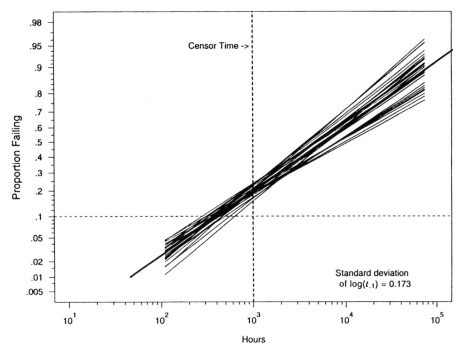

Figure 10.7. ML cdf estimates from 30 simulated samples of size $n = 208$ from a lognormal distribution with $\mu^\square = 8.658$ and $\sigma^\square = 2.079$ (shown with the thicker, longer line).

Here $\zeta_e = [\log(t_e) - \mu]/\sigma$ and Φ is the cdf of $Z = [\log(T) - \mu]/\sigma$. Because $h(t_e; \boldsymbol{\theta}) > 0$, the sample size is obtained using the confidence interval for $\log(h)$. In this case

$$V_{\log(\widehat{h})} = \frac{1}{h^2} V_{\widehat{h}} = \frac{1}{h^2} \left[\left(\frac{\partial h}{\partial \mu}\right)^2 \frac{V_{\widehat{\mu}}}{\sigma^2} + \left(\frac{\partial h}{\partial \sigma}\right)^2 \frac{V_{\widehat{\sigma}}}{\sigma^2} + 2 \left(\frac{\partial h}{\partial \mu}\right)\left(\frac{\partial h}{\partial \sigma}\right) \frac{V_{(\widehat{\mu},\widehat{\sigma})}}{\sigma^2} \right]$$

(10.9)

and n is determined from

$$n = \frac{z_{(1-\alpha/2)}^2 V_{\log(\widehat{h})}^\square}{[\log(R_T)]^2},$$

where, as before, $V_{\log(\widehat{h})}^\square$ is obtained by evaluating $V_{\log(\widehat{h})}$ at $\mu^\square$ and $\sigma^\square$.

Figures 10.8 and 10.9 give plots of variance factors $V_{\log(\widehat{h})}$ as a function of $p_c = \Pr(T \le t_c) = \Pr(Z \le \zeta_c)$ and $p_e = \Pr(T \le t_e) = \Pr(Z \le \zeta_e)$. The plots are similar to variance factor plots for the quantile estimates.

Example 10.9 Sample Size Needed to Estimate $h(1000)$ for a Weibull Distribution Used to Describe Insulation Life. Refer to the insulation evaluation problem described in Examples 10.1, 10.5, and 10.7. Suppose that the engineers need to obtain

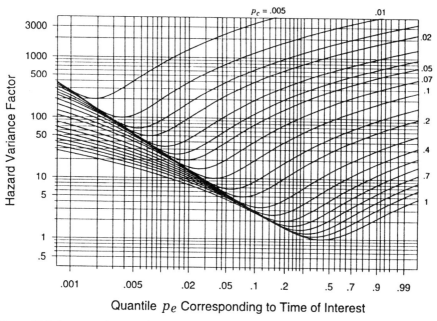

Figure 10.8. Large-sample approximate variance factor $V_{\log(\hat{h})}$ for ML estimation of the Weibull distribution hazard rate at t_e as a function of p_c (the population proportion failing by time t_c) and p_e (the population proportion failing by time t_e).

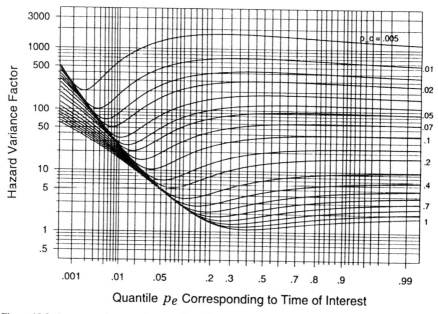

Figure 10.9. Large-sample approximate variance factor $V_{\log(\hat{h})}$ for ML estimation of the lognormal hazard rate at t_e as a function of p_c (the population proportion failing by time t_c) and p_e (the population proportion failing by time t_e).

a test plan that will estimate $h(1000)$, the Weibull hazard at 1000 hours, such that a 95% confidence interval will have endpoints that are approximately 50% away from the ML estimate of $h(1000)$ (so $R_T = 1.5$). Using $p_c^\square = .2$ and $p_e = .2$ to enter Figure 10.8 gives $V^\square_{\log[\widehat{h}(1000)]} = 10.3$. Thus

$$n = \frac{z^2_{(1-\alpha/2)} V^\square_{\log[\widehat{h}(1000)]}}{[\log(R_T)]^2} = \frac{(1.96)^2(10.3)}{[\log(1.5)]^2} \approx 240. \qquad \square$$

Example 10.10 Sample Size Needed to Estimate $h(1000)$ for a Lognormal Distribution Used to Describe Insulation Life. We use the same inputs as in Example 10.9 but assume that the underlying distribution is lognormal. Using the lognormal planning values from Example 10.1 gives $n \approx 191$. $\square$

10.6 TEST PLANS TO DEMONSTRATE CONFORMANCE WITH A RELIABILITY STANDARD

10.6.1 Reliability Demonstration Plans

It is often necessary to specify the sample size and test length for a life test that is to be used to *demonstrate*, with some level of confidence, that reliability exceeds a given standard. Often the reliability standard is specified in terms of a quantile, say, t_p. For example, a customer purchasing a product may require demonstration, by the vendor, that $t_p > t_p^\dagger$, where $t_p^\dagger$ is a specified value. In general, the demonstration that $t_p > t_p^\dagger$ will be successful at the $100(1-\alpha)\%$ level of confidence if $\utilde{t}_p > t_p^\dagger$, where $\utilde{t}_p$ is a lower $100(1-\alpha)\%$ confidence bound for t_p.

Example 10.11 Reliability Demonstration Test for a Life-Limiting Component. A relatively expensive life-limiting component is to be installed in a product with a 1-year warranty. The manufacturer of the product will purchase the component from a vendor. The vendor has been asked to demonstrate that $t_{.01}$ exceeds $24 \times 365 = 8760$ hours. Equivalently, in terms of failure probabilities the reliability requirement could be specified as

$$F(t_e) < p^\dagger,$$

which would be demonstrated if $\widetilde{F}(t_e) < p^\dagger$. For this example, $t_e = 8760$ and $p^\dagger = .01$. $\square$

10.6.2 Weibull Minimum Sample Size Reliability Demonstration Plans with Given β

Suppose that failure times have a Weibull distribution with a given shape parameter β. A *minimum sample size* test plan is one that tests n units until time t_c and the demonstration is successful if there are no failures. The particular sample size n

depends on the confidence level $1 - \alpha$, the quantile of interest p, the amount of time available for testing t_c, and the given Weibull shape parameter β. The needed sample size n is the smallest integer greater than

$$\frac{1}{k^\beta} \times \frac{\log(\alpha)}{\log(1 - p)}$$

and $k = t_c/t_p^\dagger$. This minimum sample size reliability demonstration plan (also known as a "zero-failure demonstration plan") can be justified as follows. Suppose that failure times are Weibull with a given β and there are zero failures during a test in which n units are tested until t_c. Using the results in Section 8.5.3, lower $100(1 - \alpha)\%$ confidence bounds for η and t_p are

$$\underaccent{\tilde}{\eta} = \left[\frac{2nt_c^\beta}{\chi^2_{(1-\alpha;2)}}\right]^{1/\beta} = \left[\frac{nt_c^\beta}{-\log(\alpha)}\right]^{1/\beta},$$

$$\underaccent{\tilde}{t_p} = \underaccent{\tilde}{\eta} \times \left[-\log(1-p)\right]^{1/\beta}.$$

Then using the inequality $\underaccent{\tilde}{t_p} \geq t_p^\dagger$ and solving for n gives

$$n \geq \frac{1}{k^\beta} \times \frac{\log(\alpha)}{\log(1 - p)}, \qquad (10.10)$$

where $k = t_c/t_p^\dagger$. Thus the needed minimum sample size is the smallest integer greater than or equal to the right-hand side of (10.10).

The inequality in (10.10) can also be solved for k, β, or α. For example,

$$k \geq \left\{\frac{1}{n} \times \frac{\log(\alpha)}{\log(1 - p)}\right\}^{1/\beta}. \qquad (10.11)$$

Example 10.12 Life Test to Demonstrate the Reliability of a Bearing. The manufacturer of a home food processor requires that a bearing to be used in the product have no more than 10% failing at 5 million revolutions (a conservatively high number of revolutions expected in a typical 10-year life). Thus $t_{.1}^\dagger = 5$ million revolutions. A new long-life, low-cost bearing is available from a vendor. The vendor will, however, be asked to demonstrate the specified level of reliability. Similar bearings in this kind of application have had lifetimes that could be described by a Weibull distribution with a shape parameter of $\beta = 2$ or more ($\sigma = .5$ or less). Figure 10.10 gives the needed sample size for a 99% demonstration (so $\alpha = .01$) on $t_{.1}$ (sometimes known, in the bearing industry and elsewhere, as B10 for 10% bearing life), as a function of the test-length factor $k = t_c/t_p^\dagger$ and the Weibull shape parameter β. This figure indicates that a zero-failure test on $n = 5$ units will provide the desired demonstration if there are no failures up to $3 \times 5 = 15$ million revolutions.

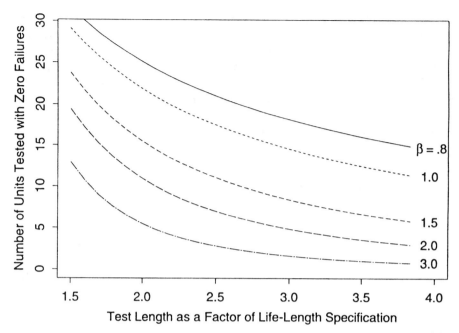

Figure 10.10. Minimum sample size for a 99% reliability demonstration for $t_{.1}$ as a function of the test-length factor k.

More precisely, substituting into (10.11) gives $k = 2.96$, which implies that the test should be run until $t_c = 2.96 \times 5 = 14.6$ million revolutions. □

For tests with $k > 1$, having a specified value of β less than the true value is conservative (in the sense that the demonstration is still valid). If, however, the specified value of β is larger than the true value, the demonstration would not be valid. The biggest danger of using such a minimum sample size zero-failure test is that defect-related failures (which might occur in only a small proportion of units) might not show up in the sample.

It is also possible to conduct minimum sample size zero-failure tests with $k < 1$. In this case having a specified value of β greater than the true value is conservative. Such tests give information quickly but require correspondingly large sample sizes. The most serious difficulty with such demonstrations is that they are based on large amounts of data early in life and if there is an unknown wearout failure mode occurring later in life that is not reflected in the given Weibull shape parameter, the results of the test could be seriously misleading.

10.6.3 Extensions for Other Reliability Demonstration Test Plans

Zero-failure test plans can be obtained for other failure-time distributions with only one unknown parameter. The ideas in this section can also be extended to test plans with one or more failures. Such test plans require more units but provide a higher probability of successful demonstration for a given $t_p^\dagger > t_p$. General reliability demonstration test plans can be obtained for any distribution, with or without specified parameter values, although, in general, it may be necessary to base the test on a large-sample approximation, corresponding to the methods used in Chapter 8. Chapter 9 of Hahn and Meeker (1991) give tables and charts for demonstration plans for $k = 1$ plans not requiring any distributional assumption or for normal/lognormal demonstration tests with no censoring. For more information on reliability demonstration tests, see Wang (1991) and Chapter 6 of Abernethy (1996).

10.7 SOME EXTENSIONS

10.7.1 Failure (Type II) Censoring

Most of this chapter has focused on planning reliability tests with time (Type I) censoring. This is because Type I censoring is most common in practice. Most reliability studies are conducted with tight time constraints. With failure (Type II) censoring, a test is stopped after a specified number (say, $r \le n$) of units have failed. The methods and figures in Section 10.5 also apply to failure censoring. In this case, $p_c = r/n$ is specified and $\zeta_c = \Phi^{-1}(p_c)$.

Although less common in practice, failure-censored tests can be useful. In particular, if one is interested in estimating the .1 quantile, then, as shown in Figures 10.5 and 10.6, there is little to be gained by continuing testing beyond the time at which about 15% of the units have failed. If there is a limited number of test positions and one needs to plan a test to estimate a specific quantile of the distribution, failure censoring provides a convenient mechanism for deciding when to replace unfailed units with new ones. In particular, if one is interested in estimating the .1 quantile of a failure-time distribution and only five test positions are available for testing, a reasonable test plan would test five units at a time, replacing the test units with a new set after the first failure in each group. Such tests are known as sudden death tests. Pascual and Meeker (1998b) describe such tests and extensions.

10.7.2 Variance Factors for Location-Scale Parameters and Multiple Censoring

Section 10.5 provides methods and easy-to-use graphs for planning life tests in which all units will be censored at the same time (single censoring). The scaled Fisher matrix elements f_{11}, f_{12}, and f_{22} (for the normal/lognormal distribution) in Appendix Table C.20 or from Algorithm LSINF from Escobar and Meeker (1994) (also for the SEV/Weibull and logistic/loglogistic distributions) can, however, also be used to compute variance factors for more complicated censoring patterns. For example, in some applications, a life test may run in groups, each group having a different censoring time (e.g., testing at two different locations or beginning times as lots of units to be tested are received). In this case it is necessary to generalize the single-censoring formula.

EXERCISES

For a life test that is to be run in k groups, let $\delta_i, i = 1, \ldots, k$ (where $\sum_{i=1}^{k} \delta_i = 1$), denote the proportion of units that will be run until standardized right-censoring time ζ_{c_i} or failure (whichever comes first). In this case,

$$\Sigma_{(\hat{\mu},\hat{\sigma})} = \frac{1}{n}\begin{bmatrix} V_{\hat{\mu}} & V_{(\hat{\mu},\hat{\sigma})} \\ V_{(\hat{\mu},\hat{\sigma})} & V_{\hat{\sigma}} \end{bmatrix} = \frac{\sigma^2}{n}\left(\frac{1}{J_{11}J_{22} - J_{12}^2}\right)\begin{bmatrix} J_{22} & -J_{12} \\ -J_{12} & J_{11} \end{bmatrix}, \quad (10.12)$$

where $J_{11} = \sum_{i=1}^{k} \delta_i f_{11}(\zeta_{c_i})$, $J_{22} = \sum_{i=1}^{k} \delta_i f_{22}(\zeta_{c_i})$, $J_{12} = \sum_{i=1}^{k} \delta_i f_{12}(\zeta_{c_i})$, and $\zeta_{c_i} = [\log(t_{c_i}) - \mu]/\sigma$, and the values of f_{11}, f_{22}, and f_{12} are the Fisher information matrix elements given by Algorithm LSINF in Escobar and Meeker (1994). Appendix Table C.20 provides these values for the lognormal distribution. Factors for the approximate variance–covariance matrix $(1/\sigma^2)V_{\hat{\mu}}$, $(1/\sigma^2)V_{\hat{\sigma}}$, and $(1/\sigma^2)V_{(\hat{\mu},\hat{\sigma})}$ depend on Φ, the standardized censoring times ζ_{c_i}, and the proportions $\delta_i, i = 1, \ldots, k$.

10.7.3 Test Planning for Distributions that Are Not Log-Location-Scale

The methods in Sections 10.1–10.4 can be applied to a much wider class of models than that treated in detail in Section 10.5. For distributions that are not log-location-scale, however, variance factors may depend on an additional shape parameter and separate graphs like those in Figures 10.5, 10.6, 10.8, and 10.9 would have to be made for each representative set of values of such shape parameters. In such cases it may be necessary to rely on a computer program to compute the Fisher information matrix or to do a simulation.

BIBLIOGRAPHIC NOTES

Escobar and Meeker (1994) describe algorithm LSINF that provides the f_{ij} values for the smallest extreme value (Weibull), normal (lognormal), and logistic (loglogistic) distributions. Meeker and Nelson (1976) present asymptotic theory, tables, and figures that can be used to plan a life test to estimate a Weibull quantile with a specified degree of precision. Figures 10.5 and 10.6 were patterned after their figure for the Weibull distribution. Meeker and Nelson (1977) present general theory and tables that can be used to choose the needed sample size for other functions of Weibull parameters. Meeker, Escobar, and Hill (1992) present asymptotic theory and figures that can be used to plan a life test to estimate a Weibull hazard function with a specified degree of precision. Figures 10.8 and 10.9 were patterned after their figure for the Weibull distribution. Escobar and Meeker (1997) show how to compute the Fisher information matrix and asymptotic variances for truncated distributions and the LFP model (Chapter 11) and regression models (Chapter 17).

EXERCISES

10.1. Use the input information in Example 10.1 to compute the planning values $\mu^\square$ and $\sigma^\square$ for both the lognormal and the Weibull distributions. Plot

10.2. Refer to the information for Example 10.3. Use (10.4) to compute the suggested n for additional values of $D_T = 20, 10, 5$. Describe the effect that changing the target precision has on the needed sample size.

▲10.3. Derive an expression for the large-sample approximate variance of the ML estimate of the logarithm of the Weibull hazard function. Start by taking the partial derivatives indicated in (10.9). This expression should depend on $\text{Avar}(\widehat{\mu})$, $\text{Avar}(\widehat{\sigma})$, and $\text{Acov}(\widehat{\mu}, \widehat{\sigma})$.

10.4. Refer to Example 10.7. Use the given inputs and interpolation in Appendix Table C.20 to compute the needed sample size assuming that the distribution is lognormal instead of Weibull.

10.5. Refer to Example 10.9. Use the given inputs to compute the needed sample size assuming that the distribution is lognormal instead of Weibull.

10.6. A reliability engineer wants to run a life test to estimate the .05 quantile of the fatigue life distribution of a metal component used in a switch. The engineer has to choose a sample size that will allow estimation to be precise enough so that the lower endpoint of a 95% confidence interval for the quantile will be about one-half of the ML estimate. It will be possible to test each specimen until about 100 thousand cycles, when it is expected that about 15% of the specimens will have failed. It is expected that about 5% will have failed after about 40 thousand cycles.

  (a) Use the information above on Weibull probability paper to obtain "planning values" for the Weibull parameters.

  (b) Determine the sample size needed to achieve the desired precision.

10.7. Consider the sample size problem in Example 10.7. Solve the same problem assuming that there is need to estimate $t_{.02}$ with $R_T = 2$.

10.8. Do the calculations for Example 10.6.

▲10.9. Section 10.5.1 shows how to use the elements of the Fisher information matrix to compute variance factors for test planning in a life test with a single censoring time. In Appendix Table C.20, use the values of f_{11}, f_{22}, and f_{12} in the row corresponding to $\zeta_c = 1$, and show how to compute the following quantities in that row:

$$\frac{1}{\sigma^2}V_{\widehat{\mu}}, \quad \frac{1}{\sigma^2}V_{\widehat{\sigma}}, \quad \frac{1}{\sigma^2}V_{(\widehat{\mu},\widehat{\sigma})}, \quad \rho_{(\widehat{\mu},\widehat{\sigma})}, \quad \frac{1}{\sigma^2}V_{\widehat{\mu}|\sigma}, \quad \text{and} \quad \frac{1}{\sigma^2}V_{\widehat{\sigma}|\mu}.$$

EXERCISES 253

▲10.10. For Example 10.3, compute $\Pr(D > D_T)$, the probability that the actual half-width will be greater than the target half-width $D_T = 10$ for samples of size $n = 200, 300,$ and $400,$ and sketch a graph of $\Pr(D > D_T)$ versus n.

▲10.11. For Example 10.4, derive an approximate expression for $\Pr(R > R_T)$, the probability that the actual confidence interval factor is greater than the target factor $R_T = 1.5$. Evaluate this expression for samples sizes ranging between $n = 30$ and 100. Make a plot of $\Pr(R > R_T)$ versus n.

▲10.12. Refer to Example 10.7. Derive an approximate expression for $\Pr(R > R_T)$, based on a large-sample approximate distribution for the random variable $\widehat{V}_{\log(\hat{t}_p)}$.

▲10.13. Refer to Example 10.4. Show that, for the exponential distribution,

$$E\left[-\frac{\partial^2 \mathcal{L}(\theta)}{\partial \theta^2}\right] = \frac{n}{\theta^2}\left[1 - \exp\left(-\frac{t_c}{\theta}\right)\right].$$

▲10.14. Refer to Section 10.5. Show how to compute the no-censoring ($\zeta_c \to \infty$) asymptotic values of $f_{11} = 1$, $f_{12} = 0$, and $f_{22} = 2$ for the normal distribution.

▲10.15. Show that, for the Weibull distribution, (10.9) reduces to

$$V_{\log(\hat{h})} = \left[V_{\hat{\mu}} + (\zeta_e + 1)^2 V_{\hat{\sigma}} + 2(\zeta_e + 1)V_{(\hat{\mu},\hat{\sigma})}\right]/\sigma^2.$$

10.16. Verify equation (10.12).

10.17. In some cases (e.g., for the lognormal distribution), planning values will be specified in terms of the shape parameter σ and a particular quantile, say, t_{p_1} for a specified p_1. Given these values, derive an expression for μ and for t_{p_2} for a given p_2.

10.18. Refer to Appendix Table C.20. What can you say about the effect that censoring has on the correlation between ML estimators $\hat{\mu}$ and $\hat{\sigma}$?

CHAPTER 11

Parametric Maximum Likelihood: Other Models

Objectives

This chapter explains:

- ML estimation for the gamma and the extended generalized gamma (EGENG) distributions.
- ML estimation for the Birnbaum–Saunders (BISA) and the inverse Gaussian (IGAU) distributions.
- ML estimation for the limited failure population (LFP) model.
- How truncation arises in reliability data applications and ML estimation for truncated data (or data from truncated distributions).
- ML estimation for distributions with a threshold parameter like the three-parameter lognormal and the three-parameter Weibull distributions [using the generalized threshold-scale (GETS) distribution].
- Potential difficulties involved in using distributions with threshold parameters and how to avoid them.

Overview

This is an advanced chapter that can be skipped without loss of continuity. Section 11.1 describes the extension of ML methods, introduced in Chapter 8, to distributions that are not location-scale or log-location-scale. Chapter 5 provides background and technical details for the non-location-scale distributions used in this chapter. Section 11.2 illustrates ML methods for the gamma distribution and Section 11.3 shows how to fit the extended generalized gamma distribution. Section 11.4 uses ML methods to fit and compare the Birnbaum–Saunders and inverse Gaussian distributions. Section 11.5 shows how to fit and interpret the limited failure population (LFP) distribution, a kind of mixture model. Section 11.6 describes various applications where truncated data arise and illustrates methods for analyzing such data. Section 11.7 illustrates methods for fitting distributions with threshold parameters, showing how to avoid potential difficulties.

11.1 INTRODUCTION

The methods presented in Chapter 8 were based on fitting log-location-scale distributions. This chapter describes the fitting of some important distributions that are not log-location-scale. For such distributions, things may be nearly as simple, but they can also be considerably more complicated.

11.1.1 Likelihood for Other Distributions and Models

General likelihood principles for fitting distributions and models are as described in Chapters 2 and 7. For most distributions, with data consisting of independent observations with exact failures and right-censored observations, the standard density-approximation form of the likelihood

$$L(\boldsymbol{\theta}) = \prod_{i=1}^{n} L_i(\boldsymbol{\theta}; \text{data}_i) = \prod_{i=1}^{n} [f(t_i; \boldsymbol{\theta})]^{\delta_i} [1 - F(t_i; \boldsymbol{\theta})]^{1-\delta_i} \quad (11.1)$$

works well. Here, as in previous chapters, $\text{data}_i = (t_i, \delta_i)$,

$$\delta_i = \begin{cases} 1 & \text{if } t_i \text{ is an exact failure} \\ 0 & \text{if } t_i \text{ is a right-censored observation,} \end{cases}$$

and $F(t; \boldsymbol{\theta})$ and $f(t; \boldsymbol{\theta})$ are the cdf and pdf, respectively, of the specified distribution.

The likelihood can be adapted easily to accommodate left-censored and interval-censored observations, as described in Chapter 2. As mentioned in Section 7.2.2, operationally, it is usually the log likelihood that is computed as the sum of log likelihoods for individual independent observations. As illustrated in Section 11.7.3, for some non-location-scale distributions (e.g., GETS) the density approximation breaks down and one should use instead the actual interval probability or "correct likelihood" given in (7.2).

In some cases, the likelihood function can be poorly behaved and, particularly in unfamiliar data/model situations, it is important to investigate $L(\boldsymbol{\theta})$ graphically. Graphical exploration of $L(\boldsymbol{\theta})$ [or $R(\boldsymbol{\theta}) = L(\boldsymbol{\theta})/L(\widehat{\boldsymbol{\theta}})$] is simple to do when $\boldsymbol{\theta}$ has length 1 or 2. When the length of $\boldsymbol{\theta}$ is 3 or more it is useful to view one- (and two-) dimensional "profiles" of $L(\boldsymbol{\theta})$. The needed computations can, however, become quite demanding.

11.1.2 Confidence Intervals for Other Distributions and Models

Methods for computing confidence intervals and confidence regions can be used in a manner that is similar to that used for log-location-scale distributions in Chapters 8. Normal-approximation confidence intervals (using the delta method and appropriate transformations) are simple and are adequate in large samples. In other situations these easy-to-compute intervals provide quick analyses, but approximations can be a bit rough. Profile likelihoods provide useful insight into the information available about a particular parameter or function of a distribution's parameters. Confidence limits based on the profile likelihood, as well as bootstrap and simulation-based intervals described in Chapter 9, generally provide confidence intervals with reasonably good approximations to nominal coverage probabilities. The approximations will be adequate even with moderately small samples (say, samples large enough to yield at least 10–15 failures). Such intervals will, however, require more computer time (and may not be available in commercial software).

11.2 FITTING THE GAMMA DISTRIBUTION

The gamma distribution, introduced in Section 5.2, is a commonly used failure-time distribution. The gamma distribution likelihood is obtained by substituting (5.2) and (5.1) into (11.1). For given DATA, the likelihood is a function of the scale parameter θ and the shape parameter κ.

Example 11.1 Gamma Distribution Fit to the Ball Bearing Fatigue Data. This example uses the ball bearing data from Example 1.1. Figure 11.1 shows a lognormal probability plot of the bearing failure data, comparing ML estimates of the gamma, lognormal, and Weibull distributions. The gamma ML estimates are $\widehat{\theta} =$

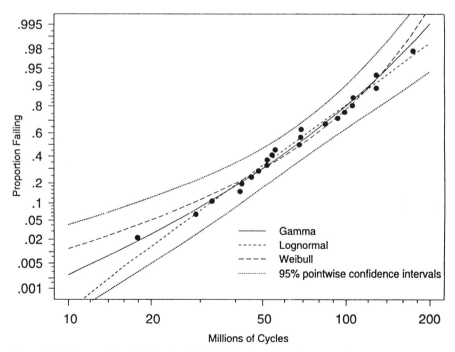

Figure 11.1. Lognormal probability plot of the ball bearing failure data, comparing gamma, lognormal, and Weibull ML estimates. Approximate 95% pointwise confidence intervals for the gamma $F(t)$ are also shown.

17.94 and $\widehat{\kappa} = 4.025$. Figure 11.1 also shows approximate 95% pointwise confidence intervals for $F(t)$, based on the gamma ML estimates. The gamma distribution fits well. Within the range of the data, however, there is very little difference among these three different distributions. □

11.3 FITTING THE EXTENDED GENERALIZED GAMMA DISTRIBUTION

The extended generalized gamma (EGENG) distribution, introduced in Section 5.3, includes the gamma, generalized gamma, Weibull, exponential, and lognormal distributions as special cases. As such, it provides a flexible distribution structure for modeling data and comparing among these widely used distributions. The EGENG distribution likelihood is obtained by substituting (5.6) and (5.5) into (11.1). For given DATA, the likelihood is a function of the parameters μ, σ, and λ.

Example 11.2 EGENG Distribution Fit to the Ball Bearing Fatigue Data. This example uses the ball bearing data from Examples 1.1 and 11.1. Figure 11.2 is a Weibull probability plot of the bearing failure data showing exponential, Weibull, lognormal, and EGENG ML estimates of $F(t)$. The EGENG ML estimates are $\widehat{\mu} =$

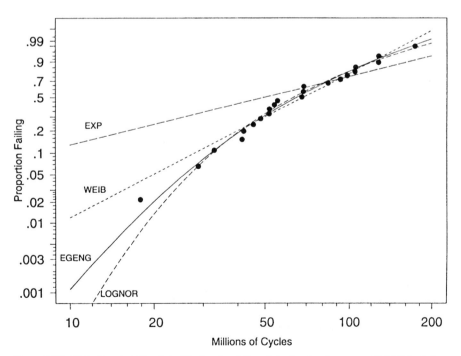

Figure 11.2. Weibull probability plot of the bearing failure data showing the exponential, Weibull, lognormal, and EGENG ML estimates of $F(t)$.

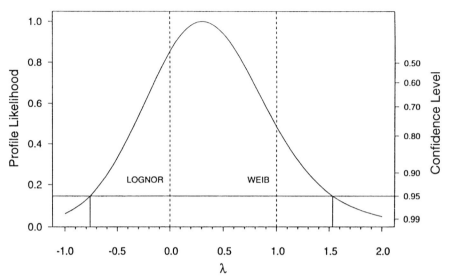

Figure 11.3. Profile likelihood plot for the EGENG shape parameter λ for the bearing failure data. The Weibull and lognormal distributions are shown as special cases.

4.23, $\hat{\sigma} = .51$, and $\hat{\lambda} = .3076$. Figure 11.2 illustrates that the EGENG distribution provides a compromise between the lognormal and Weibull distributions. Figure 11.3 is a profile likelihood plot for the EGENG shape parameter λ for the bearing failure data. The Weibull and lognormal distributions are shown as special cases. This figure shows that the lognormal relative likelihood is slightly higher than the Weibull. The data, however, do not indicate a strong preference for one or the other of these distributions. □

The previous example illustrated the fitting of the EGENG distribution to a moderate-size sample with *no* censoring. The EGENG distribution can also be fit to censored data, but the fitting can be more delicate because of difficulty statistically separating the three different parameters in (5.5). Using a robust optimization algorithm and/or a reparameterization to stable parameters will, however, allow this distribution to be fit to censored data, even with very heavy censoring (see the comments in the bibliographic notes at the end of this chapter).

Example 11.3 EGENG Distribution Fit to the Fan Data. This example fits the EGENG distribution to the fan data from Example 1.4. During the period of observation there were 12 failures out of 70 units. Due to multiple censoring, however, the nonparametric estimate extends to .29. Figure 11.4 shows a lognormal probability plot of the fan failure data showing EGENG ML estimates and corresponding 95% pointwise confidence intervals for $F(t)$. The EGENG ML estimates are $\hat{\mu} = 9.332$, $\hat{\sigma} = 2.375$, and $\hat{\lambda} = -1.764$. The pointwise confidence intervals for the EGENG

FITTING THE EXTENDED GENERALIZED GAMMA DISTRIBUTION 259

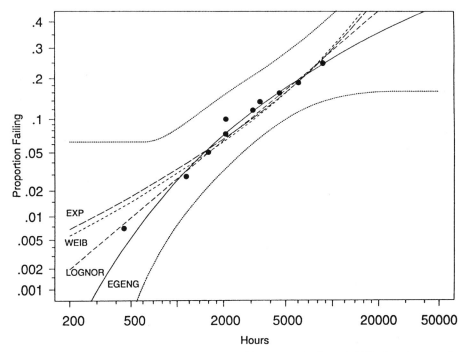

Figure 11.4. Lognormal probability plot of the fan failure data showing EGENG ML estimates and corresponding 95% pointwise confidence intervals for $F(t)$. Exponential, Weibull, and lognormal ML estimates of $F(t)$ are also shown.

$F(t)$ become extremely wide outside the range of the data. Exponential, Weibull, and lognormal ML estimates of $F(t)$ are also shown. These distributions, and especially the lognormal distribution, also fit the data reasonably well. Note that the EGENG departs quite strongly from the other distributions outside the range of the data. Except for missing the first point, the EGENG estimate also goes well with the nonparametric estimate. To assess the strength of the evidence in the data for choosing among these distributions, consider the profile likelihood plot for EGENG λ for the fan failure data in Figure 11.5. The EGENG has a larger likelihood than the other distributions, but the difference is statistically unimportant (the approximate 95% likelihood-based confidence interval endpoints for λ ranges from something less than -8 to something greater than 2, as shown in Figure 11.5). Because of the small number of failures, fitting a three-parameter distribution to these data could be considered to be "overfitting." Generally one should use the simplest model that provides an adequate fit to the data. Overfitting such as this, however, is helpful in demonstrating uncertainty in distribution choice when the decision is to be based on data alone. □

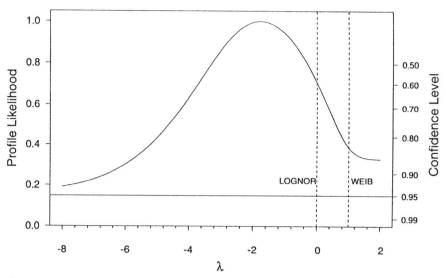

Figure 11.5. Profile likelihood plot for the EGENG shape parameter λ for the fan failure data showing the Weibull and lognormal distributions as special cases.

11.4 FITTING THE BISA AND IGAU DISTRIBUTIONS

Fitting the Birnbaum–Saunders (BISA) or the inverse Gaussian (IGAU) distributions is very similar to fitting the gamma distribution (all of these distributions have one scale and one shape parameter). The likelihood function for the BISA and IGAU distributions can be obtained by substituting the corresponding pdf and cdf from Sections 5.6.2 or 5.7.2 into (11.1). As described in Section 5.7.5, the BISA and IGAU distributions were motivated by particular degradation models.

Example 11.4 BISA and IGAU Distributions Fit to Fatigue-Fracture Data. Yokobori (1951) describes a fatigue-fracture test on .41% carbon steel cylindrical specimens, tested at $\pm 37.1\,\text{kg/mm}^2$ stress amplitude. The data are given on pages 224–225 of Bogdanoff and Kozin (1985). Figure 11.6 is a lognormal probability plot of Yokobori's fatigue failure data. The plot also shows ML estimates of $F(t)$ for the lognormal, BISA, and IGAU distributions. The ML estimates for the BISA distribution are $\widehat{\theta} = 109.24$ and $\widehat{\beta} = 1.129$. The ML estimates for the IGAU distribution are $\widehat{\theta} = 179.9$ and $\widehat{\beta} = .595$. The ML estimates for the lognormal distribution are $\widehat{\mu} = 4.72$ and $\widehat{\sigma} = 1.026$. The closeness of the $F(t)$ estimates from these three different distributions in Figure 11.6 is striking. This is not too surprising, given the correspondence between the hazard shapes that can be seen in Figures 4.3, 5.2, and 5.3. □

Figure 11.7 is a plot of cdfs of the BISA distribution with $\theta = 1$ and different shape parameters on lognormal probability paper. The approximate linearity of the cdfs, especially for small β (small coefficient of variation), suggests that the log-

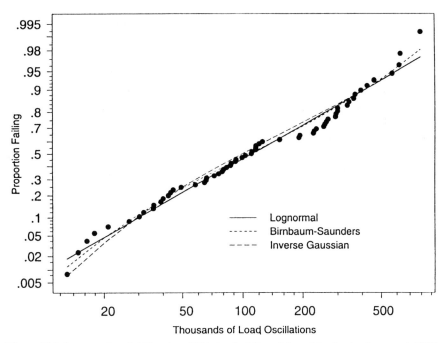

Figure 11.6. Lognormal probability plot of Yokobori's fatigue failure data showing lognormal, BISA, and IGAU distribution ML estimates.

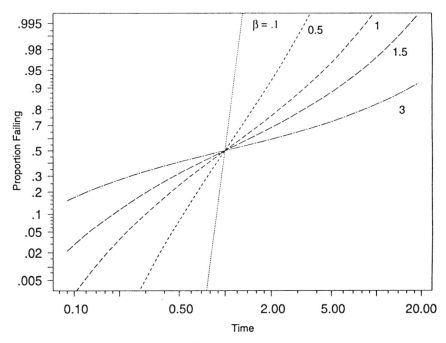

Figure 11.7. Comparison of BISA cdfs on lognormal probability paper.

261

262   PARAMETRIC MAXIMUM LIKELIHOOD: OTHER MODELS

normal and BISA distributions will often give very similar results in the center of the distribution. It was noted in Section 4.6 that the lognormal distribution is widely used to describe time to fracture from fatigue crack growth in metals. The similarity of the lognormal and BISA distributions and the fatigue-fracture justification for the BISA distribution (see Section 5.7) suggest why the lognormal distribution has been found to be a useful model for fatigue-fracture data. Figure 11.7 indicates that the lognormal distribution, when used to extrapolate into the lower tail of a distribution, will give more conservative (i.e., smaller) estimates of distribution quantiles. Such extrapolation is common, for example, when it is desired to estimate $t_{.001}$ on the basis of tests on 300 specimens.

11.5 FITTING A LIMITED FAILURE POPULATION MODEL

11.5.1 Example and Data

Example 11.5 The IC Failure Time Data. We now return to the IC failure-time data from Example 1.2, given in Table 1.2. Figure 11.8 is a Weibull probability plot of the right-censored failure data. The last failure occurred at 593 hours and

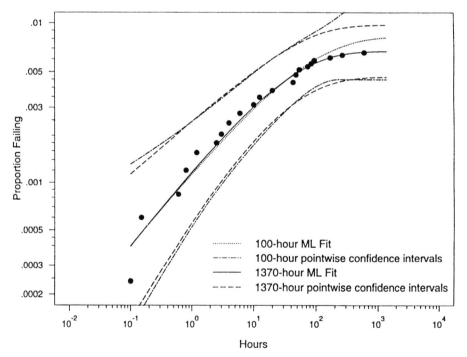

Figure 11.8. A Weibull probability plot of integrated circuit failure time data with ML estimates of the Weibull/LFP model after 1370 hours and after 100 hours of testing. The asymptotes for the ML fits in the plot correspond to the ML estimates for p, the proportion in the process susceptible to failure.

the test was stopped at 1370 hours. We see that the points on the probability plot are leveling off, apparently to something less than 1% failing. The failures were caused by manufacturing defects that could not be detected without a life test. The reliability engineers responsible for this product wanted to estimate p, the proportion of defects being manufactured by the process, in its current state. Moreover, they were attempting to make improvements to the process and wondered if informative life tests could, in the future, be run without waiting so long. □

11.5.2 The Limited Failure Population Model

The limited failure population (LFP) model has a proportion p of units from a population or process that is defective and will fail according to a distribution $F(t; \mu, \sigma)$; the remaining proportion $(1 - p)$ will never fail. This model has been found to be useful for modeling the reliability of integrated circuit infant mortality. Its use will be illustrated with the IC life test data from Example 11.5. If $F(t; \mu, \sigma)$ is Weibull, then the LFP failure-time model is

$$\Pr(T \leq t) = G(t; \mu, \sigma, p) = pF(t; \mu, \sigma) = p\Phi_{\text{sev}}\left[\frac{\log(t) - \mu}{\sigma}\right]. \quad (11.2)$$

Note that as $t \to \infty$, $G(t) \to p$. The lognormal LFP model is obtained by using Φ_{nor} instead of Φ_{sev} in (11.2).

11.5.3 The Likelihood Function and Its Maximum

The likelihood function for the Weibull LFP model is

$$L(\mu, \sigma, p) = \prod_{i=1}^{n} \left\{ \frac{p}{t_i \sigma} \phi_{\text{sev}}\left[\frac{\log(t_i) - \mu}{\sigma}\right] \right\}^{\delta_i} \left\{ 1 - p\Phi_{\text{sev}}\left[\frac{\log(t_i) - \mu}{\sigma}\right] \right\}^{1-\delta_i},$$

$$(11.3)$$

where the notation is similar to that used in Section 8.2.1. In some situations it will be difficult to estimate the parameters of this model. In particular, if the censoring time is before the nonparametric estimate of $G(t; \mu, \sigma, p)$ begins to level off, one cannot tell the difference between a population with $p = 1$ in which defective units fail slowly and a population with small p in which defective units fail rapidly.

Example 11.6 *Comparison of IC Failure-Time Data Analyzed at 100 and at 1370 Hours.* Table 11.1 summarizes and compares the results of the analyses for the data that were available at 1370 hours and the data that would have been available after only 100 hours. Figure 11.8 shows the ML estimates and 95% pointwise normal-approximation confidence intervals for $G(t)$. As might be expected, there is close agreement until approximately 100 hours, when the estimates of $G(t)$ begin to differ importantly. The upper bounds of the pointwise normal-approximation confidence intervals for $G(t)$ are larger for $t > 100$. As suggested below, however,

Table 11.1. Comparison of LFP Model Integrated Circuit Failure Data Analyses

	Analysis with Test Run Until	
	1370 Hours	100 Hours
ML estimate $\widehat{\mu}$	3.34	4.05
Standard error $\widehat{\text{se}}_{\widehat{\mu}}$	.41	1.70
Approximate 95% Confidence Intervals for μ		
Based on the likelihood	[2.50, 4.20]	[2.43, 24.99]
Based on $Z_{\widehat{\mu}} \stackrel{.}{\sim} \text{NOR}(0, 1)$	[2.55, 4.12]	[.72, 7.38]
ML estimate $\widehat{\sigma}$	2.02	2.12
Standard error $\widehat{\text{se}}_{\widehat{\sigma}}$	.31	.55
Approximate 95% Confidence Intervals for σ		
Based on the likelihood	[1.53, 2.82]	[1.40, 3.96]
Based on $Z_{\log(\widehat{\sigma})} \stackrel{.}{\sim} \text{NOR}(0, 1)$	[1.50, 2.71]	[1.28, 3.51]
Based on $Z_{\widehat{\sigma}} \stackrel{.}{\sim} \text{NOR}(0, 1)$	[1.42, 2.62]	[1.05, 3.19]
ML estimate $\widehat{p}$	.00674	.00827
Standard error $\widehat{\text{se}}_{\widehat{p}}$	.00127	.00380
Approximate 95% Confidence Intervals for p		
Based on the likelihood	[.00455, .00955]	[.00463, 1.0000]
Based on $Z_{\text{logit}(\widehat{p})} \stackrel{.}{\sim} \text{NOR}(0, 1)$	[.00466, .00975]	[.0033, .0203]
Based on $Z_{\widehat{p}} \stackrel{.}{\sim} \text{NOR}(0, 1)$	[.00426, .00923]	[.00081, .0157]

similar confidence intervals for the 100-hour data, computed with the likelihood-based method, would be *much* wider.

The nonparametric estimate of $G(t; \mu, \sigma, p)$ in Figure 11.8 (i.e., the plotted points) portends the LFP model estimation difficulties with the 100-hour data. Note that the curvature in the Weibull probability plot becomes much more pronounced after 100 hours. □

11.5.4 Profile Likelihood Functions and Likelihood-Based Confidence Intervals for μ, σ, and p

Example 11.7 Likelihood-Based Confidence Intervals for the IC Data. Table 11.1 gives numerical values for the likelihood-based confidence intervals for μ, σ, and p based on these and other profiles (not shown here). The table also gives confidence

FITTING A LIMITED FAILURE POPULATION MODEL 265

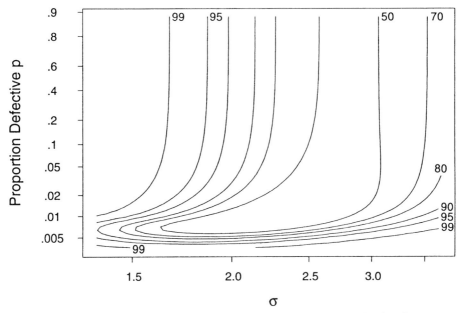

Figure 11.9. Approximate joint confidence regions for the LFP parameters p and σ based on a two-dimensional profile likelihood after 100 hours of testing.

intervals based on the normal approximation. Figure 11.9 shows approximate joint confidence regions for p and σ based on a two-dimensional profile likelihood for p and σ for the 100-hour data. Figure 11.10 provides a comparison of the one-dimensional profiles for p for the 1370- and the 100-hour data. The results of this comparison show that for the 1370-hour data, the log likelihood is approximately quadratic and the different methods of computing confidence intervals give similar results. For the 100-hour data, however, the situation is quite different. In particular, the leveling off of the 100-hour profile likelihood for p tells us that the data available after 100 hours could reasonably have come from a population with $p = 1$. That is, the 100-hour data do not allow us to clearly distinguish between a situation where there are many defectives failing slowly and a situation with just a few defectives failing rapidly. The 1370-hour data, however, allow us to say with a high degree of confidence that p is small.

For the 100-hour data, the likelihood and normal-approximation confidence intervals for p are vastly different. This is because the log likelihood is not well approximated by a quadratic function over the range of the confidence interval. The approximate confidence intervals based on the likelihood can be expected to provide coverage probabilities closer to the nominal values. □

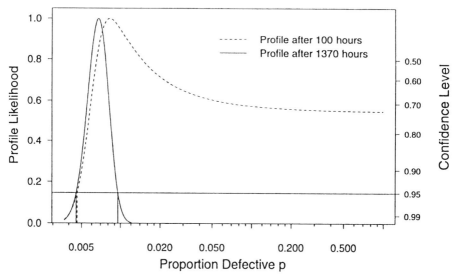

Figure 11.10. Comparison of profile likelihoods for p, the LFP proportion defective after 1370 and 100 hours of testing.

11.6 TRUNCATED DATA AND TRUNCATED DISTRIBUTIONS

It is important to distinguish between truncated data and censored data. They are sometimes confused. Censoring occurs when there is a bound on an observation (lower bound for observations censored on the right, upper bound for observations censored on the left, and both upper and lower bounds for observations that are interval censored). Truncation, however, arises when even the *existence* of a potential observation would be unknown if its value were to lie in a certain range. Usually, truncation occurs to the left of a specified point τ^L or to the right of a specified point τ^U.

11.6.1 Examples of Left Truncation

Example 11.8 Ultrasonic Inspection of Material. Ultrasonic inspection is used to detect flaws in titanium alloys during several stages of manufacturing of jet engine turbine disks. Undetected flaws in such parts could cause early initiation of a fatigue crack and thus increase the risk of failure. Ultrasonic signal amplitude is generally positively correlated with flaw size. Thus the distribution of signal amplitudes reflected from flaws provides information on the distribution of flaw sizes. Titanium is a "noisy" material. Titanium grain boundaries reflect about as well as small flaws. Thus below a specified threshold τ^L, it is impossible to be sure whether a signal is from a flaw or a grain boundary. Suppose that interest centers on the distribution of signal strengths (including signals below τ^L that would be observed in the absence of material noise). Consider the following two possibilities.

TRUNCATED DATA AND TRUNCATED DISTRIBUTIONS   267

- In a laboratory test of the inspection process, specimens with seeded flaws of known size are inspected. In this case when a reading is taken, it is known that a flaw is present. The signal's amplitude, however, is measured only when the amplitude is above the threshold τ^L. In some applications, τ^L will change from time to time (depending on the local material noise level), but generally the value τ^L can be recorded. Then the number of signals that were below τ^L is known, and these are left-censored observations.

- In an operating inspection process, a flaw is not detected when the signal's amplitude lies below the threshold τ^L. Then we observe only the signals that are greater than τ^L in amplitude. The number of flaws that were present with signals below τ^L is unknown. The observations recorded as being above τ^L are known as left-truncated observations, or observations from a left-truncated distribution (in the case where τ^L is the same for all readings). □

If all units below τ^L in a population or process are screened out before observation, the remaining data are from a "left-truncated" distribution. Depending on the application, interest could center on either the original untruncated distribution or on the truncated distribution. For most problems the additional information provided by the proportion of observations truncated (either in the population or the sample) would lead to censoring instead of truncation and importantly improve estimation precision of the original (unconditional) distribution's parameters.

Example 11.9   Life Data with Pretest Screening.   Table 1.3, described in Example 1.3, gives the number of observed failures from a field-tracking study of circuit packs. The Vendor 2 units had already seen 1000 hours of burn-in testing at the manufacturing plant, but no information was available on the number of units that had failed in that test. Thus the Vendor 2 circuit packs are left-truncated. If the number of circuit packs that failed in the burn-in period were known, then the data could be treated as censored. □

Example 11.10   Distribution of Brake Pad Life from Observational Data.
Kalbfleisch and Lawless (1992) give data on brake pad life from a study of automobiles. For each automobile in the study (where i is used to index the individual automobiles), the number of kilometers (v_i) driven and a wear measurement (w_i) were taken at a point in time when the automobiles were being serviced (for something other than brake problems). The wear measurement is such that $w_i = 0$ represents no wear and $w_i = 1$ is the level of brake wear that requires replacement of the pads. Suppose, for a given automobile, that wear is proportional to accumulated driven kilometers, which suggests a brake pad life estimate of $t_i = v_i/w_i$. The important assumption being used here is that the main source of variability in brake pad life is the automobile-to-automobile variability in the t_i values and that failure time could be predicted accurately from these observed ratios. Automobiles that had previously had brake pads replaced were not included in the study. For this reason, high-wear-rate automobiles are under represented in the study. To correct for this, Kalbfleisch

and Lawless treat the life prediction t_i as left-truncated at $\tau_i^L = v_i$, the number of kilometers of service at the time of the prediction. The idea is that if the wear rate had been high enough to cause failure before the regularly scheduled service call at τ_i^L, the automobile would not have been included in the study. □

11.6.2 Likelihood with Left Truncation

Following the general development in Section 2.4.5, if a random variable T_i is truncated on the left at τ_i^L, then the likelihood (probability) of an observation in the interval $(t_i^L, t_i]$ is the conditional probability

$$L_i(\boldsymbol{\theta}) = \Pr(t_i^L < T_i \leq t_i \mid T_i > \tau_i^L) = \frac{F(t_i; \boldsymbol{\theta}) - F(t_i^L; \boldsymbol{\theta})}{1 - F(\tau_i^L; \boldsymbol{\theta})}, \quad t_i > t_i^L \geq \tau_i^L.$$

For an observation reported as an exact failure at time t_i, the corresponding density-approximation form of the likelihood is

$$L_i(\boldsymbol{\theta}) = \frac{f(t_i; \boldsymbol{\theta})}{1 - F(\tau_i^L; \boldsymbol{\theta})}, \quad t_i > \tau_i^L. \tag{11.4}$$

It is possible to have either right or left censoring when sampling from a left-truncated distribution. The recorded censoring time will exceed τ_i^L. As in Table 2.3, to obtain $L_i(\boldsymbol{\theta})$ for a censored observation, one simply replaces the numerator in (11.4) by $F(t_i; \boldsymbol{\theta}) - F(\tau_i^L; \boldsymbol{\theta})$ for an observation that is left-censored at $t_i > \tau_i^L$ and by $1 - F(t_i; \boldsymbol{\theta})$ for an observation that is right-censored at $t_i > \tau_i^L$.

11.6.3 Nonparametric Estimation with Left Truncation

Section 3.10 presents a general method, due to Turnbull (1976), for nonparametric estimation that can be extended to the analysis of arbitrarily censored and truncated data. When there is left truncation, Turnbull's method provides a nonparametric estimate of the conditional distribution

$$F_C(t) = \Pr(T \leq t \mid T > \tau_{\min}^L) = \frac{\Pr(\tau_{\min}^L < T \leq t)}{\Pr(T > \tau_{\min}^L)} = \frac{F(t) - F(\tau_{\min}^L)}{1 - F(\tau_{\min}^L)}, \quad t > \tau_{\min}^L, \tag{11.5}$$

where $\tau_{\min}^L$ is the smallest left-truncation time in the sample. Without a parametric assumption, the data contain no information about $F(t)$ below $\tau_{\min}^L$.

For purposes of probability plotting to assess the adequacy of a parametric assumption for $F(t; \boldsymbol{\theta})$, one must have, instead, an estimate of the unconditional distribution of T. In this case we use a parametric model to estimate $\Pr(T > \tau_{\min}^L)$ and then compute a parametrically adjusted nonparametric estimate of $F(t)$. Let $\widehat{F}_{NPC}(t)$ denote the nonparametric estimate of the conditional distribution $F_C(t)$. Then a parametrically adjusted unconditional nonparametric estimate of $F(t)$ will be denoted by $\widehat{F}_{NPU}(t)$. This estimate is obtained by substituting $\widehat{F}_{NPC}(t)$ for $F_C(t)$ and $F(\tau_{\min}^L; \widehat{\boldsymbol{\theta}})$ for $F(\tau_{\min}^L)$

in (11.5) and solving for $F(t)$. This gives

$$\widehat{F}_{NPU}(t) = F(\tau_{\min}^L; \widehat{\boldsymbol{\theta}}) + \widehat{F}_{NPC}(t)\left[1 - F(\tau_{\min}^L; \widehat{\boldsymbol{\theta}})\right], \quad t > \tau_{\min}^L, \qquad (11.6)$$

where $\widehat{\boldsymbol{\theta}}$ is a parametric estimate of the parameters in $F(t; \boldsymbol{\theta})$.

11.6.4 ML Estimation with Left-Truncated Data

Example 11.11 Analysis of Life Data with Pretest Screening. As described in Example 11.9, because they had already seen 1000 hours of burn-in testing, the Vendor 2 data in Table 1.3 are left-truncated at 1000 hours. Figure 11.11 is a lognormal probability plot of the Vendor 2 failure data, parametrically adjusted with a lognormal ML fit to the data. The plot also shows the lognormal (straight line) and Weibull (curved line) ML estimates, and approximate 95% pointwise lognormal confidence intervals. In this example, the estimated proportion truncated on the left of 1000 hours appears to be (through parametric estimation) very small $[\widehat{F}(1000) = .95 \times 10^{-5}$ for the lognormal and $.38 \times 10^{-4}$ for the Weibull distribution] and so the parametric adjustment to the nonparametric estimate used in making the probability plot was extremely small. If the failure mode before 1000 hours was different from that after 1000 hours, however, this kind of extrapolation into the lower tail of the distribution

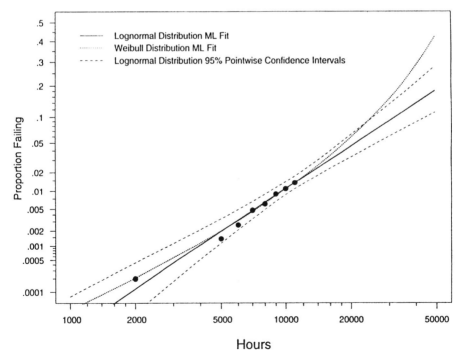

Figure 11.11. Parametrically adjusted lognormal probability plot of Vendor 2 circuit pack data comparing Weibull and lognormal ML estimates.

would give a misleading view of what happened up to 1000 hours (analysis of the Vendor 1 data described in Exercise 11.10 and Section 22.3 suggests that there well might have been such a change). In any case, the available data can be used to estimate failure probabilities beyond 1000 hours, conditional on the burn-in test results.

For extrapolation beyond the range of the data, the lognormal distribution gives much more optimistic (i.e., lower) estimates of failure probabilities than the Weibull distribution. This lognormal/Weibull contrast is common and leads to the question: "Which distribution is most appropriate for extrapolation?" The data have little to say about differentiating between these two distributions. If there is an answer to the question it would be in knowledge about the physics of failure for the observed failure mode(s). In the absence of such knowledge, reported uncertainty would have to encompass not only the sampling variability quantified in the confidence intervals, but also the distribution uncertainty. As illustrated in Section 11.3, fitting the extended generalized gamma distribution will help to quantify and illustrate such distributional uncertainty. □

11.6.5 Examples of Right Truncation

Right truncation is similar to left truncation and occurs when the values in the upper tail of the distribution are removed.

Example 11.12 Screening Out Units with Large Flaws. Degree of porosity is an important quality metric in casting processes. The distribution of pore sizes is closely related to the failure time of a component in a particular application. If large pores or other voids occur inside a casting, fatigue cracks will initiate more rapidly, leading to premature failure. In a particular manufacturing process, castings for automobile engine mounts are inspected by x-ray to make sure that there are no large internal voids or pores. Pores larger than 10 microns can be detected with high probability. The casting process output has a distribution of pore sizes. The inspection process truncates off the upper tail of the pore-size distribution (i.e., units containing pores greater than 10 microns are eliminated). Thus the distribution of pore size in units passing inspection could be described by a right-truncated distribution. □

Example 11.13 Warranty Data with Limited Information for Unfailed Units.
A particular home appliance, after purchase, is either used regularly or not at all. The percentage of units actually put into regular use is unknown. During a particular production period, an incorrect component (i.e., one that did not have the specified power rating) was installed in all of the units that were produced. When failures occur among these regularly used units, the units are returned to the manufacturer for repair or replacement under a long-term warranty program. The manufacturer learns about failures from this group of units only if the unit is actually put into service and if the unit fails before the analysis time. In this case, the observed failure times can be viewed as a sample from a distribution right-truncated at a time equal to the difference between the time of analysis and the time at which the unit was put into service. □

TRUNCATED DATA AND TRUNCATED DISTRIBUTIONS   271

Example 11.14 Limited Failure Population Data. Similar to the situation in Example 11.13, the 28 IC failure times in Example 11.5 can be viewed as a sample from a distribution right-truncated at $\tau^U = t_c = 1370$ hours. Intuitively this is because, although 4156 ICs were tested, the number of potential susceptible units in the population was unknown. Susceptible units become known as such only if they failed within the 1370-hour-long life test. Exercise 11.4 provides technical justification for this. □

As shown in the analysis of the LFP data in Section 11.5 it will be impossible to estimate the unconditional failure-time distribution from right-truncated data unless a parametric form is specified for the distribution. Even with a specified failure-time distribution, there are serious estimability problems (leading to wide confidence intervals on quantities of interest) unless the proportion truncated is very small (say, less than 5%). This problem is also described in Kalbfleisch and Lawless (1988).

11.6.6 Likelihood with Right (and Left) Truncation

If the random variable T_i is truncated when it lies above τ_i^U then the likelihood (probability) of an interval observation is

$$L_i(\boldsymbol{\theta}) = \Pr(t_i^L < T_i \leq t_i \mid T_i \leq \tau_i^U) = \frac{F(t_i; \boldsymbol{\theta}) - F(t_i^L; \boldsymbol{\theta})}{F(\tau_i^U; \boldsymbol{\theta})}, \quad 0 \leq t_i^L < t_i \leq \tau_i^U.$$

For an observation reported as an exact failure at time t_i, the corresponding density-approximation form of the likelihood is

$$L_i(\boldsymbol{\theta}) = \frac{f(t_i; \boldsymbol{\theta})}{F(\tau_i^U; \boldsymbol{\theta})}.$$

As with left truncation, it is possible to have either left or right censoring when sampling from the right-truncated distribution. With both left and right truncation, the appropriate likelihood for an interval-censored observation is

$$\begin{aligned} L_i(\boldsymbol{\theta}) &= \Pr(t_i^L < T_i \leq t_i \mid \tau^L \leq T < \tau^U) \\ &= \frac{F(t_i; \boldsymbol{\theta}) - F(t_i^L; \boldsymbol{\theta})}{F(\tau_i^U; \boldsymbol{\theta}) - F(\tau_i^L; \boldsymbol{\theta})}, \quad \tau_i^L \leq t_i^L < t_i \leq \tau_i^U. \end{aligned}$$

The likelihood for censored observations is obtained in a manner similar to that described in Section 11.6.2.

11.6.7 Nonparametric Estimation with Right (and Left) Truncation

Section 11.6.3 showed how to parametrically adjust a truncated-data nonparametric estimator so that it could be used for making a probability plot. This approach can be extended to work in situations with right or both left and right truncation. In particular, the nonparametric estimate is for the conditional probability

$$F_C(t) = \Pr(T \leq t \mid \tau_{\min}^L \leq T < \tau_{\max}^U) = \frac{F(t) - F(\tau_{\min}^L)}{F(\tau_{\max}^U) - F(\tau_{\min}^L)}, \quad \tau_{\min}^L < t \leq \tau_{\max}^U,$$

(11.7)

where $\tau_{\max}^U$ is the largest right-truncation time in the sample. Then, as in (11.6), a parametrically adjusted unconditional nonparametric estimate of $F(t)$ is obtained from

$$\widehat{F}_{NPU}(t) = F(\tau_{\min}^L; \widehat{\boldsymbol{\theta}}) + \widehat{F}_{NPC}(t) \left[F(\tau_{\max}^U; \widehat{\boldsymbol{\theta}}) - F(\tau_{\min}^L; \widehat{\boldsymbol{\theta}}) \right], \quad \tau_{\min}^L < t \leq \tau_{\max}^U.$$

(11.8)

Example 11.15 Using Right Truncation to Estimate the Failure-Time Distribution from Limited Failure Population Data. Following Example 11.14, Figure 11.12 is a lognormal probability plot of the 25 pre-100-hour IC failure times from Table 1.2. In contrast to the probability plot in Figure 11.8, the 4131 units that were unfailed (censored) at 100 hours are ignored. The plotted estimate in Figure 11.12, as described in Section 11.6.7, is really an estimate of $\Pr(T \leq t \mid T \leq 100)$. Treat-

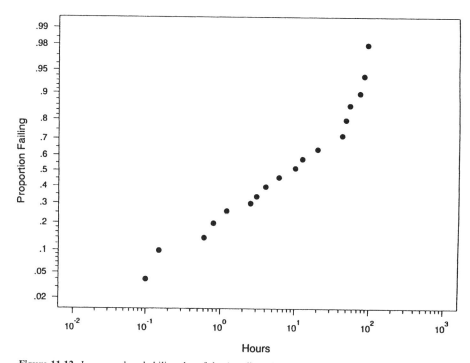

Figure 11.12. Lognormal probability plot of the (unadjusted) nonparametric estimate of the IC failure-time distribution, conditional on failure in the first 100 hours.

FITTING DISTRIBUTIONS THAT HAVE A THRESHOLD PARAMETER 273

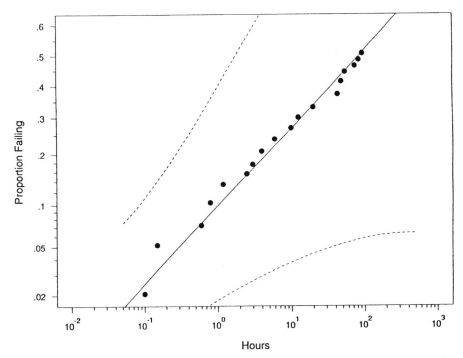

Figure 11.13. Lognormal probability plot of the lognormal-adjusted (unconditional) nonparametric estimate of the IC failure-time distribution with pointwise 95% normal-approximation confidence intervals for the lognormal $F(t)$.

ing the 25 failures as right-truncated at 100 hours (or estimating the parameters of a lognormal distribution truncated at 100 hours) gives $\hat{\mu} = 4.44$ and $\hat{\sigma} = 3.44$. From this, $F(100; \hat{\mu}, \hat{\sigma}) = \Pr(T \leq 100) = \Phi[(\log(100) - 4.44)/3.44] = .5191$. This can be used in (11.8) to parametrically adjust the nonparametric estimate of $\Pr(T \leq t \mid T \leq 100)$ for purposes of probability plotting. The parametrically adjusted probability plot is shown in Figure 11.13 along with the ML estimate $F(t; \hat{\mu}, \hat{\sigma})$ based on the truncated data. The pointwise confidence intervals are very wide because of the important amount of information lost by not having direct information about $\Pr(T \leq 100)$. If we knew the actual number of defectives in the sample of 4156 ICs, and used this to determine the number of censored susceptible ICs, the intervals would be much tighter (see Exercise 11.2). □

11.7 FITTING DISTRIBUTIONS THAT HAVE A THRESHOLD PARAMETER

In many areas of application, analysts wish to fit distributions with a "threshold parameter" γ that shifts the distribution of a positive random variable (usually to the right) by an amount γ. Recall from Section 5.10.1 that the three-parameter lognormal cdf and pdf can be written as

$$F(t; \mu, \sigma, \gamma) = \Phi_{\text{nor}}\left[\frac{\log(t-\gamma) - \mu}{\sigma}\right],$$

$$f(t; \mu, \sigma, \gamma) = \frac{1}{\sigma(t-\gamma)} \phi_{\text{nor}}\left[\frac{\log(t-\gamma) - \mu}{\sigma}\right], \quad t > \gamma.$$

Threshold versions of any other distribution with support on $[0, \infty)$ (e.g., exponential, Weibull, loglogistic, gamma, inverse Gaussian, and Birnbaum–Saunders) can be defined in this manner. Although the discussions in the following sections use the three-parameter lognormal and Weibull distributions, the basic ideas hold also for the other threshold distributions. In some physical applications it makes sense to constrain $\gamma > 0$, but there is no theoretical need to do this.

11.7.1 Estimation with a Given Threshold Parameter

If the threshold parameter γ is given, one can subtract γ from all reported failure, inspection, and censoring times and then use the simpler methods for the base distribution without the threshold parameter (e.g., the one-parameter exponential distribution methods in Chapter 7 and the two-parameter Weibull and lognormal distribution methods in Chapter 8). Of course one needs to adjust inferences accordingly. For example, the given value of γ must be added back into estimates of quantiles and one must subtract γ from times before computing failure probabilities, hazard function values, or other functions of time. Using a specified value of γ that is seriously incorrect can lead to seriously incorrect conclusions.

11.7.2 Probability Plotting Methods

With three-parameter log-location-scale distributions, $\log(T - \gamma)$ has a location-scale distribution with parameters μ and σ. For such distributions, there are two different methods of probability plotting. These methods do not compete, but rather complement each other.

- One can make log-location-scale distribution probability plots (e.g., standard Weibull or lognormal probability plots) using $T - \gamma$ over a range of different γ values. Choosing a value that linearizes the probability plot (as discussed in Section 6.2.4) provides a graphical estimate of γ. Then, conditional on the fixed value of γ, one can obtain graphical estimates of μ and σ.
- As illustrated in Section 6.5, with a specified value for the shape parameter σ, a log-location-scale distribution with a threshold parameter can be treated as a location-scale distribution with a location parameter γ and scale parameter $\exp(\mu)$. One can either use ML to estimate the shape parameter σ or try different values of σ to find one that provides a reasonably straight probability plot. Conditional on the fixed value of σ, one can obtain graphical estimates of γ and $\exp(\mu)$.

Both of these approaches are easy to implement with flexible computer programs.

11.7.3 Likelihood Methods

The likelihood for a three-parameter log-location-scale distribution using the "density approximation" for exact failures, and allowing for right-censored observations, has the form

$$
\begin{aligned}
L(\mu, \sigma, \gamma) &= \prod_{i=1}^{n} L_i(\mu, \sigma, \gamma; \text{data}_i) \\
&= \prod_{i=1}^{n} \{f(t_i; \mu, \sigma, \gamma)\}^{\delta_i} \{1 - F(t_i; \mu, \sigma, \gamma)\}^{1-\delta_i} \\
&= \prod_{i=1}^{n} \left\{ \frac{1}{\sigma(t_i - \gamma)} \phi\left[\frac{\log(t_i - \gamma) - \mu}{\sigma}\right] \right\}^{\delta_i} \left\{ 1 - \Phi\left[\frac{\log(t_i - \gamma) - \mu}{\sigma}\right] \right\}^{1-\delta_i},
\end{aligned}
$$

where $\text{data}_i = (t_i, \delta_i)$ is defined as in (11.1). This is a classic example for which density approximation defined in equation (7.13) can cause serious numerical and statistical problems in the application of ML estimation. The problem is that when this approximation is used, there can be, for some model/data combinations, a path in the parameter space for which the likelihood goes to ∞. In particular, when $\gamma \to t_{(1)}$ (the smallest observation) and $\sigma \to 0$, the likelihood $L(\mu, \sigma, \gamma)$ approaches ∞. The likelihood approaches ∞ *not* necessarily because the probability of the data is large in that region of the parameter space, but rather because of a breakdown in the density approximation in (7.13). For some (but not all) data sets there is a local maximum for $L(\mu, \sigma, \gamma)$, corresponding to the maximum of the correct likelihood (probability of the data). Although it has been suggested that one could ignore the part of the likelihood surface where $L(\mu, \sigma, \gamma) \to \infty$ and use the local maximum to provide the ML estimates, this practice can lead to numerical difficulties and it is possible for the local maximum to be masked by the breakdown of the density approximation. A better solution is to use the correct likelihood contributions,

$$L_i(\mu, \sigma, \gamma; \text{data}_i) = [F(t_i + \Delta_i; \boldsymbol{\theta}) - F(t_i - \Delta_i; \boldsymbol{\theta})] \qquad (11.9)$$

(based on small intervals implied by the data's precision) instead of the density approximation. The correct likelihood will always be bounded (because probabilities can be no larger than 1). Using the correct likelihood eliminates the problem of an unbounded likelihood and helps simplify the process of finding the ML estimates. The values of Δ_i should be chosen to reflect the round-off in the data (which often depends on the magnitude of the observations within a data set). Generally the shape and position of the likelihood (and thus the ML estimates) are not very sensitive to the value of Δ_i used here.

Numerical problems with fitting threshold parameter distributions can also arise from the embedded distributions problem described in Section 5.10.3. In order to

Table 11.2. Alloy-C Strength Data

Strength (ksi) Interval Endpoint		Number of Failures
Lower	Upper	
79	80	1
80	81	0
81	82	4
82	83	4
83	84	9
84	85	25
85	86	21
86	87	18
87	88	2

circumvent this problem, one can fit, instead, the generalized threshold-scale (GETS) distribution introduced in Section 5.11. When $\hat{\sigma} > 0$, the ML estimates for this model are equivalent to what one would get with the corresponding threshold parameter distribution. When $\hat{\sigma} < 0$, it is an indication that the ML estimates for the corresponding threshold-parameter distribution would be on the boundary of the parameter space (i.e., the limiting embedded distribution) and careful consideration should be given to using the GETS or some other alternative to the corresponding threshold-parameter distribution.

Example 11.16 Fitting the Three-Parameter Weibull Distributions to the Alloy-C Strength Data. Table 11.2 gives interval data for tensile strength (in ksi) from a sample of 84 specimens of Alloy-C. The test was run to obtain information on the strength of the alloy when produced with a modified process. Figure 11.14 gives a histogram of the data. The distribution is skewed to the left and the lowest values of strength seem far from the origin, relative to the spread in the data. This

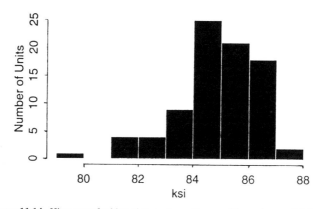

Figure 11.14. Histogram for binned strength readings on 84 specimens of Alloy-C.

FITTING DISTRIBUTIONS THAT HAVE A THRESHOLD PARAMETER 277

figure indicates that a threshold-parameter distribution could provide an adequate description for the data.

Figure 11.15 shows a sequence of fitted three-parameter Weibull distributions in which γ was fixed at a set of values between -20 and 79. The probability plot in which $\gamma = 0$ illustrates the fitting of a two-parameter Weibull distribution. As γ gets larger than 70, the likelihood drops off rapidly and, simultaneously, the fit in the probability plots becomes poorer. We see, however, that as γ gets smaller, the profile levels off and the fit remains good. This is an example in which the three-parameter Weibull is approaching the embedded SEV distribution. The scales on the probability plot suggest an intuitive explanation for the embedding behavior. As γ becomes smaller and smaller, we are adding a larger and larger number to the original strength values. With the plot running over a range of strength values that is relatively small, the log axis on the strength scale is approximately linear and further shifting will act as a location shift and thus have little effect on the fit of the distribution to the data.

Figure 11.16 shows a SEV probability plot of the Alloy-C strength data nonparametric estimate of $F(t)$ along with straight lines depicting ML estimates of the three-parameter Weibull (SEV-GETS with $\sigma > 0$) and the SEV distributions (corresponding to the limiting embedded SEV distribution with $\gamma = -\infty$, where $\hat{\mu} = 85.512$ and $\hat{\sigma} = 1.159$). Either distribution fits the data very well. The SEV-GETS ML estimates are $\hat{\alpha} = 85.503$, $\hat{\sigma} = .015$, and $\hat{\varsigma} = 1.166$. Then using the relationships in Section 5.11.4, with $\hat{\sigma} > 0$, yields the three-parameter Weibull distribution ML estimates as $\hat{\gamma} = \hat{\alpha} - \hat{\varsigma}/\hat{\sigma} = 7.77$ and $\hat{\mu} = \log(\hat{\varsigma}/\hat{\sigma}) = 4.35$. This plot confirms that the embedded SEV distribution fits well, but there may be an objection to the use of such a distribution because there is a positive probability (albeit extremely small) of a negative strength. In this example, the two-parameter Weibull fit, illustrated in the $\gamma = 0$ plot in Figure 11.15, is indistinguishable over the range of the data from the three-parameter Weibull fit in Figure 11.16. Thus the two-parameter Weibull fit provides a physically reasonable, parsimonious description of the data.

□

Example 11.17 Fan Data and the Three-Parameter Lognormal Distribution. This example fits the three-parameter lognormal distribution to the fan data that were introduced in Example 1.4 and are given in Appendix Table C.1. The profile plot in Figure 11.17 shows the breakdown in the density approximation for the likelihood. As $\gamma \to x_{(1)}$ (the smallest observation), the profile likelihood for γ blows up. To eliminate this problem, use the correct likelihood. For these data we use the likelihood based on recognizing that data were recorded to a precision of ± 5 hours and choose $\Delta_i = 5$ in (11.9). Figure 11.18 shows that, with the correct likelihood, the profile plot is well behaved with a clear maximum at a value of γ that is a little less than 400. Figure 11.19 is a lognormal probability plot of the fan data comparing the three-parameter lognormal and the three-parameter Weibull distributions with the two-parameter lognormal distribution. The NOR-GETS ML estimates in this example are $\hat{\alpha} = 41261.2588$, $\hat{\sigma} = 2.27164389$, and $\hat{\varsigma} = 92853.2782$, which, because $\hat{\sigma} > 0$, yields the three-parameter lognormal distribution ML estimates as $\hat{\gamma} = \hat{\alpha} - \hat{\varsigma}/\hat{\sigma} = 386.33$ and $\hat{\mu} = \log(\hat{\varsigma}/\hat{\sigma}) = 10.618$ and $\hat{\sigma}$ as above.

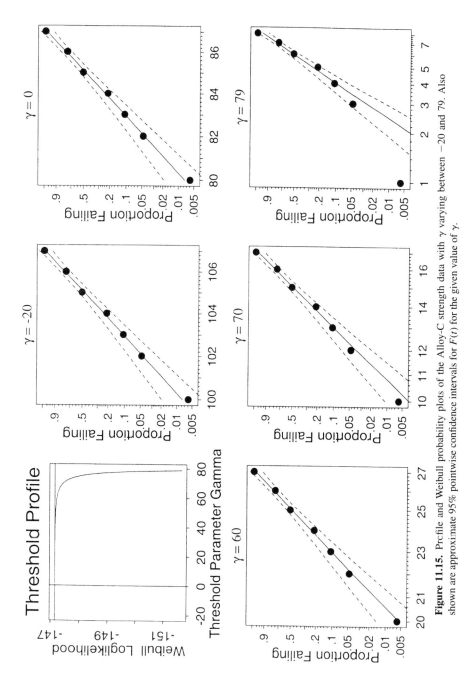

Figure 11.15. Profile and Weibull probability plots of the Alloy-C strength data with γ varying between -20 and 79. Also shown are approximate 95% pointwise confidence intervals for $F(t)$ for the given value of γ.

FITTING DISTRIBUTIONS THAT HAVE A THRESHOLD PARAMETER   279

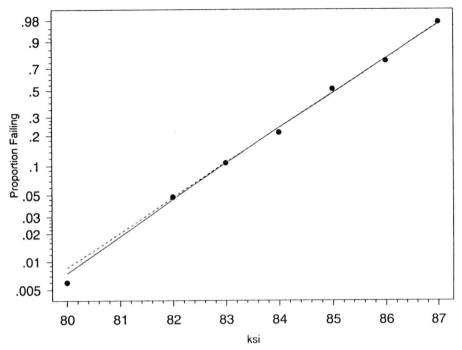

Figure 11.16. Smallest extreme value probability plot of the Alloy-C strength data showing the three-parameter Weibull (i.e., SEV-GETS distribution with $\sigma > 0$) (solid line) along with the SEV distribution (dotted line).

These analyses give a strong indication that fitting a threshold-parameter distribution to these data would be overfitting, unless there were strong physical reasons to suggest that such a threshold exists. Visually, the two-parameter lognormal distribution ($\mu = 10.14$, $\sigma = 1.68$) provides a reasonably adequate and parsimonious fit to the data. □

Example 11.18 Fitting the Threshold-Parameter Distributions to the Alloy T7987 Data. Example 6.7 suggested that a threshold-parameter distribution might be appropriate for the Alloy T7987 data given in Table 6.1. Figure 11.20 is a Weibull probability plot comparing the three-parameter lognormal and three-parameter Weibull distributions for the Alloy T7987 data. The SEV-GETS ML estimates are $\widehat{\alpha} = 186.3$, $\widehat{\sigma} = .76$, and $\widehat{\varsigma} = 70.64$, which, because $\widehat{\sigma} > 0$, yields the three-parameter Weibull distribution ML estimates as $\widehat{\gamma} = \widehat{\alpha} - \widehat{\varsigma}/\widehat{\sigma} = 92.99$ and $\widehat{\mu} = \log(\widehat{\varsigma}/\widehat{\sigma}) = 4.54$. The NORMAL-GETS ML estimates are $\widehat{\alpha} = 162.25$, $\widehat{\sigma} = .613$, and $\widehat{\varsigma} = 55.28$. Using the relationships in Section 5.11.4, with $\sigma > 0$, yields the three-parameter lognormal distribution ML estimates as $\widehat{\gamma} = \widehat{\alpha} - \widehat{\varsigma}/\widehat{\sigma} = 72.03$ and $\mu = \log(\widehat{\varsigma}/\widehat{\sigma}) = 4.50$. In this case there is strong evidence that the threshold parameter is important for describing the failure-time distribution. The data do not

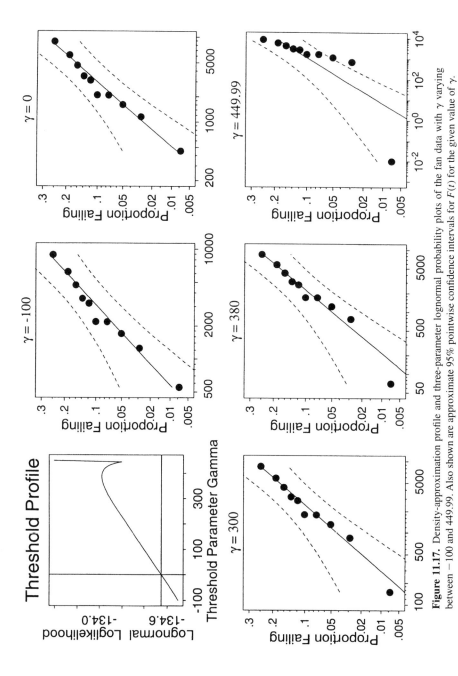

Figure 11.17. Density-approximation profile and three-parameter lognormal probability plots of the fan data with γ varying between -100 and 449.99. Also shown are approximate 95% pointwise confidence intervals for $F(t)$ for the given value of γ.

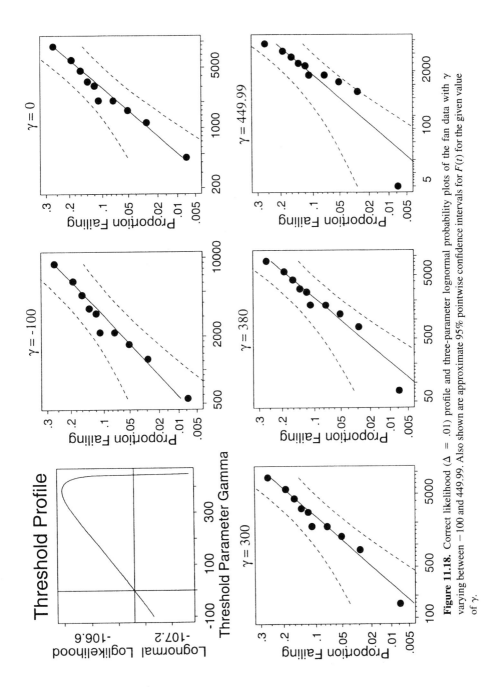

Figure 11.18. Correct likelihood ($\Delta = .01$) profile and three-parameter lognormal probability plots of the fan data with γ varying between -100 and 449.99. Also shown are approximate 95% pointwise confidence intervals for $F(t)$ for the given value of γ.

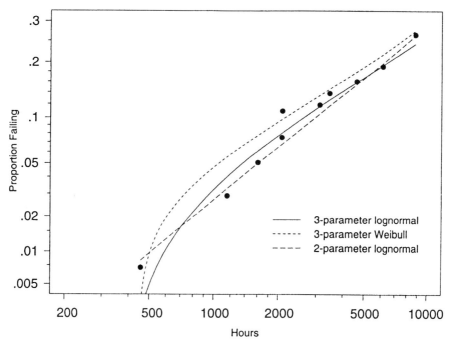

Figure 11.19. Lognormal probability plot of the fan data comparing the three-parameter lognormal and Weibull distributions with the two-parameter lognormal distribution.

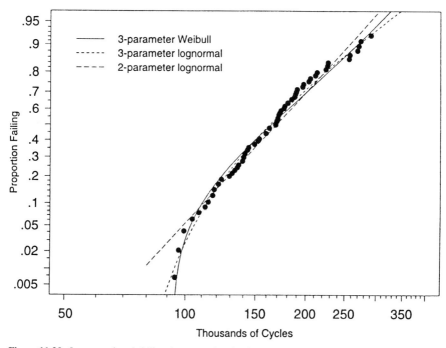

Figure 11.20. Lognormal probability plot comparing the three-parameter lognormal and three-parameter Weibull distributions and the two-parameter Weibull distribution for the Alloy T7987 data.

suggest a preference for either the three-parameter lognormal or three-parameter Weibull distribution. □

11.7.4 Summary of Results of Fitting Models to Skewed Distributions

The results in this section indicate some general guidelines for fitting parametric distributions to skewed data. We have seen that as $\gamma \to -\infty$ and $\sigma \to 0$, the shape of the three-parameter threshold distribution approaches that of the underlying embedded distribution. The Weibull approaches the form of a smallest extreme value (which is left-skewed) and the lognormal approaches the form of a normal distribution (symmetric). These results indicate that, for purposes of fitting parametric distributions to data:

- If data are left-skewed, even if far from the origin, it is generally possible to fit a three-parameter Weibull distribution and achieve a good fit to the data. In many cases, however, it will be possible to fit the simpler two-parameter Weibull or smallest extreme value distributions and get, effectively, the same results. The typical profile likelihood shape for such data is shown in Figure 11.15.
- If data are approximately symmetric, one can generally fit a three-parameter Weibull, a three-parameter lognormal model, or two-parameter versions of the distribution and get a reasonably good fit to the data. In many cases, however, it will be possible to fit the simpler two-parameter normal or logistic distributions (depending on the heaviness of the tails) and get, effectively, the same results. Often, unless there is a large amount of data (hundreds of observations), it will be difficult to distinguish among these alternative distributions.
- If the data are right-skewed, it is often possible to fit either the three-parameter Weibull or the three-parameter lognormal distribution and get a good fit to the data.
- The use of a threshold parameter can be viewed from two different directions. Sometimes it might be viewed as a physical parameter—a time before which probability of failure is zero or a threshold strength. In such cases it may be important to constrain $\gamma > 0$ or some other number. In other cases, γ is one of several parameters of a curve being fit to data. In such cases, the ML estimate of γ may be negative, yielding a positive probability of negative failure time or strength. Generally, however, the estimated probability of such small events is small enough to ignore.

When one of the simpler distributions (e.g., two-parameter lognormal or two-parameter Weibull) fits one's data well, the simpler description will be preferred to a threshold distribution, especially when the amount of data available is limited. When one of the simpler distributions does not fit, however, using a threshold parameter may provide an important improvement in data description. As usual, however, it is important to be especially cautious when making inferences outside the range of one's data, especially when the fitted distribution is chosen purely on the basis of its fit to the data.

BIBLIOGRAPHIC NOTES

Farewell and Prentice (1977) show that a judicious choice of parameters for the GENG model can make an important difference in one's ability to apply ML methods. They suggest the parameterization given in Section 5.4. Lawless (1982, Chapter 5) shows how to use likelihood-based methods with the generalized gamma distribution to assess and compare results from the special case distributions. Liu, Meeker, and Escobar (1998) suggest and illustrate the use of an EGENG parameterization that is stable even for heavy censoring.

Engelhardt, Bain, and Wright (1981) describe methods for ML estimation for the Birnbaum–Saunders distribution. Cohen and Whitten (1988) describe estimation methods, including ML, for a wide variety of life distributions including the Weibull, lognormal, inverse Gaussian, gamma, and generalized gamma distributions.

Meeker (1987) provides a more complete analysis and more technical details for the LFP model described in Section 11.5. Also, Monte Carlo simulation showed that the likelihood-based confidence intervals provide a much better approximation to the nominal confidence levels over a wide range of parameter values for the LFP model. Trindade (1991) also uses this model in a similar application.

In general, fitting mixture distributions with maximum likelihood presents some difficult and challenging issues. See, for example, Day (1969), Falls (1970), Hosmer (1973), Titterington, Smith, and Makov (1985), and Gelman, Carlin, Stern, and Rubin (1995, Chapter 16). Also Everitt and Hand (1981) review the pertinent literature.

Turnbull (1976) presents a generalization of the Kaplan–Meier estimate for arbitrarily censored and truncated data. Kalbfleisch and Lawless (1988) describe examples of field reliability data that can be analyzed using truncated data methods. Nelson (1990b) describes examples of truncated reliability data and shows how to adapt the method of hazard plotting to such data. Kalbfleisch and Lawless (1992) provide further examples and methods. Woodroofe (1972, 1974), Schneider (1986), and Cohen (1991) are useful references for the theoretical aspects of truncated data and estimating the parameters of truncated distributions. Escobar and Meeker (1998d) show how to compute the Fisher information matrix and asymptotic variances for truncated distributions and the LFP model.

Serious numerical and statistical problems can arise when estimating the parameters of threshold-parameter distributions, especially when using the density approximation for the likelihood contributions of observations reported as exact failures (see Kempthorne and Folks, 1971; Giesbrecht and Kempthorne 1976; and Cheng and Iles, 1987, for more details and other references). Griffiths (1980) and Smith and Naylor (1987) describe likelihood-based inferences for the three-parameter lognormal and Weibull distributions, respectively. Also, the asymptotic ML theory for this approach is complicated (e.g., see Smith, 1985) and, arguably, inappropriate for data with finite precision. Cheng and Iles (1990) noted that the smallest extreme value (normal) distribution is a limiting case of the three-parameter Weibull (lognormal) distribution when the threshold parameter $\to -\infty$. Hirose and Lai (1997) use an example to illustrate the problems in inference created by embedded models when fitting a threshold Weibull model with binned data. They propose a solution to those problems by embedding the Weibull model in SEV-GETS family and using ML methods.

EXERCISES

11.1. Wilk, Gnanadesikan, and Huyett (1962b) give the number of weeks until failure for a sample of 34 transistors subjected to accelerated conditions. The reported times, with the number of ties shown in parentheses, were 3, 4, 5, 6(2), 7, 8(2), 9(3), 10(2), 11(3), 13(5), 17(2), 19(2), 25, 29, 33, 42(2), 52. The other 3 transistors had not failed at the end of 52 weeks.
  (a) Use ML to estimate the parameters of the BISA, IGAU, gamma, and lognormal distributions to these data using a discrete-data likelihood. Plot all of these estimates on lognormal probability paper and compare the different estimates. Describe any important differences that you see in the estimates.
  (b) Redo the gamma distribution analyses assuming that the failures occurred at exactly the reported time. Are the differences of practical importance in this example?

11.2. The engineers who collected the IC data from Example 11.14 felt that if the life test had been extended for another 50,000 hours (corresponding to the technological life of the application system), only another 2 or 3 failures might have been observed. Use this information to construct several Weibull and lognormal probability plots for the failure-time distribution for the defective subpopulation.

11.3. For the Weibull LFP (limited failure population) model with cdf given in (11.2), a proportion p of the units in the population is defective and will eventually fail (according to a Weibull distribution) and all other units are immune to failure.
  (a) What is the practical interpretation of the parameter μ in this model?
  (b) Generally in a life test of a limited failure population, if the test is stopped far before all of the units in the population have failed, the parameters μ and p will be highly correlated. Give an intuitive explanation for this.
  (c) The expected proportion of units failing in a life test of length t_c might be a more "stable" parameter to estimate. Explain why. Write down the reparameterized model, and describe the steps that you would use to find the ML estimates of this new parameter as well as μ and σ.

11.4. Refer to Exercise 11.3. Show that the likelihood for the LFP model with a single censoring time can be factored into two components, one a binomial with parameter $\pi = p \times \Phi\{[\log(t_c) - \mu]/\sigma\}$ and another consisting of a right-truncated failure-time distribution with truncation time $\tau_i^U = t_c$.

▲11.5. Comment on the statistical implications of the factoring in Exercise 11.4.

11.6. Suppose that T has a WEIB(μ, σ) distribution. Show the following:
   (a) The cdf of the left-truncated Weibull distribution is
   $$F(t; \tau^L; \mu, \sigma) = 1 - \exp\left[(\tau^L/\eta)^\beta - (t/\eta)^\beta\right], \quad t > \tau^L \geq 0,$$
   where $\eta = \exp(u)$ and $\beta = 1/\sigma$.
   (b) The cdf of the right-truncated Weibull distribution is
   $$F(t; \tau^U; \mu, \sigma) = \frac{1 - \exp\left[-(t/\eta)^\beta\right]}{1 - \exp\left[-(\tau^U/\eta)^\beta\right]}, \quad 0 < t \leq \tau^U.$$
   (c) The cdf of the left-truncated and right-truncated Weibull distribution is
   $$F(t; \tau^L, \tau^U; \mu, \sigma) = \frac{\exp\left[-(\tau^L/\eta)^\beta\right] - \exp\left[-(t/\eta)^\beta\right]}{\exp\left[-(\tau^L/\eta)^\beta\right] - \exp\left[-(\tau^U/\eta)^\beta\right]}, \quad \tau^L < t \leq \tau^U.$$

11.7. Refer to Exercise 11.6, part (a). What is the cdf of the truncated distribution when $\sigma = 1$?

▲11.8. Derive the expressions given in parts (a) and (b) of Exercise 11.6 as limiting cases of the expression in part (c) when $\tau^L \to 0$ and $\tau^U \to \infty$, respectively.

▲11.9. Consider ML estimation from a random sample of size n from an EXP(θ) distribution. Denote by $\widehat{\theta}_n$, $\widehat{\theta}_c$, and $\widehat{\theta}_t$ the ML estimates of θ from complete, right-censored, and right-truncated samples, respectively. Define the asymptotic relative efficiency of the estimator $\widehat{\theta}_a$ with respect to the estimator $\widehat{\theta}_b$ as RE$(\widehat{\theta}_a, \widehat{\theta}_b)$ = Avar$(\widehat{\theta}_b)$/Avar$(\widehat{\theta}_a)$.
   (a) Write the Fisher information matrix for the estimators and show that
   $$\text{Avar}(\widehat{\theta}_n) = \theta^2/n,$$
   $$\text{Avar}(\widehat{\theta}_c) = \text{Avar}(\widehat{\theta}_n)/(1 - p_c),$$
   $$\text{Avar}(\widehat{\theta}_t) = \frac{\text{Avar}(\widehat{\theta}_n)}{(1 - p_t)\{1 - p_t[\log(p_t)]^2/(1 - p_t)^2\}},$$
   where $p_c = \exp(-t_c/\theta)$ is the proportion of right censoring and $p_t = \exp(-\tau^U/\theta)$ is the proportion of right truncation.
   (b) Use the results in part (a) with $p_t = p_c$ to show that
   $$\text{RE}(\widehat{\theta}_c, \widehat{\theta}_n) = (1 - p_c),$$
   $$\text{RE}(\widehat{\theta}_t, \widehat{\theta}_n) = (1 - p_t)\left\{1 - p_t[\log(p_t)]^2/(1 - p_t)^2\right\},$$
   $$\text{RE}(\widehat{\theta}_t, \widehat{\theta}_c) = 1 - p_t[\log(p_t)]^2/(1 - p_t)^2.$$

EXERCISES

When $p_t = p_c = .1$ evaluate the relative efficiencies and show that $\text{RE}(\widehat{\theta}_t, \widehat{\theta}_n) = .31$ and $\text{RE}(\widehat{\theta}_c, \widehat{\theta}_n) = .90$. Comment on these efficiencies.

(c) Again suppose that $p_t = p_c$. In the same plot draw $\text{RE}(\widehat{\theta}_c, \widehat{\theta}_n)$ and $\text{RE}(\widehat{\theta}_t, \widehat{\theta}_n)$ as a function of p_t. Comment on the effect of right censoring a proportion p of units when compared with right truncating the same proportion of units.

11.10. Consider the data in Examples 1.3, 11.9, and 11.11. Truncate the Vendor 1 data at 1000 hours and compare the resulting fit with that for Vendor 2. Can you detect any difference that might be considered to be important?

11.11. Consider the results from either Exercise 11.9 or from Exercise 11.14. Provide an intuitive explanation for the reason that precision from the censored distribution is much better than from the truncated distribution.

11.12. During a single month a company sold 2341 modems. These modems have a 36-month warranty. During the first 24 months, 75 of these modems had been returned for warranty service. Suppose that it is reasonable to assume that this is the number failing out of the 2341 modems that were sold. From previous experience with similar products, it is known that a Weibull distribution with a shape parameter of $\beta = .85$ provides an adequate description of the failure-time distribution.

(a) Show that the conditional probability of failing in the third year of life, given survival up to 2 years, can be expressed as a truncated distribution.

(b) Although the times to failure for the returned modems were not available, it is still possible to compute an estimate of the Weibull cdf using the given value of $\beta = .85$. Show how to do this.

(c) Use the estimate from part (b) to compute an estimate for the number of units that will fail in the third year of operation.

(d) Suppose that you had learned that 2 years after being sold, between 5% and 10% of the purchased modems had never been put into service. How would you do part (b)?

▲**11.13.** Consider the discrete-data likelihood

$$L(\boldsymbol{\theta}) = \prod_{i=1}^{m} [F(t_i + \Delta_i; \eta, \sigma, \zeta) - F(t_i - \Delta_i; \eta, \sigma, \zeta)]^{d_i},$$

where $F(t; \eta, \sigma, \zeta)$ is the GETS cdf in (5.13).

(a) Suppose that the GETS cdf is Weibull based (i.e., $\Phi = \Phi_{\text{sev}}$) and that $\sigma > 0$. Show that the derivatives of the log likelihood with respect to the parameters are discontinuous at t_0 when $t_0 = (\eta/\zeta - 1/\sigma)$ and $\sigma > 1$.

(b) Suppose that the GETS cdf is Weibull based with $\sigma < 0$. Show that the derivatives of the log likelihood with respect to the parameters are discontinuous at t_m when $t_m = (\eta/\zeta - 1/\sigma)$ and $|\sigma| > 1$.

(c) Show that if the GETS cdf is lognormal based, then the derivatives of the log likelihood with respect to the parameters are continuous everywhere.

◆11.14. Conduct the following simulation to compare the effects of censoring and truncation on estimation precision.

(a) Generate a sample from a Weibull distribution with $\eta = 100$ hours and $\beta = 2$.

(b) Find the ML estimates of the parameters and $t_{.1}$, treating any observations beyond 150 hours as right-censored.

(c) Find the ML estimates of the parameters and $t_{.1}$, truncating observations beyond 150 hours.

(d) Repeat the simulation 500 times. Make appropriate plots of the sample estimates (including scatter plots to see correlation). Compute and use histograms or other graphical displays to compare the estimates from the censored and the truncated samples. Also compute the sample variances for the parameter estimates and for the estimates of $t_{.1}$.

(e) What can you conclude from this simulation experiment?

CHAPTER 12

Prediction of Future Random Quantities

Objectives

This chapter explains:

- Important reliability-related applications of prediction.
- The difference between probability prediction and statistical prediction.
- Methods for computing predictions and prediction bounds for future failure times.
- Methods for computing predictions and prediction bounds for the number of failures in a future time interval.

Overview

This chapter describes methods to construct prediction bounds or intervals for future random quantities. Both new-sample prediction (using data from a previous sample to make predictions on a future unit or sample of units) and within-sample prediction problems (predicting future events in a sample based on early data from the sample) are considered. To illustrate new-sample prediction we show how to construct a prediction interval for a single future observation from a previously sampled population/process (motivated by a customer's request for an interval to contain the life of a purchased product). To illustrate within-sample prediction, we show how to compute a prediction interval for the number of future failures in a specified period beyond the observation period (motivated by a warranty prediction problem). A third example requires more general methods to deal with complicated censoring arising because units enter service at different points in time (staggered entry). Section 12.2 describes "probability prediction intervals" for a completely specified distribution. Probability prediction intervals, when computed on the basis of estimates from limited data, are sometimes called "naive prediction intervals," and they can serve as a basis for developing more commonly needed statistical prediction intervals. Section 12.3 describes coverage probability concepts and other basic ideas pertaining to statistical prediction intervals. Section 12.4 introduces the basic ideas of pivotal methods for complete and Type II censored data from log-location-scale distributions and shows how they can be extended to obtain approximate prediction intervals for the more commonly used Type I censoring. Section 12.5 describes some simple special case prediction interval procedures that can be implemented with simple hand computations. Section 12.6 presents a more general approach of calibrating naive statistical prediction intervals and shows how these methods are related to the pivotal-like methods. Section 12.7 shows how to apply the calibration method to a commonly occurring problem of predicting future field failures on the basis of early field failures. Section 12.8 extends the field prediction problem to situations where units enter the field over a longer period of time.

12.1 INTRODUCTION

12.1.1 Motivation and Prediction Problems

Practical problems often require the computation of predictions and prediction bounds for future values of random quantities. For example:

- A consumer purchasing a refrigerator would like to have a lower bound for the failure time of the unit to be purchased (with less interest in distribution of the population of units purchased by other consumers).
- Financial managers in manufacturing companies need upper prediction bounds on future warranty costs.
- When planning life tests, engineers may need to predict the number of failures that will occur by the end of the test or the amount of time that it will take for a specified number of units to fail.

Some applications require a two-sided prediction interval $[\underline{T}, \widetilde{T}]$ that will, with a specified high degree of confidence, contain the future random variable of interest, say, T. In many applications, however, interest is focused on either an upper prediction bound or a lower prediction bound (e.g., the maximum warranty cost is more important than the minimum and the time of the early failures in a product population is more important than the last ones).

Conceptually, it is useful to distinguish between "new-sample" prediction (Figure 12.1) and "within-sample" prediction (Figure 12.2). For new-sample prediction, data from a past sample is used to make predictions on a future unit or sample of units from the same process or population. For example, based on previous (possibly censored) life test data, one could be interested in predicting the following:

- Time to failure of a new item.
- Time until k failures in a future sample of m units.
- Number of failures by time t_w in a future sample of m units.

INTRODUCTION

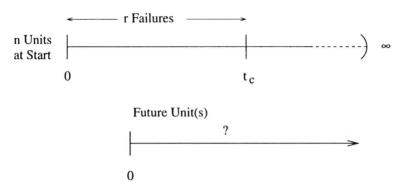

Figure 12.1. New-sample prediction.

For within-sample prediction, the problem is to predict future events in a sample or process based on early data from that sample or process. For example, if n units are followed until t_c and there are r observed failures, $t_{(1)}, \ldots, t_{(r)}$, one could be interested in predicting the following:

- Time of the next failure.
- Time until k additional failures.
- Number of additional failures in a future interval (t_c, t_w).

12.1.2 Model

In general, to predict a future realization of a random quantity one needs:

- A statistical model to describe the population or process of interest. This model usually consists of a distribution depending on a vector of parameters $\boldsymbol{\theta}$. Nonparametric new-sample prediction is also possible (Chapter 5 of Hahn and Meeker, 1991, gives examples and references).
- Information on the values of the parameters $\boldsymbol{\theta}$. This information could come from either a laboratory life test or field data.

We will assume that the failure times have a continuous distribution with cdf $F(t) = F(t; \boldsymbol{\theta})$ and pdf $f(t) = f(t; \boldsymbol{\theta})$, where $\boldsymbol{\theta}$ is a vector of parameters. Generally, $\boldsymbol{\theta}$ is unknown and will be estimated from available sample data. In such cases we will make the standard assumptions of statistical independence of failure times and that

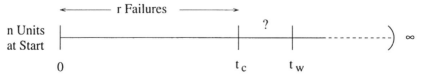

Figure 12.2. Within-sample prediction.

censoring times are independent of any future failure time that would be observed if a unit were not to be censored (as described in Section 2.3.2).

12.1.3 Data

The beginning of this chapter considers situations in which n units begin operation at time 0 and are observed until a time t_c, where the available data are to be analyzed. Failure times are recorded for the r units that fail in the interval $(0, t_c)$. Then the data consist of the r smallest order statistics $t_{(1)} < \cdots < t_{(r)} \leq t_c$ and the information that the other $n - r$ units will have failed after t_c. Section 12.8 shows how to compute prediction bounds for more complicated kinds of censored data that are frequently encountered in the analysis of field reliability data.

12.2 PROBABILITY PREDICTION INTERVALS (θ GIVEN)

With a completely specified continuous probability distribution, an exact $100(1-\alpha)\%$ "probability prediction interval" for a future observation from $F(t; \boldsymbol{\theta})$ is (ignoring any data)

$$PI(1 - \alpha) = [\utilde{T}, \; \widetilde{T}] = [t_{\alpha/2}, \; t_{1-\alpha/2}], \tag{12.1}$$

where t_p is the p quantile of $F(t; \boldsymbol{\theta})$. The probability of coverage of the interval in (12.1) is

$$\Pr[T \in PI(1 - \alpha)] = \Pr(\utilde{T} \leq T \leq \widetilde{T}) = \Pr(t_{\alpha/2} \leq T \leq t_{1-\alpha/2}) = 1 - \alpha$$

by the definition of quantiles of continuous distributions.

Example 12.1 Prediction Interval for a Completely Specified Probability Distribution. A potential customer plans to purchase a system. The system contains a bearing known to be a life-limiting component that has failed in some existing systems. The potential customer needs a lower prediction bound on T, the number of use-cycles to failure for one of these systems that is to be placed into service. Based on previous experience, the manufacturer believes that the number of cycles to failure for the bearing can be described by a lognormal cdf

$$\Pr(T \leq t) = F(t; \mu, \sigma) = \Phi_{\text{nor}}\left[\frac{\log(t) - \mu}{\sigma}\right]$$

with specified parameters $\mu = 4.0$ and $\sigma = .5$. From (12.1), a two-sided 90% probability prediction interval is

$$[\utilde{T}, \; \widetilde{T}] = [\exp(\mu + \Phi_{\text{nor}}^{-1}(.05) \times \sigma), \; \exp(\mu + \Phi_{\text{nor}}^{-1}(.95) \times \sigma)]$$
$$= [\exp(4.0 + (-1.645) \times .5), \; \exp(4.0 + 1.645 \times .5)]$$
$$= [23.93, \; 124.59].$$

Then $\Pr(\underline{T} \leq T \leq \widetilde{T}) = \Pr(23.93 \leq T \leq 124.59) = .90$. With misspecified parameters, coverage probability may not be .90. □

12.3 STATISTICAL PREDICTION INTERVAL (θ ESTIMATED)

12.3.1 Coverage Probability Concepts

Before describing methods for constructing $\boldsymbol{\theta}$-estimated prediction intervals, let us first consider methods for evaluating the properties of prediction intervals. In particular, "coverage probability" is an important property. We describe these concepts in terms of a "new-sample" prediction interval for a future observation but the ideas also hold for other new-sample prediction problems and within-sample prediction problems.

In statistical prediction, the objective is to predict the random quantity T based on "learning" sample information (denoted by DATA). Generally, with only sample data, there is uncertainty in the distribution parameters. The random DATA leads to a parameter estimate $\widehat{\boldsymbol{\theta}}$ and then to a nominal $100(1 - \alpha)\%$ prediction interval $PI(1 - \alpha) = [\underline{T}, \widetilde{T}]$. Thus $[\underline{T}, \widetilde{T}]$ and the future random variable T have a joint distribution that depends on a parameter vector $\boldsymbol{\theta}$.

There are two kinds of coverage probabilities:

- For fixed DATA (and thus fixed $\widehat{\boldsymbol{\theta}}$ and $[\underline{T}, \widetilde{T}]$) the conditional coverage probability of a particular interval $[\underline{T}, \widetilde{T}]$ is

$$\text{CP}[PI(1 - \alpha) \mid \widehat{\boldsymbol{\theta}}; \boldsymbol{\theta}] = \Pr(\underline{T} \leq T \leq \widetilde{T} \mid \widehat{\boldsymbol{\theta}}; \boldsymbol{\theta}) = F(\widetilde{T}; \boldsymbol{\theta}) - F(\underline{T}; \boldsymbol{\theta}). \quad (12.2)$$

This conditional probability is *unknown* because $F(t; \boldsymbol{\theta})$ depends on the unknown $\boldsymbol{\theta}$.

- From sample to sample, the conditional coverage probability is *random* because $[\underline{T}, \widetilde{T}]$ depends on $\widehat{\boldsymbol{\theta}}$. The unconditional coverage probability for the prediction interval *procedure* is

$$\text{CP}[PI(1 - \alpha); \boldsymbol{\theta}] = \Pr(\underline{T} \leq T \leq \widetilde{T}; \boldsymbol{\theta}) = \text{E}_{\widehat{\boldsymbol{\theta}}}\{\text{CP}[PI(1 - \alpha) \mid \widehat{\boldsymbol{\theta}}; \boldsymbol{\theta}]\} \quad (12.3)$$

where the expectation is with respect to the random $\widehat{\boldsymbol{\theta}}$. Because it can be computed (at least approximately) and can be controlled, it is this unconditional probability that is generally used to describe a prediction interval procedure. When $\text{CP}[PI(1 - \alpha); \boldsymbol{\theta}] = 1 - \alpha$ does not depend on $\boldsymbol{\theta}$, the procedure is said to be "exact." In general, $\text{CP}[PI(1 - \alpha); \boldsymbol{\theta}] \approx 1 - \alpha$ because of dependency on the unknown $\boldsymbol{\theta}$. In such cases only an approximate prediction interval procedure is available.

12.3.2 Relationship Between One-Sided and Two-Sided Prediction Intervals

Combining a one-sided lower $100(1 - \alpha/2)\%$ prediction bound and a one-sided upper $100(1 - \alpha/2)\%$ prediction bound gives an equal-tail two-sided $100(1 - \alpha)\%$ probability prediction interval. In particular, if $\Pr(\underset{\sim}{T} \leq T < \infty) = 1 - \alpha/2$ and $\Pr(0 < T \leq \widetilde{T}) = 1 - \alpha/2$, then $\Pr(\underset{\sim}{T} \leq T \leq \widetilde{T}) = 1 - \alpha$. It may be possible to find a narrower interval with unequal probabilities in the upper and lower tails, still summing to α. Use of equal-probability intervals, however, has the important advantage of providing an interval that has endpoints that can be correctly interpreted as one-sided prediction bounds (with the appropriate adjustment in the confidence level). This is important because in most applications the cost of predicting too high is different from the cost of predicting too low and two-sided prediction intervals are often reported even though primary interest is on one side or the other.

12.3.3 Naive Method for Computing a Statistical Prediction Interval

A "naive" prediction interval for continuous T is obtained by substituting the maximum likelihood (ML) estimate for $\boldsymbol{\theta}$ into (12.1), giving

$$PI(1 - \alpha) = [\underset{\sim}{T}, \ \widetilde{T}] = [\widehat{t}_{\alpha/2}, \ \widehat{t}_{1-\alpha/2}],$$

where $\widehat{t}_p = t_p(\widehat{\boldsymbol{\theta}})$ is the ML estimate of the p quantile of T. To predict a future independent observation from a log-location-scale distribution, a naive prediction interval is

$$PI(1 - \alpha) = [\underset{\sim}{T}, \ \widetilde{T}] = [\widehat{t}_{\alpha/2}, \ \widehat{t}_{1-\alpha/2}]$$
$$= [\exp(\widehat{\mu} + \Phi^{-1}(\alpha/2) \times \widehat{\sigma}), \ \exp(\widehat{\mu} + \Phi^{-1}(1 - \alpha/2) \times \widehat{\sigma})]. \quad (12.4)$$

The unconditional coverage probability for this naive procedure is approximately equal to the nominal $1 - \alpha$ with large samples sizes. For small to moderate number of units failing, however, the coverage probability may be *far* from $1 - \alpha$.

Example 12.2 Naive Prediction Interval for Predicting the Life of a Ball Bearing (Lognormal Distribution). Refer to the problem of predicting bearing life in Example 12.1, but suppose that only limited data are available to make the prediction. Figure 12.3 is a lognormal probability plot of the first 15 of 23 failures in a bearing life test described in Lawless (1982, page 228) when the data are right-censored at 80 million cycles. Failures occurred at 17.88, 28.92, 33.00, 41.52, 42.12, 45.60, 48.40, 51.84, 51.96, 54.12, 55.56, 67.80, 68.64, 68.64, and 68.88 million revolutions. The other eight bearings were treated as if they had been censored at 80 million cycles. The lognormal ML estimates are $\widehat{\mu} = 4.160$ and $\widehat{\sigma} = .5451$. From (12.4), the naive two-sided 90% prediction interval is

$$[\underset{\sim}{T}, \ \widetilde{T}] = [\exp(\widehat{\mu} + \Phi_{\text{nor}}^{-1}(.05) \times \widehat{\sigma}), \ \exp(\widehat{\mu} + \Phi_{\text{nor}}^{-1}(.95) \times \widehat{\sigma})] \quad (12.5)$$
$$= [\exp(4.160 + (-1.645) \times .5451), \ \exp(4.160 + 1.645 \times .5451)]$$
$$= [26.1, \ 157.1].$$

STATISTICAL PREDICTION INTERVAL (θ ESTIMATED)   295

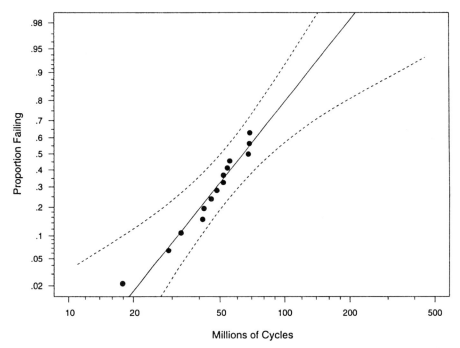

Figure 12.3. Lognormal probability plot of bearing life test data censored after 80 million cycles (with 15 of 23 units failed) with lognormal ML estimates and pointwise 95% confidence intervals.

Intervals constructed in this manner are generally too narrow and their coverage probability is below the nominal value of $1 - \alpha$ because they ignore the uncertainty in $\widehat{\mu}$ and $\widehat{\sigma}$ relative to μ and σ. $\square$

For the Weibull distribution, the prediction interval can be computed analogously by recognizing that $\log(T)$ has a smallest extreme value (SEV) distribution. In particular,

$$\Pr(T \leq t; \eta, \beta) = 1 - \exp\left[-\left(\frac{t}{\eta}\right)^{\beta}\right] = \Phi_{\text{sev}}\left[\frac{\log(t) - \mu}{\sigma}\right], \quad t > 0,$$

where $\mu = \log(\eta)$ and $\beta = 1/\sigma$.

Example 12.3 Naive Prediction Interval for Predicting the Life of a Ball Bearing (Weibull Distribution). The Weibull distribution also provides an adequate description of the censored ball bearing data. Following the approach in Example 12.2, the Weibull ML estimates are $\widehat{\mu} = 4.334$ and $\widehat{\sigma} = .4013$. From (12.4), the naive two-sided 90% prediction interval is

$$[\underline{T}, \widetilde{T}] = [\exp(\widehat{\mu} + \Phi_{\text{sev}}^{-1}(.05) \times \widehat{\sigma}), \quad \exp(\widehat{\mu} + \Phi_{\text{sev}}^{-1}(.95) \times \widehat{\sigma})] \quad (12.6)$$

$$= [\exp(4.334 + (-2.970) \times .4013), \quad \exp(4.334 + 1.097 \times .4013)]$$

$$= [23.2, \quad 118.4].$$

Note that, in comparison with the prediction interval for the lognormal distribution in Example 12.2, the Weibull prediction interval has a much smaller upper endpoint. It is typical that the lognormal distribution, when compared with the Weibull, will provide a more optimistic extrapolation into the upper tail of a fitted distribution. This is because the lognormal distribution has a much "longer" upper tail. □

12.4 THE (APPROXIMATE) PIVOTAL METHOD FOR PREDICTION INTERVALS

12.4.1 Type II (Failure) Censoring

With Type II (failure) censoring, a life test is run until a specified number of r failures, where $1 \leq r \leq n$. When T has a log-location-scale distribution and the data are complete or Type II (failure) censored, the random variable $Z_{\log(T)} = [\log(T) - \widehat{\mu}]/\widehat{\sigma}$ is pivotal. That is, the distribution of $Z_{\log(T)}$ depends only on n and r but it does not depend on μ or σ. Then

$$\Pr\left[z_{\log(T)_{(\alpha/2)}} < \frac{\log(T) - \widehat{\mu}}{\widehat{\sigma}} \leq z_{\log(T)_{(1-\alpha/2)}}\right] = 1 - \alpha,$$

where $z_{\log(T)_{(p)}}$ is the p quantile of $Z_{\log(T)}$. Thus

$$\Pr\left[\widehat{\mu} + z_{\log(T)_{(\alpha/2)}} \times \widehat{\sigma} < \log(T) \leq \widehat{\mu} + z_{\log(T)_{(1-\alpha/2)}} \times \widehat{\sigma}\right] = 1 - \alpha,$$

which suggests the prediction interval

$$[\utilde{T}, \widetilde{T}] = \left[\exp(\widehat{\mu} + z_{\log(T)_{(\alpha/2)}} \times \widehat{\sigma}), \ \exp(\widehat{\mu} + z_{\log(T)_{(1-\alpha/2)}} \times \widehat{\sigma})\right].$$

The quantiles $z_{\log(T)_{(\alpha/2)}}$ and $z_{\log(T)_{(1-\alpha/2)}}$ can be obtained from the distribution of $Z_{\log(T)}$, which can be obtained approximately (the approximation due only to Monte Carlo error) by simulating B realizations of

$$Z_{\log(T^*)} = \frac{\log(T^*) - \widehat{\mu}^*}{\widehat{\sigma}^*}. \tag{12.7}$$

The procedure is as follows:

1. Draw a sample of size n from a log-location-scale distribution with parameters $(\widehat{\mu}, \widehat{\sigma})$, censored at the rth failure.
2. Use the censored sample to compute ML estimates $\widehat{\mu}^*$ and $\widehat{\sigma}^*$.
3. Draw an additional single observation T^* from the log-location-scale distribution with parameters $(\widehat{\mu}, \widehat{\sigma})$.
4. Compute $Z_{\log(T^*)} = [\log(T^*) - \widehat{\mu}^*]/\widehat{\sigma}^*$.
5. Repeat steps 1 to 4 B times. Obtain the approximations for the quantiles $z_{\log(T)_{(\alpha/2)}}$ and $z_{\log(T)_{(1-\alpha/2)}}$ from the empirical distribution of $Z_{\log(T^*)}$.

THE (APPROXIMATE) PIVOTAL METHOD FOR PREDICTION INTERVALS

Because the quantiles of $Z_{\log(T)}$ depend only on n and r the procedure to predict T will, except for Monte Carlo error, have exactly the nominal coverage probability.

12.4.2 Type I (Time) Censoring

For single time censoring (test run until a specified censoring time t_c), the simulation procedure is modified by censoring the samples at the specified censoring time t_c. In this case, the $Z_{\log(T)}$ is only approximately pivotal and quantiles of $Z_{\log(T)}$ depend on $F(t_c, \boldsymbol{\theta})$ (the unknown expected proportion failing by time t_c) and the sample size n. Thus, with Type I censoring, the prediction interval to predict T is approximate.

Example 12.4 Approximate Prediction Interval for Predicting the Life of a Ball Bearing Based on an Approximate Pivotal (Lognormal Distribution). Figure 12.4 shows the simulated distribution of the lognormal predictive pivotal-like statistic in (12.7). The simulated values were obtained by doing 100,000 simulations of the censored life test, using the lognormal distribution and the ML estimates for bearing life from Example 12.2. The needed quantiles of the simulated distribution are $z_{\log(T^*)_{(.05)}} \approx -1.802$ and $z_{\log(T^*)_{(.95)}} \approx 1.837$. Thus a two-sided approximate 90%

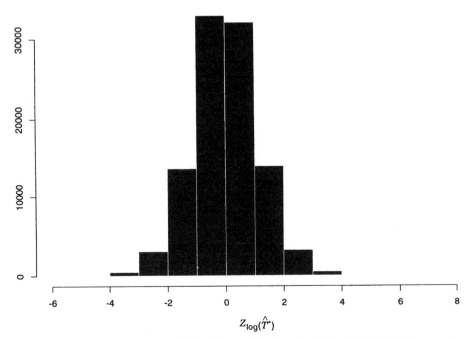

Figure 12.4. Histogram of 100,000 simulated $Z_{\log(\widehat{T}^*)}$ values, based on the bearing life test data censored after 80 million cycles.

prediction interval for lognormal T is

$$[\underline{T}, \widetilde{T}] = [\exp(\widehat{\mu} + z_{\log(T^*)_{(.05)}} \times \widehat{\sigma}), \quad \exp(\widehat{\mu} + z_{\log(T^*)_{(.95)}} \times \widehat{\sigma})]$$
$$= [\exp(4.160 + (-1.802) \times .5451), \quad \exp(4.160 + 1.837 \times .5451)]$$
$$= [24.0, \quad 174.4].$$

It is important to note that the upper prediction bound requires some extrapolation given that there were only 15 failures in the sample of 23 of the bearings. This upper bound does not account for possible model error in the unobserved upper tail of the failure-time distribution. □

Example 12.5 Approximate Prediction Interval for Predicting the Life of a Ball Bearing Based on an Approximate Pivotal (Weibull Distribution). Following the approach used in Example 12.4, a simulation for the Weibull distribution to estimate the distribution of $Z_{\log(T^*)}$ resulted in $z_{\log(T^*)_{(.05)}} \approx -3.263$ and $z_{\log(T^*)_{(.95)}} \approx 1.260$. Thus a two-sided approximate 90% prediction interval for Weibull T is

$$[\underline{T}, \widetilde{T}] = [\exp(\widehat{\mu} + z_{\log(T^*)_{(.05)}} \times \widehat{\sigma}), \quad \exp(\widehat{\mu} + z_{\log(T^*)_{(.95)}} \times \widehat{\sigma})]$$
$$= [\exp(4.334 + (-3.263) \times .4013), \quad \exp(4.334 + 1.260 \times .4013)]$$
$$= [20.6, \quad 126.4].$$

Note that, as we observed with the naive prediction intervals (Examples 12.2 and 12.3), in comparison with the prediction interval for the lognormal distribution in Example 12.4, the Weibull prediction interval has a much smaller upper endpoint. □

The (approximate) pivotal method can be extended directly to compute prediction intervals for random variables from other log-location-scale distributions. It can also be extended to other non-location-scale distributions by using a location estimate, such as a central quantile of the distribution of $\log(T)$ in place of $\widehat{\mu}$ and an estimate of the standard deviation of $[\log(T) - \widehat{\mu}]$ in place of $\widehat{\sigma}$.

12.5 PREDICTION IN SIMPLE CASES

This section describes some simple special case prediction interval procedures. The methods are based on pivotal quantities that are related to the pivotal quantity methods described in Section 12.4.1. For the simple cases presented here, the distribution of the pivotal quantities have well-known distributions (Student's t and the Snedecor F distributions) for which tables of quantiles are readily available.

12.5.1 Complete Samples from a Lognormal Distribution

Suppose that $t_1, \ldots, t_n$ is a complete sample from a LOGNOR(μ, σ) distribution and T is an independent new observation from the same distribution. Then it can be shown that $\sqrt{(n-1)/(n+1)} \times Z_{\log(T)}$ has a Student's t distribution with $n-1$ degrees of freedom. From this it follows that

$$\Pr\left[t_{(\alpha/2;n-1)} < \sqrt{\frac{n-1}{n+1}} \times \frac{\log(T) - \widehat{\mu}}{\widehat{\sigma}} \leq t_{(1-\alpha/2;n-1)}\right]$$

$$= \Pr\left[t_{(\alpha/2;n-1)} < \sqrt{\frac{n}{n+1}} \times \frac{\log(T) - \widehat{\mu}}{s} \leq t_{(1-\alpha/2;n-1)}\right] = 1 - \alpha,$$

where $s = \sqrt{n/(n-1)} \times \widehat{\sigma}$ is the sample standard deviation of the logarithms of the observed failure times and $t_{(p;\nu)}$ is the p quantile of the Student's t distribution with ν degrees of freedom. This leads to an exact $100(1 - \alpha)\%$ prediction interval for a new independent observation given by

$$[\undertilde{T}, \widetilde{T}] = [\exp(\widehat{\mu} - \omega \times s), \quad \exp(\widehat{\mu} + \omega \times s)],$$

where $\omega = t_{(1-\alpha/2;n-1)} \times \sqrt{1 + 1/n}$. This prediction interval is equivalent to the prediction interval for a new independent observation for complete data given in other books (e.g., Hahn and Meeker, 1991, page 61).

The exact prediction interval $[\undertilde{T}, \widetilde{T}]$ is wider than the naive prediction interval obtained from using the ML estimates of the quantiles of T

$$[\widehat{t}_{\alpha/2}, \widehat{t}_{1-\alpha/2}] = \left[\exp\left\{\widehat{\mu} + \Phi_{\text{nor}}^{-1}(\alpha/2) \times \widehat{\sigma}\right\}, \quad \exp\left\{\widehat{\mu} + \Phi_{\text{nor}}^{-1}(1 - \alpha/2) \times \widehat{\sigma}\right\}\right].$$

When n is large (say, $n > 30$), however, the differences are negligible because for $0 < p < 1$,

$$t_{(p;n-1)} \times \sqrt{1 + \frac{1}{n}} \times s \approx \Phi_{\text{nor}}^{-1}(p) \times \widehat{\sigma}.$$

Example 12.6 Prediction Interval for a New Observation from a Lognormal Distribution. Refer to Examples 11.4. Suppose that the analysts wanted a 95% prediction interval to contain the time to fracture of a specimen of the same type to be tested in the future. Based on a sample of size $n = 63$ specimens, the ML estimate of the parameters for the lognormal distribution are $\widehat{\mu} = 4.722$, $\widehat{\sigma} = 1.0255$, and $s = \sqrt{63/62} \times 1.0255 = 1.034$. Then, an exact 90% prediction interval for a new observation is

$$[\undertilde{T}, \widetilde{T}] = [\exp(\widehat{\mu} - \omega \times s), \quad \exp(\widehat{\mu} + \omega \times s)]$$

$$= [\exp(4.722 - 1.683 \times 1.034), \quad \exp(4.722 + 1.683 \times 1.034)]$$

$$= [19.72, \quad 640.0],$$

where $\omega = \sqrt{1 + (1/63)} \times 1.669804 = 1.683$. Thus we are 90% confident that this interval will contain the fatigue life of the specimen. □

12.5.2 Complete or Type II Censored Samples from an Exponential Distribution

If $t_1, \ldots, t_r$ is a Type II ($r < n$) censored sample or a complete sample ($r = n$) from an EXP(θ) distribution, $\widehat{\theta}$ is the ML estimator from these data, and if T is another, future independent observation from the same distribution, then $T/\widehat{\theta}$ has an F distribution with $(2, 2r)$ degrees of freedom. From this, it follows that

$$\Pr\left[\mathcal{F}_{(\alpha/2;2,2r)} < T/\widehat{\theta} \leq \mathcal{F}_{(1-\alpha/2;2,2r)}\right] = 1 - \alpha,$$

where $\mathcal{F}_{(p;\nu_1,\nu_2)}$ is the p quantile of the F distribution with (ν_1, ν_2) degrees of freedom. This leads to an exact $100(1 - \alpha)\%$ prediction interval for a new observation given by

$$[\underset{\sim}{T}, \widetilde{T}] = \left[\mathcal{F}_{(\alpha/2;2,2r)} \times \widehat{\theta}, \quad \mathcal{F}_{(1-\alpha/2;2,2r)} \times \widehat{\theta}\right],$$

where the F distribution p quantile with 2 numerator degrees of freedom can be obtained from $\mathcal{F}_{(p;2;2r)} = r \times \{\exp[-(1/r)\log(1-p)] - 1\}$. This prediction interval is wider than the naive prediction interval computed as ML estimates of quantiles of the distribution of T,

$$[\widehat{t}_{\alpha/2}, \widehat{t}_{1-\alpha/2}] = \left[\{-\log(1 - \alpha/2)\} \times \widehat{\theta}, \quad \{-\log(\alpha/2)\} \times \widehat{\theta}\right],$$

but when r is large the differences between the two intervals are negligible because for $0 < p < 1$ and large r, $\mathcal{F}_{(p;2,2r)} \approx -\log(1 - p)$.

Example 12.7 Prediction Interval for the Lifetime of an Insulation Specimen. Refer to Example 7.13, where $n = 25$ insulation specimens were tested until $r = 15$ failures had been observed. The ML estimate of the exponential mean θ is $\widehat{\theta} = 63.392$ hours. Another specimen is to be tested. An exact 90% prediction interval for the new observations is

$$[\underset{\sim}{T}, \widetilde{T}] = [\mathcal{F}_{(.05;2,30)} \times 63.392, \quad \mathcal{F}_{(.95;2,30)} \times 63.392] = [3.257, \quad 210.197].$$

Thus we are 90% confident that this interval will contain the lifetime of the specimen to be tested. □

12.6 CALIBRATING NAIVE STATISTICAL PREDICTION BOUNDS

Cox (1975) suggested a large-sample approximate method, based on maximum likelihood estimates, that can be used to calibrate or correct a naive prediction interval.

The basic idea of this approach is to calibrate the naive one-sided prediction bound by evaluating $CP[PI(1 - \alpha); \boldsymbol{\theta}]$ at $\widehat{\boldsymbol{\theta}}$ and finding a calibration value $1 - \alpha_{cl}$ such that for a one-sided *lower* prediction bound for T,

$$CP[PI(1 - \alpha_{cl}); \widehat{\boldsymbol{\theta}}] = \Pr(\underline{T} \leq T \leq \infty; \widehat{\boldsymbol{\theta}}) = \Pr(\widehat{\underline{t}}_{\alpha_{cl}} \leq T \leq \infty; \widehat{\boldsymbol{\theta}}) = 1 - \alpha. \quad (12.8)$$

Calibration for a one-sided *upper* prediction bound on T (described at the end of Section 12.6.1) is similar. For a two-sided prediction interval, the calibration is done separately such that the probability is $\alpha/2$ in each tail. In problems where $CP[PI(1 - \alpha); \boldsymbol{\theta}]$ does not depend on $\boldsymbol{\theta}$, the calibrated $PI(1 - \alpha_{cl})$ provides an exact prediction bound.

Although it is sometimes possible to do analytical calibration, operationally, the analytical approach is intractable except in the simplest of situations, where alternative, simpler methods exist (e.g., the methods in Section 12.5).

12.6.1 Calibration by Simulation of the Sampling/Prediction Process

Modern computing capabilities make it easy to use Monte Carlo methods to evaluate, numerically, quantities like (12.8), even for complicated statistical models. In particular, under the assumed model we can use ML estimates $\widehat{\boldsymbol{\theta}}$ to simulate both the sampling *and* prediction process a large number B (e.g., $B = 50,000$ or $B = 100,000$) times. Although $B = 2000$ or so is often suggested for simulation-based confidence intervals, larger values of B are usually required for prediction problems due to the added variability of the single future observation.

Conceptually, (12.8) can be evaluated as follows:

1. Choose a value of $1 - \alpha$, say, $1 - \alpha_0$.
2. Simulate $DATA_j^*$ from the assumed model with parameter values equal to the ML estimates $\widehat{\boldsymbol{\theta}}$ [i.e., from $F(t; \widehat{\boldsymbol{\theta}})$]. Use the sampling procedures and censoring that mimics the original experiment.
3. Compute simulation ML estimates $\widehat{\boldsymbol{\theta}}_j^*$ from $DATA_j^*$.
4. Compute the naive $100(1 - \alpha_0)\%$ lower prediction bound $\underline{T}_j^*$ from the simulated $DATA_j^*$. Compare $\underline{T}_j^*$ with an independent T_j^* simulated from $F(t; \widehat{\boldsymbol{\theta}})$ to see if $T_j^* > \underline{T}_j^*$.
5. Repeat steps 2 to 4 for $j = 1, 2, \ldots, B$. The proportion of the B trials having $T_j^* > \underline{T}_j^*$ gives the Monte Carlo evaluation of $CP\left[PI(1 - \alpha_0); \boldsymbol{\theta}\right]$ at $\widehat{\boldsymbol{\theta}}$, which we denote by $CP^*[PI(1 - \alpha_0); \widehat{\boldsymbol{\theta}}]$.
6. Repeat steps 2 to 5 for different values of $1 - \alpha_0$.
7. Find $1 - \alpha_{cl}$ such that $CP^*[PI(1 - \alpha_{cl}); \widehat{\boldsymbol{\theta}}] = 1 - \alpha$.

The difference between $CP[PI(1 - \alpha_0); \widehat{\boldsymbol{\theta}}]$ and $CP^*[PI(1 - \alpha_0); \widehat{\boldsymbol{\theta}}]$ is due to Monte Carlo error and can be made arbitrarily small by choosing a sufficiently large value of B. To avoid cumbersome notation we will use $\text{CP}[PI(1-\alpha_0); \widehat{\boldsymbol{\theta}}]$ even when the evaluation is done with simulation.

Operationally, for a log-location-scale distribution where $\boldsymbol{\theta} = (\mu, \sigma)$, the $\text{CP}[PI(1-\alpha); \widehat{\boldsymbol{\theta}}]$ function can be evaluated more directly by using the following procedure:

1. Use simulation to compute B realizations of the pivotal-like statistic $Z_{\log(T^*)}$, as described in Section 12.4.
2. The empirical distribution of the observed values of the random variable $P = 1 - \Phi[Z_{\log(T^*)}]$ provides a Monte Carlo evaluation of lower prediction bound $\text{CP}[PI(1-\alpha); \widehat{\boldsymbol{\theta}}]$ used in (12.8). In particular, for a lower prediction bound, $1 - \alpha_{cl}$ is the $1 - \alpha$ quantile of the distribution of the random variable $P = 1 - \Phi(Z_{\log(T^*)})$.

The naive one-sided upper prediction bound for T is calibrated by finding $1 - \alpha_{cu}$ such that

$$\text{CP}[PI(1-\alpha_{cu}); \widehat{\boldsymbol{\theta}}] = \Pr(0 \leq T \leq \widetilde{T}; \widehat{\boldsymbol{\theta}}) = \Pr(0 \leq T \leq \widehat{t}_{1-\alpha_{cu}}; \widehat{\boldsymbol{\theta}}) = 1 - \alpha. \quad (12.9)$$

For a log-location-scale distribution a Monte Carlo evaluation of the upper prediction bound $\text{CP}[PI(1-\alpha); \widehat{\boldsymbol{\theta}}]$ can be obtained from the empirical distribution of the observed values of the random variable $P = \Phi[Z_{\log(T^*)}]$. In particular, for an upper prediction bound, $1 - \alpha_{cu}$ is the $1 - \alpha$ quantile of the distribution of the random variable $P = \Phi(Z_{\log(T^*)})$.

Escobar and Meeker (1998a) provide justification for this procedure and demonstrate the equivalence of the calibration method and the approximate pivotal method from Section 12.4. For predicting random variables with distributions that are not log-location-scale, the approach is similar, as will be illustrated in Sections 12.7 and 12.8.

12.6.2 Calibration by Averaging Conditional Coverage Probabilities

As suggested by Mee and Kushary (1994), it can be much more efficient, computationally, to obtain the needed calibration curves for (12.8) and (12.9) by simulating conditional coverage probabilities like those in (12.2) and averaging these to estimate the expectation in (12.3). The procedure is similar to the one in Section 12.6.1, replacing steps 4 and 5 with the following:

4. For each simulated sample, compute the *naive* $100(1-\alpha_0)\%$ upper and lower prediction bounds $\underset{\sim}{T}$ and $\widetilde{T}$, respectively. For a log-location-scale distribution, $\underset{\sim}{T} = \exp[\widehat{\mu}^* + \Phi^{-1}(\alpha_0) \times \widehat{\sigma}^*]$ and $\widetilde{T} = \exp[\widehat{\mu}^* + \Phi^{-1}(1-\alpha_0) \times \widehat{\sigma}^*]$.
5. A Monte Carlo evaluation of the unconditional coverage probability is obtained from the average of the simulated conditional coverage probabilities $\text{CP}[PI(1-\alpha_0); \widehat{\boldsymbol{\theta}}] = \sum_{j=1}^{B} P_j/B$, where:
    (a) For the upper prediction bound calibration $P_j = \Pr(T \leq \widetilde{T})$. For a log-location-scale distribution, $P_j = \Phi[(\log(\widetilde{T}) - \widehat{\mu})/\widehat{\sigma}]$.

(b) For the lower prediction bound calibration, compute the conditional coverage probability $P_j = \Pr(T \geq \underline{T})$. For a log-location-scale distribution, $P_j = 1 - \Phi[(\log(\underline{T}) - \widehat{\mu})/\widehat{\sigma}]$.

To obtain the entire calibration curves, one would need to compute $CP[PI(1 - \alpha_0); \widehat{\boldsymbol{\theta}}]$ for a large number of different values of $1 - \alpha_0$ between 0 and 1. Operationally, to compute a one-sided prediction bound one needs only to find the appropriate $1 - \alpha_{cl}$ value. The $CP[PI(1 - \alpha_{cl}); \widehat{\boldsymbol{\theta}}]$ function is continuous and monotone increasing in $1 - \alpha_{cl}$, so the appropriate calibration value can be found by using a simple root-finding method.

The procedure for Monte Carlo evaluation of the coverage probability in Section 12.6.1 utilized the observed proportion of correct prediction intervals. The advantage of the probability-averaging procedure is that it does not include a simulation of the future random variable in the evaluation. Thus the procedure requires fewer Monte Carlo samples to get the same level of accuracy.

For either evaluation method, it is a simple matter to use standard sampling methods to quantify Monte Carlo error. For example, the standard error of the Monte Carlo evaluation of $CP[PI(1 - \alpha_0); \widehat{\boldsymbol{\theta}}]$ for any particular $1 - \alpha_0$ is

$$\sqrt{\sum_{j=1}^{B} \frac{(P_j - CP[PI(1 - \alpha_0); \widehat{\boldsymbol{\theta}}])^2}{B(B-1)}}.$$

For the probability-averaging procedure, the variability in the P_j values is related to the variability in $\widehat{\boldsymbol{\theta}}^*$. The probability-averaging procedure can provide substantial savings in computing time.

Example 12.8 Calibration of the Naive Prediction Interval for a Future Lognormal Bearing Life. Returning to Example 12.4, Figure 12.4 is a histogram of the 100,000 simulated values of $Z_{\log(T^*)}$. Figure 12.5 is a corresponding histogram of the $B = 100,000$ simulated values of $\Phi_{\text{nor}}[Z_{\log(T^*)}]$. Although the lower and upper $CP[PI(1 - \alpha); \widehat{\boldsymbol{\theta}}]$ calibration functions in Figure 12.6 could have been computed from the empirical cdfs of the simulated $1 - \Phi_{\text{nor}}[Z_{\log(T^*)}]$ and $\Phi_{\text{nor}}[Z_{\log(T^*)}]$ values, respectively, the conditional probability averaging methods with $B = 100,000$ was used instead. The simulation sample size of $B = 100,000$ was chosen to be large enough to assure that the printed calibration values are correct to the number of digits shown. Because B is so large, the differences between the calibration methods were small. With $B = 10,000$, the differences were more pronounced, but $B = 10,000$ would, for practical purposes, be large enough for the conditional probability-averaging method. Using the calibration method, a naive 96.4% lower prediction bound for T provides a calibrated approximate 95% lower prediction bound for T. Also, a naive 96.7% upper prediction bound for T provides a 95% calibrated upper prediction bound for T. Comparing to the results in Example 12.4, numerically $1 - \alpha_{cl} = .964 \approx 1 - \Phi_{\text{nor}}(-1.802)$ for an approximate 95% lower prediction bound and $1 - \alpha_{cl} = .967 \approx \Phi_{\text{nor}}(1.837)$ for an approximate 95% upper

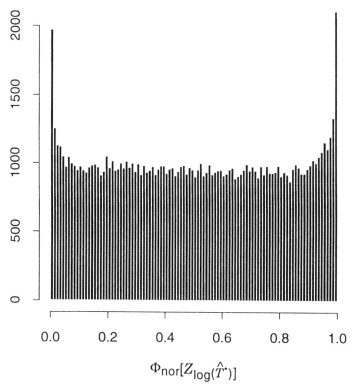

Figure 12.5. Histogram of 100,000 simulated $\Phi_{\text{nor}}[Z_{\log(\widehat{T}^*)}]$ values, based on the bearing life test data censored after 80 million cycles.

prediction bound. Differences are due to Monte Carlo error in the pivotal method. Thus substituting $\Phi_{\text{nor}}^{-1}(1 - .964)$ for $\Phi_{\text{nor}}^{-1}(.05)$ and $\Phi_{\text{nor}}^{-1}(.967)$ for $\Phi_{\text{nor}}^{-1}(.95)$ in the naive interval formula (12.5) will result in a calibrated interval from a procedure that is equivalent to the pivotal method, but with somewhat less Monte Carlo error for the same B. □

12.7 PREDICTION OF FUTURE FAILURES FROM A SINGLE GROUP OF UNITS IN THE FIELD

Consider the situation where n units are placed into service at approximately one point in time. Failures are reported until t_c, another point in time where the available data are to be analyzed. Suppose that $F(t; \boldsymbol{\theta})$ is used to describe the failure-time distribution and that $r > 0$ units have failed in the interval $(0, t_c)$. Thus there are $n - r$ unfailed units at t_c.

A common problem (e.g., in warranty exposure prediction) is the need to predict the number of additional failures K that will be reported between t_c and t_w, where

PREDICTION OF FUTURE FAILURES FROM A SINGLE GROUP OF UNITS

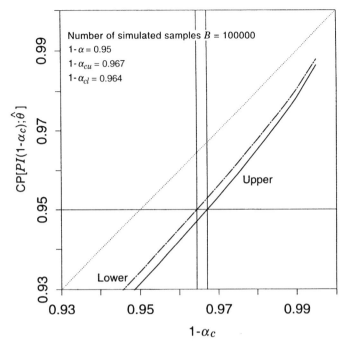

Figure 12.6. Calibration functions for predicting the failure time of a future bearing based on a lognormal distribution and life test data censored after 80 million cycles.

$t_w > t_c$. In addition, it is sometimes necessary to quantify the uncertainty in such a prediction. The upper prediction bound for K is usually of particular interest.

Conditional on the number of failures r, K follows a BINOMIAL$(n - r, \rho)$ distribution, where

$$\rho = \frac{\Pr(t_c < T \leq t_w)}{\Pr(T > t_c)} = \frac{F(t_w; \boldsymbol{\theta}) - F(t_c; \boldsymbol{\theta})}{1 - F(t_c; \boldsymbol{\theta})} \quad (12.10)$$

is the conditional probability of failing in the interval (t_c, t_w), given that a unit survived until t_c. The corresponding binomial cdf is $\Pr(K \leq k) = \text{BINCDF}(k; n - r, \rho)$.

The naive $100(1 - \alpha)\%$ upper prediction bound for K is $\widetilde{K}(1 - \alpha) = \widehat{K}_{1-\alpha}$. This upper prediction bound is computed as the smallest integer k such that $\text{BINCDF}(k; n - r, \widehat{\rho}) \geq 1 - \alpha$. The ML estimate $\widehat{\rho}$ is obtained by evaluating (12.10) at ML estimate $\widehat{\boldsymbol{\theta}}$. This upper prediction bound can be calibrated by finding $1 - \alpha_{cu}$ such that

$$\text{CP}[PI(1 - \alpha_{cu}); \widehat{\boldsymbol{\theta}}] = \Pr[K \leq \widetilde{K}(1 - \alpha_{cu})] = 1 - \alpha. \quad (12.11)$$

Then the $100(1 - \alpha)\%$ calibrated upper prediction bound would be $\widetilde{K}(1 - \alpha_{cu}) = \widehat{K}_{1-\alpha_{cu}}$.

The naive $100(1 - \alpha)\%$ lower prediction bound for K is $\underline{K}(1 - \alpha) = \widehat{K}_{\alpha}$. This lower prediction bound is computed as the largest integer k such that BINCDF$(k;$

$n - r, \widehat{\rho}) < \alpha$. This naive lower prediction bound can be calibrated by finding $1 - \alpha_{cl}$ such that

$$\text{CP}[PI(1 - \alpha_{cl}); \widehat{\boldsymbol{\theta}}] = \Pr\left[K \geq \underline{K}(1 - \alpha_{cl})\right] = 1 - \alpha \qquad (12.12)$$

and the calibrated lower prediction bound would be $\underline{K}(1 - \alpha_{cl}) = \widehat{K}_{\alpha_{cl}}$.

The needed calibration curves for (12.11) and (12.12) can be found by averaging conditional coverage probabilities obtained from Monte Carlo simulation by using the following procedure that is similar to the one in Section 12.6.2.

1. Choose a value of $1 - \alpha$, say, $1 - \alpha_0$.
2. Generate simulated samples of size n, say, DATA$^*_j$ for $j = 1, \ldots, B$ from the assumed model with parameter values equal to $\widehat{\boldsymbol{\theta}}$ and the same censoring scheme as in the original sample (leading to the same censoring pattern, except for the variability in $n - r$).
3. The jth simulated sample DATA$^*_j$ provides $n - r^*_j$, $\widehat{\boldsymbol{\theta}}^*_j$, and $\widehat{\rho}^*_j$.
4. Use the cdf BINCDF($k; n - r^*_j, \widehat{\rho}^*_j$) to compute the upper and lower *naive* prediction bounds $\widetilde{K}(1 - \alpha_0)^*_j$ and $\underline{K}(1 - \alpha_0)^*_j$.
5. For the upper prediction bound calibration, compute the conditional coverage probability $P_j = \text{BINCDF}[\widetilde{K}(1 - \alpha_0)^*_j; n - r^*_j, \widehat{\rho}]$. A Monte Carlo evaluation of the unconditional coverage probability is $\text{CP}[PI(1 - \alpha_0); \widehat{\boldsymbol{\theta}}] = \sum_{j=1}^{B} P_j/B$.
6. For the lower prediction bound calibration, compute the conditional coverage probability $P_j = 1 - \text{BINCDF}[\underline{K}(1 - \alpha_0)^*_j - 1; n - r^*_j, \widehat{\rho}]$. A Monte Carlo evaluation of the unconditional coverage probability is $\text{CP}[PI(1 - \alpha_0); \widehat{\boldsymbol{\theta}}] = \sum_{j=1}^{B} P_j/B$.

The justification for this procedure is given in the Appendix of Escobar and Meeker (1998a).

Example 12.9 Prediction Interval to Contain the Number of Future Product-A Failures. During one month, $n = 10,000$ units of Product-A (the actual name of the product is not being used to protect proprietary information) were put into service. After 48 months, 80 failures had been reported. Management requested a point prediction and an upper prediction bound on the number of the remaining $n - r = 10000 - 80 = 9920$ units that will fail during the next 12 months (i.e., between 48 and 60 months of age). The available data suggested a Weibull failure-time distribution and the ML estimates are $\widehat{\alpha} = 1152$ and $\widehat{\beta} = 1.518$. From these,

$$\widehat{\rho} = \frac{\widehat{F}(60) - \widehat{F}(48)}{1 - \widehat{F}(48)} = .003233.$$

Figure 12.7 shows the point prediction, the naive 95% upper prediction bound, and the calibrated approximate 95% upper prediction bound. The point prediction for the number failing between 48 and 60 months is $\widehat{K} = (n - r) \times \widehat{\rho} = 9920 \times .003233 = 32.07$. The naive 95% upper prediction bound on K is $\widetilde{K}(.95) = \widehat{K}_{.95} = 42$, the

PREDICTION OF FUTURE FAILURES FROM A SINGLE GROUP OF UNITS 307

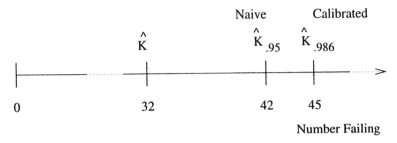

Figure 12.7. Prediction of the future number failing in the Product-A population.

smallest integer k such that $\text{BINCDF}(k; 9920, .003233) \geq .95$. The calibration curve shown in Figure 12.8 gives, for the upper prediction bound, $CP[PI(.986); \hat{\boldsymbol{\theta}}] = .95$. Thus the calibrated approximate 95% upper prediction bound on K is $\widetilde{K}(.986) = \widehat{K}_{.986} = 45$, the smallest integer k such that $\text{BINCDF}(k; 9920, .003233) \geq .986$. The naive 95% lower prediction bound on K is $\underline{K}(.95) = \widehat{K}_{.05} = 22$, the largest integer k such that $\text{BINCDF}(k; 9920, .003233) < .05$. The calibration curve shown in Figure 12.8 gives, for the lower prediction bound, $CP[PI(.981); \hat{\boldsymbol{\theta}}] = .95$. Thus the calibrated approximate 95% lower prediction bound on K is $\underline{K}(.981) = \widehat{K}_{.019} = 20$, the largest integer k such that $\text{BINCDF}(k; 9920, .003233) < 1 - .981 = .019$. □

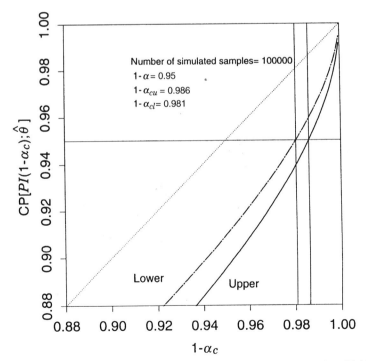

Figure 12.8. Calibration functions for upper and lower prediction bounds on the number of field failures in the next year for the Product-A population.

12.8 PREDICTION OF FUTURE FAILURES FROM MULTIPLE GROUPS OF UNITS WITH STAGGERED ENTRY INTO THE FIELD

This section describes a generalization of the prediction problem in Section 12.7. In many applications the units in the population of interest entered service over a period of time. This is called staggered entry. As in Section 12.7, the need is to use early field-failure data to construct a prediction interval for the number of future failures in some interval of calendar time, where the amount of previous operating time differs from group to group. This prediction problem is illustrated in Figure 12.9. Staggered entry failure-time data are multiply censored because of the differences in operating time. The prediction problem can be viewed as predicting the number of additional failures across the s groups during a specified period of real time. The problem is more complicated than the prediction procedure given in Section 12.7 because the age of the units, the failure probabilities, and number of units at risk to failure differ from group to group. For group i, n_i units are followed for a period of length t_{ci} and the first r_i failures were observed at times $t_{(i1)} < \cdots < t_{(ir_i)}$, $i = 1, \ldots, s$.

Conditional on $n_i - r_i$, the number of additional failures K_i from group i during interval (t_{ci}, t_{wi}) (where $t_{wi} = t_{ci} + \Delta t$) is distributed BINOMIAL$(n_i - r_i, \rho_i)$ with

$$\rho_i = \frac{\Pr(t_{ci} < T \le t_{wi})}{\Pr(T > t_{ci})} = \frac{F(t_{wi}; \boldsymbol{\theta}) - F(t_{ci}; \boldsymbol{\theta})}{1 - F(t_{ci}; \boldsymbol{\theta})}. \tag{12.13}$$

Let $K = \sum_{i=1}^{s} K_i$ be the total number of additional failures over Δt. Conditional on the DATA (and the fixed censoring times) K has a distribution that can be described by

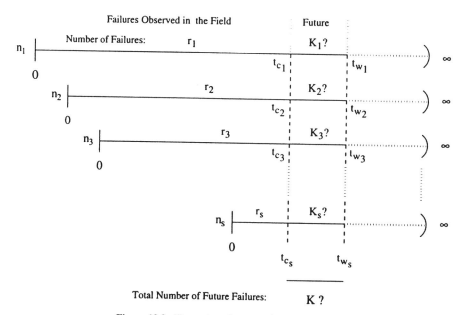

Figure 12.9. Illustration of staggered entry prediction.

PREDICTION OF FUTURE FAILURES FROM MULTIPLE GROUPS OF UNITS 309

the sum of s independent but nonidentically distributed binomial random variables with cdf denoted by $\Pr(K \leq k) = \text{SBINCDF}(k; \boldsymbol{n}-\boldsymbol{r}, \boldsymbol{\rho})$, where $\boldsymbol{n}-\boldsymbol{r} = (n_1 - r_1, \ldots, n_s - r_s)$ and $\boldsymbol{\rho} = (\rho_1, \ldots, \rho_s)$. The Appendix of Escobar and Meeker (1998a) describes methods for evaluating $\text{SBINCDF}(k; \boldsymbol{n}-\boldsymbol{r}, \boldsymbol{\rho})$.

A naive $100(1-\alpha)\%$ upper prediction bound $\widetilde{K}(1-\alpha) = \widehat{K}_{1-\alpha}$ is computed as the smallest integer k such that $\text{SBINCDF}(k; \boldsymbol{n}-\boldsymbol{r}^*, \widehat{\boldsymbol{\rho}}^*) \geq 1 - \alpha$. This upper prediction bound can be calibrated by finding $1-\alpha_{cu}$ such that

$$\text{CP}[PI(1-\alpha_{cu}); \widehat{\boldsymbol{\theta}}] = \Pr[K \leq \widetilde{K}(1-\alpha_{cu})] = 1 - \alpha.$$

A naive $100(1-\alpha)\%$ lower prediction bound $\underset{\sim}{K}(1-\alpha) = \widehat{K}_{\alpha}$ is computed as the largest integer k such that $\text{SBINCDF}(k; \boldsymbol{n}-\boldsymbol{r}^*, \widehat{\boldsymbol{\rho}}^*) < \alpha$. This lower prediction bound can be calibrated by finding $1-\alpha_{cl}$ such that

$$\text{CP}[PI(1-\alpha_{cl}); \widehat{\boldsymbol{\theta}}] = \Pr\left[K \geq \underset{\sim}{K}(1-\alpha_{cl})\right] = 1 - \alpha.$$

To calibrate these one-sided prediction bounds, one can use the same procedure outlined in Section 12.7, replacing $\text{BINCDF}(k; n-r, \widehat{\rho})$ with $\text{SBINCDF}(k; \boldsymbol{n}-\boldsymbol{r}, \widehat{\boldsymbol{\rho}})$.

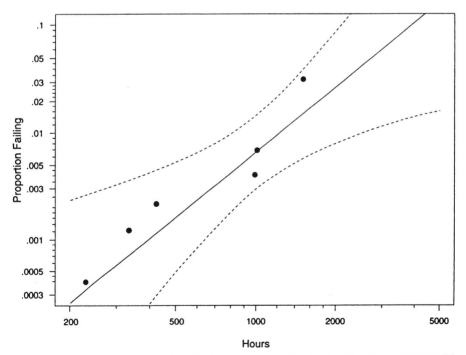

Figure 12.10. Weibull probability plot of the bearing-cage data showing the ML estimate of $F(t)$ (solid line) and a set of 95% pointwise confidence intervals (dotted lines).

Example 12.10 Prediction Interval to Contain the Number of Future Bearing-Cage Failures Abernethy, Breneman, Medlin, and Reinman (1983, pages 43–47) describe the analysis of bearing-cage failure data. Groups of bearing cages, installed in a larger system, were introduced into service at different points in time (staggered entry). Failures had occurred at 230, 334, 423, 990, 1009, and 1510 hours of service. There were 1697 other units that had accumulated various service times without failing. Figure 12.10 is a Weibull probability plot for the data. Because of an unexpectedly large number of failures in early life, the bearing cage was to be redesigned. It would, however, be some time before the design could be completed, manufacturing started, and the existing units replaced. The analysts wanted to use the initial data to predict the number of additional failures that could be expected from the population of units currently in service, during the next year, assuming that each unit will see $\Delta = 300$ hours of service during the year. Abernethy et al. (1983) computed point predictions. We will extend their results to compute a prediction interval to quantify uncertainty.

Table 12.1 is a future-failure risk analysis. This table gives, for each of the groups of units that had been put into service, the number of units installed, accumulated

Table 12.1. Bearing-Cage Data and Future-Failure Risk Analysis for the Next Year (300 Hours of Service per Unit)

Group i	Hours in Service	n_i	Failed r_i	At Risk $n_i - r_i$	$\widehat{\rho}_i$	$(n_i - r_i) \times \widehat{\rho}_i$
1	50	288	0	288	.000763	.2196
2	150	148	0	148	.001158	.1714
3	250	125	1	124	.001558	.1932
4	350	112	1	111	.001962	.2178
5	450	107	1	106	.002369	.2511
6	550	99	0	99	.002778	.2750
7	650	110	0	110	.003189	.3508
8	750	114	0	114	.003602	.4106
9	850	119	0	119	.004016	.4779
10	950	128	0	128	.004432	.5673
11	1050	124	2	122	.004848	.5915
12	1150	93	0	93	.005266	.4898
13	1250	47	0	47	.005685	.2672
14	1350	41	0	41	.006105	.2503
15	1450	27	0	27	.006525	.1762
16	1550	12	1	11	.006946	.0764
17	1650	6	0	6	.007368	.0442
18	1750	0	0	0	.007791	0
19	1850	1	0	1	.008214	.0082
20	1950	0	0	0	.008638	0
21	2050	2	0	2	.009062	.0181
Total		1703	6			5.057

Data from Abernethy, Breneman, Medlin, and Reinman (1983), pages 43–47.

PREDICTION OF FUTURE FAILURES FROM MULTIPLE GROUPS OF UNITS 311

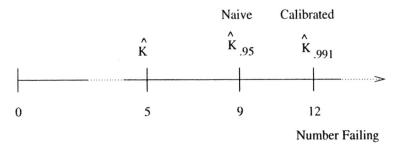

Figure 12.11. Prediction of the future number failing in the bearing-cage population.

service times, number of observed failures, estimated conditional probability of failure, and the estimated expected number failing in the 300-hour period. The sum of the estimated expected numbers failing is 5.057, providing a point prediction for the number of failures in the 300-hour period. The Poisson distribution will, in this example, provide a good approximation for the SBIN distribution of K. Figure 12.11 shows the point prediction, naive upper prediction bound, and calibrated upper prediction bound for the bearing-cage population. The naive 95% upper prediction bound on K is $\widetilde{K}(.95) = \widehat{K}_{.95} = 9$, the smallest integer k such that

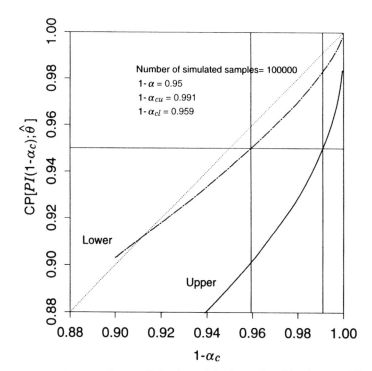

Figure 12.12. Calibration curve for a prediction interval for the number of bearing-cage failures in the next 12 months.

SBINCDF$(k; n-r, \rho) \geq .95$. The upper calibration curve shown in Figure 12.12 gives, for the upper prediction bound, CP$[PI(.991); \widehat{\theta}] = .95$. Thus the calibrated 95% upper prediction bound on K is $\widetilde{K}(.991) = \widehat{K}_{.991} = 12$, the smallest integer k such that SBINCDF$(k; n-r, \rho) \geq .991$. The naive 95% lower prediction bound on K is $\underline{K}(.95) = \widehat{K}_{.05} = 1$, the largest integer k such that SBINCDF$(k; n-r, \rho) < .05$. The lower calibration curve shown in Figure 12.12 gives CP$[PI(.959); \widehat{\theta}] = .95$. Thus the calibrated 95% lower prediction bound on K is $\underline{K}(.959) = \widehat{K}_{.041} = 1$, the largest integer k such that SBINCDF$(k; n-r, \rho) < 1 - .959 = .041$. Note that, in this particular case, the naive and the calibrated prediction bounds are the same. □

BIBLIOGRAPHIC NOTES

This chapter is based largely on Escobar and Meeker (1998a). There is a considerable amount of literature on statistical prediction. Hahn and Nelson (1973), Patel (1989), and Chapter 5 of Hahn and Meeker (1991) provide surveys of methods for statistical prediction for a variety of situations. Lawless (1973), Nelson and Schmee (1981), Engelhardt and Bain (1979), and Mee and Kushary (1994) describe exact simulation-based methods to obtain prediction intervals for Type II censored data. These methods are for log-location-scale distributions, based on the distribution of pivotal statistics. Type II censoring, however, is rare in practical application. Nelson (1995c) gives a simple procedure for computing prediction limits for the number of failures that will be observed in a future inspection, based on the number of failures in a previous inspection when the units have a Weibull failure-time distribution with a given shape parameter.

Faulkenberry (1973) suggests a method that can be applied when there is a sufficient statistic that can be used as a predictor. Cox (1975) presents a general approximate analytical approach to prediction based on the asymptotic distribution of ML estimators. Atwood (1984) used a similar approach. Beran (1990) presents the bootstrap calibration method for obtaining prediction intervals and gives theoretical results on the properties of prediction statements obtained with such calibration methods. An approximate pivotal-based approach is described in Efron and Tibshirani (1993, page 390–391). Kalbfleisch (1971) describes a likelihood-based method, Thatcher (1964) describes the relationship between Bayesian and frequentist prediction for the binomial distribution, while Geisser (1993) presents a more general overview of the Bayesian approach (see also Chapter 14).

EXERCISES

12.1. A sample of 20 aluminum specimens was tested until fatigue failure. A probability plot showed that the lognormal distribution provides an adequate description of the spread in the data. The sample mean and standard deviation of the logarithms of cycles to failure were 5.13 and .161, respectively.

(a) Compute a 95% confidence interval for the median of the cycles to failure distribution.

(b) Compute a 95% prediction interval for the number of cycles to failure for a future specimen tested in the same way. Compare this with the "naive" prediction interval computed as if the estimates are the parameters.

(c) Redo parts (a) and (b) supposing, instead, that the sample size had been 100 units. Comment on the results.

(d) Explain why there is so much difference between the confidence interval in part (a) and the prediction interval in part (b).

12.2. Show that putting together a one-sided lower and a one-sided upper $100(1 - \alpha/2)\%$ prediction bound for a future observation results in a two-sided $100(1 - \alpha)\%$ prediction interval for that observation.

▲**12.3.** Let $t_1, \ldots, t_n$ denote the r failure times and the $n - r$ censored times of a failure-censored test with EXP(θ) data. As indicated in Chapter 7, the ML estimator of θ is $\widehat{\theta} = TTT/r$ and $2(r\widehat{\theta}/\theta) \sim \chi^2_{(2r)}$. If T is a new independent observation then $2T/\theta \sim \chi^2_{(2)}$. Thus it follows that $T/\widehat{\theta} \sim \mathcal{F}_{(2,2r)}$.

(a) Show that $PI(1 - \alpha) = (0, \widetilde{T})$, where $\widetilde{T} = \mathcal{F}_{(1-\alpha;2,2r)} \times \widehat{\theta} = r[\alpha^{(-1/r)} - 1] \times \widehat{\theta}$ is an exact $100(1 - \alpha)\%$ prediction interval for T.

(b) Show that the conditional coverage probability, conditional on $\widehat{\theta}$, is

$$CP[PI(1 - \alpha) \mid \widehat{\theta}; \theta] = 1 - \exp\{-\widehat{\theta} \times r \times [\alpha^{(-1/r)} - 1]/\theta\}.$$

(c) Show that $\lim_{r \to \infty} CP[PI(1-\alpha) \mid \widehat{\theta}; \theta] = 1-\alpha$. Comment on the practical interpretation of this result.

(d) Compute the unconditional coverage probability using

$$CP[PI(1 - \alpha)] = E_{\widehat{\theta}}\{CP[PI(1 - \alpha) \mid \widehat{\theta}; \theta]\}$$
$$= E_{\widehat{\theta}}(1 - \exp\{-\widehat{\theta} \times r \times [\alpha^{(-1/r)} - 1]/\theta\})$$

and verify that it is equal to $100(1 - \alpha)\%$ for any n.

▲**12.4.** Suppose $t_1, \ldots, t_n$ is a random sample of size n from an EXP(θ). Denote the ML of θ by $\widehat{\theta}_n$ ($\widehat{\theta}_n = \bar{t}$). Consider the naive prediction interval $PI(1 - \alpha) = (0, \widetilde{T})$, where $\widetilde{T} = [-\log(\alpha)] \times \widehat{\theta}_n$ and $[-\log(\alpha)]$ is the $100(1 - \alpha)\%$ quantile of the EXP(1) distribution.

(a) Show that the conditional coverage probability is

$$CP[PI(1 - \alpha) \mid \widehat{\theta}; \theta] = 1 - \alpha^{(\widehat{\theta}/\theta)}.$$

(b) Show that $\lim_{n\to\infty} \text{CP}[PI(1-\alpha) \mid \widehat{\theta}; \theta] = 1-\alpha$. (Here you might want to use the fact that $\widehat{\theta}_n \to \theta$ in probability when $n \to \infty$.) Explain the practical implications of this result.

(c) Show that $\text{CP}[PI(1-\alpha)] = E_{\widehat{\theta}}\{\text{CP}[PI(1-\alpha) \mid \widehat{\theta}; \theta]\} = 1 - [1 - (1/n)\log(\alpha)]^{-n}$.

(d) Show that $\text{CP}[PI(1-\alpha)] < 1-\alpha$, for all $1-\alpha$ and n.

(e) For $n = 2$ draw a graph of $\text{CP}[PI(1-\alpha)]$ for values of $1-\alpha$ over the interval $[.9, 1]$ and compare the $\text{CP}[PI(1-\alpha)]$ with the nominal coverage of $100(1-\alpha)\%$. Repeat this for $n = 4, 10, 100$. Comment on the behavior of the coverage probability and the length of the naive prediction interval in large samples.

(f) Show that $\lim_{n\to\infty} \text{CP}[PI(1-\alpha)] = 1-\alpha$ and comment on the practical implications of this result.

▲**12.5.** Refer to Exercise 12.4 but now suppose that the data are failure-censored with r observed failures. Generalize all the results in that exercise to this new situation.

12.6. Suppose that $t_1, \ldots, t_n$ is a random sample from a LOGNOR(μ, σ) distribution. Suppose that σ is known and let $\widehat{\mu}_n = \bar{y} = \sum_{i=1}^{n} y_i/n$, where $y_i = \log(t_i)$. Consider the prediction of a new time to failure observation T. In this case, $[\log(T) - \widehat{\mu}_n]/\sqrt{\text{Var}(T - \widehat{\mu}_n)} = [\log(T) - \widehat{\mu}_n]/[\sigma\sqrt{1 + 1/n}] \sim \text{NOR}(0, 1)$, which suggests the prediction interval

$$PI(1-\alpha) = \left[0, \ \exp\left(\widehat{\mu}_n + \sigma\sqrt{\frac{n+1}{n}}\Phi_{\text{nor}}^{-1}(1-\alpha)\right)\right].$$

(a) Show that $\text{CP}[PI(1-\alpha)] = 1-\alpha$ for any n.

(b) Show that

$$\text{CP}[PI(1-\alpha) \mid \widehat{\mu}_n; \mu] = \Phi_{\text{nor}}\left[\frac{\widehat{\mu}_n - \mu}{\sigma} + \sqrt{\frac{n+1}{n}}\Phi_{\text{nor}}^{-1}(1-\alpha)\right].$$

(c) Show that $\lim_{n\to\infty} \text{CP}[PI(1-\alpha) \mid \widehat{\mu}_n; \mu] = 1-\alpha$.

(d) Show that $\text{CP}[PI(1-\alpha)] = E_{\widehat{\mu}_n}\{\text{CP}[PI(1-\alpha) \mid \widehat{\mu}_n; \mu]\} = 1-\alpha$. Note the complexity of this computation when compared with the computation in part (a).

(e) Derive an expression for a two-sided $100(1-\alpha)\%$ prediction interval for T.

(f) A naive prediction interval is $PI(1-\alpha) = [0, \exp(\widehat{\mu} + \Phi_{\text{nor}}^{-1}(1-\alpha) \times \sigma)]$. Show that $\text{CP}[PI(1-\alpha)] = \Phi_{\text{nor}}[\sqrt{n/(n+1)}\ \Phi_{\text{nor}}^{-1}(1-\alpha)]$. Draw a graph of this coverage probability for values of $n = 2, 4, 10$ over values of $1-\alpha$

EXERCISES

in the interval [.9, 1]. Comment on the coverage probabilities and their behavior as a function of the sample size.

▲12.7. Refer to Exercise 12.6 but suppose that σ is unknown. Define

$$\widehat{\sigma}_n = \sqrt{\frac{\sum_{i=1}^{n}(y_i - \bar{y})^2}{n}}.$$

In this case, $[\log(T) - \widehat{\mu}_n]/[\widehat{\sigma}_n\sqrt{(n+1)/(n-1)}]$ has a Student's t distribution with $(n-1)$ degrees of freedom, which suggests the prediction interval

$$PI(1-\alpha) = \left[0, \exp\left(\widehat{\mu}_n + \frac{\widehat{\sigma}_n}{\sigma}\sqrt{\frac{n-1}{n+1}} t_{(1-\alpha;n-1)}\right)\right],$$

where $t_{(1-\alpha;n-1)}$ is the $100(1-\alpha)\%$ quantile of the Student's t distribution with $(n-1)$ degrees of freedom.

(a) Show that $CP[PI(1-\alpha)] = 1 - \alpha$ for any n.

(b) Show that

$$CP[PI(1-\alpha) \mid \widehat{\mu}_n; \mu] = \Phi_{\text{nor}}\left[\frac{\widehat{\mu}_n - \mu}{\sigma} + \frac{\widehat{\sigma}_n}{\sigma}\sqrt{\frac{n-1}{n+1}} t_{(1-\alpha;n-1)}\right].$$

(c) Show that $\lim_{n\to\infty} CP[PI(1-\alpha) \mid \widehat{\mu}_n; \mu] = 1 - \alpha$.

(d) Derive the expression for a two-sided $100(1-\alpha)\%$ prediction interval for T.

(e) A naive upper prediction interval is $PI(1-\alpha) = [0, \exp(\widehat{\mu} + \Phi_{\text{nor}}^{-1}(1-\alpha) \times \widehat{\sigma})]$. Show that $CP[PI(1-\alpha)] = \Pr[X \leq \sqrt{n/(n+1)}\,\Phi_{\text{nor}}^{-1}(1-\alpha)]$, where X has a Student's t distribution with $(n-1)$ degrees of freedom. Draw a graph of this coverage probability for values of $n = 2, 4, 10$ over values of $1 - \alpha$ in the interval [.9, 1].

▲12.8. Consider the prediction of a new observation from a log-location-scale family with cdf $\Phi[(\log(t) - \mu)/\sigma]$. Let $[\widetilde{T}, \widetilde{T}] = [\widetilde{t}_{\alpha/2}, \widetilde{t}_{1-\alpha/2}]$ be the naive prediction interval computed based on a set of available data.

(a) Show that the coverage probability of the prediction interval, conditional on the ML estimates $\widehat{\theta} = (\widehat{\mu}, \widehat{\sigma})$, is given by

$$CP[PI(1-\alpha) \mid \widehat{\theta}; \theta] = \Phi\left[\frac{\widehat{\mu} - \mu}{\sigma} + \frac{\widehat{\sigma}}{\sigma} \times \Phi^{-1}(1-\alpha/2)\right]$$

$$-\Phi\left[\frac{\widehat{\mu} - \mu}{\sigma} + \frac{\widehat{\sigma}}{\sigma} \times \Phi^{-1}(\alpha/2)\right],$$

where $\widehat{\theta} = (\widehat{\mu}, \widehat{\sigma})$.

(b) Show that when the sample size increases to $+\infty$ the conditional coverage probability converges in probability to $1 - \alpha$.

CHAPTER 13

Degradation Data, Models, and Data Analysis

Objectives

This chapter explains:

- Some useful degradation models.
- The connection between degradation models and failure-time models.
- How degradation measures, when available, can be used to advantage in estimating reliability.
- Methods for data analysis and reliability inference with degradation data.
- The differences between degradation data analysis and traditional failure-time data analysis.
- A simple method for degradation analysis that can be useful in certain situations.

Overview

This chapter introduces the concepts of degradation analysis as they relate to product reliability. Many failure mechanisms can be traced to an underlying degradation process. Degradation eventually leads to a weakness that can cause failure. When it is possible to measure degradation, such measures often provide more information than failure-time data for purposes of assessing and improving product reliability. For some products direct observation of degradation level is impossible, but it may be that product performance data will be a useful substitute. This chapter, Chapter 18, and Chapter 21 provide a brief introduction to this important subject (a complete treatment would require a separate book). This chapter should be read before Chapter 21. Readers may skip this chapter if their primary interest is in failure-time data, although Section 13.2 does give an introduction to some physics-of-failure concepts that provide useful motivation for failure-time models. Section 13.3 extends ML methods from earlier chapters to deal with the more complicated degradation models. Sections 13.4, 13.5, 13.6, and 13.7 relate degradation and failure time and show how to estimate $F(t)$ from degradation data. Section 13.8 uses an example to compare degradation analysis with traditional failure-time analysis. Section 13.9 presents a simple approximate method for degradation analysis that might be appropriate in some applications.

13.1 INTRODUCTION

Design of high-reliability systems generally requires that the individual system components have extremely high reliability, even after long periods of time. With short product development times, reliability tests must be conducted with severe time constraints. Frequently no failures occur during such tests. Thus it is difficult to assess reliability with traditional life tests that record only failure time. For some components degradation measures can be taken over time. A relationship between component failure and amount of degradation makes it possible to use degradation models and data to make inferences and predictions about failure time.

13.2 MODELS FOR DEGRADATION DATA

13.2.1 Degradation Data

In some reliability studies, it is possible to measure physical degradation as a function of time (e.g., tire wear). In other applications actual physical degradation cannot be observed directly, but measures of product performance degradation (e.g., power output) may be available. Both kinds of data are generically referred to as "degradation data" and we will follow this convention. Modeling performance degradation may be useful but could be complicated because performance may be affected by more than one underlying degradation process. Depending on the application, degradation data may be available continuously or at specific points in time where measurements are taken.

In most reliability testing applications, degradation data, if available, will have important practical advantages. In particular:

- Degradation data can, especially in applications with few or no failures, provide considerably more reliability information than traditional censored failure-time data.
- Accelerated tests are commonly used to obtain reliability test information more quickly. Direct observation of the physical degradation process (e.g., tire wear) or some closely related surrogate may allow direct modeling of the failure-causing mechanism, providing more credible and precise reliability estimates and a firmer basis for often-needed extrapolation.

Example 13.1 Fatigue Crack-Size Data. Recall the Alloy-A fatigue crack-size data in Example 1.10. Figure 13.1 is similar to Figure 1.10 but includes the actual data points on crack size given in Appendix Table C.14. The initial crack size (i.e., at time 0) for each path was .9 inch, the size of the notch cut into each

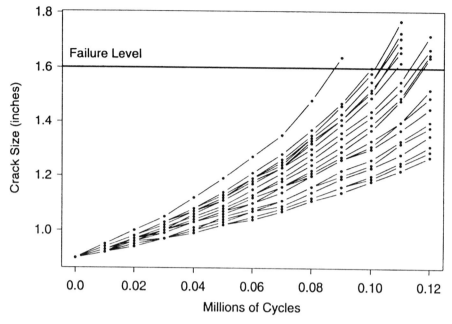

Figure 13.1. Alloy-A fatigue crack data.

specimen. Suppose that investigators wanted to estimate the material's crack growth parameters and the time (measured in number of cycles) at which 50% of the cracks would reach 1.6 inches (a size considered to be dangerous). For purposes of our degradation analysis the fatigue experiment for each specimen was terminated at the first inspection after a unit's cracks reached 1.6 inches or censored after .12 million cycles, whichever came first. □

13.2.2 Degradation Leading to Failure

Most failures can be traced to an underlying degradation process. Figure 13.2 shows examples of three general shapes for degradation curves in arbitrary units of degradation and time: linear, convex, and concave. The horizontal line at degradation level .6 represents the level or approximate level at which failure would occur. In some applications there may be more than one degradation variable or more than one underlying degradation process. The following examples, however, use only a single degradation variable.

The following examples describe some specific models for degradation curves. Engineers and physical scientists must find such models in their literature or develop them from basic principles relating to the underlying degradation process. Usually such models start with a deterministic description of the degradation process—often in the form of a differential equation or system of differential equations. Then randomness can be introduced, as appropriate, using probability distributions to describe

MODELS FOR DEGRADATION DATA

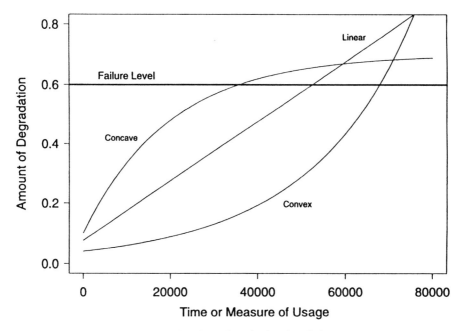

Figure 13.2. Possible shapes for univariate degradation curves.

variability in initial conditions and model parameters like rate constants or material properties.

Example 13.2 Linear Degradation. Linear degradation arises in some simple wear processes (e.g., automobile tire wear). For example, if $\mathcal{D}(t)$ is the amount of automobile tire tread wear at time t and wear rate is $d\mathcal{D}(t)/dt = C$, then $\mathcal{D}(t) = \mathcal{D}(0) + C \times t$. □

The parameters $\mathcal{D}(0)$ and C could be taken as constant for individual units, but random from unit-to-unit.

Example 13.3 Convex Degradation. Models for which the degradation rate increases with the level of degradation are, for example, used in modeling the growth of fatigue cracks. Let $a(t)$ denote the size of a crack at time t. A simple version of the deterministic Paris-rule model (e.g., Dowling, 1993),

$$\frac{d\,a(t)}{dt} = C \times \left[\Delta K(a)\right]^m, \tag{13.1}$$

provides a useful model for cracks within a certain size range. Here C and m are material properties and $\Delta K(a)$ (known as the "stress intensity range function") is a function of crack size a, the range of applied stress, part dimensions, and geometry. For example, to model a two-dimensional edge-crack in a plate with a crack that is small relative to the width of the plate (say, less than 3%), $\Delta K(a) = \text{Stress}\sqrt{\pi a}$. The deterministic solution to the resulting differential equation is

$$a(t) = \begin{cases} \left[\{a(0)\}^{1-m/2} + (1-m/2) \times C \times (\text{Stress}\sqrt{\pi})^m \times t\right]^{2/(2-m)}, & m \neq 2 \\ a(0) \times \exp\left[C \times (\text{Stress}\sqrt{\pi})^2 \times t\right], & m = 2. \end{cases} \quad (13.2)$$

This solution is illustrated for $m = 2.05$ with the convex curve in Figure 13.2. □

Example 13.4 Concave Degradation. Meeker and LuValle (1995) describe models for growth of failure-causing conducting filaments of chlorine–copper compounds in printed-circuit boards. These filaments cause failure when they reach from one copper-plated through-hole to another. In their model, $A_1(t)$ is the amount of chlorine available for reaction and $A_2(t)$ is proportional to the amount of failure-causing chlorine–copper compounds at time t. Under appropriate conditions of temperature, humidity, and electrical charge, copper combines with chlorine A_1 to produce the chlorine–copper compound A_2 with rate constant k_1. Diagrammatically,

$$A_1 \xrightarrow{k_1} A_2.$$

The rate equations for this process are

$$\frac{dA_1}{dt} = -k_1 A_1 \quad \text{and} \quad \frac{dA_2}{dt} = k_1 A_1. \quad (13.3)$$

The solution of this system of differential equations gives

$$A_1(t) = A_1(0) \times \exp(-k_1 t) \quad (13.4)$$

$$A_2(t) = A_2(0) + A_1(0) \times [1 - \exp(-k_1 t)] \quad (13.5)$$

where $A_1(0)$ and $A_2(0)$ are initial amounts. To simplify notation, let $A_2(\infty) = A_1(0) + A_2(0)$. Then if $A_2(0) = 0$, the solution for $A_2(t)$ (the quantity of primary interest) can be expressed as

$$A_2(t) = A_2(\infty) \times [1 - \exp(-k_1 t)]. \quad (13.6)$$

This function is illustrated by the concave curve in Figure 13.2. The asymptote at $A_2(\infty)$ reflects the finite amount of chlorine available for the reaction producing the harmful compounds. □

Meeker and LuValle (1995) also suggest other more elaborate models for this failure process. Carey and Koenig (1991) use similar models to describe degradation of electronic components. Chapter 18 describes the ideas behind acceleration of failure-causing processes like these.

MODELS FOR DEGRADATION DATA

13.2.3 Models for Variation in Degradation and Failure Times

If all manufactured units were identical, operated under exactly the same conditions and in exactly the same environment, and if every unit failed as it reached a particular "critical" level of degradation, then, according to the simple deterministic models above, all units would fail at exactly the same time. Of course, there is some degree of variability in all of these model factors as well as in factors that are not in the model. These factors combine to cause variability in the degradation curves and in the failure times.

Unit-to-Unit Variability

The following are examples of sources of unit-to-unit variability:

- **Initial conditions.** Individual units will vary with respect to the amount of material available to wear, initial level of degradation, amount of harmful degradation-causing material, and so on. Figure 13.3 shows the Paris model for growth of fatigue cracks, with simulated variability in the size of the initial crack, but with all other of the unit's Paris model characteristics and other factors held constant.
- **Material properties**. Figure 13.4 shows the Paris model for growth of fatigue cracks, allowing for unit-to-unit variability in the material properties parameters C and m and the size of the initial crack. In this case, as shown in the Paris model

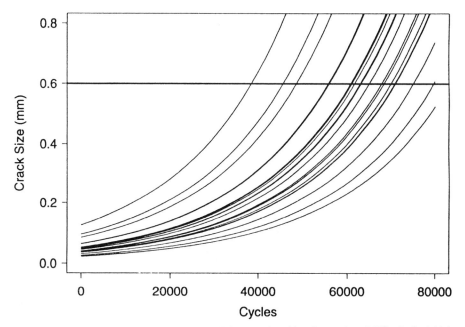

Figure 13.3. Plot of Paris model for growth of fatigue cracks with unit-to-unit variability in the initial crack size a_0 but with constant material parameters (C and m) and constant stress.

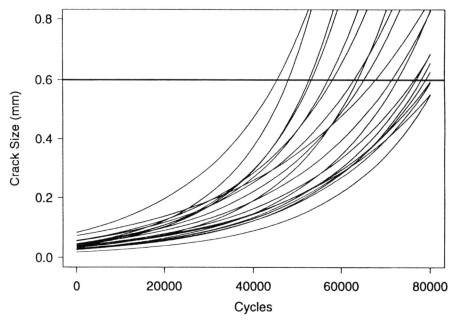

Figure 13.4. Plot of Paris model for growth of fatigue cracks with unit-to-unit variability in the initial crack size and in material parameters C and m, but with constant stress.

in (13.1), the rate of growth depends on C and m, which differ from unit to unit. This yields crossing of the crack-growth curves (typical of what is observed in actual fatigue testing).

- **Component geometry or dimensions.** Unit-to-unit variability in component geometry or dimensions can cause additional unit-to-unit variability, for example, in degradation rates [e.g., through the $\Delta K(a)$ function in (13.1)].
- **Within-unit variability.** Often there will be spatial variability in material properties within a unit (e.g., defects).

Variability Due to Operating and Environmental Conditions

Besides the material properties described above, the rate of degradation will depend on operating and environmental conditions. For example, $K(a)$ in the Paris model (13.1) depends on the amount of applied stress and the Paris parameters can depend on temperature. In laboratory fatigue tests, the stress is either fixed or changing in a systematic manner [e.g., to keep $K(a)$ nearly constant as a increases]. In actual operation of most components, stress would generally be a complicated function over time. Such variations can be described by a stochastic process model. Figure 13.5 shows the Paris model with degradation rate varying due to variations in stress that might have been caused, for example, by variation in driving conditions encountered, over time, by an automobile. In some applications, shocks or changes in environmental conditions that occur randomly in time can dominate other sources of variability in a failure-causing process.

MODELS FOR DEGRADATION DATA 323

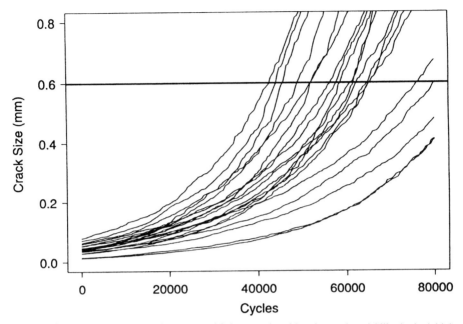

Figure 13.5. Plot of Paris model for growth of fatigue cracks with unit-to-unit variability in the initial crack size and material parameters C and m, and with a stochastic process model for the changes in stress over the life of the unit.

The models described here are simple relative to the more exact theory of failure-causing processes that almost certainly exist (but may not be known or understood). For some purposes, however, such simple first-order descriptions are useful.

13.2.4 Limitations of Degradation Data

Physical degradation or performance degradation are natural properties to measure for many testing processes (e.g., monitoring crack size of a specimen subjected to stress cycling or power output of an electronic device). Often, however, degradation measurement of a unit requires destructive inspection (e.g., destructive strength tests) or disruptive measurement (e.g., disassembly and reassembly of a motor) that has the potential to change the degradation process. In such situations one can obtain only a single measurement on each unit tested. It is possible to extract useful information from such data if a large number of units can be tested (for an example, see Nelson, 1990a, Chapter 11).

The advantages of degradation data can also be compromised when the degradation measurements are contaminated with large amounts of measurement error or when the degradation measure is not closely related to failure. For example, when the degradation measurement is on performance degradation, rather than physical degradation, failures may occur for physical reasons that are not or cannot be observed directly.

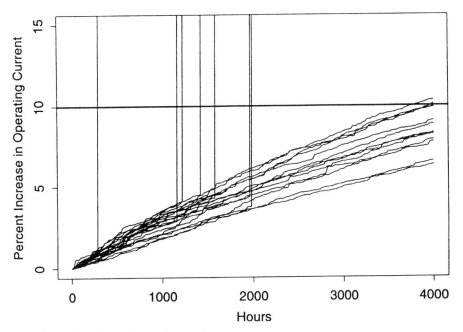

Figure 13.6. Plot of percent increase in operating current for GaAs lasers tested at 80°C.

Example 13.5 Laser Degradation and Defective Lasers. Over the life of some laser devices, degradation causes a decrease in light output. Some lasers, however, contain a feedback mechanism that will maintain nearly constant light output by increasing operating current as the laser degrades. When operating current gets too high, the device is considered to have failed. Figure 13.6 shows the increase in operating current over time for a sample of GaAs lasers tested at 80°C (this temperature, much higher than the use temperature, was used to accelerate the failure mechanism so that degradation information would be obtained more rapidly—see Chapter 21). Some of the units degrade gracefully; others fail suddenly. Sudden failures, like those in Figure 13.6, usually indicate manufacturing or other quality problems in an immature product. Such behavior is common, especially in outputs of electronic devices. Possible reasons for sudden failures include the following:

- An *unobserved* sudden change in the physical state of the unit that would lead to a subsequent increase in the degradation rate (e.g., growth of a conducting path that suddenly causes a short circuit).
- Manufacturing defects (often observed in early development of a new product).
- A different failure mode actuated only at high temperatures.
- Inadvertent shocks to units.

As a product's design, manufacturing, and testing processes mature, such problems are usually eliminated. □

MODELS FOR DEGRADATION DATA   325

When there is not a strong correlation between failure times and degradation, there may be little to be gained by using degradation data instead of traditional censored failure-time data. The limited amount of information in such degradation measurements can be the result of monitoring a *performance* variable (e.g., output voltage) rather than the actual physical degradation (e.g., amount of material displaced by electromigration). An important but difficult engineering challenge of degradation analysis is to find variables that are closely related to failure time and develop methods for accurately measuring these variables.

13.2.5 General Degradation Path Model

The actual degradation path of a particular unit over time is denoted by $\mathcal{D}(t), t > 0$. In applications, values of $\mathcal{D}(t)$ are sampled at discrete points in time $t_1, t_2, \ldots$. The observed sample degradation y_{ij} of unit i at time t_j is

$$y_{ij} = \mathcal{D}_{ij} + \epsilon_{ij}, \quad i = 1, \ldots, n, \quad j = 1, \ldots, m_i, \quad (13.7)$$

where $\mathcal{D}_{ij} = \mathcal{D}(t_{ij}, \beta_{1i}, \ldots, \beta_{ki})$ is the actual path of the unit i at time t_{ij} (the times need not be the same for all units) and $\epsilon_{ij} \sim \text{NOR}(0, \sigma_\epsilon)$ is a residual deviation for unit i at time t_j. The total number of inspections on unit i is denoted by m_i. Time t could be real-time, operating time, or some other appropriate measure of use like miles for automobile tires or cycles in fatigue tests. For the ith unit, $\beta_{1i}, \ldots, \beta_{ki}$ is a vector of k unknown parameters. Typically, sample paths have $k = 1, 2, 3$, or 4 parameters. As described in Section 13.2.3, some of the $\beta_1, \ldots, \beta_k$ parameters will be random from unit to unit. One or more of the $\beta_1, \ldots, \beta_k$ parameters could, however, be modeled as common across all units.

The scales of y and t can be chosen, as suggested by physical theory and the data, to simplify the form of $\mathcal{D}(t, \beta_1, \ldots, \beta_k)$. For example, the relationship between the logarithm of degradation and the logarithm of time might be modeled by the additive relationship in (13.7). Degradation model choice requires not only specification of the form of the $\mathcal{D}(t, \beta_1, \ldots, \beta_k)$ function, but also specification of which of the $\beta_1, \ldots, \beta_k$ are random (differing from unit to unit) and which are fixed (common to all units). Because of the flexibility in specifying the form of $\mathcal{D}(t, \beta_1, \ldots, \beta_k)$, and of the way in which the $\beta_1, \ldots, \beta_k$ come into this form, we can, for simplicity, model the unit-to-unit variability in $\beta_1, \ldots, \beta_k$ with a multivariate normal distribution with mean vector μ_β and covariance matrix Σ_β.

It is generally assumed that the random $\beta_1, \ldots, \beta_k$ are independent of the ϵ_{ij} deviations. Another common assumption is that σ_ϵ is constant. The adequacy of this assumption can be affected by transforming $\mathcal{D}(t)$. Because the y_{ij} are taken serially on a unit, however, there is potential for autocorrelation among the $\epsilon_{ij}, j = 1, \ldots, m_i$ values, especially if there are many closely spaced readings. In many practical applications involving inference on the degradation of units from a population or process, however, the correlation is weak and, moreover, dominated by the unit-to-unit variability in the $\beta_1, \ldots, \beta_k$ values and thus can be ignored. In situations where autocorrelation cannot be ignored, one can use a time series model for the residual term along with appropriate estimation methods.

13.2.6 Degradation Model Parameters

Although the values of $\beta_1, \ldots, \beta_k$ for the individual units may be of interest in some applications (e.g., to predict the future degradation of a particular unit, based on a few early readings), subsequent development in this chapter will concentrate on the use of degradation data to make inferences about the population or process or predictions about future units. In this case, the underlying model parameters are μ_β and Σ_β, as well as the residual standard deviation σ_ϵ. For shorthand, we will use $\theta_\beta = (\mu_\beta, \Sigma_\beta)$ to denote the overall population/process parameters.

13.3 ESTIMATION OF DEGRADATION MODEL PARAMETERS

The likelihood for the random-parameter degradation model in Section 13.2.5 can be expressed as

$$L(\theta_\beta, \sigma_\epsilon | \text{DATA}) = \prod_{i=1}^{n} \int_{-\infty}^{\infty} \cdots \int_{-\infty}^{\infty} \left[\prod_{j=1}^{m_i} \frac{1}{\sigma_\epsilon} \phi_{\text{nor}}(\zeta_{ij}) \right] \quad (13.8)$$
$$\times f_\beta(\beta_{1i}, \ldots, \beta_{ki}; \theta_\beta) d\beta_{1i}, \ldots, d\beta_{ki},$$

where $\zeta_{ij} = [y_{ij} - \mathcal{D}(t_{ij}, \beta_{1i}, \ldots, \beta_{ki})]/\sigma_\epsilon$ and $f_\beta(\beta_{1i}, \ldots, \beta_{ki}; \theta_\beta)$ is the multivariate normal distribution density function. Each evaluation of (13.8) will, in general, require numerical approximation of n integrals of dimension k (where n is the number of sample paths and k is the number of random parameters in each path). Maximizing (13.8) with respect to $(\mu_\beta, \Sigma_\beta, \sigma_\epsilon)$ directly, even with today's computational capabilities, is extremely difficult unless $\mathcal{D}(t)$ is a linear function. Pinheiro and Bates (1995a) describe and compare estimation schemes that provide approximate ML estimates of $\theta_\beta = (\mu_\beta, \Sigma_\beta)$ and σ_ϵ, as well as the unit-specific components in $\beta_{1i}, \ldots, \beta_{ki}, i = 1, \ldots, n$. Pinheiro and Bates (1995b) implement a modification of the method of Lindstrom and Bates (1990). The examples in this chapter were computed with the Pinheiro and Bates (1995b) program implemented in S-PLUS as function nlme.

Example 13.6 Estimates of Fatigue Data Model Parameters for Alloy-A.
Continuing with Example 13.1, we fit the model in (13.7) with $\mathcal{D}_{ij} = a(t)$ in (13.2), $a(0) = .9$, Stress $= 1$, $\beta_1 = m$, and $\beta_2 = C$, modeling (β_1, β_2) with a bivariate normal distribution. The program of Pinheiro and Bates (1995b) gives the following

13.4 MODELS RELATING DEGRADATION AND FAILURE

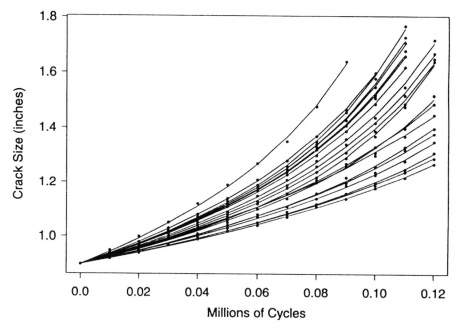

Figure 13.7. Alloy-A fatigue crack-size observations and fitted Paris-rule model.

approximate ML estimates:

$$\widehat{\boldsymbol{\mu}}_\beta = \begin{pmatrix} 5.17 \\ 3.73 \end{pmatrix}, \quad \widehat{\Sigma}_\beta = \begin{pmatrix} .251 & -.194 \\ -.194 & .519 \end{pmatrix},$$

and $\widehat{\sigma}_\epsilon = .0034$. Figure 13.7 shows the fitted Paris relationship for each of the sample paths (indicated by the points on the plot) for the Alloy-A fatigue-crack data. Figure 13.8 is a scatter plot of the estimates of the Paris relationship parameters for each of the 21 sample paths, indicating the reasonableness of the bivariate normal distribution model for this random-coefficients model. □

13.4 MODELS RELATING DEGRADATION AND FAILURE

13.4.1 Soft Failures: Specified Degradation Level

For some products there is a gradual loss of performance (e.g., decreasing light output from a fluorescent light bulb). Then failure would be defined (in a somewhat arbitrary, but purposeful, manner) at a specified level of degradation (e.g., 60% of initial output). We call this a "soft failure" definition.

A fixed value of $\mathcal{D}_f$ will be used to denote the critical level for the degradation path above (or below) which failure is assumed to have occurred. The failure time T is defined as the time when the actual path $\mathcal{D}(t)$ crosses the critical degradation level $\mathcal{D}_f$.

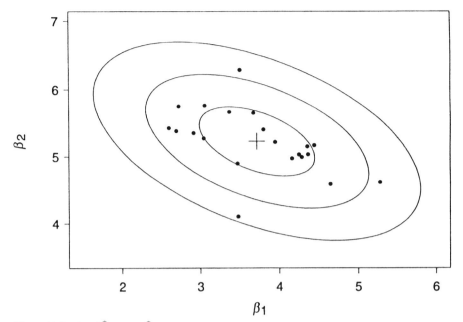

Figure 13.8. Plot of $\widehat{\beta}_{1i}$ versus $\widehat{\beta}_{2i}$ for the $i = 1, \ldots, 21$ sample paths from the Alloy-A fatigue crack-size data. The contour lines represent the fitted bivariate normal distribution for β_1 and β_2.

We use t_c to denote the planned stopping time in the experiment (as illustrated in Figure 13.1). Inferences are desired on the failure-time distribution of a particular product or material. For soft failures it may be possible to continue observation beyond $\mathcal{D}_f$.

13.4.2 Hard Failures: Joint Distribution of Degradation and Failure Level

For some products, the definition of the failure event is clear—the product stops working (e.g., when the resistance of a resistor deviates too much from its nominal value, causing the oscillator in an electronic circuit to stop oscillating or when an incandescent light bulb burns out). These are called "hard failures." With hard failures, failure times will not, in general, correspond exactly with a particular level of degradation (like the horizontal line shown in Figures 13.2–13.6). Instead, the level of degradation at which failure (i.e., loss of functionality) occurs will be random from unit to unit and even over time. This could be modeled by using a distribution to describe unit-to-unit variability in $\mathcal{D}_f$ or, more generally, the joint distribution of $\boldsymbol{\beta}$ and the stochastic behavior in $\mathcal{D}_f$.

13.5 EVALUATION OF $F(t)$

A specified model for $\mathcal{D}(t)$ and $\mathcal{D}_f$ defines a failure-time distribution. In general, this distribution can be written as a function of the degradation model parameters.

EVALUATION OF F(t)

Suppose that a unit fails at time t if the degradation level first reaches $\mathcal{D}_f$ at time t. Then

$$\Pr(T \leq t) = F(t) = F(t; \boldsymbol{\theta}_\beta) = \Pr[\mathcal{D}(t, \beta_1, \ldots, \beta_k) \geq \mathcal{D}_f]. \tag{13.9}$$

For a fixed $\mathcal{D}_f$, the distribution of T depends on the distribution of the $\beta_1, \ldots, \beta_k$, which, in turn, depends on the basic path parameters in $\boldsymbol{\theta}_\beta$. In some simple cases it is possible to write down a closed-form expression for $F(t)$. In general, however, such a closed-form expression will not exist. For most practical path models, especially when $\mathcal{D}(t)$ is nonlinear and more than one of the $\beta_1, \ldots, \beta_k$ is random, it is necessary to evaluate $F(t)$ with numerical methods.

13.5.1 Analytical Solution for F(t)

For some particularly simple path models, $F(t)$ can be expressed as a function of the basic path parameters in a closed form. Consider the following example.

Example 13.7 Linear Degradation with Lognormal Rate. Suppose the actual degradation path of a particular unit is given by

$$\mathcal{D}(t) = \beta_1 + \beta_2 t,$$

where β_1 is fixed and β_2 varies from unit to unit according to a LOGNOR(μ, σ) distribution; that is,

$$\Pr(\beta_2 \leq b) = \Phi_{\text{nor}}\left[\frac{\log(b) - \mu}{\sigma}\right].$$

The parameter β_1 represents the common initial amount of degradation of all the test units at time 0 and β_2 represents the degradation rate, random from unit to unit. Then

$$F(t; \beta_1, \mu, \sigma) = \Pr[\mathcal{D}(t) > \mathcal{D}_f] = \Pr(\beta_1 + \beta_2 t > \mathcal{D}_f) = \Pr\left(\beta_2 > \frac{\mathcal{D}_f - \beta_1}{t}\right)$$

$$= 1 - \Phi_{\text{nor}}\left(\frac{\log(\mathcal{D}_f - \beta_1) - \log(t) - \mu}{\sigma}\right)$$

$$= \Phi_{\text{nor}}\left(\frac{\log(t) - [\log(\mathcal{D}_f - \beta_1) - \mu]}{\sigma}\right), \quad t > 0.$$

This shows that T has a lognormal distribution with parameters that depend on the basic path parameters $\boldsymbol{\theta}_\beta = (\beta_1, \mu, \sigma)$ and $\mathcal{D}_f$. That is, $\exp[\log(\mathcal{D}_f - \beta_1) - \mu]$ is the lognormal median and σ is the lognormal shape parameter. □

13.5.2 Numerical Evaluation of F(t)

Algorithm 13.1 Evaluation of F(t) by Direct Integration. If (β_1, β_2) have a bivariate normal distribution with parameters $\mu_{\beta_1}, \mu_{\beta_2}, \sigma_{\beta_1}^2, \sigma_{\beta_2}^2$, and ρ_{β_1,β_2}, then

$$F(t) = P(T \le t) = \int_{-\infty}^{\infty} \Phi_{\text{nor}}\left[-\frac{g(\mathcal{D}_{\text{f}}, t, \beta_1) - \mu_{\beta_2|\beta_1}}{\sigma_{\beta_2|\beta_1}}\right] \frac{1}{\sigma_{\beta_1}} \phi_{\text{nor}}\left(\frac{\beta_1 - \mu_{\beta_1}}{\sigma_{\beta_1}}\right) d\beta_1,$$

where $g(\mathcal{D}_{\text{f}}, t, \beta_1)$ is the value of β_2 that gives $\mathcal{D}(t) = \mathcal{D}_{\text{f}}$ for specified β_1 and where

$$\mu_{\beta_2|\beta_1} = \mu_{\beta_2} + \rho\sigma_{\beta_2}\left(\frac{\beta_1 - \mu_{\beta_1}}{\sigma_{\beta_1}}\right),$$

$$\sigma^2_{\beta_2|\beta_1} = \sigma^2_{\beta_2}(1 - \rho^2).$$

In principle, this approach can be extended in a straightforward manner when there are more than two continuous random variables. The amount of computational time needed to evaluate the multidimensional integral will, however, increase exponentially with the dimension of the integral. □

13.5.3 Monte Carlo Evaluation of $F(t)$

Monte Carlo simulation, as illustrated in Figures 13.3, 13.4, and 13.5, is a particularly versatile method for evaluating $F(t)$. Evaluation is done in the following algorithm by generating a large number of random sample paths from the assumed path model, using the proportion of path crossing $\mathcal{D}_{\text{f}}$ by time t as an evaluation of $F(t)$.

Algorithm 13.2 Monte Carlo Evaluation of $F(t)$ from Degradation Model Parameters

1. Generate N simulated realizations $\check{\beta}_1, \ldots, \check{\beta}_k$ of $\beta_1, \ldots, \beta_k$ from a multivariate normal distribution with mean $\widehat{\boldsymbol{\theta}}_\beta$ and variance–covariance matrix $\widehat{\boldsymbol{\Sigma}}_\beta$, where N is a large number (e.g., $N = 100{,}000$).
2. Compute the N simulated failure times corresponding to the N realizations of $\check{\beta}_1, \ldots, \check{\beta}_k$. Conceptually this can be done by substituting the realizations of $\check{\beta}_1, \ldots, \check{\beta}_k$ into $\mathcal{D}(t, \beta_1, \ldots, \beta_k)$, finding the crossing time for each (often the crossing time can be expressed conveniently as a function of the $\beta_1, \ldots, \beta_k$; otherwise the crossing time can be found by using a numerical root-finding algorithm).
3. For any desired values of t, use

$$F(t) \approx \frac{\text{Number of Simulated First Crossing Times } \le t}{N}$$

as an evaluation of $F(t)$. □

The potential error in this Monte Carlo approximation is easy to evaluate by using the binomial distribution. The error can be made arbitrarily small by choosing N large enough. In particular, the standard deviation of the Monte Carlo error in $F(t)$ at a given point is $\sqrt{F(t)(1 - F(t))/N}$. For example, if $F(t) = .01$ and $N = 100{,}000$, the standard deviation of the Monte Carlo error is .0003.

13.6 ESTIMATION OF $F(t)$

One can estimate the failure-time distribution $F(t)$ by substituting the estimates $\widehat{\boldsymbol{\theta}}_\beta$ into (13.9) giving $\widehat{F}(t) = F(t; \widehat{\boldsymbol{\theta}}_\beta)$. This is straightforward for the case when $F(t)$ can be expressed in a closed form. When there is no closed-form expression for $F(t)$, and when numerical transformation methods are too complicated, one can use Algorithm 13.1 or 13.2 to evaluate (13.9) at $\widehat{\boldsymbol{\theta}}_\beta$.

Example 13.8   Degradation Data Estimate of $F(t)$.   Figure 13.9 shows $\widehat{F}(t)$ for the Alloy-A data, estimated with Algorithm 13.1, using the estimates $\widehat{\boldsymbol{\theta}}_\beta = (\widehat{\boldsymbol{\mu}}_\beta, \widehat{\boldsymbol{\Sigma}}_\beta)$ obtained in Example 13.6. The plotted points from Table 13.1 (computed as described in Chapter 6) provide a nonparametric estimate based on the complete crossing time data (i.e., including the eventual crossing times beyond .12 million cycles, where we assume that the test has ended). This nonparametric estimate provides a useful comparison with the parametric degradation and failure-time models based on the data available up to $t_c = .12$. The confidence intervals in this figure will be explained in the next section.   □

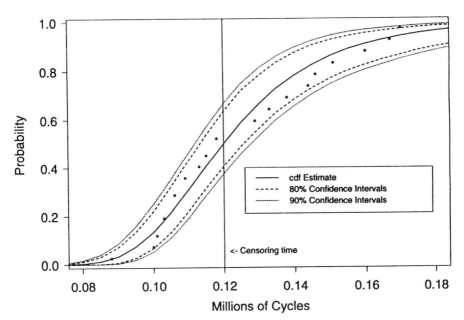

Figure 13.9. Degradation model estimate of $F(t)$ with pointwise two-sided approximate 90% and 80% bootstrap bias-corrected percentile confidence intervals, based on the Alloy-A crack-size data censored at $t_c = .12$. The dots track the nonparametric estimate of $F(t)$.

Table 13.1. Crossing Times and Plotting Positions for the Alloy-A Data

Path	Crossing Time (Million Cycles)	Plotting Positions $(i-.5)/21$
1	.088	.024
2	.100	.071
3	.101	.119
4	.103	.167
5	.103	.214
6	.106	.262
7	.106	.310
8	.109	.357
9	.113	.405
10	.115	.452
11	.118	.500
12	.118	.548
13	.129*	.595
14	.133*	.643
15	.138*	.690
16	.144*	.738
17	.146*	.786
18	.151*	.833
19	.160*	.881
20	.167*	.929
21	.170*	.976

Observations marked with an asterisk would have been censored for a test that ended at .12 million cycles.

13.7 BOOTSTRAP CONFIDENCE INTERVALS

Because there is no simple method of computing standard errors for $\widehat{F}(t)$, we use the bias-corrected percentile bootstrap method in this chapter. This method is described briefly in Section 9.6 and more fully in Efron (1985) and Efron and Tibshirani (1993). The method is implemented with the following algorithm.

Algorithm 13.3 Bootstrap Confidence Intervals from Degradation Data

1. Use the observed data from the n sample paths to compute the estimates $\widehat{\boldsymbol{\theta}}_\beta$ and $\widehat{\sigma}_\epsilon^2$.
2. Use Algorithm 13.1 or 13.2 with $\widehat{\boldsymbol{\theta}}_\beta$ as input to compute the estimate $\widehat{F}(t)$ at desired values of t.
3. Generate a large number B (e.g., $B = 4000$) of bootstrap samples that mimic the original sample and compute the corresponding bootstrap estimates $\widehat{F}^*(t)$ according to the following steps.

(a) Generate, from $\widehat{\boldsymbol{\theta}}_\beta$, n simulated realizations of the random path parameters $\beta_{1i}^*, \ldots, \beta_{ki}^*, i = 1, \ldots, n$.

(b) Using the same sampling scheme as in the original experiment, compute n simulated observed paths from

$$y_{ij}^* = \mathcal{D}(t_{ij}; \beta_{1i}^*, \ldots, \beta_{ki}^*) + \epsilon_{ij}^*$$

up to the planned stopping time t_c, where the ϵ_{ij}^* values are independent simulated residual values generated from $\mathrm{NOR}(0, \widehat{\sigma}_\epsilon)$.

(c) Use the n simulated paths to estimate parameters of the path model, giving the bootstrap estimates $\widehat{\boldsymbol{\theta}}_\beta^*$.

(d) Use Algorithm 13.1 or 13.2 with $\widehat{\boldsymbol{\theta}}_\beta^*$ as input to compute the bootstrap estimates $\widehat{F}^*(t)$ at desired values of t.

4. For each desired value of t, the bootstrap confidence interval for $F(t)$ is computed using the following steps.

(a) Sort the B bootstrap estimates $\widehat{F}^*(t)_1, \ldots, \widehat{F}^*(t)_B$ in increasing order giving $\widehat{F}^*(t)_{(b)}, b = 1, \ldots, B$.

(b) Following Efron and Tibshirani (1993), the lower and upper bounds of pointwise approximate $100(1 - \alpha)\%$ confidence intervals for the distribution function $F(t)$ are

$$\left[\underset{\sim}{F(t)}, \widetilde{F(t)}\right] = \left[\widehat{F}^*(t)_{(l)}, \widehat{F}^*(t)_{(u)}\right],$$

where

$$l = B \times \Phi_{\mathrm{nor}}\left[2\Phi_{\mathrm{nor}}^{-1}(q) + \Phi_{\mathrm{nor}}^{-1}(\alpha/2)\right],$$

$$u = B \times \Phi_{\mathrm{nor}}\left[2\Phi_{\mathrm{nor}}^{-1}(q) + \Phi_{\mathrm{nor}}^{-1}(1 - \alpha/2)\right],$$

and q is the proportion of the B values of $\widehat{F}^*(t)$ that are less than $\widehat{F}(t)$ (using $q = .5$ is equivalent to the percentile bootstrap method). □

Example 13.9 Degradation Data Bootstrap Confidence Intervals for the Alloy-A F(t). Continuing with Example 13.8, Figure 13.9 shows pointwise two-sided approximate 90% and 80% bootstrap bias-corrected percentile confidence intervals for $F(t)$, based on the crack-size data censored at $t_c = .12$. Figure 13.10 shows a subset of the bootstrap estimates that were used to compute the bias-corrected percentile confidence intervals for $F(t)$. □

13.8 COMPARISON WITH TRADITIONAL FAILURE-TIME ANALYSES

This section compares the degradation and failure-time data analyses for the Alloy-A data. Based on the degradation data and the failure-time data censored at $t_c = .12$,

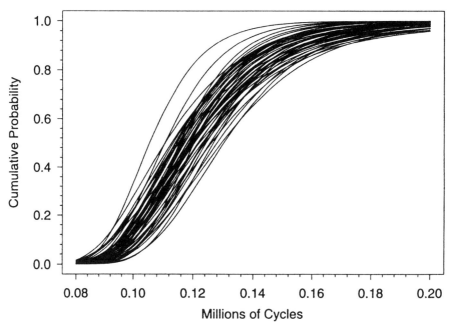

Figure 13.10. Bootstrap estimates of $F(t)$ for Alloy-A.

Figure 13.11 shows a lognormal probability plot of the nonparametric estimate of $F(t)$ (the dots) and the lognormal distribution ML estimate of $F(t)$ based on the failure-time data available at time $t_c = .12$ million cycles (the line). Figure 13.12 shows the nonparametric estimate of $F(t)$, the lognormal distribution ML estimate of $F(t)$, and the corresponding approximate 90% pointwise confidence intervals. Figure 13.13 compares the degradation data/model estimates with the ML estimates of the lognormal, normal, and Weibull failure-time distributions, based on the censored failure-time data. Figures 13.12 and 13.13 also show the nonparametric estimate of the failure-time distribution using the actual crossing times that occurred after $t_c = .12$. Some observations from these figures are as follows:

1. Figures 13.11 and 13.12 show that the lognormal distribution provides a good fit to the failure-time data up to $t_c = .12$, but not beyond.
2. Figure 13.13 shows that the other commonly used parametric models, which fit almost as well before $t_c = .12$, do not do any better beyond $t_c = .12$. The degradation analysis, however, does provide a reasonable extrapolation beyond $t_c = .12$. This is because the degradation analysis method directly models the relationship between degradation and time and takes account of the amount of degradation in the censored observations when estimating $F(t)$. See the distribution of crack lengths for the Alloy-A units that had not failed before $t_c = .12$, shown in Figure 13.1. The traditional failure-time data analysis ignores this important information.

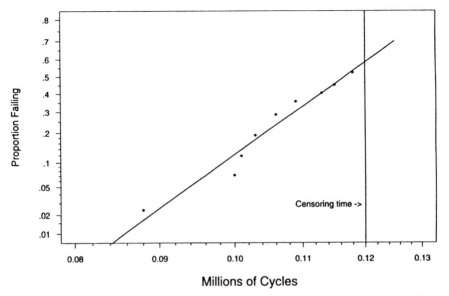

Figure 13.11. Lognormal probability plot of the nonparametric estimate (dots) and the lognormal distribution ML fit (line) based on the failure-time Alloy-A data censored at $t_c = .12$.

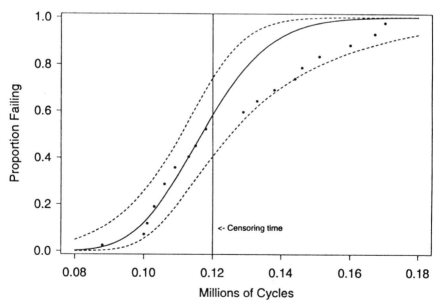

Figure 13.12. Lognormal distribution ML estimate and pointwise approximate 90% confidence intervals. The dots indicate the nonparametric estimate based on the Alloy-A *uncensored* failure-time data (i.e., based on failure times beyond the artificial censoring time used in the parametric degradation and failure-time analyses).

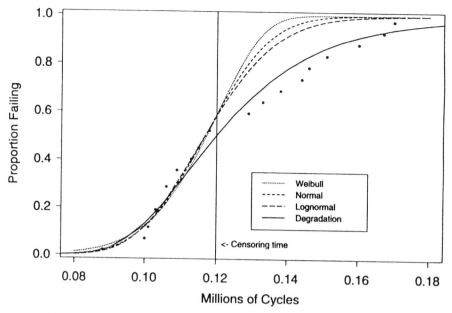

Figure 13.13. Comparison of degradation and failure-time data analyses based on the Alloy-A data censored at t_c. The dots indicate the nonparametric estimate based on the Alloy-A *uncensored* failure-time data (i.e., based on failure times beyond the artificial censoring time used in the parametric degradation and failure-time analyses).

3. Comparing Figures 13.9 and 13.12 shows that the confidence intervals based on the degradation and failure-time data have similar widths for $.10 < t < .12$. Outside this range, however, the confidence intervals are narrower for the degradation method.

13.9 APPROXIMATE DEGRADATION ANALYSIS

This section describes an alternative (but only approximately correct) method of analyzing degradation data. Consider the general degradation model given in Section 13.2.5. There are two steps in the approximate method. The first step consists of a separate analysis for each unit to predict the time at which the unit will reach the critical degradation level corresponding to failure. These times are called "pseudo failure times." In the second step, the n pseudo failure times are analyzed as a complete sample of failure times to estimate $F(t)$. Formally, the method is as follows.

- For the unit i, use the path model $y_{ij} = \mathcal{D}_{ij} + \epsilon_{ij}$ and the sample path data $(t_{i1}, y_{i1}), \ldots, (t_{im_i}, y_{im_i})$ to find the (conditional) ML estimate of $\boldsymbol{\beta}_i = (\beta_{1i}, \ldots, \beta_{ki})$, say, $\widehat{\boldsymbol{\beta}}_i$. This can be done using nonlinear least squares.
- Solve the equation $\mathcal{D}(t, \widehat{\boldsymbol{\beta}}_i) = \mathcal{D}_f$ for t and call the solution $\widehat{t}_i$.

- Repeat the procedure for each sample path to obtain the pseudo failure times $\widehat{t}_1, \ldots, \widehat{t}_n$.
- Do a single distribution analysis of the data $\widehat{t}_1, \ldots, \widehat{t}_n$ to estimate $F(t)$.

13.9.1 Simple Linear Path

For simple linear degradation processes the degradation path for a unit can be written as $\mathcal{D}(t) = \beta_1 + \beta_2 t$. In some cases log transformations on the sample degradation values or on the time scale or both will result in a simple linear-path model. In this case the pseudo failure times are obtained from

$$\widehat{t}_i = \frac{\mathcal{D}_f - \widehat{\beta}_{1i}}{\widehat{\beta}_{2i}},$$

where

$$\widehat{\beta}_{1i} = \bar{y}_i - \widehat{\beta}_{2i} \times \bar{t}_i, \qquad \widehat{\beta}_{2i} = \frac{\sum_{j=1}^{m_i}(t_{ij} - \bar{t}_i) \times y_{ij}}{\sum_{j=1}^{m_i}(t_{ij} - \bar{t}_i)^2},$$

and $\bar{t}_i$ and $\bar{y}_i$ are the means of $t_{i1}, \ldots, t_{im_i}$ and $y_{i1}, \ldots, y_{im_i}$, respectively.

13.9.2 Simple Linear Path Through the Origin

For some degradation, all paths start at the origin ($t_{i1} = 0, y_{i1} = 0$). If, in addition, the degradation rate is constant, then the degradation path has the form $\mathcal{D}(t) = \beta_2 t$. Then the pseudo failure times are obtained from

$$\widehat{t}_i = \frac{\mathcal{D}_f}{\widehat{\beta}_{2i}},$$

where

$$\widehat{\beta}_{2i} = \frac{\sum_{j=1}^{m_i} t_{ij} \times y_{ij}}{\sum_{j=1}^{m_i} t_{ij}^2}.$$

13.9.3 Comments on the Approximate Degradation Analysis

For simple problems the approximate degradation analysis is attractive because the computations are relatively simple. The approximate method is less appealing when the degradation paths are nonlinear.

The approximate method may give adequate analysis if:

- The degradation paths are relatively simple.
- The fitted path model is approximately correct.
- There are enough data for precise estimation of the $\boldsymbol{\beta}_i$ values.

- The amount of measurement error is small.
- There is not too much extrapolation in predicting the $\widehat{t}_i$ "failure times."

There are, however, potential problems with the approximate degradation analysis because of the following:

- The method ignores the prediction error in $\widehat{t}$ and does not account for measurement error in the observed sample paths.
- The distributions fitted to the pseudo failure times will not, in general, correspond to the distribution induced by the degradation model.
- For some applications, there may be sample paths that do not contain enough information to estimate all of the path parameters (e.g., when the path model has an asymptote but the sample path has not begun to level off). This might necessitate fitting different models for different sample paths in order to predict the crossing time.

Overall, extrapolation into the tails of the failure-time distribution may be more valid with the actual crossing distribution implied by the degradation model (as used in Sections 13.4–13.6) than with the empirically predicted failure times.

Example 13.10 Laser Life Analysis. The data in Figure 13.14 and Appendix Table C.17 are from a laser life test similar to the one described in Example 13.5,

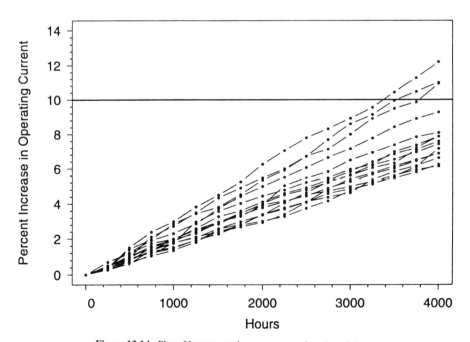

Figure 13.14. Plot of laser operating current as a function of time.

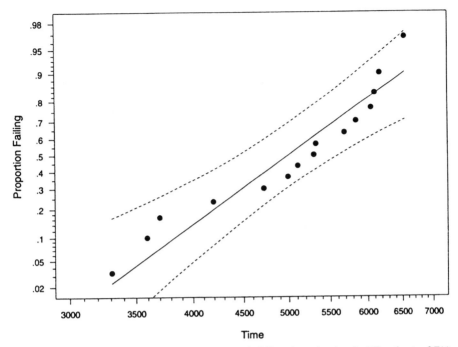

Figure 13.15. Weibull probability plot of the laser pseudo failure times showing the ML estimate of $F(t)$ and approximate 95% pointwise confidence intervals.

except that there were no early failures. For this device and the corresponding application, a $\mathcal{D}_f = 10\%$ increase in current was the specified failure level. The failure times (for paths exceeding $\mathcal{D}_f = 10\%$ increase in current before 4000 hours) and the pseudo failure times were obtained by fitting straight lines through the data for each path. These pseudo failure times are 3702, 4194, 5847, 6172, 5301, 3592, 6051, 6538, 5110, 3306, 5326, 4995, 4721, 5689, and 6102 hours. Using methods from Chapter 8, Figure 13.15 is a Weibull probability plot of the laser pseudo failure times showing the ML estimate for $F(t)$ and approximate 95% pointwise confidence intervals. □

BIBLIOGRAPHIC NOTES

Gertsbakh and Kordonsky (1969) discuss the degradation approach from an engineering point of view. They point out the value of analyzing degradation measures in terms of sample paths to assess product reliability. They present the Bernstein distribution, which describes the failure-time distribution for a simple linear model with random intercept and random slope. Knezevic (1993) presents similar probability models based on what he calls a "condition parameter." Nelson (1990a, Chapter 11) discusses a special situation in which the degradation measurement is destructive (only one measurement could be made on each item). Tomsky (1982) uses a multivariate normal regression model to evaluate component degradation. Linear degradation models were used in Suzuki, Maki, and Yokogawa (1993) to model the increase in a resistance measurement over time and in Tseng, Hamada, and Chiao (1995) to model the lumen output from fluorescent light bulbs over time. Yanagisawa (1997) fits models to degradation data for silicon solar cells where the accelerating factors were light intensity and temperature.

Murray (1993, 1994) and Murray and Maekawa (1996) describe degradation data for disk storage media (e.g., compact disk) error rates in accelerated testing. These papers and Tseng et al. (1995) and Tobias and Trindade (1995) use the approximate analysis method described in Section 13.9 to analyze their degradation data. Tseng and Yu (1997) and Yu and Tseng (1998) propose methods for choosing the time to terminate a degradation test.

Much of the material in this chapter is based on methods presented in Lu and Meeker (1993), Lu, Meeker, and Escobar (1996), and ideas in Meeker and Hamada (1995). Nelson (1995b) describes models and analysis methods for problems with random nonzero degradation initiation times. His methods assume destructive inspection so that each sample unit will provide a single (possibly censored) degradation response. It would be useful to extend this work to allow for multiple readings on individual test units. Crowder (1997) describes methods for developing component-based preventive maintenance plans. One of these methods used degradation-type data. Davidian and Giltinan (1995) provide an excellent description and development of methods for fitting statistical models to nonlinear degradation-type models (also known as "growth curves") and estimating random parameters.

EXERCISES

13.1. Show that (13.2) is the solution to (13.1).

13.2. Use the degradation equation (13.2) to obtain an expression for the time that $\mathcal{D}(t)$ crosses a specified $\mathcal{D}_f$.

13.3. Show that (13.4) is the solution to (13.3).

13.4. Use the degradation equation (13.6) to obtain an expression for the time that $A_2(t)$ crosses a specified A_{2f}.

13.5. Determine the value of N needed in Algorithm 13.2 to evaluate $F(t)$ at the point where $F(t) = .01$ so that the probability that Monte Carlo error in evaluation is less than .0001, with probability approximately .95. Use the normal distribution approximation to the binomial probability.

13.6. Discuss the advantages and disadvantages of using Algorithm 13.1 versus 13.2 when estimating $F(t)$ from degradation data.

EXERCISES

13.7. Consider the analysis of the laser degradation data in Example 13.10.
   (a) Repeat the analysis using the time at which current has increased by 5% to define failure.
   (b) Repeat the analysis using the time at which current has increased by 15% to define failure.
   (c) Compare the results in parts (a) and (b). Comment on the differences in assumptions needed to estimate these two different distributions.

▲**13.8.** Example 13.10 illustrates the simple analysis of the data in Appendix Table C.17. Do the analysis of these data using the random-coefficient degradation model.
   (a) Identify and estimate a parametric distribution for the slopes.
   (b) Derive an expression for the failure-time distribution based on the degradation model, where failure is defined as the time at which current has increased by 10%.
   (c) Plot the estimate of the failure-time distribution. Compare it with the simple estimate obtained in Example 13.10.

13.9. Appendix Table C.18 gives degradation data on block error rates (the ratio of number of bytes with errors to the total number of bytes tested) of magneto-optic data storage disks tested for 2000 hours at 80°C and 85% relative humidity. Use the simple analysis method described in Section 13.9 to estimate the failure-time distribution of these disks at 80°C and 85% relative humidity, where failure is defined as the time that it takes to reach an error rate of 50×10^{-5} (a safe level at which error-correcting codes can be expected to correct errors).

13.10. As an electronic device ages, its power output decreases. Because the degradation results from a simple one-step chemical reaction with a limited amount of harmful material available for reaction, the decrease in power can be described by the degradation model $\mathcal{D}(t) = \beta_2[1 - \exp(-\beta_1 t)]$, where $\mathcal{D}(t)$ is the power output at time t, $\beta_2 < 0$ is nearly the same for all units, and β_1 comes from a lognormal distribution. System performance degrades noticeably when $\mathcal{D}(t)$ falls below $\mathcal{D}_f$. Thus we define a failure as the point in time when $\mathcal{D}(t) < \mathcal{D}_f$.
   (a) Describe some possible physical reasons for the asymptotic behavior of $\mathcal{D}(t)$.
   (b) What happens, in the long run, if $\beta_2 > \mathcal{D}_f$?
   (c) Assuming that $\mathcal{D}_f$ is a fixed constant and that $\beta_2 < \mathcal{D}_f$, derive an expression for $F(t)$, the failure-time cdf.

13.11. Suppose the actual degradation path of a particular unit is given by $\mathcal{D}(t) = \beta_1 t$, where $-\log(\beta_1)$ varies from unit to unit according to a $\text{SEV}(\mu, \sigma)$

distribution. If failure occurs when $\mathcal{D}(t) > \mathcal{D}_f$, where $\mathcal{D}_f$ is a fixed constant, show that the failure-time distribution is Weibull.

▲13.12. Suppose the actual degradation path of a particular unit is given by $\mathcal{D}(t) = \beta_1 t$, where β_1 varies from unit to unit according to a LOGNOR(μ_1, σ_1) distribution. Also suppose that failure occurs when $\mathcal{D}(t) > \mathcal{D}_f$, and $\mathcal{D}_f$ has a LOGNOR(μ_2, σ_2) distribution. Derive an expression for $F(t)$, the failure-time cdf.

▲13.13. Suppose the actual degradation path of a particular unit is given by $\mathcal{D}(t) = \beta_1 + \beta_2 t$, where (β_1, β_2) vary from unit to unit according to a bivariate normal distribution with parameters $\mu_{\beta_1}, \mu_{\beta_2}, \sigma_{\beta_1}, \sigma_{\beta_2}$, and $\rho_{\beta_1 \beta_2}$.
(a) Assuming that failure occurs when $\mathcal{D}(t) > \mathcal{D}_f$, derive an expression for the failure-time distribution $F(t) = \Pr(T \le t) = \Pr[\mathcal{D}(t) > \mathcal{D}_f]$.
(b) Explain why $\lim_{t \to \infty} F(t) < 1$.

CHAPTER 14

Introduction to the Use of Bayesian Methods for Reliability Data

Objectives

This chapter explains:

- The use of Bayesian statistical methods to combine "prior" information with data to make inferences.
- The relationship between Bayesian methods and the likelihood methods used in earlier chapters.
- Sources of prior information.
- Useful statistical and numerical methods for Bayesian analysis.
- Bayesian methods for estimating reliability.
- Bayesian methods for prediction.
- The dangers of using "wishful thinking" or expectations as prior information.

Overview

This chapter explains basic Bayesian methods and illustrates them with some simple applications in reliability analysis. This chapter builds on material from Chapter 8. Section 14.2 shows how Bayes's rule can be used to combine prior information with data. Section 14.3 explains the different kinds of prior distributions and how they are obtained. Section 14.4 describes numerical methods for combining prior information with a likelihood and for computing marginal posterior distributions. Section 14.5 describes and gives methods for using the posterior distribution to obtain point estimates and compute Bayesian confidence and prediction intervals. Section 14.7.2 describes some of the dangers involved in using prior information in a statistical analysis.

14.1 INTRODUCTION

Bayesian methods are closely related to likelihood methods. Bayesian methods, however, allow data to be combined with "prior" information to produce a posterior distribution for a parameter or parameters. This posterior is used to quantify uncertainty about the parameters and functions of parameters, much as the likelihood was used in earlier chapters.

Combinations of extensive past experience and physical/chemical theory can provide prior information to form a framework for inference and decision making. In many applications it is necessary to combine prior information with limited additional observational or experimental data. For example, reliability engineers may know, with a high degree of certainty, that products made out of a certain alloy will eventually fail from fracture caused by repeated fatigue loading. The lognormal (base e) distribution with shape parameter σ in the interval of .5 to .7 has always provided an adequate model. To estimate the cycles-to-failure distribution of a new product made from the same alloy with needed precision might require hundreds of sample units. By incorporating the prior information about σ into the analysis, an adequate estimate of reliability might be obtained with 20 or 30 units. There are, of course, dangers involved in making strong assumptions about knowledge of model parameters. These will be described in more detail in Section 14.7.2.

14.2 USING BAYES'S RULE TO UPDATE PRIOR INFORMATION

14.2.1 Notation

To keep the presentation of the basic ideas simple, in this chapter we follow the usual convention in the Bayesian statistics literature and let the argument of pdfs and cdfs indicate the parameter for which uncertainty is being described. For example, $f(\boldsymbol{\theta})$ denotes the prior pdf of $\boldsymbol{\theta}$ and $f[\log(\boldsymbol{\theta})]$ denotes the prior pdf of $\log(\boldsymbol{\theta})$. Also $f(\boldsymbol{\theta} \mid \text{DATA})$ and $f[\log(\boldsymbol{\theta}) \mid \text{DATA}]$ will represent the posterior pdfs of $\boldsymbol{\theta}$ (distribution of $\boldsymbol{\theta}$ given the available data) and $\log(\boldsymbol{\theta})$, respectively.

In some cases it will be necessary to start with the prior pdf of one parameter and then use it to obtain the prior pdf for a function of that parameter. For example, for a scalar parameter θ, $f(\theta) = f[\log(\theta)]/\theta$. See Appendix Section B.1 for details on deriving the pdf of a function of a parameter vector $\boldsymbol{\theta}$.

14.2.2 Bayes's Rule

Bayes's rule provides a mechanism for combining *prior* information with sample data to make inferences on model parameters. This is illustrated in Figure 14.1. Analytically, for a vector parameter $\boldsymbol{\theta}$ the procedure is as follows. Prior knowledge about $\boldsymbol{\theta}$ is expressed in terms of a pdf denoted by $f(\boldsymbol{\theta})$. The likelihood for the available data and specified model is given by $L(\text{DATA} \mid \boldsymbol{\theta}) = L(\boldsymbol{\theta}; \text{DATA})$. Then, using Bayes's rule, the conditional distribution of $\boldsymbol{\theta}$ given the data provides the *posterior* pdf of $\boldsymbol{\theta}$, representing the updated state of knowledge about $\boldsymbol{\theta}$. This posterior

PRIOR INFORMATION AND DISTRIBUTIONS

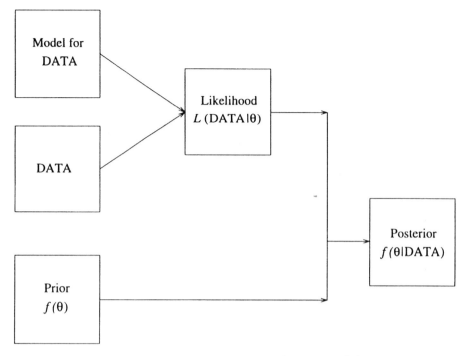

Figure 14.1. Bayesian method for making inferences or predictions.

distribution can be expressed as

$$f(\boldsymbol{\theta} \mid \text{DATA}) = \frac{L(\text{DATA} \mid \boldsymbol{\theta}) f(\boldsymbol{\theta})}{\int L(\text{DATA} \mid \boldsymbol{\theta}) f(\boldsymbol{\theta}) \, d\boldsymbol{\theta}} = \frac{R(\boldsymbol{\theta}) f(\boldsymbol{\theta})}{\int R(\boldsymbol{\theta}) f(\boldsymbol{\theta}) \, d\boldsymbol{\theta}}. \tag{14.1}$$

Here $R(\boldsymbol{\theta}) = L(\boldsymbol{\theta})/L(\widehat{\boldsymbol{\theta}})$ is the relative likelihood (introduced in Chapters 7 and 8) and the integral is computed over the region $f(\boldsymbol{\theta}) > 0$.

In general, it is impossible to compute the integral in (14.1) in closed form. Numerical methods are needed and these methods can be computationally intensive when the length of $\boldsymbol{\theta}$ is more than two or three. In the past this has been an impediment to the use of Bayesian methods. Today, however, new statistical and numerical methods that take advantage of modern computing power are making it feasible to apply Bayesian methods to a much wider range of real problems.

14.3 PRIOR INFORMATION AND DISTRIBUTIONS

It is convenient to divide available prior information about a parameter into three different categories:

1. Parameters that are given as known, leading to a degenerate prior distribution.
2. Parameters with a diffuse or approximately noninformative prior distribution.
3. An informative, nondegenerate prior distribution.

In general, there are two possible sources of prior information: (1) expert or other subjective opinion or (2) past data. The prior pdf $f(\boldsymbol{\theta})$ may be either informative or not. Loosely speaking, a noninformative[1] prior distribution is one that provides little or no information about any of the parameters in $\boldsymbol{\theta}$. Such a prior distribution is useful when it is desired to let the data speak for themselves without being influenced by previous data, expert opinion, or other available prior information.

The most important motivation for using prior information is to combine it with data to provide more and better information about model parameters of interest. An informative prior distribution is expressed in the form of a (joint) pdf for the parameter(s) for which information is available. A "proper" prior pdf $f(\boldsymbol{\theta})$ is a nonnegative function that is defined for all values of the parameters and that integrates to one. Some examples of proper prior distributions for scalar parameters include:

- Normal prior distribution with mean a and a standard deviation b so that $f(\theta) = (1/b)\phi_{\text{nor}}[(\theta - a)/b]$ for $-\infty < \theta < \infty$.
- Uniform prior distribution between a and b [denoted by $\text{UNIF}(a, b)$] so that $f(\theta) = 1/(b - a)$ for $a \leq \theta \leq b$. This prior distribution does not express strong preference for specific values of θ in the interval.
- Beta prior distribution between specified a and b with specified shape parameters (allows for a more general shape).
- Isosceles triangle prior distribution with base (range) between a and b.

For a prior pdf with a finite endpoint on one or both sides of its range, $f(\theta) = 0$ outside the specified range. For a positive parameter θ it is often more natural or convenient to specify the prior pdf in terms of $\log(\theta)$.

14.3.1 Noninformative (Diffuse) Prior Distributions

Noninformative (or approximately noninformative) prior pdfs are constant (or approximately constant) over the range(s) of the model parameter(s). Other names for noninformative prior distributions are "vague prior" and "diffuse prior" distributions.

Some noninformative prior pdfs are "improper" because they do not integrate to a finite quantity [i.e., $\int f(\boldsymbol{\theta}) d\boldsymbol{\theta} = \infty$]. For example, with an unrestricted scalar θ, a uniform distribution $f(\theta) = c$ for all $-\infty < \theta < \infty$ does not have a finite integral. For a positive scalar parameter θ the corresponding choice is $f[\log(\theta)] = c$ or $f(\theta) = c/\theta$, $0 < \theta < \infty$, and this prior pdf is also improper. Improper prior pdfs cause no difficulties as long as the resulting *posterior* pdf is "proper" (integrates to one). Whether this is so or not depends on the form of the model and the available data.

---
[1] There is a particular technical definition for a noninformative prior distribution, but we use the term loosely to indicate a prior distribution that carries little or no weight in estimation relative to the information in the available data.

PRIOR INFORMATION AND DISTRIBUTIONS

The effect of using a noninformative prior distribution for $\boldsymbol{\theta}$ can be seen as follows. For a uniform prior pdf $f(\boldsymbol{\theta})$ (possibly improper) across all possible values of $\boldsymbol{\theta}$,

$$f(\boldsymbol{\theta} \mid \text{DATA}) = \frac{R(\boldsymbol{\theta})f(\boldsymbol{\theta})}{\int R(\boldsymbol{\theta})f(\boldsymbol{\theta})\,d\boldsymbol{\theta}} = \frac{R(\boldsymbol{\theta})}{\int R(\boldsymbol{\theta})\,d\boldsymbol{\theta}}.$$

This indicates that the posterior pdf $f(\boldsymbol{\theta} \mid \text{DATA})$ is proportional to the likelihood.

It is possible, for example, to replace an improper uniform prior with a proper uniform prior by limiting the range of the pdf to a finite interval. As long as the range of the uniform prior pdf includes all values of the parameters with substantially large relative likelihood, the prior distribution will remain (approximately) uninformative. That is, with a finite-range uniform prior pdf, the posterior pdf is approximately proportional to the likelihood if the range of the uniform prior distribution is large enough so that $R(\boldsymbol{\theta}) \approx 0$, where $f(\boldsymbol{\theta}) = 0$.

14.3.2 Using Past Data to Specify a Prior Distribution

Prior distributions can also be based on available data. Combining past data with a noninformative prior distribution gives a posterior pdf that is proportional to the likelihood. This posterior pdf can then serve as a prior pdf for further updating with new data.

14.3.3 Expert Opinion and Eliciting Prior Information

The elicitation of a prior distribution for a single parameter may be straightforward if there has been considerable experience in estimating or observing estimates of that parameter in similar situations. For a vector of parameters, however, the elicitation and specification of a meaningful joint prior distribution is more difficult. In general, marginal distributions for individual parameters do not completely determine the joint distribution. Also, it is difficult to elicit opinion on dependences among parameters and then express these as a joint distribution. For example, if previous experience with integrated circuit devices is always obtained from studies with just a few percent failing, then past estimates of the lognormal parameters μ and σ would be highly correlated, implying that the prior distribution for these parameters should reflect this dependency. Also, it may not be reasonable to elicit opinion about parameters from a standard parameterization when those parameters have no physical or practical meaning. Again, for the integrated circuit devices, if only a few percent fail in studies, it might be better to elicit information about a quantile at which a few percent fail, rather than about μ, which, for the lognormal distribution, corresponds to the logarithm of the time at which 50% of the units in a population will fail.

A general approach is to elicit information about particular quantities (or parameters) that, from past experience (or data), can be specified approximately independently. For example, for a high-reliability integrated circuit, a good choice would be a quantile in the lower tail of the failure-time distribution and the lognormal shape parameter σ. Then the corresponding prior distributions for these quantities can be described as being approximately independent.

When there is useful informative prior information for a parameter, one elicits a general shape or form of the distribution and the range (or approximate range) of the distribution. For example, the uncertainty in the .01 quantile of a failure-time distribution (a positive quantity) might be described by lognormal prior distribution with 99.7% content between two specified time points (expressing the approximate range).

When specifying the prior distribution for quantities for which there is *no* prior information, no detailed elicitation is necessary, but one does have to specify the form and range of the vague prior distribution (e.g., a uniform distribution over a sufficiently wide interval, as described in Section 14.3.1, is generally satisfactory).

Example 14.1 Prior Distributions for Estimating the Bearing-Cage Life Distribution. This example revisits the bearing-cage field data that were fit to a Weibull distribution in Example 8.16. Suppose that with appropriate questioning, engineers provided the following information based on experience with previous products of the same material and knowledge of the failure mechanism. Life can be described adequately with a Weibull distribution and the Weibull shape parameter β would almost certainly be between 2 and 5 (σ between .2 and .5). Using a normal distribution to express the uncertainty in $\log(\sigma)$ gives $\log(\sigma) \sim \text{NOR}(a_0, b_0)$, where a_0 and b_0 are obtained from the specification of two extreme quantiles $\sigma_{(\gamma/2)}$ and $\sigma_{(1-\gamma/2)}$ of the prior distribution for σ. Then

$$a_0 = \log\left[\sqrt{\sigma_{(\gamma/2)} \times \sigma_{(1-\gamma/2)}}\right], \quad b_0 = \frac{\log\left[\sqrt{\sigma_{(1-\gamma/2)}/\sigma_{(\gamma/2)}}\right]}{z_{(1-\gamma/2)}}.$$

The prior (normal) pdf for $\log(\sigma)$ is

$$f[\log(\sigma)] = \frac{1}{b_0} \phi_{\text{nor}}\left[\frac{\log(\sigma) - a_0}{b_0}\right], \quad \sigma > 0.$$

The corresponding prior pdf for σ is $f(\sigma) = (1/\sigma)f[\log(\sigma)]$. Figure 14.2 shows the marginal prior pdfs for $\log(\sigma)$ and σ when $\sigma_{.005} = .2, \sigma_{.995} = .5$ and $\gamma = .01$ (corresponding to 1% probability outside the limits $\sigma_{.005} = .2, \sigma_{.995} = .5$ and giving $a_0 = -1.151$ and $b_0 = .178$).

In previous studies for similar products, censoring was heavy, and there was a much better sense of expected time for 1% failing than for 50% or more failing. Thus for small p (near the proportion failing in previous studies), t_p and σ are approximately independent (which allows for specification of independent priors). Actually, however, little was known about the Weibull .01 quantile for this particular product and, in this application, it was decided to describe the uncertainty in $\log(t_p)$ with a $\text{UNIF}[\log(a_1), \log(b_1)]$ distribution, where $a_1 = 100$ and $b_1 = 5000$. This is a wide range and thus this part of the prior distribution is not very informative. Then the prior pdf for $\log(t_p)$ is uniform:

$$f[\log(t_p)] = \frac{1}{\log(b_1/a_1)}, \quad a_1 \leq t_p \leq b_1.$$

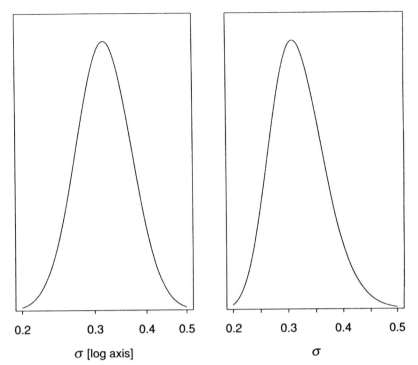

Figure 14.2. Prior pdfs for $\log(\sigma)$ and σ when the prior for σ is a lognormal distribution specified by $\sigma_{.005} = .2, \sigma_{.995} = .5$.

The corresponding prior pdf for t_p is $f(t_p) = (1/t_p)f[\log(t_p)]$. Figure 14.3 shows the marginal prior pdfs for $\log(t_p)$ and t_p. This figure shows why a prior that is noninformative (uniform) for a parameter is informative (nonuniform) for a nonlinear function of that parameter. The distribution for $t_{.01}$ is, however, approximately noninformative over the range $1000 < t_{.01} < 5000$.

Using the (approximate) independence of t_p and σ, the joint prior pdf for (t_p, σ) is

$$f(t_p, \sigma) = \frac{f[\log(t_p)]}{t_p} \times \frac{f[\log(\sigma)]}{\sigma}, \quad a_1 \le t_p \le b_1, \sigma > 0.$$

The transformation $\mu = \log(t_p) - \Phi_{\text{sev}}^{-1}(p)\sigma, \sigma = \sigma$ yields

$$f(\mu, \sigma) = \frac{f[\log(t_p)]}{t_p} \times \frac{f[\log(\sigma)]}{\sigma} \times t_p = f[\log(t_p)] \times \frac{f[\log(\sigma)]}{\sigma}$$

$$= \frac{1}{\log(b_1/a_1)} \times \frac{\phi_{\text{nor}}\{[\log(\sigma) - a_0]/b_0\}}{\sigma b_0} \qquad (14.2)$$

as the joint prior pdf for (μ, σ), where $\log(a_1) - \Phi_{\text{sev}}^{-1}(p)\sigma \le \mu \le \log(b_1) - \Phi_{\text{sev}}^{-1}(p)\sigma, \sigma > 0$ (see Appendix Section B.1). □

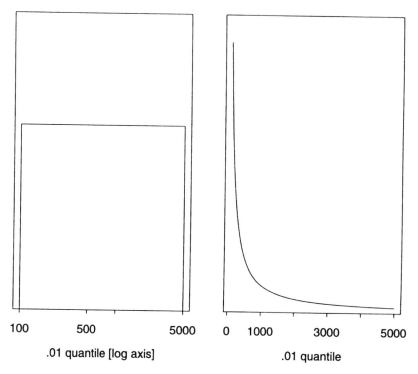

Figure 14.3. Prior pdfs for $\log(t_{.01})$ and $t_{.01}$ with $a_1 = 100, b_1 = 5000$.

14.4 NUMERICAL METHODS FOR COMBINING PRIOR INFORMATION WITH A LIKELIHOOD

Given a specified prior pdf and likelihood function, it is easy to write an expression for the posterior pdf using (14.1).

Example 14.2 Posterior for the Bearing-Cage Life Distribution. Following Section 8.2.2, the likelihood for the bearing-cage life distribution is

$$L(\text{DATA} \mid \mu, \sigma) = \prod_{i=1}^{2003} \left\{ \frac{1}{\sigma t_i} \phi_{\text{sev}} \left[\frac{\log(t_i) - \mu}{\sigma} \right] \right\}^{\delta_i}$$

$$\times \left\{ 1 - \Phi_{\text{sev}} \left[\frac{\log(t_i) - \mu}{\sigma} \right] \right\}^{1-\delta_i}, \quad (14.3)$$

where $\delta_i = 1$ ($\delta_i = 0$) if observation i is a failure (right-censored observation). Substituting the prior pdf in (14.2) and the likelihood in (14.3) into (14.1) provides an expression for the posterior pdf for the bearing-cage life distribution. In general, this posterior pdf cannot be evaluated analytically. Numerical methods must be used instead. □

14.4.1 Numerical Integration Methods for Computing the Posterior pdf of $\boldsymbol{\theta}$

For problems with one or two parameters it is reasonably easy to compute, numerically, the posterior distribution by using numerical integration. It is difficult, however, to provide general-purpose software that will work on all problems, especially with two or more parameters. Although numerical integration is generally a reasonably stable numerical procedure and there are many algorithms available for one-dimensional integration, it is possible for numerical problems to arise. To guarantee accurate results, it is generally necessary to have an idea of the shape of the function being integrated and to make sure that the algorithm performed satisfactorily over the entire relevant range of integration. Such checking becomes difficult when there is more than one variable of integration. It is relatively difficult to find good algorithms for integration over two or more variables.

14.4.2 Simulation-Based Methods for Computing the Posterior Distribution of $\boldsymbol{\theta}$

Simulation can be used to generate a sample from the posterior distribution of $\boldsymbol{\theta}$. Then this sample can be used to approximate the posterior distribution. Using a larger number of simulated points provides a better approximation and the number of points used is limited only by computing equipment and time constraints. The procedure is general and easy to apply, requiring only computable expressions of the relative likelihood and the inverse cdf of the independent marginal prior distributions. The procedure is as follows.

Algorithm 14.1 Computation of a Sample from the Posterior Distribution with Monte Carlo Simulation

1. Generate a random sample $\boldsymbol{\theta}_i$, $i = 1, \ldots, M$, from the prior $f(\boldsymbol{\theta})$ (as described in Section 4.13).
2. Retain the ith sample value, $\boldsymbol{\theta}_i$, with probability $R(\boldsymbol{\theta}_i)$. Do this by generating U_i, a random quantity from a uniform $(0, 1)$, and retain $\boldsymbol{\theta}_i$ if $U_i \leq R(\boldsymbol{\theta}_i)$. □

It can be shown that the retained sample values, say, $\boldsymbol{\theta}_1^\star, \ldots, \boldsymbol{\theta}_{M^\star}^\star$ ($M^\star \leq M$), are a random sample from the posterior pdf $f(\boldsymbol{\theta} \mid \text{DATA})$.

Example 14.3 Computation of a Sample from the Prior Distribution of the Parameters of the Bearing-Cage Life Distribution. Continuing with Example 14.2, the joint prior for $\boldsymbol{\theta} = (\mu, \sigma)$ is generated as follows. First use the inverse cdf method (see Section 4.13) to obtain a loguniform random sample for t_p from

$$(t_p)_i = a_1 \times (b_1/a_1)^{U_{1i}}, \quad i = 1, \ldots, M,$$

where $U_{11}, \ldots, U_{1M}$ is a random sample from a UNIF(0, 1) distribution. For this example we use $p = .01$ because it was thought that $t_{.01}$ and σ could be described as being approximately independent. Similarly, obtain a lognormal random sample for σ, say,

$$\sigma_i = \exp\left[a_0 + b_0 \Phi_{\text{nor}}^{-1}(U_{2i})\right], \quad i = 1, \ldots, M,$$

where $U_{21}, \ldots, U_{2M}$ is another independent random sample from a UNIF(0, 1). Figure 14.4 shows simulated points from the joint prior distribution for $t_{.01}$ and σ. Then $\boldsymbol{\theta}_i = (\mu_i, \sigma_i)$ with $\mu_i = \log[(t_p)_i] - \Phi_{\text{sev}}^{-1}(p)\sigma_i$ is a random sample from the (μ, σ) prior. Figure 14.5 shows the simulated prior, transformed from the points in Figure 14.4. Histograms of the individual (marginal) samples for μ and σ are also shown. Consider the sample from the joint distribution; the histogram of the sample from the distribution of μ shows why the prior for μ is not uniform. □

The size of the random sample $M^\star$ from the posterior is random with an expected value of

$$\mathrm{E}(M^\star) = M \int f(\boldsymbol{\theta}) R(\boldsymbol{\theta}) \, d\boldsymbol{\theta}.$$

When the prior distribution and the data (i.e., relative likelihood contours) do not agree well, $M^\star$ can be much less than M. In such cases it may be necessary to use a very large sample from the prior distribution. Operationally, one can add to the posterior

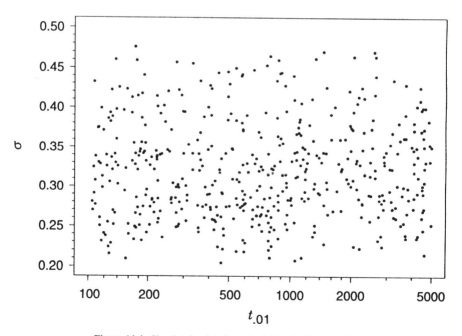

Figure 14.4. Simulated points from the joint prior for $t_{.01}$ and σ.

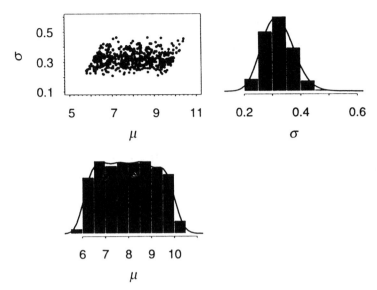

Figure 14.5. Simulated points from the joint prior and the corresponding marginal prior distributions for μ and σ.

by sequentially filtering groups of prior points until a sufficient number of random values are available in the posterior. Generally 2000–4000 points in the posterior provide sufficient accuracy to get a smooth estimate of a marginal distribution for a scalar quantity.

Example 14.4 Computation of a Sample from the Posterior Distribution of the Parameters of the Bearing-Cage Life Distribution. Continuing with Example 14.3, the joint posterior for $\boldsymbol{\theta} = (\mu, \sigma)$ is generated by using Algorithm 14.1. Figure 14.6 shows the same (μ, σ) prior given in Figure 14.5 with the bearing-cage data relative likelihood contours superimposed. Figure 14.7 shows a sample of 500 points from the (μ, σ) posterior, showing the effect of the filtering illustrated in Figure 14.6. Over 4000 points were actually computed to provide the Monte Carlo approximation to the posterior distribution. □

14.4.3 Marginal Posterior Distributions

Inferences on individual parameters are obtained by using the marginal posterior distribution of the parameter of interest. Mathematically, the marginal posterior pdf of a scalar θ_j is

$$f[\theta_j \mid \text{DATA}] = \int f(\boldsymbol{\theta} \mid \text{DATA}) \, d\boldsymbol{\theta}_{[j]},$$

where $\boldsymbol{\theta}_{[j]}$ is the vector $\boldsymbol{\theta}$ with θ_j omitted. Using the general resampling method described above, one gets a sample from the posterior distribution for $\boldsymbol{\theta}$, say, $\boldsymbol{\theta}_i^\star =$

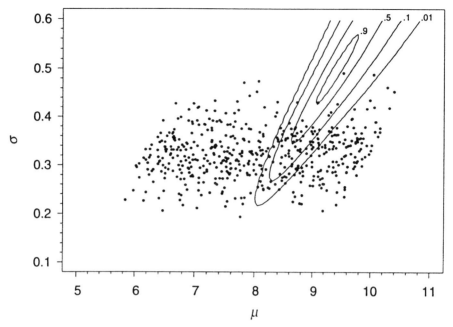

Figure 14.6. Simulated points from the joint prior distribution for μ and σ with the bearing-cage data Weibull relative likelihood contours.

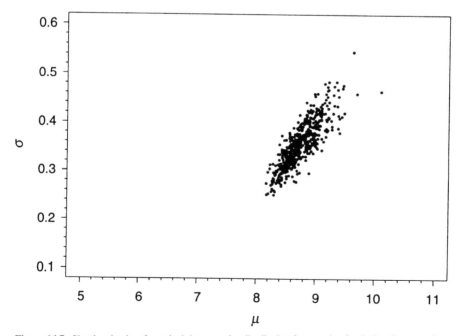

Figure 14.7. Simulated points from the joint posterior distribution for μ and σ for the bearing-cage data.

NUMERICAL METHODS FOR COMBINING PRIOR INFORMATION 355

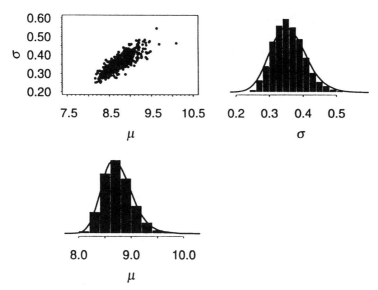

Figure 14.8. Simulated points from the joint posterior and corresponding marginal posterior distributions for μ and σ for the bearing-cage life distribution.

$(\mu_i^\star, \sigma_i^\star)$, $i = 1, \ldots, M^\star$, and uses this to approximate $f[\theta_j \mid \text{DATA}]$. Then, for example, inferences for μ or σ alone are based on the corresponding "marginal" distributions of $\mu_i^\star$ and $\sigma_i^\star$, respectively.

Example 14.5 Marginal Distribution for μ and σ for the Bearing-Cage Life Distribution. Figure 14.8 shows the same simulated points from the joint posterior distribution given in Figure 14.7, but it also provides histograms of the samples from the corresponding marginal distributions for μ and σ. □

Estimates and confidence intervals for a scalar function of the parameters $g(\boldsymbol{\theta})$ are obtained by using the marginal posterior pdf $f[g(\boldsymbol{\theta}) \mid \text{DATA}]$ and the corresponding posterior cdf $F[g(\boldsymbol{\theta}) \mid \text{DATA}]$ of the functions. Using the simulation method, $f[g(\boldsymbol{\theta}) \mid \text{DATA}]$ and $F[g(\boldsymbol{\theta}) \mid \text{DATA}]$ are approximated by using the empirical pdf and cdf of $g(\boldsymbol{\theta}^\star)$, respectively.

Example 14.6 Joint Posterior and Marginal Distributions for Functions of μ and σ for the Bearing-Cage Life Distribution. The marginal posterior distribution of t_p is used to estimate distribution quantiles. This marginal posterior is obtained from the empirical distribution of $\mu_i^\star + \Phi_{\text{sev}}^{-1}(p)\sigma_i^\star$. The marginal posterior distribution of $F(t_e)$ is used to estimate the population fraction failing at t_e. This distribution is obtained from the empirical distribution of $\Phi_{\text{sev}}(\zeta_e^\star)$, where $\zeta_e^\star = [\log(t_e) - \mu_i^\star]/\sigma_i^\star$.
□

14.5 USING THE POSTERIOR DISTRIBUTION FOR ESTIMATION

14.5.1 Bayesian Point Estimation

Bayesian inference for $\boldsymbol{\theta}$ and functions of the parameters $g(\boldsymbol{\theta})$ are entirely based, respectively, on the posterior pdfs $f(\boldsymbol{\theta} \mid \text{DATA})$ and $f[g(\boldsymbol{\theta}) \mid \text{DATA}]$. If $g(\boldsymbol{\theta})$ is a scalar, a common Bayesian estimate of $g(\boldsymbol{\theta})$ is the mean of the posterior distribution, which is given by

$$\widehat{g}(\boldsymbol{\theta}) = \text{E}[g(\boldsymbol{\theta}) \mid \text{DATA}] = \int g(\boldsymbol{\theta}) f(\boldsymbol{\theta} \mid \text{DATA}) \, d\boldsymbol{\theta}.$$

This estimate of $g(\boldsymbol{\theta})$ is the Bayesian estimate that minimizes the square error loss. Other possible choices to estimate $g(\boldsymbol{\theta})$ include the mode of the posterior pdf (which is very similar to the ML estimate) and the median of the posterior distribution. Such estimates are easy to compute from a simulated sample from a posterior. In particular,

$$\widehat{g}(\boldsymbol{\theta}) \approx \frac{1}{M^\star} \sum_{i=1}^{M^\star} g(\boldsymbol{\theta}_i^\star)$$

is the sample mean, the posterior median is the sample median of the $g(\boldsymbol{\theta}^\star)$ values, and the mode can be obtained by finding the maximum of a smooth density estimate of the distribution of $g(\boldsymbol{\theta}^\star)$ values.

14.5.2 Bayesian Interval Estimation

A $100(1-\alpha)\%$ Bayesian lower confidence bound (or credible bound) for a scalar function $g(\boldsymbol{\theta})$ is value $\underset{\sim}{g}$ satisfying $\int_{\underset{\sim}{g}}^{\infty} f[g(\boldsymbol{\theta}) \mid \text{DATA}] \, dg(\boldsymbol{\theta}) = 1-\alpha$. A $100(1-\alpha)\%$ Bayesian upper confidence bound (or credible bound) for a scalar function $g(\boldsymbol{\theta})$ is value $\widetilde{g}$ satisfying $\int_{-\infty}^{\widetilde{g}} f[g(\boldsymbol{\theta}) \mid \text{DATA}] \, dg(\boldsymbol{\theta}) = 1-\alpha$. A $100(1-\alpha)\%$ Bayesian confidence interval (or credible interval) for a scalar function $g(\boldsymbol{\theta})$ is any interval $[\underset{\sim}{g}, \widetilde{g}]$ satisfying

$$\int_{\underset{\sim}{g}}^{\widetilde{g}} f[g(\boldsymbol{\theta}) \mid \text{DATA}] \, dg(\boldsymbol{\theta}) = 1 - \alpha. \tag{14.4}$$

The two-sided interval $[\underset{\sim}{g}, \widetilde{g}]$ can be chosen in different ways:

- Combining two $100(1-\alpha/2)\%$ intervals puts equal probability in each tail (preferable when there is more concern for being incorrect in one direction than the other).
- A $100(1-\alpha)\%$ highest posterior density (HPD) confidence interval chooses $[\underset{\sim}{g}, \widetilde{g}]$ to consist of all values of g with $f(g \mid \text{DATA}) > c$, where c is chosen such that (14.4) holds. HPD intervals are similar to likelihood-based approximate confidence intervals, calibrated with a χ^2 quantile.

USING THE POSTERIOR DISTRIBUTION FOR ESTIMATION   357

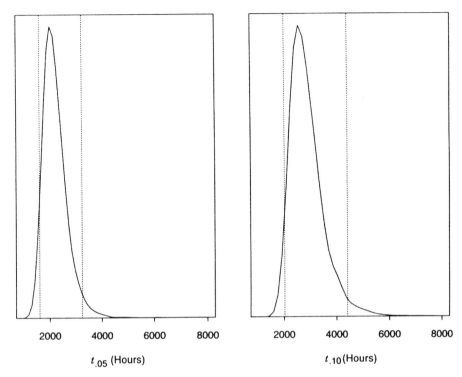

Figure 14.9. Marginal posterior distributions for $t_{.05}$ and $t_{.10}$ (quantiles) of bearing-cage life.

Example 14.7 Marginal Posterior Distributions for the $t_{.05}$ and $t_{.10}$ Quantiles of Bearing-Cage Life. Figure 14.9 shows the marginal posterior distributions for the $t_{.05}$ and $t_{.10}$ quantiles of the bearing-cage life distribution. The figure also shows 95% two-sided Bayesian confidence intervals for the quantiles. These confidence intervals were obtained by combining lower and upper one-sided 97.5% confidence bounds for each quantile. Numerically, the interval for $t_{.05}$ is [1613, 3236] hours and for $t_{.10}$, the interval is [2018, 4400] hours. □

Example 14.8 Marginal Posterior Distributions for F(t) of the Bearing-Cage Life Distribution. Similar to Example 14.7 and Figure 14.9, Figure 14.10 shows the marginal posterior distributions for the $F(2000)$ and $F(5000)$ points on the bearing-cage life distribution, along with 95% Bayesian confidence intervals. Numerically, the interval for $F(2000)$ is [.015, .097], and for $F(5000)$ the interval is [.132, .905].
□

The procedure for constructing a confidence interval for a scalar generalizes to the computation of confidence regions for vector functions $g(\boldsymbol{\theta})$ of $\boldsymbol{\theta}$. In particular, a $100(1 - \alpha)\%$ Bayesian confidence region (or credible region) for a vector-valued

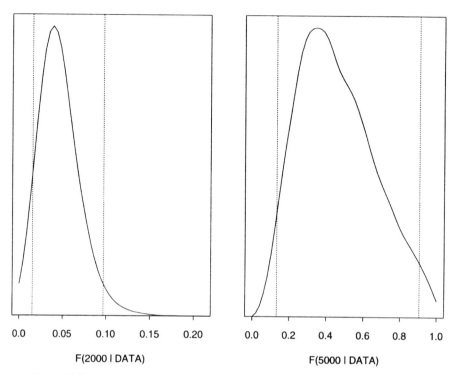

Figure 14.10. Marginal posterior distributions for $F(2000)$ and $F(5000)$ for bearing-cage life.

function $g(\boldsymbol{\theta})$ is defined as

$$\mathrm{CR_B} = \{g(\boldsymbol{\theta}) \mid f[g \mid \mathrm{DATA}] \geq c\},$$

where c is chosen such that

$$\int_{\mathrm{CR_B}} f[g(\boldsymbol{\theta}) \mid \mathrm{DATA}]\, dg(\boldsymbol{\theta}) = 1 - \alpha.$$

The presentation of the confidence region is difficult when $\boldsymbol{\theta}$ has more than two components.

14.6 BAYESIAN PREDICTION

Bayesian methods are also useful for predicting a future event like the failure of a unit from a specified population or a process. Future events can be predicted by using the Bayesian posterior predictive distribution.

14.6.1 Bayesian Posterior Predictive Distribution

If X [with pdf $f(x \mid \boldsymbol{\theta})$] represents a future random variable, then the posterior predictive pdf of X is

$$f(x \mid \text{DATA}) = \int f(x \mid \boldsymbol{\theta}) f(\boldsymbol{\theta} \mid \text{DATA}) \, d\boldsymbol{\theta} = \mathrm{E}_{\boldsymbol{\theta} \mid \text{DATA}}[f(x \mid \boldsymbol{\theta})]. \tag{14.5}$$

The corresponding posterior predictive cdf of X is

$$F(x \mid \text{DATA}) = \int_{-\infty}^{x} f(u \mid \text{DATA}) \, du = \int F(x \mid \boldsymbol{\theta}) f(\boldsymbol{\theta} \mid \text{DATA}) \, d\boldsymbol{\theta} \tag{14.6}$$

$$= \mathrm{E}_{\boldsymbol{\theta} \mid \text{DATA}} \left[F(x \mid \boldsymbol{\theta}) \right].$$

Both posterior predictive distributions are expectations computed with respect to the posterior distribution of $\boldsymbol{\theta}$.

14.6.2 Approximating Posterior Predictive Distributions

Using the simulation approach described in Section 14.4.2, the Bayesian posterior predictive pdf can be approximated by the average of the posterior pdfs $f(x \mid \boldsymbol{\theta}_i^\star)$ using

$$f(x \mid \text{DATA}) \approx \frac{1}{M^\star} \sum_{i=1}^{M^\star} f(x \mid \boldsymbol{\theta}_i^\star). \tag{14.7}$$

Similarly, the Bayesian posterior predictive cdf can be approximated by the average of the posterior cdfs $F(x \mid \boldsymbol{\theta}_i^\star)$. In particular,

$$F(x \mid \text{DATA}) \approx \frac{1}{M^\star} \sum_{i=1}^{M^\star} F(x \mid \boldsymbol{\theta}_i^\star). \tag{14.8}$$

A two-sided $100(1 - \alpha)\%$ Bayesian prediction interval for a new observation is given by the $\alpha/2$ and $(1 - \alpha/2)$ quantiles of $F(x \mid \text{DATA})$. A one-sided $100(1 - \alpha)\%$ lower (upper) Bayesian prediction bound for a new observation is given by the α quantile [or the $(1 - \alpha)$ quantile] of $F(x \mid \text{DATA})$.

14.6.3 Prediction of an Observation from a Log-Location-Scale Distribution

This section describes prediction of a future observation T that has a log-location-scale distribution (extension to other distributions with shape parameters is straightforward, but the notation becomes more complicated). Using $X = T$ and $x = t$, the pdf and cdf of the observation to be predicted (conditional on the parameters in $\boldsymbol{\theta}$) are

$$f(t \mid \boldsymbol{\theta}) = \frac{1}{\sigma t} \phi(\zeta) \quad \text{and} \quad F(t \mid \boldsymbol{\theta}) = \Phi(\zeta), \tag{14.9}$$

where $\zeta = [\log(t) - \mu]/\sigma$. These functions are then averaged over the posterior distribution of $\boldsymbol{\theta}$ using (14.5) and (14.6), giving the posterior predictive pdf and cdf. To implement with the simulation method, substitute instead into (14.7) and (14.8).

Example 14.9 Predictive Distributions for a Bearing-Cage Failure. Continuing with the bearing-cage fracture example, the Weibull distribution Monte Carlo approximations for the posterior predictive pdf and cdf are

$$f(t \mid \text{DATA}) \approx \frac{1}{M^\star} \sum_{i=1}^{M^\star} \frac{1}{\sigma_i^\star t} \phi_{\text{sev}} \left[\frac{\log(t) - \mu_i^\star}{\sigma_i^\star} \right]$$

$$F(t \mid \text{DATA}) \approx \frac{1}{M^\star} \sum_{i=1}^{M^\star} \Phi_{\text{sev}} \left[\frac{\log(t) - \mu_i^\star}{\sigma_i^\star} \right].$$

The posterior predictive pdf $f(t \mid \text{DATA})$ for the failure time of a bearing-cage is shown in Figure 14.11, along with a 95% prediction interval. Numerically, the interval is [1618, 13,500] hours. □

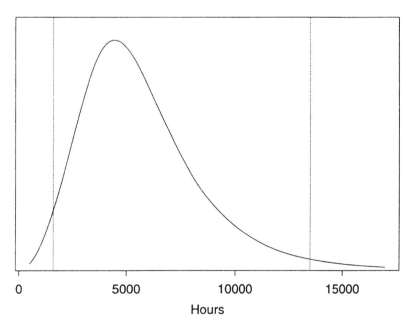

Figure 14.11. Posterior predictive pdf and Bayesian 95% prediction interval for a future observation from the bearing-cage population.

BAYESIAN PREDICTION

14.6.4 Posterior Predictive Distribution for the *k*th Failure from a Future Sample of Size *m*

It is often of interest to predict the *k*th failure (or more generally the *k*th order statistic) in a future sample of size *m*. We will illustrate how to do this when the distribution of time T has a log-location-scale distribution. We let $T_{(k)}$ denote the *k*th largest observation for a sample of size *m* from the distribution of T. In this case, $X = T_{(k)}$ and $x = t_{(k)}$. Then the pdf for $T_{(k)}$, conditional on $\boldsymbol{\theta}$, is

$$f[t_{(k)} \mid \boldsymbol{\theta}] = \frac{m!}{(k-1)!(m-k)!} \times [\Phi(\zeta)]^{k-1} \times \frac{1}{\sigma t_{(k)}} \phi(\zeta) \times [1 - \Phi(\zeta)]^{m-k}, \quad (14.10)$$

where $\zeta = [\log(t_{(k)}) - \mu]/\sigma$. The corresponding cdf of $T_{(k)}$, conditional on $\boldsymbol{\theta}$, is

$$\Pr[T_{(k)} \le t_{(k)} \mid \boldsymbol{\theta}] = F[t_{(k)} \mid \boldsymbol{\theta}]$$

$$= \sum_{j=k}^{m} \frac{m!}{j!(m-j)!} [\Phi(\zeta)]^{j} \times [1 - \Phi(\zeta)]^{m-j}. \quad (14.11)$$

Substituting (14.10) and (14.11) into (14.7) and (14.8) provides the following expressions for the posterior predictive pdf and cdf of $T_{(k)}$, respectively:

$$f[t_{(k)} \mid \text{DATA}] \approx \frac{1}{M^{\star}} \sum_{i=1}^{M^{\star}} \left\{ \frac{m!}{(k-1)!(m-k)!} \times [\Phi(\zeta_i^{\star})]^{k-1} \times \frac{1}{\sigma_i^{\star} t_{(k)}} \phi(\zeta_i^{\star}) \right.$$
$$\left. \times [1 - \Phi(\zeta_i^{\star})]^{m-k} \right\},$$

$$F[t_{(k)} \mid \text{DATA}] \approx \frac{1}{M^{\star}} \sum_{i=1}^{M^{\star}} \left\{ \sum_{j=k}^{m} \frac{m!}{j!(m-j)!} [\Phi(\zeta_i^{\star})]^{j} \times [1 - \Phi(\zeta_i^{\star})]^{m-j} \right\},$$

where $F[t_{(k)} \mid \text{DATA}] = \Pr[T_{(k)} \le t_{(k)} \mid \text{DATA}]$ and $\zeta_i^{\star} = [\log(t_{(k)}) - \mu_i^{\star}]/\sigma_i^{\star}$.

Example 14.10 Posterior Predictive Distributions for the First Failure from a Group of 50 Bearing-Cages. Fifty bearings have been put into service. A prediction interval is needed for the time (measured in hours of operation) of the first failure from this group. Using $m = 50$ and the simulated $\mu_i^{\star}$ and $\sigma_i^{\star}$ values from Example 14.4, the posterior predictive pdf for the first-order statistic is obtained as follows. First, with $k = 1$ the pdf of $T_{(k)}$ in (14.10) simplifies to

$$f[t_{(1)} \mid \boldsymbol{\theta}] = m \times \frac{1}{\sigma t_{(1)}} \phi(\zeta) \times [1 - \Phi(\zeta)]^{m-1},$$

where $\zeta = [\log(t_{(1)}) - \mu]/\sigma$. The corresponding cdf for $T_{(1)}$, the first-order statistic, is

$$\Pr[T_{(1)} \le t \mid \boldsymbol{\theta}] = F[t_{(1)} \mid \boldsymbol{\theta}] = 1 - [1 - \Phi(\zeta)]^{m}.$$

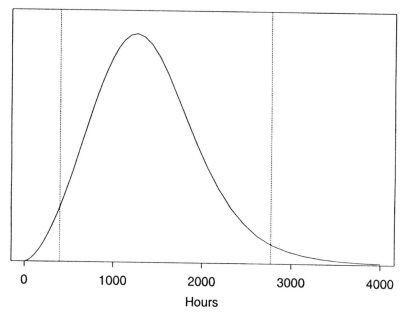

Figure 14.12. Posterior predictive pdf and Bayesian 95% prediction interval for the first failure from a future sample of 50 bearing cages.

Thus the posterior predictive pdf for $T_{(1)}$ is

$$f[t_{(1)} \mid \text{DATA}] \approx \frac{1}{M^\star} \sum_{i=1}^{M^\star} \left\{ m \times \frac{1}{\sigma_i^\star t_{(1)}} \phi\left(\zeta_i^\star\right) \times \left[1 - \Phi\left(\zeta_i^\star\right)\right]^{m-1} \right\},$$

where $\zeta_i^\star = [\log(t_{(1)}) - \mu_i^\star]/\sigma_i^\star$. The corresponding posterior predictive cdf for $T_{(1)}$ is

$$\Pr[T_{(1)} \leq t \mid \text{DATA}] = F[t_{(1)} \mid \text{DATA}] \approx \frac{1}{M^\star} \sum_{i=1}^{M^\star} \left\{ 1 - \left[1 - \Phi\left(\zeta_i^\star\right)\right]^m \right\}.$$

The posterior predictive pdf $f[t_{(1)} \mid \text{DATA}]$ for the first failure time out of a sample of size $m = 50$ bearing cages is shown in Figure 14.12, along with a 95% prediction interval. Numerically, the interval is [401, 2771] hours. □

14.7 PRACTICAL ISSUES IN THE APPLICATION OF BAYESIAN METHODS

14.7.1 Comparison Between Bayesian and Likelihood/Frequentist Statistical Methods

One of the most important differences between Bayesian methods and the likelihood method of making inferences is the manner in which nuisance parameters are handled.

Bayesian interval inference methods are based on a marginal distribution in which nuisance parameters have been integrated out and parameter uncertainty can be interpreted in terms of probabilities from the marginal posterior distribution. In the profile likelihood method, on the other hand, nuisance parameters can be maximized out, as suggested by large-sample theory. Confidence intervals based on likelihood and profile likelihood functions can be calibrated and interpreted in terms of repeated-sampling coverage probabilities (as described in Section 3.3.2). In large samples (where the likelihood and posterior are approximately symmetric), Bayesian and likelihood/frequentist confidence interval methods give very similar answers when prior information is approximately uninformative.

14.7.2 Cautions on the Use of Prior Information

In many applications, engineers really have useful, indisputable prior information (e.g., information from physical theory or past experience deemed relevant through engineering or scientific knowledge). In such cases, the information should be integrated into the analysis. Analysts and decision makers must, however, beware of and avoid the use of "wishful thinking" as prior information. The potential for generating seriously misleading conclusions is especially high when experimental data will be limited and the prior distribution will dominate in the final answers (common in engineering applications). Evans (1989) describes such concerns from an engineering point of view.

As with other analytical methods, when using Bayesian statistics, it is important to do sensitivity analyses with respect to uncertain inputs to one's model. For some model/data combinations, Bayes's estimates and confidence bounds can depend entirely on prior assumptions. This possibility can be explored by changing prior distribution assumptions and checking the effect that the changes have on final answers of interest.

BIBLIOGRAPHIC NOTES

Lindley (1972) describes the basic ideas and philosophy of Bayesian inference. Singpurwalla (1988b) explains reasons for using Bayesian methods in reliability and risk analysis applications. Box and Tiao (1973) present procedures for applying Bayesian methods to a wide range of commonly used statistical models, including analysis of variance and regression analysis. Martz and Waller (1982) apply Bayesian methods to numerous applications in reliability and risk assessment. Gelman, Carlin, Stern, and Rubin (1995) provide a comprehensive treatment of modern methods of Bayesian modeling and computation and illustrate how to apply the methods in an impressive array of applications. Singpurwalla (1988a) presents methods and a computer program for eliciting prior information from experts. Smith and Gelfand (1992) describe simple Monte Carlo methods for doing Bayesian computations, including the method used in this chapter. Gelfand and Smith (1990) describe and compare three more sophisticated Monte Carlo methods (data augmentation, Gibbs sampling, and importance sampling) that can be used to compute marginal posterior distributions. Gelfand and Smith (1992) apply the Gibbs sampling methods to constrained parameter and censored data problems. Severini (1991) describes the relationship between Bayesian and non-Bayesian confidence intervals. Smith and Naylor (1987) compare maximum likelihood and Bayesian methods for estimating the parameters of a three-parameter Weibull (see Section 11.7), providing a convincing example of the potential difference between Bayesian and likelihood methods. Hamada and Wu (1995) show how to use Bayesian methods to analyze data from fractional factorials experiments when faced with censored data. Such experiments are sometimes used in reliability improvement efforts. Geisser (1993) describes Bayesian methods for prediction.

EXERCISES

14.1. Starting with the traditional form of Bayes's rule, show that it can be expressed in terms of relative likelihoods.

▲**14.2.** The Monte Carlo approach to making Bayesian inferences is convenient, intuitive, and easy to explain.

(a) Show that the expected number of retained units using the Monte Carlo technique is given by

$$E(M^\star) = M \int f(\boldsymbol{\theta}) R(\boldsymbol{\theta}) \, d\boldsymbol{\theta}.$$

Hint: The number of retained units has a binomial distribution with parameters M and success probability equal to $\Pr(U \leq R(\boldsymbol{\theta}))$, where U and $R(\boldsymbol{\theta})$ are independent. Then the proof consists of showing that $\Pr(U \leq R(\boldsymbol{\theta})) = \int f(\boldsymbol{\theta}) R(\boldsymbol{\theta}) \, d\boldsymbol{\theta}$, which is the expected relative likelihood under the prior.

(b) Use the result in part (a) to argue that if there is agreement between the prior pdf and the likelihood then the ratio $M^\star/M$ tends to be large and the prior pdf and the posterior pdf are similar.

(c) Discuss the case in which the prior pdf and the likelihood do not agree and indicate when the prior and the posterior pdfs tend to agree (or disagree) in this case.

14.3. Explain why doing a Bayesian analysis with a specified prior distribution to estimate an unknown exponential distribution mean θ is not a model that implies that θ varies from unit to unit in the population or process being studied. *Hint*: Consider what happens to the posterior pdf when the sample size approaches infinity.

14.4. Explain how one would compute $f[g(\boldsymbol{\theta}) \mid \text{DATA}]$, the marginal posterior pdf of the function of the parameters of interest, using numerical integration. Contrast this with the simulation-based method.

14.5. Discuss the advantages and disadvantages of the Monte Carlo method of computing the posterior distribution, relative to the use of numerical integration.

14.6. A total of 100 new units have been introduced into service. It is believed that the underlying time-to-failure distribution is Weibull. The analysts have available prior information and censored data on a sample of n similar units.

  (a) Given a Monte Carlo sample of 2000 pairs of μ^*, σ^* values from the posterior pdf $f(\mu, \sigma \mid \text{DATA})$, show how to compute the marginal posterior predictive distribution for $T_{(1)}$, the time of the first failure out of the 100 units. Also show how to get a 95% lower Bayesian prediction bound for $T_{(1)}$.

  (b) Suppose that you can compute (but perhaps not write in closed form) the joint posterior pdf $f(\mu, \sigma \mid \text{DATA})$. Explain how you could, using numerical integration and other numerical (but not simulation) methods, compute the marginal posterior predictive pdf and a 95% lower Bayesian prediction bound for $T_{(1)}$.

▲**14.7.** The model for failure time T of an electronic component is $\text{LOGNOR}(\mu, \sigma)$ with μ unknown and σ known. The uncertainty in μ can be described by a $\text{NOR}(a_1, b_1)$ distribution, where the prior distribution parameters a_1 and b_1 are also known. Suppose that a sample of size n will be used, along with the prior information, to make inferences on μ (and thus other quantities of interest like quantiles of the distribution of T). The sample will provide a realization of $\text{DATA} = (T_1, \ldots, T_n)$.

  (a) Find the conditional pdf of $\bar{Y} = (1/n)\sum_{i=1}^{n} Y_i = (1/n)\sum_{i=1}^{n} \log(T_i)$ for a specified fixed value of μ (which, when viewed as a function of μ for fixed DATA, is also the likelihood).

  (b) Combining the variability in $\bar{Y}$ with the uncertainty in μ, find the joint pdf of μ and $\bar{Y}$.

  (c) Show that the marginal distribution of $\bar{Y}$ is $\text{NOR}\left(a_1, \sqrt{\sigma^2/n + b_1^2}\right)$. What is the practical interpretation of this distribution, relative to this application?

  (d) Show that, for a given DATA realization, the posterior distribution of μ is normal with mean and variance

  $$E(\mu \mid \text{DATA}) = \frac{\sigma^2/n}{b_1^2 + \sigma^2/n} \times a_1 + \frac{b_1^2}{b_1^2 + \sigma^2/n} \times \bar{Y},$$

  $$\text{Var}(\mu \mid \text{DATA}) = \frac{\sigma^2/n}{b_1^2 + \sigma^2/n} \times b_1^2.$$

  (e) Consider what happens to the expressions in part (d) as $n \to \infty$.

(f) Consider the expression for Var(μ | DATA) in part (d). Use this expression to explain how prior information might be related quantitatively to information from a previous sample of a certain size (note that this pseudosample size does not have to be an integer).

14.8. Refer to Exercise 14.7. Derive a simple expression for the posterior predictive pdf of T given DATA.

14.9. Consider the following quantities used in Exercise 14.7: $\bar{Y}$, μ, a_1, and b_1.
(a) Use the setting of Exercise 14.7 and these quantities to explain the important differences between variability and uncertainty in physical processes and statistical inference.
(b) Explain how one could generalize the model in Exercise 14.7 to allow for batch-to-batch variability in the reliability of the electronic component and describe a sampling plan that could be used to combine prior information with data to estimate the parameters of this more general model.

▲**14.10.** When the prior pdf for $\log(t_p)$ is uniform,

$$f[\log(t_p)] = \frac{1}{\log(b_1/a_1)}, \quad a_1 \le t_p \le b_1.$$

When the prior pdf for $\log(\sigma)$ is triangular,

$$f[\log(\sigma)] = \begin{cases} \dfrac{4[\log(\sigma/a_2)]}{[\log(b_2/a_2)]^2} & \text{if } \quad a_2 \le \sigma \le \sqrt{a_2 b_2} \\ \dfrac{4[\log(b_2/\sigma)]}{[\log(b_2/a_2)]^2} & \text{if } \quad \sqrt{a_2 b_2} < \sigma \le b_2. \end{cases}$$

Under these specifications:
(a) Show that the prior pdf for σ is $f(\sigma) = (1/\sigma) f[\log(\sigma)]$.
(b) Show that the joint pdf for (t_p, σ) is

$$f(t_p, \sigma) = \frac{f[\log(t_p)] f[\log(\sigma)]}{t_p \sigma}, \quad a_1 \le t_p \le b_1, \ a_2 \le \sigma \le b_2.$$

(c) Use the transformation $\mu = \log(t_p) - \Phi_{\text{sev}}^{-1}(p)\sigma, \sigma = \sigma$ to show that the joint prior pdf for (μ, σ) is

$$f(\mu, \sigma) = f[\log(t_p)] \frac{f[\log(\sigma)]}{\sigma}$$

$$= \frac{1}{\log(b_1/a_1)} \frac{f[\log(\sigma)]}{\sigma},$$

where $\log(a_1) - \Phi_{\text{sev}}^{-1}(p)\sigma \le \mu \le \log(b_1) - \Phi_{\text{sev}}^{-1}(p)\sigma$, and $a_2 \le \sigma \le b_2$.

(d) Show that the region in which $f(\mu, \sigma) > 0$ is southwest to northeast oriented.

▲14.11. For the prior pdf for (t_p, σ) given in Exercise 14.10, show the following:
(a) The prior cdf for t_p is

$$F(t_p) = \frac{\log(t_p/a_1)}{\log(b_1/a_1)}, \quad a_1 \le t_p \le b_1.$$

(b) The prior cdf for σ is

$$F(\sigma) = \begin{cases} \dfrac{2[\log(\sigma/a_2)]^2}{[\log(b_2/a_2)]^2} & \text{if} \quad a_2 \le \sigma \le \sqrt{a_2 b_2} \\ 1 - \dfrac{2[\log(b_2/\sigma)]^2}{[\log(b_2/a_2)]^2} & \text{if} \quad \sqrt{a_2 b_2} < \sigma \le b_2. \end{cases}$$

(c) Invert these cdfs to verify that the the following formulas provide random numbers from the prior (t_p, σ):

$$(t_p)_i = a_1 \times (b_1/a_1)^{U_{1i}}, \quad i = 1, \ldots, M,$$

where $U_{11}, \ldots, U_{1M}$ is a random sample from a UNIF(0, 1).
(d) For σ,

$$\sigma_i = \begin{cases} a_2 \times (b_2/a_2)^{\sqrt{U_{2i}/2}} & \text{if} \quad U_{2i} \le 1/2 \\ b_2 \times (a_2/b_2)^{\sqrt{(1-U_{2i})/2}} & \text{if} \quad U_{2i} > 1/2 \end{cases} \quad i = 1, \ldots, M,$$

where $U_{21}, \ldots, U_{2M}$ is another independent random sample from a UNIF(0, 1).
(e) Then $\boldsymbol{\theta}_i = (\mu_i, \sigma_i)$ with $\mu_i = \log[(t_p)_i] - \Phi_{\text{sev}}^{-1}(p)\sigma_i$ is a random sample from the (μ, σ) prior distribution.

▲14.12. Consider the prior pdfs $f[\log(\sigma)]$ and $f(\sigma)$.
(a) Plot the pdf $f[\log(\sigma)]$ in a log-scale for the following choices of the parameters: $[a_2, b_2] = [1, 7], [1, 10]$.
(b) For the same choices of $[a_2, b_2]$ plot the pdf $f(\sigma)$ in an arithmetic scale.
(c) Show analytically that the mode of $f[\log(\sigma)]$ occurs at $\sigma = \sqrt{a_2 b_2}$.
(d) Show, however, that the mode of $f[\log(\sigma)]$ occurs at $\sigma = \sqrt{a_2 b_2}$ if $\exp(1) \ge \sqrt{b_2/a_2}$ and at $\sigma = a_2 \exp(1) < \sqrt{a_2 b_2}$ otherwise.
(e) Indicate the modes for the graphs in parts (a) and (b) above. Explain what you observe on the plots.

▲14.13. Consider the pdf $f[t_{(k)} \mid \boldsymbol{\theta}]$ and cdf $F[t_{(k)} \mid \boldsymbol{\theta}]$ for the kth largest observation given in Section 14.6.4.

(a) Integrate the pdf $f[t_{(k)} \mid \boldsymbol{\theta}]$ to obtain the cdf $F[t_{(k)} \mid \boldsymbol{\theta}]$. *Hint*: Use integration by parts.

(b) Use the general formula given in Section 14.6.4 for $F[t_{(k)} \mid \boldsymbol{\theta}]$ to derive the predictive cdf when $k = 1$. *Hint*: The general formula is a sum of binomial probabilities.

14.14. Show that for the special case $k = m = 1$, (14.10) and (14.11) reduce to the pdf and cdf given in (14.9).

CHAPTER 15

System Reliability Concepts and Methods

Objectives

This chapter explains:

- Important system reliability concepts like system structure, redundancy, nonrepairable and repairable systems, and maintainability and availability.
- Basic concepts of system reliability modeling.
- The distribution of system failure time as a function of individual component failure-time distributions.
- Simple methods for using component test data to estimate system reliability.
- Analysis of data with more than one failure mode.

Overview

This chapter describes and illustrates some basic ideas behind system reliability analysis. Section 15.2 describes some simple system structures and shows how to compute system reliability as a function of component reliability. These simple structures can be used as building blocks to compute system reliability for more complicated systems. Section 15.3 shows how to estimate and compute confidence intervals for system reliability from limited component data. Section 15.4 explains methods of analyzing failure-time data with more than one cause of failure and how to use such data to estimate the distribution corresponding to the individual failure modes or the overall system. Section 15.5 provides a brief overview of other topics and references related to system reliability.

15.1 INTRODUCTION

A system is a collection of components interconnected to perform a given task. Component state (e.g., working or not working) and system structure determine whether a system is working or not. System structure is described by a logic diagram illustrating the relationship between components and satisfactory system performance. Ultimately, interest centers on the reliability of specific systems. Assessing and improving system reliability generally requires consideration of system structure and component reliability. Some systems are replaced upon failure. Many systems, however, are maintained (e.g., replacing worn components before they fail) and/or repaired after failure. For repairable systems, availability (the fraction of time that a system is available for use) may be the appropriate metric. This leads to consideration of maintainability (e.g., improvement of reliability through inspection and/or preventive maintenance) and repairability (characterized by the distribution of time to do a repair). In general, availability can be increased by increasing reliability or by improving maintainability and repairability.

15.2 SYSTEM STRUCTURES AND SYSTEM FAILURE PROBABILITY

System failure probability, $F_T(t; \boldsymbol{\theta})$, is the probability that the system fails before t. The failure probability of the system is a function of time in operation t (or other measure of use), the system structure, reliability of system components, interconnections, and interfaces (including, e.g., human operators).

This section describes several simple system structures. Not all systems fall into one of these categories, but the examples provide a collection of building blocks to illustrate the basics of system structure. Complicated system structures can generally be decomposed into collections of the simpler structures presented here. The methods for evaluation of system reliability can be adapted to more complicated structures.

15.2.1 Time Dependency of System Reliability

For a new system (i.e., all components starting a time 0) with s independent components, the cdf for component i is denoted by $F_i = F_i(t; \boldsymbol{\theta}_i)$. The corresponding survival probability (reliability) for component i is $S_i = S_i(t; \boldsymbol{\theta}_i) = 1 - F_i(t; \boldsymbol{\theta}_i)$. The $\boldsymbol{\theta}_i$ vectors may have some elements in common. We let $\boldsymbol{\theta}$ denote the unique elements in $(\boldsymbol{\theta}_1, \ldots, \boldsymbol{\theta}_s)$. The cdf for the system is denoted by $F_T = F_T(t; \boldsymbol{\theta})$. This cdf is determined by the component F_i functions and the system structure. Then $F_T(t; \boldsymbol{\theta}) = g[F_1(t; \boldsymbol{\theta}_1), \ldots, F_s(t; \boldsymbol{\theta}_s)]$. To simplify the presentation, time (and parameter) dependency will usually be suppressed in this chapter. Then this function can also be expressed in one of the simpler forms $F_T(\boldsymbol{\theta}) = g[F_1(\boldsymbol{\theta}_1), \ldots, F_s(\boldsymbol{\theta}_s)]$ or $F_T = g(F_1, \ldots, F_s)$.

15.2.2 Systems with Components in Series

A series structure with s components works if and only if all the components work. Examples of systems with components in series include chains, high-voltage multicell batteries, inexpensive computer systems, and inexpensive decorative tree lights using low-voltage bulbs. For a system with two independent components in series,

Figure 15.1. A system with two components in series.

illustrated in Figure 15.1, the cdf is

$$F_T(t) = \Pr(T \le t) = 1 - \Pr(T > t) = 1 - \Pr(T_1 > t \cap T_2 > t) \quad (15.1)$$
$$= 1 - \Pr(T_1 > t)\Pr(T_2 > t) = 1 - (1 - F_1)(1 - F_2) = F_1 + F_2 - F_1 F_2.$$

For s independent components $F_T(t) = 1 - \prod_{i=1}^{s}(1 - F_i)$ and for s iid components ($F = F_i$, $i = 1, \ldots, s$), $F_T(t) = 1 - (1 - F)^s$. This is the same as the minimum-type distributions discussed in Section 5.12.3. For a series system of independent components, the system hazard function is the sum of the component hazard functions

$$h_T(t) = \sum_{i=1}^{s} h_i(t). \quad (15.2)$$

Figure 15.2 shows the relationship between system reliability $1 - F_T(t)$ and individual component reliability $1 - F(t)$ for different numbers of identical independent components in series. This figure shows that extremely high component reliability is needed to maintain high system reliability, particularly if the system has many components in series.

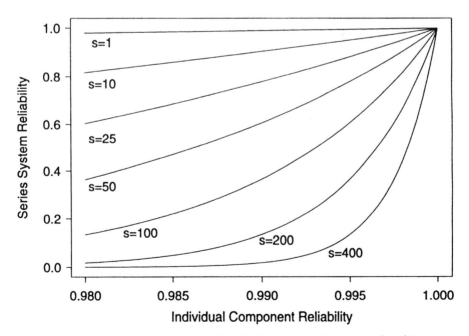

Figure 15.2. Reliability of a system with s identical independent components in series.

Importance of Part Count in Product Design

An important rule of thumb in reliability engineering design practice is "keep the part count small," meaning keep the number of individual parts (or components) in a system to a minimum. Besides the cost of purchase and handling of additional individual parts, there is also an important reliability motivation for having a smaller number of parts at risk of failure in a product. Of course, this rule of thumb holds when the reliability of the individual parts in the design with a smaller number of parts is the same or similar to the reliability of the parts in the design with a larger number of parts.

Example 15.1 Effect of Part-Count Reduction on Modem Reliability. The design for a new computer modem uses a higher level of microelectronic integration and requires only 20 discrete parts instead of the 40 parts required in the previous generation. For a series system of parts with independent failure times, the hazard function of the system can be obtained by summing the hazards for the individual parts. This is particularly simple if a constant hazard rate (or, equivalently, an exponential time-to-failure distribution) provides an adequate model for part life. As a rough approximation, suppose that all failures are due to part failures, and that all of the parts have the same hazard function. Then the population of modems produced with the new design with only 20 parts will experience only half the failures when compared to the old design. Allowing that failures can occur at interfaces and interconnections between parts with the same frequency in the new and old designs would widen the reliability gap because of the larger number of such interfaces with a higher number of parts. With a nonconstant hazard function (more common in practice) the idea is similar. □

Series System of Independent Components Having Weibull Distributions with the Same Shape Parameter

Recall from Section 4.8 that the Weibull hazard function can be written as

$$h(t) = \frac{\beta}{\eta} \left(\frac{t}{\eta}\right)^{\beta-1}, \quad t > 0.$$

For a series system of s independent components having a Weibull distribution with the same shape parameter β but possibly differing η values, the system failure-time distribution is also Weibull. The system hazard function is

$$h_T(t) = \sum_{i=1}^{s} \frac{\beta}{\eta_i} \left(\frac{t}{\eta_i}\right)^{\beta-1} = \frac{\beta}{\eta_T} \left(\frac{t}{\eta_T}\right)^{\beta-1}, \quad t > 0,$$

where

$$\eta_T = \left(\sum_{i=1}^{s} \frac{1}{\eta_i^\beta}\right)^{-1/\beta}.$$

If all of the s components have the same η, then this simplifies to $\eta_T = \eta/s^{1/\beta}$.

Example 15.2 Reliability of a Chain. A particular kind of chain link can fail from growth of fatigue cracks and eventual fracture. The life distribution of a single link has a Weibull distribution with $\eta = 100$ thousand use-cycles and a shape parameters $\beta = 2.3$. A chain of 75 links can be viewed as a series system. When the first link breaks, the chain fails and has to be replaced. In the application, all links in the chain are subject to the same level of stress. If the cycles to failure for the individual links are independent, then the chain has a Weibull time-to-failure distribution with $\eta_T = 100/(75)^{1/2.3} = 15.30$ thousand cycles and a shape parameter $\beta = 2.3$. □

Effect of Positive Dependency in a Two-Component Series System

If a series system contains two components with dependent failure times, then the first line of (15.1) still gives $F_T(t)$, but the evaluation has to be done with respect to the bivariate distribution of T_1 and T_2. More generally, for a system with s components in series, the system $F_T(t)$ would have to be computed with respect to the underlying s-variate distribution. Such computations are, in general, difficult. If the correlation among the s series components is positive, then the assumption of independence is conservative in the sense that the actual $F_T(t)$ is smaller than that predicted by the independent-component model. For a simple two-component series system, Figure 15.3 shows the reliability $1 - F_T(t)$ of a two-component series system as a function of the reliability $1 - F(t)$ of the individual components that have positive correlation. For this example, the distributions of log failure times for the individual components is bivariate normal with the same (arbitrary) mean and standard deviation for both

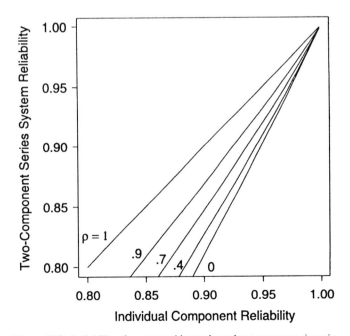

Figure 15.3. Reliability of a system with two dependent components in series.

components and correlation ρ. When $\rho = 1$ (so the two components are perfectly dependent and will fail at exactly the same time), the curve is the same as the $s = 1$ (single-component) system shown in Figure 15.2. When $\rho = 0$ (so the two components are independent), the curve would correspond to an $s = 2$ curve in Figure 15.2. Figure 15.3 shows that when there is positive correlation between the failure times of the individual components, the actual reliability of the system exceeds that predicted by the independent-component series system. The multivariate generalization of this result is important in reliability modeling applications.

Example 15.3 Reliability of a Jet Engine Turbine Disk. The primary threat for a jet engine turbine disk failure is the initiation and growth of a fatigue crack. Generally it is not economically desirable to test more than one or two jet engine turbine disks. Additionally, even with realistic continuous accelerated testing, no failures would be expected for years. Instead, disk reliability is predicted by using a model. The reliability model for the disk is obtained by dividing the disk into a large number of small "elements" that are first modeled individually. Accelerated tests on material specimens provide information to predict the life of an element as a function of temperature and stress. The overall reliability of the system can then be modeled as a series system of independent components. Modeling the individual elements' failure-time distribution as a function of temperature and stress (which depend on position within the disk) improves the adequacy of the independence assumption. Still, however, one would expect the initiation and growth of cracks to be positively correlated from element to element within a disk, especially among elements that are close together. When the failure times of a series system's components have positive association (nonnegative correlation between all pairs), the independence model provides a conservative prediction of the system's overall reliability. Theoretical justification for this result is given in Chapter 2 of Barlow and Proschan (1975). □

Of course, if the lifetimes of components in a series system have negative association, then the reliability predicted with the independence model will be anticonservative. We do not give details for this situation because it is not common in physical systems.

15.2.3 Systems with Components in Parallel

A parallel structure with s components works if at least one of the components works. Examples of systems with components in parallel include automobile headlights, RAID computer disk array systems, stairwells with emergency lighting, overhead projectors with backup bulb, and multiple light banks in classrooms. For two independent parallel components, illustrated in Figure 15.4,

$$F_T(t) = \Pr(T \leq t) = \Pr(T_1 \leq t \cap T_2 \leq t) \tag{15.3}$$
$$= \Pr(T_1 \leq t) \Pr(T_2 \leq t) = F_1 F_2.$$

SYSTEM STRUCTURES AND SYSTEM FAILURE PROBABILITY

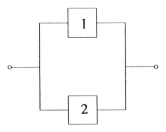

Figure 15.4. A system with two components in parallel.

For s independent components $F_T(t) = \prod_{i=1}^{s} F_i$ and for s iid components ($F_i = F$, $i = 1, \ldots, s$), $F_T(t) = F^s$.

Figure 15.5 shows the relationship between system reliability $1 - F_T(t)$ and individual component reliability $1 - F(t)$ for different numbers of identical independent components in parallel. The figure shows the dramatic effect that parallel redundancy can have on the reliability of the system or subsystem. If the components are not independent, then the first line of (15.3) still gives $F_T(t)$, but the evaluation has to be done with respect to the bivariate distribution of T_1 and T_2.

Effect of Positive Dependency in a Two-Component Parallel-Redundant System
For a simple two-component parallel system, Figure 15.6 shows the effect that positive dependency between the failure times of the two components has on system

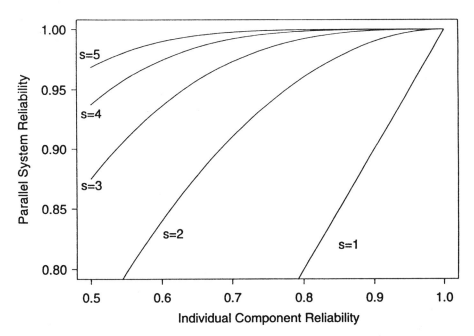

Figure 15.5. Reliability of a system with s iid components in parallel.

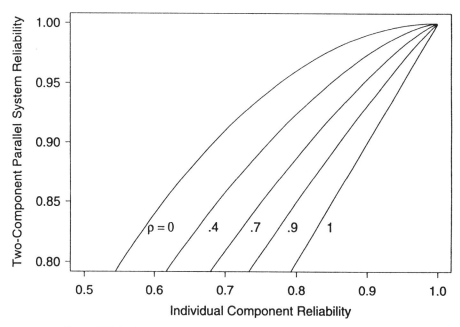

Figure 15.6. Reliability of a system with two dependent components in parallel.

reliability $1 - F_T(t)$. For this example, the distributions of log failure times for the individual components is bivariate normal with the same (arbitrary) mean and standard deviation for both components and correlation ρ. With $\rho = 0$ (so the components are independent), the curve is the same as the $s = 2$ curve shown in Figure 15.5. When $\rho = 1$ (so the components are perfectly dependent and will fail at exactly the same time), the curve is the same as the $s = 1$ (single-component) curve shown in Figure 15.5. The advantages of redundancy can be degraded seriously when the failure times of the individual components have positive dependence.

15.2.4 Systems with Components in Series–Parallel

Methods for evaluating the reliability of structures with components in both series and parallel provide the basis for evaluating more complicated structures that use redundancy to increase system reliability. There are two types of simple (i.e., rectangular) series–parallel structures: series–parallel with system-level redundancy and series–parallel with component-level redundancy.

Series–Parallel System Structure with System-Level Redundancy

In some applications it is more cost effective to achieve higher reliability by using two or more copies of a series system rather than having to improve the reliability of the single system itself. Series–parallel system structures with system-level redundancy are used in applications like parallel central processors for a system-critical

SYSTEM STRUCTURES AND SYSTEM FAILURE PROBABILITY

communications switching system, spacecraft or aircraft fly-by-wire computer control systems, automobile brake system (hydraulic and mechanical), and multiple trans-Atlantic transmission cables.

A $r \times k$ series–parallel system-level redundancy structure has r parallel sets, each of k components in series. For a 2×2 structure with independent components, illustrated in Figure 15.7,

$$F_T(t) = \Pr(T \le t) = \Pr[\text{``series 1 failed''} \cap \text{``series 2 failed''}] \quad (15.4)$$
$$= [1 - (1 - F_{11})(1 - F_{12})][1 - (1 - F_{21})(1 - F_{22})]$$
$$= [F_{11} + F_{12} - F_{11}F_{12}][F_{21} + F_{22} - F_{21}F_{22}],$$

where F_{ij}, $j = 1, 2$, are the cdfs for the series subsystem i. For a $r \times k$ structure with independent components $F_T(t) = \prod_{i=1}^{r}\left[1 - \prod_{j=1}^{k}(1 - F_{ij})\right]$ and for a $r \times k$ parallel–series structure with iid components $F_T(t) = [1 - (1 - F)^k]^r$. If the system components are not independent, then the first line of (15.4) still gives $F_T(t)$, but the evaluation has to be done with respect to the multivariate distribution of the component failure times.

Series-Parallel System Structure with Component-Level Redundancy

Component redundancy is an important method for improving system reliability. Series–parallel system structures with component-level redundancy are found in numerous applications, including parallel dual repeaters in undersea fiber-optic data transmission systems, and the human body (lungs, kidneys). A $k \times r$ component-level redundant structure has k series structures, each one made of r components in parallel. If it is necessary to have only one path through the system, such a structure is, for a given number of identical components, more reliable than the series–parallel system-level redundancy. For a 2×2 series–parallel system with independent components, illustrated in Figure 15.8,

$$F_T(t) = 1 - \Pr(T > t) = 1 - \Pr[\text{``parallel 1 works''} \cap \text{``parallel 2 works''}]$$
$$= 1 - (1 - F_{11}F_{21})(1 - F_{12}F_{22})$$
$$= F_{11}F_{21} + F_{12}F_{22} - F_{11}F_{21}F_{12}F_{22}, \quad (15.5)$$

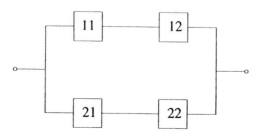

Figure 15.7. A series–parallel system structure with system-level redundancy.

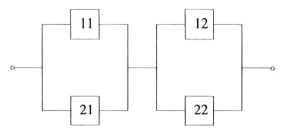

Figure 15.8. A simple series–parallel system structure with component-level redundancy.

where F_{ij}, $i = 1, 2$, are the cdfs for parallel subsystem j. For a $k \times r$ series–parallel system with independent components,

$$F_T(t) = 1 - \prod_{i=1}^{k}\left(1 - \prod_{j=1}^{r} F_{ij}\right).$$

When all of the system's components are iid $F_T(t) = 1 - (1 - F^r)^k$. If the system components are not independent, then the first line of (15.5) still gives $F_T(t)$, but the evaluation has to be done with respect to the multivariate distribution of the component failure times.

15.2.5 Bridge-System Structure

Bridge-structure systems provide another useful structure for improving the reliability of certain systems. Bridge structures are common in computer and electric power-distribution networks. Figure 15.9 illustrates a simple bridge-structure system. Note that if component 3 is *not* working, the bridge system has the same structure as Figure 15.7. If component 3 is working, the bridge system has the same structure as Figure 15.8. In many practical situations a bridge such as the one at component 3 in Figure 15.9 can be installed at little extra cost but provides a potentially important improvement in reliability when compared to a simple series–parallel system (see Exercise 15.1).

The relationship between the bridge structure and the two different series–parallel structures provides a method to compute the bridge-structure system reliability. We

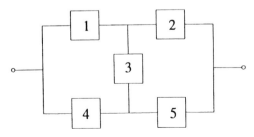

Figure 15.9. A bridge-system structure.

use A_3 (A_3^c) to denote the event that component 3 is working (is not working). Then similar to (15.5) $\Pr(T \leq t \mid A_3) = F_1F_4 + F_2F_5 - F_1F_2F_4F_5$, and similar to (15.4) $\Pr(T \leq t \mid A_3^c) = [F_1 + F_2 - F_1F_2][F_4 + F_5 - F_4F_5]$; thus

$$F_T(t) = \Pr(T \leq t \cap A_3) + \Pr(T \leq t \cap A_3^c)$$
$$= \Pr(A_3)\Pr(T \leq t \mid A_3) + \Pr(A_3^c)\Pr(T \leq t \mid A_3^c)$$
$$= (1 - F_3)(F_1F_4 + F_2F_5 - F_1F_2F_4F_5)$$
$$+ F_3(F_1 + F_2 - F_1F_2)(F_4 + F_5 - F_4F_5).$$

15.2.6 *k*-Out-of-*s* System Structure

Some systems work if at least *k* out of *s* components work, but not otherwise. Special cases include the 1 of *s* parallel structure and the *s* of *s* series structure. Other examples of *k*-out-of-*s* system structures include a satellite battery system in which the system will continue to operate as long as 6 of 10 batteries continue to operate correctly, or computer storage disks that continue to provide service by blocking out bad sectors up to a certain limit, and heat exchangers that continue to operate even when a certain small proportion of their tubes have been plugged (see Example 1.5).

Figure 15.10 illustrates a logic diagram for a system requiring that at least two out of three components work. Note that the system structure diagram does not reflect physical layout, but rather paths through the system that will allow operation of the system. Computationally, for a two-out-of-three system of independent components,

$$F_T(t) = \Pr(T \leq t)$$
$$= \Pr(\text{"exactly two fail"}) + \Pr(\text{"exactly three fail"})$$
$$= [F_1F_2(1 - F_3) + F_1F_3(1 - F_2) + F_2F_3(1 - F_1)] + F_1F_2F_3$$
$$= F_1F_2 + F_1F_3 + F_2F_3 - 2F_1F_2F_3.$$

For *k*-out-of-*s* independent components,

$$F_T(t) = \sum_{j=k}^{s} \left\{ \sum_{\delta \in A_j} \left[\prod_{i=1}^{s} F_i^{\delta_i}(1 - F_i)^{(1-\delta_i)} \right] \right\}, \tag{15.6}$$

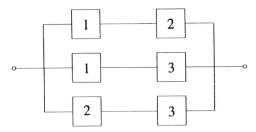

Figure 15.10. A *k*-out-of-*s* system structure.

where $\boldsymbol{\delta} = (\delta_1, \ldots, \delta_s)$ with $\delta_i = 1$ indicating failure of component i by time t and $\delta_i = 0$ otherwise and A_j is the set of all $\boldsymbol{\delta}$ such that $\sum_{i=1}^{s} \delta_i = j$. For identically distributed components ($F = F_i, i = 1, \ldots, s$) $F_T(t) = \sum_{j=k}^{s} \binom{s}{j} F^j (1-F)^{s-j}$, a binomial distribution.

Example 15.4 **Spacecraft Power System.** A spacecraft power system uses 10 rechargable batteries in parallel. The capacity of the power system is equal to the sum of the power provided by each of the batteries. The system can continue to operate at design specifications as long as at least 7 of the 10 batteries are functional. □

15.3 ESTIMATING SYSTEM RELIABILITY FROM COMPONENT DATA

15.3.1 Computing System Reliability from Component Reliability

To compute the system cdf, one can use $F_T = g(F_1, \ldots, F_s)$ when g is known from the system structure. If g cannot be expressed in closed form or is otherwise difficult to compute, one can use a computer simulation of the system based on the F_i and the system structure. When the F_i are unknown, an estimate of the system F_T can be obtained by evaluating F_T at the ML estimates of the needed F_i values.

15.3.2 Sources of Reliability Data

Laboratory tests are used widely, especially to test new materials and components where there is little past experience. Such testing is generally expensive and may have limited ability to predict product field reliability. Special care must be taken to assure that test conditions can be related accurately to actual field conditions (as described in Chapters 18–21, laboratory tests are often accelerated with the goal of getting component reliability information more quickly). Carefully collected field data, when available, provide the most accurate information about how components and systems behave in the field. Field data collection, however, is also expensive. Warranty data often have serious deficiencies. For example, warranty data often contain no information on units that do not fail (see Robinson and McDonald, 1991, for further discussion of this and other related issues). Ireson (1996) describes some general issues relating to the collection and storage of reliability data. Reliability handbooks and data banks can be useful (e.g., Klinger, Nakada, and Menendez, 1990, and MIL-HDBK 217E, 1986). One common complaint about such handbooks, however, is that data become obsolete by the time they are published and that reported hazard rates and failure probabilities may be off by an order of magnitude or more. Technology, in many areas, is moving faster than accurate traditional reliability data can be obtained. Expert knowledge is often used when no other source of information is available. Unless data are collected from carefully conducted statistical studies, quantifying uncertainty may be impossible.

15.3.3 Maximum Likelihood Estimation of System Reliability

Suppose that sample data are available to estimate the failure-time distributions of the system's individual components. For example, data on component i for $i = 1, \ldots, s$ can be used to estimate $\boldsymbol{\theta}_i, i = 1, \ldots, s$, providing estimates $\widehat{F}_1, \ldots, \widehat{F}_s$, respectively. These cdf estimates are functions of time, as described in Section 15.2.1. Then the system cdf (or other related functions) can be estimated as functions of $\widehat{F}_1, \ldots, \widehat{F}_s$, the function being determined from the system's structure, as described in Sections 15.2.2–15.2.6. Let $\widehat{\boldsymbol{\theta}}$ be the ML estimate of $\boldsymbol{\theta}$ (the unique parameters describing the components' cdfs) and $\widehat{\Sigma}_{\widehat{\boldsymbol{\theta}}}$ the ML estimate of $\Sigma_{\widehat{\boldsymbol{\theta}}}$ obtained from the component data. Then using the same methods as in previous chapters, $\widehat{F}_T = F_T(\widehat{\boldsymbol{\theta}}) = g[F_1(\widehat{\boldsymbol{\theta}}_1), \ldots, F_s(\widehat{\boldsymbol{\theta}}_s)]$. The variance of $\widehat{F}_T$ can be computed by using the delta method (Appendix Section B.2) as $\widehat{\text{Var}}(\widehat{F}_T) = (\partial F_T/\partial \boldsymbol{\theta})' \widehat{\Sigma}_{\widehat{\boldsymbol{\theta}}} (\partial F_T/\partial \boldsymbol{\theta})$, where the derivatives are evaluated at $\widehat{\boldsymbol{\theta}}$.

Example 15.5 Maximum Likelihood Estimation for a Simple Parallel-Structure System. For a parallel structure with s iid components

$$\widehat{F}_T = [\widehat{F}]^s = [F(\widehat{\boldsymbol{\theta}})]^s,$$

$$\widehat{\text{Var}}(\widehat{F}_T) = \left(\frac{\partial F_T}{\partial \boldsymbol{\theta}}\right)' \widehat{\Sigma}_{\widehat{\boldsymbol{\theta}}} \left(\frac{\partial F_T}{\partial \boldsymbol{\theta}}\right) = \left(\frac{\partial F_T}{\partial F} \frac{\partial F}{\partial \boldsymbol{\theta}}\right)' \widehat{\Sigma}_{\widehat{\boldsymbol{\theta}}} \left(\frac{\partial F_T}{\partial F} \frac{\partial F}{\partial \boldsymbol{\theta}}\right)$$

$$= \left(s\widehat{F}^{s-1} \frac{\partial F}{\partial \boldsymbol{\theta}}\right)' \widehat{\Sigma}_{\widehat{\boldsymbol{\theta}}} \left(s\widehat{F}^{s-1} \frac{\partial F}{\partial \boldsymbol{\theta}}\right).$$

Then $\widehat{\text{se}}_{\widehat{F}_T} = \sqrt{\widehat{\text{Var}}(\widehat{F}_T)}$. □

15.3.4 Normal-Approximation Confidence Intervals for System Reliability

A normal approximation $100(1 - \alpha)\%$ confidence interval for $F_T(t; \boldsymbol{\theta})$ based on $Z_{\text{logit}(\widehat{F}_T)} \overset{.}{\sim} \text{NOR}(0, 1)$ is

$$[\underset{\sim}{F_T}, \; \widetilde{F}_T] = \left[\frac{\widehat{F}_T}{\widehat{F}_T + (1 - \widehat{F}_T) \times w}, \; \frac{\widehat{F}_T}{\widehat{F}_T + (1 - \widehat{F}_T)/w}\right],$$

where $w = \exp\{z_{(1-\alpha/2)} \widehat{\text{se}}_{\widehat{F}_T} / [\widehat{F}_T(1 - \widehat{F}_T)]\}$.

15.3.5 Bootstrap Approximate Confidence Intervals for System Reliability

Bootstrap confidence intervals can be used to improve upon the simple normal-approximation method in Section 15.3.4. One iteration of the bootstrap procedure requires the computation of bootstrap estimates $\widehat{F}_i^*, i = 1, \ldots, s$ (as in Chapter 9), for each of the s components. These lead to the system bootstrap estimate $\widehat{F}_T^* = g(\widehat{F}_1^*, \ldots, \widehat{F}_s^*)$. The procedure is repeated B times and, as in Section 9.4.2, an approximate $100(1-\alpha)\%$ confidence interval for $F_T(t)$ based on $Z_{\text{logit}(\widehat{F}_T)} \stackrel{.}{\sim} Z_{\text{logit}(\widehat{F}_T^*)}$ and the B bootstrap samples is

$$[\underset{\sim}{F_T}, \widetilde{F}_T] = \left[\frac{\widehat{F}_T}{\widehat{F}_T + (1-\widehat{F}_T) \times \underset{\sim}{w}}, \frac{\widehat{F}_T}{\widehat{F}_T + (1-\widehat{F}_T) \times \widetilde{w}}\right].$$

Here $\underset{\sim}{w} = \exp\{z_{\text{logit}(\widehat{F}^*)_{(1-\alpha/2)}} \widehat{\text{se}}_{\widehat{F}_T} / [\widehat{F}_T(1-\widehat{F}_T)]\}$ and $\widetilde{w} = \exp\{z_{\text{logit}(\widehat{F}^*)_{(\alpha/2)}} / [\widehat{F}_T(1-\widehat{F}_T)]\}$, where $z_{\text{logit}(\widehat{F}^*)_{(1-\alpha/2)}}$ and $z_{\text{logit}(\widehat{F}^*)_{(\alpha/2)}}$ are obtained from the quantiles of the B bootstrap estimates $\widehat{F}_T^*$, as described in Section 9.4.2.

15.4 ESTIMATING RELIABILITY WITH TWO OR MORE CAUSES OF FAILURE

15.4.1 Products with Two or More Causes of Failure

Many systems, subsystems, and components (which we generically refer to as "units") have more than one cause of failure. In some applications and for some purposes it is important to distinguish between the different failure causes (sometimes referred to as "failure modes"). For purposes of improving reliability, it is essential to identify the cause of failure down to the component level and, in many applications, down to the actual physical cause of failure.

Multiple failure modes should be distinguished from population mixtures (e.g., Section 5.12.1). Population mixtures divide a population into different mutually exclusive groupings of units. Such subpopulations result from differences in the manufacture or use of the product. Multiple failure modes, on the other hand, are the different ways in which a particular unit might fail.

Failure time of a system with two or more failure modes can be modeled with a series-system or competing risk model. Each risk is like a component in a series system. When one component fails, the system (i.e., product) fails. Each unit has a potential failure time associated with each failure mode. The observed failure time is the minimum of these individual potential failure times.

15.4.2 Estimation with Two or More Causes of Failure

This section explores applications in which failed units are replaced rather than repaired after failure (repairable system data analysis is described in Chapter 16). Some life tests result in failure-time data that have more than one cause of failure. Most field data could have both failure-time and failure-cause information reported for each failure (although failure-cause information is often expensive or otherwise difficult to obtain and is therefore often *not* reported). Warranty data have potential problems of bias and limited information about surviving units. As described in Example 11.13, it may be necessary to conduct a survey to get information about the status of units that have not been reported as failing (e.g., if and when units have been retired and the number of use-cycles for units still in service). Typically, many

ESTIMATING RELIABILITY WITH TWO OR MORE CAUSES OF FAILURE 383

units reported as failing in the warranty period are units that have been subjected to the harshest use conditions.

Field-tracking studies will follow, more carefully, a group of units in service (or simulated service). Such studies are more expensive but provide better information about field reliability. See Amster, Brush, and Saperstein (1982) for more information on planning field-tracking studies. For some applications it is possible to test in an "accelerated" field environment where failures could be expected to occur more rapidly than in typical service applications (providing time to make corrections in customer units before serious problems might arise).

Example 15.6 Estimation of Device-G $F_T(t)$ Using Failure Mode Information.
Table 15.1 gives times of failure and running times for a sample of devices from a field-tracking study of a larger system. At a certain point in time, 30 units were installed in typical service environments. Cause of failure information was determined for each unit that failed. Mode S failures were caused by an accumulation of randomly occurring damage from power-line voltage spikes during electric storms, resulting in failure of a particular unprotected electronic component. These failures predominated early in life. Mode W failures, caused by normal product wear, began to appear after 100 thousand cycles of use. The NW corner of Figure 15.11 displays the results of a Weibull analysis of the Mode S failures only (failures due to Mode W were treated as censored at the time of the Mode W failure—all we know is that the unobserved Mode S failure time would have been sometime after the observed Mode W failure). Similarly, the NE corner of Figure 15.11 displays the results of a Weibull analysis of the Mode W failures only. The results for these two analyses are also summarized in Table 15.2. In both cases, the Weibull distribution provides a good fit to the data. The SW corner shows the results of a Weibull analysis ignoring the cause of failure information. Looking carefully we note evidence of a change in the slope of the plotted

Table 15.1. Device-G Failure Times and Cause of Failure for Devices that Failed and Running Times for Units that Did Not Fail

Thousands of Cycles	Failure Mode	Thousands of Cycles	Failure Mode	Thousands of Cycles	Failure Mode
275	W	106	S	88	S
13	S	300	—	247	S
147	W	300	—	28	S
23	S	212	W	143	S
181	W	300	—	300	—
30	S	300	—	23	S
65	S	300	—	300	—
10	S	2	S	80	S
300	—	261	S	245	W
173	S	293	W	266	W

W indicates a wearout failure, S indicates an electrical surge failure, and — indicates a unit still operating after 300 thousand cycles.

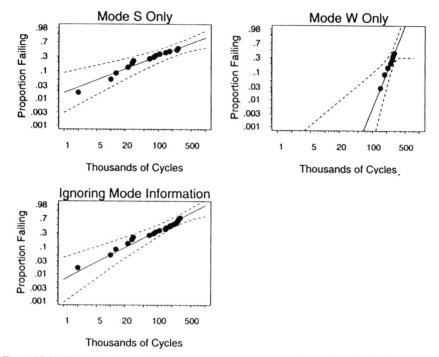

Figure 15.11. Weibull analyses of Device-G data estimating time to failure Mode S only, failure Mode W only, and ignoring the cause of failure.

Table 15.2. Device-G Field-Tracking Data Weibull ML Estimation Results for the Electric Surge (S) and Wearout (W) Failure Modes

Mode	Parameter	ML Estimate	Standard Error	Approximate 95% Confidence Interval	
				Lower	Upper
S	μ_S	6.11	.427	5.27	6.95
	σ_S	1.49	.35	.94	2.36
W	μ_W	5.83	0.11	5.62	6.04
	σ_W	.23	.08	.12	.44
S and W	μ_{SW}	5.49	0.23	5.04	5.94
	σ_{SW}	1.08	.21	.74	1.57

For Mode S alone, $\mathcal{L}_S = -101.36$; for Mode W alone, $\mathcal{L}_W = -47.16$; and for both modes together, $\mathcal{L}_{SW} = -142.62$.

ESTIMATING RELIABILITY WITH TWO OR MORE CAUSES OF FAILURE

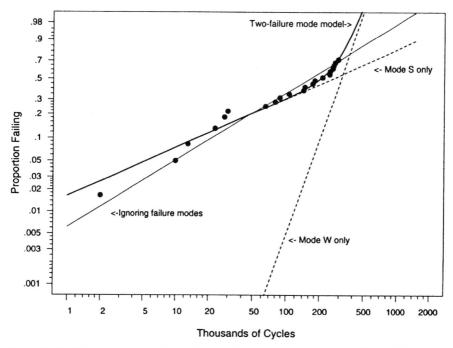

Figure 15.12. Weibull analyses of Device-G data estimating time to failure Mode S only, failure Mode W only, and distribution to the minimum of Mode S and Mode W.

points, indicating a gradual shift from one failure mode to another. The dotted lines in Figure 15.12 show the estimated ML lines for the two individual failure modes. The thin, straight, solid line is the ML line estimating the Weibull $F_T(t)$ obtained from ignoring the cause of failure information (i.e., using both failure modes together in the analysis). The curved line is the series-system estimate of $F_T(t)$ for the two failure modes acting together. This estimate was computed, under the assumption of independence of T_1 and T_2, as $\widehat{F}_T(t) = 1 - [1 - \widehat{F}_1(t)] \times [1 - \widehat{F}_2(t)]$. The two estimates diverge rapidly after 100 thousand cycles. Estimates of the mean time to failure were computed from $\widehat{\mathrm{MTTF}} = \int_0^\infty [1 - \widehat{F}_T(t)]\, dt$ in (4.3) and were 251.3 and 196.0 thousand cycles, respectively, for the models ignoring and using the failure mode information. The difference between these estimates would have been greater if the censoring had been heavier (implying more extrapolation in time). □

15.4.3 Estimation of Multiple Failure Mode Distributions when Only Some Failure Modes are Identified in the Data

When the failure modes are not identified or are only partially identified, it is sometimes possible to estimate the individual $F_i(t)$ distributions by using maximum likelihood. This may, however, be difficult because the analyses are no longer separable, and failure-time distributions must be estimated simultaneously even if the modes act independently. Also, the parameter estimates for the distribution for one mode will be correlated with those of the other modes (Friedman and Gertsbakh, 1980). In practice, one is likely to analyze the data as if there were only a single mode. This can result in the pitfalls described earlier, especially when the shapes of the distributions of the individual failure modes are not similar or if censoring is somehow linked to one or more of the failure modes. Guess, Usher, and Hodgson (1991), for example, describe maximum likelihood methods. Determining the appropriate likelihood is straightforward. Determining if there is enough information in the data to estimate all of the model parameters can be problematic. Also see the example in Section 22.3.

15.5 OTHER TOPICS IN SYSTEM RELIABILITY

15.5.1 Other System Structures

For standby redundancy (also known as passive redundancy), a redundant unit is activated only when another unit fails and the redundant unit is needed to keep the system working. There are many variations of this including cold standby and partially loaded redundancy. Also it is necessary, in some systems, to consider the reliability of component and subsystem interfaces as well as the switching mechanism that activates the standby units. This can be done by including such interfaces and the switching mechanism into the overall system structure.

15.5.2 Dependency Among Components

The common assumption of components with independent failures is sometimes unrealistic. For example, it is possible that failure of one component either improves or degrades the reliability of other system components (leading to either negative or positive correlation between failure times in different components). Another common source of dependency is "common cause of failure," which occurs when an external force causes failure of more than one component. Figure 15.3 (Figure 15.6) showed the effect of dependency on a simple two-component series (parallel) system. The same effect would be amplified in a multicomponent redundant system.

15.5.3 Systems with Repair

Many systems are repaired after failure. Questions of repairability, maintainability, and availability generally depend on knowledge of a repair time distribution. The methods used in this book can also be used to estimate such distributions.

Analyses involving time dependence of system failure/repair cycles generally use models different from those used in the previous chapters in this book. Chapter 16 describes simple methods for analyzing system repair-history and other recurrence-type data.

15.5.4 FMEA/FMECA

Products and systems often have complicated designs that are the result of efforts of one or more design teams. Management for system reliability requires a global process to assure that the product/system reliability will meet customer requirements.

Failure modes and effect analysis (FMEA) is a systematic, structured method for identifying system failure modes and assessing the effects or consequences of the identified failure modes. Failure modes and effect criticality analysis (FMECA) considers, in addition, the criticality (or importance) of identified failure modes, with respect to safety, successful completion of system mission, or other criteria. The goal of FMEA/FMECA is to identify all possible failure modes at a specified level of system architecture. These methods are used typically in product/system design review processes. The use of FMEA/FMECA typically begins in the early stages of product/system conceptualization and design. Then the FMEA/FMECA evolves over time along with changes in the product/system design and accumulation of information about product/system performance in preproduction testing and field experience. FMEA/FMECA is used during the product/system design phase to help guide decision making. FMEA/FMECA is also used to develop product/system guidelines for system repair and maintenance procedures, to make statements about system safety, and to provide direction for reliability improvement efforts.

Operationally, FMEA/FMECA begins by defining the scope of the analysis, specified by the system level at which failures are to be considered. FMEA/FMECA can be conducted at various different levels in a product or a system. FMEA/FMECA might be done initially for individual subsystems. Then the results can be integrated to provide an FMEA/FMECA for an entire system comprised of many subsystems. For example, an FMEA to study the reliability of a telecommunications relay repeater might consider, as basic components, each discrete device in the electronic circuit (e.g., ICs, capacitors, resistors, diodes). At another level, an FMEA for a large telecommunications network might consider as components all of the network nodes and node interconnections (ignoring the electronic detail within each node).

The next step in the FMEA/FMECA process is the identification of all components that are subject to failure. This is followed by identification of all component interfaces or connections between components that might fail. In many applications, environmental and human-factor-related failures are considered in defining failure modes. Finally, the effects of the identified failure modes are delineated. Determining the effect of failure modes and combinations of failure modes uses the detailed specification of the relationship among the product/system components (system structure). Special worksheets and/or computer software can be used to organize all of the information.

15.5.5 Fault Trees

The FMEA/FMECA process described above is sometimes referred to as the "bottom-up" approach to reliability modeling. Fault tree analysis, on the other hand, quantifies system failure using a "top-down" approach. First, one or more critical "top-events" (such as loss of system functionality) are defined. Then in a systematic manner, the combination (or combinations) of conditions required for that event to occur is delineated. Generally this is done by identifying how failure-related events at a higher level are caused by lower level "primary events" (e.g., failure of an individual component) and "intermediate events" (e.g., failure of a subsystem). Information from an FMEA analysis might be used as input to this step. The information is organized in the form of a "fault-tree diagram" with the top event at the top of the diagram. Events at different levels of the tree are connected by logic gates defined on a system of Boolean logic (e.g., AND, OR, Exclusive OR gates).

A complete fault tree can be used to model the probability of critical system events. Additional inputs required for this analysis include probability or conditional probabilities of the primary events. With this information and the detailed system structure specification provided by the fault tree, it is possible to compute critical event probabilities.

Fault tree diagrams are, in one sense, similar to the reliability block diagrams presented earlier in this section. It is generally possible to translate from one to the other. Fault tree analysis differs in its basic approach to system reliability. Reliability block diagrams are structured around the event that the system does *not* fail. Fault tree analysis, however, provides focus on the critical failure-causing top-events like loss of system functionality or other safety-critical events. The tree shows, directly, the root causes of these top-events, and other contributing events, at all levels within the scope of the analysis. The structure and logic of the fault tree itself provide not only a mechanism for quantitative reliability assessment but also clearer insight into possible approaches for reliability improvement.

15.5.6 Component Importance

A component's importance to overall system reliability depends on the reliability of the component and the component's position in the system structure. Measures of component importance with respect to reliability provide information that is needed to develop effective strategies to improve system reliability. In particular, such measures suggest which components should get attention in reliability improvement efforts. For example, one particularly simple measure of component reliability, motivated by traditional sensitivity analysis, is the partial derivative of overall system reliability with respect to the individual component's reliability. Chapter 5 of Høyland and Rausand (1994) provides details on a number of other useful measures of component importance.

15.5.7 Markov and Other State-Space Reliability Models

System models described up to this point have had only two states of interest: failed or not failed. A state-space model can be used to allow for a richer formulation of system behavior. For example, in a parallel system, the system state might describe the number of failed components. A state-space model would describe the different system states, possible transitions from one state to another, and probability distributions describing how the system goes from one state to another (transition probabilities).

A Markov model is a special case of a state-space model requiring (1) a memoryless property that future events depend only on the current state and not on the manner in which the system arrived at that state and (2) a stationarity property that transition probability distributions do not change with time. Although somewhat restrictive, it is possible to use Markov models to describe many kinds of practical systems with useful approximations. If a model with a limited number of states does not have the memoryless property, it might be possible to reformulate the state definition, add some more states, and find a structure that does meet (at least approximately) the required conditions.

Markov models are useful for handling dependencies among system components, complicated repair policies, common-cause failures, and other system complexities. With large, complicated systems, however, the number of states can be large, leading to computational difficulties. Also, because of the memoryless property, Markov models are limited to exponential distributions for life and repair distributions.

Non-Markovian generalizations of state-space models are possible, but there are few analytical results available and numerical computations become exceedingly difficult when dealing with nontrivial system structures. Analyses of such models are generally done by using simulation methods.

BIBLIOGRAPHIC NOTES

Barlow and Proschan (1975) is the classic reference outlining the mathematical theory of system reliability. Kozlov and Ushakov (1970), Høyland and Rausand (1994), Ushakov (1994), and Gnedenko and Ushakov (1995) provide detailed coverage of many different kinds of system reliability models. O'Connor (1985) and Lewis (1996) provide engineering-oriented descriptions of system reliability concepts. Gertsbakh (1989) also describes a number of important system reliability concepts and methods.

Chapter 5 of Nelson (1982) provides theory and applications of multiple failure mode (competing risk) methods for series systems of independent components. David and Moeschberger (1978) and Birnbaum (1979) provide theory, some applications, and numerous important references for such models (for a single subpopulation).

O'Connor (1985), Sundararajan (1991), Høyland and Rausand (1994), and Lewis (1996) provide more details, examples, and references for fault tree methods. MIL-STD-1629A (1980) and books like Høyland and Rausand (1994), Klion (1992), and Sundararajan (1991) outline in more detail and provide examples for the procedures for performing FMEA/FMECA analyses. Høyland and Rausand (1994) also list several computer programs designed to facilitate FMEA/FMECA analyses.

EXERCISES

15.1. Consider the bridge-system structure in Figure 15.9. Assume that components 1, 2, 4, and 5 all have the same cdf $F(t)$. Plot $F_T(t)$ versus $0 < F(t) < 1$ with a separate line for each value of $\Pr(A_3) = 0, .25, .5, .75,$ and 1. Comment on the results relative to the reliability of the two different series–parallel structures in Section 15.2.4.

15.2. Revisit the shock absorber data introduced in Example 3.8 and given in Appendix Table C.2. There were actually two different causes of failure for this product. The analyses in Chapters 3 and 8 ignored the different "failure modes" in the analysis. An alternative analysis (important for some purposes) would take into consideration the different failure modes. Such an analysis is straightforward when the failure times to the two different failure modes are statistically independent.

(a) Assuming that the failure modes are independent, fit Weibull distributions separately to estimate the time to failure for each failure mode. Also fit other parametric distributions to the individual failure modes to see if there are other distributions that fit better than the Weibull.

(b) Suppose that the shock absorber is a series system that fails as soon as there is a failure due to one mode or the other. Combine the results from part (a) to obtain an estimate of the cdf for the shock absorbers with both modes acting together. Plot these on Weibull probability paper and compare the results with the analysis that ignores the differences between the two different failure modes.

(c) Provide an intuitive explanation for the result from part (b).

(d) Explain why the agreement between the two methods of analysis is so good in this case (as compared with Example 15.6). In general, when would you expect to see more important differences between the analysis that accounts for the different failure modes and the analysis that does not account for the different failure modes?

15.3. Consider the following system diagram. Derive an expression for the cdf of the system as a function of F_1, F_2, and F_3 under the assumption that component failure times are statistically independent.

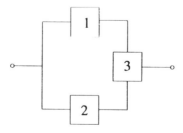

15.4. Consider the following system diagram. Derive an expression for the cdf of the system as a function of $F_1, \ldots, F_6$ under the assumption that component failure times are statistically independent.

EXERCISES

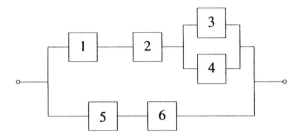

15.5. Consider the following system diagram. Derive an expression for the cdf of the system as a function of $F_1, \ldots, F_6$ under the assumption that component failure times are statistically independent.

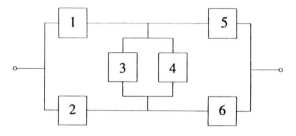

▲**15.6.** Show that for a series system of s independent components, the system hazard function is the sum of the component hazard functions: that is, $h_T(t) = \sum_{i=1}^{s} h_i(t)$.

15.7. A new computer-based home entertainment system will require 16 megabytes of RAM to run the operating system, store needed information for quick access, and perform other tasks. The product designers can use a single 16-megabyte chip or four 4-megabyte chips, the latter option being 30% less expensive. The manufacturers of the memory claim that the average fit rate for 2 years at normal operating conditions is 10 fits/per chip for all of their memory chips. Compare the reliability of the memory system for the two different design options.

◆**15.8.** Consider the model used in Example 15.6, with the generalization that only a proportion ξ of the units in the population are susceptible to failure Mode S.
   (a) Write down expressions for the cdf and pdf of the overall failure-time distribution.
   (b) Write a computer program to compute and plot the hazard function for this model. Find combinations of the five parameters that give a bathtub-shaped hazard function.
   See Section 22.3 for an example using this model.

▲15.9. Consider a parallel system with s independent components. Show the following.
 (a) The pdf for the system is
$$f_T(t) = F_T(t) \times \sum_{j=1}^{s} \frac{f_j(t)}{F_j(t)}.$$
 (b) The hazard function is
$$h_T(t) = \frac{F_T(t)}{1 - F_T(t)} \times \sum_{j=1}^{s} \left[\frac{1 - F_j(t)}{F_j(t)}\right] h_j(t).$$

▲15.10. Suppose that a series system has s iid components. If the life of each component is modeled with a WEIB(μ, σ) then:
 (a) Show that the cdf for the system is WEIB[$\mu - \sigma \log(s), \sigma$].
 (b) Using the cdf obtained in part (a) compute the hazard function for the system. Verify that the same answer can be obtained using the formula for the hazard function given in Section 15.2.
 (c) Compute the reliability and hazard of the components at $t = 1$ month when $\sigma = .5$ and $\mu = 2.3$.
 (d) Use the information in part (c) to compute the reliability and hazard of the system when $s = 10$.

▲15.11. For a series system with s iid components each having a failure-time cdf $F = F(t)$, show that
$$\widehat{\text{Var}}(\widehat{F}_T) = \left[s(1 - \widehat{F})^{s-1} \frac{\partial F}{\partial \boldsymbol{\theta}}\right]' \widehat{\Sigma}_{\hat{\boldsymbol{\theta}}} \left[s(1 - \widehat{F})^{s-1} \frac{\partial F}{\partial \boldsymbol{\theta}}\right].$$

▲15.12. Beginning with the general formula given in (15.6) for F_T in a k-out-of-s system:
 (a) Show that with identically distributed components,
$$F_T(t) = \sum_{j=k}^{s} \binom{s}{j} F^j (1 - F)^{s-j}.$$
 (b) Use the result in part (a) to obtain $F_T(t)$ when $k = 2$ and $s = 3$.

CHAPTER 16

Analysis of Repairable System and Other Recurrence Data

Objectives

This chapter explains:

- Typical data from repairable systems and other applications that have recurrence data.
- Simple nonparametric graphical methods for presenting recurrence data.
- Simple parametric models for recurrence data.
- The combined use of simple parametric and nonparametric graphical methods for drawing conclusions from recurrence data.
- A method of simulating recurrence data.
- Some basic ideas of software reliability modeling.

Overview

This chapter describes methods for analyzing recurrence data where the recurrence times may not be statistically independent. A primary application for such methods is in the analysis of system repair data. The methods are useful for empirically quantifying overall system reliability, for monitoring and predicting repair cost, and for checking to see if the times between repairs of individual components in a system can be treated as being independent or not. Section 16.2 describes nonparametric graphical methods to estimate mean cumulative recurrence rates and, when appropriate, confidence intervals to quantify sampling uncertainty. Section 16.3 gives nonparametric two-sample comparison methods. These nonparametric methods require few assumptions. Section 16.4 describes some simple point-process models that are useful for describing repairable system data for a single system. Section 16.5 gives methods for checking point-process model adequacy and Section 16.6 shows how to use ML to fit parametric models to recurrence data for a single system. Section 16.7 gives methods for simulating data from a nonhomogeneous Poisson process while Section 16.8 explains some of the basic ideas of software reliability.

16.1 INTRODUCTION

16.1.1 Repairable System Reliability Data and Other Recurrence Data

Recurrence data arise frequently in reliability applications. The stochastic model for recurrence data is called a "point-process" model. An important application is system repair data. A repair process for a single system can be viewed as a sequence of repair times $T_1, T_2, \ldots$. In the following discussion of system reliability, the term "system repair" describes a general event of interest. In particular applications, however, the event may be a failure, replacement, adjustment, and so on.

Generally, repair times are measured in terms of system age or time since some well-defined specific event in the system's history. Repairs are typically observed over a fixed observation interval (t_0, t_a), where, typically, $t_0 = 0$. In some cases the number of repairs in each of a set of smaller intervals is reported (e.g., because problems are detected and repairs are initiated at fixed times of inspection) and in other cases, exact times of repairs are recorded.

Some applications have data on only one system. In other applications there may be data from a sample or other collection of systems. When data from such a collection are combined into data to form a single process, the resulting process is known as the superposition of several point-processes or a superimposed point-process. For some applications, cause of failure and/or cost of repair may also be recorded. For some purposes, it is necessary to consider that the population contains a mixture of systems operating in different environments.

Repairable system data are collected to estimate or predict quantities like:

- The distribution of the times between repairs, $\tau_j = T_j - T_{j-1}$ ($j = 1, 2, \ldots$), where $T_0 = 0$.
- The cumulative number of repairs in the interval $(0, t]$ as a function of system age t.
- The expected time between failures (also known as mean time between failures or MTBF).
- The expected number of repairs in the interval $(0, t]$ as a function of t.
- The repair rate as a function of t.
- Average repair cost as a function of t.

Example 16.1 Unscheduled Maintenance Actions for the U.S.S. Grampus Number 4 Diesel Engine. Table 16.1 gives the times (in thousands of operating hours) of unscheduled maintenance actions for the number 4 diesel engine of the U.S.S. *Grampus*, up to 16 thousand hours of operation. This is an example of data on a single system. The unscheduled maintenance actions were caused by system failure or imminent failure. Such maintenance actions are inconvenient and expensive. We will use the data to assess if the system was deteriorating (i.e., maintenance actions occurring more frequently as the system ages) and whether the occurrence of unscheduled maintenance actions could be modeled by a homogeneous Poisson process (discussed in Section 16.4.2). □

INTRODUCTION

Table 16.1. Times (in Thousands of Operating Hours) of Unscheduled Maintenance Actions for the U.S.S. *Grampus* Number 4 Main Propulsion Diesel Engine

.860	1.258	1.317	1.442	1.897	2.011	2.122	2.439
3.203	3.298	3.902	3.910	4.000	4.247	4.411	4.456
4.517	4.899	4.910	5.676	5.755	6.137	6.221	6.311
6.613	6.975	7.335	8.158	8.498	8.690	9.042	9.330
9.394	9.426	9.872	10.191	11.511	11.575	12.100	12.126
12.368	12.681	12.795	13.399	13.668	13.780	13.877	14.007
14.028	14.035	14.173	14.173	14.449	14.587	14.610	15.070
16.000							

Data from Lee (1980).

Example 16.2 Times of Replacement of Diesel Engine Valve Seats. For a fleet of 41 diesel engines, Appendix Table C.8 gives engine age (in days) at the time of a valve seat replacement. These data on a *sample* of systems appeared in Nelson and Doganaksoy (1989) and also in Nelson (1995a). Questions to be answered by these data include:

- Does the replacement rate increase with age?
- How many replacement valves will be needed in a future period of time?
- Can valve life in these systems be modeled as a superimposed renewal process? (If so, the methods in Chapters 3–15 can be used to model the data.)

Figure 16.1 is an event plot of the valve seat repair data showing the observation period and the reported repair times for a subset of 22 diesels. □

16.1.2 A Nonparametric Model for Recurrence Data

For a single system, recurrence data can be expressed as $N(s, t)$, the cumulative number of recurrences in the system age interval $(s, t]$. To simplify notation, $N(t)$ is used to represent $N(0, t)$. The corresponding model, used to describe a population of systems, is based on the mean cumulative function (MCF) at system age t. The population MCF is defined as $\mu(t) = E[N(t)]$, where the expectation is over the variability of each system and the unit-to-unit variability in the population. Assuming that $\mu(t)$ is differentiable,

$$\nu(t) = \frac{d\,E[N(t)]}{dt} = \frac{d\mu(t)}{dt}$$

defines the recurrence rate per system for the population.

Although data on the number of repairs (or other recurrent events related to reliability) are common in practice, the methods in this chapter can be used to model other quantities accumulating in time, including continuous variables like cost. Then,

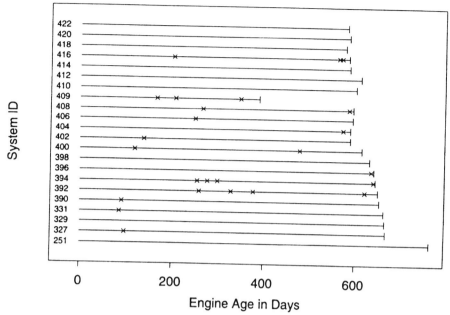

Figure 16.1. Valve seat event plot showing engine age at time of replacement and period of observation for a subset of 22 observed engines.

for example, $\mu(t) = E[C(t)]$ is the mean cumulative cost per system in the time interval $(0, t]$.

16.2 NONPARAMETRIC ESTIMATION OF THE MCF

This section describes a simple method for estimating the MCF.

16.2.1 Nonparametric Model Assumptions

Suppose that an observed collection of $n \geq 1$ systems is an entire population of interest or a sample from a larger population of systems. Then the method described here can be used to estimate the population MCF. The method is nonparametric in the sense that it does not use a parametric model for the population MCF. The method requires minimal assumptions. The basic assumption is that there exists a population of cumulative functions (one for each system in the population) from which a sample has been observed. Randomness in the sample is due to the random sampling of cumulative functions from the population. The method also assumes that the time that observation of a system is terminated does not depend on the system's history. Biased MCF estimators will result, for example, if units follow a staggered scheme of entry into service (e.g., some units put into service each month) and the recurrence rate $\nu(t)$ is increasing in real time due to external events affecting all systems simultaneously. Then newer systems that have a more stressful life will be censored earlier, causing an overly optimistic estimate of the recurrence rate. The nonparametric estimate, however, does not require that the sampled systems be statistically independent.

16.2.2 Point Estimate of the MCF

Let $N_i(t)$ denote the cumulative number of system recurrences for system i before time t, and let t_{ij}, $j = 1, \ldots, m_i$, be the recurrence times for system i. A simple naive estimator of the population MCF at time t would be the sample mean of the available $N_i(t)$ values for the systems still operating at time t. This estimator is simple, but appropriate only if all systems are still operating at time t. Nelson (1988) provided an appropriate unbiased estimator, allowing for different lengths of observation among systems. Nelson's estimate of the population MCF can be computed by using the following algorithm.

Algorithm 16.1 Computation of the MCF Estimate

1. Order the unique recurrence times t_{ij} among all of the n systems. Let m denote the number of unique times. These ordered unique times are denoted by $t_1 < \cdots < t_m$.
2. Compute $d_i(t_k)$, the total number of recurrences for system i at t_k.
3. Let $\delta_i(t_k) = 1$ if system i is still being observed at time t_k and $\delta_i(t_k) = 0$ otherwise.
4. Compute

$$\widehat{\mu}(t_j) = \sum_{k=1}^{j} \left[\frac{\sum_{i=1}^{n} \delta_i(t_k) d_i(t_k)}{\sum_{i=1}^{n} \delta_i(t_k)} \right] = \sum_{k=1}^{j} \frac{d.(t_k)}{\delta.(t_k)} = \sum_{k=1}^{j} \bar{d}(t_k) \quad (16.1)$$

for $j = 1, \ldots, m$, where $d.(t_k) = \sum_{i=1}^{n} \delta_i(t_k) d_i(t_k)$, $\delta.(t_k) = \sum_{i=1}^{n} \delta_i(t_k)$, and $\bar{d}(t_k) = d.(t_k)/\delta.(t_k)$.

Note that $d.(t_k)$ is the total number of system recurrences at time t_k, $\delta.(t_k)$ is the size of the risk set at t_k, and $\bar{d}(t_k)$ is the average number of recurrences per system at t_k (or proportion of systems with recurrences if individual systems have no more than one recurrence at a point time). Thus the estimator of the MCF is obtained by accumulating the mean number (across systems) of recurrences per system in each time interval. □

Like the nonparametric estimate of a cdf (see Chapter 3), the estimate $\widehat{\mu}(t)$ is a step function, with jumps at recurrence times, but constant between the recurrence times. To provide better visual perceptions of shape, one might plot $\widehat{\mu}(t)$ as a piecewise linear function.

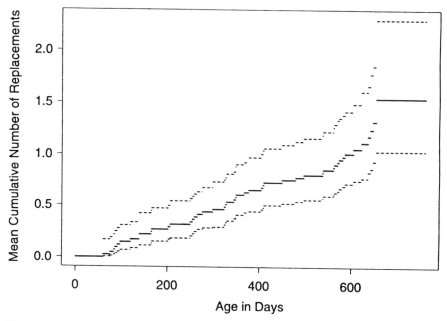

Figure 16.2. Estimate of the mean cumulative number of valve seat replacements for all 41 engines and a set of pointwise approximate 95% confidence intervals.

Example 16.3 MCF Estimate for the Valve Seat Replacements. Figure 16.2 shows the estimate of the valve seat MCF as a function of engine age in days. The estimate increases sharply between 620 and 650 days, but it is important to recognize that this part of the estimate is based on only a small number of systems (i.e., 10 operating at 650 hours). The estimate is flat after 653 hours, but this is largely because there were only two systems being observed between 667 and 759 hours and neither failed during this period. The uncertainty in the estimate for longer times, due to the small number of systems, is reflected in the width of the confidence intervals (to be explained in Section 16.2.3). □

16.2.3 Standard Errors and Nonparametric Confidence Intervals for the MCF

For a random sample of $n \geq 2$ cumulative functions (from a finite or infinite population of systems), there is a simple formula for the true $\text{Var}[\widehat{\mu}(t_j)]$. A corresponding estimate of this variance can be computed from the sample data.

Let $d(t_k)$ denote the random number of recurrences at t_k for a cumulative function drawn at random from the population of cumulative functions. Then, as shown in Nelson (1995a), the true variance of $\widehat{\mu}(t_j)$ for a large population of cumulative functions can be expressed as

NONPARAMETRIC ESTIMATION OF THE MCF

$$\text{Var}[\widehat{\mu}(t_j)] = \sum_{k=1}^{j} \text{Var}[\bar{d}(t_k)] + 2 \sum_{k=1}^{j-1} \sum_{v=k+1}^{j} \text{Cov}[\bar{d}(t_k), \bar{d}(t_v)]$$

$$= \sum_{k=1}^{j} \frac{\text{Var}[d(t_k)]}{\delta.(t_k)} + 2 \sum_{k=1}^{j-1} \sum_{v=k+1}^{j} \frac{\text{Cov}[d(t_k), d(t_v)]}{\delta.(t_k)}. \quad (16.2)$$

To estimate $\text{Var}[d(t_k)]$, we use the assumption that $d_i(t_k)$, $i = 1, \ldots, n$, is a random sample from the population of $d(t_k)$ values. Moment estimators for the variances and covariances on the right-hand side of (16.2) are

$$\widehat{\text{Var}}[d(t_k)] = \sum_{i=1}^{n} \frac{\delta_i(t_k)}{\delta.(t_k)} [d_i(t_k) - \bar{d}(t_k)]^2, \quad (16.3)$$

$$\widehat{\text{Cov}}[d(t_k), d(t_v)] = \sum_{i=1}^{n} \frac{\delta_i(t_v)}{\delta.(t_v)} [d_i(t_v) - \bar{d}(t_v)] d_i(t_k) \quad (16.4)$$

for $t_v > t_k$. Substituting these into (16.2) and simplifying gives

$$\widehat{\text{Var}}[\widehat{\mu}(t_j)] = \sum_{k=1}^{j} \frac{\widehat{\text{Var}}[d(t_k)]}{\delta.(t_k)} + 2 \sum_{k=1}^{j-1} \sum_{v=k+1}^{j} \frac{\widehat{\text{Cov}}[d(t_k), d(t_v)]}{\delta.(t_k)}$$

$$= \sum_{i=1}^{n} \left\{ \sum_{k=1}^{j} \frac{\delta_i(t_k)}{\delta.(t_k)} [d_i(t_k) - \bar{d}(t_k)] \right\}^2. \quad (16.5)$$

Nelson (1995a) presented an unbiased estimator of $\text{Var}[\widehat{\mu}(t_j)]$. We use the estimator in (16.5), suggested by Lawless and Nadeau (1995), because it is always greater than or equal to zero.

Example 16.4 *Computation of the MCF Estimate and its Variance for Simulated Data.* To illustrate the computations of $\widehat{\mu}(t_j)$ and $\widehat{\text{Var}}[\widehat{\mu}(t_j)]$, we use the simple 3-system simulated data shown in Table 16.2. Suppose that the three cumulative functions are a random sample from a large population of cumulative functions. The unique system repair times are $t_1 = 1$, $t_2 = 5$, $t_3 = 8$, and $t_4 = 16$ days. Table 16.3 summarizes the computations for $\widehat{\mu}(t_j)$. Substituting results of Table 16.3

Table 16.2. Simulated System Repair Times

System Number	System Age at Time of Repair	System Age at End of Observation
1	5, 8	12
2		16
3	1, 8, 16	20

Table 16.3. Sample MCF Computations for Simulated System Repair Times

j	t_j	δ_1	δ_2	δ_3	d_1	d_2	d_3	$\delta.$	$d.$	$\bar{d}$	$\widehat{\mu}(t_j)$
1	1	1	1	1	0	0	1	3	1	1/3	1/3
2	5	1	1	1	1	0	0	3	1	1/3	2/3
3	8	1	1	1	1	0	1	3	2	2/3	4/3
4	16	0	1	1	0	0	1	2	1	1/2	11/6

into (16.5) gives

$$\widehat{\text{Var}}[\widehat{\mu}(t_1)] = [(1/3)\times(0-1/3)]^2 + [(1/3)\times(0-1/3)]^2 + [(1/3)\times(1-1/3)]^2 = 6/81.$$

Similar computations yield $\widehat{\text{Var}}[\widehat{\mu}(t_2)] = 6/81 = .0741$, $\widehat{\text{Var}}[\widehat{\mu}(t_3)] = 24/81 = .296$, and $\widehat{\text{Var}}[\widehat{\mu}(t_4)] = 163/216 = .755$. □

Pointwise normal-approximation confidence intervals for the population MCF at a specified time t can be computed following the general approach used in Chapters 7 and 8. In particular, a normal-approximation $100(1-\alpha)\%$ confidence interval based on $Z_{\widehat{\mu}(t)} = [\widehat{\mu}(t) - \mu(t)]/\widehat{\text{se}}_{\widehat{\mu}(t)} \sim \text{NOR}(0, 1)$ is

$$[\underline{\mu}(t),\ \widetilde{\mu}(t)] = \widehat{\mu}(t) \pm z_{(1-\alpha/2)}\widehat{\text{se}}_{\widehat{\mu}(t)}, \tag{16.6}$$

where $\widehat{\text{se}}_{\widehat{\mu}(t)} = \sqrt{\widehat{\text{Var}}[\widehat{\mu}(t)]}$. When $\mu(t)$ is positive (which is common in applications), an alternative interval based on $Z_{\log[\widehat{\mu}(t)]} \sim \text{NOR}(0, 1)$ is

$$[\underline{\mu}(t),\ \widetilde{\mu}(t)] = [\widehat{\mu}(t)/w,\ \widehat{\mu}(t) \times w], \tag{16.7}$$

where $w = \exp[z_{(1-\alpha/2)}\widehat{\text{se}}_{\widehat{\mu}(t)}/\widehat{\mu}(t)]$. Intervals constructed using (16.7) will always have positive endpoints and, for some positive processes, can be expected to have coverage probabilities closer to the nominal confidence level.

Example 16.5 MCF Estimate for the Cylinder Replacements. Cylinders in a type of diesel engine can develop leaks or have low compression for some other reason. Each engine has 16 cylinders. Cylinders are inspected at times of convenience, along with other usual engine maintenance operations. Faulty cylinders are replaced by a rebuilt cylinder. More than one cylinder could be replaced at an inspection. Nelson and Doganaksoy (1989) present data on replacement times for 120 engines. We take these engines to be a sample from a larger population of engines. Management needed to know if the company should perform preventive replacement of cylinders before they develop low compression.

Figure 16.3 displays cylinder replacement times for a subset of 31 of the engines. Except for one outlying replacement at 568 days of service, no replacements occurred until after 847 days. Figure 16.4 shows the MCF plot for the cylinder replacements for all 120 engines. The estimate is close to 0 until about 800 days, after which the

NONPARAMETRIC ESTIMATION OF THE MCF

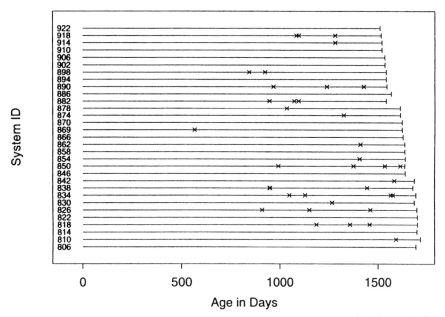

Figure 16.3. Cylinder replacement event plot showing replacement times and period of observation for a subset of 31 observed engines.

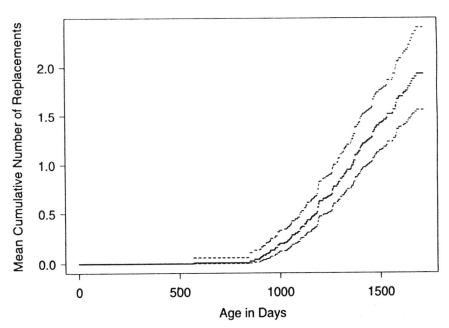

Figure 16.4. MCF estimate for the cylinder replacements for the 120 engines and a set of pointwise approximate 95% confidence intervals.

MCF estimate increases rapidly. Interestingly, the sample MCF is nearly linear after about 1200 days. A possible explanation for this behavior in the sample MCF will be given in Section 16.4.5. □

Finite Population Correction
When the number of cumulative functions sampled is more than 5% or 10% of the population, finite population methods should be used for estimating standard errors. In this case, the following should be substituted into (16.2):

$$\widehat{\text{Var}}[d(t_k)] = \left[1 - \frac{\delta.(t_k)}{N}\right] \sum_{i=1}^{n} \frac{\delta_i(t_k)}{\delta.(t_k)} \left[d_i(t_k) - \bar{d}(t_k)\right]^2,$$

$$\widehat{\text{Cov}}[d(t_k), d(t_v)] = \left[1 - \frac{\delta.(t_v)}{N}\right] \sum_{i=1}^{n} \frac{\delta_i(t_v)}{\delta.(t_v)} \left[d_i(t_v) - \bar{d}(t_v)\right] d_i(t_k),$$

where N is the total number of cumulative functions in the population of interest. The factors $[1 - \delta.(t_v)/N]$ are known as finite population correction factors.

Nonparametric Estimation with a Single System
When there is a single system the point estimate $\hat{\mu}(t)$ is the number of system recurrences up to t. Due to the limited information (a sample of size one at each recurrence time), the quantities (16.2) and (16.3) cannot be computed for single systems.

16.2.4 Adequacy of Normal-Approximation Confidence Intervals

The adequacy of the normal-approximation confidence interval procedures in (16.6) and (16.7) depends on the number of sample cumulative functions (or sample systems) at risk to failure *and* on the shape of the distribution of the cumulative function levels at the point in time where the interval is to be constructed.

As mentioned in Sections 7.5 and 8.4.4, the normal-approximation intervals like those in (16.6) can be improved by using $t_{(p;v)}$ instead of $z_{(p)}$. When the number of sample systems at risk is small (say, less than 30), using $t_{(p;v)}$ instead of $z_{(p)}$ can provide important improvements in confidence interval accuracy. If the cumulative function at a point in time has a normal distribution and if all units are still under observation at that point, then using $t_{(p;v)}$ instead of $z_{(p)}$ and substituting $[n/(n-1)]^{1/2} \text{se}_{\hat{\mu}}$ for $\text{se}_{\hat{\mu}}$ in (16.6) provides an exact interval for two or more systems. For a counting process like the Poisson processes described in Section 16.4, the distribution of the level of a sample cumulative function at a point in time can be described adequately by a normal distribution when the expected level of the cumulative function is 30 or more at that point in time.

Example 16.6 Maintenance Costs for an Earth-Moving Machine. A construction company owns 23 large earth-moving machines that were put into service over a period of time. At intervals of approximately 300–400 hours of operation, these

NONPARAMETRIC ESTIMATION OF THE MCF

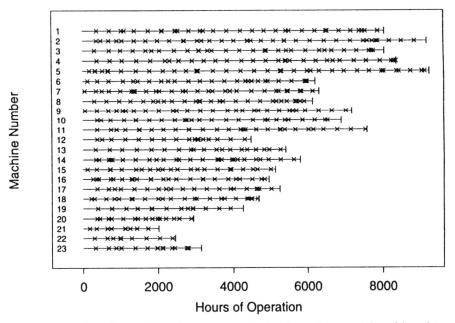

Figure 16.5. Earth-moving machine maintenance event plot showing machine age at time of the maintenance action for the 23 machines.

machines receive scheduled preventive maintenance. Unscheduled maintenance actions are also needed from time to time. Additionally, these machines require major (and costly) overhaul, usually every 2000–3000 hours of operation. The event plot in Figure 16.5 shows times of the maintenance actions for the 23 machines. Here time is taken to be the machine's age in hours operated. The cost of each maintenance action was also recorded in terms of the number of hours of labor that were required for the maintenance work. Figure 16.6 gives the estimate of the mean cumulative number of hours of labor for the earth movers as a function of hours of operation and corresponding pointwise normal-approximation confidence intervals. The periodicity of the early scheduled maintenance actions can be seen in the first 1500 hours or so. After that, the randomness in the intervals averages out over time, reducing the amplitude of the periodicity. The slope of the MCF is a bit larger over some intervals of time. These are intervals in which more of the machines were required to have major overhauls (after 3000 and 5000 hours). The confidence intervals have no real meaning relative to the 23 machines owned by the company. If, on the other hand, the 23 machines were being viewed as a random sample from a much larger population (e.g., of other similar machines in other parts of the company or of similar machines to be purchased in the future), the confidence intervals would quantify the uncertainty in the mean of that larger population. The intervals are wider later in time because there are fewer machines with that much exposure time. □

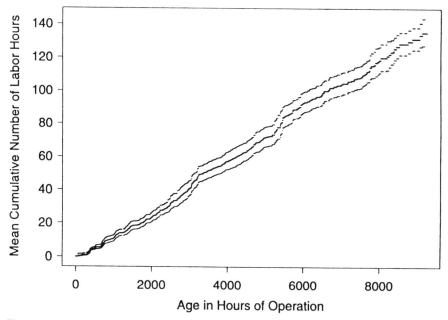

Figure 16.6. Mean cumulative number of hours required for maintenance for earth-moving machines and approximate 95% confidence intervals, as a function of machine age.

16.3 NONPARAMETRIC COMPARISON OF TWO SAMPLES OF RECURRENCE DATA

Decisions often need to be made on the basis of a comparison between two manufacturers, product designs, environments, and so on. This section describes methods for comparing recurrence-data samples from two different groups of systems.

Example 16.7 Replacement Times of Locomotive Braking Grids. A particular type of locomotive has six braking grids. Appendix Table C.9 shows data on locomotive ages when braking grids were replaced and the largest observed age for each locomotive. The data are from Doganaksoy and Nelson (1991). A comparison between two different production batches of the braking grids is desired. □

Suppose that there are two populations or processes with mean cumulative functions $\mu_1(t)$ and $\mu_2(t)$, respectively. Let $\Delta_\mu(t) = \mu_1(t) - \mu_2(t)$ represent the difference in the mean cumulative functions at time t. Based on independent samples from the two populations, a nonparametric estimator of $\Delta_\mu(t)$ is

$$\widehat{\Delta}_\mu(t) = \widehat{\mu}_1(t) - \widehat{\mu}_2(t).$$

If $\widehat{\mu}_1(t)$ and $\widehat{\mu}_2(t)$ are independent, an estimate of $\text{Var}[\widehat{\Delta}_\mu(t)]$ is

$$\widehat{\text{Var}}[\widehat{\Delta}_\mu(t)] = \widehat{\text{Var}}[\widehat{\mu}_1(t)] + \widehat{\text{Var}}[\widehat{\mu}_2(t)]$$

and $\widehat{\text{se}}_{\widehat{\Delta}_\mu} = \sqrt{\widehat{\text{Var}}[\widehat{\Delta}_\mu(t)]}$. An approximate $100(1 - \alpha)\%$ confidence interval for $\Delta_\mu(t)$ based on $Z_{\widehat{\Delta}_\mu} = [\widehat{\Delta}_\mu(t) - \Delta_\mu(t)]/\widehat{\text{se}}_{\widehat{\Delta}_\mu} \dot\sim \text{NOR}(0, 1)$ is

$$\left[\utilde{\Delta}_\mu, \widetilde{\Delta}_\mu\right] = \left[\widehat{\Delta}_\mu - z_{(1-\alpha/2)}\widehat{\text{se}}_{\widehat{\Delta}_\mu}, \; \widehat{\Delta}_\mu + z_{(1-\alpha/2)}\widehat{\text{se}}_{\widehat{\Delta}_\mu}\right].$$

Example 16.8 Comparison of Two Production Batches of the Locomotive Braking Grids. Figure 16.7 shows the sample MCFs for the braking grids from production Batches 1 and 2. This figure shows that the sample MCF for Batch 2 is greater than that for Batch 1. Figure 16.8 plots the nonparametric estimate and confidence intervals for the population $\Delta_\mu(t)$. This figure indicates that there is a statistically significant difference between the MCFs over almost the entire span of locomotive age. □

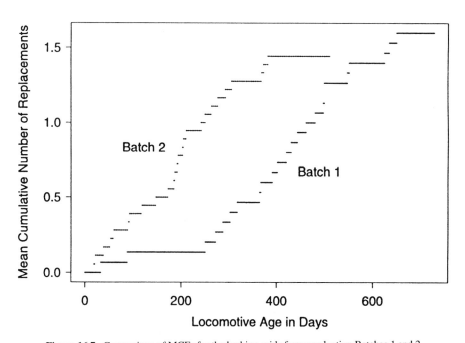

Figure 16.7. Comparison of MCFs for the braking grids from production Batches 1 and 2.

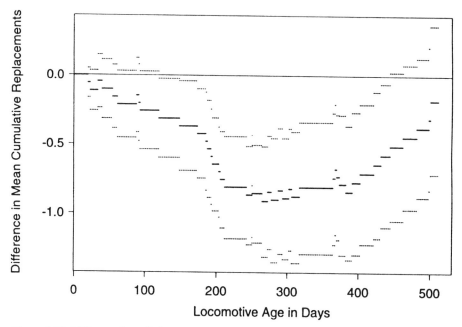

Figure 16.8. Difference $\hat{\mu}_1 - \hat{\mu}_2$ between sample MCFs for production Batches 1 and 2 and a set of pointwise approximate 95% confidence intervals for the population difference.

16.4 PARAMETRIC MODELS FOR RECURRENCE DATA

Parametric point-process models are also useful for recurrence data. The most commonly used models for recurrence data are Poisson processes (homogeneous and nonhomogeneous), renewal processes, and superimposed versions of these processes. The most common application is for monitoring and modeling recurrences, such as repairs, from a *single system*.

16.4.1 Poisson Process

The Poisson process is a simple parametric model that is commonly used for the analysis of certain kinds of recurrence data. An integer-valued point process on $[0, \infty)$ is said to be a Poisson process if it satisfies the following three conditions:

- $N(0) = 0$.
- The numbers of recurrences in disjoint time intervals are statistically independent. A process with this property is said to have "independent increments."
- The process recurrence rate $\nu(t)$ is positive and $\mu(a,b) = \mathrm{E}[N(a,b)] = \int_a^b \nu(u)\,du < \infty$, when $0 \leq a < b < \infty$.

PARAMETRIC MODELS FOR RECURRENCE DATA

It follows, for a Poisson process, that $N(a, b)$, has a Poisson distribution with pdf

$$\Pr[N(a,b) = d] = \frac{[\mu(a,b)]^d}{d!} \exp[-\mu(a,b)], \ d = 0, 1, 2, \ldots.$$

16.4.2 Homogeneous Poisson Process

A homogeneous Poisson process (HPP) is a Poisson process with a *constant* recurrence rate, say, $\nu(t) = 1/\theta$. In this case:

- $N(a, b)$ has a Poisson distribution with parameter $\mu(a, b) = (b - a)/\theta$.
- The expected number of recurrences in $(a, b]$ is $\mu(a, b)$. Equivalently, the expected number of recurrences per unit time over $(a, b]$ is constant and equal to $1/\theta$. This property is called "stationary increments."
- The interrecurrence times (the times between recurrences), $\tau_j = T_j - T_{j-1}$, are independent and identically distributed (iid), each with an EXP(θ) distribution. This follows directly from the relationship

$$\Pr(\tau_j > t) = \Pr\left[N(T_{j-1}, T_{j-1} + t) = 0\right] = \exp(-t/\theta).$$

Thus the steady-state mean time between recurrences for an HPP is equal to θ. For a failure process one would say that the mean time between failures is MTBF = θ.
- The time $T_k = \tau_1 + \cdots + \tau_k$ to the kth recurrence has a GAM(θ, k) distribution.

16.4.3 Nonhomogeneous Poisson Process

A nonhomogeneous Poisson process (NHPP) model is a Poisson process model with a nonconstant recurrence rate $\nu(t)$. In this case the interrecurrence times are neither independent nor identically distributed. The expected number of recurrences per unit time over $(a, b]$ is

$$\frac{\mu(a,b)}{b-a} = \frac{1}{b-a} \int_a^b \nu(t)\, dt.$$

An NHPP model is often specified in terms of the recurrence rate $\nu(t)$. To specify an NHPP model we use $\nu(t) = \nu(t; \boldsymbol{\theta})$, a function of an unknown vector of parameters $\boldsymbol{\theta}$. For example, the power-model recurrence rate is

$$\nu(t; \beta, \eta) = \frac{\beta}{\eta}\left(\frac{t}{\eta}\right)^{\beta-1}, \quad \beta > 0, \eta > 0.$$

The corresponding mean cumulative number of recurrences over $(0, t]$ is $\mu(t; \beta, \eta) = (t/\eta)^\beta$. When $\beta = 1$, this model reduces to the HPP model. The loglinear model recurrence rate is $\nu(t; \gamma_0, \gamma_1) = \exp(\gamma_0 + \gamma_1 t)$. The corresponding mean cumulative number of recurrences over $(0, t]$ is $\mu(t; \gamma_0, \gamma_1) = [\exp(\gamma_0)][\exp(\gamma_1 t) - 1]/\gamma_1$. When $\gamma_1 = 0$, $\nu(t; \gamma_0, 0) = \exp(\gamma_0)$, which is an HPP.

16.4.4 Renewal Processes

A sequence of system recurrences at system ages $T_1, T_2, \ldots$ is a renewal process if the interrecurrence times $\tau_j = T_j - T_{j-1}$, $j = 1, 2, \ldots$ ($T_0 = 0$), are iid. The MCF for a renewal process is also known as the "renewal function." If a renewal process provides an adequate model for interrecurrence times, one can use the single distribution statistical techniques, as described in Chapters 3–11, to display, model, and draw conclusions from the data. Note that an HPP is a renewal process [and that interrecurrence times $\tau_j \sim \text{EXP}(\theta)$ with $\theta = 1/\nu$] but the NHPP is not.

Before using a renewal process model, it is important to check for departures from the model such as trend and nonindependence of interrecurrence times (note that, in general, independent increments, as defined in Section 16.4.2, and independent interrecurrence times are not the same). The methods in Section 16.5 can be used for this purpose. If trend and nonindependence tests suggest a renewal process, then one might use methods in the earlier chapters of this book to describe the distribution of interrecurrence times (e.g., times between repairs, failures).

Renewal process characteristics that are typically of interest include:

- The distribution of the τ_j values.
- Thus the steady-state mean time between recurrences for a renewal process is $E(\tau)$. Again, for a failure renewal process one would say that MTBF $= E(\tau)$.
- The distribution of the time until the kth recurrence for a system, $k = 1, 2, \ldots$.
- The recurrence (or renewal) rate.
- The number of recurrences that will be observed in a given future time interval.

16.4.5 Superimposed Renewal Processes

The point process arising from the aggregation of renewals from a group of n independent renewal processes operating simultaneously is known as a superimposed renewal process (SRP). Unless the individual renewal processes are HPP, an SRP is not a renewal process. Drenick's theorem (Drenick, 1960) says, however, that when the number of systems n is large and the systems have run long enough to eliminate transients, an SRP behaves as an HPP. This result is rather like a central limit theorem for renewal processes. This result is sometimes used to justify the use of the exponential distribution to model interrecurrence times in a large population of systems. This result, for example, provides a possible explanation for why the MCF for the cylinder replacement data in Example 16.5 is nearly linear after about 1200 days.

Depending on the shape of the underlying distributions in the individual superimposed renewal processes, large samples and long times may be needed for the HPP approximation to be adequate (see Blumenthal, Greenwood, Herbach 1973, 1976). In practice, it is straightforward to check the adequacy of the approximation by using simulation.

16.5 TOOLS FOR CHECKING POINT-PROCESS ASSUMPTIONS

This section describes graphical and analytical tools for checking point-process model assumptions.

16.5.1 Tests for Recurrence Rate Trend

The simplest plot for recurrence data of a single system shows the cumulative number of system recurrences versus time (a special case of an MCF used with multiple systems). Nonlinearity in this plot indicates that the interrecurrence times are not identically distributed. For Poisson processes, this implies a nonconstant recurrence rate. An HPP should have an MCF plot that is approximately linear (but a linear MCF alone does not imply an HPP). A plot of interrecurrence times versus system age or a "time series plot" showing interrecurrence times versus recurrence number will allow discovery of trends or cycles that would suggest that the interrecurrence times are not identically distributed.

Several formal tests for trend are available. The "Military Handbook" test (so known because it appears in MIL-HDBK-189) tests for $\beta = 1$ (implying HPP and thus no trend) in the power NHPP model. The statistic

$$X^2_{\text{MHB}} = -2 \sum_{j=1}^{r} \log(t_j/t_a)$$

approximately has a $\chi^2_{(2r)}$ distribution under the HPP model. Thus values of X^2_{MHB} greater than $\chi^2_{(1-\alpha;2r)}$ provide evidence of a nonconstant recurrence rate $\nu(t)$ at the $100\alpha\%$ level of significance. This is a powerful tool for testing HPP versus NHPP with a power-model $\nu(t)$.

The well-known Laplace test has a similar basis for testing for trend in the loglinear NHPP model. In this case if the underlying process is HPP ($\gamma_1 = 0$), the test statistic

$$Z_{\text{LP}} = \frac{\sum_{j=1}^{r} t_j/t_a - r/2}{\sqrt{r/12}}$$

has, approximately, a NOR(0, 1) distribution. Thus values of Z_{LP} in excess of $z_{(1-\alpha/2)}$ provide evidence of a nonconstant $\nu(t)$ at the $100\alpha\%$ level of significance. This is a powerful tool for testing HPP versus NHPP with a loglinear $\nu(t)$.

Both the Military Handbook test and the Laplace test can give misleading conclusions for situations where there is no trend but the underlying process is a renewal process other than HPP. The Lewis–Robinson test for trend uses

$$Z_{\text{LR}} = Z_{\text{LP}} \times \frac{\bar{\tau}}{S_\tau}, \tag{16.8}$$

where $\bar{\tau}$ and S_τ are, respectively, the sample mean and standard deviation of the interrecurrence times. The fraction on the right-hand side of (16.8) is the reciprocal of the sample coefficient of variation (the population coefficient of variation is defined in

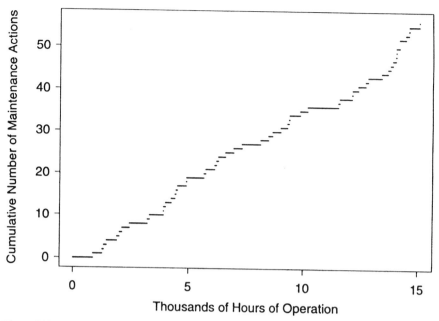

Figure 16.9. Cumulative number of unscheduled maintenance actions versus age in operating hours for a U.S.S. *Grampus* diesel engine.

Section 4.2). In large samples, Z_{LR} follows approximately a NOR(0, 1) distribution if the underlying process is a renewal process (renewal processes, in general, have no trend). The statistic Z_{LR} was derived from heuristic arguments to allow for nonexponential interrecurrence times by adjusting for a different coefficient of variation (the exponential distribution has a coefficient of variation equal to 1). Results in Lawless and Thiagarajah (1996) indicate that Z_{LR} is preferable to Z_{LP} as a general test of trend in point-process data.

Example 16.9   Initial Graphical Analysis of the Grampus Diesel Engine Data.
Continuing from Example 16.1, Figure 16.9 shows the cumulative number of unscheduled maintenance actions for a U.S.S. *Grampus* diesel engine versus operating hours. The plot is nearly linear, indicating that the recurrence rate $v(t)$ is nearly constant (as in the HPP model). Figure 16.10 shows the times between unscheduled maintenance actions plotted against maintenance action number. This plot indicates that there has not been a discernible trend over time (again consistent with the HPP model) but there is one large outlier (which might, e.g., have been caused by failing to report a single maintenance action). □

16.5.2   Test for Independent Interrecurrence Times

When assessing the adequacy of a renewal process model, it is also necessary to check if the model assumption of independent interrecurrence times is consistent with the

TOOLS FOR CHECKING POINT-PROCESS ASSUMPTIONS   411

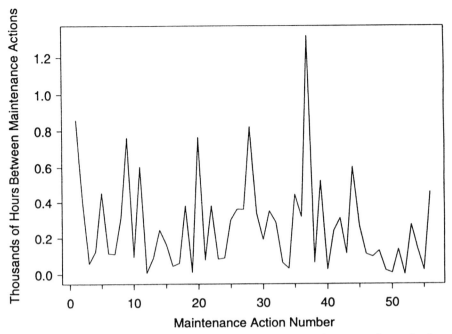

Figure 16.10. Times between unscheduled maintenance actions versus maintenance action number for a U.S.S. *Grampus* diesel engine.

data. To do this we consider the serial correlation in the sequence of interrecurrence times. Plotting interrecurrence times versus lagged interrecurrence times (τ_i versus τ_{i+k}) provides a graphical representation of serial correlation (the correlation between adjacent interrecurrence times). The serial correlation coefficient of lag-k is

$$\rho_k = \frac{\text{Cov}(\tau_j, \tau_{j+k})}{\text{Var}(\tau_j)}.$$

First-order serial correlation ($k = 1$) is typically the most important lag to consider.

If $\tau_1, \ldots, \tau_r$ are observed interrecurrence times, then the sample serial correlation coefficient is

$$\widehat{\rho}_k = \frac{\sum_{j=1}^{r-k}(\tau_j - \bar{\tau})(\tau_{j+k} - \bar{\tau})}{\sqrt{\sum_{j=1}^{r-k}(\tau_j - \bar{\tau})^2 \sum_{j=1}^{r-k}(\tau_{j+k} - \bar{\tau})^2}},$$

where $\bar{\tau} = \sum_{j=1}^{r} \tau_j / r$. When $\rho_k = 0$ and r is large, $\sqrt{r-k} \times \widehat{\rho}_k \overset{\cdot}{\sim} \text{NOR}(0, 1)$. This approximate distribution can be used to assess if ρ_k is different from zero.

Example 16.10 Checking for Independent Times Between Maintenance Actions in the U.S.S. Grampus Diesel Engine Data. Continuing from Examples 16.1 and 16.9, Figure 16.11 plots the times between unscheduled maintenance actions versus lagged times between unscheduled maintenance actions for the U.S.S. *Grampus*

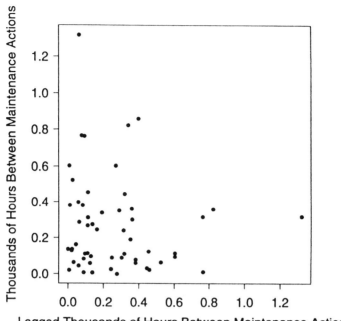

Figure 16.11. Plot of τ_{j+1} versus τ_j for the U.S.S. *Grampus* diesel engine.

diesel engine. Although at first glance there might seem to be some negative correlation, ignoring the one outlying large time between unscheduled maintenance actions (which appears twice in Figure 16.11) suggests that there is no discernible serial correlation. Numerically, including the outlier, $\widehat{\rho}_1 = -.094$ and $\sqrt{r-1} \times \widehat{\rho}_1 = -.70$ (with an approximate *p*-value of .48). □

16.6 MAXIMUM LIKELIHOOD FITTING OF POISSON PROCESS

16.6.1 Poisson Process Likelihood

For a system that has been observed for a period $(0, t_a]$, the data are the number of recurrences $d_1, \ldots, d_m$ in the nonoverlapping intervals $(t_0, t_1], (t_1, t_2], \ldots, (t_{m-1}, t_m]$ (with $t_0 = 0$, $t_m = t_a$). The likelihood for the NHPP model is

$$L(\boldsymbol{\theta}) = \Pr[N(t_0, t_1) = d_1, \ldots, N(t_{m-1}, t_m) = d_m] = \prod_{j=1}^{m} \Pr[N(t_{j-1}, t_j) = d_j]$$

$$= \prod_{j=1}^{m} \frac{[\mu(t_{j-1}, t_j; \boldsymbol{\theta})]^{d_j}}{d_j!} \exp[-\mu(t_{j-1}, t_j; \boldsymbol{\theta})]$$

$$= \prod_{j=1}^{m} \frac{[\mu(t_{j-1}, t_j; \boldsymbol{\theta})]^{d_j}}{d_j!} \times \exp[-\mu(t_0, t_a; \boldsymbol{\theta})].$$

MAXIMUM LIKELIHOOD FITTING OF POISSON PROCESS 413

As the number of intervals m increases and the size of the intervals approaches zero, there are *exact* reported recurrence times at $t_1 \leq \cdots \leq t_r$ (here $r = \sum_{j=1}^{m} d_j$, $t_0 \leq t_1, t_r \leq t_a$), then using a limiting argument it follows that the likelihood in terms of the density approximation is

$$L(\boldsymbol{\theta}) = \prod_{j=1}^{r} \nu(t_j; \boldsymbol{\theta}) \times \exp[-\mu(0, t_a; \boldsymbol{\theta})].$$

16.6.2 Superimposed Poisson Process Likelihood

Suppose that data are available from n independent systems with the same intensity function $\nu(t)$ and system i is observed in the interval $(0, t_{a_i}]$, $i = 1, \ldots, n$, and the system i recurrence times are denoted by $t_{i1}, \ldots, t_{ir_i}$. Then the NHPP likelihood is simply the product of the individual system likelihoods

$$L(\boldsymbol{\theta}) = \prod_{i=1}^{n}\prod_{j=1}^{r_i} \nu(t_{ij}; \boldsymbol{\theta}) \times \exp\left[-\sum_{i=1}^{n} \mu(0, t_{a_i}; \boldsymbol{\theta})\right].$$

The assumption that all systems have the same $\nu(t)$ is a strong assumption that was not required for the nonparametric estimation method in Section 16.2. This assumption will often be inappropriate in practical applications. Generalizations (e.g., use of explanatory variables to account for system-to-system differences) are possible but are beyond the scope of this book.

16.6.3 ML Estimation for the Power NHPP

For the NHPP model with a power $\nu(t)$, the single-system likelihood for exact recurrence times is

$$L(\beta, \eta) = \left(\frac{\beta}{\eta^\beta}\right)^r \times \prod_{j=1}^{r} t_j^{\beta-1} \times \exp[-\mu(t_a; \beta, \eta)].$$

The ML estimates of the parameters are $\widehat{\beta} = r/\sum_{j=1}^{r} \log(t_a/t_j)$ and $\widehat{\eta} = t_a/r^{1/\widehat{\beta}}$. The relative likelihood is

$$R(\beta, \eta) = \left(\frac{\beta}{\widehat{\beta}} \times \frac{\widehat{\eta}^{\widehat{\beta}}}{\eta^\beta}\right)^r \times \left(\prod_{j=1}^{r} t_j\right)^{\beta-\widehat{\beta}} \exp[r - \mu(t_a; \beta, \eta)].$$

16.6.4 ML Estimation for the Loglinear NHPP

With a loglinear $\nu(t)$ and exact recurrence times, the single-system likelihood is

$$L(\gamma_0, \gamma_1) = \exp\left(r\gamma_0 + \gamma_1 \sum_{j=1}^{r} t_j\right) \times \exp[-\mu(t_a; \gamma_0, \gamma_1)].$$

The ML estimates $\widehat{\gamma}_0$ and $\widehat{\gamma}_1$ are obtained by solving

$$\sum_{j=1}^{r} t_j + \frac{r}{\gamma_1} - \frac{r t_a \exp(\gamma_1 t_a)}{\exp(\gamma_1 t_a) - 1} = 0 \quad \text{and} \quad \gamma_0 = \log\left(\frac{r \gamma_1}{\exp(t_a \gamma_1) - 1}\right).$$

The relative likelihood is

$$R(\gamma_0, \gamma_1) = \exp\left[r(\gamma_0 - \widehat{\gamma}_0) + (\gamma_1 - \widehat{\gamma}_1) \sum_{j=1}^{r} t_j\right] \times \exp\{r - \mu(t_a; \gamma_0, \gamma_1)\}.$$

Example 16.11   U.S.S. Grampus Diesel Engine Data.   Figure 16.12 shows the fitted $\mu(t)$ for both the power and loglinear NHPP models. There is very little difference between the two NHPP models and both seem to fit the data very well. For the power NHPP model, $\widehat{\beta} = 1.22$ and $\widehat{\eta} = .553$. For the loglinear NHPP model, $\widehat{\gamma}_0 = 1.01$ and $\widehat{\gamma}_1 = .0377$. The Lewis–Robinson test gave $Z_{LR} = 1.02$ with p-value $= .21$. The MIL-HDBK-189 test gave $X^2_{MHB} = 92$ with p-value $= .08$. These results are consistent with a renewal process. Exponential and Weibull probability plots (details requested in Exercise 16.7) strongly suggested that the times between unscheduled maintenance actions could be described by an exponential distribution. Thus these data seem to be consistent with the HPP model.   □

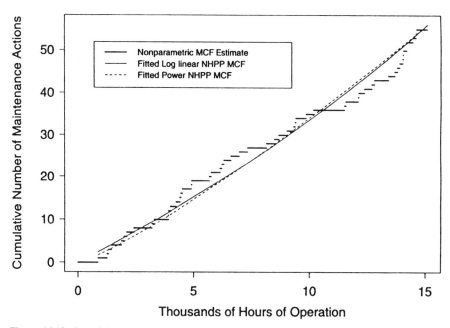

Figure 16.12. Cumulative number of unscheduled maintenance actions for a U.S.S. *Grampus* diesel engine versus operating hours with fitted power and loglinear NHPP models.

MAXIMUM LIKELIHOOD FITTING OF POISSON PROCESS

Table 16.4. Engine Age at Time of Unscheduled Maintenance Actions (in Thousands of Hours of Operation) for the U.S.S. *Halfbeak* Number 4 Main Propulsion Diesel Engine

1.382	2.990	4.124	6.827	7.472	7.567	8.845	9.450
9.794	10.848	11.993	12.300	15.413	16.497	17.352	17.632
18.122	19.067	19.172	19.299	19.360	19.686	19.940	19.944
20.121	20.132	20.431	20.525	21.057	21.061	21.309	21.310
21.378	21.391	21.456	21.461	21.603	21.658	21.688	21.750
21.815	21.820	21.822	21.888	21.930	21.943	21.946	22.181
22.311	22.634	22.635	22.669	22.691	22.846	22.947	23.149
23.305	23.491	23.526	23.774	23.791	23.822	24.006	24.286
25.000	25.010	25.048	25.268	25.400	25.500	25.518	

Data from Ascher and Feingold (1984, page 75). Reprinted with permission, copyright Marcel Dekker.

Example 16.12 U.S.S. Halfbeak Diesel Engine Data. Table 16.4 gives times of unscheduled maintenance actions for the U.S.S. *Halfbeak* number 4 main propulsion diesel engine over 25,518 operating hours. As with the *Grampus* data, questions to be answered were: (1) Was the system deteriorating (i.e., are unscheduled maintenance actions occurring more rapidly as the system ages)? and (2) Can unscheduled maintenance actions be modeled by an HPP? Figure 16.13 shows the cumulative number of unscheduled maintenance actions versus operating hours with fitted $\mu(t)$ for both the power and loglinear NHPP models. Both NHPP models roughly follow

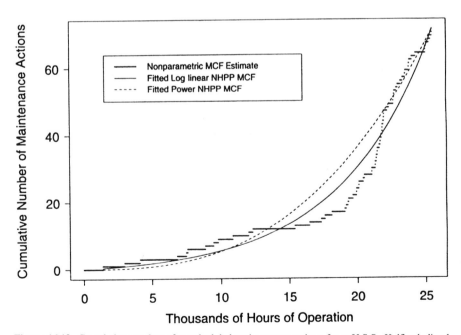

Figure 16.13. Cumulative number of unscheduled maintenance actions for a U.S.S. *Halfbeak* diesel engine versus operating hours with fitted power and loglinear NHPP models.

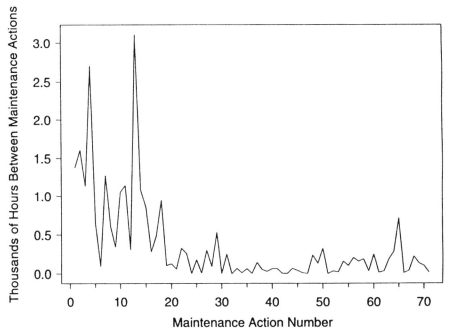

Figure 16.14. Times between unscheduled maintenance actions for a U.S.S. *Halfbeak* diesel engine versus maintenance action number.

the pattern in the data (nonparametric estimate), but the loglinear $\nu(t)$ seems to fit a little better than the power $\nu(t)$. For the power model, $\hat{\beta} = 2.76$ and $\hat{\eta} = 5.45$. For the loglinear model, $\hat{\gamma}_0 = -1.43$ and $\hat{\gamma}_1 = .149$. Figure 16.14 shows a strong downward shift in the times between unscheduled maintenance actions after maintenance action number 17, suggesting a change in the distribution of times between maintenance actions. Confirming this, the Lewis–Robinson test gave $Z_{LR} = 4.70$ with p-value $= 2.5 \times 10^{-6}$. The MIL-HDBK-189 test gave $X^2_{MHB} = 51$ with p-value ≈ 0. Figure 16.15 plots the times between unscheduled maintenance actions versus lagged times between unscheduled maintenance actions. This plot indicates a strong positive correlation. Numerically, $\hat{\rho}_1 = .43$ and $\sqrt{r-1} \times \hat{\rho}_1 = 3.58$. In contrast to the *Grampus* data, there is strong evidence against the HPP model for these data. □

16.6.5 Confidence Intervals for Parameters and Functions of Parameters

Confidence intervals for NHPP parameters or functions of the parameters can be computed using the general ideas developed in Chapters 7 and 8.

16.6.6 Prediction of Future Recurrences with a Poisson Process

The expected number of recurrences in an interval $[a, b]$ is $\int_a^b \nu(u, \boldsymbol{\theta}) \, du$. The corresponding ML point prediction is $\int_a^b \nu(u, \hat{\boldsymbol{\theta}}) \, du$. A point prediction for the future

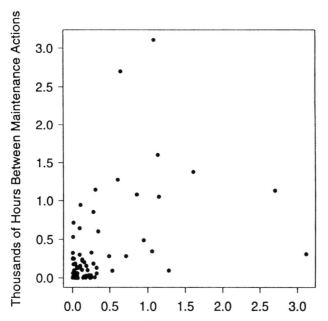

Figure 16.15. Times between unscheduled maintenance actions versus lagged times between unscheduled maintenance actions for U.S.S. *Halfbeak* diesel engine.

number of recurrences using the power NHPP model is

$$\int_a^b \nu(u, \widehat{\boldsymbol{\theta}}) \, du = \left(\frac{1}{\widehat{\eta}}\right)^{\widehat{\beta}} (b^{\widehat{\beta}} - a^{\widehat{\beta}}).$$

Similarly, a point prediction for the future number of recurrences using the loglinear NHPP model is

$$\int_a^b \nu(u, \widehat{\boldsymbol{\theta}}) \, du = \frac{\exp(\widehat{\gamma}_0)}{\widehat{\gamma}_1} \left[\exp(\widehat{\gamma}_1 b) - \exp(\widehat{\gamma}_1 a)\right].$$

Prediction intervals could be computed using bootstrap/simulation methods like those used in Chapter 12.

16.7 GENERATING PSEUDORANDOM REALIZATIONS FROM AN NHPP PROCESS

This section shows how to generate a pseudorandom realization $T_1, T_2, \ldots$ from an NHPP process. Such realizations are useful for checking the adequacy of large-sample approximations and for implementing bootstrap methods like those described in Chapter 9.

16.7.1 General Approach

Using the properties of the NHPP model, it is easy to show (see Exercise 16.14) that for a monotone increasing $\mu(t)$, the random variables $\mu(T_{i-1}, T_i)$, $i = 1, \ldots$, are iid, each with an EXP(1) (where $T_0 = 0$). Suppose that U_i, $i = 1, \ldots, r$, is a pseudorandom sample from a UNIF(0, 1). Then using $\mu(T_{i-1}, T_i) = \mu(T_i) - \mu(T_{i-1})$ and solving sequentially for $T_1, T_2, \ldots$ gives the equations

$$\mu(T_1) = -\log(U_1),$$
$$\mu(T_2) - \mu(T_1) = -\log(U_2),$$
$$\vdots$$
$$\mu(T_r) - \mu(T_{r-1}) = -\log(U_r).$$

Then solving for $T_1, \ldots, T_r$ gives r recurrence times from the NHPP with recurrence rate $\nu(t)$ [or MCF $\mu(t)$]. If one wants a realization in an interval $(0, t_a]$, then r is random and the sequential process is stopped when $T_i > t_a$, which implies that the number of recurrences is $r = i - 1$.

The general solution can be expressed as

$$T_j = \mu^{-1}\left[-\sum_{i=1}^{j} \log(U_i)\right], \quad j = 1, \ldots.$$

Sometimes it is more convenient to express the general solution recursively as

$$T_j = \mu^{-1}[\mu(T_{j-1}) - \log(U_j)], \quad j = 1, \ldots,$$

where $T_0 = 0$. The following subsections give explicit formulas for two cases of interest.

16.7.2 NHPP with a Power Recurrence Rate

In this case $\mu(t) = (t/\eta)^\beta$ and $\mu^{-1}(s) = \eta \times s^{(1/\beta)}$. Then using the general solution, one gets

$$T_j = \eta \times \left[-\sum_{i=1}^{j} \log(U_i)\right]^{1/\beta} = \left[T_{j-1}^\beta - \eta^\beta \times \log(U_j)\right]^{1/\beta}, \quad j = 1, \ldots.$$

16.7.3 NHPP with a Loglinear Recurrence Rate

In this case $\mu(t) = [\exp(\gamma_0)][\exp(\gamma_1 t) - 1]/\gamma_1$ and

$$\mu^{-1}(s) = (1/\gamma_1)\log[\gamma_1 \exp(-\gamma_0)s + 1].$$

Again using the general solution, one gets

$$T_j = \frac{1}{\gamma_1} \times \log\left[-\gamma_1 \times \exp(-\gamma_0) \times \sum_{i=1}^{j} \log(U_i) + 1\right]$$

$$= \frac{1}{\gamma_1} \times \log[\exp(\gamma_1 T_{j-1}) - \gamma_1 \times \exp(-\gamma_0) \times \log(U_j)], \quad j = 1,\ldots.$$

16.8 SOFTWARE RELIABILITY

State-of-the-art reservation, banking, billing, accounting, and other financial and business systems depend on complicated software systems. Additionally, modern hardware systems of all kinds, from automobiles and televisions to communications networks and spacecraft, contain complicated electronic circuitry. Most of these electronic systems depend heavily on software for important functionality and flexibility. For many systems, software reliability has become an important limiting factor in system reliability. The Institute of Electrical and Electronic Engineers defines software reliability as "the probability that software will not cause a system failure for a specified time under specified conditions." This is very similar to the general definition of reliability given in Section 1.1.1.

Software reliability differs from hardware reliability in at least one important way. Hardware failures can generally be traced to some combination of a physical fault and a physical/chemical degradation that progresses over time, perhaps accelerated by stress, shocks, or other environmental or operating conditions. Software failures, on the other hand, are generally caused by inherent faults (or "bugs") in the software that are usually present all along (although a new fault can be introduced during the process of fixing an existing fault). Actual failure may not occur until a particular set of inputs is used or until a particular system state or level of system load is reached. The state of the software itself does not change without intervention.

Software errors differ in their criticality. Those who work with personal computers know that from time to time the system will stop functioning for reasons that are unknown. The cause is often software related (i.e., would not have occurred if the software had been designed to anticipate the conditions that caused the problem). Restarting the computer and the application will seem to make the problem disappear. Important data in the application being used at the time of the failure may or may not have been lost. Future versions of the operating system or the application software may reduce the probability of such problems. In safety-critical systems (e.g., medical, air-traffic control, or military weapons-control systems) software failures can, of course, have much more serious (e.g., life-threatening) consequences.

For some purposes, statistical methods for software reliability are similar to those used in monitoring a repairable system or another sequence of recurrences. Software reliability data often consist of a sequence of times of failures (or some other specific event of interest) in the operation of the software system. Software reliability data are collected for various reasons, including assessment of the distribution of times between failures, tracking the effect of continuing efforts to find and correct software errors, making decisions on when to release a software product, and assessing the effect of changes to improve the software development process.

Numerous special models have been suggested and developed to model software reliability data. The simplest of these describe the software failure rate as a smooth function of time of the service and other factors, such as system load and amount of testing or use to which the system has been exposed. The models and data analysis methods presented in this chapter are useful for software data analysis. In an attempt to be more mechanistic and to incorporate information from the fix process directly into the software reliability model, many of these models have a parameter corresponding to the number of faults remaining in the system. In some models, the failure rate would be proportional to the number of faults. When a "repair" is made, there is some probability that the fault is fixed and, perhaps, a probability that a new fault is introduced. For more information on software reliability and software reliability models, see Musa, Iannino, and Okumoto (1987), Shooman (1983), Neufelder (1993), Chapter 6 of Pecht (1995), or Azem (1995).

BIBLIOGRAPHIC NOTES

Ascher and Feingold (1984) provide a comprehensive review of the important ideas for modeling a single repairable system. Thompson (1981) outlines and explains important aspects of probabilistic models and metrics relating to reliability of replaceable units and repairable systems. Snyder (1975), Cox and Isham (1980), and Thompson (1988) present theory and methods for general point-process models. Nelson (1988) describes simple graphical methods for the analysis of system repair data. Nelson (1995a) extends these results, showing how to compute confidence limits for the MCF, and references available computer programs. Our presentation of methods for nonparametric estimation depends heavily on the results in these papers, but the formulas given in this chapter are closer to those in Lawless and Nadeau (1995). Lawless and Nadeau (1995) also present models and methods for covariate adjustment and for comparing point processes. Robinson (1995) derived the finite-sample variances given in Section 16.2.3. Doganaksoy and Nelson (1991) explain how to compare two samples of repair data and provide a computer program. Cox and Lewis (1966) provide an outline of methods for analyzing point-process data. Cox (1962) gives special attention to renewal processes. Crowder, Kimber, Smith, and Sweeting (1991), Ansell and Phillips (1994), and Høyland and Rausand (1994) contain useful chapters on counting processes, including Poisson process models, and corresponding methods for data analysis. Lee (1980) presents methods for testing between HPP with NHPP alternative models. Lawless and Thiagarajah (1996) present a general model that connects NHPP and renewal processes and allows for adjustment for covariates. Crow (1982) uses the power NHPP model to describe reliability growth of a product. Kuo and Yang (1996) describe Bayesian computations for NHPP models with applications to software reliability.

EXERCISES

16.1. A small unmonitored computer laboratory contains 10 networked microcomputers. Users who notice a hardware or software problem with a computer are supposed to report the problem to a technician who will fix the problem. The following table gives, for each computer, the days in which trouble calls were received. Most of the trouble reports were easy to address (replace a defective mouse, reboot the computer, remake the computer's file system from the server, remove stuck floppy disk, tighten loose connector, etc.). Calls reporting network problems or problems in the remote file server (which would affect all of the computers in the laboratory) were eliminated from the database. All of the computers were in operation for the entire semester (day 1 through 105).

Unit Number	Day Trouble Reported
401	18, 22, 45, 52, 74, 76, 91, 98, 100, 103
402	11, 17, 19, 26, 27, 38, 47, 48, 53, 86, 88
403	2, 9, 18, 43, 69, 79, 87, 87, 95, 103, 105
404	3, 23, 47, 61, 80, 90
501	19, 43, 51, 62, 72, 73, 91, 93, 104, 104, 105
502	7, 36, 40, 51, 64, 70, 73, 88, 93, 99, 100, 102
503	28, 40, 82, 85, 89, 89, 95, 97, 104
504	4, 20, 31, 45, 55, 68, 69, 99, 101, 104
601	7, 34, 34, 79, 82, 85, 101
602	9, 47, 78, 84

(a) Plot the trouble reports on a time–event chart.

(b) Compute an estimate of the mean cumulative number of trouble reports, as a function of days.

(c) What do you notice about the pattern of trouble reports over the semester? What could explain this pattern?

(d) What information could be added to the table above to make the data on computer trouble reports more informative and more useful?

(e) Could these data be used to predict the number of trouble reports next semester for the same lab?

(f) Could these data be used to predict the number of trouble reports next semester for a different computer lab in the same building? Explain.

▲**16.2.** Consider the NHPP model with a power $\nu(t)$ given in Section 16.6.3.

(a) Verify the formulas given there for the ML estimates of the parameters β and η.

(b) Verify the expression given for the relative likelihood $R(\beta, \eta)$.

(c) Show that the profile likelihood for β has the form

$$R(\beta) = \left(\frac{\beta}{\hat{\beta}}\right)^r \times \left[\prod_{j=1}^{r}\left(\frac{t_j}{t_a}\right)\right]^{\beta-\hat{\beta}}.$$

▲16.3. Consider the NHPP model with a loglinear $v(t)$ given in Section 16.6.4.
  (a) Do parts (a) and (b) as in Exercise 16.2, but using the loglinear NHPP model.
  (b) Show that the profile likelihood for γ_1 is

$$R(\gamma_1) = \left[\frac{\gamma_1(\exp(t_a\hat{\gamma}_1) - 1)}{\hat{\gamma}_1(\exp(t_a\gamma_1) - 1)}\right]^r \times \exp\left[(\gamma_1 - \hat{\gamma}_1)\sum_{j=1}^{r} t_j\right].$$

▲16.4. A repair process has been observed for a period $(t_0, t_a]$ and the data are the number of repairs $d_1, \ldots, d_m$ in the nonoverlapping intervals $(t_0, t_1], (t_1, t_2], \ldots, (t_{m-1}, t_m]$ (with $t_m = t_a$). Suppose that the process can be described with the loglinear $v(t)$ NHPP model.
  (a) Show that the ML estimates $\hat{\gamma}_0$ and $\hat{\gamma}_1$ are the solution to the equations

$$\sum_{j=1}^{m} d_j \left[\frac{t_j \exp(\gamma_1 t_j) - t_{j-1}\exp(\gamma_1 t_{j-1})}{\exp(\gamma_1 t_j) - \exp(\gamma_1 t_{j-1})}\right]$$

$$- r\left[\frac{t_m \exp(\gamma_1 t_m) - t_0 \exp(\gamma_1 t_0)}{\exp(\gamma_1 t_m) - \exp(\gamma_1 t_0)}\right] = 0,$$

$$\gamma_0 = \log\left[\frac{r\gamma_1}{\exp(\gamma_1 t_m) - \exp(\gamma_1 t_0)}\right],$$

where $r = \sum_{i=1}^{m} d_j$.
  (b) Show that when the length of the intervals $t_{i+1} - t_i$ approaches zero (exact recurrence times reported), then the formulas in part (a) simplify to the formulas given in Section 16.6.4.

16.5. Show that substituting (16.3) and (16.4) into (16.2) and simplifying gives (16.5).

16.6. Consider the power-loglinear recurrence rate given by

$$v(t; \gamma_0, \gamma_1, \gamma_2, \gamma_3) = \frac{\gamma_3}{\gamma_2}\left(\frac{t}{\gamma_2}\right)^{\gamma_3-1} \exp(\gamma_0 + \gamma_1 t).$$

  (a) Show that this model includes as special cases the power and loglinear models.

(b) Show that under the NHPP model, the mean cumulative number of system events over $(0, t]$ is

$$\mu(t; \gamma_0, \gamma_1, \gamma_2, \gamma_3) = \gamma_3 \left(\frac{t}{\gamma_2}\right)^{\gamma_3} \exp(\gamma_0) \left[\frac{1}{\gamma_3} + \sum_{i=1}^{\infty} \frac{(\gamma_1 t)^i}{(\gamma_3 + i) \times i!}\right].$$

(c) Derive expressions for $\mu(t)$ for the power and loglinear models from the general expression given above.

16.7. Example 16.11 showed that the times between unscheduled maintenance actions were consistent with a renewal process. Use exponential and Weibull probability plots to explore the form of the distribution of the times between unscheduled maintenance actions.

16.8. Verify the computation of $\widehat{\text{Var}}[\widehat{\mu}(t_2)]$ and $\widehat{\text{Var}}[\widehat{\mu}(t_3)]$ given in Example 16.4.

▲**16.9.** A company manufactures systems. The number of faults in a new system, say, X, has a Poisson distribution with mean γ. Each fault has associated with it a time of occurrence. These times can be modeled as being independent and distributed with a pdf $f(t; \boldsymbol{\theta})$. Consider the counting process $N(t)$ giving the number of failures at time t (see Kuo and Yang, 1996, for a more detailed description and applications for this model).

(a) Show that, conditional on fixed number of faults $X > 0$, the distribution of $N(t)$ is binomial with probability of success equal to $F(t)$.
(b) Show that $N(t)$ is an NHPP with $\nu(t) = \gamma f(t; \boldsymbol{\theta})$.
(c) Derive an expression for the MCF $\mu(t)$.
(d) Show that $\lim_{t \to \infty} \mu(t) = \gamma < \infty$, and give an intuitive explanation for this result.
(e) Derive an expression for the probability of zero failures in the interval $(0, t]$. Use this expression to show that there is always a positive probability of zero failures even for large t.
(f) Suppose that $f(t; \boldsymbol{\theta})$ is an exponential density with mean θ.
  (i) Show that in this case the recurrence rate $\nu(t)$ is a loglinear function and express β_0 and β_1 as a function of θ and γ.
  (ii) Show that there is a restriction on the sign of β_1 imposed by the model.
(g) Derive expressions for $\nu(t)$ and $\mu(t)$ when $f(t; \boldsymbol{\theta})$ is a Weibull density.

16.10. A company that runs a fleet of passenger automobiles would like to do a retrospective study to compare two different brands of automobile batteries. The fleet contains 55 automobiles. All of the automobiles started with Brand B, but during the life of the automobiles, when a battery failed, it was replaced with either Brand A or Brand B. The available data show the date of purchase of the original automobile and the date and brand of replacement for batteries that were replaced.

(a) How would you organize the data for analysis? What kind of questions could you answer from an analysis focusing on the mean cumulative number of failures?

(b) Describe some of the potential pitfalls involved in drawing inferences from a retrospective study like this. What advantages would there be for doing a prospective study where the failure times of future batteries would be studied?

(c) What assumptions would you have to make about the battery failures in order to be able to use the times between battery failures to estimate a failure-time distribution for the batteries? How could you use the available data to check the assumptions?

16.11. Suppose that n repairable systems were put into service at the same time and that, up to time t_j, all systems are still being monitored.

(a) Show that, in this case, the estimator of the MCF at t_j in (16.1) reduces to the sample mean of the $N_i(t_j)$ values:

$$\widehat{\mu}(t_j) = \bar{N}_i(t_j) = \frac{\sum_{i=1}^{n} N_i(t_j)}{n}.$$

(b) Show that the estimator of $\text{Var}[\widehat{\mu}(t_j)]$ in (16.3) reduces to

$$\widehat{\text{Var}}[\widehat{\mu}(t_j)] = \frac{\widehat{\text{Var}}[N(t_j)]}{n} = \frac{\sum_{i=1}^{n}[N_i(t_j) - \bar{N}_i(t_j)]^2}{n^2}.$$

(c) Show why $\widehat{\text{Var}}[\widehat{\mu}(t_j)] = 0$ when $N_1(t_j) = N_2(t_j) = \cdots = N_n(t_j)$.

(d) Give the formula for the unbiased estimator of $\text{Var}[\widehat{\mu}(_j)]$.

▲**16.12.** As in Exercise 16.11 suppose that n repairable systems are being observed but that at time t_j some of the systems have been censored.

(a) Show the result in Exercise 16.11 part (a) does not hold.

(b) Show that the result in Exercise 16.11 part (c) is still true.

▲**16.13.** Consider a sequence of recurrence times $T_1, T_2, \ldots$ from an NHPP with recurrence rate $\nu(t)$.

(a) Show that the cdf of T_i conditional on T_{i-1} is

$$F(t \mid T_{i-1}) = \Pr(T_i \leq t \mid T_{i-1}) = 1 - \exp\left[-\int_{T_{i-1}}^{t} \nu(u)\, du\right].$$

Hint: Note that $\Pr(T_i > t \mid T_{i-1}) = \Pr[N(t) - N(T_{i-1}) = 0]$ and use the properties of the NHPP.

(b) Using the method given in Section 4.13, show that

$$F(T_i \mid T_{i-1}) \sim \text{UNIF}(0, 1).$$

(c) Suppose that $T_0 = 0$. Show that pseudorandom observations of T_i are obtained from the sequence $T_1 = F^{-1}(U_1), T_2 = F^{-1}(U_2 \mid T_1), \ldots, T_i = F^{-1}(U_i \mid T_{i-1})$ where $U_1, \ldots, U_i$ is a pseudorandom sample from a UNIF(0, 1).

(d) For the power model, show that simulated values of the NHPP can be obtained from the sequence,

$$T_1 = \left[-\eta^\beta \log(1 - U_1)\right]^{1/\beta},$$

$$T_2 = \left[T_1^\beta - \eta^\beta \log(1 - U_2)\right]^{1/\beta},$$

$$\vdots$$

$$T_i = \left[T_{i-1}^\beta - \eta^\beta \log(1 - U_i)\right]^{1/\beta}.$$

Also show that the recursive formula can be expressed as

$$T_1 = \eta\left[-\log(1 - U_1)\right]^{1/\beta},$$

$$T_2 = \eta\left[-\sum_{j=1}^{2} \log(1 - U_j)\right]^{1/\beta},$$

$$\vdots$$

$$T_i = \eta\left[-\sum_{j=1}^{i} \log(1 - U_j)\right]^{1/\beta}.$$

(e) Show why and how one can replace $1 - U_j$ with $U_j, j = 1, \ldots$, in part (d), and still the expressions will provide the desired sequence of NHPP recurrence times.

(f) Derive the corresponding recursive formulas for the NHPP model with loglinear recurrence rate function.

16.14. Consider an NHPP model with a strictly positive recurrence rate $\nu(t)$ [i.e., $\nu(t) > 0$ for all t in $[0, \infty)$]. Let $T_1, T_1, \ldots$ be the random times at which the recurrent events occur in the stochastic process. Let $W_1 = \mu(T_1), W_2 = \mu(T_2)$ denote transformed times, where $\mu(t)$ is the process MCF.

(a) Show that the time transformation $w = \mu(t)$ is monotone increasing. This implies that the inverse transformation $t = \mu^{-1}(w)$ is well defined.

(b) Let $N_W(a)$ be the number of events in the interval $[0, a)$ in the W time scale. Show that events in nonoverlapping intervals are independent and that

$$N_W(0) = 0$$
$$\Pr[N_W(a,b) = d] = \Pr\{N[\mu^{-1}(a), \mu^{-1}(b)] = d\}$$
$$= \frac{(b-a)^d}{d!} \exp[-(b-a)], \, d = 0, 1, 2, \ldots.$$

This implies that in the W time scale, the point-process is HPP with a constant recurrence rate equal to 1.

(c) Using the result in part (b), show that the interrecurrence transformed times $W_i - W_{i-1} = \mu(T_{i-1}, T_i)$, $i = 1, 2, \ldots$ (where $W_0 = 0$), are iid with an EXP(1) distribution.

◆16.15. The first confidence interval procedure given in Section 16.2.3 is based on the large-sample approximation that $Z_{\hat{\mu}(t)} = [\hat{\mu}(t) - \mu(t)]/\widehat{se}_{\hat{\mu}(t)} \overset{\sim}{\cdot} \text{NOR}(0, 1)$. The adequacy of this approximation depends on the number of system in the sample, and the distribution of the cumulative function levels at the point in time of interest. The adequacy of the approximation can be checked by doing a Monte Carlo simulation.

(a) Show that if the underlying processes generating repairs for refrigerators can be described by an NHPP (with possibly different recurrence rate functions from refrigerator to refrigerator), the cumulative number of repairs for a sample of such refrigerators at a particular point in time (assuming no censoring) has a Poisson distribution.

(b) Simulate 1000 samples, each giving the total number of repairs from $n = 5$ systems. Suppose, as suggested in part (a), that $\sum_{i=1}^{n} N_i(t)$, the cumulative number of repairs at time t for all n systems, has a Poisson distribution with mean $\text{E}\left[\sum_{i=1}^{n} N_i(t)\right] = 30$. For each sample, compute (again, assuming no censoring) $\hat{\mu}(t)$, $\widehat{se}_{\hat{\mu}(t)}$, and $Z_{\hat{\mu}(t)}$. Make a normal probability plot of the $Z_{\hat{\mu}(t)}$. What do you conclude?

(c) Repeat part (a), but use samples of size $n = 10, 25, 50$, and 100. What do you conclude?

(d) Explain how you could use the bootstrap methods in Chapter 9 to obtain better approximate confidence intervals for $\mu(t)$.

◆16.16. Redo Exercise 16.15 with $\text{E}\left[\sum_{i=1}^{n} N_i(t)\right] = 5, 10, 20, 50$, and 100. Comment on the results.

CHAPTER 17

Failure-Time Regression Analysis

Objectives

This chapter explains:

- Modeling life as a function of explanatory variables.
- Graphical methods for displaying censored regression data.
- Time-scaling transformation functions and other forms of relationships between life and explanatory variables.
- Simple regression models to relate life to explanatory variables.
- Likelihood methods to draw conclusions from censored regression data.
- How to detect departures from an assumed regression model.
- Extensions to more elaborate nonstandard regression models that can be used to relate life to explanatory variables, including quadratic regression, models in which σ depends on explanatory variables, and proportional hazards models.

Overview

This chapter builds on the material in Chapter 8 and other earlier chapters. It shows how to model time as a function of explanatory variables. Section 17.2 introduces the idea of time acceleration and related failure-time regression models and contrasts them with other regression models. Section 17.3 uses a simple regression example to explain and illustrate the most important concepts in this chapter. Section 17.4 shows how to extend methods from Chapter 8 to compute confidence intervals for model parameters and functions of parameters (e.g., quantiles and failure probabilities). Sections 17.5, 17.7, and 17.8 describe applications requiring other, more complicated, regression models. Section 17.6 shows how to adapt traditional regression diagnostics to nonnormal distributions and problems involving censored data. Section 17.9 describes the Cox proportional hazard model, which has been used occasionally for analyzing field-failure data.

17.1 INTRODUCTION

Chapters 3–11 presented and illustrated the use of models to describe a single failure-time distribution. In this chapter we present and illustrate methods for including explanatory variables in failure-time models. Some of the material in this chapter will be familiar to those who have previously studied traditional statistical regression analysis. There the mean of a normal distribution is modeled as a linear function of one or more (possibly transformed) explanatory variables. The more general regression methods presented here were needed to solve practical problems in reliability testing and data analysis.

Example 17.1 Computer Program Execution Time Data. Appendix Table C.11 and Figure 17.1 give the amount of time it took to execute a particular computer program, on a multiuser computer system, as a function of the system load (obtained with the Unix uptime command) at the point in time when execution was beginning. The figure shows that, as one might expect, it takes longer to execute a program when the system is more heavily loaded. The figure also indicates that there is more variability at higher levels of system load. Execution-time predictions are useful for scheduling subsequent steps in a multistep series–parallel computing problem. □

Example 17.2 Low-Cycle Fatigue Data on a Nickel-Base Superalloy. Appendix Table C.12 and Figure 17.2 give low-cycle fatigue life data for a strain-controlled test on 26 cylindrical specimens of a nickel-base superalloy. The data were

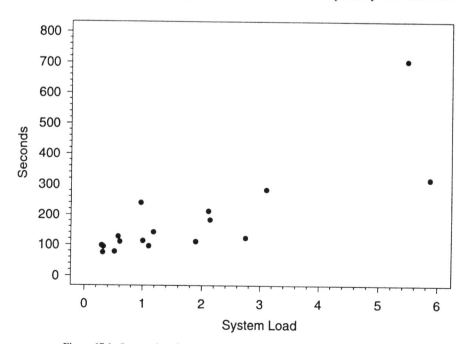

Figure 17.1. Scatter plot of computer program execution time versus system load.

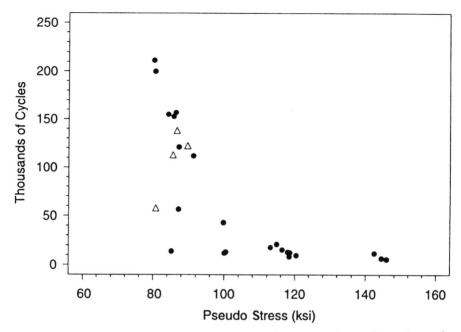

Figure 17.2. Scatter plot of low-cycle superalloy fatigue life versus pseudostress for specimens of a nickel-base superalloy. Points marked with △ are right-censored observations; others are failures.

originally described and analyzed in Nelson (1984; 1990a, page 272). Four of the specimens were removed from test before failure. These censored observations are indicated by a △ in Figure 17.2. In addition to number of cycles, each specimen has a level of pseudostress (Young's modulus times strain). The purpose of Nelson's analysis was to estimate the curve giving the number of cycles at which .1% of the population of such specimens would fail, as a function of the pseudostress. Figure 17.2 shows that (as expected) failures occur sooner at high pseudostress. Also, at lower stress there is more spread in the failure times. The following sections will explore several models for the relationship between cycles to failure and pseudostress. □

17.2 FAILURE-TIME REGRESSION MODELS

A model with explanatory variables sometimes explains or predicts why some units fail quickly and other units survive a long time. Also, if important explanatory variables are ignored in an analysis, it is possible that resulting estimates of quantities of interest (e.g., distribution quantiles or failure probabilities) could be biased seriously. In reliability studies possible explanatory variables include:

- Continuous variables like stress, temperature, voltage, and pressure.
- Discrete variables like number of hardening treatments or number of simultaneous users of a system.
- Categorical variables like manufacturer, design, and location.

The general idea of a regression model is to express the failure-time distribution as a function of k explanatory variables denoted by $x = (x_1, \ldots, x_k)$. For example,

$$\Pr(T \le t; x) = F(t; x) = F(t).$$

In some cases, to simplify notation, we will suppress the dependency of $F(t)$ on x. Regression models can come from physical/chemical theory, curve fitting to empirical observations, or some combination of theory and empiricism. In science and engineering, new knowledge and theory are often developed and refined through iterative experimentation.

17.2.1 Parameters as Functions of Explanatory Variables

An important class of regression models allows one or more of the elements of the model parameter vector $\boldsymbol{\theta} = (\theta_1, \ldots, \theta_r)$ to be a function of the explanatory variables. Generally, one employs a function with a specified form with one or more unknown parameters that need to be estimated from data. This is a generalization of statistical regression analysis in which the most commonly used models have the mean of the normal distribution depending linearly on a vector x of explanatory variables. For example, if x_i is a scalar explanatory variable for observation i, then the normal distribution mean is

$$\mu_i = \beta_0 + \beta_1 x_i.$$

In this case we think of x_i as being a fixed part of data$_i$. When the x_i values are themselves random, standard statistical methods and models provide inferences that are conditional on the fixed, observed values of these explanatory variables. Then the unknown regression model coefficients (β_0 and β_1) replace the model parameter μ_i in a new definition of $\boldsymbol{\theta}$. In some situations, there may be more than one model parameter that will depend on one or more explanatory variables.

17.2.2 The Scale-Accelerated Failure-Time Model

The scale-accelerated failure-time (SAFT) model is commonly used to describe the effect that explanatory variables x have on time. This model has a simple time-scaling acceleration factor that is a function of x and is defined by

$$T(x) = T(x_0)/\mathcal{AF}(x), \quad \mathcal{AF}(x) > 0, \quad \mathcal{AF}(x_0) = 1, \tag{17.1}$$

where $T(x)$ is the time at conditions x and $T(x_0)$ is the corresponding time at some "baseline" conditions x_0. Some commonly used forms for the time-scale factor $\mathcal{AF}(x)$ include the following log linear relationships [assuring that $\mathcal{AF}(x) > 0$]:

- For a scalar x, $\mathcal{AF}(x) = 1/\exp(\beta_1 x)$ with $x_0 = 0$.
- For a vector $x = (x_1, \ldots, x_k)$, $\mathcal{AF}(x) = 1/\exp(\beta_1 x_1 + \cdots + \beta_k x_k)$ with $x_0 = \mathbf{0}$.

FAILURE-TIME REGRESSION MODELS

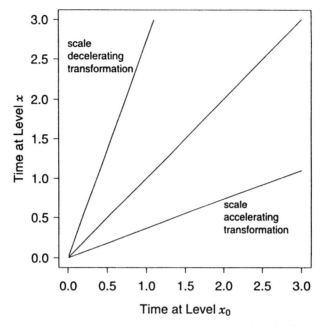

Figure 17.3. SAFT models illustrating acceleration and deceleration.

When $\mathcal{AF}(x) > 1$, the model accelerates time in the sense that time moves more quickly at x than at x_0 so that $T(x) < T(x_0)$. When $0 < \mathcal{AF}(x) < 1$, $T(x) > T(x_0)$, and time at x is decelerated relative to time at x_0, but we still refer to the model as SAFT. Figure 17.3 illustrates that scale failure-time transformation functions are straight lines starting at the origin:

- Lines *below* the diagonal accelerate time relative to time at x_0.
- Lines *above* the diagonal decelerate time relative to time at x_0.

Figure 17.3 and equation (17.1) describe the effect that $\mathcal{AF}(x)$ has on time. In terms of cdfs, with baseline cdf $F(t; x_0)$, $F(t; x) = F[\mathcal{AF}(x) \times t; x_0]$. This shows that if $\mathcal{AF}(x) \neq 1$, the cdfs $F(t; x)$ and $F(t; x_0)$ do not cross each other. In terms of distribution quantiles, $t_p(x) = t_p(x_0)/\mathcal{AF}(x)$. Then taking logarithms, and isolating the term $\log[\mathcal{AF}(x)]$, gives

$$\log[t_p(x_0)] - \log[t_p(x)] = \log[\mathcal{AF}(x)]. \tag{17.2}$$

Thus, as shown in Figure 17.4, in a cdf plot or a probability plot with a log data scale, $F(t, x)$ is a translation of $F(t, x_0)$ along the $\log(t)$ axis.

SAFT models are relatively simple to interpret. Also they are often suggested by the physical theory of some simple failure mechanisms (as shown in Chapter 18). They do *not*, however, hold universally (as shown in Sections 17.5.2 and 18.3.5).

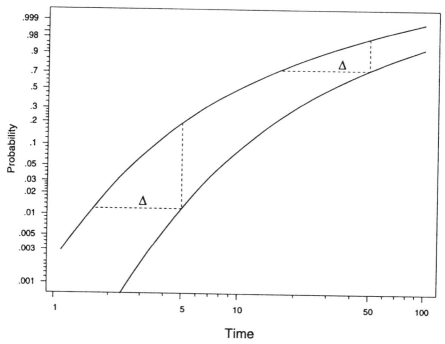

Figure 17.4. Weibull probability plot showing that two lognormal cdfs with a scale-accelerated failure-time (SAFT) regression relationship are equidistant along the log-time axis.

17.3 SIMPLE LINEAR REGRESSION MODELS

This section describes and illustrates the use of simple linear regression models (i.e., models with a single explanatory variable) based on location-scale distributions, including the normal, logistic, and smallest extreme value. These models are closely related with the traditional linear regression model. Maximum likelihood (ML) estimation is used instead of least squares, however, to handle censored data and nonnormal distributions. Because of the relationships described in Sections 4.6, 4.8, and 4.11, location-scale models can be used also to fit log-location-scale distributions (e.g., Weibull, lognormal, and loglogistic) regression models. The methods developed in this and some of the following sections in this chapter can be viewed as a generalization of the material in Chapter 8, allowing μ (and later σ) to depend on explanatory variables. Although it would be possible to use notation like $\mu(x)$ to make this dependency explicit, we will keep the notation simple, as the dependency will be clear from the context of the problem.

We will first illustrate the methods for the lognormal distribution. The methods apply, however, directly to other log-location-scale distributions or other distributions that can be transformed into the location-scale form. Sections 17.5–17.10 illustrate various extensions of the simple regression model. These extensions include

SIMPLE LINEAR REGRESSION MODELS

quadratic relationships, models with more than one explanatory variable, a model for product comparison, and relationships based on more general time-transformation functions.

17.3.1 Location-Scale Regression Model and Likelihood

With only one explanatory variable, the location-scale simple regression model (including the normal, logistic, and smallest extreme value distributions as special cases) is

$$\Pr(Y \leq y) = F(y; \mu, \sigma) = F(y; \beta_0, \beta_1, \sigma) = \Phi\left(\frac{y - \mu}{\sigma}\right), \quad (17.3)$$

where $\mu = \beta_0 + \beta_1 x$ and σ does not depend on the explanatory variable x. The quantile function for this model

$$y_p(x) = \mu + \Phi^{-1}(p)\sigma = \beta_0 + \beta_1 x + \Phi^{-1}(p)\sigma \quad (17.4)$$

is linear in x. As with the Chapter 8 models, choosing Φ determines the shape of the distribution for a particular value of x. One uses Φ_{nor} for normal, Φ_{logis} for logistic, and Φ_{sev} for the smallest extreme value distribution. In this model β_0 can be interpreted as the value of μ when $x = 0$ (when this has meaning) and β_1 is the change in μ [or $y_p(x)$] for a one-unit increase in x.

The likelihood for a sample of n independent observational units with right-censored and exact-failure observations has the form

$$L(\beta_0, \beta_1, \sigma) = \prod_{i=1}^{n} L_i(\beta_0, \beta_1, \sigma; \text{data}_i)$$

$$= \prod_{i=1}^{n} \left[\frac{1}{\sigma} \phi\left(\frac{y_i - \mu_i}{\sigma}\right)\right]^{\delta_i} \left[1 - \Phi\left(\frac{y_i - \mu_i}{\sigma}\right)\right]^{1-\delta_i}, \quad (17.5)$$

where $\mu_i = \beta_0 + \beta_1 x_i$, $\delta_i = 1$ for an exact failure time, and $\delta_i = 0$ for a right-censored observation. Similar terms could be added for left-censored or interval-censored data, as described in Section 2.4.3. As in Chapter 8, ML estimates are obtained by finding the values of β_0, β_1, and σ that maximize (17.5) or, equivalently, the corresponding log likelihood (preferred for numerical reasons). With an underlying normal distribution and complete data, there are simple closed-form equations to compute the ML estimates. Generally, it is necessary to use numerical optimization methods to maximize the log likelihood.

17.3.2 Log-Location-Scale Regression Model and Likelihood

Following the development in Section 17.3.1, the log-location-scale distribution simple regression model (including the lognormal, Weibull, and loglogistic distributions as special cases) is

$$\Pr(T \leq t) = F(t; \mu, \sigma) = F(t; \beta_0, \beta_1, \sigma) = \Phi\left[\frac{\log(t) - \mu}{\sigma}\right], \quad (17.6)$$

where $\mu = \beta_0 + \beta_1 x$ and σ does not depend on x. The quantile function for this model

$$\log[t_p(x)] = y_p(x) = \mu + \Phi^{-1}(p)\sigma = \beta_0 + \beta_1 x + \Phi^{-1}(p)\sigma \quad (17.7)$$

is linear in x. Such a relationship between $t_p(x)$ and x is sometimes known as a "loglinear relationship." Choosing Φ determines the shape of the distribution for a particular value of x. One uses Φ_{nor} for lognormal, Φ_{logis} for loglogistic, and Φ_{sev} for the Weibull distribution. Again, β_1 can be interpreted as the change in μ (or in $\log[t_p(x)]$) for a one-unit increase in x. Relatedly, because the response is on the log scale, $100\beta_1$ can be interpreted as the approximate percent increase in $t_p(x)$ for a one-unit increase in x. Reexpressing the quantile function as

$$t_p(x) = \exp[y_p(x)] = \exp(\beta_1 x) \, t_p(0)$$

shows that this regression model is a SAFT model with $\mathcal{AF}(x) = 1/\exp(\beta_1 x)$.

The likelihood for a combination of n independent exact-failure and right-censored observations is

$$L(\beta_0, \beta_1, \sigma) = \prod_{i=1}^{n} L_i(\beta_0, \beta_1, \sigma; \text{data}_i)$$

$$= \prod_{i=1}^{n} \left\{ \frac{1}{\sigma t_i} \phi\left[\frac{\log(t_i) - \mu_i}{\sigma}\right] \right\}^{\delta_i} \left\{ 1 - \Phi\left[\frac{\log(t_i) - \mu_i}{\sigma}\right] \right\}^{1-\delta_i}, \quad (17.8)$$

where $\mu_i = \beta_0 + \beta_1 x_i$, $\delta_i = 1$ for an exact failure time, and $\delta_i = 0$ for a right-censored observation. Similar terms could be added for left-censored or interval-censored data, as described in Section 2.4.3. As in Chapter 8, ML estimates are obtained by finding the values of β_0, β_1, and σ that maximize (17.8) or the corresponding log likelihood. Also, as described in Section 8.2.2, some computer programs omit the $1/t_i$ factor in the density part of the likelihood and therefore caution should be used when comparing values of log likelihoods computed with different software.

Example 17.3 *Loglinear Lognormal Regression Model for the Computer Program Execution Times.* Continuing with Example 17.1, Figure 17.5 shows the lognormal simple regression model fit to the computer program execution-time data. Table 17.1 summarizes the numerical results. Fitting this model is equivalent to fitting a simple linear relationship to the logarithms of the times. The estimated densities in Figure 17.5 are normal distribution densities because they are plotted on a log-time scale. This loglinear model provides a useful description of the data. □

STANDARD ERRORS AND CONFIDENCE INTERVALS FOR REGRESSION MODELS

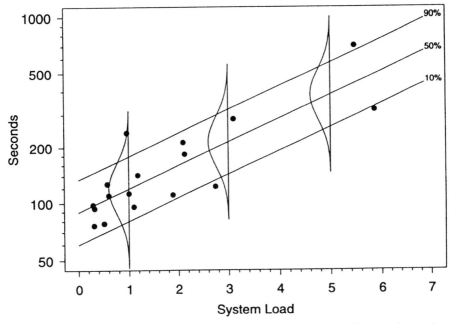

Figure 17.5. Computer program execution time versus system load with fitted lognormal regression model. The lines show ML estimates of the .1, .5, and .9 lognormal distribution quantiles as a function of system load.

Table 17.1. ML Estimates for the Computer Execution-Time Example

Parameter	ML Estimate	Standard Error	Approximate 95% Confidence Interval	
			Lower	Upper
β_0	4.49	.11	4.28	4.71
β_1	.290	.05	.20	.38
σ	.312	.05	.22	.44

The log likelihood is $\mathcal{L} = -89.50$. The confidence intervals for β_0, β_1, and $\log(\sigma)$ are based on the normal-approximation method.

17.4 STANDARD ERRORS AND CONFIDENCE INTERVALS FOR REGRESSION MODELS

This section shows how to compute estimates of standard errors and confidence intervals for parameters and functions of parameters. We use the computer execution-time data and the simple regression model in Section 17.3 as an example. The general theory is in Appendix Section B.6.7. Other examples in this chapter, in the exercises, and in subsequent chapters illustrate how the general ideas apply directly to more complicated models. We will focus primarily on normal-approximation confidence intervals. Likelihood and bootstrap confidence intervals described in Chapters 8 and 9 can also be used for these models.

Standard errors are provided by most computer software packages. Normal-approximation confidence intervals may or may not be provided but are easy to compute given the ML estimates and standard errors. Although they are superior, likelihood and simulation (bootstrap) intervals typically require specialized procedures not commonly available in today's commercial software.

17.4.1 Standard Errors and Confidence Intervals for Parameters

Appendix Section B.6.7 describes the general theory for computing normal-approximation confidence intervals that can be applied to regression models. These confidence intervals are based on the large-sample approximate normal distribution of the ML estimators. These intervals employ an estimate of the variance–covariance matrix for the ML estimates of the model parameters $\boldsymbol{\theta} = (\beta_0, \beta_1, \sigma)$. Extending the ideas presented in Section 8.4.1, the "local" estimate $\widehat{\Sigma}_{\widehat{\boldsymbol{\theta}}}$ of $\Sigma_{\widehat{\boldsymbol{\theta}}}$ is the inverse of the "observed" information matrix, namely,

$$\widehat{\Sigma}_{\widehat{\boldsymbol{\theta}}} = \begin{bmatrix} \widehat{\text{Var}}(\widehat{\beta}_0) & \widehat{\text{Cov}}(\widehat{\beta}_0, \widehat{\beta}_1) & \widehat{\text{Cov}}(\widehat{\beta}_0, \widehat{\sigma}) \\ \widehat{\text{Cov}}(\widehat{\beta}_1, \widehat{\beta}_0) & \widehat{\text{Var}}(\widehat{\beta}_1) & \widehat{\text{Cov}}(\widehat{\beta}_1, \widehat{\sigma}) \\ \widehat{\text{Cov}}(\widehat{\sigma}, \widehat{\beta}_0) & \widehat{\text{Cov}}(\widehat{\sigma}, \widehat{\beta}_1) & \widehat{\text{Var}}(\widehat{\sigma}) \end{bmatrix} \quad (17.9)$$

$$= \begin{bmatrix} -\dfrac{\partial^2 \mathcal{L}(\beta_0,\beta_1,\sigma)}{\partial \beta_0^2} & -\dfrac{\partial^2 \mathcal{L}(\beta_0,\beta_1,\sigma)}{\partial \beta_0 \partial \beta_1} & -\dfrac{\partial^2 \mathcal{L}(\beta_0,\beta_1,\sigma)}{\partial \beta_0 \partial \sigma} \\ -\dfrac{\partial^2 \mathcal{L}(\beta_0,\beta_1,\sigma)}{\partial \beta_1 \partial \beta_0} & -\dfrac{\partial^2 \mathcal{L}(\beta_0,\beta_1,\sigma)}{\partial \beta_1^2} & -\dfrac{\partial^2 \mathcal{L}(\beta_0,\beta_1,\sigma)}{\partial \beta_1 \partial \sigma} \\ -\dfrac{\partial^2 \mathcal{L}(\beta_0,\beta_1,\sigma)}{\partial \sigma \partial \beta_0} & -\dfrac{\partial^2 \mathcal{L}(\beta_0,\beta_1,\sigma)}{\partial \sigma \partial \beta_1} & -\dfrac{\partial^2 \mathcal{L}(\beta_0,\beta_1,\sigma)}{\partial \sigma^2} \end{bmatrix}^{-1},$$

where the partial derivatives are evaluated at $\widehat{\beta}_0, \widehat{\beta}_1, \widehat{\sigma}$.

Example 17.4 Parameter Variance–Covariance Matrix Estimate for the Computer Execution-Time Example. For the fitted computer execution-time model, the estimate of the variance–covariance matrix for the ML estimates $\widehat{\boldsymbol{\theta}} = (\widehat{\beta}_0, \widehat{\beta}_1, \widehat{\sigma})$ is

$$\widehat{\Sigma}_{\widehat{\boldsymbol{\theta}}} = \begin{bmatrix} .012 & -.0037 & 0 \\ -.0037 & .0021 & 0 \\ 0 & 0 & .0029 \end{bmatrix}. \quad (17.10)$$

These quantities will be used in subsequent numerical examples. □

Normal-approximation confidence intervals for regression model parameters can be computed using the methods described in Section 8.4.2.

STANDARD ERRORS AND CONFIDENCE INTERVALS FOR REGRESSION MODELS 437

Example 17.5 Normal-Approximation Confidence Interval for the Regression Slope in the Computer Execution-Time Example. For the computer time data, $\widehat{\text{se}}_{\widehat{\beta}_1} = \sqrt{.0021} = .046$ [see (17.10) and Table 17.1] and

$$[\underset{\sim}{\beta_1}, \widetilde{\beta}_1] = .2907 \pm 1.960 \times .0459 = [.20, \quad .38]$$

provides an approximate 95% confidence interval for β_1. This interval is based on the approximation $Z_{\widehat{\beta}_1} = (\widehat{\beta}_1 - \beta_1)/\widehat{\text{se}}_{\widehat{\beta}_1} \overset{\cdot}{\sim} \text{NOR}(0, 1)$. This interval implies that we are 95% confident that a one-unit increase in load will increase quantiles of the running time distribution by an amount between 20% and 38%. □

As described in Section 8.4.2, the log transformation generally improves normal-approximation confidence intervals for positive parameters like σ. In particular, using (8.8) provides an approximate $100(1 - \alpha)\%$ confidence interval for σ based on the assumption that $Z_{\log(\widehat{\sigma})} = [\log(\widehat{\sigma}) - \log(\sigma)]/\widehat{\text{se}}_{\log(\widehat{\sigma})}$ can be approximated by a $\text{NOR}(0, 1)$ distribution.

Example 17.6 Normal-Approximation Confidence Intervals for σ in the Computer Execution-Time Example. For the computer execution-time example $\widehat{\text{se}}_{\widehat{\sigma}} = \sqrt{.0029} = .054$. An approximate 95% confidence interval for σ is

$$[\underset{\sim}{\sigma}, \widetilde{\sigma}] = [.3125/1.400, \quad .3125 \times 1.400] = [.22, \quad .44],$$

where $w = \exp[1.96(.05359)/.3125] = 1.400$. □

17.4.2 Standard Errors and Confidence Intervals for Distribution Quantities at Specific Explanatory Variable Conditions

Appendix Section B.6.7 describes the general theory for computing standard errors and normal-approximation confidence intervals for functions of model parameters. This general theory can be used to estimate distribution quantities (such as quantiles, failure probabilities, or hazard function values) at specified levels of the explanatory variables. This section illustrates an equivalent two-stage approach that will simplify the presentation of the method (and computations). The presentation and examples are for simple regression models with log-location-scale distributions, but the ideas apply more generally.

As expressed in (17.6) and (17.7), there are unknown values of μ and σ at each level of x (or combination of levels of x in the case of multiple explanatory variables). For specified values of the explanatory variables, one can compute the ML estimates $(\widehat{\mu}, \widehat{\sigma})$ of these parameters and an estimate of the corresponding variance–covariance matrix. Then the methods given in Sections 8.4.2 and 8.4.3 can be applied directly to compute standard errors and confidence intervals for quantities of interest at specified values of the explanatory variable(s).

The general formulas needed to compute an estimate of the variance–covariance matrix of $(\widehat{\mu}, \widehat{\sigma})$ are (B.3) and (B.4) in Appendix Section B.2. For the simple linear regression model, at a particular value of x, $\widehat{\mu} = \widehat{\beta}_0 + \widehat{\beta}_1 x$, σ does not depend on x, and using a special case of (B.10),

$$\widehat{\Sigma}_{\widehat{\mu},\widehat{\sigma}} = \begin{bmatrix} \widehat{\mathrm{Var}}(\widehat{\mu}) & \widehat{\mathrm{Cov}}(\widehat{\mu},\widehat{\sigma}) \\ \widehat{\mathrm{Cov}}(\widehat{\mu},\widehat{\sigma}) & \widehat{\mathrm{Var}}(\widehat{\sigma}) \end{bmatrix} \qquad (17.11)$$

is obtained from $\widehat{\mathrm{Var}}(\widehat{\mu}) = \widehat{\mathrm{Var}}(\widehat{\beta}_0) + 2x\widehat{\mathrm{Cov}}(\widehat{\beta}_1,\widehat{\beta}_0) + x^2\widehat{\mathrm{Var}}(\widehat{\beta}_1)$ and $\widehat{\mathrm{Cov}}(\widehat{\mu},\widehat{\sigma}) = \widehat{\mathrm{Cov}}(\widehat{\beta}_0,\widehat{\sigma}) + x\widehat{\mathrm{Cov}}(\widehat{\beta}_1,\widehat{\sigma})$. Then standard errors for functions of $(\widehat{\mu},\widehat{\sigma})$ can be computed by using the delta method, as in (8.10). The following examples show how the the general approach, given in Section 8.4.3, can be used to find confidence intervals for a function of regression model parameters.

Example 17.7 Confidence Interval for $t_p(x)$ in the Computer Execution-Time Example. Refer to Figure 17.5 and the results from Example 17.3. For scheduling purposes, it was necessary to estimate and obtain a 95% confidence interval for $t_{.9}(5)$, the time at which 90% of jobs running at a system load of 5 will have finished. From Table 17.1,

$$\widehat{\mu} = \widehat{\beta}_0 + \widehat{\beta}_1 x = 4.494 + .2907 \times 5 = 5.947,$$

$$\widehat{t}_{.9}(5) = \exp(\widehat{\mu} + \Phi_{\mathrm{nor}}^{-1}(p)\widehat{\sigma}) = \exp(5.947 + 1.2816 \times .3125) = 571.2.$$

From (17.10) computing the elements in (17.11) gives

$$\widehat{\Sigma}_{\widehat{\mu},\widehat{\sigma}} = \begin{bmatrix} \widehat{\mathrm{Var}}(\widehat{\mu}) & \widehat{\mathrm{Cov}}(\widehat{\mu},\widehat{\sigma}) \\ \widehat{\mathrm{Cov}}(\widehat{\mu},\widehat{\sigma}) & \widehat{\mathrm{Var}}(\widehat{\sigma}) \end{bmatrix} = \begin{bmatrix} .0277 & 0 \\ 0 & .00287 \end{bmatrix}.$$

Substituting into (8.12) gives

$$\widehat{\mathrm{se}}_{\widehat{t}_{.9}(5)} = 571.2[.0277 + 2 \times (1.2816) \times 0 + (1.2816)^2 \times .00287]^{1/2} = 102.9.$$

Then an approximate 95% confidence interval for $t_{.9}(5)$ based on $Z_{\log[\widehat{t}_{.9}(5)]} \stackrel{\cdot}{\sim}$ NOR(0, 1) is obtained by substituting into (8.11), giving

$$[\underaccent{\tilde}{t}_{.9},\ \widetilde{t}_{.9}] = [571.2/1.423,\ \ 571.2 \times 1.423] = [401,\ \ 813],$$

where $w = \exp(1.96 \times 102.9/571.2) = 1.423$. Thus we are 95% confident that when the system load is 5, $t_{.9}$ lies between 401 and 813 seconds. □

Normal-approximation confidence intervals for other quantities (like failure probabilities or hazard values) at this or other levels of system load can be found similarly using the methods of Section 8.4.3. More accurate confidence intervals for regression models can be based on the likelihood ratio approach used in Chapter 8 or on the bootstrap methods described in Chapter 9. Implementation here is a straightforward extension of the methods presented there and is not described in detail here.

17.5 REGRESSION MODEL WITH QUADRATIC μ AND NONCONSTANT σ

This section describes and illustrates the use of regression models that extend the simple linear model in Section 17.3.

17.5.1 Quadratic Regression Relationship for μ and a Constant Spread Parameter

Consider the log-quadratic relationship that uses (17.6) with $\mu = \beta_0 + \beta_1 x + \beta_2 x^2$ and σ does not depend on x. Then

$$\log[t_p(x)] = y_p(x) = \beta_0 + \beta_1 x + \beta_2 x^2 + \Phi^{-1}(p)\sigma$$

is quadratic in x. The quantile function can be written as

$$t_p(x) = \exp[y_p(x)] = \exp(\beta_1 x + \beta_2 x^2) t_p(0).$$

Thus this is a SAFT model with $\mathcal{AF}(x) = 1/\exp(\beta_1 x + \beta_2 x^2)$. Substituting $\mu_i = \beta_0 + \beta_1 x_i + \beta_2 x_i^2$ into (17.8) gives a likelihood having the form $L(\beta_0, \beta_1, \beta_2, \sigma)$. In the Weibull model $\beta = 1/\sigma$ is the Weibull shape parameter. On the log-time scale (corresponding to the smallest extreme value distribution), σ is a scale parameter. Generally we will refer to σ as a "spread" parameter.

Example 17.8 *Model 1: Log-Quadratic Weibull Regression Model for Superalloy Fatigue Data.* Continuing from Example 17.2, we will fit a curvilinear regression model to the superalloy fatigue data. Figure 17.6 shows the log-quadratic Weibull regression model with constant σ, fit to the superalloy fatigue data with $x = \log(\text{pseudostress})$. This model provides a reasonable fit to the data, but there is, even on the log scale, some evidence that the spread in the data is greater at lower levels of stress. The estimated densities in Figure 17.6 are smallest extreme value densities because they are plotted on a log-cycles scale. Table 17.2 summarizes the numerical results of this and another model. The quadratic model has to be used with caution. As can be visualized in Figure 17.6, extrapolation to levels of pseudostress beyond 160 ksi would lead to nonsensical estimates of longer life. Section 22.4 suggests an alternative model for these data. □

17.5.2 Quadratic Regression Model with Nonconstant Spread Parameter (σ)

This section illustrates the fitting of a regression model in which both μ and σ depend on an explanatory variable. The model is given by (17.6) with $\mu = \beta_0^{[\mu]} + \beta_1^{[\mu]} x + \beta_2^{[\mu]} x^2$ and $\log(\sigma) = \beta_0^{[\sigma]} + \beta_1^{[\sigma]} x$. The log-quantile function for this model is

$$\log[t_p(x)] = y_p(x) = \beta_0^{[\mu]} + \beta_1^{[\mu]} x + \beta_2^{[\mu]} x^2 + \Phi^{-1}(p) \exp\left(\beta_0^{[\sigma]} + \beta_1^{[\sigma]} x\right), \quad (17.12)$$

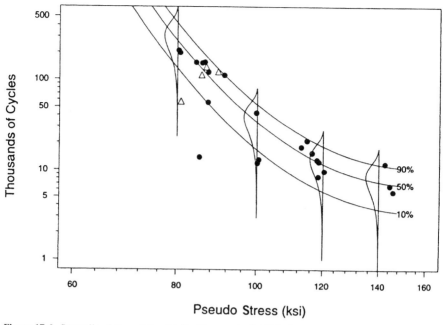

Figure 17.6. Superalloy fatigue data with fitted log-quadratic Weibull regression model with constant σ, plotted on log–log axes.

Table 17.2. ML Estimates for Life Versus Stress Weibull Regression Relationships for Nickel-Base Superalloy Specimens

	Parameter	ML Estimate	Standard Error	Approximate 95% Confidence Intervals	
				Lower	Upper
Model 1	β_0	217.61	62.1	95.9	339.3
	β_1	−85.52	26.53	−137.5	−33.53
	β_2	8.48	2.83	2.93	14.03
	σ	.375	.067	.26	.53
Model 2	$\beta_0^{[\mu]}$	243.2	58.12	129.3	357.1
	$\beta_1^{[\mu]}$	−96.54	24.73	−145.0	−48.07
	$\beta_2^{[\mu]}$	9.67	2.63	4.52	14.8
	$\beta_0^{[\sigma]}$	4.47	4.17	−3.71	12.6
	$\beta_1^{[\sigma]}$	−1.18	.89	−2.93	.58

The log likelihoods for Model 1 and Model 2 are, respectively, $\mathcal{L}_1 = -93.38$ and $\mathcal{L}_2 = -92.58$.

which is *not* quadratic in x. Also, because $t_p(x)/t_p(0) = \exp[y_p(x) - y_p(0)]$ depends on p, this model is *not* a SAFT model. Substituting $\mu_i = \beta_0^{[\mu]} + \beta_1^{[\mu]} x_i + \beta_2^{[\mu]} x_i^2$ and $\sigma_i = \exp\left(\beta_0^{[\sigma]} + \beta_1^{[\sigma]} x_i\right)$ into (17.8) gives a likelihood having the form $\mathcal{L}(\beta_0^{[\mu]}, \beta_1^{[\mu]}, \beta_2^{[\mu]}, \beta_0^{[\sigma]}, \beta_1^{[\sigma]})$.

Example 17.9 Model 2: Weibull Log-Quadratic Regression Model with Nonconstant σ for the Superalloy Fatigue Data. Continuing from Examples 17.2 and 17.8, Figure 17.7 shows the log-quadratic Weibull regression model with nonconstant σ fit to the superalloy fatigue data. The numerical results are summarized in Table 17.2. This model seems to provide a reasonably good fit to these data. At first sight, the failure at 13,949 cycles with pseudostress equal to 85.2 ksi appears to be an outlier. Relative to the long lower tail of the smallest extreme distribution, however, the observation is not surprisingly extreme (see Exercise 17.8). Comparing Model 1 and Model 2 [e.g., $-2(\mathcal{L}_1 - \mathcal{L}_2) = 1.6 < \chi^2_{(.90;1)} = 2.71$] indicates that the evidence for nonconstant σ in the data is not strong. On the other hand, having σ decrease with stress or strain is typical in fatigue data and this is what we see in data points plotted in Figure 17.6. Thus it would be reasonable to use a model with decreasing σ in this case, even in the absence of "statistical significance." □

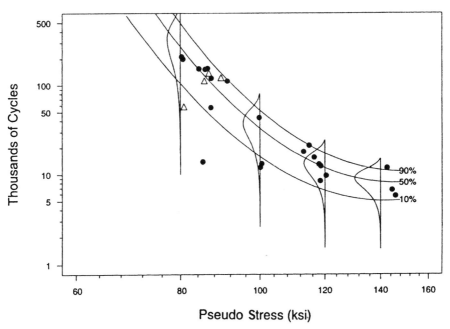

Figure 17.7. Superalloy fatigue data with fitted log-quadratic Weibull regression model with nonconstant σ.

17.5.3 Further Comments on the Use of Empirical Regression Models

As described in Example 17.8, the quadratic relationship should not be used for extrapolation. The fatigue-data models used in Examples 17.8 and 17.9 are purely empirical, without any physical basis. In general this is true of other quadratic and higher order polynomial relationships. Such relationships can be useful, providing a smooth curve to describe a population or a process *within the range of the data*, but should not be used to extrapolate outside the range of one's data.

The nonconstant-σ model also has a potential extrapolation pitfall (even if the relationship for μ is linear). Refer to Figure 17.7. Because of the potentially longer lower tail of the distribution at low levels of stress, depending on the values of the parameters of the model, it is possible to have the lower-tail quantile of the cycles to failure distribution decrease as one moves to lower levels of pseudostress, predicting shorter life at lower stress, leading physically nonsensical extrapolations (see Exercise 17.9). Section 22.4 suggests an alternative model for these data.

17.5.4 Comment on Numerical Methods

For certain model/data combinations, the shape of the likelihood can be such that the commonly used optimization software will find it difficult or impossible to find the maximum of the function. This is a problem that is analogous to the numerical problem of finding regression coefficients when there is a strong degree of linear dependence (multicollinearity) among the explanatory variables. High-quality software for least squares regression does not use the standard textbook formulas based on matrix inversion. Similarly, software developers for ML iterations need to consider carefully the numerical analysis aspects of the calculations to be programmed. Good software will deal with these numerical issues in a manner that is transparent to the user. That is, when a reparameterization or reformulation is used, this need not be brought to the user's attention (but such information should, perhaps, be available as an option).

The superalloy regression models are a case in point. Nelson (1984) centers the x variables, which alleviates the collinearity problem in polynomial regression. If, however, the different explanatory variables are close to being linearly dependent (multicollinear) or if a model without an intercept term is required, then a different approach is needed. Escobar and Meeker (1998c) describe an appropriate reparameterization for multiple linear regression with censored data. In general, having robust ML iterations (i.e., having a high probability of finding the maximum, given that one exists) requires the following:

- A parameterization that makes the likelihood have a shape that is not too different from a quadratic with major axes corresponding to the transformed model parameters (resulting in transformed parameter estimates that are approximately uncorrelated), and
- Starting values that are not too far from the maximum.

For further discussion of numerical methods for ML, see Section 8.6 of Nelson (1982), Ross (1990), and Escobar and Meeker (1998c).

CHECKING MODEL ASSUMPTIONS 443

17.6 CHECKING MODEL ASSUMPTIONS

An important part of any statistical analysis is diagnostic checking for departures from model assumptions. In conducting a failure-time regression analysis we recommend the use of graphical methods, using generalizations of usual regression diagnostics (including residual analysis). These diagnostic methods can be used in a manner that is similar to their use in ordinary regression analysis, except that interpretation is often complicated by the censoring. The analysis can also be complicated when fitting underlying nonnormal distributions.

17.6.1 Definition of Residuals

Consider a set of independent (possibly censored) observations y_i, $i = 1, \ldots, n$, from location-scale distributions with cdfs $\Phi[(y - \mu_i)/\sigma_i]$, where μ_i and σ_i may be functions of regression parameters and explanatory variables x_i. This would include data from normal, logistic, or smallest extreme value regression models. A natural and commonly used definition of standardized residuals for this model is

$$\widehat{\epsilon}_i = \frac{y_i - \widehat{\mu}_i}{\widehat{\sigma}_i}, \qquad (17.13)$$

where $\widehat{\mu}_i$ and $\widehat{\sigma}_i$ are the ML estimates of μ_i and σ_i. Then, under the assumed regression model, these residuals should look like a (possibly censored) random sample from a standardized (i.e., $\mu = 0$ and $\sigma = 1$) location-scale distribution (e.g., standard normal, smallest extreme value, or logistic). When y_i is a censored observation, the corresponding residual is also censored. For example, if y_i is a right-censored observation, the corresponding $\widehat{\epsilon}_i$ is also right-censored (we only know that the actual residual would have been larger than the censored residual).

For regression data from log-location-scale distributions with cdfs $\Phi\{[\log(t) - \mu_i]/\sigma_i\}$ (e.g., regression with the lognormal, loglogistic, or Weibull data) a natural extension of the definition of standardized residuals is

$$\widehat{\epsilon}_i = \exp\left[\frac{\log(t_i) - \widehat{\mu}_i}{\widehat{\sigma}_i}\right] = \left[\frac{t_i}{\exp(\widehat{\mu}_i)}\right]^{1/\widehat{\sigma}_i}. \qquad (17.14)$$

Again, when t_i is a censored observation, the corresponding residual is also censored. Under the assumed regression model, these residuals should look like they came from a standardized (i.e., $\mu = 0$ and $\sigma = 1$) log-location-scale distribution. Adequacy of the fitted distribution can be assessed by making a probability plot of the (possibly censored) residuals using the methods in Chapter 6. We will refer to (17.13) and (17.14) as "censored Cox–Snell" residuals because they are special cases of general Cox–Snell (Cox and Snell, 1968) residuals as applied to censored data. Nelson (1973) shows how to use such residuals to detect departures from distributional assumptions and check for other kinds of model inadequacies. These ideas are illustrated in Section 17.6.2.

Cox–Snell residuals can also be used for checking model assumptions for non-location-scale distributions. The general definition of Cox–Snell residuals is as follows. Consider observed failure times t_i corresponding to random variables T_i

($i = 1, \ldots, n$). Suppose that the T_i are functions of explanatory variables x_i, model parameters $\boldsymbol{\theta}$, and a set of iid random deviations ϵ_i ($i = 1, \ldots, n$) having a distribution with no unknown parameters. This function defines the assumed model for T. Then if there is a function $w_i(T; x, \boldsymbol{\theta})$ such that

$$\epsilon_i = w_i(T_i; x_i, \boldsymbol{\theta}),$$

the Cox–Snell residuals are defined as and can be computed from

$$\widehat{\epsilon}_i = w(t_i; x_i, \widehat{\boldsymbol{\theta}}).$$

Here $\widehat{\boldsymbol{\theta}}$ is the ML estimate of $\boldsymbol{\theta}$ obtained using the assumed model, the data (t_i, x_i) ($i = 1, \ldots, n$), and the censoring information. For example, for the lognormal loglinear regression a choice is $w_i(t; x_i, \boldsymbol{\theta}) = \exp\{[\log(t) - \mu_i]/\sigma_i\}$, which yields the residuals in (17.14).

For a given model and data set, Cox–Snell residuals are not uniquely defined because any parameter-free one-to-one transformations of the $w_i(t; x_i; \widehat{\boldsymbol{\theta}})$ values would also satisfy the Cox–Snell definition [with a corresponding change in the definition of the deviations ϵ_i ($i = 1, \ldots, n$) in the assumed model]. For example, for the lognormal regression above, one can choose $w_i(t; x_i, \boldsymbol{\theta}) = [\log(t) - \mu_i]/\sigma_i$, which provides Cox–Snell residuals that are unrestricted in sign in contrast with the residuals in (17.14), which are all positive.

In general, when the observations are independent (possibly censored) and the T_i have strictly increasing cdfs $F(t; x_i, \boldsymbol{\theta})$, a natural choice is $w_i(t; x_i, \boldsymbol{\theta}) = F(t; x_i, \boldsymbol{\theta})$, which provides Cox–Snell residuals on $(0, 1)$ given by

$$\widehat{u}_i = F(t_i; x_i, \widehat{\boldsymbol{\theta}}). \qquad (17.15)$$

Because $F(T_i; x_i, \boldsymbol{\theta})$ has a UNIF(0, 1) distribution (see Exercise 2.16), these residuals should look like a (possibly censored) random sample from a UNIF(0, 1) distribution. For example, with the log-location-scale regression model above with $F(t; x_i, \boldsymbol{\theta}) = \Phi\{[\log(t) - \mu_i]/\sigma_i\}$, a set of Cox–Snell residuals in $(0, 1)$ are

$$\widehat{u}_i = F(t_i; x_i, \widehat{\boldsymbol{\theta}}) = \Phi\left[\frac{\log(t_i) - \widehat{\mu}_i}{\widehat{\sigma}_i}\right].$$

These residuals are related to the residuals in (17.14) through the transformation $\widehat{\epsilon}_i = \exp[\Phi^{-1}(\widehat{u}_i)]$. Thus the information in the $\widehat{u}_i$ residuals is equivalent to that in the $\widehat{\epsilon}_i$ residuals.

To assess the adequacy of the distributional assumption, one can use a P-P plot (e.g., page 66 of Crowder et al., 1991) of the residuals defined in (17.15). To do this, obtain the nonparametric estimate of the cdf of the residuals using the methods in Chapter 3. Then obtain probability plotting positions $p_{\widehat{u}}$ at each point $\widehat{u}$ at which the nonparametric estimate jumps (see Section 6.4). The P-P plot is obtained by plotting $\{\widehat{u}$ versus $p_{\widehat{u}}\}$ on linear axes. Strong departures from linearity in this plot indicate a departure from the assumed model. Note that the P-P plot can be viewed as a probability plot corresponding to a UNIF(0, 1) distribution.

CHECKING MODEL ASSUMPTIONS

Alternatively, one can transform the (possibly censored) $\widehat{u}$ values to $-\log(1-\widehat{u})$, which, under the assumed model, should look like a (possibly censored) sample from an EXP(1) distribution. In this case, the adequacy of the distributional assumption can be checked by making an exponential or a Weibull probability plot of the $-\log(1-\widehat{u})$ values, using the methods in Chapter 6. The Weibull probability plot provides much better resolution in the lower tail of the distribution.

Lawless (1982, pages 280–281) describes the use of probability plots for Cox–Snell residual analysis and Collett (1994, page 158) describes closely related Weibull Q-Q plots for Cox–Snell residuals.

17.6.2 Regression Diagnostics

Some suggested regression model diagnostics include:

- **Plot of standardized residuals versus fitted values.** As mentioned in Section 17.6.1, fitted values can be defined in several different ways. For any of the suggested definitions, plotting the residuals versus fitted values can help detect nonlinearity not modeled in the underlying relationship or nonconstant variability in life. Heavy censoring can, however, make such plots difficult to interpret.
- **Probability plot of standardized residuals.** When the data at different levels of the explanatory variables are censored at a common censoring time, it is possible that the computed residuals will be multiply censored. In such cases, one can use the methods in Chapters 3 and 6 to produce the probability plot.
- **Other residual plots.** Residuals can be plotted in a variety of other ways. For example, one might plot residuals against other potential explanatory variables not in the model to see if they provide any explanatory power. If data were collected sequentially over time, then plotting residuals versus observation order can help to detect process trends and cycles.
- **Influence (or sensitivity) analysis.** It is important to assess the degree to which estimates depend on model assumptions and other uncertain inputs to the data analysis process. In some cases, simple reanalysis under alternative model assumptions will suffice. It is, however, possible to systematize the process by, for example, dropping out one observation at a time and refitting the model to detect highly influential observations.

Additionally, most analytical tests commonly used to detect departures from an assumed model can be suitably generalized, at least approximately, for censored data (especially using likelihood ratio tests).

Example 17.10 Residual Plots for Model 1 and the Superalloy Fatigue Data.
Figure 17.8 shows the standardized residuals from (17.14) versus the fitted values for Model 1. This figure gives a strong indication that there is more variability among the observations with larger fitted values. Figure 17.9 is a Weibull probability plot of the standardized residuals, revealing the early outlying observation. Similar plots

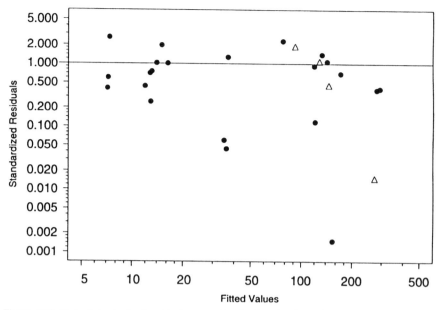

Figure 17.8. Plot of standardized residuals versus fitted values for the Weibull regression model with log-quadratic μ and constant σ, fit to the superalloy data on log–log axes.

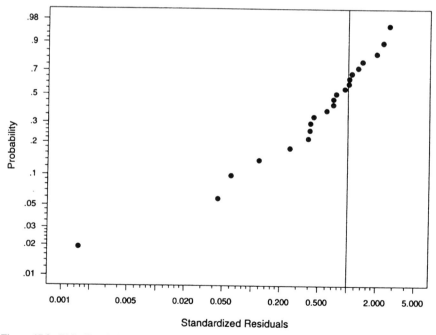

Figure 17.9. Weibull probability plot of the standardized residuals from the Weibull regression model with log-quadratic μ and constant σ, fit to the superalloy data.

for Model 2 (not shown here) are better behaved, but still show the early outlying observation. □

17.7 MODELS WITH TWO OR MORE EXPLANATORY VARIABLES

This section illustrates the use of regression models having two different explanatory variables. The approach generalizes to problems involving more than two explanatory variables.

17.7.1 Model-Free Graphical Analysis of Two-Variable Regression Data

As in simple regression, it is useful to view two-variable regression data graphically before making assumptions about the relationship between life and the explanatory variables. Probability plots and scatter plots are useful tools for doing this.

Example 17.11 Effect of Voltage and Temperature on Glass Capacitor Life.
Table 17.3 gives data from a factorial experiment on the life of glass capacitors as a function of voltage and operating temperature. The data were originally analyzed in

Table 17.3. Glass Capacitor Life Test Failure Times and Weibull ML Estimates for Each Temperature/Voltage Combination

Temperature			Applied Voltage			
			200	250	300	350
170°C	Hours to Failure		439	572	315	258
			904	690	315	258
			1092	904	439	347
			1105	1090	628	588
		$\widehat{\mu}$	7.13	7.10	6.57	6.54
		$\widehat{se}_{\widehat{\mu}}$	.15	.16	.21	.26
		$\widehat{\sigma}$	.26	.28	.37	.46
		$\widehat{se}_{\widehat{\sigma}}$	.13	.13	.17	.21
		max $\mathcal{L}(\mu, \sigma)$	−31.8	−31.7	−30.2	−30.3
180°C	Hours to Failure		959	216	241	241
			1065	315	315	241
			1065	455	332	435
			1087	473	380	455
		$\widehat{\mu}$	7.01	6.28	6.00	6.24
		$\widehat{se}_{\widehat{\mu}}$	.21	.16	.09	.17
		$\widehat{\sigma}$	.04	.28	.17	.30
		$\widehat{se}_{\widehat{\sigma}}$	.02	.13	.08	.14
		max $\mathcal{L}(\mu, \sigma)$	−24.8	−28.4	−26.0	−28.4

There were eight capacitors tested at each combination of temperature and voltage. Testing at each combination was terminated after the fourth failure, yielding failure-censored (Type II) data. The data are from Zelen (1959).

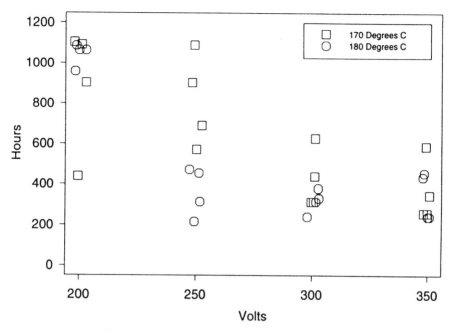

Figure 17.10. Scatter plot of glass capacitor life test data.

Zelen (1959), using a two-parameter exponential distribution. Figure 17.10 is a scatter plot of life in hours versus voltage, with different symbols for different temperatures. A small amount of "jitter" was introduced into the voltage variable before plotting so that the graph would separate the tied values. Figure 17.11, using methods from Chapters 6 and 8, contains Weibull probability plots for each of the eight individual test conditions. The straight lines on these plots are the ML estimates of Weibull cdfs. Table 17.3 summarizes numerical results. □

17.7.2 Two-Variable Regression Model Without Interaction

The log-location-scale two-variable regression model uses (17.6) with $\mu = \beta_0 + \beta_1 x_1 + \beta_2 x_2$. When σ does not depend on $x = (x_1, x_2)$, the p quantile of the life distribution at a specified x is

$$\log[t_p(x)] = y_p(x) = \beta_0 + \beta_1 x_1 + \beta_2 x_2 + \Phi^{-1}(p)\sigma. \qquad (17.16)$$

Re-expressing the quantile function gives $t_p(x) = \exp[y_p(x)] = \exp(\beta_1 x_1 + \beta_2 x_2) t_p(0)$, showing that this is a SAFT model with $\mathcal{AF}(x) = 1/\exp(\beta_1 x_1 + \beta_2 x_2)$. Substituting $\mu_i = \beta_0 + \beta_1 x_{1i} + \beta_2 x_{2i}$ into (17.8) gives a likelihood having the form $L(\beta_0, \beta_1, \beta_2, \sigma)$. The model generalizes easily when there are more than two explanatory variables. In the model, β_0 is the value of μ when $x_1 = x_2 = 0$ (if this makes sense physically), β_1 is the change in μ for a one-unit change in x_1 (holding x_2 constant), and β_2 is the change in μ for a one-unit change in x_2 (holding x_1 constant). In

MODELS WITH TWO OR MORE EXPLANATORY VARIABLES   449

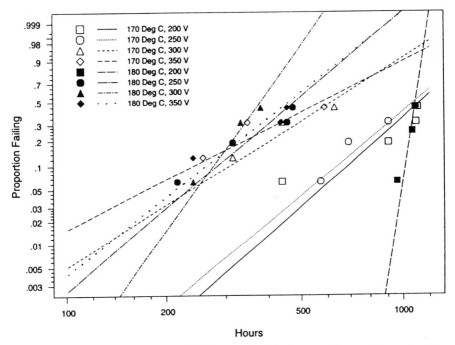

Figure 17.11. Weibull probability plots of glass capacitor life data at each temperature and voltage condition, each with its Weibull ML estimate of $F(t)$ plotted as a straight line.

particular, the effect of changing x_1 does not depend on the level of x_2 and vice versa. This model is known as the "additive" or "no-interaction" two-variable regression model.

17.7.3 Two-Variable Regression Model with Interaction

The log-location-scale two-variable regression model *with* interaction uses (17.6) with $\mu = \beta_0 + \beta_1 x_1 + \beta_2 x_2 + \beta_3 x_1 x_2$. If σ does not depend on $\boldsymbol{x}$, the log-quantile function is

$$\log[t_p(\boldsymbol{x})] = y_p(\boldsymbol{x}) = \beta_0 + \beta_1 x_1 + \beta_2 x_2 + \beta_3 x_1 x_2 + \Phi^{-1}(p)\sigma.$$

The distribution quantile can be reexpressed as $t_p(\boldsymbol{x}) = \exp[y_p(\boldsymbol{x})] = \exp(\beta_1 x_1 + \beta_2 x_2 + \beta_3 x_1 x_2) t_p(\boldsymbol{0})$. Thus this is a SAFT model with $\mathcal{AF}(\boldsymbol{x}) = 1/\exp(\beta_1 x_1 + \beta_2 x_2 + \beta_3 x_1 x_2)$. Substituting $\mu_i = \beta_0 + \beta_1 x_{1i} + \beta_2 x_{2i} + \beta_3 x_{1i} x_{2i}$ into (17.8) gives a likelihood having the form $L(\beta_0, \beta_1, \beta_2, \beta_3, \sigma)$. In this model $\beta_1 + \beta_3 x_2$ is the change in μ for a one-unit change in x_1. Similarly, $\beta_2 + \beta_3 x_1$ is the change in μ for a one-unit change in x_2.

Example 17.12   Regression Models for Glass Capacitor Life. The top half of Table 17.4 provides information on the ML fit of Model 1, the two-variable no-interaction regression model with $x_1 =$ temperature in °C and $x_2 =$ voltage. The

Table 17.4. Glass Capacitor Life Test ML Estimates for the Weibull Regression Model

	Parameter	ML Estimate	Standard Error	Approximate 95% Confidence Interval	
				Lower	Upper
Model 1	β_0	13.41	2.30	8.90	17.91
	β_1	−.02	.01	−.05	−.004
	β_2	−.006	.001	−.008	−.004
	σ	0.36	.055	.27	.49
Model 2	β_0	9.41	10.5	−11.2	30.1
	β_1	−.0062	.060	−.12	.11
	β_2	.0086	.037	−.065	.082
	β_3	.000082	.00021	−.00050	.00034
	σ	.362	.055	.27	.49

The log likelihoods for Models 1 and 2 are, respectively, $\mathcal{L}_1 = -244.24$ and $\mathcal{L}_2 = -244.17$.

bottom half of Table 17.4 provides information on the ML estimates for Model 2. This model includes the interaction term $x_3 = x_1 x_2$. Comparing the log-likelihood values for the two models indicates that the interaction term has not improved the model's ability to explain variability in the failure times. In particular, the log-likelihood ratio statistic $-2 \times (\mathcal{L}_1 - \mathcal{L}_2) = .14$ is small relative to $\chi^2_{(.75;1)} = 1.32$ (from standard chi-square quantile tables). Figure 17.12 is similar to Figure 17.11, but in this case, the parallel lines are the Model 2 ML estimates for each of the eight test conditions. The lines are parallel because σ in the model does not depend on $x = (x_1, x_2)$.

The plotted points in Figure 17.13 are the individual estimates of $t_{.5}$ for each of the eight combinations of the glass capacitor test conditions (computed from the ML estimates in Table 17.3). The lines in Figure 17.13 give the Model 2 fitted relationship between life and voltage for temperatures of 170°C and 180°C. The Model 1 plot (not shown here) was similar, but the 170°C and 180°C lines were exactly parallel. Figure 17.14 is a plot of the Model 2 regression residuals. There does not seem to be any serious departure from the fitted model or the Weibull distribution assumptions in the model. □

17.8 PRODUCT COMPARISON: AN INDICATOR-VARIABLE REGRESSION MODEL

This section illustrates methods for comparing samples from two different populations or processes, which we generically call "groups." In the first analysis (Examples 17.13 and 17.14) the samples are simply analyzed separately to make a comparison. The second analysis (Example 17.15) uses an indicator-variable regression model to compare the samples under the assumption that the spread parameter σ is the same for both groups.

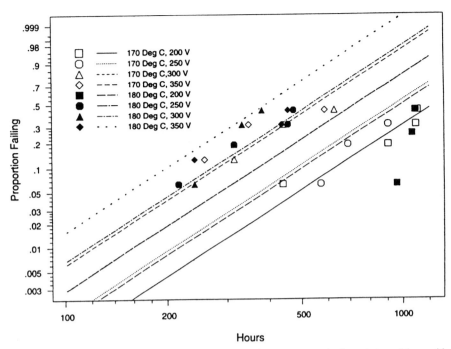

Figure 17.12. Weibull probability plots of data at individual temperature and voltage test conditions, with the Weibull regression Model 2 ML estimates of $F(t)$ at each set of conditions for the glass capacitor data.

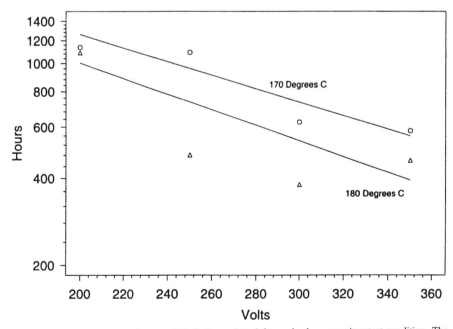

Figure 17.13. Individual estimates of Weibull $t_{.5}$ plotted for each glass capacitor test condition. The straight lines depict the with-interaction (Model 2) Weibull regression model estimate of $t_{.5}$ versus voltage for 170°C and 180°C.

451

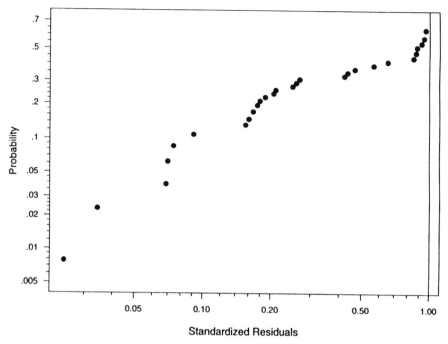

Figure 17.14. Weibull probability plots of the residuals from Model 2 fit to the glass capacitor data.

17.8.1 Comparison of Groups Using Separate Analyses

The simplest method for comparing two groups is to analyze the groups separately.

Example 17.13 Comparison of Snubber Designs—Separate Analyses. Appendix Table C.4 gives data from Nelson (1982, page 529), from a life test to compare two different snubber designs. A snubber is a component in an electric toaster. Following the analysis used by Nelson, Figure 17.15 shows the results of fitting separate normal distributions to the data from each snubber design. The different slopes reflect the different standard deviation estimates. The numerical results are summarized in Table 17.5, under Model 1. ☐

As illustrated in Figure 17.15, comparing distributions may not be straightforward in applications where the cdfs from the two groups cross. If we took the lines in Figure 17.15 to be true cdfs, the implication would be that the old design is better up until about 550 cycles, after which the new design is better. The ambiguity arises because the estimate for σ from the new design is larger. It is, however, possible to compare particular points on the cdfs.

Example 17.14 Comparison of Snubber Designs—Comparing Quantiles. To make a quantitative comparison between the two designs, we will estimate the difference between the values of $y_{.5}$(new) and $y_{.5}$(old) (we use y instead of t here because

PRODUCT COMPARISON: AN INDICATOR-VARIABLE REGRESSION MODEL

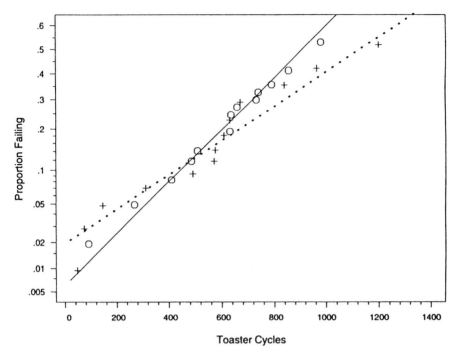

Figure 17.15. Normal probability plot showing separate analyses comparing the old and new snubber designs (old design "○" and the new design "+"). The solid (dotted) line is the ML estimate of the normal distribution cdf for the old (new) design.

Table 17.5. ML Estimates for the Snubber Life Test Data

	Parameter	ML Estimate	Standard Error	Approximate 95% Confidence Interval	
				Lower	Upper
Model 1	μ_{old}	908	76.2	759	1057
	σ_{old}	362	63.4	238	487
	μ_{new}	1126	123	885	1368
	σ_{new}	546	100	351	741
Model 2	β_0	975	89.1	800	1149
	β_1	86.7	114	−137	311
	σ	459	57.7	346	572

For Model 1, $\mathcal{L}_{old} = -138.6$ for the old design and $\mathcal{L}_{new} = -146.8$ for the new design (totaling $\mathcal{L}_1 = \mathcal{L}_{old} + \mathcal{L}_{new} = -285.4$). For Model 2, $\mathcal{L}_2 = -286.7$.

we are fitting a normal distribution, which has a theoretical range extending to negative numbers). For the normal distribution, this is the same as comparing the means of the two groups. Using the results for Model 1 in Table 17.5,

$$\widehat{\mu}_{\text{new}} - \widehat{\mu}_{\text{old}} = 1126 - 908 = 218,$$

$$\widehat{\text{se}}_{\widehat{\mu}_{\text{new}} - \widehat{\mu}_{\text{old}}} = \sqrt{\widehat{\text{se}}^2_{\widehat{\mu}_{\text{new}}} + \widehat{\text{se}}^2_{\widehat{\mu}_{\text{old}}}}$$
$$= \sqrt{(76.2)^2 + (123)^2} = 144.7,$$

and an approximate 95% confidence interval for $\delta = \mu_{\text{new}} - \mu_{\text{old}}$ is

$$[\utilde{\delta}, \widetilde{\delta}] = \widehat{\mu}_{\text{new}} - \widehat{\mu}_{\text{old}} \pm z_{(1-\alpha/2)}\widehat{\text{se}}_{\widehat{\mu}_{\text{new}} - \widehat{\mu}_{\text{old}}}$$
$$= 218 \pm 1.96 \times 144.7 = [-66, \quad 501].$$

Because this interval contains 0, we conclude that there is not a convincing difference between the mean cycles to failure for the two designs. □

17.8.2 Comparison of Groups Using Combined Analyses

In some situations it might be reasonable to use a model in which σ is the same across the groups, with differences only in the values of μ. The analysis can be made with a simple regression relationship using $\mu = \beta_0 + \beta_1 x$, where $x = 0$ for one group and $x = 1$ for the other. Then substituting x into the model gives $\mu(0) = \beta_0$ and $\mu(1) = \beta_0 + \beta_1$. Furthermore $\delta = t_p(1) - t_p(0) = \mu(1) - \mu(0) = \beta_1$, making comparisons with a common σ less ambiguous because, in this model, δ does not depend on which quantile is compared.

Example 17.15 Comparison of Snubber Designs—Common σ Analysis. Figure 17.16 displays estimates of the cdfs for the new and old designs. Because in this model σ is the same for both designs, the fitted lines are parallel. Table 17.5 gives corresponding numerical results. The dotted curves on Figure 17.16 are pointwise approximate 95% confidence intervals for $F(t)$ for the old design. Although these intervals do not answer precisely the question of interest, they do indicate that there is considerable variability in the estimates, relative to the difference between the point estimates of the distributions for the old and the new designs. An approximate 95% confidence interval for $\delta = t_p(\text{new}) - t_p(\text{old}) = \mu_{\text{new}} - \mu_{\text{old}} = \beta_1$ is

$$[\utilde{\delta}, \widetilde{\delta}] = [\utilde{\beta}_1, \widetilde{\beta}_1] = \widehat{\beta}_1 \pm z_{(1-\alpha/2)}\widehat{\text{se}}_{\widehat{\beta}_1}$$
$$= 86.7 \pm 1.96 \times 114 = [-137, \quad 311].$$

Because this interval contains zero, we conclude, as before, that there is not a convincing difference between the new and old designs. □

THE PROPORTIONAL HAZARDS FAILURE-TIME MODEL

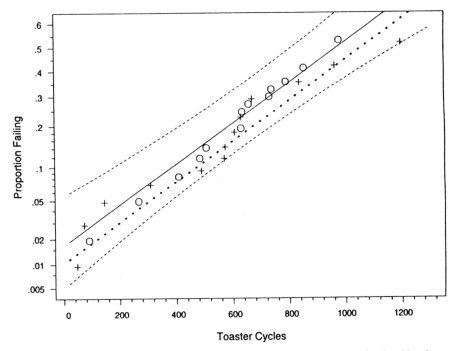

Figure 17.16. Normal probability plot summarizing the fitted common-σ models for the old and new snubber designs. Observations from the old design are indicated by a "○" and observations from the new design are indicated by a "+." The solid (dotted) line is the ML estimate of the normal distribution cdf for the old (new) design. The dashed lines are approximate 95% pointwise confidence intervals for the old design.

17.9 THE PROPORTIONAL HAZARDS FAILURE-TIME MODEL

17.9.1 Proportional Hazards Relationships

The proportional hazards (PH) model relates the hazard functions at conditions x and baseline conditions x_0 by

$$h(t; x) = \Psi(x) h(t; x_0) \tag{17.17}$$

for all $t > 0$, where $\Psi(x)$, like $\mathcal{AF}(x)$ in Section 17.2.2, is a positive function with $\Psi(x_0) = 1$. The proportional hazards model can also be written as

$$S(t; x) = [S(t; x_0)]^{\Psi(x)} \tag{17.18}$$

or

$$F(t; x) = 1 - [1 - F(t; x_0)]^{\Psi(x)}. \tag{17.19}$$

Again, if $\Psi(x) \neq 1$, then $F(t;x)$ and $F(t;x_0)$ do not cross. When $\Psi(x) > 1$, the model accelerates time in the sense that $F(t;x) > F(t;x_0)$ for all t. When $\Psi(x) < 1$, the model decelerates time in the sense that $F(t;x) < F(t;x_0)$ for all t.

Reexpressing (17.19) leads to $1 - F(t;x) = [1 - F(t;x_0)]^{\Psi(x)}$ and taking logs (twice) gives

$$\log\{-\log[1 - F(t;x)]\} - \log\{-\log[1 - F(t;x_0)]\} = \log[\Psi(x)]. \qquad (17.20)$$

Thus when $F(t;x)$ and $F(t;x_0)$ are related by a PH model, they are vertically equidistant at any given t on Weibull probability paper, as shown in Figure 17.17. This graphical relationship is useful for assessing the reasonableness of a PH regression model.

A proportional hazards model can also be expressed as a failure time transformation model. In particular, if $T(x_0) \sim F(t;x_0)$ and if $T(x)$ and $T(x_0)$ are related by the time transformation function

$$T(x) = F^{-1}\left(1 - \{1 - F[T(x_0);x_0]\}^{1/\Psi(x)}; x_0\right) \qquad (17.21)$$

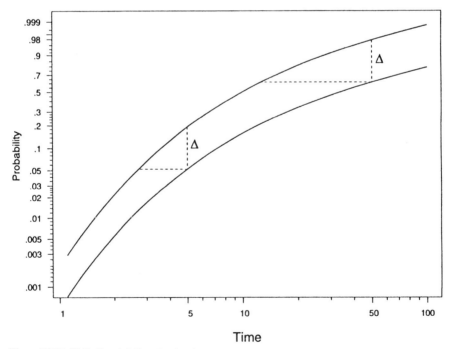

Figure 17.17. Weibull probability plot showing a proportional hazards accelerated failure time regression relationship with a lognormal baseline distribution (lower line). The upper line corresponds to a cdf from the power-lognormal distribution.

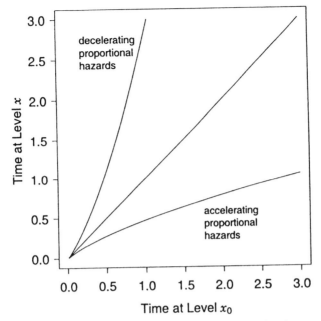

Figure 17.18. Proportional hazards model with a lognormal baseline hazard function expressed as a time transformation function.

then it can be shown that $T(x)$ and $T(x_0)$ have the PH relationship in (17.17). This time transformation function is illustrated in Figure 17.18. In this example, the amount of acceleration (or deceleration), $T(x_0)/T(x)$, depends on the position in time.

17.9.2 The Weibull Proportional Hazards Model

For the Weibull distribution (and only the Weibull distribution), a PH regression model is also a SAFT regression model. This can be seen by noting that the Weibull distribution is the only distribution in which both (17.2) and (17.20) hold. Relatedly, Weibull probability plots of Weibull cdfs with the same σ are translations of each other in *both* the probability and the log(t) scale. Thus the Weibull cdfs at x and x_0 are parallel straight lines on Weibull probability paper, as shown in Figure 17.19.

17.9.3 Other Proportional Hazards Models

Except in the case of the Weibull distribution, the PH regression relationship changes the shape of the underlying distribution as a function of the explanatory variables x. That is, in general, the PH model does not preserve the form of baseline distribution. For example, if $T(x_0)$ has a lognormal distribution then $T(x)$ has a power-lognormal distribution (Section 5.12.3).

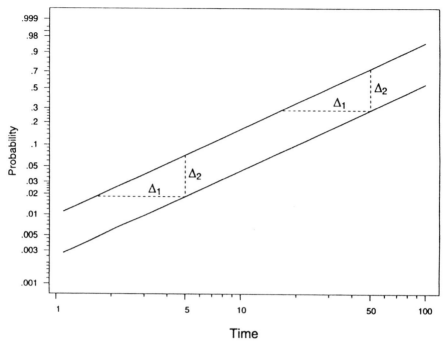

Figure 17.19. Weibull probability plot of two Weibull distributions. The parallel straight lines here come from a regression relationship that is both accelerated failure time and proportional hazards.

17.9.4 The Semiparametric (Cox) Proportional Hazards Model

In its semiparametric form, the PH model in (17.17) is known as the Cox proportional hazards model. In this model, the form of the hazard function $h(t; x_0)$ is unspecified. One can estimate the regression coefficients [the parametric part of the model $\Psi(x)$] and obtain nonparametric estimates of the hazard (or cdf or survival) functions at any specified values of the explanatory variable(s). The Cox PH model is widely used in biomedical applications and especially in the analysis of clinical-trial data.

17.9.5 PH Model Applications in Reliability

Because models of failure based on physics and chemistry (see Chapter 18) generally suggest a SAFT model (or other non-PH models), the main area for potential application of PH models would appear to be in the analysis of field reliability data for which it is necessary to adjust for covariates like operating environment, use-rate, and so on. Bendell (1985) and Dale (1985) describe applications of semiparametric proportional hazards modeling to the analysis of reliability data.

Landers and Kolarik (1987) describe an application where a parametric PH model was used in the analysis of field reliability data. They, however, used a Weibull baseline distribution and so their model was really the same as the SAFT Weibull regression model used in Sections 17.2–17.5.

17.10 GENERAL TIME TRANSFORMATION FUNCTIONS

In Sections 17.2 and 17.9 we expressed both SAFT and PH models as special time transformation functions. Time transformation functions provide a general model for relating time at one level of x with time at another level of x. This can be expressed as

$$T(x) = Y[T(x_0), x],$$

where x_0 are again *baseline* conditions. To be a time transformation, the function $Y(t, x)$ must have the following properties:

- For any x, $Y(0, x) = 0$, as in Figure 17.20.
- $Y(t, x)$ is nonnegative, that is, $Y(t, x) \geq 0$ for all t and x.
- For fixed x, $Y(t, x)$ is monotone increasing in t.
- When evaluated at x_0, the transformation is the identity transformation [i.e., $Y(t, x_0) = t$ for all t].

A quantile of the distribution of $T(x)$ can be determined as a function of the corresponding quantile of the distribution of $T(x_0)$ and x. In particular, $t_p(x) = Y[t_p(x_0), x]$

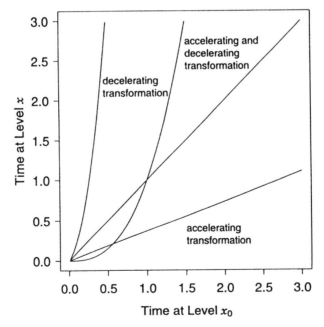

Figure 17.20. General time transformation model.

for $0 \leq p \leq 1$. As shown in Figure 17.20, a plot of $T(x_0)$ versus $T(x)$ can imply a particular class of transformation functions. In particular:

- $T(x)$ entirely below the diagonal line implies acceleration.
- $T(x)$ entirely above the diagonal line implies deceleration.
- $T(x)$ can cross the diagonal, in which case the transformation is accelerating over some times and decelerating over other times. In this case the cdfs of $T(x)$ and $T(x_0)$ cross.

BIBLIOGRAPHIC NOTES

Seber (1977) and Neter, Kutner, Nachtsheim, and Wasserman (1996) are useful references on linear regression analysis for complete (uncensored) data. Seber and Wild (1989) describe methods for nonlinear regression. Lawless (1982), Cox and Oakes (1984), and Nelson (1990a) describe applications of failure-time regression models and illustrate fitting methods for censored data. Nelson (1990a) describes failure time regression modeling, analysis, and test plans with applications to accelerated life testing, a topic covered in Chapters 18 and 19 of this book. The ideas behind Figures 17.4, 17.18, and 17.19 came from Lawless (1986). Nelson (1984) analyzed the superalloy fatigue data used in Examples 17.2, 17.8, and 17.9. Nelson (1984) also used a model in which both μ and σ (parameters of the lognormal distribution) depend on a stress variable and outlined the pitfalls of using this model. Our development has followed this work. Escobar and Meeker (1992) give expressions for computing the elements in (17.9) and present methods for doing influence analysis with censored data. Meeker and LuValle (1995) describe a regression model in which two different chemical reaction rate constants depend on temperature, leading to a non-SAFT model. Martin (1982) describes the use of general time transformation functions used in life testing. General time transformation functions are discussed, in the context of accelerated testing, in Chapter 17 of Ushakov (1994).

Nelson (1973) describes methods for residual analysis with censored data. Schmee and Hahn (1979) present an iterative least squares method of finding estimates for regression parameters with censored data. The method is useful for finding starting values for ML estimation.

Kalbfleisch and Prentice (1980), Lawless (1982), and Cox and Oakes (1984) provide detailed treatments of the important facets of the Cox proportional hazards model. Bagdonavičius and Nikulin (1995, 1997) describe general classes of semiparametric failure-time regression methods, methods of estimation, and asymptotic theory for the estimators.

EXERCISES

17.1. The confidence intervals for Example 17.3 given in Table 17.1 were computed by using the normal-approximation method that is commonly used for censored data problems. For this data set, with the lognormal model, it is

EXERCISES

possible to use standard "exact" methods to compute confidence intervals for these parameters, based on a simple ordinary least squares regression relating the logarithms of execution time to system load. Formulas for these exact methods are available in almost any elementary statistics text covering regression analysis. Use a computer program to do the necessary least squares computations and compute, by hand, the "exact" confidence intervals. Compare the estimates and confidence intervals. What is your conclusion?

17.2. Refer to the loglinear model in (17.7). Show why $100\beta_1$ can be interpreted as the approximate percent increase in $t_p(x)$ for a one-unit increase in x.

17.3. McCool (1980) gives the results of a life test on rolling contact fatigue of ceramic ball bearings. Ten specimens were tested at each of four levels of stress. The ordered failure times are given in the following table. McCool indicates that it is customary to model such data with the two-parameter Weibull distribution with a shape parameter that does not depend on stress.

Stress (10^6 psi)	Ordered Lifetimes (10^6 revolutions)
.87	1.67, 2.20, 2.51, 3.00, 3.90, 4.70, 7.53, 14.70, 27.80, 37.40
.99	.80, 1.00, 1.37, 2.25, 2.95, 3.70, 6.07, 6.65, 7.05, 7.37
1.09	.012, .18, .20, .24, .26, .32, .32, .42, .44, .88
1.18	.073, .098, .117, .135, .175, .262, .270, .350, .386, .456

(a) Plot the failure times versus stress on log–log axes (or, alternatively, take logs and plot on linear axes).

(b) It is often suggested that median failure time is proportional to a power transformation of stress: that is, $t_{.5} = e^{\beta_0} \times (\text{stress})^{\beta_1}$ or $\log(t_{.5}) = \beta_0 + \beta_1 \log(\text{stress})$. Is the suggestion reasonable in this case? Plot the sample medians on the graph in part (a) to help answer this question.

(c) Use a hand-drawn line through sample medians versus stress points to obtain a graphical estimate of the exponent (or slope) β_1.

(d) Make separate Weibull probability plots for the data at each level of stress, plotting them all on the same graph. What does this plot suggest about the customary assumption that the Weibull shape parameter is the same for all levels of stress? Provide possible explanations for the observed differences in the estimates of the Weibull shape parameter at each level of stress.

17.4. Figure 17.2 shows that there is more spread (variability) in the observed log failure times at low stress, as compared with high stress. Provide a simple, nontechnical, intuitive explanation for this common behavior of fatigue data.

▲17.5. Refer to Exercise 17.3. Suppose that log life can be described adequately by a normal distribution. Consider the regression model $\mu = \beta_0 + \beta_1 \log(\text{stress})$ with constant σ, where σ is the standard deviation of log life, the same for any fixed levels of stress.

(a) Use ordinary least squares to fit this model to the rolling contact fatigue data. Compare your answer to the graphical estimate from part (c) of Exercise 17.3.

(b) Use a computer program that does maximum likelihood estimation and fit the lognormal regression model (17.7) to the rolling contact fatigue data.

(c) Compute and compare estimates from these two different methods of estimation for the median time to failure at 1.05×10^6 psi, the .01 quantile at 1.05×10^6 psi, and the .01 quantile at $.85 \times 10^6$ psi. Comment on the results of this comparison.

17.6. Use the results in Table 17.2 for Example 17.8 to compute an 80% normal-approximation confidence interval for σ. Explain the interpretation of this interval.

17.7. Use the results in Table 17.2 for Example 17.9 to do the following:

(a) Compute a 90% normal-approximation confidence interval for $\beta_1^{[\sigma]}$.

(b) Explain how this confidence interval can be used to judge whether σ depends on the level of pseudostress or not.

(c) Do a likelihood ratio test to determine whether or not there exists strong evidence that σ depends on the level of pseudostress.

(d) Explain the steps you would follow to compute a likelihood-ratio-based confidence interval for $\beta_1^{[\sigma]}$.

▲17.8. Refer to Example 17.9 and Figure 17.7. It has been suggested that the failure at 13,949 cycles with pseudostress equal to 85.2 ksi appears to be an outlier.

(a) Use the ML estimates for Model 2 in Table 17.2 as if they were the actual parameter values and compute the probability that one would have a failure before 13,949 cycles when running at a pseudostress equal to 85.2 ksi.

(b) Following the approach in part (a), if 26 units were to be tested at a pseudostress equal to 85.2 ksi, what is the probability that the *earliest* failure would occur before 13,949 cycles?

(c) There were 26 observations in the superalloy fatigue example. Explain why the probability in part (b) is the relevant probability to consider when judging whether the single observation in the example departs importantly from the assumed model or not.

17.9. Consider the model used in Example 17.9 and the corresponding ML estimates for Model 2 in Table 17.2.

(a) Show that it is possible to have $t_p(x_1) < t_p(x_2)$ when $x_1 < x_2$.

(b) Explain why the relationship in part (a) is physically unreasonable.

▲17.10. Follow the general approach outlined in Section 17.4.2 and Example 17.7 to do the following for Model 2 used in Example 17.9:

(a) Using the results in Table 17.2, compute estimates $(\widehat{\mu}, \widehat{\sigma})$ for pseudostress 100 ksi.

(b) Derive the expressions needed to compute $\widehat{\Sigma}_{\widehat{\mu},\widehat{\sigma}}$ as a function of the elements of the variance–covariance matrix of the regression parameters in Model 2.

▲17.11. Refer to Exercise 17.3. There is some evidence that the Weibull shape parameter depends on stress, but it might be argued that the observed differences are due to random variation in the data.

(a) Fit separate two-parameter Weibull distributions to the data at each level of stress.

(b) Fit a regression model with an indicator variable allowing for a different Weibull scale parameter $\eta = \exp(\mu)$ at each level of stress. Hold $\sigma = 1/\beta$ (the reciprocal of the Weibull shape parameter β) constant over all levels of stress.

(c) Use (a) and (b) to do a likelihood ratio test to see if there is evidence that the values of σ differ with stress.

17.12. Return to the capacitor life test data in Example 17.11. Physical theory suggests that the Weibull shape parameter $\beta = 1/\sigma$ will depend on temperature if there is unit-to-unit variability in both the initial level of dielectric degradation and in dielectric degradation rate. Insulation engineers were interested in using these data to see if there was evidence for a temperature or voltage effect on $\beta = 1/\sigma$.

(a) Use the fitted model in Table 17.3 giving the Weibull distribution estimates at each combination of the individual experimental conditions. Compute the sum of the log likelihoods for all of the conditions.

(b) Plot, on Weibull probability paper, the estimates of the Weibull cdfs for each test condition, as done in Figure 17.11.

(c) Fit an indicator-variable regression model to the capacitor life test data such that there is a constant value of σ but that a separate value of μ is estimated at each test condition.

(d) Compare the sum of the likelihoods from part (a) with the likelihood obtained from the model in part (c). Use these results to assess the evidence for nonconstant σ in these data. What do you conclude?

▲17.13. Write a time transformation function corresponding to model (17.12).

▲17.14. Show that (17.17) implies (17.18).

17.15. Refer to Example 17.12 and Figure 17.13. The two points corresponding to the subexperiments at 250 and 300 volts and 180°C seem to be out of line from the other points. Use the individual subexperiment results given in Table 17.3 to compute a confidence interval for $t_{.5}$ at 300 volts and 180°C. What does this suggest about the two outlying points?

17.16. Refer to Example 17.12 and the numerical results in Table 17.4. For both models, compute the ML estimates of $t_{.5}$ for the eight factor-level combinations and plot these on a graph like that in Figure 17.13. What do you conclude?

17.17. Consider the two-variable regression model with interaction given in Section 17.7.3. Show that $\beta_1 + \beta_3 x_2$ is the change in μ for a one-unit change in x_1.

17.18. In some reliability applications it is common to use transformations of explanatory variables in regression modeling (models involving such transformations are described in Chapter 18). Also, it might be possible to find a distribution other than the Weibull that will provide a better fit to the data. Use the glass capacitor data in Examples 17.11 and 17.12, and fit the following alternative models.

   (a) In the two-variable regression model, use the transformations $x_1 = 1/(\text{temp}°C + 273.15)$ (known as the Arrhenius relationship) and $x_2 = \log(\text{voltage})$ (known as the inverse power relationship). Compare estimates of $t_{.5}$ at the different levels of temperature and voltage used in the life test. Is there evidence that these factors affect life? Explain.

   (b) Repeat the analysis in part (a), using the normal distribution instead of the Weibull distribution. Again, compare estimates of $t_{.5}$ at the different levels of temperature and voltage used in the life test.

17.19. The Weibull SAFT regression model is also a PH model. Thus fitting a parametric PH model with a Weibull baseline distribution is equivalent to fitting a Weibull regression model in which $\mu = \log(\eta)$ is a function of explanatory variables and $\sigma = 1/\beta$ is constant.

   (a) Why is the Cox PH model called "semiparametric"?

   (b) What are some advantages of using the semiparametric Cox PH model, when compared to using the parametric Weibull model?

   (c) What are some advantages of using the parametric Weibull model, when compared to using the semiparametric Cox PH model?

▲**17.20.** Show that a parametric SAFT model is also a parametric PH model if and only if the underlying distribution is a Weibull. *Hint:* Divide the proof in the following steps:

(a) (Sufficient Condition). Assume that $T(x_0)$ has a Weibull distribution. Then show that $T(x) = T(x_0)/\mathcal{AF}(x)$ has Weibull distribution and that $h(t;x) = \Psi(x)h(t;x_0)$.

(b) (Necessary Condition). Assume that $T(x) = T(x_0)/\mathcal{AF}(x)$ and that $h(t;x) = \Psi(x)h(t;x_0)$. First show that $S[\mathcal{AF}(x)t;x_0] = [S(t;x_0)]^{\Psi(x)}$. Then argue that this can only be true if a Weibull probability plot of $F(t;x_0)$ is a straight line.

CHAPTER 18

Accelerated Test Models

Objectives

This chapter explains:

- Motivation and applications of accelerated testing.
- Connections between degradation and physical failure.
- Models for temperature acceleration.
- Models for voltage and pressure acceleration.
- How to compute time-acceleration factors.
- Other accelerated test models and their assumptions.

Overview

This chapter describes models used for accelerated tests and introduces concepts of physics of failure. Some aspects of the models introduced here follow from the degradation models introduced in Chapter 13. The acceleration models described here are fitted to data in Chapter 19 (accelerated life tests) and Chapter 21 (accelerated degradation tests). Section 18.1 motivates and describes the general methods for accelerating reliability tests. Sections 18.2, 18.3, and 18.4 describe, respectively, use-rate, temperature, and voltage acceleration. Section 18.5 describes some models with a combination of accelerating variables.

18.1 INTRODUCTION

18.1.1 Motivation

Today's manufacturers face strong pressure to develop new, higher technology products in record time, while improving productivity, product field reliability, and overall quality. This has motivated the development of methods like concurrent engineering and encouraged wider use of designed experiments for product and process improvement. The requirements for higher reliability have increased the need for more *up-front* testing of materials, components, and systems. This is in line with the modern quality philosophy for producing high-reliability products: achieve high reliability by improving the design and manufacturing processes; move away from reliance on inspection (or screening) to achieve high reliability.

Estimating the failure-time distribution or long-term performance of components of *high-reliability* products is particularly difficult. Most modern products are designed to operate without failure for years, decades, or longer. Thus few units will fail or degrade appreciably in a test of practical length at normal use conditions. For example, the design and construction of a communications satellite may allow only 8 months to test components that are expected to be in service for 10 or 15 years. For such applications, Accelerated Tests (ATs) are used widely in manufacturing industries, particularly to obtain timely information on the reliability of simple components and materials. There are difficult practical and statistical issues involved in accelerating the life of a complicated product that can fail in different ways. Generally, information from tests at high levels of one or more accelerating variables (e.g., use-rate, temperature, voltage, or pressure) is extrapolated, through a physically reasonable statistical model, to obtain estimates of life or long-term performance at lower, normal levels of the accelerating variable(s). In some cases, stress is increased or otherwise changed during the course of a test (step-stress and progressive-stress ATs). AT results are used in the reliability-design process to assess or demonstrate component and subsystem reliability, to certify components, to detect failure modes so that they can be corrected, to compare different manufacturers, and so forth. ATs have become increasingly important because of rapidly changing technologies, more complicated products with more components, higher customer expectations for better reliability, and the need for rapid product development.

18.1.2 Different Types of Acceleration

The term "acceleration" has many different meanings within the field of reliability, but the term generally implies making "time" (on whatever scale is used to measure device or component life) go more quickly, so that reliability information can be obtained more rapidly. Different types of reliability tests and screens are used in different phases of the product development/production processes. Table 18.1 outlines the kinds of reliability tests done at different stages of product design and production.

Table 18.1. Reliability Tests at Different Product Stages

Product Design		Product Production
Qualification Testing of Materials and Components	Prototype Testing of Systems and Subsystems	Production and Postproduction Screens for Systems and Subsystems
Accelerated Degradation Tests Accelerated Life Tests	Robust Design Test STRIFE Tests Test-and-Fix for Reliability Growth	Component Certification Burn-in Environmental Stress Screening

Generally, there is a need to do all of these tests as quickly as possible, and methods have been developed to accelerate all of these different types of tests and screens. The main focus of this chapter and Chapters 19, 20, and 21, and Section 22.5 will be Accelerated Life Tests and Accelerated Degradation Tests that are done during product design to assess reliability and qualify the use of proposed materials and components. A section at the end of Chapter 19 provides definitions and further discussion of the statistical aspects of the other types of acceleration as well as references to sources of further information.

18.1.3 Types of Responses

It is useful to distinguish between ATs on the basis of what is observed.

- **Accelerated Life Tests (ALTs).** One obtains information on the failure time (actual failure time or an interval containing the failure time) for units that fail and lower bounds for the failure time (also known as the running time or runout time) for units that do not fail.
- **Accelerated Degradation Tests (ADTs).** As described in Chapter 13, one observes, at one or more points in time, the amount of degradation for a unit (perhaps with measurement error).

Many of the underlying physical model assumptions, concepts, and practices are the same for ALTs and ADTs. In some cases, analysts use degradation-level data to define failure times, turning ADT data into ALT data (generally simplifying analysis, but often sacrificing useful information). There are close relationships between ALT and ADT models. Because of the different types of response, however, the actual models fitted to the data and methods of analysis differ. Analyses of ALT and ADT data are covered in Chapters 19 and 21, respectively. An important characteristic of both types of ATs is the need to extrapolate outside the range of available data: tests are done at accelerated conditions, but estimates are needed at use conditions. Such extrapolation requires strong model assumptions.

18.1.4 Methods of Acceleration

There are three different methods of accelerating a reliability test:

- Increase the use-rate of the product. Consider the reliability of a toaster, which is designed for a median lifetime of 20 years, assuming a usage rate of twice each day. If, instead, we test the toaster 365 times each day, we could reduce the median lifetime to about 40 days. Also, because it is not necessary to have all units fail in a life test, useful reliability information could be obtained in a matter of days instead of months.
- Increase the aging-rate of the product. For example, increasing the level of experimental variables like temperature or humidity can accelerate the chemical processes of certain failure mechanisms, such as chemical degradation (resulting

INTRODUCTION

in eventual weakening and failure) of an adhesive mechanical bond or the growth of a conducting filament across an insulator (eventually causing a short circuit).
- Increase the level of stress (e.g., temperature cycling, voltage, or pressure) under which test units operate. A unit will fail when its *strength* drops below applied stress. Thus a unit at a high stress will generally fail more rapidly than it would have failed at low stress.

Combinations of these methods of acceleration are also employed. Variables like voltage and temperature cycling can both increase the rate of an electrochemical reaction (thus accelerating the aging-rate) and increase stress relative to strength. In such situations, when the effect of an accelerating variable is complicated, there may not be enough physical knowledge to provide an adequate physical model for acceleration (and extrapolation). Empirical models may or may not be useful for extrapolation to use conditions.

18.1.5 Acceleration Models

Interpretation of accelerated test data requires models that relate accelerating variables like temperature, voltage, pressure, and size to time acceleration. For testing over some range of accelerating variables, one can fit a model to the data to describe the effect that the variables have on the failure-causing processes. The general idea is to test at high levels of the accelerating variable(s) to speed up failure processes and then to extrapolate to lower levels of the accelerating variable(s). For some situations, a physically reasonable statistical model may allow such extrapolation.

Physical Acceleration Models

For well-understood failure mechanisms, one may have a model based on physical/chemical theory that describes the failure-causing process over the range of the data and provides extrapolation to use conditions. The relationship between accelerating variables and the actual failure mechanism is usually extremely complicated. Often, however, one has a simple model that adequately describes the process. For example, failure may result from a complicated chemical process with many steps, but there may be one rate-limiting (or dominant) step and a good understanding of this part of the process may provide a model that is adequate for extrapolation.

Empirical Acceleration Models

When there is little understanding of the chemical or physical processes leading to failure, it may be impossible to develop a model based on physical/chemical theory. An empirical model may be the only alternative. An empirical model may provide an excellent fit to the available data but may provide nonsense extrapolations (e.g., the quadratic models used in Section 17.5). In some situations there may be extensive empirical experience with particular combinations of variables and failure mechanisms and this experience may provide the needed justification for extrapolation to use conditions.

18.2 USE-RATE ACCELERATION

Increasing use-rate will, for some components, accelerate failure-causing wear and degradation. Examples include:

- Running automobile engines, appliances, and similar products continuously or with higher than usual use-rates.
- Higher than usual cycling rates for relays and switches.
- Increasing the cycling rate (frequency) in fatigue testing.

There is a basic assumption underlying simple use-rate acceleration models. Useful life must be adequately modeled by cycles of operation and cycling rate (or frequency) should not affect the cycles-to-failure distribution. This is reasonable if cycling simulates actual use and if the cycling frequency is low enough that test units return to steady state after each cycle (e.g., cool down).

Example 18.1   Increased Cycling Rate for Low-Cycle Fatique Tests.   Fatigue life is typically measured in cycles to failure. To estimate low-cycle fatigue life of metal specimens, testing is done using cycling rates typically ranging between 10 and 50 Hz (where 1 Hz is one stress cycle per second), depending on material type and available test equipment. At 50 Hz, accumulation of 10^6 cycles would require about 5 hours of testing. Accumulation of 10^7 cycles would require about 2 days and 10^8 about 20 days. Higher frequencies are used in the study of high-cycle fatigue.   □

Testing at higher frequencies could shorten test times but could also affect the cycles-to-failure distribution due to specimen heating or other effects. In some complicated situations, wear rate or degradation rate depends on cycling frequency. Also, a product may deteriorate in stand-by as well as during actual use.

For a certain type of fatigue test, notched test specimens are subjected to cyclic loading, as shown in Figure 18.1. Because the notch is a point of highest stress, a crack will initiate and grow out of it. Cycling rates in such tests are generally increased to a point where crack growth or cycles to failure (two common responses) can still be measured without distortion. There is a danger, however, that increased temperature

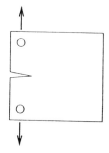

Figure 18.1.   Fatigue test notched "compact" specimen to which cyclic stress will be applied.

due to cycling rate will affect crack growth. This is especially true if there are effects like creep–fatigue interaction (see Dowling, 1993, page 706, for further discussion). In another example, there was concern that changes in cycling rate would affect the distribution of lubricant on a rolling bearing surface.

18.3 TEMPERATURE ACCELERATION

It is sometimes said that high temperature is the enemy of reliability. Increasing temperature is one of the most commonly used methods to accelerate a failure mechanism.

Example 18.2 Resistance Change of Carbon-Film Resistors. Appendix Table C.3 and Figure 18.2 show the percent increase in resistance over time for a sample of carbon-film resistors. These data were previously analyzed by Shiomi and Yanagisawa (1979) and Suzuki, Maki, and Yokogawa (1993). Samples of resistors were tested at each of three levels of temperature. At standard operating temperature (e.g., 50°C), carbon-film resistors will degrade slowly. Changes in resistance can cause reduced product performance or even system failures. The test was run at high levels of temperature to accelerate the chemical degradation process and obtain degradation data more quickly. Figure 18.2 shows that the resistors degrade more rapidly at high temperature. □

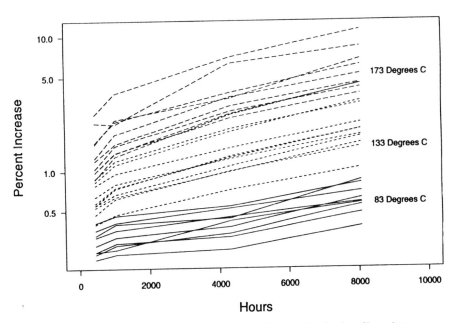

Figure 18.2. Percent increase in resistance over time for a sample of carbon-film resistors.

18.3.1 Arrhenius Relationship Time-Acceleration Factor

The Arrhenius relationship is a widely used model describing the effect that temperature has on the rate of a simple chemical reaction. This relationship can be written as

$$\mathcal{R}(\texttt{temp}) = \gamma_0 \exp\left(\frac{-E_a}{k_B \times \texttt{temp K}}\right) = \gamma_0 \exp\left(\frac{-E_a \times 11605}{\texttt{temp K}}\right) \quad (18.1)$$

where $\mathcal{R}$ is the reaction rate and $\texttt{temp K} = \texttt{temp }°C + 273.15$ is temperature in the absolute Kelvin scale, $k_B = 8.6171 \times 10^{-5} = 1/11605$ is Boltzmann's constant in electron volts per °C, and E_a is the activation energy in electron volts (eV). The parameters E_a and γ_0 are product or material characteristics. The Arrhenius acceleration factor is

$$\mathcal{AF}(\texttt{temp}, \texttt{temp}_U, E_a) = \frac{\mathcal{R}(\texttt{temp})}{\mathcal{R}(\texttt{temp}_U)} = \exp\left[E_a\left(\frac{11605}{\texttt{temp}_U \text{ K}} - \frac{11605}{\texttt{temp K}}\right)\right]. \quad (18.2)$$

When $\texttt{temp} > \texttt{temp}_U$, $\mathcal{AF}(\texttt{temp}, \texttt{temp}_U, E_a) > 1$. When $\texttt{temp}_U$ and E_a are understood to be, respectively, product use temperature and reaction-specific activation energy, $\mathcal{AF}(\texttt{temp}) = \mathcal{AF}(\texttt{temp}, \texttt{temp}_U, E_a)$ will be used to denote a time-acceleration factor. Figure 18.3 gives the acceleration factor

$$\mathcal{AF}(\texttt{temp}_{\text{High}}, \texttt{temp}_{\text{Low}}, E_a) = \exp(E_a \times \text{TDF})$$

as a function of E_a and the temperature differential factor (TDF) values

$$\text{TDF} = \left(\frac{11605}{\texttt{temp}_{\text{Low}} \text{ K}} - \frac{11605}{\texttt{temp}_{\text{High}} \text{ K}}\right)$$

given in Table 18.2.

The Arrhenius relationship does not apply to all temperature-acceleration problems and will be adequate over only a limited temperature range (depending on the particular application). Yet it is satisfactorily and widely used in many applications. Nelson (1990a, page 76) comments that "in certain applications (e.g., motor insulation), if the Arrhenius relationship... does not fit the data, the data are suspect rather than the relationship."

Example 18.3 *Arrhenius Time-Acceleration Factor for a Metallization Failure Mode.* An accelerated life test will be used to study a metallization failure mechanism for a transistor. Experience with this type of failure mechanism suggests that the activation energy should be in the neighborhood of $E_a = 1.2$. The usual operating junction temperature for the transistor is 90°C. To determine the acceleration factor for testing at 160°C, enter Table 18.2 with these temperatures and read TDF = 5.16. Then enter Figure 18.3 with this figure on the bottom and read up to the line with $E_a = 1.2$ eV, giving an acceleration factor of approximately 4.9×10^2 [or computed more precisely using (18.2) as 491]. □

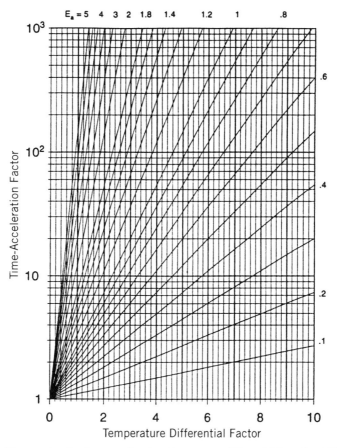

Figure 18.3. Time-acceleration factor as a function of temperature differential factor from Table 18.2 and activation energy E_a.

18.3.2 Eyring Relationship Time-Acceleration Factor

The Arrhenius relationship (18.1) was obtained through empirical observation. Eyring (e.g., Eyring, Gladstones, and Laidler, 1941, or Eyring, 1980) gives physical theory describing the effect that temperature has on a reaction rate. Written in terms of a reaction rate,

$$\mathcal{R}(\texttt{temp}) = \gamma_0 \times A(\texttt{temp}) \times \exp\left(\frac{-E_a}{k_B \times \texttt{temp K}}\right),$$

where $A(\texttt{temp})$ is a function of temperature depending on the specifics of the reaction dynamics and γ_0 and E_a are again constants (e.g., Weston and Schwarz, 1972, provide more detail). Applications in the literature have typically used $A(\texttt{temp}) = (\texttt{temp K})^m$ with a fixed value of m ranging between $m = 0$ (Boccaletti et al., 1989, page 379), $m = .5$ (Klinger, 1991a), to $m = 1$ (Nelson, 1990a, page 100; Mann, Schafer, and Singpurwalla, 1974, page 436).

Table 18.2. Temperature Differential Factors (TDFs) from the Arrhenius Time-Acceleration Model

Higher Temperature (°C)	Lower Temperature (°C)							
	30	40	50	60	70	80	90	100
80	5.42	4.20	3.05	1.97	0.96	0.00		
85	5.88	4.66	3.51	2.43	1.42	0.46		
90	6.32	5.10	3.96	2.88	1.86	0.90	0.00	
95	6.76	5.54	4.39	3.31	2.30	1.34	0.43	
100	7.18	5.96	4.81	3.73	2.72	1.76	0.86	0.00
105	7.59	6.37	5.22	4.14	3.13	2.17	1.27	0.41
110	7.99	6.77	5.62	4.55	3.53	2.57	1.67	0.81
115	8.38	7.16	6.01	4.94	3.92	2.96	2.06	1.20
120	8.76	7.54	6.39	5.32	4.30	3.34	2.44	1.58
125	9.13	7.91	6.76	5.69	4.67	3.71	2.81	1.95
130	9.49	8.27	7.13	6.05	5.03	4.08	3.17	2.31
135	9.85	8.63	7.48	6.40	5.39	4.43	3.52	2.67
140	10.19	8.97	7.82	6.74	5.73	4.77	3.87	3.01
145	10.53	9.31	8.16	7.08	6.07	5.11	4.20	3.35
150	10.86	9.63	8.49	7.41	6.39	5.44	4.53	3.67
155	11.18	9.95	8.81	7.73	6.71	5.76	4.85	3.99
160	11.49	10.27	9.12	8.04	7.03	6.07	5.16	4.31
165	11.79	10.57	9.43	8.35	7.33	6.37	5.47	4.61
170	12.09	10.87	9.72	8.65	7.63	6.67	5.77	4.91
175	12.39	11.16	10.02	8.94	7.92	6.97	6.06	5.20
180	12.67	11.45	10.30	9.22	8.21	7.25	6.35	5.49
185	12.95	11.73	10.58	9.50	8.49	7.53	6.63	5.77
190	13.22	12.00	10.85	9.78	8.76	7.80	6.90	6.04
195	13.49	12.27	11.12	10.04	9.03	8.07	7.17	6.31
200	13.75	12.53	11.38	10.31	9.29	8.33	7.43	6.57
205	14.01	12.79	11.64	10.56	9.55	8.59	7.69	6.83
210	14.26	13.04	11.89	10.81	9.80	8.84	7.94	7.08
215	14.51	13.28	12.14	11.06	10.04	9.09	8.18	7.33
220	14.75	13.53	12.38	11.30	10.29	9.33	8.42	7.57
225	14.98	13.76	12.62	11.54	10.52	9.56	8.66	7.80
230	15.22	13.99	12.85	11.77	10.75	9.80	8.89	8.03
235	15.44	14.22	13.07	12.00	10.98	10.02	9.12	8.26
240	15.67	14.44	13.30	12.22	11.20	10.25	9.34	8.48
245	15.88	14.66	13.51	12.44	11.42	10.46	9.56	8.70
250	16.10	14.88	13.73	12.65	11.64	10.68	9.77	8.92

TDF = $(11605/\text{temp}_{\text{low}} \text{ K}) - (11605/\text{temp}_{\text{high}} \text{ K})$ used as input to Figure 18.3.

TEMPERATURE ACCELERATION

The Eyring relationship temperature-acceleration factor is

$$\mathcal{AF}_{\text{Ey}}(\text{temp}, \text{temp}_U, E_a) = \left(\frac{\text{temp K}}{\text{temp}_U \text{ K}}\right)^m \times \mathcal{AF}_{\text{Ar}}(\text{temp}, \text{temp}_U, E_a),$$

where $\mathcal{AF}_{\text{Ar}}(\text{temp}, \text{temp}_U, E_a)$ is the Arrhenius acceleration factor from (18.2). For use over practical ranges of temperature acceleration, and when m is close to 0, the factor outside the exponential has relatively little effect on the acceleration factor and the additional term is often dropped in favor of the simpler Arrhenius relationship.

Example 18.4 Eyring Acceleration Factor for a Metallization Failure Mode. Returning to Example 18.3, the Eyring acceleration factor, using $m = 1$, is

$$\mathcal{AF}_{\text{Ey}}(160, 90, 1.2) = \left(\frac{160 + 273.15}{90 + 273.15}\right) \times \mathcal{AF}_{\text{Ar}}(160, 90, 1.2)$$
$$= 1.1935 \times 491 = 586,$$

where $\mathcal{AF}_{\text{Ar}}(160, 90, 1.2) = 491$ from Example 18.3. We see that, for a *fixed* value of E_a, the Eyring relationship predicts, in this case, an acceleration that is 19% greater than the Arrhenius relationship. As explained below, however, this figure exaggerates the practical difference between these models. □

When fitting models to limited data, the estimate of E_a depends strongly on the assumed value for m (e.g., 0 or 1). This dependency will compensate for and reduce the effect of changing the assumed value of m. Only with extremely large amounts of data would it be possible to adequately separate the effects of m and E_a using data alone. If m can be determined accurately on the basis of physical considerations, the Eyring relationship could lead to better low-stress extrapolations. With $m > 0$ the Eyring acceleration factor is larger than the Arrhenius acceleration factor. One argument in favor of the Arrhenius relationship (and perhaps a reason for its more common use) is that extrapolation to use-levels of temperature will be more conservative (i.e., predicting shorter life) than with the Eyring relationship with $m > 0$.

18.3.3 Reaction-Rate Acceleration for a Nonlinear Degradation Path Model

Some simple chemical degradation processes might be described by the following path model (previously used in Section 13.2.2):

$$\mathcal{D}(t; \text{temp}) = \mathcal{D}_\infty \times \left\{1 - \exp\left[-\mathcal{R}_U \times \mathcal{AF}(\text{temp}) \times t\right]\right\}, \tag{18.3}$$

where $\mathcal{R}_U$ is the reaction rate at use temperature temp_U, $\mathcal{R}_U \times \mathcal{AF}(\text{temp})$ is the reaction rate at a general temperature temp, and for $\text{temp} > \text{temp}_U$, $\mathcal{AF}(\text{temp}) > 1$. Figure 18.4 shows this function for fixed $\mathcal{R}_U$, E_a, and $\mathcal{D}_\infty$, but at different temperatures. Note from (18.3) that when $\mathcal{D}_\infty > 0$, $\mathcal{D}(t)$ is increasing and failure occurs when $\mathcal{D}(t) > \mathcal{D}_f$. For the example in Figure 18.4, however, $\mathcal{D}_\infty < 0$, $\mathcal{D}(t)$ is decreasing,

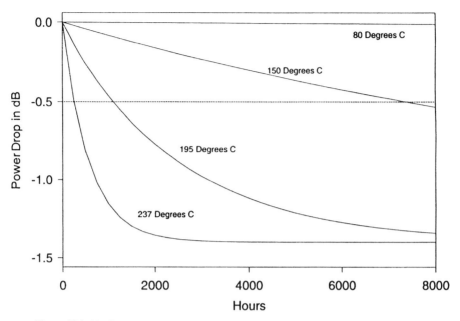

Figure 18.4. Nonlinear degradation paths at different temperatures with a SAFT relationship.

and failure occurs when $\mathcal{D}(t) < \mathcal{D}_f$. In either case, equating $\mathcal{D}(T; \texttt{temp})$ to $\mathcal{D}_f$ and solving for failure time gives

$$T(\texttt{temp}) = \frac{T(\texttt{temp}_U)}{\mathcal{AF}(\texttt{temp})}, \qquad (18.4)$$

where $T(\texttt{temp}_U) = -(1/\mathcal{R}_U)\log(1 - \mathcal{D}_f/\mathcal{D}_\infty)$ is failure time at use conditions. Faster degradation shortens time to any particular definition of failure (e.g., crossing $\mathcal{D}_f$ or some other specified level) by a *scale factor* that depends on temperature. Thus changing temperature is similar to changing the units of time. Consequently, the failure-time distributions at $\texttt{temp}_U$ and $\texttt{temp}$ are related by

$$\Pr[T(\texttt{temp}_U) \le t] = \Pr[T(\texttt{temp}) \le t/\mathcal{AF}(\texttt{temp})]. \qquad (18.5)$$

Equations (18.4) and (18.5) are forms of the scale-accelerated failure-time (SAFT) model introduced in Section 17.2.2.

With a SAFT model, for example, if $T(\texttt{temp}_U)$ (time at use or some other baseline temperature) has a log-location-scale distribution with parameters μ_U and σ, then

$$\Pr[T \le t; \texttt{temp}_U] = \Phi\left[\frac{\log(t) - \mu_U}{\sigma}\right].$$

At any other temperature,

$$\Pr[T \leq t; \texttt{temp}] = \Phi\left[\frac{\log(t) - \mu}{\sigma}\right],$$

where

$$\mu = \mu(x) = \mu_U - \log[\mathcal{AF}(\texttt{temp})] = \beta_0 + \beta_1 x, \qquad (18.6)$$

$x = 11605/(\texttt{temp K})$, $x_U = 11605/(\texttt{temp}_U \text{ K})$, $\beta_1 = E_a$, and $\beta_0 = \mu_U - \beta_1 x_U$. This is the same regression model used in Section 17.3.2 (e.g., for the lognormal, Weibull, and loglogistic distributions). LuValle, Welsher, and Svoboda (1988) and Klinger (1992) describe more general degradation model characteristics needed to assure that the SAFT property holds. Figure 18.5 shows a typical example of an Arrhenius relationship between life and temperature. Using an Arrhenius temperature axis and a log-life axis, the relationship plots as a family of straight lines. Because of the SAFT relationship, the logarithms of different lognormal distribution quantile lines all have the same slope.

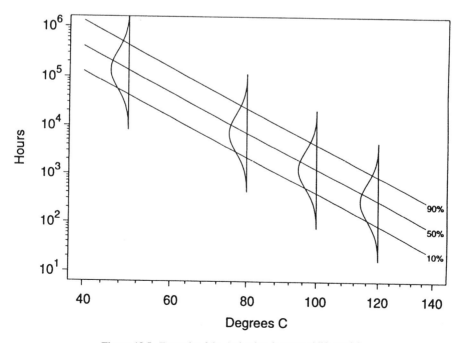

Figure 18.5. Example of the Arrhenius–lognormal life model.

18.3.4 Acceleration for a Linear Degradation Path Model

If $\mathcal{R}_U \times \mathcal{AF}(\texttt{temp}) \times t$ in (18.3) is small so that $\mathcal{D}(t)$ is small relative to $\mathcal{D}_\infty$, then

$$\mathcal{D}(t; \texttt{temp}) = \mathcal{D}_\infty \times \left\{1 - \exp\left[-\mathcal{R}_U \times \mathcal{AF}(\texttt{temp}) \times t\right]\right\}$$
$$\approx \mathcal{D}_\infty \times \mathcal{R}_U \times \mathcal{AF}(\texttt{temp}) \times t = \mathcal{R}_U^+ \times \mathcal{AF}(\texttt{temp}) \times t \quad (18.7)$$

is approximately linear in t. This is apparent when comparing the early-time behavior in Figure 18.4 with Figure 18.6. Also some degradation processes (e.g., automobile tire wear) are approximately linear in time. In this case, if $\mathcal{D}(0; \texttt{temp}) = 0$,

$$\mathcal{D}(t; \texttt{temp}) = \mathcal{R}_U^+ \times \mathcal{AF}(\texttt{temp}) \times t.$$

Again, $\mathcal{R}_U^+$ is the degradation rate at use conditions and $\mathcal{R}_U^+ \times \mathcal{AF}(\texttt{temp})$ is the degradation rate at general temperature $\texttt{temp}$. Failure occurs when $\mathcal{D}(t)$ crosses $\mathcal{D}_f$. Equating $\mathcal{D}(T; \texttt{temp})$ to $\mathcal{D}_f$ and solving for the failure time gives

$$T(\texttt{temp}) = \frac{T(\texttt{temp}_U)}{\mathcal{AF}(\texttt{temp})},$$

where $T(\texttt{temp}_U) = \mathcal{D}_f/\mathcal{R}_U^+$ is the failure time at use conditions. This is also a SAFT model. If $T(\texttt{temp})$ has a log-location-scale distribution, the parameters of the distribution can, as in Section 18.3.3, be expressed as $\mu = \beta_0 + \beta_1 x$ and a constant σ.

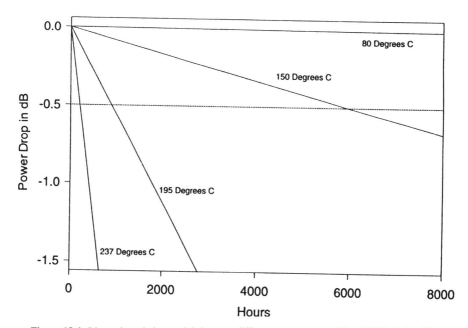

Figure 18.6. Linear degradation model shown at different temperatures with a SAFT relationship.

18.3.5 Acceleration of Parallel Chemical Reactions

Consider the more complicated chemical degradation path model having two separate reactions contributing to failure and described by

$$\mathcal{D}(t; \texttt{temp}) = \mathcal{D}_{1\infty} \times \left\{ 1 - \exp\left[-\mathcal{R}_{1U} \times \mathcal{AF}_1(\texttt{temp}) \times t\right] \right\}$$
$$+ \mathcal{D}_{2\infty} \times \left\{ 1 - \exp\left[-\mathcal{R}_{2U} \times \mathcal{AF}_2(\texttt{temp}) \times t\right] \right\}.$$

Here $\mathcal{R}_{1U}$ and $\mathcal{R}_{2U}$ are the use-condition rates of the two parallel reactions contributing to failure. Suppose that the Arrhenius relationship can be used to describe temperature dependence for these rates, providing acceleration functions $\mathcal{AF}_1(\texttt{temp})$ and $\mathcal{AF}_2(\texttt{temp})$. Then, unless $\mathcal{AF}_1(\texttt{temp}) = \mathcal{AF}_2(\texttt{temp})$ for all $\texttt{temp}$, this degradation model does *not* lead to a SAFT model. Intuitively, this is because temperature affects the two degradation processes differently, inducing a nonlinearity into the acceleration function relating times at two different temperatures. To obtain useful extrapolation models it is, in general, necessary to have adequate models for the important individual degradation processes.

In some situations (e.g., when the individual processes can be observed) it may be possible to use such a model by estimating the effect that temperature (or other accelerating variable) has on both $\mathcal{R}_{1U}$ and $\mathcal{R}_{2U}$.

18.4 VOLTAGE AND VOLTAGE-STRESS ACCELERATION

Increasing voltage or voltage stress (electric field) is another commonly used method to accelerate failure of electrical materials and components like light bulbs, capacitors, transformers, heaters, and insulation. Voltage is defined as the difference in electrical potential between two points. Physically it can be thought of as the amount of pressure behind an electrical current. Voltage stress across a dielectric is measured in units of volts/thickness (e.g., V/mm or kV/mm).

Example 18.5 Accelerated Life Test of a Mylar–Polyurethane Insulation. Appendix Table C.13 and Figure 18.7 show data from an ALT on a special type of mylar–polyurethane insulation used in high-performance electromagnets. The data, from Kalkanis and Rosso (1989), give time to dielectric breakdown of units tested at 100.3, 122.4, 157.1, 219.0, and 361.4 kV/mm. The purpose of the experiment was to evaluate the reliability of the insulating structure and to estimate the life distribution at system design voltages. The figure shows that failures occur much sooner at high voltage stress. Except for the data at 361.4 kV/mm, the relationship between log life and log voltage appears to be approximately linear. □

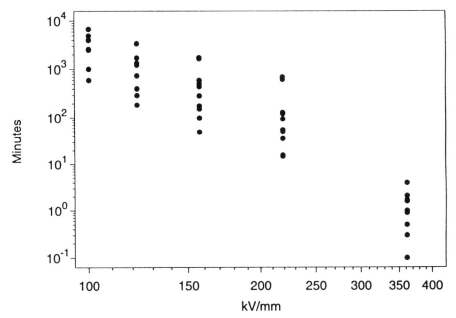

Figure 18.7. Times to dielectric breakdown of mylar–polyurethane insulation tested at 100.3, 122.4, 157.1, 219.0, and 361.4 kV/mm.

18.4.1 Voltage-Acceleration Mechanisms

Depending on the failure mode, raising voltage can:

- Increase the voltage-stress level relative to dielectric strength of a specimen. The dielectric strength of certain types of insulation will decline over time from chemical degradation.
- Increase the strength of the electric field, thereby accelerating some failure-causing electrochemical reactions or accelerating the growth of failure-causing discontinuities in the dielectric material.

Sometimes one or the other of these effects will be the primary cause of failure. In other cases, both effects will be important.

18.4.2 The Inverse Power Relationship

The most commonly used model for voltage acceleration is the "inverse power relationship" (also known as the "inverse power rule" and the 'inverse power law"). Let $T(\text{volt})$ and $T(\text{volt}_U)$ be the failure times that would result for a particular unit tested at increased voltage and use-voltage conditions, respectively. Then the inverse power relationship is

$$T(\text{volt}) = \frac{T(\text{volt}_U)}{\mathcal{AF}(\text{volt})} = \left(\frac{\text{volt}}{\text{volt}_U}\right)^{\beta_1} T(\text{volt}_U), \tag{18.8}$$

which is a SAFT model. The relationship in (18.8) is known as the inverse power relationship because, generally, $\beta_1 < 0$.

The inverse power relationship voltage-acceleration factor can be expressed as

$$\mathcal{AF}(\text{volt}) = \mathcal{AF}(\text{volt}, \text{volt}_U, \beta_1) = \frac{T(\text{volt}_U)}{T(\text{volt})} = \left(\frac{\text{volt}}{\text{volt}_U}\right)^{-\beta_1}. \quad (18.9)$$

When $\text{volt} > \text{volt}_U$, and $\beta_1 < 0$, $\mathcal{AF}(\text{volt}, \text{volt}_U, \beta_1) > 1$. When volt_U and β_1 are understood to be, respectively, product use (or other baseline) voltage and the material-specific exponent, $\mathcal{AF}(\text{volt}) = \mathcal{AF}(\text{volt}, \text{volt}_U, \beta_1)$ denotes the acceleration factor. Figure 18.8 gives $\mathcal{AF}$ as a function of the stress ratio (e.g., $\text{volt}_{\text{High}}/\text{volt}_{\text{Low}}$) and β_1.

If the model for $T(\text{volt})$ is a log-location-scale distribution, its parameters can be expressed as $\mu = \beta_0 + \beta_1 x$ with constant σ, where $x = \log(\text{volt})$ and β_0 is the

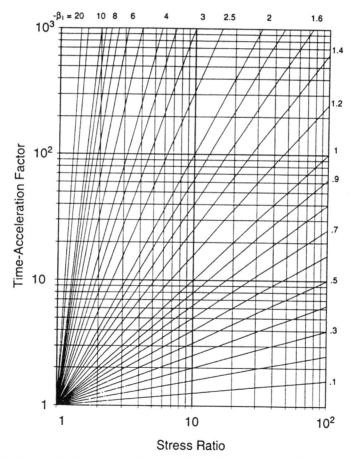

Figure 18.8. Time-acceleration factor as a function of stress ratio and exponent $-\beta_1$ for the inverse power relationship.

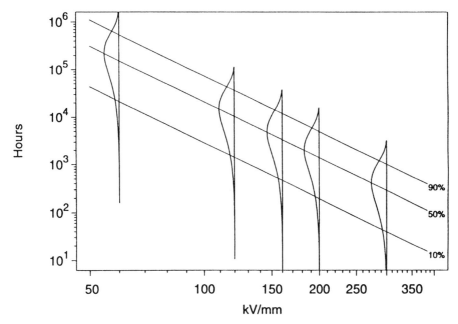

Figure 18.9. Example of the inverse power relationship/Weibull life model.

value of μ at `volt` $= 1$. Then $\log(t_p) = \beta_0 + \beta_1 x + \Phi^{-1}(p)\sigma$. The parameters β_0, β_1, and σ are product or material characteristics. Figure 18.9 shows a typical example of an inverse-power relationship between quantiles of a Weibull life distribution and voltage. Using log axes for time and voltage, the relationship plots as a family of straight lines. Because of the SAFT relationship, the quantile lines all have the same slope.

Example 18.6 Time-Acceleration Factor for Glass Capacitors. From extensive experience with glass capacitors, it is known that the power parameter in the inverse power relationship model is in the neighborhood of $\beta_1 = -2$. For capacitors rated at 100 volts, testing at 300 volts should provide an acceleration factor of $(300/100)^2 = 9$. This can be seen directly from Figure 18.8 entering with $300/100 = 3$ and reading $\mathcal{AF} \approx 9$ from the $\beta_1 = -2$ line. □

18.4.3 Physical Motivation for the Inverse Power Relationship for Voltage-Stress Acceleration

The inverse power relationship is generally considered to be an empirical model for the relationship between life and the level of certain accelerating variables and especially those that are pressure-like stresses. This section presents a simple physical motivation for the inverse power relationship for voltage-stress acceleration under constant-temperature situations. Section 18.5.2 describes a more general model for voltage acceleration involving a combination of temperature and voltage acceleration.

This discussion is for insulation. The ideas extend, however, to other dielectric materials, products, and devices like insulating fluids, transformers, and capacitors. In applications, an insulation should not conduct an electrical current. An insulation has a characteristic dielectric strength which can be expected to be random from unit to unit. The dielectric strength of an insulation specimen operating in a specific environment at a specific voltage may degrade with time. Figure 18.10 shows a family of simple curves to model unit-to-unit variability and degradation in dielectric strength over time. The unit-to-unit variability could be caused, for example, by materials or manufacturing variability. The horizontal lines represent voltage-stress levels that might be present in actual operation or in an accelerated test. When a specimen's dielectric strength falls below the applied voltage stress, there will be flash-over, a short circuit, or other failure-causing damage to the insulation. Analytically, suppose that degrading dielectric strength at age t can be expressed as

$$\mathcal{D}(t) = \delta_0 \times t^{1/\beta_1}.$$

Here, as in Section 18.3.3, failure occurs when $\mathcal{D}(t)$ crosses $\mathcal{D}_f$, the applied voltage stress, denoted by `volt`. In Figure 18.10, the unit-to-unit variability is in the δ_0

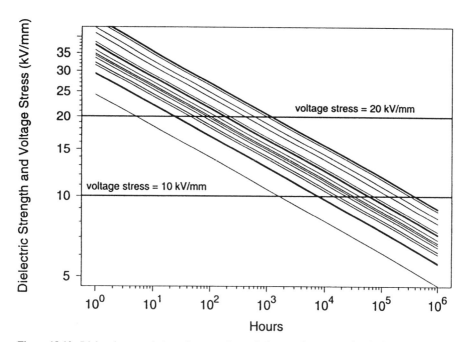

Figure 18.10. Dielectric strength degrading over time, relative to voltage-stress levels (horizontal lines).

parameter. Equating $\mathcal{D}(T)$ to volt and solving for failure time T gives

$$T(\text{volt}) = \left(\frac{\text{volt}}{\delta_0}\right)^{\beta_1}.$$

Then the acceleration factor for volt versus volt_U is

$$\mathcal{AF}(\text{volt}) = \mathcal{AF}(\text{volt}, \text{volt}_U, \beta_1) = \frac{T(\text{volt}_U)}{T(\text{volt})} = \left(\frac{\text{volt}}{\text{volt}_U}\right)^{-\beta_1},$$

which is an inverse power relationship, as in (18.9).

To extend this model, suppose that higher voltage also leads to an increase in the degradation rate and that this increase is described with the degradation model

$$\mathcal{D}(t) = \delta_0 \left[\mathcal{R}(\text{volt}) \times t\right]^{1/\gamma_1},$$

where

$$\mathcal{R}(\text{volt}) = \gamma_0 \exp[\gamma_2 \log(\text{volt})].$$

Suppose failure occurs when $\mathcal{D}(t)$ crosses $\mathcal{D}_f$, the applied voltage stress, denoted by volt. Then equating $\mathcal{D}(T)$ to volt and solving for failure time T gives the failure time

$$T(\text{volt}) = \frac{1}{\mathcal{R}(\text{volt})} \left(\frac{\text{volt}}{\delta_0}\right)^{\gamma_1}.$$

Then the ratio of failure times at volt_U versus volt is the acceleration factor

$$\mathcal{AF}(\text{volt}) = \frac{T(\text{volt}_U)}{T(\text{volt})} = \left(\frac{\text{volt}}{\text{volt}_U}\right)^{\gamma_2 - \gamma_1},$$

which is again an inverse power relationship with $\beta_1 = \gamma_1 - \gamma_2$.

18.4.4 Other Inverse Power Relationships

The inverse power relationship is also commonly used for other accelerating variables including pressure, cycling rate, electric current, stress, and humidity. Some examples are given in the next section.

18.5 ACCELERATION MODELS WITH MORE THAN ONE ACCELERATING VARIABLE

Some accelerated tests use more than one accelerating variable. Such tests might be suggested when it is known that two or more potential accelerating variables contribute to degradation and failure. Using two or more variables may provide needed time acceleration without requiring levels of the individual accelerating variables to be too high. Some accelerated tests include engineering variables that are not accelerating variables. Examples include material type, design, and operation.

18.5.1 Generalized Eyring Relationship

The generalized Eyring relationship extends the Eyring relationship in Section 18.3.2, allowing for one or more nonthermal accelerating variables (such as humidity or voltage). For one additional nonthermal accelerating variable X, the model, in terms of reaction rate, can be written as

$$\mathcal{R}(\text{temp}, X) = \gamma_0 \times (\text{temp K})^m \times \exp\left(\frac{-\gamma_1}{k_B \times \text{temp K}}\right)$$

$$\times \exp\left(\gamma_2 X + \frac{\gamma_3 X}{k_B \times \text{temp K}}\right), \qquad (18.10)$$

where X is a function of the nonthermal stress. The parameters $\gamma_1 = E_a$ (activation energy) and γ_0, γ_2, γ_3 are characteristics of the particular physical/chemical process. Additional factors like the one on the right of (18.10) can be added for other nonthermal accelerating variables.

The following sections, following common practice, set $(\text{temp K})^m = 1$, using what is essentially the Arrhenius temperature-acceleration relationship. They describe some important special-case applications of this more general model. If the underlying model relating the degradation process to failure is a SAFT model, then, as in Section 18.3.1, the generalized Eyring relationship can be used to describe the relationship between times at different sets of conditions temp and X. In particular, the acceleration factor relative to use conditions temp_U and X_U is

$$\mathcal{AF}(\text{temp}, X) = \frac{\mathcal{R}(\text{temp}, X)}{\mathcal{R}(\text{temp}_U, X_U)}.$$

The same approach used in Section 18.3.3 shows the effect of accelerating variables on time to failure. For example, suppose that $T(\text{temp}_U)$ (time at use or some other baseline temperature) has a log-location-scale distribution with parameters μ_U and σ. Then $T(\text{temp})$ has the same log-location-scale distribution with

$$\mu = \mu_U - \log[\mathcal{AF}(\text{temp}, X)] = \beta_0 + \beta_1 x_1 + \beta_2 x_2 + \beta_3 x_1 x_2, \qquad (18.11)$$

where $\beta_1 = E_a$, $\beta_2 = -\gamma_2$, $\beta_3 = -\gamma_3$, $x_1 = 11605/(\text{temp K})$, $x_2 = X$, and $\beta_0 = \mu_U - \beta_1 x_{1U} - \beta_2 x_{2U} - \beta_3 x_{1U} x_{2U}$.

18.5.2 Temperature–Voltage Acceleration

Example 17.11 describes an analysis of the Zelen (1959) data from a life test of glass capacitors at higher than usual levels of temperature and voltage. That example used only simple linear relationships between log-life and the accelerating variables. McPherson and Baglee (1985) used accelerated life test data to model the joint effect of thermal and electrical accelerating variables for failure of thin-gate 100-Å oxides. Boyko and Gerlach (1989) investigate the effect of temperature and strength of electrical field on the time to the generalized Eyring relationship.

To put the Eyring/Arrhenius temperature–voltage acceleration model in the form of (18.11), let $x_1 = 11605/\text{temp K}$, $x_2 = \log(\text{volt})$, and $x_3 = x_1 x_2$. The terms with x_1 and x_2 correspond, respectively, to the Arrhenius and the power relationship acceleration models. The term with x_3, a function of both temperature and voltage, is an interaction suggesting that the temperature-acceleration factor depends on the level of voltage. Similarly, a voltage–temperature interaction suggests that the voltage-acceleration factor depends on the level of temperature. Klinger (1991a) suggests an alternative physically motivated model for the Boyko–Gerlach data with second-order terms involving both temperature and voltage stress.

The dynamic voltage-stress/dielectric-strength model introduced in Section 18.4.3 can be generalized to provide motivation for the generalized Eyring relationship in (18.10), where $X = \log(\text{volt})$. In particular, for the degradation path model

$$\mathcal{D}(t) = \delta_0 [\mathcal{R}(\text{temp}, \text{volt}) \times t]^{1/\gamma_1}.$$

Then, from (18.10),

$$\mathcal{R}(\text{temp}, \text{volt}) = \gamma_0 \times (\text{temp K})^m \times \exp\left(\frac{-E_a}{k_B \times \text{temp K}}\right)$$

$$\times \exp\left(\gamma_2 \log(\text{volt}) + \frac{\gamma_3 \log(\text{volt})}{k_B \times \text{temp K}}\right).$$

Failure occurs when $\mathcal{D}(t)$ crosses $\mathcal{D}_f =$ applied voltage stress, denoted by volt. Equating $\mathcal{D}(T)$ to volt and solving for failure time T gives

$$T(\text{temp}, \text{volt}) = \frac{1}{\mathcal{R}(\text{temp}, \text{volt})} \left(\frac{\text{volt}}{\delta_0}\right)^{\gamma_1}.$$

Then the ratio of failure times at $(\text{temp}_U, \text{volt}_U)$ versus $(\text{temp}, \text{volt})$ is the acceleration factor

$$\mathcal{AF}(\text{temp}, \text{volt}) = \frac{T(\text{temp}_U, \text{volt}_U)}{T(\text{temp}, \text{volt})}$$

$$= \exp[E_a(x_{1U} - x_1)] \times \left(\frac{\text{volt}}{\text{volt}_U}\right)^{\gamma_2 - \gamma_1}$$

$$\times \{\exp[x_1 \log(\text{volt}) - x_{1U} \log(\text{volt}_U)]\}^{\gamma_3},$$

where $x_{1U} = 11605/(\text{temp}_U \text{ K})$ and $x_1 = 11605/(\text{temp K})$. For the special case when $\gamma_3 = 0$ (no interaction), $\mathcal{AF}(\text{temp}, \text{volt})$ is composed of separate factors for temperature and voltage acceleration. In this case the voltage-acceleration factor (holding temperature constant) does not depend on the temperature level used in the acceleration.

18.5.3 Temperature–Current Density Acceleration

Accelerated tests for electromigration typically use temperature and current density as the accelerating variables. To put the Eyring/Arrhenius temperature–current density acceleration model in the form of (18.11), let $x_1 = 11605/\text{temp}$ K, $x_2 = \log(\text{current})$, and $x_3 = x_1 x_2$. The terms with x_1 and x_2 correspond to Arrhenius and the power relationship acceleration models. When the interaction term is omitted (i.e., β_3 assumed to be 0), this is known as "Black's equation" (described in Black, 1969).

18.5.4 Temperature–Humidity Acceleration

Humidity is another commonly used accelerating variable, particularly for failure mechanisms involving corrosion and certain kinds of chemical degradation.

Example 18.7 Accelerated Life Test of a Printed Wiring Board. Example 1.8 introduced data, shown in Figure 1.9, from an ALT of printed circuit boards. It illustrates the use of humidity as an accelerating variable. This is a subset of the larger experiment described by LuValle, Welsher, and Mitchell (1986), involving acceleration with temperature, humidity, and voltage. The figure shows clearly that failures occur earlier at higher levels of humidity. □

A variety of different humidity models (mostly empirical but a few with some physical basis) have been suggested for different kinds of failure mechanisms. Much of this work has been motivated by concerns about the effect of environmental humidity on plastic-packaged electronic devices. Humidity is also an important factor in the service-life distribution of paints and coatings. In most test applications where humidity is used as an accelerating variable, it is used in conjunction with temperature. For example, Peck (1986) presents data and models relating life of semiconductor electronic components to humidity and temperature. Gillen and Mead (1980) describe kinetic models for accelerated aging that include humidity terms. See also Peck and Zierdt (1974) and Joyce et al. (1985). LuValle, Welsher, and Mitchell (1986) describe the analysis of time-to-failure data on printed circuit boards that have been tested at higher than usual temperature, humidity, and voltage. They suggest ALT models based on the physics of failure. Chapter 2 of Nelson (1990a) and Boccaletti et al. (1989) review and compare a number of different humidity models.

The Eyring/Arrhenius temperature–humidity acceleration relationship in the form of (18.11) uses $x_1 = 11605/\text{temp}$ K, $x_2 = \log(\text{RH})$, and $x_3 = x_1 x_2$, where RH is relative humidity, expressed as a proportion. An alternative humidity relationship suggested by Klinger (1991b), on the basis of a simple kinetic model for corrosion, uses the term $x_2 = \log[\text{RH}/(1 - \text{RH})]$ (a logistic transformation) instead.

18.6 GUIDELINES FOR THE USE OF ACCELERATION MODELS

Because most applications of accelerated testing involve extrapolation, users must exercise caution in planning tests (accelerated test planning is described in Chapter 20 and Section 22.5) and interpreting the results of data analyses (accelerated test data analysis is described in Chapters 19 and 21). Some guidelines for the use of acceleration models include:

- Accelerating variables should be chosen to correspond with variables that cause actual failures.
- It is useful to investigate previous attempts to accelerate failure mechanisms similar to the ones of interest. There are many research reports and papers that have been published in the physics of failure literature.
- Accelerated tests should be designed, as much as possible, to minimize the amount of extrapolation required (see Chapter 20 and Section 22.5). High levels of accelerating variables can cause extraneous failure modes that would never occur at use-levels of the accelerating variables. If extraneous failures are not recognized and properly handled, they can lead to seriously incorrect conclusions. Also, the relationship may not be accurate enough over a wide range of acceleration.
- Generally, accelerated tests are used to obtain information about one particular, relatively simple failure mechanism (or corresponding degradation measure). If there is more than one failure mode, it is possible that the different failure mechanisms will be accelerated at different rates. Then, unless this is accounted for in the modeling and analysis, estimates could be seriously incorrect when extrapolating to lower use-levels of the accelerating variables.
- In practice, it is difficult or impractical to verify acceleration relationships over the entire range of interest. Of course, accelerated test data should be used to look for departures from the assumed acceleration model. It is important to recognize, however, that the available data will generally provide very little power to detect anything but the most serious model inadequacies. Typically, there is no useful diagnostic information about possible model inadequacies at accelerating variable levels close to use conditions.
- Simple models with the right shape have generally proved to be more useful than elaborate multiparameter models.
- Sensitivity analysis should be used to assess the effect of perturbing uncertain inputs (e.g., inputs related to model assumptions).
- Accelerated test programs should be planned and conducted by teams including individuals knowledgeable about the product and its use environment, the physical/chemical/mechanical aspects of the failure mode, and the statistical aspects of the design and analysis of reliability experiments.

BIBLIOGRAPHIC NOTES

Nelson (1990a) provides an extensive and comprehensive source for background material, practical methodology, basic theory, and examples for accelerated testing models. See Smith (1996), Chapter 7 of Tobias and Trindade (1995), Chapters 2 and 9 of Jensen (1995), and Klinger, Nakada, and Menendez (1990) for additional discussion of these topics. Thomas (1964) describes some practical aspects of accelerated testing and describes what we have called SAFT as "true acceleration." Harter (1977) provides a detailed review of the literature on the effect that size has on reliability. Starke et al. (1996) describe the use of long-term elevated temperature exposure and the prospects for the use of accelerated aging of materials and structures. Feinberg and Windom (1995) describe the reliability physics of thermodynamic aging and its relationship to device reliability. Fukuda (1991) describes degradation models for lasers and LEDs. Howes and Morgan (1981), Hakim (1989), Pollino (1989), and Christou (1992, 1994a,b) describe degradation models for microelectronic devices. Gillen and Clough (1985) describe a kinetic model for predicting oxidative degradation rates in combined radiation–thermal environments.

LuValle (1990) and LuValle and Hines (1992) show how to use step-stress methods to extract information about the kinetics of failure processes. Drapella (1992) provides a mathematical model illustrating how a failure process with a kinetic model more complicated than first order can lead to a breakdown of the commonly used Arrhenius relationship. Costa and Mercer (1993) describe degradation models for corrosion. Cragnolino and Sridhar (1994) present a collection of papers describing the use of accelerated corrosion tests for the prediction of service life. Bro and Levy (1990) describe kinetic models for degradation of batteries. Starke et al. (1996) describe issues relating to aging of materials and structures. Bayer (1994) describes models for prediction and prevention of wear. Castillo and Galambos (1987) derive a regression model for fatigue failure, based on established physical models. The annual *Proceedings of the International Reliability Physics Symposium*, sponsored by the IEEE Electron Devices Society and the IEEE Reliability Society, contain numerous articles describing physical models for acceleration and failure.

Often accelerated test models are derived or specified through a system of differential equations. When, as is often the case, no closed-form solution is available, it becomes necessary to use numerical solutions. Nash and Quon (1996) describe software for fitting differential equation models to data.

Evans (1977) makes the important point that the need to make rapid reliability assessments and the fact that accelerated tests may be "the only game in town" are not sufficient to *justify* the use of the method. Justification must be based on physical models or empirical evidence. Evans (1991) describes difficulties with accelerated testing and suggests the use of sensitivity analysis. He also comments that acceleration factors of 10 "are not unreasonable" but that "factors much larger than that tend to be figments of the imagination and lots of correct but irrelevant arithmetic."

EXERCISES

18.1. For the toaster example in Section 18.1.4, toasters were cycled 365 times per day to get reliability information more quickly. Discuss the practical limitations of increasing the cycling frequency to get information even more quickly.

18.2. Based on previous experience with similar products, the failure time of a particular field effect transistor on a monolithic microwave GaAs integrated circuit, operating at 100°C (channel temperature), is expected to have a lognormal distribution with a median time to failure of 30 years. The primary failure mode is caused by a chemical reaction that has an activation energy of $E_a = .6$ eV. The value of the lognormal scale parameter for this failure mode is expected to be $\sigma = .7$.

(a) What are the lognormal parameters μ and σ if time is recorded in hours?

(b) For operation at 100°C channel temperature, what is the time at which 5% of the units would fail? 10%? 90%?

(c) Calculate the time-acceleration factor for testing at 250°C, 200°C, and 150°C channel temperature. Use Table 18.2 and Figure 18.3 and check with equation (18.2).

(d) Obtain an expression for the temperature at which $100p\%$ of tested units would be expected to fail in a 6000-hour test. Use this expression to compute the temperatures at which 90% and 10% would be expected to fail.

18.3. Refer to Exercise 18.2. Obtain an expression for the (average) FIT rate (in standard units of failures per hour in parts per billion) for the first 10 years of operation at 100°C? How much would this improve if the operating channel temperature is changed to only 90°C?

18.4. A mechanical adhesive has been designed for 10-year life at 60°C ambient temperature. Over time, the bond will degrade chemically and will eventually fail. The rate of the chemical reaction can be increased by testing at higher levels of temperature. Using an activation energy of $E_a = 1.2$ eV and the Arrhenius relationship, calculate the time-acceleration factors for testing at 120°C, 90°C, and 80°C.

18.5. Time to failure of incandescent light bulbs can be described accurately with a lognormal distribution. A test engineer claims that a 10% increase in voltage decreases life by approximately 50%. A particular brand of 100-watt bulb has a median life of 1200 hours at 110 volts.

(a) Give an inverse power relationship expression for the life of such light bulbs as a function of voltage.

(b) Calculate the time-acceleration factors for operating the light bulb at 120 volts and 130 volts?

18.6. Refer to Exercise 8.7.

(a) Assume that the activation energy of the observed failure mode is $E_a = 1.2$ eV and that the Arrhenius relationship provides an adequate description of the effect of temperature on the reaction rate. Compute estimates of the lognormal distribution parameters at 50°C, 80°C, and 120°C.

(b) On lognormal probability paper, plot the estimate of the life distributions at 50°C, 80°C, and 120°C.

(c) Repeat part (a), using an activation energy of $E_a = .7$ eV and also plot these results.

(d) Explain the effect that an incorrect assumption about activation energy could have on estimates of life at low temperature.

18.7. A certain kind of capacitor has an exponential life distribution with a median life of 10 thousand hours at operating voltage of 400 volts. The relationship between life and voltage can be described by the inverse power relationship with an exponent $\beta = -10$. Determine the time-acceleration factors for accelerated testing of these capacitors at 500 volts, 600 volts, and 800 volts.

18.8. A particular type of integrated circuit is thought to have a dominant failure mode with an activation energy of $E_a = 1.2$ eV. This circuit is designed to operate at 50°C. If a 1000-hour life test is conducted at 120°C, under the Arrhenius relationship, what is the equivalent amount of operating time for this failure mode?

18.9. Consider a failure mechanism modeled with an underlying degradation path model in (18.3). Suppose that the reaction activation energy is $E_a = 1.8$ eV. For $\mathcal{R}_U = .2$ and $\mathcal{D}_\infty = .6$, compute the crossing times for all combinations of $t = 1000, 2000$ hours and $\mathcal{D}_f = .5, 1$. Use these results to verify that the SAFT property holds in this case.

18.10. Table 18.2 and Figure 18.3 can be used together to determine time-acceleration factors for different levels of use and test temperatures. Create a similar table and figure that can be used to obtain time-acceleration factors for the logit-transformation relative humidity model described in Section 18.5.4.

18.11. Equation (18.6) gives the Arrhenius relationship between the location parameter of the log-life distribution and temperature. When needed, suppose that $\beta_0 = -17$, $\beta_1 = .86$, and $\sigma = 1.2$. Using this model:

(a) Give an expression for t_p, the p quantile as a function of temperature in °C.

(b) Show that the relationship between $\log(t_p)$ versus $1/(\text{temp K})$ is linear.

(c) Starting with linear graph paper, make an "Arrhenius plot" for this model, similar to Figure 18.5. Start by plotting the linear relationship in part (b) on linear paper over the range of interest. Use a range of temperatures running from 50°C to 140°C. Note that in order to have the slope of the plotted line decreasing with temperature, it will be necessary to have the $1/(\text{temp K})$ axis running left to right from largest to smallest values of $1/(\text{temp K})$ (instead of the customary increasing axis). Then, finally, add in new (nonlinear) axes for Time and °C. Generally, it is most convenient to do this on the axis opposite to the corresponding linear axis. On these axes, use tick and tick labels at major points on the scale (e.g., for temperature at $50, 60, \ldots, 140$, corresponding to $1/(\texttt{temp K}) = .003094538, .003001651, \ldots, .002420428$).

18.12. Show why the relationships between the TDF and $\mathcal{AF}$ in Figure 18.3 plot as straight lines.

18.13. Show why the relationships between the voltage ratio and $\mathcal{AF}$ in Figure 18.8 plot as straight lines.

CHAPTER 19

Accelerated Life Tests

Objectives

This chapter explains:

- Nonparametric and graphical methods for presenting and analyzing accelerated life test (ALTs) data.
- Likelihood methods for analyzing right-censored data from an ALT with a single accelerating factor.
- Analysis of other kinds of ALT experiments including experiments yielding interval data and experiments with two accelerating variables.
- Some other common forms of accelerated testing.
- Some potential pitfalls of accelerated testing.

Overview

This chapter describes and illustrates some basic data analysis methods for accelerated life tests (ALTs). These tests are used to characterize durability properties or the life distribution of materials or simple components. The presentation employs the ALT models described in Chapter 18. Section 19.2 explains and illustrates basic important ideas for a single-variable ALT with right-censored data and exact failure times. Section 19.3 presents several other important examples with special models or data features. Section 19.4 gives some suggestions and cautions for drawing conclusions from AT data. Section 19.5 briefly describes other kinds of "accelerated tests." Section 19.6 outlines a number of potential pitfalls that can arise in the application of accelerated testing.

19.1 INTRODUCTION

This chapter shows how to apply regression methods from Chapter 17 to the analysis of accelerated life test (ALT) data.

Example 19.1 Temperature-Accelerated Life Test on Device-A. Hooper and Amster (1990) analyze the temperature-accelerated life test data on a particular kind

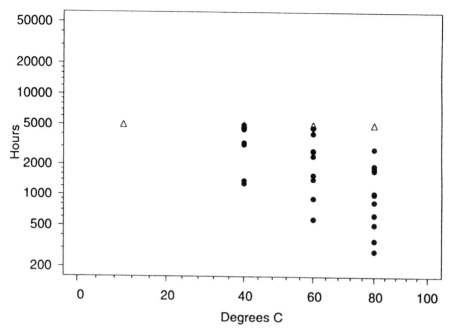

Figure 19.1. Scatter plot of temperature-accelerated life test data for Device-A. Censored observations are indicated by △. The number of censored/tested units were 30/30, 90/100, 11/20, 1/15 at 10, 40, 60, and 80°C, respectively.

of device. Because they do not identify the particular device, we will refer to it as Device-A. The data are given in Appendix Table C.10 and Figure 19.1. The purpose of the experiment was to determine if Device-A would meet its failure rate objective through 10,000 hours and 30,000 hours at its operating ambient temperature of 10°C. In this context, failure rate is usually taken to mean the proportion failing over the specified time interval [see (2.4) in Section 2.1.1]. In the following sections we will show how to fit an accelerated life regression model to these data to answer this and other questions. □

19.1.1 Accelerated Life Test Models

Most parametric ALT models have the following two components:

1. A parametric distribution for the life of a population of units at a particular level(s) of an experimental variable or variables. It might be possible to avoid this parametric assumption for some applications, but when appropriate, parametric models (e.g., Weibull and lognormal) provide important practical advantages for most applications.
2. A relationship between one (or more) of the distribution parameters and the acceleration or other experimental variables. Such a relationship models the effect that variables like temperature, voltage, humidity, and specimen or unit

ANALYSIS OF SINGLE-VARIABLE ALT DATA

size will have on the failure-time distribution. As described in Chapter 18, this part of the accelerated life model should be based on a physical model such as one relating the accelerating variable to degradation, on a well-established empirical relationship, or some combination.

The examples in this section use the log-location-scale regression models described in Section 17.3.2 and illustrated with other examples in Chapter 17. The relationships between parameters and accelerating variables come from considerations like those described in Chapter 18.

19.1.2 Strategy for Analyzing ALT Data

This section outlines and illustrates a strategy that is useful for analyzing ALT data consisting of a number of groups of specimens, each having been run at a particular set of conditions. The basic idea is to start by examining the data graphically. Use probability plots to analyze each group separately and explore the adequacy of candidate distributions. Then fit a model that describes the relationship between life and the accelerating variable(s). Briefly, the strategy is to:

1. Examine a scatter plot of failure time versus the accelerating variable.
2. Fit distributions individually to the data at separate levels of the accelerating variable. Plot the fitted ML lines on a multiple probability plot along with the individual nonparametric estimates at each level of the accelerating variable. Use the plotted points and fitted lines to assess the reasonableness of the corresponding life distribution and the constant-σ assumption. Repeat with probability plots for different assumed failure-time distributions.
3. Fit an overall model with the proposed relationship between life and the accelerating variable.
4. Compare the combined model from Step 3 with the individual analyses in Step 2 to check for evidence of lack of fit for the overall model.
5. Perform residual analyses and other diagnostic checks of the model assumptions.
6. Assess the reasonableness of the ALT data to make the desired inferences.

The first examples in this chapter have just one accelerating variable (the simplest and most common type of ALT). Section 19.3.3 shows how to apply the same general strategy to an ALT with two or more accelerating variables.

19.2 ANALYSIS OF SINGLE-VARIABLE ALT DATA

This section describes methods for analyzing ALT data with a single accelerating variable. The subsections illustrate, in sequence, the steps in the strategy described in Section 19.1.2.

19.2.1 Scatter Plot of ALT Data

Start by examining a scatter plot of failure-time data versus the accelerating-variable data. A different symbol should be used to indicate censored observations.

Example 19.2 Scatter Plot of the Device-A Data. Figure 19.1 is a scatter plot of the Device-A ALT data introduced in Example 19.1. As expected, units fail sooner at higher levels of temperature. The heavy censoring (e.g., note that there were no failures at 10°C) makes it difficult to see the form of the life/accelerating variable relationship from this plot. □

19.2.2 Multiple Probability Plot of Nonparametric cdf Estimates at Individual Levels of the Accelerating Variable

To make a multiple probability plot, first compute nonparametric estimates of the failure-time distribution for each group of specimens tested at the same level of the accelerating variable. Then plot these on probability paper. Use the plot to assess the distributional model for the different levels of the accelerating variable (or variable-level combinations).

Example 19.3 Multiple Probability Plot of the Device-A Data. Figure 19.2 is a Weibull multiple probability plot of the Device-A data. Figure 19.3 is a corresponding

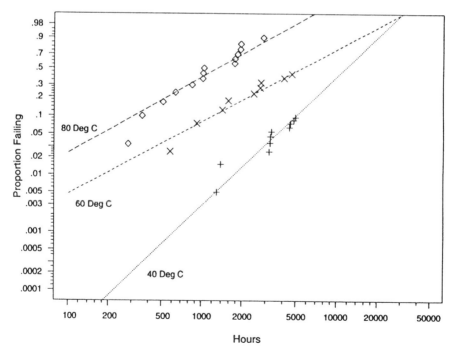

Figure 19.2. Weibull Multiple probability plot with individual Weibull ML fits for each temperature for the Device-A data.

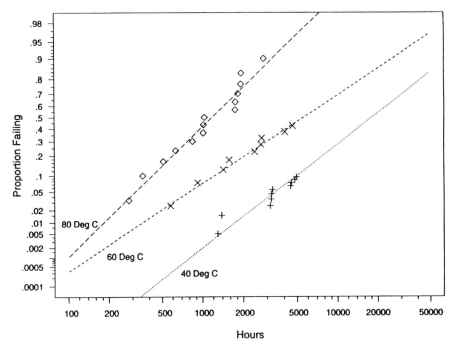

Figure 19.3. Lognormal multiple probability plot with lognormal ML fits for each temperature for the Device-A data.

lognormal probability plot. Comparison of these plots indicates that both the Weibull and the lognormal distributions provide a reasonable fit to the failure data at the different levels of temperature, but that the lognormal distribution provides a better fit to the individual temperature groups. □

19.2.3 Adding ML Estimates at Each Level of the Accelerating Variable(s) to a Multiple Probability Plot

If a suitable parametric distribution can be found, then ML estimates of the cdf at each level of the accelerating variable should be computed and put on the probability plot along with the corresponding nonparametric cdf estimates. This plot is useful for assessing the commonly used assumptions that distribution shape does not depend on the level of the accelerating variable and that the accelerating variable only affects the distribution scale parameter. The slopes of the lines are related to the distribution shape parameter values. Thus we can assess graphically the assumption that temperature has no effect on distribution shape.

Example 19.4 ML Estimates of Device-A Life at 40, 60, and 80°C. The straight lines on Figures 19.2 and 19.3 depict, respectively, individual Weibull and lognormal ML estimates of the cdfs at the different levels of temperature. Table 19.1

Table 19.1. Device-A ALT Lognormal ML Estimation Results at Individual Temperatures

	Parameter	ML Estimate	Standard Error	95% Approximate Confidence Intervals	
				Lower	Upper
40°C	μ	9.81	.42	8.9	10.6
	σ	1.0	.27	.59	1.72
60°C	μ	8.64	.35	8.0	9.3
	σ	1.19	.32	.70	2.0
80°C	μ	7.08	.21	6.7	7.5
	σ	.80	.16	.55	1.17

The individual log likelihoods were $\mathcal{L}_{40} = -115.46$, $\mathcal{L}_{60} = -89.72$, and $\mathcal{L}_{80} = -115.58$. The confidence intervals are based on the normal-approximation method.

summarizes the lognormal ML estimation results. In Figure 19.3, there are small differences among the slopes, but this could be due to sampling error. This can be seen, informally, from the overlapping confidence intervals for σ (this issue is addressed more formally in Example 19.6). Although both distributions fit reasonably well, tradition and physical theory (see Section 4.6) suggest the lognormal distribution to describe the failure-time distribution for such devices. Subsequent Device-A examples will also use the lognormal distribution. □

19.2.4 Multiple Probability Plot of ML Estimates with a Fitted Acceleration Relationship

In order to draw conclusions about life at low levels of accelerating variables, one needs to use a life/accelerating variable relationship to tie together results at the different levels of the accelerating variable. The cdfs estimated from the model fit can also be plotted on a probability plot along with the data to assess how well the life/accelerating variable model fits the data. Extrapolations to other levels of the accelerating variable can also be plotted.

Example 19.5 *ML Estimates of Device-A Data for the Arrhenius–Lognormal Model.* The Arrhenius–lognormal regression model, described in Section 18.3.3, is

$$\Pr[T \le t; \texttt{temp}] = \Phi_{\text{nor}}\left[\frac{\log(t) - \mu}{\sigma}\right],$$

where $\mu = \beta_0 + \beta_1 x$, $x = 11605/(\texttt{temp K})$, and $\beta_1 = E_a$ is the activation energy. Table 19.2 contains ML estimates and other information. The estimate of the variance-

ANALYSIS OF SINGLE-VARIABLE ALT DATA

Table 19.2. ML Estimates for the Device-A Data and the Arrhenius-Lognormal Regression Model

Parameter	ML Estimate	Standard Error	95% Approximate Confidence Intervals	
			Lower	Upper
β_0	−13.5	2.9	−19.1	−7.8
β_1	.63	.08	.47	.79
σ	.98	.13	75	1.28

The log likelihood is $\mathcal{L} = -321.7$. The confidence intervals are based on the normal-approximation method.

covariance matrix for the ML estimates $\widehat{\theta} = (\widehat{\beta}_0, \widehat{\beta}_1, \widehat{\sigma})$ is

$$\widehat{\Sigma}_{\widehat{\theta}} = \begin{bmatrix} 8.336 & -.239 & -.195 \\ -.239 & .0069 & .0059 \\ -.195 & .0059 & .0176 \end{bmatrix}. \tag{19.1}$$

These quantities will be used in subsequent numerical examples. Figure 19.4 is a lognormal probability plot showing the Arrhenius–lognormal model fit to the Device-A

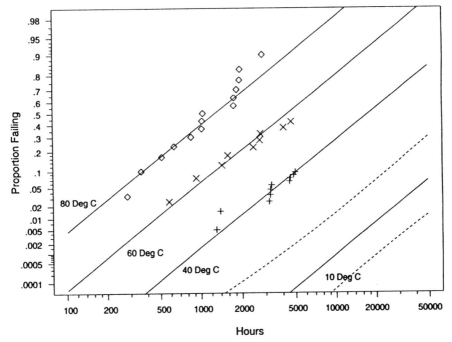

Figure 19.4. Lognormal multiple probability plot depicting the Arrhenius–lognormal regression model ML fit to the Device-A ALT data.

ALT data. The solid line at the bottom of the graph is the ML estimate of the cdf at 10°C, extrapolated from the ML fit.

The dotted curves are a set of pointwise 95% normal-approximation confidence intervals. They reflect the random "sampling uncertainty" arising from the limited sample data. The necessary computations are illustrated in Example 19.8. It is important to note that these intervals do *not* reflect model-specification and other errors (and we know that the model is only an approximation for the exact relationship). Figure 19.5 shows directly the fitted life/accelerating variable relationship and the estimated densities at each level of temperature, and lines indicating ML estimates of percent failing as a function of temperature. The density estimates are normal densities because time is plotted on a log scale. □

It is useful to compare individual analyses with model analyses. This can be done both graphically and analytically. A likelihood-ratio test provides an analytical assessment about whether observed deviations between the individual model fit and the overall life/accelerating variable relationship can be explained by random variability or not.

Example 19.6   Analytical Comparison of Individual Lognormal Fits and Arrhenius-Lognormal Model Fit to the Device-A Data. Fitting individual lognormal distributions (Table 19.1 and Figure 19.3) estimates μ and σ at each level of temperature without any constraints. Fitting the Arrhenius–lognormal model

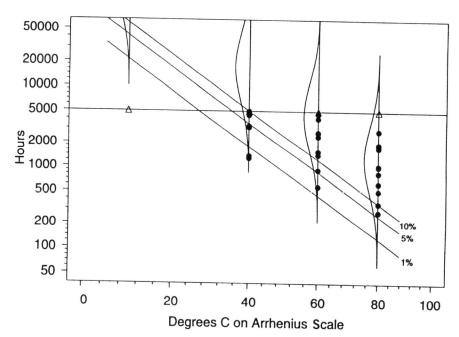

Figure 19.5. Scatter plot showing hours to failure versus °C (on an Arrhenius scale) and the Arrhenius–lognormal regression model fitted to the Device-A data. Censored observations are indicated by $\triangle$.

ANALYSIS OF SINGLE-VARIABLE ALT DATA

(Table 19.2 and Figure 19.4) estimates μ and σ at each level of temperature with μ constrained to be a linear function of $x = 11605/(\text{temp K})$ and σ constrained to be the same for all temperatures. The total likelihood for the unconstrained Arrhenius–lognormal model will always be larger than the likelihood for the constrained model. If the total likelihood for the unconstrained model is *much* larger than the total likelihood for the constrained model, there is evidence of lack of fit for the constrained Arrhenius–lognormal model. These two approaches for fitting the data can be compared with an "omnibus" likelihood-ratio test. From Table 19.1, for the unconstrained model, $\mathcal{L}_{\text{unconst}} = \mathcal{L}_{40} + \mathcal{L}_{60} + \mathcal{L}_{80} = -320.76$ and from Table 19.2, for the constrained model, $\mathcal{L}_{\text{const}} = -321.7$. If the constrained model is "correct" then the test statistic $Q = -2(\mathcal{L}_{\text{const}} - \mathcal{L}_{\text{unconst}})$ has a χ^2_3 distribution (see Appendix Section B.6.5). In this case the 3 degrees of freedom is the difference between the 6 parameters in the unconstrained model and the 3 parameters in the constrained model. Thus $Q = -2(-321.7 + 320.76) = 1.88 < \chi^2_{(.75;3)} = 4.1$, indicating that there is no evidence of inadequacy of the constrained model, relative to the unconstrained model. □

19.2.5 Checking Other Model Assumptions

Before drawing conclusions from a set of data, it is important to check, as carefully as possible, model assumptions by using residual analysis and other model diagnostics, as explained in Section 17.6.1. A probability plot of residuals is useful for assessing the overall adequacy of a fitted distribution. A standard plot of residuals versus fitted values (e.g., $\widehat{\mu}$) can also be useful for identifying other departures from the fitted model. Heavy censoring, however, makes such plots difficult to interpret. Without censoring, the residuals should not show any structure that might have been explained by a more elaborate model. In interpreting such a plot with censoring, one has to make allowances for the predictable patterns that censoring will induce.

Example 19.7 Residual Analysis for the Arrhenius–Lognormal Model Fit to the Device-A Data. Figure 19.6 is a lognormal probability plot of the (censored) standardized residuals from the life/temperature model. In this case the deviation from linearity is not strong and can be attributed to randomness in the data. Thus the lognormal distribution appears to be adequate.

Figure 19.7 is a plot of the (censored) standardized residuals (as defined in Section 17.6) versus the Arrhenius–lognormal model fitted values (as in other plots of failure times, Δ represents right-censored residuals). The heavy censoring makes this plot more difficult to interpret. All that we know about the residuals marked with a Δ is that they are larger than the plotted points. The appearance of a downward sloping trend is mostly due to the right-censored observations. The smallest observations also appear to slope downward, but there are only a few observations involved in this trend. Although this might, at first, suggest that there are outliers or some kind of other model departure, there is no evidence of outliers in Figure 19.6 and the observed pattern could be a result of randomness from the Arrhenius–lognormal model. □

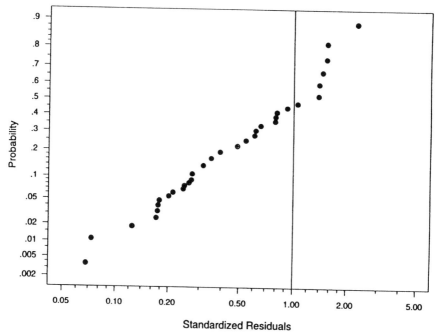

Figure 19.6. Lognormal probability plot of the standardized residuals from the Arrhenius–lognormal model fit to the Device-A data.

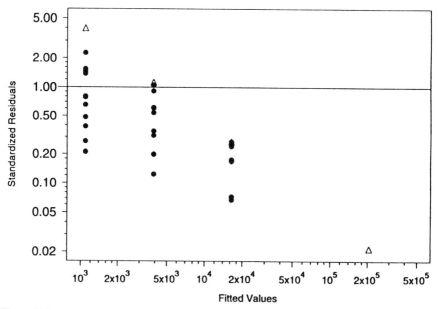

Figure 19.7. Plot of standardized residuals versus fitted values for the Arrhenius–lognormal regression model fitted to the Device-A data. Censored residuals are indicated by Δ.

19.2.6 Estimation at Use Conditions

The methods presented in Section 17.4.2 can also be used to compute estimates and confidence intervals for quantities of interest at use conditions. The following example gives details on how to compute an estimate of $F(t)$ but, as explained in Section 17.4.2, the same ideas can be applied to compute estimates of other quantities of interest such as distribution quantiles or hazard function values.

Example 19.8 Confidence Interval for the Device-A Lognormal Distribution $F(30000)$ and $F(10000)$ at $10°C$. As mentioned in Example 19.1, the purpose of the ALT was to estimate the proportion failing at 30,000 and 10,000 hours. To illustrate the methods we use $t_e = 30{,}000$ hours. Using methods described in Sections 8.4.3 and 17.4.2, and the numerical results in Table 19.2, simple computations give, at $10°C$,

$$\hat{\mu} = \hat{\beta}_0 + \hat{\beta}_1 x$$
$$= -13.469 + .6279 \times 11605/(10 + 273.15) = 12.2641,$$
$$\hat{\zeta}_e = [\log(t_e) - \hat{\mu}]/\hat{\sigma} = [\log(30000) - 12.2641]/.9778 = -2.000,$$
$$\hat{F}(30000) = \Phi_{\text{nor}}(\hat{\zeta}_e) = \Phi_{\text{nor}}(-2.000) = .02281.$$

The estimated covariance matrix for $\hat{\mu}$ and $\hat{\sigma}$ at $10°C$ can be computed as shown in equation (17.11), using (19.1). The result is

$$\hat{\Sigma}_{\hat{\mu},\hat{\sigma}} = \begin{bmatrix} \widehat{\text{Var}}(\hat{\mu}) & \widehat{\text{Cov}}(\hat{\mu},\hat{\sigma}) \\ \widehat{\text{Cov}}(\hat{\mu},\hat{\sigma}) & \widehat{\text{Var}}(\hat{\sigma}) \end{bmatrix} = \begin{bmatrix} .287 & .048 \\ .048 & .0176 \end{bmatrix}.$$

The off-diagonal elements are zero for complete data but are nonzero here because of the right censoring. Then, following the general approach in Section 8.4.3, equation (8.14) gives

$$\widehat{\text{se}}_{\hat{F}} = \frac{\phi(\hat{\zeta}_e)}{\hat{\sigma}} \left[\widehat{\text{Var}}(\hat{\mu}) + 2\hat{\zeta}_e \widehat{\text{Cov}}(\hat{\mu},\hat{\sigma}) + \hat{\zeta}_e^2 \widehat{\text{Var}}(\hat{\sigma}) \right]^{1/2}$$
$$= \frac{\phi(-2.000)}{.9778} \left[.287 + 2 \times (-2.000) \times .048 + (-2.000)^2 \times .0176 \right]^{1/2}$$
$$= .0225.$$

From this, the 95% normal-approximation confidence interval for $F(30000)$, using (8.15), is

$$[\underset{\sim}{F}(t_e),\ \widetilde{F}(t_e)] = \left[\frac{\hat{F}}{\hat{F} + (1 - \hat{F}) \times w},\ \frac{\hat{F}}{\hat{F} + (1 - \hat{F})/w} \right]$$
$$= \left[\frac{.02281}{.02281 + (1 - .02281) \times w},\ \frac{.02281}{.02281 + (1 - .02281)/w} \right]$$
$$= [.0032,\ .14],$$

where

$$w = \exp\{z_{(1-\alpha/2)}\widehat{se}_{\widehat{F}}/[\widehat{F}(1-\widehat{F})]\}$$
$$= \exp\{1.96 \times .0225/[.02281(1-.02281)]\} = 7.232.$$

This interval is based on the assumption that $Z_{\text{logit}(\widehat{F})} \stackrel{\sim}{\cdot} \text{NOR}(0, 1)$. The interval is wide (also see Figure 19.4) but properly reflects the sampling uncertainty when the activation energy is unknown. Additionally, it is important to note that the interval does not reflect model uncertainty. On the other hand, if one assumed that the activation energy is known, this and other confidence intervals would be much narrower. A similar approximate 95% confidence interval for $F(10000)$ (computations left as an exercise) is [.00006, .013]. □

Conclusions based on unverified assumptions are subject to error. Although confidence intervals provide an assessment of sampling uncertainty, they do not reflect possible model deviations. When model assumptions are uncertain, repeating computations with alternative assumptions provides informative sensitivity analyses. The following example illustrates this by changing the fitted distribution for the Device-A data.

Example 19.9 *Confidence Interval for the Weibull Distribution $F(30000)$ and $F(10000)$ for Device-A at $10°C$.* For the Weibull distribution model (with other details omitted to save space), an approximate 95% confidence interval for $F(30000)$ is [.0092, .126]. An approximate 95% confidence interval for $F(10000)$ is [.0021, .027]. The differences between the lognormal and Weibull confidence intervals is not that large, relative to width of these intervals. Without some physical basis for a choice between these two distributions, both sets of intervals would have to be taken into consideration. In this case, however, the lognormal was favored on physical grounds, and the lognormal distribution also provided a better fit. □

19.3 FURTHER EXAMPLES

19.3.1 Voltage Acceleration

This section illustrates statistical methods for fitting a model to and making inferences from voltage-accelerated life data. Section 18.4 describes the inverse power relationship for voltage acceleration used here.

Example 19.10 *Accelerated Life Test of a Mylar–Polyurethane Insulating Structure.* Returning to the data introduced in Example 18.5, it was clear from the scatter plot in Figure 18.7 that the linear relationship for log life versus log voltage relationship implied by the inverse power relationship Section 18.4.2) did not hold for the data at 361.4 kV/mm. This suggests that the failure mechanism might be different at 361.4 kV/mm. In this kind of situation, particularly when primary interest is in extrapolating to lower ranges of voltage, it is appropriate to drop the data at 361.4 kV/mm and analyze the remaining data.

FURTHER EXAMPLES

Figure 19.8 is a lognormal probability plot of the data at each of the five different levels of voltage stress along with individual lognormal ML estimates. Although the σ estimates differ across voltage-stress levels, the differences are small and consistent with ordinary random variability. A similar Weibull plot (not shown here) also provided a reasonable fit to the data at the different levels of voltage stress but, overall, the lognormal seemed to fit better. Figure 19.9 shows the inverse power relationship–lognormal model fitted to the mylar–polyurethane data. The model is

$$\Pr[T \leq t; \text{volt}] = \Phi_{\text{nor}}\left[\frac{\log(t) - \mu}{\sigma}\right],$$

where $\mu = \beta_0 + \beta_1 x$, $x = \log(\text{volt})$, and volt is voltage stress in kV/mm. Numerical estimates are summarized in Table 19.3. The solid line in the SE part of Figure 19.9 is the ML estimate of the cdf at 50 kV/mm, extrapolated from the ML fit. As in Example 19.5, the dotted lines are pointwise 95% normal-approximation confidence intervals, reflecting sampling uncertainty. Figure 19.10 is a lognormal probability plot of the residuals from the inverse power relationship–lognormal model (Section 18.4.2) fitted to the mylar–polyurethane data. This plot does not suggest any important departure from the assumed lognormal distribution and is closer to linear than the corresponding plot for the inverse power relationship–Weibull model (not shown here). Figure 19.11 shows the fitted model and the original data (showing the

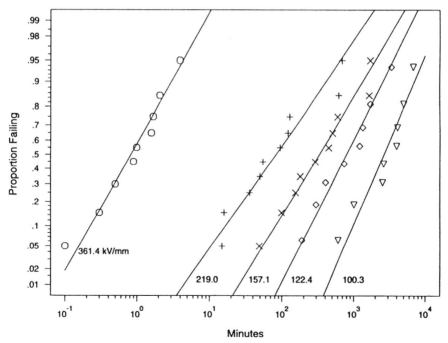

Figure 19.8. Lognormal multiple probability plot and ML fit for each voltage in the mylar–polyurethane ALT.

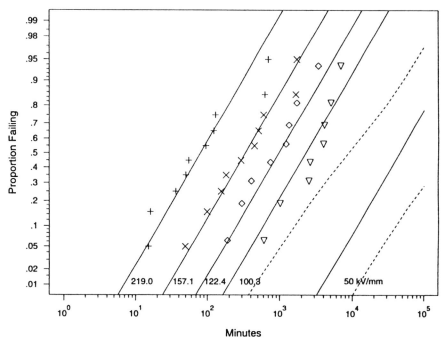

Figure 19.9. Lognormal multiple probability plot and ML fit of the inverse power relationship–lognormal model to the mylar–polyurethane data with the 361.4-kV/mm data omitted.

data at 361.4 kV/mm even though these data were not used in the model fitting). This plot indicates that there is a substantial probability of failure before 10,000 hours [$F(10000) = .076$] for this insulating structure at 50 kV/mm. A 95% confidence interval for $F(10000)$ is [.0058, .54]. Incorrectly including the 361.4-kV/mm data in the analysis (details not shown here, but see Exercise 19.17) changes the confidence interval to [.00012, .064], resulting in a very optimistic and misleading impression. □

Table 19.3. Inverse Power Relationship–Lognormal Model ML Estimates for the Mylar-Polyurethane Data

Parameter	ML Estimate	Standard Error	95% Approximate Confidence Intervals	
			Lower	Upper
β_0	27.5	3.0	21.6	33.4
β_1	−4.29	.60	−5.46	−3.11
σ	1.05	.12	.83	1.32

The log likelihood is $\mathcal{L} = -271.4$. The confidence intervals are based on the normal-approximation method.

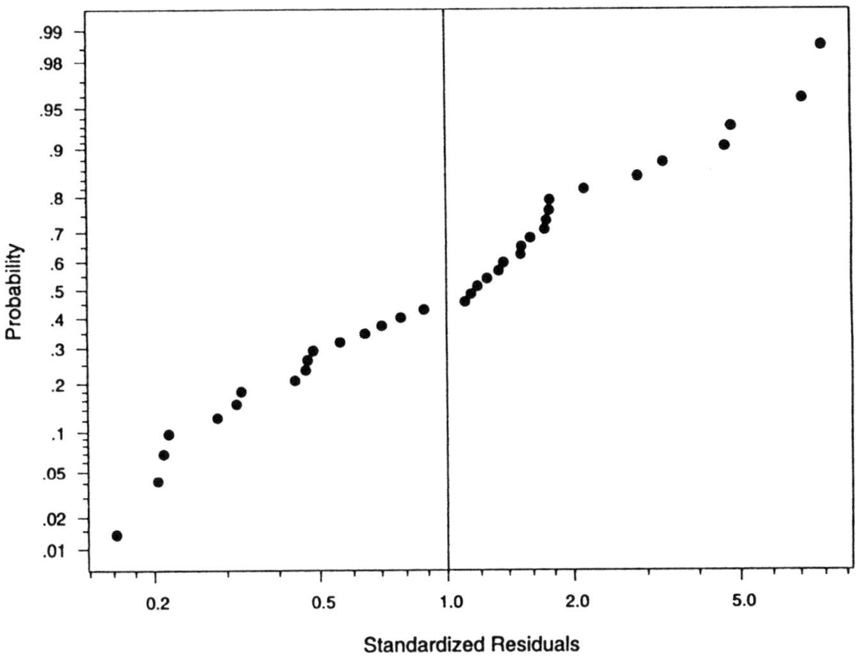

FURTHER EXAMPLES

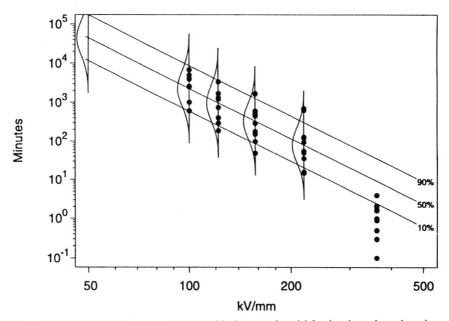

Figure 19.10. Lognormal probability plot of the residuals from the inverse power relationship–lognormal model fitted to the mylar–polyurethane data.

Figure 19.11. Plot of the inverse power relationship–lognormal model fitted to the mylar–polyurethane data (also showing 361.4-kV/mm data omitted from the ML estimation).

19.3.2 Analysis of Interval ALT Data

Previous examples have shown that interval (or read-out) data occur frequently in reliability studies. Such data also arise in accelerated tests. With a complicated evaluation process and limited resources, it is often possible to do only a few inspections on each unit.

Example 19.11 Analysis of ALT Data on a New-Technology IC Device. Appendix Table C.15 gives data from an accelerated life test on a new-technology integrated circuit (IC) device. The device inspection involved an expensive electrical diagnostic test. Thus only a few inspections could be conducted on each device. One common method of planning the times for such inspections is to choose a first inspection time and then space the inspections such that they are equally spaced on a log axis. In this case, the first inspection was after one day with subsequent inspections at two days, four days, and so on (except for one day when the person doing the inspection had to leave early). Tests were run at 150, 175, 200, 250, and 300°C. Failures had been found only at the two higher temperatures. After an initial analysis based on early failures at 250°C and 300°C, there was concern that no failures would be observed at 175°C before the time at which decisions would have to be made. Thus the 200°C test was started later than the others to assure some failures and only limited running time on these units had been accumulated by the time of the analysis.

The developers were interested in estimating the activation energy of the failure mode and the long-life reliability of the ICs. Initially engineers asked about "MTTF" at use conditions of 100°C junction temperature. After recognizing that the estimate of the mean would be on the order of 6 million hours (more than 700 years), it was decided that this would not be a useful reliability metric. Subsequently they decided that the average hazard rate or the proportion that would fail by 100 thousand hours (about 11 years) would be more useful for decision-making purposes.

Figure 19.12 is a lognormal probability plot of the failures at 250°C and 300°C along with the ML estimates of the individual lognormal cdfs. Table 19.4 summarizes the individual lognormal ML estimates. The different slopes in the plot suggest that the lognormal shape parameter σ changes from 250 to 300°C. Such a change could be caused by the occurrence of a different failure mode at high temperatures, casting doubt on the simple first-order Arrhenius model. Failure modes with a higher activation energy, that might never be seen at low levels of temperature, can appear at higher temperatures (or other accelerating variables). A 95% confidence interval for the ratio $\sigma_{250}/\sigma_{300}$ is [1.01, 3.53] (calculations requested in Exercise 19.18), suggesting that there could be a real difference. These results also suggested that detailed physical failure mode analysis should be done for at least some of the failed units and that the accelerated test should be extended until some failures are observed at lower levels of temperature. Even if the Arrhenius model is questionable at 300°C, it could be adequate below 250°C.

Table 19.5 gives Arrhenius–lognormal model ML estimates for the new-technology IC device. Figure 19.13 is a lognormal probability plot showing the Arrhenius–lognormal model fit to the new-technology IC device ALT data. This figure shows

FURTHER EXAMPLES 509

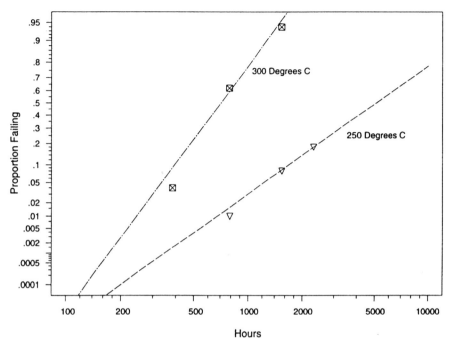

Figure 19.12. Lognormal probability plot of the failures at 250°C and 300°C for the new-technology integrated circuit device ALT experiment.

lognormal cdf estimates for all of the test levels of temperature as well as the use condition of 100°C.

Following the approach employed in Example 19.6, we use the "omnibus" test to compare the constrained and unconstrained models for estimating the lognormal distributions at the different levels of temperature. From Table 19.4, for the unconstrained model, $\mathcal{L}_{\text{unconst}} = \mathcal{L}_{250} + \mathcal{L}_{300} = -86.01$ and from Table 19.5, for the constrained model, $\mathcal{L}_{\text{const}} = -88.36$. In this case the comparison has just 1 degree of

Table 19.4. Individual Lognormal ML Estimation Results for the New-Technology IC Device

	Parameter	ML Estimate	Standard Error	95% Approximate Confidence Intervals	
				Lower	Upper
250°C	μ	8.54	.33	7.9	9.2
	σ	7.87	.26	.48	1.57
300°C	μ	6.56	.07	6.4	6.7
	σ	.46	.05	.36	.58

The log-likelihood values were $\mathcal{L}_{250} = -32.16$ and $\mathcal{L}_{300} = -53.85$. The confidence intervals are based on the normal-approximation method.

Table 19.5. Arrhenius–Lognormal Model ML Estimation Results for the New-Technology IC Device

Parameter	ML Estimate	Standard Error	95% Approximate Confidence Intervals	
			Lower	Upper
β_0	−10.2	1.5	−13.2	−7.2
β_1	.83	.07	.68	.97
σ	.52	.06	.42	.64

The log likelihood is $\mathcal{L} = -88.36$. The confidence intervals are based on the normal-approximation method.

freedom (i.e., dof $= 4 - 3 = 1$) and the test statistic is $Q = -2(-88.36 + 86.01) = 4.7 > \chi^2_{(.95;1)} = 3.84$. This indicates that there is some lack of fit in the constant-σ Arrhenius–lognormal model. Note that there were no failures at 150, 175, or 200°C. These results were implicit in the likelihood. Because the result of 0 failures at these temperatures is consistent with the model and the other data, the computed likelihood at these three temperatures would be very close to 1 and the computed values of $\mathcal{L}$ are therefore very close to 0. Thus the results have no direct effect on the ML fit. A result of 0 failures at any temperature greater than 250°C combined with the rest of these data would, however, have had a strong effect on the fit.

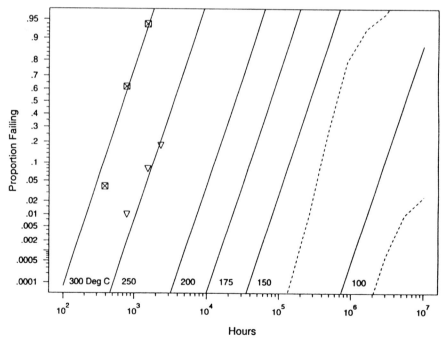

Figure 19.13. Lognormal probability plot showing the ML fit of the Arrhenius–lognormal model for the new-technology IC device.

FURTHER EXAMPLES 511

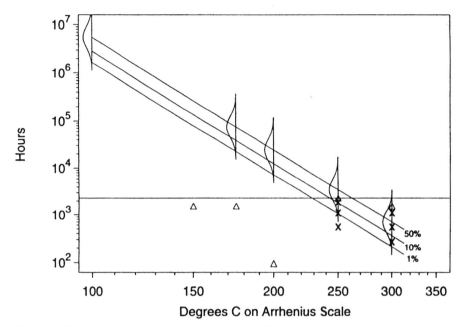

Figure 19.14. Arrhenius plot showing the new-technology IC device data and the Arrhenius–lognormal model ML estimates. Censored observations are indicated by △.

Figure 19.14 is an Arrhenius plot of the Arrhenius–lognormal model fit to the IC new-technology device data. This plot shows the rather extreme extrapolation needed to estimate the failure-time distribution at the use condition of 100°C. If the projections are close to the truth, it appears unlikely that there will be any failures below 200°C during the remaining 3000 hours of testing and, as mentioned before, this was the reason for starting some units at 200°C. ☐

In some applications, temperature-accelerated life tests are run with only one level of temperature. Then a given value of activation energy is used to compute an acceleration factor to estimate life at use temperature. Resulting confidence intervals are generally unreasonably precise because activation energy is generally not known exactly. For example, MIL-STD-883 provides reliability demonstration tests based on a given value of E_a.

Example 19.12 Analysis of New-Technology IC Device ALT Data with Given Activation Energy. Figure 19.15, similar to Figure 19.12, shows the effect of assuming that $E_a = .8\text{eV}$ and having to estimate only β_0 and σ from the limited data. Using a given E_a results in a set of approximate 95% confidence intervals for $F(t)$ at 100°C that are unrealistically narrow. See Section 22.2 for alternative analyses of these data. ☐

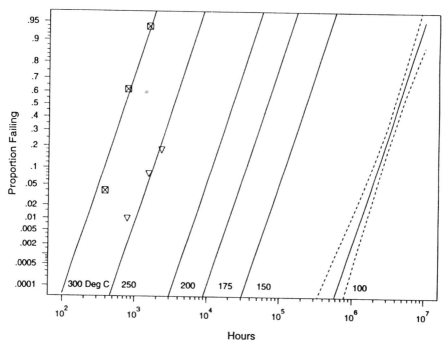

Figure 19.15. Lognormal probability plot showing the Arrhenius–lognormal model ML estimates and 95% confidence intervals for $F(t)$ at $100°C$ for the new-technology IC device with given $E_a = .8$.

19.3.3 Analysis of a Two-Variable ALT

This section shows how to analyze ALT data with two experimental variables. In this case, both variables were thought to be accelerating. The methods illustrated apply to experiments with any number of variables.

Appendix Table C.16 contains temperature/voltage ALT data on tantalum electrolytic capacitors. These data come from Singpurwalla, Castellino, and Goldschen (1975). Tests were conducted at temperature/voltage combinations that were nonrectangular and with unequal allocations of units. Figure 19.16 is a scatter plot of hours to failure versus voltage with temperature indicated by different symbols in the plot (with some jitter used in voltage to help in viewing ties in the data). The amount of censoring is indicated in the top margin. There were various censoring times that are given in Table C.16.

Figure 19.17 is a multiple Weibull probability plot for the individual combinations of voltage and temperature for the tantalum capacitor data. The plot also shows individual ML estimates of Weibull cdfs for those combinations having more than one failure. The line for 85°C and 46.5 volts is much steeper than the others, but this line results from only 2 out of 50 capacitors failing. Thus this deviation in the slopes could be due to random variability. The Weibull distribution seems to provide a reasonable model for the failure-time distribution at those conditions with enough

FURTHER EXAMPLES

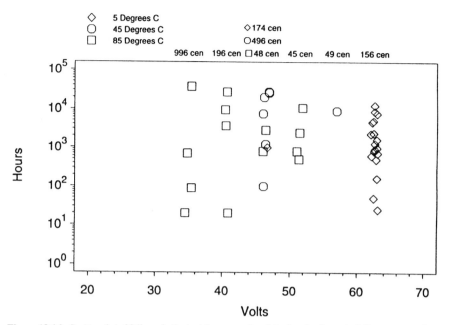

Figure 19.16. Scatter plot of failures in the tantalum capacitor data showing hours to failure versus voltage with temperature indicated by different symbols.

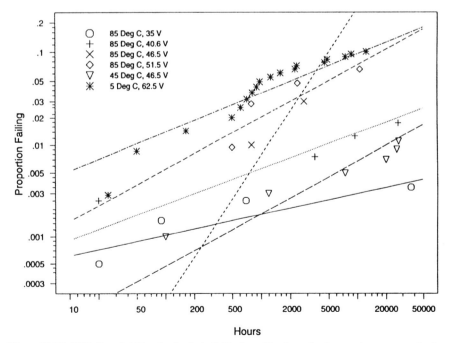

Figure 19.17. Weibull probability plot for the individual combinations of voltage and temperature for the tantalum capacitor data, along with ML estimates of Weibull cdfs.

failures to make a judgment. Figure 19.6 shows that units fail more rapidly at high voltage. Any possible temperature effect is not as strong.

The Arrhenius–inverse power relationship–Weibull model (Section 18.5.2), both without the interaction term (Model 1) and with the interaction term (Model 2), was fitted to these data. In particular, the fitted relationships were

$$\text{Model 1:} \quad \mu = \beta_0 + \beta_1 x_1 + \beta_2 x_2,$$

$$\text{Model 2:} \quad \mu = \beta_0 + \beta_1 x_1 + \beta_2 x_2 + \beta_3 x_1 x_2,$$

where $x_1 = 11605/(\text{temp K})$, $x_2 = \log(\text{volt})$, and $\beta_1 = E_a$. The results for Model 1 are depicted in Figure 19.18 and the results for Models 1 and 2 are summarized numerically in Table 19.6. Comparing the log-likelihood values in Table 19.6 indicates that the interaction term in Model 2 is not helpful in explaining variability in these data (correspondingly, the confidence interval for β_3 contains zero). There is strong evidence for an important voltage effect on life. There is also some evidence for a temperature effect, but the evidence is not strong. Physical theory, however, predicts the positive coefficient β_1; the lack of strong evidence could be the result of a small number of failures at most variable-level combinations and the odd-shaped experimental region.

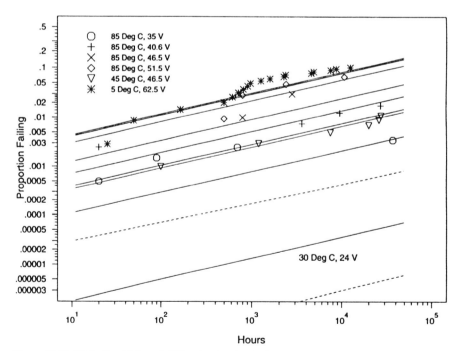

Figure 19.18. Weibull multiple probability plot showing the fitted Arrhenius–inverse power relationship–Weibull model (with no interaction) for the tantalum capacitor data. The dotted lines are 95% confidence limits for $F(t)$ at use conditions.

Table 19.6. Tantalum Capacitor Weibull–Inverse Power Relationship Regression ML Estimation Results

	Parameter	ML Estimate	Standard Error	95% Approximate Confidence Intervals	
				Lower	Upper
Model 1	β_0	84.4	13.6	57.8	111.
	β_1	.33	.19	−.04	.69
	β_2	−20.1	4.4	−28.8	−11.4
	σ	2.33	.36	1.72	3.16
Model 2	β_0	−78.6	109.0	−292.3	135.1
	β_1	5.13	3.3	−1.35	11.6
	β_2	19.9	26.7	−32.5	72.35
	β_3	−1.17	.80	−2.8	.40
	σ	2.33	.36	1.72	3.16

The log likelihoods for Models 1 and 2 are, respectively, $\mathcal{L}_1 = -539.63$ and $\mathcal{L}_2 = -538.40$. The confidence intervals are based on the normal-approximation method.

In Table 19.6, it is interesting that the coefficient estimates of the regression model are highly sensitive to whether the interaction term is included in the model or not. This is due, in part, to the highly unbalanced allocation in the experiment. Figure 19.19 is a Weibull probability plot of the regression residuals for Model 1. The Weibull distribution appears to provide a reasonable description of the variability in these data. Figure 19.20 shows Model 1 estimates of the .01 quantile of the life distribution of the tantalum capacitors as a function of voltage, for the three temperatures in the data (5, 45, and 85°C). The plotted points give the individual ML estimates of the .01 quantile at the temperature/voltage combinations that had more than one failure. This figure shows the relatively strong effect of voltage, relative to the effect of temperature. The lines in this plot are parallel because there is no interaction term in Model 1.

19.4 SOME PRACTICAL SUGGESTIONS FOR DRAWING CONCLUSIONS FROM ALT DATA

Due to their extrapolative nature, drawing conclusions from ALT data can be difficult. This section describes some cautions.

19.4.1 Predicting Product Performance

It is particularly difficult to use AT data to predict actual product performance. Because most products tend to be complicated combinations of materials and components, AT data are mostly successful for estimating life distributions of simple components and materials. Extrapolation is needed to make predictions on the life

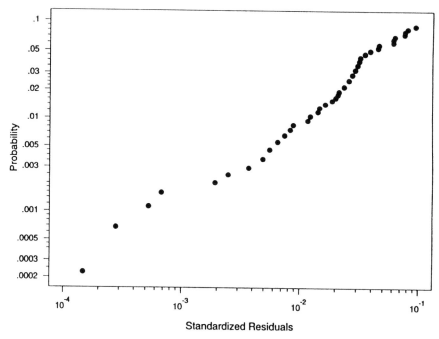

Figure 19.19. Weibull probability plot of the residuals from the Arrhenius–inverse power relationship–Weibull model (with no interaction) for the tantalum capacitor data.

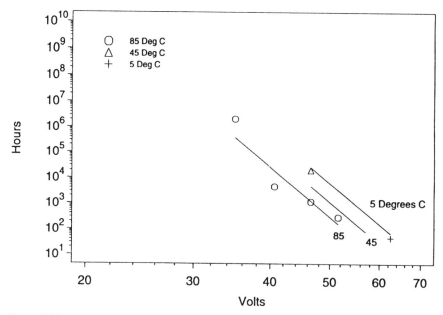

Figure 19.20. Log–log plot of the ML estimates of the .01 quantile of tantalum capacitor life based on data fitted to the Arrhenius–inverse power relationship–Weibull model with no interaction.

distribution of a product in the field environment (which, itself, may be difficult to predict). Extrapolation is difficult or impossible to justify completely.

ALT experiments should be planned and executed with a great deal of care. Inferences and predictions should be made with a great deal of caution. Some particular suggestions for doing this are as follows:

- Use previous experience with similar products and materials.
- Conduct initial studies (pilot experiments) to evaluate the effect that the accelerating variable or variables will have on degradation and the effect that degradation will have on life or performance. Information from preliminary tests provides useful input for planning ALTs (as described in Chapter 20).
- Use failure mode analysis and physical/chemical theory to improve or develop physical understanding to provide a better physical basis for ALT models.
- Limit, as much as possible, the amount of extrapolation (in time and in the accelerating variable). Methods for doing this are described in Chapter 20.

19.4.2 Drawing Conclusions from Accelerated Test Experiments

A typical AT is an extreme example of what Deming (1975) calls an "analytic study." As such, when making predictions from an AT, one has to question the reasonableness of using the AT manufacturing/testing process to represent the actual manufacturing/use process and the adequacy of the life–acceleration model. Typically, confidence intervals based on analytic studies provide, at best, only a lower bound on the total variability and uncertainty; see Hahn and Meeker (1991, Chapter 1) for more detail. For example, such intervals do *not* account for model inadequacy. Extrapolation will amplify model errors, often dominating other sources of uncertainty.

19.5 OTHER KINDS OF ACCELERATED TESTS

The first parts of this chapter have discussed traditional ATs that are used to obtain timely information on a product life distribution at use conditions by testing units at one or more higher levels of accelerating variables. Usually such tests are done on materials, devices, other components, or relatively simple subsystems or systems where the focus is one or a small number of known failure modes. The primary purpose of these tests has been to assess the failure-time distribution of the failure mode(s) of interest. Such information is then used to characterize product life at use conditions and to make product design, warranty coverage, and other decisions.

There are several other important types of accelerated reliability testing, particularly in the electronics industry. These additional types have also been called "accelerated tests." Their purpose is generally other than direct estimation of reliability. Nevertheless, these other tests do generate data that can, in some cases, provide useful information for reliability estimation.

19.5.1 Continuous Operation Product Accelerated Testing

Assuring the reliability of individual components is generally not sufficient to assure the reliability of the larger product or other system within which the components are installed. In particular, it is necessary to assure that components, subsystems, and various interfaces work together. Acceleration is more difficult when testing a complete system and acceleration factors will generally be small (i.e., 2–10). During a system test, there are more potential failure modes. Relative to materials or component tests, there generally has to be more stringent limitations on how much accelerating variables can be increased. One must be careful not to cause damage to the system (e.g., melting of components, overheating causing a rapid change in material properties, or other damage that would not accurately simulate degradation at use conditions). Usually acceleration is achieved by running the system more or less continuously. For example, a refrigerator might be tested, with focus on the compressor, by running the cooling system continuously. Similar tests are used for systems like washing machines and automobile engines (where the purpose of the test might be to accelerate the life of either the engine itself or the engine's lubricant).

Usually accelerated system tests are started on prototype units or early-production units, before the product is introduced into the field. For some products, however, tests will continue for some period of time after units have been introduced into the field (to protect against failure modes that will not show up until later in life). If a design change is being contemplated new tests may be needed.

Sometimes a manufacturer will design an accelerated system test to be run on an audit basis to check the output of the production process over time. For example, two units might be selected from production at the beginning and end of each week. These units would be run at stressful conditions for some period of time (e.g., 1 month) to see if any early failures occur (which might indicate that there is a problem in the manufacturing process).

Manufacturers have developed a number of strategies for accelerated testing of complete systems. The characteristics of system accelerated tests are often product and failure mode specific. In addition to testing units at higher than usual use-rates, systems might be placed in a more stressful environment (e.g., high temperature and humidity, or higher or lower voltage). A manufacturer of electric garbage disposals tests units by having them "chew" on hard plastic cubes to provide a more stressful kind of "garbage." If the motor withstands an equivalent amount of service with the more stressful garbage, then, in this case, the reliability engineers believe that the product will not experience motor or bearing wearout problems in actual use.

To test the starting system of an automobile, the engine could be started and stopped with high frequency. Such a start–stop test would also tend to put special stresses on the engine (or lubricant) itself and might be used to study different failure modes than the continuous test. The manufacturer of laser printers has the printer print a ream of paper, rest 2 minutes, and then continue printing. A sump pump needs to be tested in humid environments. Continuous tests will track some kinds of failure modes quite well. One does, however, run the risk that continuous operation can actually inhibit some failure-causing processes.

19.5.2 Highly Accelerated Life Tests

When planning ALTs to make projections about life at use conditions, Meeker and Hahn (1985) suggest that tests should be planned to minimize the amount of extrapolation in both the accelerating variable and in time. In other applications, where life information at use conditions is not needed, Highly Accelerated Life Tests (HALTs) may be useful. Confer, Canner, and Trostle (1991) discuss the use of extremely high temperature (up to 150°C) and voltage (8 times rating) to achieve acceleration factors of up to 2555 times. They suggest that such tests can be used as (1) a means of sampling inspection for incoming component lots and (2) as burn-in screening test (in this application there is danger that high levels of the accelerating variable will damage units that are to be put into service). Other applications include (3) pilot tests to get information needed for planning a more extensive ALT at lower levels of the accelerating variable and (4) experiments to obtain information on the relevance of failure modes discovered in STRIFE testing (see next section). When using very high levels of the accelerating variable(s), one must watch for failure modes that would never occur at use conditions and have concern for the adequacy of the model (Chapter 7 of Nelson, 1990a, shows how modeling might be used to deal with multiple failure modes in ALTs).

19.5.3 Environmental Stress Testing

The pressure to quickly develop new, high-reliability products has motivated the development of new product testing methods. The purpose of these testing methods is to quickly identify and eliminate potential reliability problems early in product development. One such testing method is known as STRIFE (STRESS-LIFE) testing. The basic idea of STRIFE testing is to aggressively stress and test prototype or early-production units to force failures. It is common to test only one or two units, but more test units can provide important additional information, for example, on unit-to-unit variation. Typical STRIFE tests use combinations of temperature and vibration cycling. The amplitude of the cycling is increased continuously until the end of the test. When possible, use-rate may also be increased. Such a test could be run for days, provided appropriate fixes for detected failure modes can be effected without long delays that might be needed for a complicated redesign (as opposed to a simple part substitution). Bailey and Gilbert (1981) report an example in which the complete STRIFE test and improvement program was successfully completed in 3 weeks. Nelson (1990a, pages 37–39) describes environmental stress tests as "elephant tests" and describes some important issues. The *Proceedings of the Institute of Environmental Sciences* often contain papers on this subject, as do various journals on reliability of electronics systems.

Generally, failures in STRIFE testing can be due to product or process design flaws. When there is a failure in a STRIFE test it is necessary to find and carefully study the failure's root cause. First it is necessary to assess whether the failure could occur in actual use or not. Knowledge and physical/chemical modeling of the particular failure mode are useful in making this assessment. Nelson (1990a, page 38) describes an example where a costly effort was made to remove a high-stress-induced failure mode that never would have occurred in actual use. The occurrence of such failures might indicate that the test is using some combinations of accelerating variables that are too high to be useful. When it is determined that a failure could occur in actual use, it is necessary to change the product design or manufacturing process to eliminate that cause of failure. In some cases, the fix is obvious. In other cases additional focused research and experimentation at the component level or at component interfaces may be required.

Because the results of STRIFE testing are used to make changes on the product design and manufacturing process, it is difficult, or at the very least very risky, to use the test data to predict what will happen in normal use. Even so, ideas from statistical experimental design and models relating stress to life could be useful in choosing stresses, stress ramp speed, and other aspects of the test.

In other testing programs, the goal is to test prototype or early-production units at somewhat accelerated conditions to obtain information on field performance of a product. For example, a newly designed automobile engine may be run continuously at high rpm to simulate rapidly 50,000 miles of service. The danger in interpreting the results of such tests is that the assumed acceleration factors can be seriously inaccurate. Different failure modes have different acceleration factors. For example, such an accelerated test may accurately predict a wear mechanism, but not even discover a corrosion mechanism. In another accelerated test application, humidity was used to accelerate a corrosion mechanism, but had the unexpected effect of reducing wear rate. For these reasons, it is important to have a good understanding of the physics and/or chemistry of possible failure mechanisms.

Schinner (1996) describes and gives examples of system-level and subsystem-level accelerated tests such as STRIFE.

19.5.4 Burn-in

The most common reliability problem for manufacturers and consumers of electronic equipment has been early (infant mortality) failures. Such failures are typically caused by manufacturing defects, which often appear in only a small proportion of the manufactured units. In electronic manufacturing such defective components are called "freaks." The problem of early failure also arises in other kinds of products. Such problems are sometimes referred to as a "quality problem." Often such problems are cured as a product's design and manufacturing process matures. Manufacturers would prefer to "build in" reliability by eliminating all manufacturing defects from the start of production. With rapidly changing technology, however, it has been difficult to achieve a goal of zero defects, particularly early in the product development cycle. To achieve sufficiently high reliability, particularly in critical applications (e.g., space and undersea systems), it has been common practice to use burn-in of components and systems to screen out the units that would otherwise fail early in life. For components like integrated circuits it is common to do burn-in at high humidity and temperature. For system burn-in, it is generally necessary to avoid the use of high levels of the accelerating variables to avoid damaging sensitive components. In either case, burn-in may also involve continuous operational exercising and monitoring of the units. Such

OTHER KINDS OF ACCELERATED TESTS   521

burn-in is useful for detecting intermittent failure modes that have a low probability of being detected in ordinary testing.

Burn-in can be viewed as a type of 100% inspection or screening of the product population to eliminate or reduce the number of defective items going to customers. Burn-in may be necessary if the output of production does not meet reliability specifications. It is important that the burn-in stress or temperature not be so high as to damage the good units. Burn-in is expensive and thus the length of the burn-in is typically limited. The decision on how long to run a burn-in can be based on the desired level of reliability and the distribution of the observed failures during the screening. For a stable production process, which has been characterized adequately, this can be set in advance. Otherwise a sequential or dynamic stopping rule may be needed. Kuo (1984) provides and applies a cost-optimization model for important burn-in decisions. He also reviews previous literature on this subject. Most of the literature deals with burn-in at use conditions. Kuo, Chien, and Kim (1998) describe methods for implementing and optimizing the use of burn-in for electronic components and systems.

Jensen and Petersen (1982) provide an engineering approach to this subject. Also see Nelson (1990a, page 43). Statistical methods can and have been useful for helping to choose stress levels, length of burn-in, and in using burn-in data to assess the state of the production process and the likely field reliability of a product going into service.

19.5.5 Environmental Stress Screening

Environmental Stress Screening (ESS) was developed as an improvement over traditional burn-in methods. ESS provides a more economical and more effective means of removing defective units from a product population when testing units at the system or subsystem (e.g., circuit board) level. Because systems and subsystems cannot tolerate high levels of stress for long periods of time, ESS uses mild, but more complicated, stressing. High levels of temperature and humidity at the component level are replaced by more moderate temperature cycling, physical vibration, and perhaps stressful operational regimes (e.g., running computer CPU chips at higher than usual clock speeds and lower than usual voltages) to help identify the defective units. These tests can be viewed as generalizations of step-stress tests (but their purposes are much different). The tests are sometimes called "shake and bake" tests. Again, the goal is to screen out, as effectively and as quickly as possible, the defective items without otherwise doing harm to the product.

Numerous articles on ESS testing appear each year in the *Proceedings of the Institute of Environmental Sciences*. Tustin (1990) provides a motivational description of the methodology and several references. Nelson (1990a, page 39) gives additional references, including military standards. Kececioglu and Sun (1995) provide a comprehensive description of ESS methods, including optimization and management of ESS programs. MIL-STD-2164 describes standard procedures for ESS of electronic equipment.

Statisticians have had little impact in the development of ESS methods because they are mostly based on engineering knowledge. There are, however, some areas where statistical methods could have an impact on ESS. Some of these will require the development of better models to relate the effect that complicated stressing has on the life distribution of both the defective and the nondefective units. For example, advanced application of the statistical principles of experimental design, modeling, and data analysis can be used to help:

- Choose stress conditions that are best for weeding out manufacturing defects but minimize the chance of doing damage to good units.
- Design screens to provide feedback that can be used to improve product design or the manufacturing process by reducing the frequency of manufacturing defects or eliminating them.
- Assess information that would allow prediction of field reliability. The typically complicated stress patterns make it difficult or impossible to use ESS data directly to make predictions about field performance. However, with enough screening data, correlated with field data, one could find relationships that would allow such predictions.

The development of physical/statistical models to describe the effect of ESS stress patterns on product life would be useful. This task, however, will not be easy. LuValle and Hines (1992) report experimental evidence indicating that varying several accelerating variables in typical ESS procedures leads to complicated effects on life. There can be interactions among the variables and there may be "memory" of past stress patterns in that, from a particular point in time, future degradation may depend not only on the current amount of degradation but also on how (i.e., under which stress patterns) that degradation has accrued. This is in contrast to traditional step-stress failure models (e.g., Chapter 10 of Nelson, 1990a).

Like burn-in, ESS is an inspection/screening scheme. In line with the modern quality precept of eliminating reliance on mass inspection, most manufacturers would prefer not to use burn-in or ESS. They are expensive and may not be totally effective. By improving the reliability through continuous improvement of the product design and the manufacturing process, it is often possible to reduce or eliminate reliance on screening tests except, perhaps, in the most critical applications. Some companies apply ESS only on an audit basis to monitor production quality on an ongoing basis.

19.6 POTENTIAL PITFALLS OF ACCELERATED LIFE TESTING

As described earlier in this chapter, accelerated life testing can be a useful tool for obtaining timely information about materials and products. There are, however, a number of important potential pitfalls that could cause an ALT to lead to seriously incorrect conclusions. Users of ALTs should be careful to avoid these pitfalls.

POTENTIAL PITFALLS OF ACCELERATED LIFE TESTING   523

19.6.1   Pitfall 1: Multiple (Unrecognized) Failure Modes

High levels of accelerating variables like temperature or voltage can induce failure modes that would not be observed at normal operating conditions. In some cases new failure modes result from a fundamental change in the way that the material or component degrades or fails at high levels of the accelerating variable(s). For example, instead of simply accelerating a failure-causing chemical process, increased temperature may actually change certain material properties (e.g., cause melting). In less extreme cases, high levels of an accelerating variable will change the relationship between life and the accelerating variable (e.g., life at high temperatures may not be linear in inverse absolute temperature, as predicted by the Arrhenius relationship).

If other failure modes are caused at high levels of the accelerating variables and this is recognized, it can be accounted for in the data analysis by treating the failure for the new failure modes as a censored observation (as long as the new failure mode does not completely dominate the failure mode(s) of interest). Chapter 7 of Nelson (1990a) gives several examples. In this case, however, such censoring can severely limit the information available on the failure mode of interest. If other failure modes are present but not recognized in data analysis, seriously incorrect conclusions are likely.

19.6.2   Pitfall 2: Failure to Properly Quantify Uncertainty

It is important to recognize that there is uncertainty in statistical estimates. Basing decisions on point estimates alone can, in many applications, be seriously misleading. Standard statistical confidence bounds quantify uncertainty arising from limited data. For example, Figure 19.13 shows an enormous amount of uncertainty in life at 100°C, due to the small number of failures and the large amount of extrapolation in temperature. The corresponding analysis depicted in Figure 19.15 uses a given value of activation energy for the life–temperature relationship. Because the activation energy is not known exactly, the precision exhibited in this plot is too small and potentially misleading. For many applications, neither of these extremes would provide a proper quantification of uncertainty. Section 22.2 describes an appropriate compromise for situations where there is useful information about activation energy.

It is also important to remember that statistical confidence bounds do not account for model uncertainty (which can be tremendously amplified by extrapolation in accelerated testing). In general, performing sensitivity analysis is an important step in any quantitative analysis involving uncertainty and is particularly useful for assessing the effects of model uncertainty. For example, one can rerun analyses under different assumed models to see the effect that different model assumptions have on important conclusions.

19.6.3   Pitfall 3: Multiple Time Scales and Degradation Affected by More than One Accelerating Variable

Section 1.3.4 described issues relating to time scales. These issues become even more important with accelerated testing, and particularly when there is more than one failure-causing mechanism that might be accelerated. Standard acceleration methods generally will not accelerate all time scales in the same manner. A serious pitfall of accelerated testing is to assume a simple relationship between life and the accelerating variables when the actual relationship if really very complicated. Consider the following examples.

- In an accelerated test to estimate the lifetime characteristics of a composite material, chemical degradation over time changes material ductility. Failures, however, are actually caused by stress cycles during use, leading to initiation and growth of cracks, and eventually to fracture. Thus there are two failure-causing mechanisms. The acceleration model would be complicated because the effect of cycling depends on the material ductility and because increasing temperature would affect the time scales of both mechanisms.
- An incandescent light bulb usually fails when its filament breaks. During burn time the bulb's filament will go through an evaporation process, eventually leading to failure. There are, however, other variables that can shorten a bulb's life. In particular, on–off cycles can induce both thermal and mechanical shocks that can, over time, lead to the growth of fatigue cracks in the filament. Thus the on–off frequency can have an effect on bulb life. Accelerating only the burn time (e.g., by testing at higher voltage) may give misleading predictions of life in an environment with many on–off cycles. Relatedly, light bulbs operated in an environment with physical vibration (e.g., in automobiles, on large ships, or in a motorized appliance) will often exhibit shorter lifetimes, depending on the frequency and amplitude of the vibrations as well as the bulb's design.
- The degradation of coatings like paint depends on a number of different variables relating to time scales. Most coatings degrade chemically over time. UV light accelerates the degradation process of many kinds of coatings, as does high temperature and humidity. The *number* of wet–dry and thermal cycles is also important to coating life, but generally relates to a separation or peeling failure mechanism that is different from (but perhaps related to) the chemical degradation mechanism. Each of these variables and each failure mode has its own underlying time scale.

Generally, there will be a distribution of product-use conditions in the field: for example, the number of fatigue cycles as a function of the changing ductility of the composite material over time or the ratio giving the number of on–off cycles per hour of burn time for an incandescent light bulb. Similarly, some automobiles are driven in the North and some in the South; some spend substantial time in direct sunlight, while others do not. In these situations, product-use environment plays an important, but complicated, role in planning and making life predictions from accelerated tests.

In simple situations where the ratio of the time scales for different mechanisms is known and reasonably constant in the product population, an accelerated test could be conducted to simulate life in that ratio. When the ratio has a known distribution in the product population, tests can be conducted over an appropriate range of the ratio. In other applications, it will be necessary to use an accelerated test in which accelerating variables (e.g., temperature, humidity, and UV exposure simultaneously) are varied simultaneously. To predict life at specified use conditions, one needs an adequate physical model to describe the relationship among these variables, the different degradation scales, and the definition of failure.

19.6.4 Pitfall 4: Masked Failure Mode

Figure 19.21 shows a graph of what might illustrate the results of a typical accelerated life test if there were just a single failure mode and if increased temperature accelerated that failure mode in a simple manner, described by the Arrhenius relationship. It is possible that such an accelerated test, while focusing on one known failure mode, may mask another! This is illustrated in Figure 19.22. Moreover, as shown in Figure 19.22, it is often the masked failure mode that is the first one to show up in the field. In such cases, the masked failure modes often dominate among reported field failures.

19.6.5 Pitfall 5: Faulty Comparison

It is sometimes claimed that accelerated testing is not really useful for predicting reliability but is useful for comparing alternatives (e.g., alternative designs, vendors). Consider comparing similar products from two different vendors. The thought behind this claim is that laboratory accelerated tests generally cannot be expected to adequately approximate actual use conditions, but that if Vendor 1 is better than Ven-

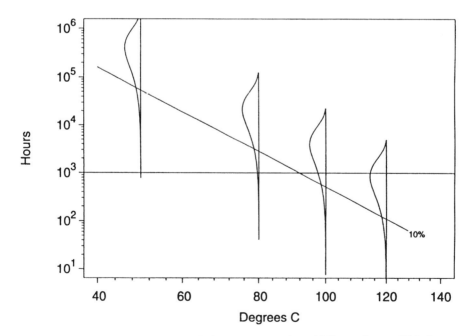

Figure 19.21. Possible results for a typical temperature-accelerated failure mode on an IC device.

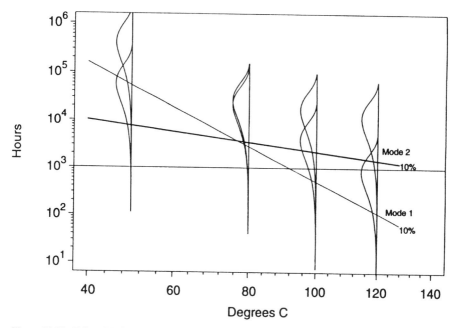

Figure 19.22. Failure Mode 2 with lower activation energy, masked at high temperature and unmasked at low temperature.

dor 2 in an accelerated test, then the same would be true in field use, as illustrated in Figure 19.23. Comparisons based on ALTs, however, are subject to some of the same difficulties as other ALTs. In particular, consider the results depicted in Figure 19.24. In this case, Vendor 1 had longer life at both of the accelerated test conditions, but the prediction at use conditions suggested that Vendor 2 would have higher reliability. An important decision on the basis of limited results in this ALT would be, at best, difficult to justify. It would be most important to find out why the slopes are different and to understand the life-limiting failure modes at use conditions. If the failures at the use conditions are not the same as those at the accelerated conditions, then the ALT results would be wrong. Also, it might be possible that the early failures for Vendor 2 are masking the failure mode that we see in Vendor 1's test results. One cannot use an ALT to compare products that have different kinds of failure modes.

19.6.6 Pitfall 6: Accelerating Variables Can Cause Deceleration!

In some cases it is possible that increasing what is thought to be an accelerating variable will actually cause deceleration! For example, increased temperature in an "accelerated" circuit-pack reliability audit predicted few field failures. The number of failures in the field was much higher than predicted because the increased temperature resulted in lower humidity in the "accelerated" test and the primary failure mode in the field was caused by corrosion that did not occur at high temperature and low

POTENTIAL PITFALLS OF ACCELERATED LIFE TESTING 527

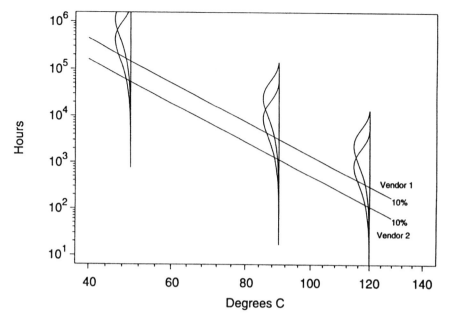

Figure 19.23. Well-behaved comparison of two products.

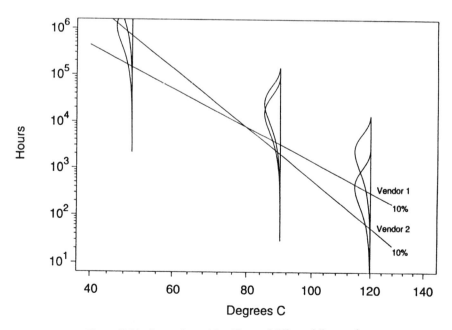

Figure 19.24. Comparison with evidence of different failure modes.

humidity. It is for this reason that in most accelerated tests of electronic equipment, both temperature and humidity need to be controlled.

In another similar application, a higher than usual use-rate for a mechanical device in an accelerated test inhibited a corrosion failure mechanism. That corrosion mechanism eventually caused a serious field problem that was not predicted by the accelerated test.

In an accelerated test of a newly designed automobile air conditioner, reliability, based on a series of constant-run accelerated life tests, was predicted to be very high over a 5-year period. However, after 2 years, a substantial number of the in-service air conditioners failed due to a "drying out" material degradation. These failures were caused by lack of use in winter and had never been seen in the continuous accelerated testing.

19.6.7 Pitfall 7: Untested Design/Production Changes

A new electromechanical device was to be used in a system designed for 20 years of service in a protected environment. An accelerated test of the device was conducted and this test "demonstrated" 20-year life (no more than 10% failing) under normal operating conditions (typical use-rate). After the accelerated test, and as the product was going to production, a material change was made by the device vendor. The change led to a material-degradation failure mode that caused (or would have caused) all in-service units to fail within 10 years. Eventually, all installed devices had to be replaced.

19.6.8 Pitfall 8: Drawing Conclusions on the Basis of Specially Built Prototype Test Units

Seriously incorrect conclusions can result from an accelerated life test if test units will differ importantly from actual production units. For example, factory manufacturing conditions are different from those in a laboratory. Cleanliness and care in building prototype versus production units may differ substantially. Material and parts in prototype units might differ from those that will be used in production. Highly trained technicians may build prototype units that are importantly different from units that would be made in the factory.

As much as possible, test units for an accelerated test should be manufactured under actual production conditions, using raw materials and parts that are the same as or as close as possible to those that will be used in actual manufacturing of units. As much as possible, the test units should reflect variabilities that will be present in actual production.

In one situation, an accelerated test was conducted on 12 prototype units. The units contained epoxy that had to be cured in an oven for a specified amount of time. The product passed its accelerated test with a safe margin. In actual manufacturing operations, however, the curing process was not well controlled. Uncured epoxy can be highly reactive. For this product, a substantial proportion of installed units eventually failed due to corrosion caused by the improperly controlled curing.

BIBLIOGRAPHIC NOTES

Nelson (1990a) is an extensive and comprehensive source for further material, practical methodology, basic theory, and examples for accelerated testing. Viertl (1988) provides a briefer (and more academic) overview of the available statistical methods for ALTs, with more focus than Nelson (1990a) on a large class of statistical methods that, for a variety of practical reasons, seem not to have been used widely in practice. These methods include nonparametric and semiparametric statistical methods. Viertl (1988) also discusses Bayesian methods for ALT planning and analysis. Nelson (1990a) contains 431 references; Viertl (1988) contains 208. The intersection is only 55 references. Mann, Schafer, and Singpurwalla (1974, Chapter 9) overview ALT methods available at the time. Derringer (1982) describes some important practical considerations in the planning and analysis of ALTs. Meeker and Escobar (1993) survey important areas of research in accelerated testing. LuValle (1993) illustrates the use of graphical methods that can be used to detect departures from the SAFT model.

Nelson (1975a; 1975b; 1990a, Chapter 7) describes graphical and ML methods for analyzing ALT data with competing failure modes.

EXERCISES

19.1. Explain the importance of having understanding of the physics or chemistry of failure mechanisms when one is doing accelerated life testing.

19.2. The following table contains data from an accelerated life test on Device-C, an integrated circuit. Failures were caused by a chemical reaction inside the circuit package. Reliability engineers tested 10 circuits at each temperature over a period of 3000 hours. The purpose of the experiment was to estimate the activation energy of the failure-causing reaction and to obtain an estimate of the integrated circuit life distribution at 80°C junction temperature.

Junction Temperature	# of Units Tested	Recorded Failure Times in Thousands of Hours
80°C	10	None by 3000 hours
125°C	10	None by 3000 hours
150°C	10	2.35, 2.56, 2.98
175°C	10	.80, 1.13, 1.21, 1.31, 1.35, 1.35, 1.37, 1.42, 1.77, 1.96
200°C	10	.22, .25, .28, .33, .37, .38, .46, .46, .51, .61

(a) For each temperature with failures, plot the ordered failure time $t_{(i)}$ versus $(i - .5)/10$ on lognormal probability paper.
(b) Repeat part (a), but use Weibull probability paper.
(c) Make a judgment as to whether the lognormal or the Weibull distribution is a more adequate distribution for these data.

(d) Using the lognormal probability plot, draw a set of parallel straight lines through the plotted points, one line for each temperature having failures.

(e) Use each line on the probability plot to obtain graphical estimates of the .5 quantiles at the corresponding temperatures.

(f) Use each line on the probability plot to obtain graphical estimates of the .01 quantiles at the corresponding temperatures. Describe the nature of the extrapolation in these estimates.

(g) Plot the estimates of the .5 quantile versus temperature on Arrhenius paper. Draw a straight line to estimate the relationship between life and temperature. Do the same with the graphical estimates of the .01 quantile.

(h) Graphically estimate the slope of the lines drawn in part (g), and use these to obtain a graphical estimate of the failure mode's activation energy.

(i) Use the lines drawn in part (g) to obtain estimates of the .01 and .5 quantiles of the life distribution of Device-C at 80°C. Describe the nature of the extrapolation in these estimates.

(j) Predict the effect on the estimates of the Arrhenius relationship if the data at 80°C and 125°C were to be omitted from the analysis. Explain.

19.3. Provide a list of the different things that one can learn from plotting individual nonparametric estimates and parametric ML estimates on Weibull probability paper (such as Figure 19.8).

19.4. Suppose that failure time $T \sim \text{LOGNOR}(\mu, \sigma)$ at a given level of temperature and that the Arrhenius model can be used to get a temperature/time acceleration factor as in (18.2). Then show that the logarithm of quantiles of the failure-time distribution will be a linear function of $1/(\text{temp °C} + 273.15)$.

19.5. Suppose that failure time $T \sim \text{WEIB}(\mu, \sigma)$ at a given level of voltage and that the inverse power relationship can be used to get a voltage/time acceleration factor as in (18.9). Then show that the logarithm of quantiles of the failure-time distribution will be a linear function of $\log(\text{Voltage})$.

19.6. An analyst has fit the Arrhenius–lognormal model

$$\Pr[T \le t; \text{temp}] = \Phi_{\text{nor}}\left[\frac{\log(t) - \mu}{\sigma}\right],$$

where $\mu = \beta_0 + \beta_1 x$, σ is constant, and $x = 1/(\text{temp K})$. Show how β_1 in this model is related to activation energy E_a in the Arrhenius relationship.

19.7. An analyst has fit the Arrhenius–lognormal model

$$\Pr[T \le t; \text{temp}] = \Phi_{\text{nor}}\left[\frac{\log_{10}(t) - \mu}{\sigma}\right],$$

where $\mu = \beta_0 + \beta_1 x$, σ is constant, and $x = 11605/(\text{temp K})$. This differs from the model presented in Example 19.5 because base-10 logarithms have

EXERCISES   531

been used (something done quite commonly, and usefully, in engineering and other nonmathematical disciplines). Show how estimates of β_0, β_1, σ, and t_p in this model relate to estimates in the traditional base-e lognormal distribution.

19.8. Refer to Example 19.8. Write down expressions that could be used to compute a normal-approximation confidence interval for the p quantile of the life distribution at a specified level of temperature.

19.9. A particular type of IC has two different failure modes, both of which can be accelerated by increasing temperature. The random failure time of Mode 1 is T_{M_1} and the random failure time due to Mode 2 is T_{M_2}. In a unit, suppose that the IC fails at $T = \min(T_{M_1}, T_{M_2})$. Failure mode M_1 has an activation energy of $E_a = 1.4$ electron volts and failure mode M_2 has an activation energy of $E_a = .7$. The physical nature of the failure mechanisms suggests that T_{M_1} and T_{M_2} are independent. The engineers involved believe that both T_{M_1} and T_{M_2} have a lognormal distribution. At the proposed highest test temperature of 120°C, with failure time measured in hours, assume that $\mu_1 = 6.9$, $\sigma_1 = .6$, $\mu_2 = 9.0$, and $\sigma_2 = .8$.

(a) Assuming an Arrhenius ALT model, plot the median of the failure-time distribution versus temperature for failure mode M_1 and also for failure mode M_2. Plot using temperatures between the use temperature of 40°C and 120°C.

(b) Plot the .1 and .9 quantiles of the mode M_1 failure-time distribution and of the mode M_2 failure-time distribution, as a function of temperature.

(c) Use the plots to help describe potential dangers of using a temperature-accelerated life test on a component, ignoring failure mode differences, when the different failure modes have vastly different activation energies.

▲**19.10.** Refer to Exercise 19.9. Compute and plot the median, as well as the .1 and .9 quantiles, of the failure-time distribution for the IC, as a function of temperature, when both failure modes are active. That is, use the distribution of $T = \min(T_{M_1}, T_{M_2})$. What do these results suggest about appropriate methods for dealing with test acceleration when there are two or more failure modes? That is, suggest how a useful ALT can be conducted and how the data could be analyzed when there are two or more failure modes.

▲**19.11.** To describe the failure-time distribution of specimens of an insulating material at use operating conditions temp_U, use a Weibull distribution with cdf $F(t) = 1 - \exp[-(t/\eta_U)^\beta]$. Suppose that the Arrhenius model applies. Then the model is SAFT and at some high temperature temp_H, $\eta_U = \mathcal{AF}(\text{temp}_H)\eta_H$.

(a) Show that $\log(t_p)$, the logarithm of the Weibull quantile, is a linear function of $1/(\text{temp °C} + 273.15)$.

(b) Derive an expression for the pdf at temp_H.

(c) Derive an expression for the hazard function at temp_H.

(d) Show that the ratio of the hazard function at temp_U and at temp_H does not depend on time. This implies that the Weibull SAFT is also a proportional hazards model.

19.12. For a particular kind of insulating material, life can be described by the inverse power model

$$\Pr[T \leq t; \texttt{volt}] = \Phi_{\text{sev}}\left[\frac{\log(t) - \mu}{\sigma}\right],$$

where σ is constant, $\mu = \beta_0 + \beta_1 x$, and $x = \log(\texttt{volt})$.

(a) Show that $\log(t_p)$, the logarithm of the Weibull quantile, is a linear function of $\log(\texttt{volt})$.

(b) If it is known that $100p_1\%$ would fail by time t_c at voltage $\texttt{volt}_1$ and $100p_2\%$ would fail by the same time at voltage $\texttt{volt}_2$, derive an expression for the proportion failing at $\texttt{volt}_3$.

(c) Derive expressions for the inverse power relationship parameters β_0 and β_1 as functions of p_1, p_2, and σ.

19.13. Example 19.6 showed how to use an omnibus likelihood-ratio test to compare fitting individual lognormal distributions at each level of temperature and fitting the Arrhenius–lognormal ALT model. This provides an overall test for model adequacy.

(a) Explain, precisely, the hypothesis or hypotheses being tested in Example 19.6.

(b) Explain which constrained and unconstrained models you would fit in order to test the assumption of a common σ at all levels of temperature. Do not assume anything about the temperature–life relationship.

(c) Explain possible likelihood-ratio tests on whether σ is constant, under the assumption that μ is linearly related to $11605/\texttt{temp}$ K.

(d) Explain how to do a likelihood-ratio test to check the assumption that μ is related linearly to $11605/\texttt{temp}$ K versus an alternative relationship with curvature, assuming that σ does not depend on temperature.

19.14. Refer to Example 19.8 for the Device-A data. Compute a confidence interval for the lognormal distribution $F(10000)$ at $10°C$.

19.15. The ALT data on the mylar–polyurethane insulating structure in Appendix Table C.13 are complete data (no censored observations). Thus it is possible to fit a lognormal distribution to these data using the standard least squares regression analysis procedure in a standard statistical package. Use these data, dropping the 361.4-kV/mm observations, to do the following.

EXERCISES

    (a) Use such a statistical package to compute the least squares estimates for the inverse power relationship–lognormal model.

    (b) Plot the regression estimates of median of the failure-time distribution on log–log axes. Also plot the sample median at each level of voltage. What conclusions can you draw from this?

    (c) Compare the least squares estimates with estimates obtained by using ML estimation with the inverse power relationship–lognormal model. What differences do you notice in the parameter estimates?

19.16. Refer to Exercise 19.15. Use a maximum likelihood program to fit the inverse power relationship–Weibull model to these data. Compare Weibull and lognormal estimates of the .1 quantile of failure at 50 kV/mm. What do you conclude about the importance of the distribution used for this estimate?

19.17. Redo the analyses in Exercise 19.15 but include the 361.4-kV/mm observations. What do you conclude?

19.18. For Example 19.11, using the results in Table 19.4, compute a 95% normal-approximation confidence interval for the ratio $\sigma_{250}/\sigma_{300}$. Base the interval on the large-sample distribution of $Z_{\log(\hat{\sigma}_{250}/\hat{\sigma}_{300})}$.

19.19. In Example 19.11, there is evidence that σ differs from one level of temperature to the other.

    (a) Suppose that σ is really changing (as suggested by the point estimates). Describe the effect that using a constant-σ model would have on estimates of the life distribution at 100°C. You do not need to do any numerical computations; answer using intuition or analytical arguments.

    (b) With reference to the simple Arrhenius–lognormal failure-time model described in Sections 18.3.1 and 18.3.3, suggest possible physical reasons (i.e., deviations from the simple model) that might cause σ to change as a function of temperature.

▲**19.20.** In Example 19.11, the likelihood-ratio test for comparing the constrained and unconstrained models has only 1 degree of freedom. In Example 19.6, however, the comparison test has 3 degrees of freedom.

    (a) Explain why there is a difference.

    (b) If the 200°C subexperiment had run to 10^5 hours without failure, would the number of degrees of freedom in the test change? Why or why not?

CHAPTER 20

Planning Accelerated Life Tests

Objectives

This chapter explains:

- Criteria for planning accelerated life tests (ALTs).
- Simulation and analytical methods for evaluating proposed ALT plans.
- The value and limitations of theoretically optimum ALT plans.
- Compromise accelerated test plans that have good statistical properties and, at the same time, meet practical constraints.
- How to extend methods for planning single-variable ALT plans to planning ALTs with more than one variable.

Overview

Chapter 10 described methods for evaluating proposed test plans for estimating a single failure-time distribution. These methods led to some simple formulas that provide a means of choosing test length and sample size to control estimation precision. This chapter describes methods for planning ALTs and for evaluating the precision of estimates illustrated in Chapter 19. Familiarity with the ideas from Sections 10.1 and 10.2 is important for understanding the underlying technical methods. A good understanding of the basic data analysis material in Chapter 19 is also helpful.

Section 20.1 uses a test-planning example to introduce some of the concepts involved in planning an ALT. Section 20.2 describes how to evaluate the properties of a specified ALT plan. The ability to evaluate the properties of these plans allows one to choose a plan that will optimize according to particular criteria. Section 20.3 provides details and examples on how to plan a single-variable ALT. Section 20.4 extends these methods to two-variable ALTs. Section 20.5 describes how to extend the concepts to ALTs with more than two variables.

20.1 INTRODUCTION

Usually ALTs need to be conducted within stringent cost and time constraints. Careful planning is essential. Resources need to be used efficiently and the amount of extrapolation should be kept to a minimum. During the test-planning phase, experimenters should be able to explore the kind of results that they might obtain as a function of the specified model and proposed test plan.

Example 20.1 *Reliability Estimation of an Adhesive Bond.* The following example comes from Meeker and Hahn (1985). The engineers responsible for the adhesive bond reliability needed to estimate the .1 quantile of the failure-time distribution at the usual operating temperature of 50°C. The .1 quantile was expected to be more than 10 years, but this needed to be demonstrated. There were 300 units, but only 6 months (183 days) available for testing. If testing had been done at 50°C, no failures would be expected. But no failures in 6 months would not provide the needed degree of assurance that the .1 quantile is at least 10 years. An ALT was proposed to make the required demonstration. □

20.1.1 Planning Information

The properties of ALT plans depend on the underlying model and the parameters of that model. The form of the underlying model and at least some of the parameters are generally unknown. To evaluate and compare alternative test plans, it is necessary to have some planning information about the model. Sources of such planning information include previous experience with similar products and failure modes, expert opinion, and other engineering information or judgment. There are a number of different questions that one could ask to get the required information.

Example 20.2 *Planning Values for the Adhesive Bond ALT.* Failure is thought to be caused by an unobservable simple chemical degradation process, leading to weakening and eventual failure of the bond. The engineers feel that the rate of the chemical reaction can be modeled with the Arrhenius relationship over some reasonable range of higher temperatures. This would suggest a SAFT model (see Section 18.3), implying that the form and shape of the failure-time distributions are the same at all levels of temperature. The engineers felt that something like .1% of the bonds might fail in 6 months at 50°C, but that something like 90% would fail in 6 months at 120°C. Additionally, the Weibull distribution had been used successfully in the past to model data from similar adhesive bonds. This information, alone, allows one to obtain algebraically the failure probability in 6 months at any level of temperature (deriving such a formula is left as an exercise). The Weibull shape parameter was thought to be near $\beta^{\square} = 1.667$ (or $\sigma^{\square} = 1/\beta^{\square} = .6$). With this additional information, the other Weibull regression model parameters are defined. In particular, for time measured in days, $\beta_0^{\square} = -16.733$ and $\beta_1^{\square} = .7265$. □

As in Chapter 10 the superscript $\square$ is used to denote a planning value of a population or process quantity needed to plan the ALT. Functions of the planning values (e.g., failure probabilities and, indeed, the plan and characteristics of the plan), will not be encumbered with this symbol.

In applications, it is important to assess the sensitivity of ALT plans to misspecifications of the unknown inputs. Generally this is done by developing a test plan with the given planning values and then, at the end, doing some sensitivity analysis to assess the effect that changes have on the suggested plan.

20.1.2 Model Assumptions

The model assumptions used in this chapter parallel those introduced and used in Chapters 17, 18, and 19. The presentation in this chapter continues to use log-location-scale distributions. Most of the general ideas however, can be applied (with a higher level of technical difficulty) to other parametric distributions like the ones in Chapter 5.

As described in Chapters 18 and 19, most common parametric ALT models use a log-location-scale distribution to describe the variability in failure times. The cdf for failure time T is

$$\Pr(T \leq t) = F(t; \mu, \sigma) = \Phi \left[\frac{\log(t) - \mu}{\sigma} \right], \tag{20.1}$$

where $\mu = \mu(x)$, the location parameter for $\log(T)$, is a function of the accelerating variable and σ is constant. For the Arrhenius relationship between life and temperature, $\mu(x) = \beta_0 + \beta_1 x$ and $x = 11605/(\text{temp }°C + 273.15)$. Here units are tested simultaneously until censoring time t_c. Care in testing must be taken to assure that failure times are independent from unit to unit.

Again, most of the important ideas in this chapter can be applied, with appropriate modifications, to other testing situations (e.g., other failure-time distributions, acceleration relationships, types of censoring).

20.1.3 Traditional Test Plans

Traditional test plans use equally spaced levels of the accelerating variable(s) and equal allocation of test units to those levels.

Example 20.3 Engineers' Originally Proposed Test Plan for the Adhesive Bond ALT. The engineers responsible for the adhesive-bonded power element reliability had developed a preliminary ALT plan, given in Table 20.1 and shown graphically in Figure 20.1. This traditional plan used equal spacing and equal allocation of units to the different levels of temperature but had some deficiencies. In particular, there was concern about the large amount of extrapolation in temperature (to 50°C) and the Arrhenius relationship was in doubt at temperatures above 120°C. The engineers had proposed testing at the very high levels of temperature under the mistaken belief that it would be necessary to have all or almost all of the test units fail before the end of the test. As shown in Chapter 19, however, ML methods can be used to estimate the parameters of the ALT model, even if data are censored.

INTRODUCTION

Table 20.1. Engineers' Preliminary ALT Plan with a Maximum Test Temperature of 150°C

Level TEMPC	Failure Probability p_i	Allocation Proportion π_i	Allocation Number n_i	Expected Number Failing $E(r_i)$
50	.001			
110	.59	1/3	100	59
130	1.00	1/3	100	100
150	1.00	1/3	100	100

Moreover, because interest centered on the lower tail of the distribution at 50°C, data in the upper tail of the distribution would be of limited value (and could even be a source of bias if the fitted distribution were not adequate there). It would be more appropriate to test at lower more realistic temperatures (even if only a small fraction of units will fail) and to allocate more units to lower temperatures. Intuitively, this is because such units would be closer to the use conditions and because, with smaller failure probability at low temperature, more units need to be tested to have assurance that at least a few units will fail. Such a test plan is shown in Table 20.2. □

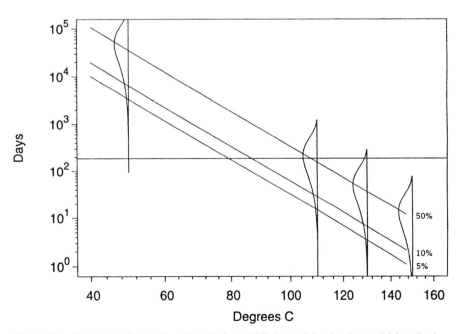

Figure 20.1. Illustration of the engineers' preliminary ALT plan and the planning model for adhesive-bonded power elements on Arrhenius paper.

Table 20.2. Engineers' Modified Traditional ALT Plan with a Maximum Test Temperature of 120°C

Level TEMPC	Failure Probability p_i	Allocation		Expected Number Failing $E(r_i)$
		Proportion π_i	Number n_i	
50	.001			
80	.04	1/3	100	4
100	.29	1/3	100	29
120	.90	1/3	100	90

For this plan, $\text{Ase}[\log(\hat{t}_{.1}(50))] = .4167$ for the Weibull–Arrhenius model.

20.2 EVALUATION OF TEST PLANS

20.2.1 Evaluation Using Monte Carlo Simulation

As described in Section 10.1.2, simulation provides a powerful, insightful tool for planning experiments. For a specified model and planning values for the model parameters, it is possible to use a computer to simulate ALT experiments to see the kind of data that will be obtained and to visualize the variability from trial to trial. Such simulations provide an assessment of sampling uncertainty that will result from using a limited number of test specimens.

Example 20.4 Simulation Evaluation of the Modified Traditional Test Plan for the Adhesive Bond. The planning values from Example 20.2 define the complete ALT model for adhesive bond failure. Based on the engineers' modified test plan in Table 20.2, Figure 20.2 shows 50 ML estimate lines for the .1 quantile, simulated from 50 ALTs. This figure illustrates the amount of variability expected if the ALT experiment were to be repeated over and over, assuming the Arrhenius relationship to be correct over the entire range of temperature. The figure also clearly shows the deteriorating effect that extrapolating to 50°C has on precision, even assuming that the Arrhenius relationship is correct. Approximate precision is also reflected in the sample standard deviations given directly in Figure 20.2 (the approximation here is due to using only 500 simulations). The sample standard deviation $\text{SD}[\log(\hat{t}_{.1})] = .427$ agrees well with the large-sample approximate standard error $\text{Ase}[\log(\hat{t}_{.1}(50))] = .4167$ from Table 20.2. Relative to the plan from Example 20.3, the engineers felt more comfortable with the degree of extrapolation in this plan even though this modified plan would provide less precision, due to the smaller range of test temperatures and the larger proportion of units that would be censored. □

20.2.2 Evaluation Using Large-Sample Approximations

Section 10.2 provides motivation for and shows how to compute approximate standard errors of sample estimates for a given model and test plan. The general formulas

EVALUATION OF TEST PLANS   539

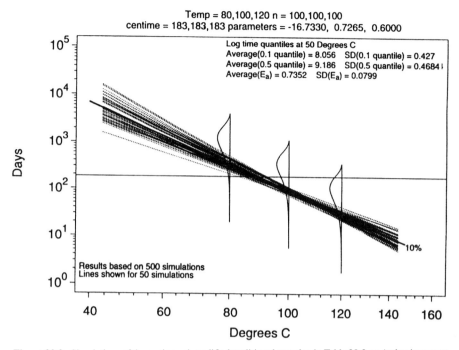

Figure 20.2. Simulations of the engineers' modified traditional test plan in Table 20.2 on Arrhenius paper.

given there and in Section 10.5.1 and Appendix Section B.6.1 can be used to compute approximate standard errors and other properties for ALT plans and these can be used as an aid in comparing and designing ALT plans. The references in the bibliographic notes at the end of this chapter give references for particular models considered here. For a specified model, planning values for the model parameters, and test plan, these methods allow one to compute the large-sample approximate variance–covariance matrix of the ML estimators of the model parameters $\boldsymbol{\theta}$. Using this matrix it is easy to compute large-sample approximate standard errors of ML estimates and these easy-to-compute quantities are useful for comparing different test plans. For example, with the simple linear regression model used in Sections 19.2.4, 19.3.1, and 19.3.2, the large-sample approximate variance–covariance matrix for $\widehat{\boldsymbol{\theta}} = (\widehat{\beta}_0, \widehat{\beta}_1, \widehat{\sigma})$ is

$$\Sigma_{(\widehat{\beta}_0, \widehat{\beta}_1, \widehat{\sigma})} = \begin{bmatrix} \mathrm{Avar}(\widehat{\beta}_0) & \mathrm{Acov}(\widehat{\beta}_0, \widehat{\beta}_1) & \mathrm{Acov}(\widehat{\beta}_0, \widehat{\sigma}) \\ \mathrm{Acov}(\widehat{\beta}_0, \widehat{\beta}_1) & \mathrm{Avar}(\widehat{\beta}_1) & \mathrm{Acov}(\widehat{\beta}_1, \widehat{\sigma}) \\ \mathrm{Acov}(\widehat{\beta}_0, \widehat{\sigma}) & \mathrm{Acov}(\widehat{\beta}_1, \widehat{\sigma}) & \mathrm{Avar}(\widehat{\sigma}) \end{bmatrix}.$$

The following is similar to the development for (10.8). The ML estimator of the p quantile of $\log(T)$ at transformed accelerating variable x is $\log(\widehat{t}_p) = \widehat{\beta}_0 + \widehat{\beta}_1 x + \Phi^{-1}(p)\widehat{\sigma} = \widehat{\mu} + \Phi^{-1}(p)\widehat{\sigma}$. As a special case of (B.9) from Appendix Section B.6.3,

$$\mathrm{Avar}[\log(\widehat{t}_p)] = \mathrm{Avar}(\widehat{\mu}) + \left[\Phi^{-1}(p)\right]^2 \mathrm{Avar}(\widehat{\sigma}) + 2\Phi^{-1}(p)\mathrm{Acov}(\widehat{\mu}, \widehat{\sigma}), \quad (20.2)$$

where $\text{Avar}(\widehat{\mu}) = \text{Avar}(\widehat{\beta}_0) + 2 \times x \times \text{Acov}(\widehat{\beta}_0, \widehat{\beta}_1) + x^2 \times \text{Avar}(\widehat{\beta}_1)$ and $\text{Acov}(\widehat{\mu}, \widehat{\sigma}) = \text{Acov}(\widehat{\beta}_0, \widehat{\sigma}) + x \times \text{Acov}(\widehat{\beta}_1, \widehat{\sigma})$. Then $\text{Ase}[\log(\widehat{t}_p)] = \sqrt{\text{Avar}[\log(\widehat{t}_p)]}$. Large-sample approximate standard deviations of other quantities of interest can be computed in a similar manner. Appendix Section B.2 gives more details.

20.3 PLANNING SINGLE-VARIABLE ALT EXPERIMENTS

This section develops some further concepts relating to planning ALTs.

20.3.1 Specifying the ALT Plan

To plan an ALT one needs to:

- Specify the experimental range(s) of the accelerating (or experimental) variable(s).
- Choose levels of the accelerating variable(s).
- Choose the number of test units to allocate to each level of the accelerating variable.

This section describes analytical methods for making these decisions.

Examples 20.3 and 20.4 illustrated traditional ALT plans with equal allocation of units to equally spaced levels of temperature. Because of censoring and extrapolation, traditional test plans may not be the best alternative. It is possible to choose levels of accelerating variables and the corresponding allocation of test units to minimize the large-sample approximate variance of the ML estimator of a quantity of interest. Plans developed in this way are called "optimum plans." Although optimum plans may be best in terms of estimation precision, they generally have practical deficiencies. This leads to compromise plans that are optimized subject to practical constraints. This section will show how to construct and compare such plans.

The Experimental Region. In theory, testing over a wider range of an accelerating variable provides higher degrees of precision. The highest level of the accelerating variable, however, has to be constrained to prevent testing beyond the range where the acceleration model is adequate (the problem with the plan in Example 20.3). On the other hand, testing at levels of the accelerating variable that are too low result in few or no failures during the time available for testing. These and perhaps other constraints define the range(s) of the accelerating variable(s).

Levels of the Accelerating Variable. The ALT plans developed here use either two or three levels of the accelerating variable. Let x_H denote the highest allowable level of transformed experimental variable and let x_U denote the use-condition. For a three-level test plan, x_L and x_M will denote the low and middle levels of the (transformed) accelerating variable, respectively. To describe the ALT plan accelerating variable levels independent of specific test situations and units, it is convenient to use a standardized acceleration level $\xi_i = (x_i - x_U)/(x_H - x_U)$ so that $\xi_U = 0$, $\xi_H = 1$, and other values of $0 < \xi_i < 1$ represent the fraction of the distance between x_H and x_U. A negative value of ξ_i implies a level of an experimental variable that is less than the use condition x_U. If there is a specified lower limit for an accelerating variable, it will be denoted by x_A and ξ_A for the corresponding standardized level.

Allocation of Test Units. To describe ALT plan allocations of test units or specimens independent of the total number of units to be tested, we will allocate units by proportion, using π_i as the allocation to x_i (or standardized level ξ_i).

Standardized Censoring Times. The standardized censoring time at the ith set of experimental conditions x_i is defined as $\zeta_i = [\log(t_c) - \mu(x_i)]/\sigma = \Phi^{-1}(p_i)$, where $\mu(x_i)$ is the location parameter at x_i and p_i is the expected proportion failing at x_i. Because $p_i = \Phi(\zeta_i)$, ζ_i can be used as a surrogate for p_i. In situations where p_i is very close to 0 or 1, it is more convenient to specify ζ_i.

Testing Units at Use Conditions. Some experimenters, when conducting an ALT, choose to test a small number of units at use conditions. These "insurance" units are typically tested to watch for evidence of other potential modes, especially when it is possible to take degradation measurements (or other parametric measurements) over time. Such units would not be expected to fail in the accelerated test and therefore will have no noticeable effect on estimates. For this reason, decisions about allocation of the other units in the test can be made independently of decisions on the insurance units.

20.3.2 Planning Criteria

The appropriate criteria for choosing a test plan depend on the purpose of the experiment. In some cases, optimizing under one criterion will result in a plan with poor properties under other criteria and it is useful to evaluate the trade-offs to obtain a satisfactory practical plan. In developing test plans, the following figures of merit are useful:

- A common purpose of an ALT experiment is to estimate a particular quantile, t_p, in the lower tail of the failure-time distribution at use conditions. Thus a natural criterion is to minimize $\text{Ase}[\log(\widehat{t_p})]$, the large-sample approximate standard error of $\log(\widehat{t_p})$, the ML estimator of the target quantile at use conditions x_U.
- Some experiments have more general purposes with corresponding overall interest in estimation precision for the parameters in $\boldsymbol{\theta}$. Let $I_{\boldsymbol{\theta}}$ denote the Fisher information matrix for the model parameters. A useful secondary criterion is to maximize $|I_{\boldsymbol{\theta}}|$, the determinant of $I_{\boldsymbol{\theta}}$. This criterion is motivated because the volume of an approximate joint confidence region for all of the model parameters in $\boldsymbol{\theta}$ is inversely proportional to an estimate of $\sqrt{|I_{\boldsymbol{\theta}}|}$.
- To assess robustness to departures from the fitted model it is useful to evaluate test properties under alternative, typically more general, models. For example, if one is planning a single-variable experiment under a linear model, it is useful to evaluate test plan properties under a quadratic model. Also, when planning

a two-variable experiment under the assumption of a linear model with no interaction, it is useful to evaluate test plan properties under a linear model with an interaction term.

- To have a useful amount of precision in one's estimates, it is necessary to have more than the minimum number of failures needed for the ML estimates to exist (e.g., at least four or five or more, depending on the desired precision) at two and, preferably, three or four levels of the accelerating variable. Thus it is important to evaluate the expected number of failures at each test condition.

These first three figures of merit depend on $\Sigma_{(\widehat{\beta}_0,\widehat{\beta}_1,\widehat{\sigma})}$, the large-sample approximate covariance matrix of the ML estimators of the model parameters. All of these figures of merit are easy to evaluate with a computer program. It is important to recognize that, generally, all of the evaluation criteria depend on unknown parameter values. Because these parameters are unknown, we use the planning values in their place. It is important to do sensitivity analysis over the plausible range of parameter values.

20.3.3 Statistically Optimum Test Plans

For a specified model and planning values for the model parameters, it is possible to find an optimum test plan that will, for example, minimize $\text{Ase}(\widehat{t}_p)$ (or, equivalently, $\text{Ase}[\log(\widehat{t}_p)]$) at a specified value of x, the transformed accelerating variable. With a linear relationship between log life and x for a log-location-scale distribution, a statistically optimum plan will:

- Test units at only two levels of x (denoted by x_L and x_H).
- Choose the highest level of x to be as high as possible. Using a larger value of x_H increases precision and statistical efficiency. The highest level of x should not, however, be chosen to be so high that it actuates new failure modes or that it otherwise causes the relationship between the accelerating variable and life to be inadequate.
- Optimize the location of x_L (the lowest level of x) and π_L (the allocation to x_L).

Example 20.5   Optimum Test Plan for the Adhesive Bond ALT.   Continuing with Examples 20.1–20.4, it will be interesting to consider a statistically optimum plan with $x_H = 120°\text{C}$. The optimization criterion is to minimize the large-sample approximate standard error of $\log(\widehat{t}_{.1})$, the ML estimator of the logarithm of the .1 quantile of the adhesive bond log-life distribution at the use conditions of $50°\text{C}$. The large-sample approximate variance $\text{Avar}[\log(\widehat{t}_{.1})]$ is a function of ξ_L and π_L. Figure 20.3 is a contour plot of log base 10 of the variance $\text{Avar}[\log(\widehat{t}_{.1})]$, relative to the minimum variance. Thus points on the contour marked 1 (marked 2) have a variance that is 10 (100) times that at the minimum. The actual optimum plan is indicated by the "+" at the minimum of the variance surface in Figure 20.3 and given numerically in Table 20.3. Figure 20.4 indicates the temperature levels for the optimum plan and shows ML estimate lines from 50 simulated ALTs from this test plan. As with

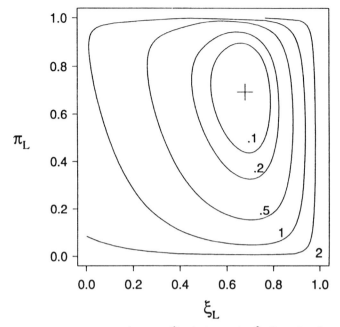

Figure 20.3. Contour plot showing $\log_{10}\{\text{Avar}[\log(\hat{t}_{.1})]/\min \text{Avar}[\log(\hat{t}_{.1})]\}$ as a function of ξ_L and π_L being varied to find the optimum ALT plan.

Figure 20.2 from Example 20.5, Figure 20.4 illustrates the kind of variability expected if the ALT experiment were to be repeated over and over, assuming the Arrhenius relationship to be correct over the entire range of temperature. The plot and the corresponding sample standard deviations given directly in the figure show that precision is somewhat better than that with the traditional plan in Example 20.4 and Figure 20.2. The sample standard deviation $\text{SD}[\log(\hat{t}_{.1})] = .3848$ agrees well with the large-sample approximate standard error $\text{Ase}[\log(\hat{t}_{.1}(50))] = .3794$ from Table 20.3.

The optimum plan, however, has some serious deficiencies. In particular, the plan caused an uncomfortable feeling that, for estimating life at 50°C, there was (relative

Table 20.3. Statistically Optimum ALT Plan to Estimate $t_{.1}$

Condition i	Level TEMPC	Standardized Level ξ_i	Time ζ_i	Failure Probability p_i	Allocation Proportion π_i	Number n_i	Expected Number Failing $E(r_i)$
Use	50	.000	−6.91	.001			
Low	95	.687	−1.59	.18	.71	212	38
High	120	1.000	.83	.90	.29	88	79

For this plan, $\text{Ase}[\log(\hat{t}_{.1}(50))] = .3794$ for the Weibull–Arrhenius model.

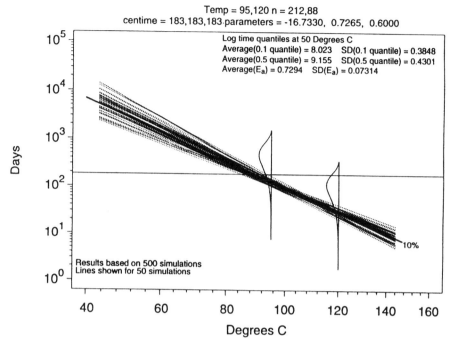

Figure 20.4. Simulations of the statistically optimum ALT plan on Arrhenius paper for the Weibull–Arrhenius model.

to the plan in Figure 20.2) too much temperature extrapolation. Also, the optimum plan uses only two levels of temperature, providing no ability to detect departures from the Arrhenius relationship and no insurance in case something goes wrong at one of the temperature levels (e.g., no failures at the lower level). Also, the optimum Weibull and lognormal plans were quite different (95°C and 120°C for Weibull versus 70°C and 120°C for lognormal). This was disconcerting because there were no strong feelings about which model would be used in the end and both models had, in the past, provided reasonable descriptions of adhesive bond failure times. In general, optimum plans tend not to be robust to model departures and deviations between the planning values and the actual model parameters. □

The main reason for consideration of the optimum plan is to provide a "best case" bench mark (i.e., the best that one could do if model assumptions were known to be correct), insight into possible good design practices (e.g., the optimum plan suggests testing more units at lower levels of the accelerating variable and that we have to constrain the highest level of the accelerating variable), and a starting point leading to a compromise plan that has good statistical properties and that meets necessary practical constraints (including robustness to departures to unknown inputs).

20.3.4 Compromise Test Plans

Real applications require a test plan that meets practical constraints, has intuitive appeal, is robust to deviations from specified inputs, and has reasonably good statistical properties. Compromise test plans that use three or four levels of the accelerating variable have somewhat reduced statistical efficiency but provide important practical advantages. They tend to be more robust to misspecification of unknown inputs and they allow one to estimate model parameters even if there are no failures at one level of the accelerating variable. The traditional plans address some of these concerns.

There are two basic issues in finding a good compromise test plan:

Basic Issue 1: Choose Levels of the Accelerating Variable. In choosing levels of the accelerating variable it is necessary to balance extrapolation in the accelerating variable (e.g., the fitted temperature–time relationship) with extrapolation in time (fitted failure-time distribution). Consider the distribution at 78°C in Figure 20.5. Moving the 78°C test to a higher temperature would provide a higher proportion of failures. But this would also increase the degree of extrapolation down to 50°C and reduce the resolution needed to estimate the slope precisely. Moving the 78°C test to a lower temperature would reduce extrapolation in temperature and increase the resolution to estimate the slope, except that it would also increase extrapolation in time (with the expected number of failures becoming smaller).

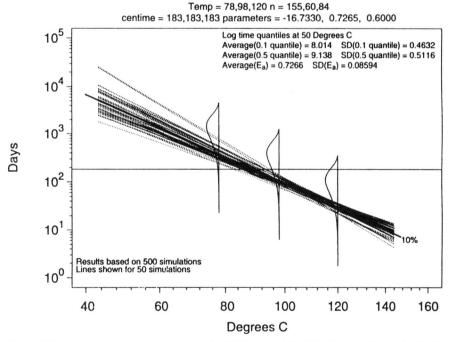

Figure 20.5. Simulations of the 20% compromise ALT plan and the Weibull–Arrhenius model for the adhesive-bonded power elements on Arrhenius paper.

Generally, at the middle and high levels of the accelerating variable we would have enough failures to interpolate in time in order to estimate quantiles in the lower tail of the distribution. For example, if interest is in the .1 quantile, we would try to test at conditions where more than 10% would be expected to fail. At the lower level of the accelerating variable we would often expect to extrapolate in time. That is, if interest is in the .1 quantile, we might have to test at conditions where somewhat less than 5% of the test units would be expected to fail.

Basic Issue 2: Allocation of Units to the Accelerating Variable Levels. As suggested by optimum test plans, one should allocate more test units to the lower accelerating variable level than to the high accelerating variable levels. This compensates for the small proportion failing at low levels of the accelerating variable. Also, testing more units near the use conditions is intuitively appealing because more testing is being done closer to the use conditions. In trying to optimize allocation, it is necessary to constrain a certain percentage of units to the middle level of the accelerating variable. Otherwise optimizing a three-level plan will result in the three-level plan degenerating to a two-level plan.

Generally, it is sufficient to use three or four levels of an accelerating variable. It is always necessary to limit the highest level of an accelerating variable to the maximum reasonable condition. Optimization of the position of the lowest level of the accelerating variable (constraining the middle level to be halfway between) often leads to an intolerable degree of extrapolation. In this case, reduce the lowest level of the accelerating variable (to minimize extrapolation)—subject to the expectation of seeing a minimum four or five failures. After deciding on some candidate plans, they can be evaluated using either large-sample approximations or simulation methods.

Example 20.6 Evaluation of a Compromise Plan for the Adhesive Bond ALT. Table 20.4 shows a compromise plan in which tests are run at 78, 98, and 120°C. Relative to optimum plans, this compromise plan increases the large-sample approximate standard deviation of the ML estimator of the .1 quantile at 50°C by 15% (if assumptions are correct). However, it reduces the low test temperature to 78°C (from 95°C) and uses three levels of the accelerating variable, instead of two levels. It is also more robust to departures from assumptions and uncertain inputs.

Table 20.4. Compromise ALT Plan for the Adhesive Bond

Condition i	Level TEMPC	Standardized Level ξ_i	Standardized Time ζ_i	Failure Probability p_i	Allocation Proportion π_i	Allocation Number n_i	Expected Number Failing $E(r_i)$
Use	50	.000	−6.91	.001			
Low	78	.448	−3.44	.03	.52	156	5
Mid	98	.726	−1.28	.24	.20	60	14
High	120	1.000	.83	.90	.28	84	76

For this plan, $\text{Ase}[\log(\widehat{t}_{.1}(50))] = .4375$ for the Weibull–Arrhenius model.

Figure 20.5 shows the results obtained by simulating from this proposed compromise test plan. The sample standard deviation $\text{SD}[\log(\widehat{t}_{.1})] = .4632$ agrees well with the large-sample approximate standard error $\text{Ase}[\log(\widehat{t}_{.1}(50))] = .4375$ from Table 20.4.

□

20.4 PLANNING TWO-VARIABLE ALT EXPERIMENTS

This section describes some of the basic ideas for planning ALTs with two variables. The discussion extends the material in Sections 20.1–20.3 and uses the same general setting and model assumptions, except that the regression model allows for two experimental variables affecting the scale parameter of the log-location-scale distribution (location parameter of the location-scale distribution). Most of the ideas can be extended to ALTs with more than two experimental variables.

20.4.1 Two-Variable ALT Model

For a log-location-scale distribution, the two-variable ALT model is similar to the model in (20.1) except that μ is a linear function of two experimental variables. Specifically,

$$\mu = \mu(x_1, x_2) = \beta_0 + \beta_1 x_1 + \beta_2 x_2,$$

where x_1 and x_2 are the (possibly transformed) levels of the accelerating or other experimental variables. For some underlying failure processes, it is possible for the underlying accelerating variables to "interact." For example, in the model

$$\mu = \mu(x_1, x_2) = \beta_0 + \beta_1 x_1 + \beta_2 x_2 + \beta_3 x_1 x_2$$

the transformed variables x_1 and x_2 interact in the sense that the effect of changing x_1 depends on the level of x_2 and vice versa. As before, the simple ALT models (i.e., SAFT models) assume that σ does not depend on the experimental variables. β and σ are unknown parameters that are characteristics of the material or product being tested and they are to be estimated from the available ALT data.

20.4.2 Examples

Example 20.7 Voltage-Stress/Thickness ALT for an Insulation. Nelson (1990a, page 349) describes the design of a complicated ALT with several experimental variables. To provide input for the design of a product, reliability engineers needed a rapid assessment of insulation life at use conditions. They also wanted to estimate the effect of insulation thickness on life, and to compare different conductors in the insulation. For purposes of test planning, Nelson (1990a) used the standard Weibull regression model in which log hours has a smallest extreme value distribution with location

$$\mu = \beta_0 + \beta_1 \log(\texttt{vpm}) + \beta_2 \log(\texttt{thick})$$

and a σ that does not depend on the accelerating or other experimental variables. Here vpm is voltage stress in volts/mm of insulation thickness and thick is insulation thickness in cm. Nelson (1990a, page 352) gives "planning" values $\beta_0^{\square} = 67.887$, $\beta_1^{\square} = -12.28$, $\beta_2^{\square} = -1.296$, and $\sigma^{\square} = .6734$. Nelson considered test plans with vpm ranging between $\text{vpm}_A = 120$ volts/mm and $\text{vpm}_H = 200$ volts/mm and thick between $\text{thick}_A = .163$ cm and $\text{thick}_H = .355$ cm. The variable levels at use conditions were $\text{vpm}_U = 80$ volts/mm for voltage stress and $\text{thick}_U = .266$ cm for thickness. Voltage stress was the accelerating variable in the experiment. Thickness was an ordinary experimental variable; its levels were chosen because they were of interest to the engineers. For purposes of illustration, we follow Nelson (1990a) and plan 1000-hour ALTs using $n = 170$ insulation specimens. The traditional plan in Figure 20.6 was obtained by choosing the lowest level of vpm to minimize $\text{Ase}[\log(\widehat{t_p})]$ with the middle vpm constrained to lie halfway between the high and the lower levels. The slanted lines in Figure 20.6 are lines of experimental variable combinations that yield equal probability of failing during the 1000-hour life test (these probabilities are indicated by a "$p =$" in the neighborhood of the lines). The slope of these lines show that the effect of changing vpm is much stronger than that of changing thick. As with the one-variable ALT, testing at combinations of the accelerating variables with small p will result in few failures and little information. On the other hand, we have a need to spread out the test conditions to get a better estimate of μ over the experimental region. $\square$

In contrast to the previous example, the following example uses two different *accelerating* experimental variables.

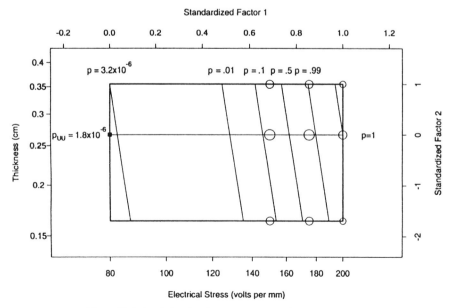

Figure 20.6. Insulation 3×3 (vpm $\times$ thick) factorial ALT plan.

Example 20.8 Voltage-Stress/Temperature ALT for an Insulation. This example is a modification of Example 20.7 in which thick will be held constant at its use conditions of $\text{thick}_U = .266$ cm and insulation life will be accelerated by using levels of voltage stress and temperature that are higher than use conditions. The Weibull regression model will again be used for purposes of test planning, with

$$\mu = \beta_0 + \beta_1 \log(\text{vpm}) + \beta_2[11605/(\text{temp } °C + 273.15)]$$

and constant σ. Again vpm is voltage stress in volts/mm of insulation thickness and temp is temperature in °C. The primary purpose of the test is to estimate $t_{.001}$ at the use conditions of $\text{vpm}_U = 80$ volts/mm and $\text{temp}_U = 120°C$. Both vpm and temp are accelerating variables in this experiment. The highest level of the variables should be no more than $\text{vpm}_H = 200$ volts/mm and $\text{temp}_H = 260°C$. Lower limits on testing are $\text{vpm}_A = 80$ volts/mm and $\text{temp}_A = 120°C$. The lower limit on temperature was chosen as the use temperature because it is generally not economical to *lower* temperatures for reliability testing.

The "planning values" $\beta_1^{\square} = -12.28$ and $\sigma^{\square} = .6734$ carry over from the previous example. The value $\beta_2^{\square} = .3878$ was chosen, by reviewing examples in Nelson (1990a), as a typical activation energy for temperature-accelerated insulation ALTs. Then $\beta_0^{\square} = 58.173$ was chosen to give the probability $p_{UU} = 1.82 \times 10^{-6}$ (as in the previous example) at use conditions $\text{vpm}_U = 80$, $\text{thick}_U = .266$, and $\text{temp}_U = 120°C$. Then $\zeta_U = -13.216$. For purposes of illustration we will again plan a 1000-hour ALT using a total of $n = 170$ insulation specimens.

The slanted lines in Figure 20.7 are also lines of equal probability failing during the 1000-hour life test. The slopes of these lines show that both vpm and temp will

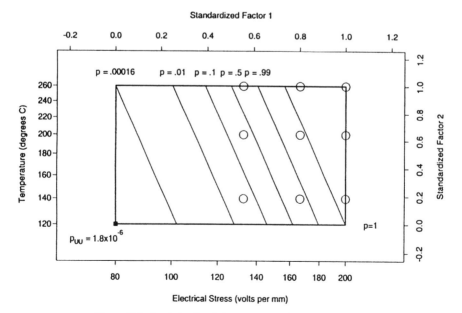

Figure 20.7. Insulation 3 × 3 (vpm × temp) factorial ALT plan.

have a strong effect on life. The circles in Figure 20.7 correspond to a traditional test plan using equally spaced levels of the accelerating variables temperature and voltage stress and equal allocation of specimens to the nine different test conditions. □

20.4.3 Two-Variable ALT Plans

As with the one-variable ALT plans, it is convenient to use standardized units for the accelerating (or other experimental) variables. The standardized variable for level i of variable j is defined as $\xi_{ji} = (x_{ji} - x_{jU})/(x_{jH} - x_{jU})$. Then $\xi_{jU} = 0$ and $\xi_{jH} = 1$, ($j = 1, 2$). This implies that, at use conditions, $\boldsymbol{\xi}_U = (0, 0)$ and at the highest levels of the experimental variables, $\boldsymbol{\xi}_H = (1, 1)$.

In the two-variable setup, there are several distinctly different kinds of test plans that provide estimates of any specified quantile t_p at $\boldsymbol{\xi}_U$.

- Test all units at $\boldsymbol{\xi}_U$. This is a degenerate test plan (because it does not allow the estimation of all of the regression model parameters) that is capable of estimating the failure-time distribution at $\boldsymbol{\xi}_U$. When the quantile of interest p and p_{UU} are not too small (e.g., $p = .01$ and $p_{UU} > .1$), concentrating all test units at $\boldsymbol{\xi}_U$ can actually minimize $\mathrm{Avar}[\log(\widehat{t}_p)]$ (see the figures in Meeker and Nelson, 1975).

- Test at any two (or more) combinations of variable levels on a line that passes through $\boldsymbol{\xi}_U$. See, for example, the circles on the dashed lines in Figures 20.8

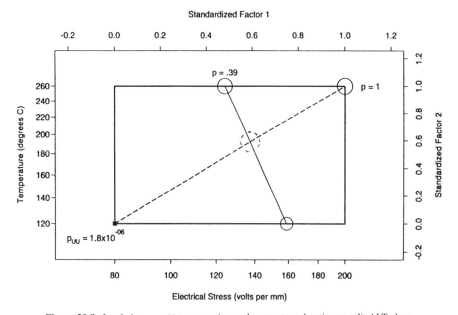

Figure 20.8. Insulation vpm × temp optimum degenerate and optimum split ALT plans.

PLANNING TWO-VARIABLE ALT EXPERIMENTS

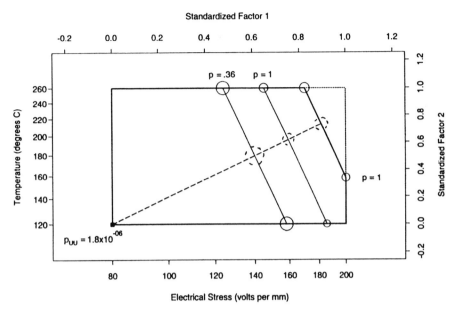

Figure 20.9. Insulation vpm × temp 20% compromise ALT plans with ζ^* constraint.

and 20.9. Such plans are also degenerate but allow estimation of t_p at $\boldsymbol{\xi}_U$ (or any other point on the line).

- Test at three (or more) noncollinear combinations of the experimental variables in the plane. This is the type of plan that one would use in practice.

Degenerate plans would be an unlikely choice in practice. They are, however, useful for developing more reasonable optimum and compromise test plans. In the next section we show how to obtain an optimum (or compromise) two-variable ALT plan by first finding a *degenerate* optimum (or compromise) plan that yields a particular Avar[log($\widehat{t}_p$)]. Then we show how to "split" this degenerate plan into an optimum (or compromise) two-variable plan that gives the same Avar[log($\widehat{t}_p$)] and that has other desirable properties. The choice among possible split plans allows us to also optimize with respect to secondary criteria and to evaluate trade-offs among these criteria.

20.4.4 Optimum Two-Variable ALT Plans

When testing is allowed anywhere in the square defined by the limits on the individual variables, an optimum degenerate plan is on the line ℓ going through $\boldsymbol{\xi}_U$ and $\boldsymbol{\xi}_H$. This optimum plan corresponds to the optimum test plan for a one-variable testing situation specified by $\zeta_U = \zeta_{UU}$ (standardized censoring time at use conditions $\boldsymbol{\xi}_U$) and $\zeta_H = \zeta_{HH}$ (standardized censoring time at conditions $\boldsymbol{\xi}_H$). The single-variable optimum plan provides the optimum ζ_L (standardized censoring time at the optimum lowest test variable level) and π_L (allocation to this variable level). The optimum degenerate two-variable plan allocates π_L to the diagonal point $\boldsymbol{\xi}_L = (\xi_{1L}, \xi_{2L})$ on the line ℓ having $\zeta_{LL} = \zeta_L$ (or, equivalently, so that $p_{LL} = p_L$) and $\pi_H = 1 - \pi_L$ to the point $\boldsymbol{\xi}_H$. This diagonal point has components

$$\xi_{1L} = \xi_{2L} = \frac{\sigma(\zeta_U - \zeta_L)}{(x_{1H} - x_{1L})\beta_1 + (x_{2H} - x_{2L})\beta_2}. \tag{20.3}$$

Figure 20.8 shows the degenerate optimum plan along the dashed line from $\boldsymbol{\xi}_H$ to $\boldsymbol{\xi}_U$. The centers of the circles on this line indicate the variable-level combination and the areas of the circles are proportional to the allocations to the different variable-level combinations.

Although a degenerate ALT plan may not be directly useful in practice, it does provide a means for finding nondegenerate optimum two-variable ALT plans. In particular, it is possible to "split" a degenerate plan into a nondegenerate optimum test plan (maintaining optimum Avar[log($\hat{t}_p$)]). Thus it is possible to use some secondary criteria to chose a "best" split plan. A reasonable strategy for many testing situations is to split the degenerate plan points into two points that extend along the equal-probability line to reach the boundary of the experimental region. As shown in Escobar and Meeker (1995), a two-variable degenerate optimum test plan having $\boldsymbol{\xi}_L = (\xi_{1L}, \xi_{2L})$ with allocation π_L can be split into two points on the same equal-probability line

$$\boldsymbol{\xi}_{L1} = (\xi_{1L1}, \xi_{2L1}) \quad \text{with allocation } \pi_{L1},$$

$$\boldsymbol{\xi}_{L2} = (\xi_{1L2}, \xi_{2L2}) \quad \text{with allocation } \pi_{L2},$$

where $\pi_L = \pi_{L1} + \pi_{L2}$. To maintain the optimality, the split allocations are chosen such that

$$\pi_{L1} \xi_{1L1} + \pi_{L2} \xi_{1L2} = \pi_L \xi_{1L}, \tag{20.4}$$

$$\pi_{L1} \xi_{2L1} + \pi_{L2} \xi_{2L2} = \pi_L \xi_{2L}.$$

Depending on the value of ζ_L, the point $\boldsymbol{\xi}_{L1}$ will be either on the North boundary or the West boundary of the experimental region. Similarly, the point $\boldsymbol{\xi}_{L2}$ will be either on the South boundary or the East boundary of the experimental region.

When $\zeta_L \geq \zeta_{LH}$, $\boldsymbol{\xi}_{L1} = (\xi_{1L1}, 1)$ will be on the North boundary of the experimental region, with

$$\xi_{1L1} = \frac{\sigma(\zeta_U - \zeta_L) - (x_{2H} - x_{2L})\beta_2}{(x_{1H} - x_{1L})\beta_1}. \tag{20.5}$$

When $\zeta_L < \zeta_{LH}$, $\boldsymbol{\xi}_{L1} = (\xi_{1A}, \xi_{2L1})$ will be on the West boundary of the experimental region, with

$$\xi_{2L1} = \frac{\sigma(\zeta_U - \zeta_L) - (x_{1H} - x_{1L})\beta_1 \xi_{1A}}{(x_{2H} - x_{2L})\beta_2}. \tag{20.6}$$

When $\zeta_L \leq \zeta_{LH}$, $\boldsymbol{\xi}_{L2} = (\xi_{1L2}, \xi_{2A})$ will be on the South boundary of the experimental region, with

$$\xi_{1L2} = \frac{\sigma(\zeta_U - \zeta_L) - (x_{2H} - x_{2L})\beta_2 \xi_{2A}}{(x_{1H} - x_{1L})\beta_1}. \tag{20.7}$$

Finally, when $\zeta_L > \zeta_{HL}$, $\boldsymbol{\xi}_{L2} = (\xi_{1A}, \xi_{2L2})$ will be on the East boundary of the experimental region, with

$$\xi_{2L2} = \frac{\sigma(\zeta_U - \zeta_L) - (x_{1H} - x_{1L})\beta_1}{(x_{2H} - x_{2L})\beta_2}. \tag{20.8}$$

Example 20.9 Voltage-Stress/Temperature Optimum ALT for an Insulation.
Figure 20.8 shows the optimum plan for the voltage-stress/temperature-accelerated test on the insulation. Using (20.3), the diagonal point $\boldsymbol{\xi}_L$ has components

$$\xi_{1L} = \xi_{2L} = \frac{.6734 \times [-13.216 - (-0.71)]}{.9163 \times (-12.28) + (-7.7511) \times .3878} = .591.$$

Thus the degenerate optimum is $\boldsymbol{\xi}_H = (1, 1)$ with allocation $\pi_H = .386$, and $\boldsymbol{\xi}_L = (.591, .591)$ with allocation $\pi_L = .614$, which are the entries for the optimum degenerate plan given in Table 20.5.

To split the degenerate optimum plan, use (20.5) and (20.7) with $\xi_{1A} = 0$, giving

$$\xi_{1L1} = \frac{.6734 \times [-13.216 - (-0.71)] - (-7.7511) \times .3878}{.9163 \times (-12.28)} = .481,$$

$$\xi_{1L2} = \frac{.6734 \times [-13.216 - (-0.71)]}{.9163 \times (-12.28)} = .748.$$

Table 20.5. Accelerating Variable Levels and Allocations for the Optimum Degenerate and Optimum Split Test Plans to Estimate $t_{.001}$ at Use Conditions

Point i	Levels		Standardized Levels		Time	p_i	Allocation		
	VPM	TEMPC	ξ_{1i}	ξ_{2i}	ζ_i		π_i	n_i	$E(r_i)$
Optimum Degenerate									
Use	80	120	.000	.000	−13.22	1.8×10^{-6}			
Low	137	192	.591	.591	−.71	.387	.614	104	40
High	200	260	1.000	1.000	7.95	1.000	.386	66	66
Optimum Split									
Use	80	120	.000	.000	−13.22	1.8×10^{-6}			
Low$_1$	124	260	.481	1.000	−.71	.387	.363	62	24
Low$_2$	159	120	.748	.000	−.71	.387	.251	42	16
High	200	260	1.000	1.000	7.95	1.000	.386	66	66

Table 20.6. Comparison of vpm × temp ALT Plans to Estimate $t_{.001}$

Plan	Figure	ζ^*	No Interaction Model		Interaction Model	
			$\dfrac{n}{\sigma^2}\text{Avar}[\log(\hat{t}_p)]$	$\dfrac{\sigma^2}{n}\|F\|$	$\dfrac{n}{\sigma^2}\text{Avar}[\log(\hat{t}_p)]$	$\dfrac{\sigma^2}{n}\|F\|$
3 × 3 Factorial adapted from Nelson (1990a)	20.7	—	77.3	1.7×10^{-3}	349	2.7×10^{-6}
Optimum split	20.8	—	50.5	1.3×10^{-3}	∞	0.0
20% Compromise with no ζ^* constraint	—	—	54.7	2.0×10^{-3}	430	3.0×10^{-6}
20% Compromise split with $\zeta^* = 5.0$ constraint	20.9	5.0	77.7	1.2×10^{-3}	324	1.7×10^{-6}

Consequently, the standardized levels for the split optimum plan are $\boldsymbol{\xi}_H = (1,1)$, $\boldsymbol{\xi}_{L1} = (.481, 1)$, and $\boldsymbol{\xi}_{L2} = (.748, 0)$. The allocation at $\boldsymbol{\xi}_H$ is $\pi_H = .386$. From the second equation in (20.4), it follows that the allocation at $\boldsymbol{\xi}_{L1}$ is $\pi_{L1} = \pi_L \times .591 = .614 \times .591 = .363$. Similarly, $\pi_{L2} = .251$.

Table 20.6 compares this optimum plan with the traditional plan shown in Figure 20.6. The variance of the optimum plan is about 35% smaller. The two-variable optimum plan, however, like the one-variable optimum plan, has some deficiencies. Tests are run at only three combinations of temperature and voltage stress and the degree of extrapolation is, perhaps, rather large. The optimum plan has no ability to estimate the parameters of the model with interaction. □

20.4.5 Splitting Degenerate Compromise Plans

A degenerate compromise plan can also be split to develop a nondegenerate compromise plan. Thus one can find a one-variable degenerate compromise plan using the methods in Section 20.3.4 and then split the plan in a manner that is analogous to that used in Section 20.4.4. For example, a two-variable degenerate compromise test plan having a middle accelerating variable level $\boldsymbol{\xi}_M = (\xi_{1M}, \xi_{2M})$ with allocation π_M can be split into two points,

$$\boldsymbol{\xi}_{M1} = (\xi_{1M1}, \xi_{2M1}) \quad \text{with allocation } \pi_{M1},$$

$$\boldsymbol{\xi}_{M2} = (\xi_{1M2}, \xi_{2M2}) \quad \text{with allocation } \pi_{M2},$$

on the same equal-probability line, where $\pi_M = \pi_{M1} + \pi_{M2}$. To maintain the optimality, the split allocations are chosen such that

$$\pi_{M1} \xi_{1M1} + \pi_{M2} \xi_{1M2} = \pi_M \xi_{1M}, \qquad (20.9)$$

$$\pi_{M1} \xi_{2M1} + \pi_{M2} \xi_{2M2} = \pi_M \xi_{2M}.$$

To obtain ξ_{M1} and ξ_{M2}, one uses (20.5)–(20.8) with ζ_L replaced by ζ_M.

When there is a ζ^* constraint [or a p^* probability constraint where $p^* = \Phi(\zeta^*)$] on the NE corner of the experimental region, there are multiple degenerate compromise plans with the same Avar[log($\hat{t}_p$)]. To specify a degenerate plan one chooses a line ℓ passing through ξ_U and intersecting the ζ^* constraint line at any point within the experimental region. Then $\xi_L, \xi_M,$ and ξ_H are determined by the intersection of ℓ and the failure probability lines defined by $\zeta_L, \zeta_M,$ and ζ_H. Such ζ^* constraint compromise plans can also be split in the same way [i.e., ξ_H is split in a manner that is analogous to that used for ξ_M in (20.9)]. The freedom of choosing the slope of ℓ allows optimization on another criterion. A desirable property of the split compromise plan is that π_{1L} and π_{2L} be equal if possible or nearly equal otherwise.

Example 20.10   Voltage-Stress/Temperature 20% Compromise ALT Plans for an Insulation.   The circles on the dashed lines in Figure 20.9 show a degenerate 20% compromise test plan with a $\zeta^* = 5.0$ constraint. Figure 20.9 also shows the split plan that maintains the Avar[log($\hat{t}_p$)] while splitting all three points to the boundaries of the experimental region. Table 20.7 shows the numerical values of the standardized

Table 20.7. Accelerating Variable Levels and Allocations for 20% Compromise Degenerate and 20% Compromise Split Plans with a $\zeta^* = 5.0$ NE Corner Constraint

Point i	Levels		Standardized Levels		Time		Allocation		
	VPM	TEMPC	ξ_{1i}	ξ_{2i}	ζ_i	p_i	π_i	n_i	$E(r_i)$
			20% Compromise Degenerate						
Use	80	120	.000	.000	−13.22				
Low	140	179	.610	.500	−.79	.364	.535	91	33
Mid	159	196	.752	.617	2.10	1.000	.200	34	34
High	182	214	.895	.733	5.00	1.000	.265	45	45
			20% Compromise Split						
Use	80	120	.000	.000	−13.22				
Low$_1$	124	260	.477	1.000	−.79	.364	.268	46	17
Mid$_1$	145	260	.650	1.000	2.10	1.000	.123	21	21
High$_1$	170	260	.823	1.000	5.00	1.000	.158	27	27
Low$_2$	158	120	.744	.000	−.79	.364	.268	46	17
Mid$_2$	185	120	.917	.000	2.10	1.000	.077	13	13
High$_2$	200	158	1.000	.339	5.00	1.000	.107	18	18

and actual levels of the experimental variables and the corresponding allocations. The starting point for obtaining the two-variable plan in this table is the computation of the standardized one-variable compromise test plan specified (for a three-level plan) in terms of the standardized censoring times $\zeta_L, \zeta_M, \zeta_H = \zeta^*$ and corresponding allocations π_L, π_M, π_H.

The criterion $\pi_{1L} = \pi_{2L}$ was used to determine the slope of the dashed line in Figure 20.9. Using (20.5) and (20.7) with $\zeta_L = -.79$, and the other inputs as before, gives $\boldsymbol{\xi}_{L1} = (.477, 1)$ and $\boldsymbol{\xi}_{L2} = (.744, 0)$. To obtain $\boldsymbol{\xi}_L$ observe that when $\pi_{L1} = \pi_{L2}$, (20.4) implies $\boldsymbol{\xi}_L = (\boldsymbol{\xi}_{L1} + \boldsymbol{\xi}_{L2})/2$. In this case, $(\boldsymbol{\xi}_{L1} + \boldsymbol{\xi}_{L2})/2 = (.610, .5)$ and the line ℓ through $\boldsymbol{\xi}_U$ and $(.610, .5)$ crosses the ζ^* constraint line within the experimental region. Then we can choose $\boldsymbol{\xi}_L = (.610, .5)$ and the corresponding line ℓ with slope equal to $.5/.610$. The intersections of ℓ with the failure probability lines with $\zeta_M = 2.10$ and $\zeta_H = 5.00$ determine the points $\boldsymbol{\xi}_M = (.752, .617)$ and $\boldsymbol{\xi}_H = (.895, .733)$. To obtain $\boldsymbol{\xi}_{M1}$ and $\boldsymbol{\xi}_{M2}$, we use formulas analogous to (20.5) and (20.7), respectively. In particular, $\boldsymbol{\xi}_{M1} = (.650, 1)$, where

$$\xi_{1M1} = \frac{\sigma(\zeta_U - \zeta_M) - (x_{2H} - x_{2L})\beta_2}{(x_{1H} - x_{1L})\beta_1}$$

$$= \frac{.6734 \times (-13.216 - 2.10) - (-7.7511) \times .3878}{.9163 \times (-12.28)} = .650.$$

Once the design points have been determined, it is simple to find the allocations. In this case, by choice $\pi_{L1} = \pi_{L2} = .535/2 = .268$. The allocations π_{M1} and π_{M2} are obtained using (20.9) and the allocations π_{H1} and π_{H2} are obtained using an equation similar to (20.9) in which the index M is replaced by an H.

As shown in Table 20.6 the 20% split compromise plan with no constraint on the NE corner of the experimental region provides considerably more precision for estimating $t_{.1}$ than the traditional factorial plan. Introducing the constraint allocates more experimental resources to lower levels of the accelerating variables, reducing precision somewhat, but also reducing extrapolation to use conditions. The variance under the interaction model is smaller for the constrained compromise plan, indicating a useful degree of robustness. The constrained compromise plan has properties that are comparable to the traditional factorial plan, but the compromise plan requires less extrapolation. □

20.4.6 Another Example

Example 20.11 Test Plans for a Voltage-Stress/Thickness ALT for an Insulation. This example returns to the setting described in Example 20.7, where there is one accelerating variable and another experimental variable that is not expected to be accelerating. For this experimental setting, the 3×3 factorial with unequal allocations (illustrated in Figure 20.6) provides a test plan with $\text{Avar}[\log(\widehat{t}_p)] = 144$ and good statistical properties across all of the evaluated criteria (see Table 20.8).

In Figure 20.10, the circles on the dashed lines show a degenerate 20% compromise test plan for which the experimental region was extended slightly from a maximum

Table 20.8. Comparison of vpm × thick **ALT Plans to Estimate** $t_{.001}$

Plan	Figure	ζ^*	No Interaction Model		Interaction Model	
			$\dfrac{n}{\sigma^2}\text{Avar}[\log(\widehat{t}_p)]$	$\dfrac{\sigma^2}{n}\lvert F \rvert$	$\dfrac{n}{\sigma^2}\text{Avar}[\log(\widehat{t}_p)]$	$\dfrac{\sigma^2}{n}\lvert F \rvert$
3 × 3 Factorial from Nelson (1990a)	20.6	—	144	2.4×10^{-3}	145	1.2×10^{-5}
Optimum split	—	—	80.1	7.3×10^{-4}	∞	0.0
Optimum split with $\zeta^* = 2.55$ constraint	—	2.5454	131	1.6×10^{-3}	138	1.7×10^{-5}
20% Compromise with $\zeta^* = 4.04$ constraint	20.10	4.04	96.1	7.0×10^{-3}	102	1.2×10^{-4}

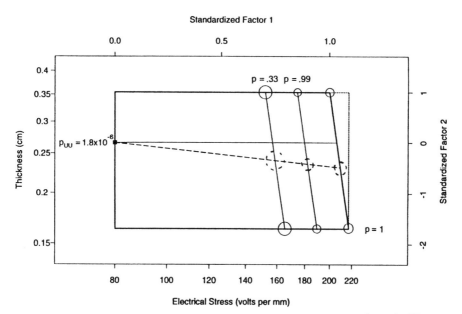

Figure 20.10. Insulation vpm × thick 20% compromise ALT plan with a constraint on the NE corner of the experimental region.

of 200 to 217 VPM, but using a standardized censoring time constraint $\zeta^* = 4.04$ (the highest value of ζ in the original experimental region used by Nelson, 1990a) on the NE corner of the experimental region. Figure 20.10 also shows the split plan that maintains the $\text{Avar}[\log(\widehat{t}_p)]$ while splitting all points to the boundaries of the experimental region. As in Example 20.10 (where there was a ζ^* constraint), the slope s of the degenerate plan was chosen to equalize the allocations at conditions corresponding to the censoring standardized time ζ_L. □

20.5 PLANNING ALT EXPERIMENTS WITH MORE THAN TWO EXPERIMENTAL VARIABLES

In some applications it is necessary or useful to conduct an accelerated test with more than two experimental variables. In such situations the ideas presented in this chapter can be extended and combined with traditional experimental design concepts. Some general ideas along these lines are as follows:

- Consider planning an experiment with just a single accelerating variable but with one or more other nonaccelerating experimental variables for which the effect on life is expected to be small or for which the direction of possible effects is unknown (i.e., the best guess of the regression coefficients for these variables is 0). Such a plan was illustrated in the vpm × thick example. In such situations, a reasonable plan would replicate a single-variable ALT experiment at various combinations of the nonaccelerating variables (see Figure 20.10). The move away from equal allocation can be viewed as a generalization of the traditional factorial plan.
- When the number of nonaccelerating variables is more than two or three, complete factorial designs may lead to an unreasonably large number of variable-level combinations. In this case, a reasonable strategy would be to use a standard fraction of a factorial design for the nonaccelerating variables and to run a single-variable ALT compromise plan (providing appropriate variable levels and allocations for the accelerating variable) at each of the combinations in the fraction.

In such situations, as with the examples in this chapter, the ideas of evaluation of test plan properties *before* running the experiment are extremely important. As with the simpler ALTs, evaluation of test plan properties is recommended and can be done using either large-sample approximations or simulation methods.

BIBLIOGRAPHIC NOTES

Chapter 6 of Nelson (1990a) reviews much of the literature and provides an overview and illustration of the most important methods for planning ALTs. Nelson (1998) provides an extensive list of references on accelerated test plans. Nelson and Kielpinski (1976) and Nelson and Meeker (1978) develop theory for optimum ALTs. Meeker (1984) compares optimum and compromise ALT plans. Meeker and Hahn (1985) describe practical aspects of ALT planning and provide tables that allow those planning tests to develop and compare alternative test plans. Jensen and Meeker (1990) describe corresponding software. Escobar and Meeker (1995) give technical details and examples of planning ALTs with two variables. They describe statistically optimum plans and show how these can be used to develop more practical compromise plans. Meeter and Meeker (1994) give references and develop methods for planning one-variable ALTs when life has a log-location-scale distribution and both μ and $\log(\sigma)$ can be written as linear functions of (transformed) accelerating variables. Escobar and Meeker (1998d) provide the extension to regression models with more than one variable. Chaloner and Larntz (1992) show how to use a prior distribution in place of particular planning values for model parameters when planning an ALT.

EXERCISES

20.1. Consider the ALT described in Exercise 19.2. The reliability engineers who ran that accelerated test want to run another accelerated test on a similar device. They have asked you to help them evaluate the properties of some alternative test plans.

(a) Relative to the plan used in Exercise 19.2, what modifications would you suggest for evaluation?

(b) List the criteria that you would use to compare the plans and make a recommendation on how to conduct the accelerated test.

20.2. In general, planning values are needed to do test planning and to determine the sample size needed to provide a specified degree of precision.

(a) Explain why such planning values are needed.

(b) Product or reliability engineers may be able to provide some useful information, but they cannot be expected to provide accurate planning values (otherwise they would have no reason to run the test!). What can be done to protect against the use of potentially misspecified planning values?

20.3. In planning an ALT, the large-sample approximate variance of the ML estimator of a particular quantile of the failure-time distribution at use conditions is often used to judge the precision that one could expect from a proposed test plan. Suppose that the Arrhenius/lognormal model will provide an adequate description of the relationship between life and temperature. As described in Appendix Section B.6.2, there are computer algorithms that can be used to compute the large-sample approximate covariance matrix $\Sigma_{\widehat{\theta}}$ of the ML estimates $\widehat{\beta}_0$, $\widehat{\beta}_1$, and $\widehat{\sigma}$. In this case, $\Sigma_{\widehat{\theta}}$ will be a function of the proposed test plan and the parameters β_0, β_1, and σ.

(a) Given the individual elements of $\Sigma_{\widehat{\theta}}$, provide an expression for the large-sample approximate variance of the ML estimator of the p quantile at use temperature temp$_U$. Do not use matrix algebra.

(b) Approximate standard errors of ML estimators from a proposed test plan can also be obtained by using Monte Carlo simulation. Explain the advantages and disadvantages of this approach relative to using the large-sample approximate variance.

20.4. An ALT is going to be conducted to investigate the effect of size on the life of an insulating material. The accelerating variable will be voltage. Based on previous experience, for purposes of planning the experiment, use the model

$$\Pr[T \leq t; \text{thick}, \text{vpm}] = \Phi_{\text{sev}}\left[\frac{\log(t) - \mu(\text{thick}, \text{vpm})}{\sigma}\right],$$

where $\mu = \beta_0 + \beta_1 \text{thick} + \beta_2 \text{vpm}$, thick is the specimen size in cm, vpm is voltage stress in volts/mm, and σ is constant.

(a) What would be the model relating μ to thick and voltage (instead of volts/mm)?

(b) What is an important advantage of modeling and experimenting in terms of vpm and thick rather than voltage and thick?

20.5. Consider the planning values given in Example 20.2.

(a) Compute the proportion failing after 6 months at 80°C.

(b) Suppose that there is a simple chemical degradation process that causes the adhesive to degrade over time. Compute the implied activation energy for this degradation process assuming that the Weibull shape parameter is $\beta = 1/\sigma = 2$, and 3.

◆**20.6.** Refer to Exercise 19.9. At use conditions, Failure Mode 2 will be dominant. We would not expect to see Failure Mode 1, except at higher levels of stress. Using the parameter values given in Exercise 19.9 as planning values, and assuming that 4000 hours of test time will be available, suggest an appropriate test plan that could be used to estimate $F(10000)$ at a use temperature of 40°C.

(a) Write a simulation program to evaluate alternative test plans and to answer the following questions. Use the simulation to evaluate quantities like $\text{Var}(\widehat{t}_{.1})$ at 40°C and the expected number of failures at the different levels of temperature.

(b) What is an appropriate highest level of temperature for such a test?

(c) What other levels of temperature would you recommend?

EXERCISES 561

(d) Suppose that 100 units are available for an ALT. How would you allocate these units to the different levels of temperature?

▲20.7. Refer to Exercise 20.6. Develop formulas that would allow easy evaluation of the large-sample approximate variance of $\widehat{F}(10000)$ and the expected number of failures at the different levels of temperature, as a function of the test plan and the planning values.

▲20.8. Use the ML estimates for Device-A given in Table 19.1 as planning values to design an ALT for a similar device with a similar failure mode. All of the large-sample approximate variances in this problem will depend on the values of these planning values. As in the original test plan, 80°C will be the highest temperature.

(a) For a test plan having three levels of temperature, write down an expression for the large-sample approximate variance of $\log(\widehat{t}_p)$ at the use temperature of 10°C as a function of the standardized levels of test temperature and the proportion of units allocated to the different temperatures.

(b) An optimum test plan for this problem will have only two levels of temperature. For this plan, write the large-sample approximate variance of $\log(\widehat{t}_p)$ as a function of ξ_L, the standardized location of the lowest temperature, and π_L, the allocation to this temperature.

◆20.9. Refer to Exercise 20.8.

(a) Write a computer program to compute the large-sample approximate variance in part (a) of Exercise 20.8. To do this you will need access to the LSINF algorithm in Escobar and Meeker (1994) (which is available from Statlib at FTP site `lib.stat.cmu.edu`).

(b) Use the computer program requested in part (a) to compute the large-sample approximate variance of $\log(\widehat{t}_{.1})$ for a grid (say, 11 by 11) of values with ξ_L and π_L ranging between 0 and 1. Plot these with a contour plot. What does this plot suggest for an "optimum" test plan?

(c) Redo the computation in part (b), now allowing the highest temperature to be at 90°C. What effect does this have on the test plan and the variance of $\log(\widehat{t}_{.1})$?

▲20.10. Here we consider planning of an ALT with a single accelerating variable ξ, right-censored at t_c for which failure time $T \sim \text{EXP}(\theta)$ distribution with $\theta = \exp(\beta_0 + \beta_1 \xi)$. The testing will be at two levels of the accelerating variable ξ_L and ξ_H. The use condition is $\xi_U = 0$.

(a) Show that the total Fisher information matrix, $\mathcal{I}_\theta$, for $\boldsymbol{\theta} = (\beta_0, \beta_1)$ is $\mathcal{I}_\theta = nF = n\left[\pi_L p_L \mathcal{I}_L + \pi_H p_H \mathcal{I}_H\right]$, where π_L is the proportion of units allocated at ξ_L, $\pi_H = 1 - \pi_L$, $p_i = 1 - \exp(-t_c/\theta_i)$ is the expected proportion of failure at ξ_i, and

$$\mathcal{I}_i = \begin{bmatrix} 1 & \xi_i \\ \xi_i & \xi_i^2 \end{bmatrix}.$$

(b) Show that if the goal is to minimize the large-sample approximate variance of the ML estimators of the logarithm of a particular quantile of the life distribution at use conditions, then it suffices to find a plan that minimizes the large-sample approximate variance of $\widehat{\beta}_0$.

(c) Show that

$$\text{Avar}(\widehat{\beta}_0) = \frac{\pi_L p_L \xi_L^2 + \pi_H p_H \xi_H^2}{(\pi_L p_L \xi_L^2 + \pi_H p_H \xi_H^2)(\pi_L p_L + \pi_H p_H) - (\pi_L p_L \xi_L + \pi_H p_H \xi_H)^2}.$$

(d) Suppose that the probabilities of failing at $\xi_U = 0$ and $\xi_H = 1$ are p_U and p_H, respectively. Then show that the proportion of failures at any value of ξ is

$$p = 1 - \exp\{-[-\log(1 - p_U)]^{1-\xi} \times [-\log(1 - p_H)]^{\xi}\}.$$

(e) For fixed values of $p_U < p_H$, $0 < \pi_L < 1$, $0 \le \xi_L < 1$, draw plots of $\text{Avar}(\widehat{\beta}_0)$ as a function of ξ_H. Observe that the large-sample approximate variance is a decreasing function of ξ_H.

(f) In practice it is necessary to bound the highest level of stress, say, at $\xi_H = 1$. Then

$$\text{Avar}(\widehat{\beta}_0) = \frac{\pi_L p_L \xi_L^2 + \pi_H p_H}{(\pi_L p_L \xi_L^2 + \pi_H p_H)(\pi_L p_L + \pi_H p_H) - (\pi_L p_L \xi_L + \pi_H p_H)^2}.$$

Use the particular fixed values of $p_U = .0001$, $p_H = .9$ and construct a contour plot of $\text{Avar}(\widehat{\beta}_0)$ as a function of π_L and ξ_L. Do this for other practical choices of p_U and p_H.

(g) For $p_U = .0001$ and $p_H = .9$, verify that the optimum test plan is $\pi_L = .795$, $\xi_L = .711$, $\pi_H = .205$, and $\xi_H = 1$.

CHAPTER 21

Accelerated Degradation Tests

Objectives

This chapter explains:

- How accelerated degradation tests can be used to assess and improve product reliability.
- Models for accelerated degradation tests.
- How to analyze accelerated degradation data.
- How accelerated degradation test methods compare with traditional accelerated life test methods.
- A simple approximate method that can be used for some accelerated degradation data analyses.

Overview

This chapter explains and illustrates the use of acceleration models from Chapter 18 with the degradation analysis methods in Chapter 13. Both of these chapters are important in the understanding of the material in this chapter. The comparison of accelerated degradation tests with accelerated life tests also depends on the material in Chapter 19. Section 21.2 introduces an example and describes a model for accelerated degradation data. Section 21.3 shows how to estimate the parameters of this model. Section 21.4 applies methods from Section 13.6 to estimate the failure-time distribution corresponding to the degradation model. Section 21.5 shows how to apply the methods in Section 13.7 to obtain bootstrap confidence intervals for a failure-time distribution. Section 21.6 compares the accelerated degradation analysis for the example with a corresponding accelerated life test analysis. Section 21.7 describes and illustrates the use of a simpler approximate method for accelerated degradation analysis.

21.1 INTRODUCTION

The degradation analysis methods described in Chapter 13 can provide useful information for reliability studies, even when failures are not observed. For some products, however, degradation rates at use conditions are so low that appreciable degradation will not be observed during usual tests. In such cases, it might be possible to accelerate the degradation process. For example, raising temperature will often accelerate the rate of a chemical degradation process.

Example 21.1 Device-B Power Output Degradation. Figure 21.1 shows the decrease in power, over time, for a sample of integrated circuit devices called "Device-B." Samples of devices were tested at each of three levels of junction temperature. Based on a life test lasting about 6 months, design engineers needed an assessment of the proportion of these devices that would "fail" before 15 years (about 130 thousand hours) of operation at 80°C. This assessment would be used to determine the amount of redundancy required in the full system. Failure for an individual device was defined as power output more than .5 decibels (dB) below initial output. At standard operating temperatures (e.g., 80°C), the devices will degrade too slowly to provide useful information in 6 months. Because units at low temperature degrade more slowly, they had to be run for longer periods of time to accumulate appreciable degradation. Because of severe limitations in the number of test positions, fewer units were run at lower temperatures. The original data from this

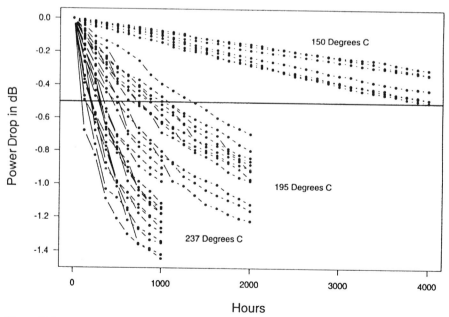

Figure 21.1. Accelerated degradation test results giving power drop in Device-B output for a sample of units tested at three levels of junction temperature.

experiment are proprietary. The data shown in Figure 21.1 were actually simulated from a model suggested by physical theory and limited real data available at the time the more complete experiment was being planned. □

21.2 MODELS FOR ACCELERATED DEGRADATION TEST DATA

Section 13.2 presents models for degradation data. Generally these models describe the behavior, over time, of a particular degradation or product performance measure (such as crack size, resistance, or power output) as well as the unit-to-unit variability in degradation paths. For accelerated degradation data, the model also describes the relationship between degradation or performance and the accelerating variable or variables (e.g., voltage or temperature).

As in Chapter 13, the observed sample degradation y_{ij} of unit i at time t_{ij} is a unit's actual degradation plus measurement error and is given by

$$y_{ij} = \mathcal{D}_{ij} + \epsilon_{ij}, \quad i = 1,\ldots,n, \quad j = 1,\ldots,m_i, \tag{21.1}$$

where $\mathcal{D}_{ij} = \mathcal{D}(t_{ij}, \beta_{1i}, \ldots, \beta_{ki})$ is the actual path of the unit i at time t_{ij} (the times need not be the same for all units) and $\epsilon_{ij} \sim \text{NOR}(0, \sigma_\epsilon)$ is a residual deviation for unit i at time t_j. The total number of inspections on unit i is denoted by m_i. Typically, a path model will have $k = 1, 2, 3,$ or 4 parameters. As described in Section 13.2.3, some of the $\beta_1, \ldots, \beta_k$ parameters will be random from unit to unit. One or more of the $\beta_1, \ldots, \beta_k$ parameters could, however, be modeled as common across all units.

The simple chemical degradation path model from Example 13.4, rewritten in the generic notation and with a temperature acceleration variable affecting the rate of the reaction, is

$$\mathcal{D}(t; \text{temp}) = \mathcal{D}_\infty \times \{1 - \exp[-\mathcal{R}_U \times \mathcal{AF}(\text{temp}) \times t]\}. \tag{21.2}$$

Here $\mathcal{R}_U$ is the rate reaction at use temperature temp_U, $\mathcal{R}_U \times \mathcal{AF}(\text{temp})$ is the rate reaction at temperature temp, and $\mathcal{D}_\infty$ is the asymptote. For $\mathcal{D}_\infty < 0$, we specify that failure occurs at the smallest t such that $\mathcal{D}(t) < \mathcal{D}_f$.

Following from (18.2), the Arrhenius acceleration factor

$$\mathcal{AF}(\text{temp}, \text{temp}_U, E_a) = \exp\left[E_a\left(\frac{11605}{\text{temp}_U + 273.15} - \frac{11605}{\text{temp} + 273.15}\right)\right] \tag{21.3}$$

depends only on the two temperature levels and the activation energy E_a. If $\text{temp} > \text{temp}_U$, then $\mathcal{AF}(\text{temp}, \text{temp}_U, E_a) > 1$. For simplicity, we use the notation $\mathcal{AF}(\text{temp}) = \mathcal{AF}(\text{temp}, \text{temp}_U, E_a)$ when temp_U and E_a are understood to be, respectively, product use (or other specified baseline) temperature and a reaction-specific activation energy.

21.2.1 Accelerated Degradation Model Parameters

In general, rate-acceleration parameters are unknown fixed-effect parameters (e.g., the Arrhenius model suggests no unit-to-unit variability in activation energy). As described in Section 13.2.5, fixed-effect parameters are included, notationally, in the parameter vector $\boldsymbol{\beta}$ introduced in Section 13.2.5. Thus for the single-step chemical reaction models in Sections 18.3.3 and 18.3.4, we have one additional parameter to estimate. The total number of parameters in $\boldsymbol{\beta}$ is still denoted by k.

The values of $\boldsymbol{\beta}$ corresponding to an individual unit may be of interest in some applications (e.g., predict the future degradation of a particular unit, based on a few early readings). Subsequent development in this chapter, however, will concentrate on the use of degradation data to make inferences about the population or process from which the sample units were obtained or predictions about the life distribution of a population of units at specific levels of the accelerating variable (e.g., temperature). In this case, the underlying model parameters are $\boldsymbol{\mu}_\beta$ and Σ_β, as well as the residual standard deviation σ_ϵ. Again, the appropriate rows and columns in Σ_β, corresponding to the fixed parameters in $\boldsymbol{\beta}$, contain zeros. For shorthand, we will use $\boldsymbol{\theta}_\beta = (\boldsymbol{\mu}_\beta, \Sigma_\beta)$ to denote the parameters of the overall degradation population or process.

Example 21.2   Device-B Power Output Degradation Model Parameterization. For the Device-B power-drop data in Example 21.1, the scientists responsible for the product were confident that degradation was caused by a simple one-step chemical reaction that could be described by the model in Example 13.4. Thus for the data in Figure 21.1, we will use the accelerated degradation model in (21.2), assuming that $\mathcal{R}_U$ and $\mathcal{D}_\infty$ are random from unit to unit. Then a possible parameterization would be $(\beta_1, \beta_2, \beta_3) = [\log(\mathcal{R}_U), \log(-\mathcal{D}_\infty), E_a]$, where the first two parameters are random effects and the activation energy E_a is a fixed effect. That is, E_a is regarded as a material property that does not depend on temperature and that is constant from unit to unit. □

21.3 ESTIMATING ACCELERATED DEGRADATION TEST MODEL PARAMETERS

The likelihood for the random-parameter degradation model is the same as that given in (13.8) and the methods of estimation described there can be applied directly to the accelerated degradation model.

Example 21.3   Estimates of the Device-B Model Parameters. Continuing with Example 21.2, we fit the mixed-effect model (21.2) to the Device-B data. In order to improve the stability and robustness of the approximate ML algorithm, it is important to keep the correlation between the estimates of E_a and the parameters relating to reaction rate $\mathcal{R}$ small. This can be done by estimating the distribution of $\mathcal{R}$ at some stress that is central to the experimental temperatures, rather than the use-temperature. Thus we parameterize with $\beta_1 = \log[\mathcal{R}(195)]$, $\beta_2 = \log(-\mathcal{D}_\infty)$, and

ESTIMATION OF FAILURE PROBABILITIES

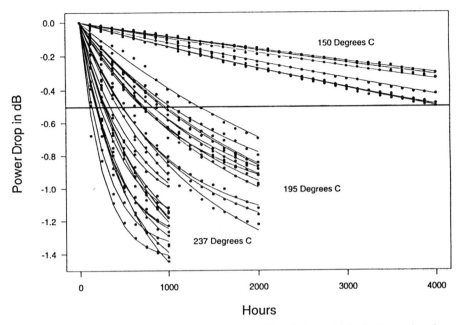

Figure 21.2. Device-B power-drop observations and fitted degradation model for the 34 sample paths.

$\beta_3 = E_a$, where $\mathcal{R}(195) = \mathcal{R}_U \times \mathcal{AF}(195)$ is the reaction rate at 195°C. Our model uses a bivariate normal distribution to describe unit-to-unit variability in (β_1, β_2). Also, activation energy $\beta_3 = E_a$ is a constant, but unknown, material property. The nonlinear mixed-effects computer program of Pinheiro and Bates (1995b) gives the following approximate ML estimates of the model parameters:

$$\widehat{\boldsymbol{\mu}}_\beta = \begin{pmatrix} -7.572 \\ .3510 \\ .6670 \end{pmatrix}, \quad \widehat{\boldsymbol{\Sigma}}_\beta = \begin{pmatrix} .15021 & -.02918 & 0 \\ -.02918 & .01809 & 0 \\ 0 & 0 & 0 \end{pmatrix}, \quad (21.4)$$

and $\widehat{\sigma}_\epsilon = .0233$. The lines in Figure 21.2 show the fitted model (21.2) for each of the sample paths (indicated by the points on the plot) for the Device-B degradation data. Figure 21.3 plots the estimates of the β_1 and β_2 parameters for each of the 34 sample paths, indicating the reasonableness of the bivariate normal distribution model for this random-coefficient model. □

21.4 ESTIMATION OF FAILURE PROBABILITIES, DISTRIBUTION QUANTILES, AND OTHER FUNCTIONS OF MODEL PARAMETERS

One can estimate the failure-time distribution $F(t)$ by substituting the estimates $\widehat{\boldsymbol{\theta}}_\beta$ into (13.9), giving $\widehat{F}(t) = F(t; \widehat{\boldsymbol{\theta}}_\beta)$. This is straightforward when $F(t)$ can be

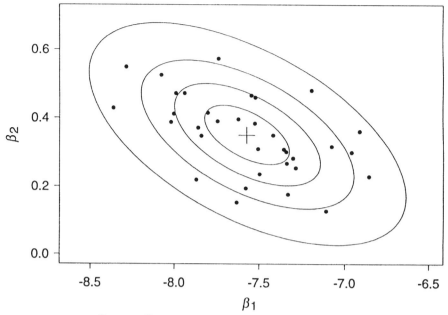

Figure 21.3. Plot of $\widehat{\beta}_{1i}$ versus $\widehat{\beta}_{2i}$ for the $i = 1,\ldots,34$ sample paths from Device-B, also showing contours corresponding to the fitted bivariate normal distribution. The + marks the estimates of the means μ_{β_1} and μ_{β_2}.

expressed in closed form. When there is no closed-form expression for $F(t)$, and when numerical transformation methods are too complicated, one can use either Algorithm 13.1 or 13.2, to evaluate (13.9) at $\widehat{\boldsymbol{\theta}}_\beta$.

Example 21.4 Device-B Degradation Data Estimate of F(t). Figure 21.4 shows $\widehat{F}(t)$ for Device-B based on the power-drop data with failure defined as a power drop of $\mathcal{D}_f = -.5$ dB. Estimates are shown for 195, 150, 100, and 80°C. These estimates were computed with Algorithm 13.1, using the estimates of the model parameters $\widehat{\boldsymbol{\theta}}_\beta = (\widehat{\boldsymbol{\mu}}_\beta, \widehat{\Sigma}_\beta)$ from Example 21.3. □

21.5 CONFIDENCE INTERVALS BASED ON BOOTSTRAP SAMPLES

Because there is no simple method of computing standard errors for $\widehat{F}(t)$, we use a simulation of the sampling/degradation process and the bias-corrected percentile bootstrap method to obtain parametric bootstrap confidence intervals for quantities of interest. This method is described in Section 9.6 and more fully in Efron (1985) and Efron and Tibshirani (1993). The method is a straightforward implementation of Algorithm 13.3, described in Section 13.7.

For a SAFT model, once $\widehat{F}^*(t)$ has been computed in step 4 of Algorithm 13.3 for one set of accelerating variable conditions, it is possible to obtain $\widehat{F}^*(t)$ for other

COMPARISON WITH TRADITIONAL ACCELERATED LIFE TEST METHODS 569

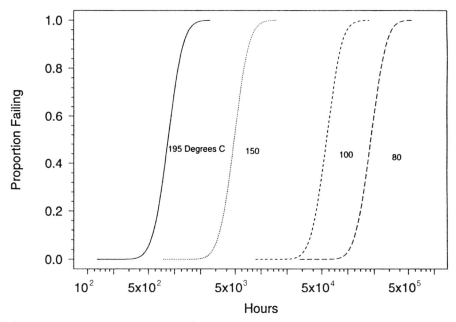

Figure 21.4. Estimates of the Device-B failure-time distributions at 80, 100, 150, and 195°C, based on the degradation data.

accelerating variable conditions by simply scaling times. Otherwise the bootstrap estimates from step 3 need to be reused in step 4 to recompute the $\widehat{F}^*(t)$ values for each new set of conditions.

Example 21.5 Degradation-Data Bootstrap Confidence Intervals for the Device-B F(t) at 80°C. Continuing with Example 21.4, Figure 21.5 shows the point estimate and a set of pointwise two-sided approximate 90% and 80% bootstrap bias-corrected percentile confidence intervals for $F(t)$ at 80°C, based on the IC power-drop data with failure defined as a power drop of $\mathcal{D}_f = -.5$ dB. The bootstrap confidence intervals were computed by using Algorithm 13.1 and Algorithm 13.3 to evaluate $\widehat{F}^*(t)$. Specifically, the point estimate for $F(t)$ at 130 thousand hours is .14 and the approximate 90% confidence interval is [.005, .64]. The extremely wide interval is due to the small number of units tested a 150°C and the large amount of extrapolation required to estimate to $F(t)$ at 80°C. It is important to recognize that this interval does not reflect possible deviations from the assumed model. □

21.6 COMPARISON WITH TRADITIONAL ACCELERATED LIFE TEST METHODS

This section compares accelerated degradation and accelerated life test analyses. With failure defined as power drop below −.5 dB, there were no failures at 150°C.

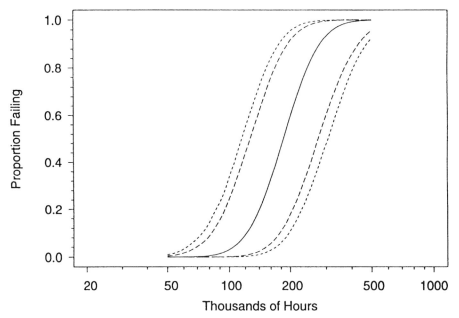

Figure 21.5. Estimates of the Device-B failure-time distribution at 80°C with approximate 80% and 90% pointwise two-sided bootstrap confidence intervals based on the power-drop data with failure defined as a power drop of $\mathcal{D}_f = -.5$ dB.

Although it is possible to fit a model to the resulting failure-time data, the degree of extrapolation with no failures at 150°C would be, from a practical point of view, unacceptable. The comparison will be useful for showing one of the main advantages of degradation analysis—the ability to use degradation data for units that have not failed. Degradation data provide important information at lower levels of stress where few, if any, failures will be observed, thus reducing the degree of extrapolation.

Figure 21.6 shows a scatter plot of the failure-time data. These failure-time data were obtained from the degradation data in Figure 21.1. All seven units tested at 150°C were right-censored. Figure 21.7 is a lognormal multiple probability plot with the straight lines showing individual lognormal distributions fitted to the samples at 237°C and 195°C. This figure shows that the lognormal distributions provide a good fit at both temperatures. Figure 21.8 is also a lognormal multiple probability plot for the individual samples at 237°C and 195°C. In this case, however, the superimposed lines show the fitted lognormal–Arrhenius model relating the failure-time distributions to temperature. This is a commonly used accelerated life test model for electronic components, as described in Chapters 18 and 19. The lognormal–Arrhenius model assumes that log failure time has a normal distribution with mean

$$\mu = \beta_0 + \beta_3 \left(\frac{11605}{\texttt{temp} + 273.15} \right)$$

COMPARISON WITH TRADITIONAL ACCELERATED LIFE TEST METHODS

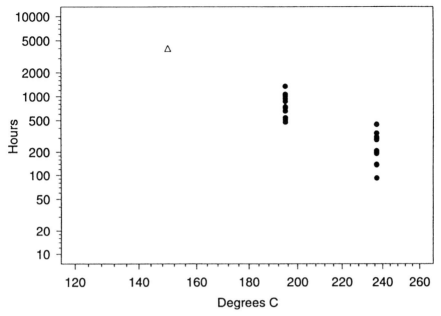

Figure 21.6. Scatter plot of Device-B failure-time data with failure defined as power drop below $-.5$ dB. The symbol Δ indicates the seven units that were tested at 150°C and had not failed at the end of 4000 hours.

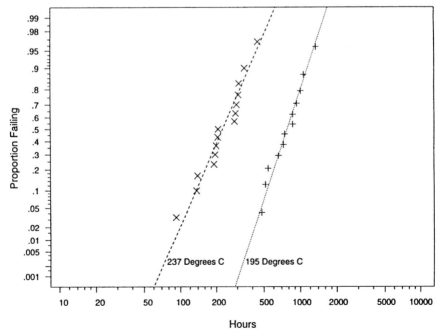

Figure 21.7. Individual lognormal probability plots of the Device-B failure-time data with failure defined as power drop below $-.5$ dB.

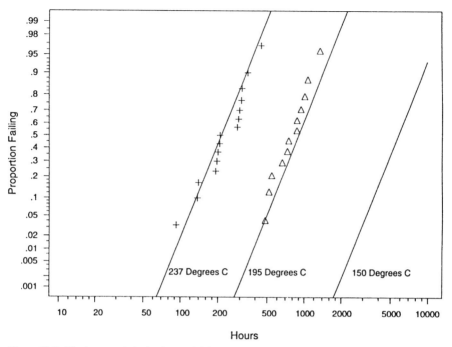

Figure 21.8. The lognormal–Arrhenius model fit to the Device-B failure-time data with failure defined as power drop below −.5 dB.

and constant standard deviation σ. In relation to the lognormal–Arrhenius failure-time model described in Section 18.3.3, the slope $\beta_3 = E_a$ is the activation energy and the intercept is

$$\beta_0 = \mu_U - \beta_3 \left(\frac{11605}{\text{temp}_U + 273.15} \right).$$

The estimated failure-time lognormal cdfs in Figure 21.8 are parallel because of the constant-σ assumption. This plot shows some deviations from the assumed model. These deviations, however, are within what could be expected from random variability alone (a likelihood-ratio test comparing the model depicted in Figure 21.8 with independent ML fits at each level of temperature, shown on Figure 21.7, had a p-value of .052).

Figure 21.9 shows the same lognormal–Arrhenius model fit given in Figure 21.8 with an extrapolated estimate of the cdf at 80°C. The dotted lines on this figure are the degradation-model-based estimates of the failure-time distributions shown in Figure 21.4. There are small differences between the lognormal and the degradation models at 237°C and 195°C. The differences at 150°C and 80°C have been amplified by extrapolation. The degradation estimate would have more credibility because it makes full use of the information available at 150°C.

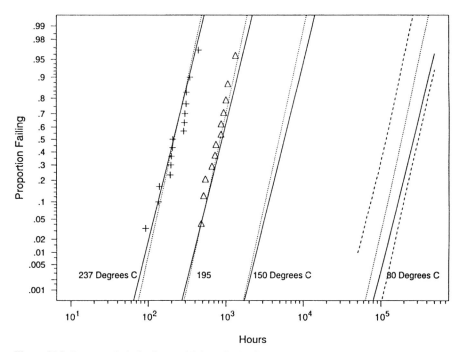

Figure 21.9. Lognormal–Arrhenius model fit to the Device-B failure-time data with failure defined as power drop below $-.5$ dB (solid lines) compared with the corresponding degradation model estimates (dotted lines). Also shown is the set of pointwise approximate 90% confidence intervals for $F(t)$ at 80°C (dashed lines), based on a bootstrap of the degradation analysis.

The overall close agreement between the degradation model and the lognormal failure-time model can be explained by referring to the models introduced in Section 18.3.3. There we showed that failure time will have a lognormal distribution if $T(\text{temp}_U) = -(1/\mathcal{R}_U)\log\left(1 - \mathcal{D}_f/\mathcal{D}_\infty\right)$ has a lognormal distribution. In our degradation model, $\log(\mathcal{R}_U)$ and $\log(-\mathcal{D}_\infty)$ [and thus $\log(\mathcal{D}_f/\mathcal{D}_\infty)$] are assumed to have a joint normal distribution. If $\mathcal{D}_f/\mathcal{D}_\infty$ is small relative to 1 (as in this example), then $\log(1 - \mathcal{D}_f/\mathcal{D}_\infty) \approx -\mathcal{D}_f/\mathcal{D}_\infty$ and thus $T(\text{temp}_U)$ is approximately the ratio of two lognormal random variables, and the ratio of two lognormal random variables also has a lognormal distribution.

Figure 21.10 is similar to Figure 21.9 with a fitted Weibull distribution for failure time. Comparing Figures 21.9 and 21.10, the lognormal ALT and degradation models provide a somewhat better fit to the data. As explained above and in Section 4.6, experience and physical theory also favor the lognormal distribution in this application.

574 ACCELERATED DEGRADATION TESTS

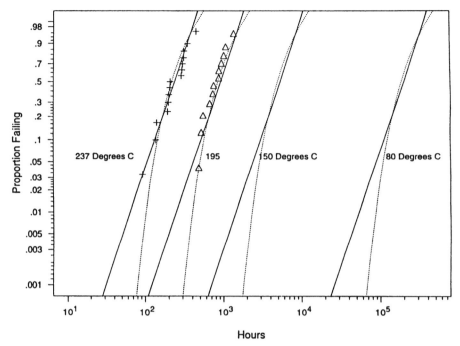

Figure 21.10. The Weibull–Arrhenius model fit to the Device-B failure-time data (solid lines) compared with the degradation-model estimates (dotted lines).

21.7 APPROXIMATE ACCELERATED DEGRADATION ANALYSIS

The simple method for degradation data analysis explained in Section 13.9 extends directly to accelerated degradation analysis. In particular, one can use the algorithm described there to predict the failure time for each sample path. Then these data can be analyzed using the methods from Chapter 19, as shown in the following example. It is important to remember, however, that such an analysis has the same limitations described in Section 13.9.

Example 21.6 Sliding Metal Wear Data Analysis. An experiment was conducted to test the wear resistance of a particular metal alloy. The sliding test was conducted over a range of different applied weights in order to study the effect of weight and to gain a better understanding of the wear mechanism. The data are given in Appendix Table C.19. Figure 21.11 shows the resulting degradation data. The same data are given in Figure 21.12, plotted on log–log axes. The predicted pseudo failure times were obtained by using ordinary least squares to fit a line through each sample path on the log–log scale (Figure 21.12) and extrapolating to the time at which the scar width would be 50 microns. These predicted pseudo failure times are given in Table 21.1. Figure 21.13 plots the pseudo failure times (on a log axis) versus applied weight. This plot also shows a fitted linear relationship between log cycles to 50

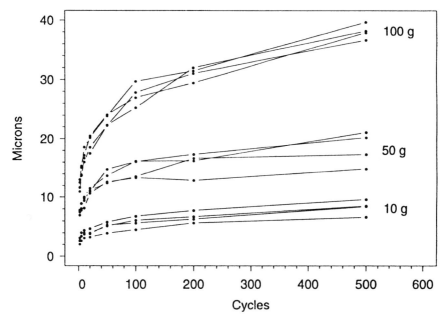

Figure 21.11. Scar width resulting from a metal-to-metal sliding test for different applied weights.

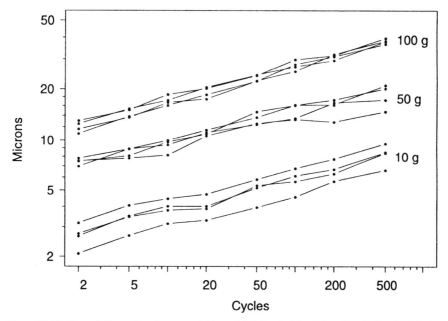

Figure 21.12. Scar width resulting from a metal-to-metal sliding test for different applied weights (using log-log axes).

Table 21.1. Metal-Wear "Failure" Times in Cycles

Grams	Pseudo Failure Times			
100	724	718	659	677
50	3216	1729	2234	1689
10	3981	4600	5718	4487

microns and applied weight. The variability at 100 grams is much smaller than at the other two weights, but with the small sample sizes involved, it is possible that this could be due to variability in the data. Figure 21.14 is a lognormal probability plot for the data at the three different levels of weight. The plot shows the smaller amount of variability at 100 grams (indicated by the steeper slope in the fitted line). The lognormal distribution fits quite well at all levels of weight (the normal and Weibull distributions did not fit as well as the lognormal distribution). Figure 21.15 is a lognormal probability plot depicting the lognormal regression model. Looking at the points relative to the fitted model suggests that this is a plausible model for the data. However, because the model is purely empirical, it would be risky to extrapolate to lower levels of weight. □

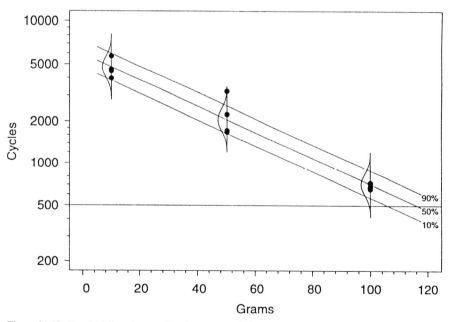

Figure 21.13. Pseudo failure times to 50 microns scar width versus applied weight for the metal-to-metal sliding test.

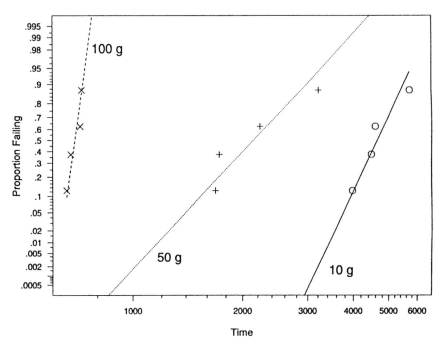

Figure 21.14. Lognormal probability plot showing the ML estimates of time to 50 microns width for each weight.

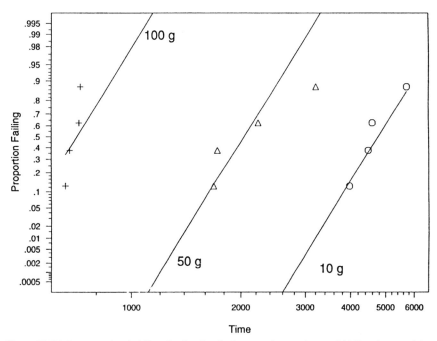

Figure 21.15. Lognormal probability plot showing the lognormal regression model ML estimates of time to 50 microns width for each weight.

577

BIBLIOGRAPHIC NOTES

Much of the material in this chapter has been taken from Meeker, Escobar, and Lu (1998). Overall, the literature describing the application of accelerated degradation methods is limited. The following is a brief summary of some available references. Beckwith (1979, 1980) describes methods of evaluating the decrease in strength of an adhesive over time. Amster and Hooper (1983) propose a simple degradation model for single-, multiple-, and step-stress life tests. They show how to use this model to estimate the central tendency of the failure-time distribution. Lu and Pantula (1989) use a repeated-measures model to analyze accelerated test degradation data from silicon devices. Nelson (1981; 1990a, Chapter 11) reviews the degradation literature, surveys applications, describes basic ideas on accelerated test degradation models, and, using a specific example, shows how to analyze degradation data with only one degradation reading per unit. Carey and Tortorella (1988) describe a Markov process model for degradation data and give methods of estimating parameters and testing goodness of fit. Similar results are given in Carey (1989). Carey and Koenig (1991) describe an application of the Carey and Tortorella (1988) methods of accelerated degradation analysis in the assessment of the reliability of a logic devices that are components in a new generation of submarine cables. Chan, Boulanger, and Tortorella (1994) illustrate the use of the simple approximate linear regression methods in Section 21.7 for analyzing degradation data. Tobias and Trindade (1995) use similar methods.

Murray (1993, 1994) and Murray and Maekawa (1996) describe accelerated degradation test data for data-storage disk error rates. These papers, the papers by Carey and Tortorella (1988), and Tobias and Trindade (1995) use the approximate analysis method described in Section 21.7 to analyze their degradation data. Tseng and Wen (1997) describe the use of step-stress ADTs for assessing the reliability of light-emitting diodes (LEDs). Chang (1992) analyzes ADT data from a test on power supplies.

Boulanger and Escobar (1994) describe methods for planning accelerated degradation tests for an important class of degradation models. Chow and Liu (1995, Chapter 9) describe applications of accelerated degradation testing for estimating the shelf life of pharmaceuticals.

EXERCISES

21.1. Appendix Table C.3 gives ADT data on the increase in resistance over time of carbon-film resistors tested at three different levels of temperature. Suppose that failure is defined as the time at which the resistance has increased by 5%. Use the approximate method of analysis described in Sections 13.9 and 21.7 to analyze these data. In particular:

(a) Make a plot of the degradation versus time for each of the sample paths. Use both log and linear axes.

(b) Choose an appropriate transformation scale for the data to make the paths approximately linear. Fit a separate linear regression model to the

EXERCISES 579

appropriate subset of the data in each sample path. Plot the estimated regression estimates in various ways.

(c) Explain the interpretation of the parameters of the regression model for the individual paths. In what sense are these regression parameters random?

(d) Compute the average of the slope estimates within each temperature group. Plot these averages versus $11605/(\texttt{temp}\ °C+273.15)$ (or, equivalently, plot on Arrhenius scales, as described in Exercise 18.11). Assess the adequacy of a linear relationship. What measure could you use, in an informal way, to help in this assessment?

(e) Write down the temperature/degradation model implied by the relationship graphed in part (d). Obtain a graphical estimate of the activation energy for this degradation process.

(f) Compute the pseudo failure time for each of the sample paths and use these to do a life data analysis and to estimate the failure-time distribution at 50°C.

21.2. Refer to Example 21.6.

(a) Repeat the analysis using 30 microns as the definition of failure.

(b) Repeat the analysis using 100 microns as the definition of failure.

(c) Compare the results in parts (a) and (b). Comment on the differences in assumptions needed to estimate these two different distributions.

◆**21.3.** Refer to Example 21.6. Repeat the analysis using the approximate ML method like that used in Examples 21.2–21.5. Explain the reason(s) for any differences in the analyses.

21.4. When iterative techniques are used for maximum likelihood estimation, there is always some chance that the iterations will not converge to the actual maximum. As described in Section 17.5.4, two precautions that will improve the probability of success are to use a parameterization that does not result in highly correlated parameter estimates and to have good starting values for the iterations. In most cases, the best way to get good starting values is to find simple graphical or moment-based estimates (e.g., estimates based on sample means and variances). For the accelerated degradation model described by (21.2) and (21.3), and the corresponding data from Example 21.1, suggest expressions that can be used to obtain starting values for the approximate ML estimation algorithm.

21.5. The relationship graphed in Exercise 21.1(d) seems to provide an adequate description of the available data. Describe the risks of using these data to predict life at lower levels of temperature.

21.6. Extend the analysis done in Exercise 13.9. The block error rate data were obtained by testing at higher than usual conditions. Suppose that the activation energy for the degradation process in $E_a = .9$. Use this to obtain an estimate of disk life at 50°C and 85% relative humidity.

21.7. An alternative to the simple graphical/ordinary least squares estimation method described in Exercise 21.1 is to do "full maximum likelihood" (or a close approximation to full ML), as described in Section 13.3 and illustrated in Examples 21.2–21.5. Explain the reason(s) for any differences in the analyses. Explain the trade-offs between these two different approaches.

21.8. Refer to the sliding metal wear data in Appendix Table C.19, also used in Example 21.6. Data were collected by testing four specimens at three different levels of applied weight (10, 50, and 100 grams). Scar depth was measured at 2, 5, 10, 20, 50, 100, 200, and 500 cycles. Comment on the practical value and the potential cost of the additional information that would be obtained by:

(a) Sampling wear at 2, 5, 10, 15, 20, 25, ..., 495, 500 cycles.
(b) Sampling wear at 2, 5, 10, 20, 50, 100, 200, 500, 1000, 2000, 5000 cycles.
(c) Testing eight units each at 10, 50, and 100 grams, using the original sampling rate in time.
(d) Testing three units each at 10, 30, 60, and 100 grams, using the original sampling rate in time.
(e) Testing three units each at 5, 20, 40, and 50 grams, using the original sampling rate in time.
(f) Testing three units each at 20, 60, 200, and 400 grams, using the original sampling rate in time.

21.9. Refer to Figure 21.1. Comment on the possible loss of information that would result from using the censored time to failure data instead of the degradation data to make inferences on the failure-time distribution.

21.10. It has sometimes been suggested that one can (or even should) use degradation data to obtain failure-time (or crossing time) data to be used in analysis. Is this a good thing to do? What are the trade-offs? How should one handle observations that have not yet crossed the boundary, but are close? Refer to the data in Figure 21.1 to help formulate your answer.

21.11. Design a computer program that can be used to simulate the results of an ADT experiment, using the model defined by (21.1), (21.2), and (21.3). Write down all of the needed inputs. Then outline each step of the process, including the formulas that you would use to generate the needed random numbers, assuming that you have access to a uniform random number generator. Note that there are two stages of randomness in this "mixed-effect" model.

◆**21.12.** Use a programming language to implement the algorithm described in Exercise 21.11. This will be much simpler if you use a high-level language like S-PLUS, Matlab, or Gauss rather than a low-level language like Fortran or C. Test the program by using the ML estimates in (21.4) to replace the model parameters. Plot the set of sample paths for simulated experiment. Compare with the sample paths in Figure 21.1. Comment on the results.

◆**21.13.** If your simulation program for Exercise 21.12 is written in S-PLUS (version 3.4 or later) you can use the nlme() function to compute approximate ML estimates of the parameters of your simulated ADT experiments. Implement this along with your data-simulation program. Again, test the program by using the ML estimates in (21.4) to replace the model parameters. Repeat the simulation 100 times. Plot the resulting estimates of the model parameters in various ways. Comment on the results of this simulation.

21.14. Refer to Exercise 21.11. Explain how you could use such a simulation program to help plan an ADT like the one in Example 21.1.

▲**21.15.** Give a detailed justification of the claim made in Section 21.6 about the approximate lognormal distribution of the random variable $T(\text{temp}_U) = -(1/\mathcal{R}_U) \log\left(1 - \mathcal{D}_\text{f}/\mathcal{D}_\infty\right)$.

CHAPTER 22

Case Studies and Further Applications

Objectives

This chapter:

- Describes additional applications of the reliability data analysis methods in this book.
- Shows how to extend the general methods covered in the earlier chapters to handle other special models and applications.
- Provides additional discussion of some important practical aspects of reliability data analysis applications.

Overview

This chapter presents several case studies that illustrate some additional important concepts and pitfalls of reliability data analysis, shows how to integrate ideas taken from several different places in the book, and presents some important additional examples.

Section 22.1 describes a serious problem that can arise when different cohorts of units arc ccnsorcd uncqually, having the potential to lead to misleading conclusions about product life. Section 22.2 applies Bayesian methods from Chapter 14 to an accelerated testing example that was first introduced in Chapter 19. This example shows how the introduction of prior information can importantly improve the precision with which one can estimate a failure-time distribution with an accelerated life test. Section 22.3 illustrates the use of a model that can be used to describe the failure-time distribution of a product that has both infant mortality and wearout failure causes. Section 22.4 suggests a physically motivated model that nicely describes the features and complicated relationship between fatigue life and applied stress or strain. Finally, Section 22.5 shows how to use simulation methods to plan an accelerated degradation test.

22.1 DANGERS OF CENSORING IN A MIXED POPULATION

The life distribution of a product can change from one production period to the next. Changes in the design or method of manufacture may improve product reliability. This is especially likely for new products undergoing reliability improvement efforts. In other cases, reliability may deteriorate due to the adverse consequences of a cost reduction, change in raw materials, or a relaxation of process monitoring standards. Thus field-tracking life data usually involve a mixture of failure-time distributions.

22.1.1 A Conceptual Example

Suppose that a product had been manufactured in equal quantities in each of two short production periods, one year apart. Units from the two periods had exponential life distributions (constant hazard rate) with mean times to failure of $\theta_1 = 1$ year (constant hazard rate $\lambda_1 = 1/\theta_1 = 1$) and $\theta_2 = 5$ years ($\lambda_2 = .2$), respectively. Due to their earlier availability, the units made in period 1 were put into service approximately one year earlier than those from period 2. An analysis of the failure-time data is performed two years after the first group (or, equivalently, one year after the second group) was put into service.

For the first year of operation, based on combining the units from both production periods, the average failure rate would be approximately .6 [i.e., $(\lambda_1 + \lambda_2)/2 = (1 + .2)/2$]. For the second year of operation, based on the units only from production period 1, the failure rate would be approximately 1. Thus because the production period 2 units with the lower failure rate are mixed with production period 1 units with the higher failure rate for the first year of operation (but not the second), there is an incorrect indication of an increasing failure rate. This is so despite the fact that a population consisting of a mixture of two different exponential distributions has a decreasing failure rate (see Exercises 5.1 and 5.4).

22.1.2 A Numerical Example

To illustrate the dangers of censoring in this simple setting, we simulated data from two populations with known characteristics. We used samples of 1000 units each from exponential distributions with $\theta_1 = 1$ year (production period 1) and $\theta_2 = 5$ years (production period 2), respectively. The interval data are given in Table 22.1. The data available after two years from the start of production (one year in service for the second group) are plotted on Weibull paper in Figure 22.1 both separately and combined for the two groups. Weibull probability plots of the data after two years of operation for both groups are given in Figure 22.2. This is the plot that would have been obtained after two years if both groups had been put into service at the same time. Curvature in the combined-sample plots in Figures 22.1 and 22.2 suggests some deviation from a Weibull distribution. Because the data were generated from a mixture of exponentials, a single Weibull distribution is not strictly correct in either case.

Table 22.1. Simulated Data from Two Different Production Periods

Time to Failure (years)	Production Period 1 $\theta = 1$ year Data After 2 Years of Operation	Production Period 2 $\theta = 5$ years Data After	
		1 Year of Operation	2 Years of Operation
0.0–0.2	185	33	33
0.2–0.4	163	33	33
0.4–0.6	134	35	35
0.6–0.8	90	39	39
0.8–1.0	83	45	45
1.0–1.2	58		34
1.2–1.4	42		33
1.4–1.6	44		31
1.6–1.8	36		24
1.8–2.0	35		31
>1.0		815	
>2.0	130		662

The total sample size for each production period was 1000 units. Data from Hahn and Meeker (1982b).

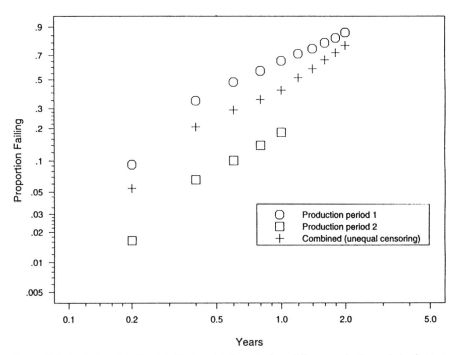

Figure 22.1. Weibull probability plot of failure data for units from different production periods after two years since production startup (but only one year of operation for the second production period).

DANGERS OF CENSORING IN A MIXED POPULATION   585

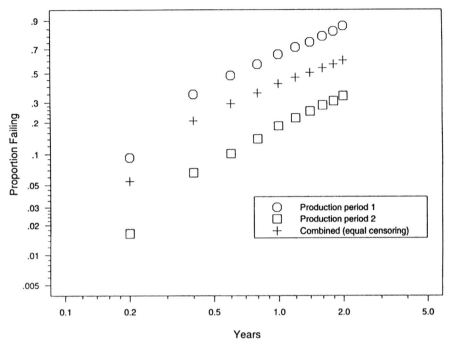

Figure 22.2. Weibull probability plot of failure data for units from different production periods after two years operating time for both production periods.

Analysis 1 is based on the combined data after two years since production startup (one year of operation for units from production period 2), that is, the crosses in Figure 22.1. Analysis 2 is based on the combined data after two years of operation for the units from both production periods, that is, the crosses in Figure 22.2.

- **Analysis 1.** A Weibull distribution is fit to the data with unequal right censoring for the two production periods. A 95% confidence interval for the Weibull distribution shape parameter is [1.02,   1.15], incorrectly indicating an increasing hazard rate with time (as expected from the conceptual example).
- **Analysis 2.** A Weibull distribution is fit to the data with equal right censoring at two years for both production periods. A 95% confidence interval for the Weibull distribution shape parameter is [.82,   .92]. This indicates a decreasing hazard rate with time, as expected from theory (see Proschan, 1963).

An appropriate analysis of the unequally censored data would fit separate exponential distributions for the two production periods. This analysis (left as an exercise) gives estimates of failure probabilities and quantiles that agree well with the true model from which the data were simulated.

22.1.3 Analyzing Data from Different Production Periods

Because reliability can change over time, it is advisable, when possible, to conduct separate analyses for each production period, to compare the results, and to combine them only if this seems appropriate. Separate analyses, however, are not possible in the following circumstances.

- A unit's production period is not known.
- Production periods are not well defined. For example, when production is continuous, there may not be well-defined points in time where the process has changed.
- The data are too scanty for reasonable dissection.

In most practical data analysis problems, available data could be viewed as having come from two or more populations. Analyses are most often done with the pooled data. This is appropriate when interest centers on failure-time distribution of the mixture and either (1) there are only small differences among the populations or (2) the amount of censoring is approximately the same over the different populations.

The effect of fitting a simple distribution to mixtures of two (or more) populations with unequal censoring depends on the degree of dissimilarity between or among the subpopulations and on the relative number of units produced in the two periods. The example in this section illustrates the desirability of doing separate analyses for units from different production periods, especially when field exposure periods differ for the different groups. If separate analyses are not possible because of the sparsity of the data or limitations in identifying the production period, one needs to recognize that a seriously incorrect model can give misleading results.

22.2 USING PRIOR INFORMATION IN ACCELERATED TESTING

This section uses an extension of the Bayesian methods presented in Chapter 14, to reanalyze the data from Example 19.11. Lerch and Meeker (1998) present similar examples. The computational methods used here, like those used in Chapter 14, follow Smith and Gelfand (1992).

Example 19.11 illustrated the analysis of accelerated life test data on a new-technology IC device. As a contrast, Example 19.12 showed how much smaller the confidence intervals on $F(t)$ would be if the Arrhenius activation energy were known. Generally it is unreasonable to assume that a parameter like activation energy is known exactly. For some applications, however, it may be useful or even important to bring outside knowledge into the analysis. Otherwise it would be necessary to spend scarce resources to conduct experiments to learn what is already known. In some applications, knowledgeable reliability engineers can, for example, specify the approximate activation energy for different expected failure modes. Translating the information about activation energy into a prior distribution will allow the use of Bayesian methods like those introduced in Chapter 14. This section shows how to incorporate prior information on the activation energy for a failure mode into an analysis of the new-technology IC device data.

22.2.1 Prior Distributions

Section 14.3 describes different kinds of prior information. This section reanalyzes the new-technology IC device ALT data in order to compare:

- A diffuse (wide uniform) prior distribution for activation energy E_a.
- A given value (degenerate prior distribution) for E_a.
- The engineers' prior information, converted into an informative prior distribution for E_a.

On the basis of previous experience with a similar failure mode, the engineers responsible for this device felt that it would be safe to presume that, with a "high degree of certainty," the activation energy E_a is somewhere in the interval .80 to .95. They also felt that a normal distribution could be used to describe the uncertainty in E_a. We use normal distribution 3-SD limits (i.e., mean ± three standard deviations) to correspond to an interval with a high degree of certainty, corresponding to about 99.7% probability. This is an informative prior distribution for E_a. The engineers did not have any firm information about the other parameters of the model. To specify prior distributions for the other parameters, it is then appropriate to choose a diffuse prior. A convenient choice is a UNIF distribution that extends far beyond the range of the data and physical possibility. As described in Chapter 14, the parameters used to specify the joint posterior distributions should be given in terms of parameters that can be specified somewhat independently and conveniently. For this example, the prior distribution for σ was specified as UNIF(.2, .9) and the prior distribution for $t_{.1}$ at 250°C was specified to be UNIF(500, 7000) hours. Comparison with the ML estimates from Example 19.11 shows that the corresponding joint uniform distribution is relatively diffuse.

Figure 22.3 compares a NOR prior distribution with a 3-SD range of (.80, .95) and a UNIF(.4, 1.4) (diffuse) prior for E_a. The corresponding marginal posterior distributions for E_a are also shown. The center of the marginal posterior distribution for E_a corresponding to the informative NOR prior is very close to that of the prior itself. This is mostly because the prior is strong relative to the information in the data. Figure 22.3 also shows a posterior distribution corresponding to the uniform (diffuse) prior. The corresponding joint posterior is approximately proportional to the profile likelihood for E_a. The uniform prior has had little effect on the posterior and therefore is approximately noninformative.

In addition to activation energy E_a, the reliability engineers also wanted to estimate life at 100°C. Figure 22.4 shows different posterior distributions for $t_{.01}$ at 100°C. The plots on the top row of Figure 22.4 compare posteriors computed under the informative and diffuse prior distributions for E_a. This comparison shows the strong effect of using the prior information in this application.

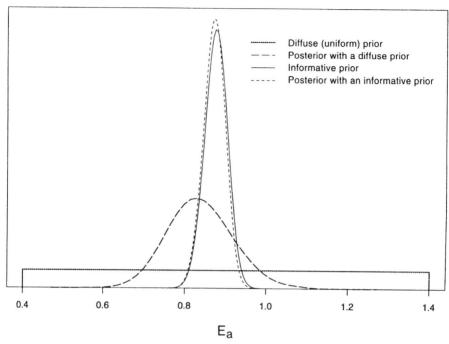

Figure 22.3. Plot of diffuse and informative prior distributions for the new-technology device activation energy E_a along with corresponding posterior distributions.

Recall from Example 19.11 that there was some concern (because of the different slopes in Figure 19.12) about the possibility of a new failure mode occurring at 300°C. Sometimes physical failure mode analysis is useful for assessing such uncertainties. In this application the information was inconclusive.

When using ML estimation or when using Bayesian methods with a diffuse prior for E_a, it is necessary to have failures at two or more levels of temperature in order to be able to extrapolate to 100°C. With a given value of E_a or an informative prior distribution on E_a, however, it is possible to use Bayesian methods to estimate $t_{.01}$ at 100°C with failures at only one level of temperature. The posterior distributions in the bottom row of Figure 22.4 assess the effect of dropping the 300°C data, leaving failures only at 250°C. Comparing the graphs in the NW and SW corners, the effect of dropping the 300°C data results in a small leftward shift in the posterior. Relative to the confidence intervals, however, the shift is small. Comparing the two plots in the bottom row suggests that using a given value of $E_a = .8$ results in an interval that is probably unreasonably narrow and potentially misleading. If the engineering information and previous experience used to specify the informative prior on E_a is credible for the new device, then the SW analysis provides an appropriate compromise between the commonly used extremes of assuming nothing about E_a and assuming that E_a is known.

If one tried to compute the posterior after dropping the 300°C, using a uniform prior distribution on E_a, the posterior distribution would be strongly dependent on

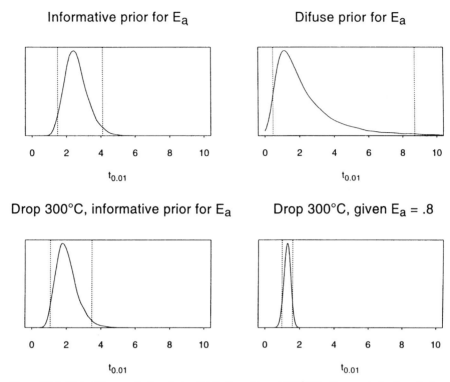

Figure 22.4. Plot of the marginal posterior distribution of $t_{.01}$ at 100°C for the new-technology device, based on different assumptions. NW corner: all data and an informative prior for E_a. NE corner: all data and a diffuse prior for E_a. SW corner: drop 300°C data and an informative prior for E_a. SE corner: drop 300°C data and given $E_a = .8$. The vertical lines are two-sided 95% Bayesian confidence intervals for $t_{.01}$ at 100°C.

the range of the uniform distribution. This is because with failures only at 250°C, there is no information on how large E_a might be. In this case there would be no approximately uninformative prior distribution.

To put the meaning of the results in perspective, the analysis based on the informative prior distribution for E_a after dropping the suspect data at 300°C would be more credible than the alternatives. The 95% Bayesian confidence intervals for $t_{.01}$ at 100°C for this analysis are [.6913, 2.192] million hours or [79, 250] years. This does not imply that the devices will last this long (we are quite sure that they will not!). Instead, the results of the analysis suggest that, if the Arrhenius model is correct, this particular failure mode is unlikely to occur until far beyond the technological life of the system into which the IC would be used. It is likely, however, that there are other failure modes (perhaps with smaller E_a) that will be observed, particularly at lower levels of temperature (see also the discussion of failure mode masking in Section 19.6.4).

590 CASE STUDIES AND FURTHER APPLICATIONS

22.3 AN LFP/COMPETING RISK MODEL

Chan and Meeker (1998) describe a model that combines components from the LFP model (for infant mortality, as described in Section 11.5.2) with a competing risk model (for longer-term wearout, as described in Section 15.4.2). This model is called the Generalized Limited Failure Population (GLFP) model. This section briefly describes the GLFP model and results of using maximum likelihood to estimate the parameters of the model.

22.3.1 Background

Consider the Vendor 1 data in Examples 1.3, 11.9, and 11.11. Most of the early failures were known to have been caused by defective integrated circuits (Mode 1). Only a small proportion of the circuit packs would contain an integrated circuit (IC) with such a defect (something like 1% or 2% was expected for this particular technology). After about 2000–4000 hours, however, the failure rate began to increase and there was some evidence (both in the data and some limited physical failure analysis) that the latter failures were being caused by a combination of a corrosion and another chemical degradation failure mode to which all units would eventually succumb (Mode 2). This can be seen in the Weibull probability plot shown in Figure 22.5,

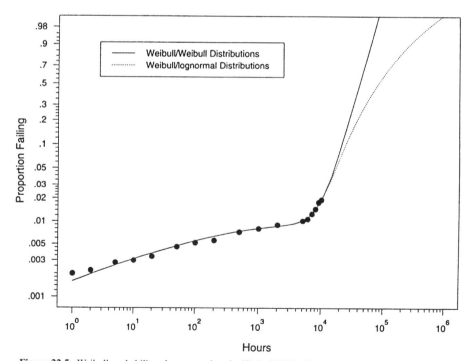

Figure 22.5. Weibull probability plot comparing the Weibull/Weibull and the Weibull/lognormal GLFP competing risk $F(t; \hat{\boldsymbol{\theta}})$ ML estimates for the Vendor 1 circuit pack failure data.

where the plotted points change direction after 2000 hours. Although something like this secondary failure mode had been expected, the managers responsible for the operation of the system in which the circuit packs were to be deployed were concerned at how early such failures were beginning to appear. They were interested in obtaining a prediction for the proportion of units that would fail in the first 5 years (43,800 hours) of operation (approximate technological life of the system).

22.3.2 The GLFP Model

Let T_1 denote the "infant mortality" failure time for a unit. If a unit is not defective, then $T_1 = \infty$. As with the LFP model described in Section 11.5.2, the cdf for T_1, conditional on the unit being defective, is $F_1(t; \boldsymbol{\theta}_1)$, where $\boldsymbol{\theta}_1$ is a vector of unknown parameters. The unconditional cdf of T_1 is $pF_1(t; \boldsymbol{\theta}_1)$, where p is the proportion of defective units in the population. Similarly, let T_2 denote the unit's wearout failure time and let $F_2(t; \boldsymbol{\theta}_2)$ denote the cdf of T_2. The unit's actual failure time is $T = \min(T_1, T_2)$. As in (15.1), if T_1 and T_2 are independent, the cdf of failure time T is

$$F_T(t; \boldsymbol{\theta}) = \Pr(T \leq t) = 1 - [1 - pF_1(t; \boldsymbol{\theta}_1)][1 - F_2(t; \boldsymbol{\theta}_2)], \quad (22.1)$$

where $\boldsymbol{\theta} = (\boldsymbol{\theta}_1, \boldsymbol{\theta}_2)$. The pdf of T is then given by

$$f_T(t; \boldsymbol{\theta}) = \frac{dF_T(t; \boldsymbol{\theta})}{dt} = pf_1(t; \boldsymbol{\theta}_1)[1 - F_2(t; \boldsymbol{\theta}_2)] + f_2(t; \boldsymbol{\theta}_2)[1 - pF_1(t; \boldsymbol{\theta}_1)], \quad (22.2)$$

where $f_1(t; \boldsymbol{\theta}_1)$ and $f_2(t; \boldsymbol{\theta}_2)$ are the pdfs of T_1 and T_2, respectively. Chan and Meeker (1998) used Weibull and lognormal distributions for $F_1(t; \boldsymbol{\theta}_1)$ and $F_2(t; \boldsymbol{\theta}_2)$. Other distributions could, however, be substituted without difficulty.

22.3.3 Likelihood Contributions

The likelihood for the circuit pack data can be written using the general form for independent right-censored and interval-censored observations given in (7.2). The contributions for the individual observations for the GLFP model depend, however, on whether the cause of failure is known or not and, if so, on which type of failure occurred.

If the cause of failure is *known*, then either T_1 or T_2 is also known. Otherwise if failure cause is *unknown*, then only T is known. If $T_1 < T_2$ the unit fails from a defective IC and T_2 is not observed. Similarly, if $T_2 < T_1$, the unit fails from a wearout mode and T_1 is not observed.

If unit i is known to have failed between times t_{i-1} and t_i from failure Mode 1, the probability of the observation is

$$\begin{aligned} L_i(\boldsymbol{\theta}) &= \Pr[(t_{i-1} < T \leq t_i) \cap (T_1 < T_2)] \\ &= \Pr[(t_{i-1} < T_1 \leq t_i) \cap (T_1 < T_2)] \end{aligned}$$

$$= \int_{t_{i-1}}^{t_i} \int_s^\infty p f_1(s; \boldsymbol{\theta}_1) f_2(v; \boldsymbol{\theta}_2) \, dv \, ds$$

$$= \int_{t_{i-1}}^{t_i} p f_1(s; \boldsymbol{\theta}_1)[1 - F_2(s; \boldsymbol{\theta}_2)] \, ds,$$

where, as before, $T = \min(T_1, T_2)$. Similarly, if the cause of failure is Mode 2,

$$L_i(\boldsymbol{\theta}) = \int_{t_{i-1}}^{t_i} f_2(s; \boldsymbol{\theta}_2)[1 - p F_1(s; \boldsymbol{\theta}_1)] \, ds.$$

If the cause of failure is not known, then

$$L_i(\boldsymbol{\theta}) = \Pr(t_{i-1} < T \leq t_i)$$
$$= F_T(t_i; \boldsymbol{\theta}) - F_T(t_{i-1}; \boldsymbol{\theta})$$
$$= [1 - p F_1(t_{i-1}; \boldsymbol{\theta}_1)][1 - F_2(t_{i-1}; \boldsymbol{\theta}_2)] - [1 - p F_1(t_i; \boldsymbol{\theta}_1)][1 - F_2(t_i; \boldsymbol{\theta}_2)].$$

For a right-censored observation at time t_i (i.e., the failure time is after t_i),

$$L_i(\boldsymbol{\theta}) = \Pr(T > t_i)$$
$$= [1 - p F_1(t_i; \boldsymbol{\theta}_1)][1 - F_2(t_i; \boldsymbol{\theta}_2)].$$

Chan and Meeker (1998) also give expressions for left-censored and exact failure-time observations.

22.3.4 ML Estimates

To develop predictions for the proportion failing in 5 years, we fit the Weibull/Weibull and Weibull/lognormal models to the available data. We did this by supposing, on the basis of engineering judgment, that the failures before 200 hours were due to defective components and that failures after 5000 hours were due to a wearout mechanism. No assumption was used for the failures between 200 and 5000 hours. Figure 22.5 shows a Weibull probability plot comparing the Weibull/Weibull and the Weibull/lognormal GLFP competing risk models for the Vendor 1 circuit pack failure data. Although the two different distributions provide excellent agreement within the range of the data, they differ importantly in extrapolation. It is interesting to note that, as we have seen in other similar examples, the lognormal extrapolation provides a much more optimistic (smaller) prediction of the future proportion of units that will fail. Figure 22.6 shows the ML estimate of the Weibull/lognormal GLFP competing risk cdf along with pointwise approximate 95% confidence intervals for the Vendor 1 circuit pack failure data. At 10^4 hours, the confidence interval is narrow because of the large sample size ($n = 4993$). At 10^5 hours, the confidence interval is wide, ranging from about .3 to .85. This is due to the large amount of extrapolation. It is important to note, however, that the width of this confidence interval does *not* reflect the deviation (which almost certainly exists) from the assumed Weibull/lognormal

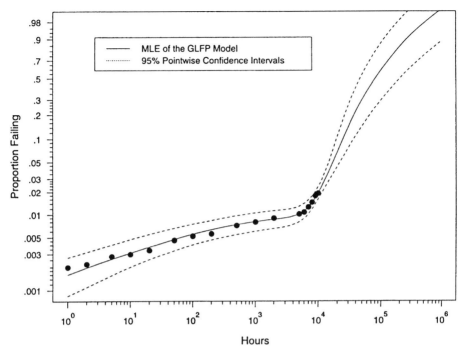

Figure 22.6. Weibull probability plot showing the ML estimate of the Weibull/lognormal GLFP competing risk cdf with pointwise approximate 95% confidence intervals for the Vendor 1 circuit pack failure data.

GLFP distribution. Indeed, as seen in Figure 22.5, the Weibull/Weibull ML estimate at 10^5 hours is almost 1.

Figure 22.7 shows a plot of the corresponding Weibull/Weibull and the Weibull/lognormal GLFP competing risk hazard function ML estimates. The hazard function was computed as $h_T(t) = f_T(t; \boldsymbol{\theta})/[1 - F_T(t; \boldsymbol{\theta})]$, where $F_T(t; \boldsymbol{\theta})$ and $f_T(t; \boldsymbol{\theta})$ are defined in (22.1) and (22.2), respectively. The plot shows the hazard decreasing until the corrosion/degradation failure mode becomes active, at which time the hazard increases markedly. We also fit lognormal/Weibull and the lognormal/lognormal GLFP models, but the estimates of the cdf and hazard function estimates were, for all practical purposes, the same as the Weibull/Weibull and Weibull/lognormal models, respectively. The shape of $h_T(t)$ results from adding the decreasing hazard of T_1 to the increasing hazard for T_2.

22.4 FATIGUE-LIMIT REGRESSION MODEL

Section 17.5 illustrated the fitting of a quadratic regression model to nickel-base superalloy fatigue data from Nelson (1984). Fitting a quadratic function is relatively easy to do and may be satisfactory for some purposes. Alternative functional forms,

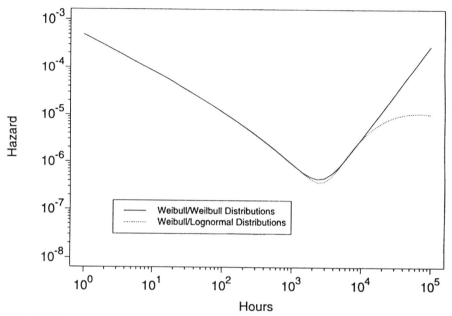

Figure 22.7. Plot comparing the Weibull/Weibull and the Weibull/lognormal GLFP competing risk mode hazard function estimates for the Vendor 1 circuit pack failure data.

however, may provide a better description of the data. Pascual and Meeker (1998a) use ML methods to fit an alternative regression model that contains a fatigue-limit parameter. The material in this section has been adapted from this paper.

Under the fatigue-limit model, specimens operated at levels of stress below the fatigue limit will never fail. The fatigue-limit model also allows the standard deviation of fatigue life to be a function of stress. The purpose of the analysis is to obtain an estimate of the small quantiles of the fatigue-life distribution.

22.4.1 The Fatigue-Limit Model

Let $x_1, \ldots, x_n$ denote pseudostress levels of n specimens and let $t_1, \ldots, t_n$ be actual failure times or censoring times. Censoring times may vary from specimen to specimen. Let γ be the fatigue limit. At each pseudostress level with $x_i > \gamma$, fatigue life t_i is modeled with a lognormal distribution; that is, the cumulative proportion failing function and its derivative are given by

$$\Pr(T \leq t) = F[t; \mu(x), \sigma(x)] = \Phi\left[\frac{\log(t) - \mu(x)}{\sigma(x)}\right] \quad (22.3)$$

$$f[t; \mu(x), \sigma(x)] = \frac{1}{\sigma(x)t}\phi\left[\frac{\log(t) - \mu(x)}{\sigma(x)}\right], \quad t > 0.$$

FATIGUE-LIMIT REGRESSION MODEL

For example, using Φ_{nor} and ϕ_{nor} implies that $\log(T)$ is modeled with a normal distribution with mean $\mu(x)$ and standard deviation $\sigma(x)$. These parameters are related to stress according to

$$\mu(x) = \mathrm{E}[\log(T)] = \beta_0^{[\mu]} + \beta_1^{[\mu]} \log(x - \gamma), \quad x > \gamma, \tag{22.4}$$

$$\sigma(x) = \sqrt{\mathrm{Var}[\log(T)]} = \exp\left[\beta_0^{[\sigma]} + \beta_1^{[\sigma]} \log(x)\right], \quad x > \gamma, \tag{22.5}$$

where $\beta_0^{[\mu]}$, $\beta_1^{[\mu]}$, $\beta_0^{[\sigma]}$, $\beta_1^{[\sigma]}$, and γ are unknown parameters to be estimated from data. If x_{minf} is the smallest observed stress level that yields a failure, then γ must be in the interval $[0, x_{\text{minf}})$.

Note that when $\beta_1^{[\sigma]} = 0$, the model has a constant standard deviation. In most fatigue data, the standard deviation decreases as stress increases, which corresponds to $\beta_1^{[\sigma]} < 0$. The scatter plot of the superalloy data (and the fitted model) in Figure 22.8 has cycles to failure on the horizontal axis, as is commonly done in the fatigue literature. This scatter plot indicates more scatter at the lower stress levels and less at the higher stress levels.

The value of γ determines the amount of curvature present in the plotted S-N curve for values of stress that are not far from x_{minf}. When γ is close to zero, the S-N curve is close to linear. Larger values of γ result in more curvature in the plot. When $\gamma = 0$, the model is equivalent to the simple linear regression model used in Section 17.3. Curvature in Figure 22.8 suggests the inclusion of a fatigue limit

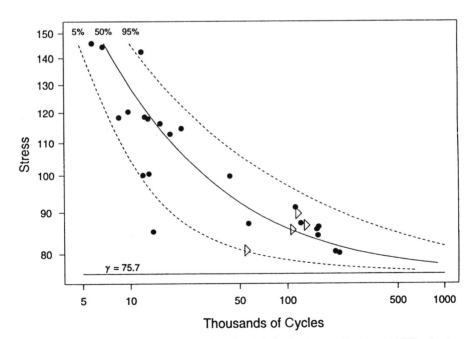

Figure 22.8. Log–log S-N (stress versus number of cycles) plot for the superalloy data with ML estimates of the .05, .5, and .95 quantiles from the fatigue-limit model (• failure, ▷ censored).

γ in the model. Although a fixed fatigue limit may be unrealistic for describing a population of specimens, the fatigue limit provides a physically appealing alternative to the quadratic term in the $\mu(x)$ relationship used in Section 17.5 for describing S-N curvature.

The maximum likelihood methods described in the next section use the following assumptions: (1) specimens are tested independently and (2) for $x > \gamma$ the times at which observations were censored are independent of actual failure times that would be observed if the experiment were to be run until failure.

22.4.2 Maximum Likelihood Estimation

The parameters of the fatigue-limit model can be estimated by using the method of maximum likelihood in a manner that is very similar to that described in Chapter 17. As before, we use $\boldsymbol{\theta} = (\beta_0^{[\mu]}, \beta_1^{[\mu]}, \beta_0^{[\sigma]}, \beta_1^{[\sigma]}, \gamma)$ to denote the vector of model parameters. The log likelihood function is

$$\mathcal{L}(\boldsymbol{\theta}) = \log[L(\boldsymbol{\theta})] = \sum_{i=1}^{n} \mathcal{L}_i(\boldsymbol{\theta}),$$

where

$$\mathcal{L}_i(\boldsymbol{\theta}) = \delta_i\{\log[\phi(z_i)] - \log[\sigma(x_i)t_i)]\} + (1 - \delta_i)\log[1 - \Phi(z_i)],$$

where $\delta_i = 1$ ($\delta_i = 0$) if observation i is a failure (right-censored observation) and $z_i = [\log(t_i) - \mu(x_i)]/\sigma(x_i)$.

The ML estimate $\boldsymbol{\theta}$ is the set of parameter values that maximize $\mathcal{L}(\boldsymbol{\theta})$. Table 22.2 gives the ML estimates of all model parameters resulting from fitting the fatigue-limit model to the data. This table also shows normal-approximation and likelihood confidence intervals for the parameters. Figure 22.8 shows curves of the ML estimates of the .05, .5, and .95 quantiles of fatigue life.

Nelson (1984, pages 72–73) comments that the quadratic fatigue life models produce quantiles larger at an intermediate stress than at a lower stress. Such a relationship is physically implausible. Although such behavior is also theoretically possible for the fatigue-limit model, it seems to be less of a problem and does not occur within the range of interest for these data.

Table 22.2. Maximum Likelihood Results for the Superalloy Data

Parameter	Estimate	Approximate 95% Confidence Interval	
		Normal-Theory	Likelihood-Ratio
$\beta_0^{[\mu]}$	14.75	(12.06, 17.44)	(12.90, 21.45)
$\beta_1^{[\mu]}$	−1.39	(−2.02, −.76)	(−2.81, −.92)
$\beta_0^{[\sigma]}$	10.97	(3.82, 18.12)	(3.22, 17.90)
$\beta_1^{[\sigma]}$	−2.50	(−4.04, −.96)	(−3.98, −.81)
γ	75.71	(67.35, 84.06)	(49.98, 79.79)

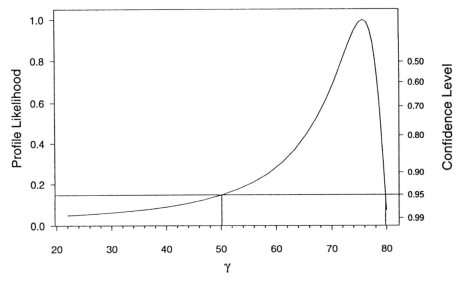

Figure 22.9. Profile likelihood plot for the fatigue limit γ for the superalloy data.

22.4.3 Profile Likelihoods and Likelihood-Ratio-Based Confidence Regions

The profile likelihood for γ is defined by

$$R(\gamma) = \max_{\boldsymbol{\theta}_{[\gamma]}} \left[\frac{L(\boldsymbol{\theta}_{[\gamma]}, \gamma)}{L(\widehat{\boldsymbol{\theta}})} \right]$$

and is shown, for the superalloy data, in Figure 22.9. Here $\boldsymbol{\theta}_{[\gamma]}$ is the vector $\boldsymbol{\theta}$ with γ removed. The likelihood confidence interval for γ in Table 22.2 is indicated by the vertical lines where $R(\gamma)$ intersects the horizontal critical level. Note that the upper bound of the normal-approximation interval exceeds $x_{\text{minf}} = 80.3$. This will never happen with the likelihood interval.

The confidence intervals in Table 22.2 indicate that the parameters $\beta_1^{[\mu]}$, $\beta_1^{[\sigma]}$, and γ are different from zero. The confidence intervals for $\beta_1^{[\sigma]}$ indicate that the standard deviation of fatigue life depends on the stress level and, moreover, that the standard deviation decreases as stress increases, a commonly observed phenomenon in metal fatigue data. The confidence intervals for γ support the inclusion of a fatigue limit as suggested by the curvature in Figure 22.8. Similar confidence intervals could also be computed for functions of the parameters, following the methods described in Chapter 8.

22.5 PLANNING ACCELERATED DEGRADATION TESTS

Morse and Meeker (1998) describe the use of simulation methods to help plan accelerated degradation tests. This section has been adapted from their work.

The motivating problem arose from an extension of the accelerated degradation analysis described in Example 21.1. The Device-B accelerated test had been designed as an accelerated life test. The engineers were going to define failure time to be the time at which power output first dropped .5 decibels (dB) below initial output. After learning about the accelerated degradation analysis methods, the engineers wanted to know how to design an accelerated test with degradation analysis in mind and how to assess the potential advantage of using degradation methods. In particular, they were interested in seeing whether, with a different test plan, they could expect to obtain better precision for estimating $F(130000)$, the proportion failing at 130 thousand hours at 80°C junction temperature.

22.5.1 Experimental Design Parameters and Test Constraints

The evaluation of alternative test plans will be based on the information obtained in the initial study (see Examples 21.1–21.3). The test plans will be evaluated under the following constraints:

- The test will use three levels of accelerated temperature.

    The highest test temperature will be 237°C.
    The middle temperature will be halfway between 237°C and the low temperature (on the Arrhenius scale).

- Twenty percent of the units on test will be allocated to the middle level of temperature.
- There is a constraint on the overall number of test positions and the length of the test. To get useful information, it is necessary to test units at lower temperatures for longer periods of time. The time to leave a unit on test at a particular level of temperature (censoring time) will be found through a censoring function:

    Censoring time $= -15621 + 730 \times 11605/(273.15 + \text{temp} \,°C)$ hours. This was the approximate censoring function used in the original study.

    A total of 67,000 test position-hours will be available for testing, as in the original study (an average of about 22 test positions over 3000 hours).

- The parameters of the study that will be varied are:

    The low level of test-plan temperature, denoted by temp_L.
    The proportion of units allocated to temp_L, denoted by π_L.

Because the highest accelerated temperature is $\text{temp}_H = 237°C$ and the middle temperature is halfway between the low and the high, specifying the low level fixes all three accelerated temperatures. Likewise, because 20% of the units will always be allocated to the middle temperature ($\pi_M = .2$), specifying the proportion allocated to the low temperature fixes the allocations to all temperatures.

22.5.2 Evaluation of Test Plan Properties

Chapters 10 and 20 showed how to evaluate the properties of proposed test plans by:

- Using large-sample approximations (easy to compute for simple problems but depend on an approximation that might not be adequate with small samples).
- Using Monte Carlo simulation (requires much more computing time but does not rely on large-sample approximations).

For the nonlinear accelerated degradation models, easy-to-compute large-sample approximations have not been derived. To answer the engineers' questions quickly, Monte Carlo simulation was used.

22.5.3 Simulation Procedure

This section shows how to simulate the accelerated degradation experiment. Model parameter estimates from the original study were used as planning values (see Example 21.3). Test plan properties can be computed for test plans specified by different combinations of temp_L and π_L. These evaluations can be used to determine if a different test plan would yield more precise estimates of the failure-time distribution. The procedure for doing this is as follows:

1. Fix the test plan by specifying temp_L and π_L from a candidate list of combinations.
2. Randomly generate sample degradation paths for each of the n units according to the specified test plan, using parameter estimates from the original study as planning values. Add simulated measurement error.
3. Fit the degradation model to the n simulated paths and compute approximate ML estimates of the model parameters.
4. Compute $\widehat{F}(130000)$ at $80°C$ as a function of the parameter estimates.
5. Repeat steps 2 to 4 N times, obtaining $\widehat{F}_1, \ldots, \widehat{F}_N$.
6. Compute some statistic quantifying the variation in these point estimates (e.g., a sample standard deviation).
7. Return to step 1 until the list of temp_L and π_L combinations has been exhausted.

For example, Figure 22.10 shows $\widehat{F}(t)$ curves computed from 20 simulated degradation analysis experiments, using a low accelerated temperature of $\text{temp}_L = 130°C$ and allocation $\pi_L = .1$. Overall the $\widehat{F}$ estimates in Figure 22.10 show a considerable amount of variability. At 130 thousand hours, however, most of the simulated $\widehat{F}$ estimates are close to 0. Of course, if the specified planning values are optimistic [in terms of having $\widehat{F}(130000)$ be much less than the true $F(130000)$], these calculations could be misleading. Thus one should use sensitivity analysis to investigate the effect that deviations from the assumed (uncertain) inputs will have on conclusions.

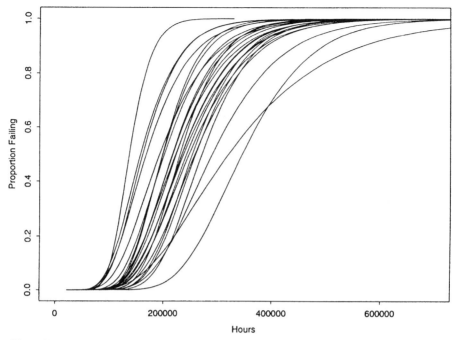

Figure 22.10. An example of variation in failure-time distributions for $(\text{temp}_L, \pi_L) = (130°\text{C}, .1)$ test parameter combination.

22.5.4 Evaluation of Test Plans Over a Grid of (temp_L, π_L) Values

Test plan simulations were run at each combination of (temp_L, π_L) values for $\text{temp}_L = 130(5)160$ and $\pi_L = .05(.05).30$. At each point in the grid, 2000 tests were simulated. For 32 out of the 42 combinations of (temp_L, π_L) values, all 2000 samples converged. For the combination $(\text{temp}_L, \pi_L) = (135, .25)$, about 6% of the simulated tests resulted in samples that did not converge and about 1% failed to converge at $(\text{temp}_L, \pi_L) = (130, .20)$ and $(130, .30)$. In the other 7 cells with some convergence difficulties, the percentage was on the order of .5% or less. The nonconverging samples were omitted from summarizing computations.

At each point the sample standard deviation of the $\widehat{F}(130000)$ values was computed. Figure 22.11 is a contour plot of the results. The original test plan used $\text{temp}_L = 150°\text{C}$ and $\pi_L = .2$. The orientation of the contour lines suggests that higher precision can be obtained for estimating $F(130000)$ by using tests having lower levels of temp_L. At lower ranges of temperature, precision seems not to be highly dependent on π_L. The final recommendation was to choose a plan with $(\text{temp}_L, \pi_L) = (130, .15)$. In comparison with the original test plan, this would reduce the amount of extrapolation in temperature, provide a test plan with somewhat more precision, and be safely away from the points that had convergence difficulties.

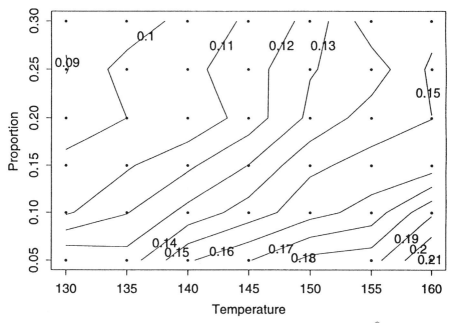

Figure 22.11. Contour plot of sample standard deviations of the simulated degradation $\widehat{F}(130000)$ values for different proposed compromise accelerated degradation plans.

BIBLIOGRAPHIC NOTES

The example in Section 22.1 was adapted from Hahn and Meeker (1982b). In addition to the unequal censoring example presented in Section 22.1, Hahn and Meeker (1982a, b) describe several other potential pitfalls that arise in the analysis of life data. The material in Section 22.2 builds on work done in Lerch and Meeker (1998). The example in Section 22.3 was extracted from Chan and Meeker (1998), who also describe in detail a Monte-Carlo-based method for finding likelihood-based confidence intervals for functions of model parameters when one is faced with a model with many parameters. The material in Section 22.4 was adapted from Pascual and Meeker (1997). In addition to analyzing the superalloy data, Pascual and Meeker (1997) also investigate the effect that censoring has on the ability to estimate the fatigue-limit parameter. Pascual and Meeker (1998a) describe a related regression model for fatigue data, suggested in Nelson (1990a, page 93). In this model, the fatigue-limit parameter is allowed to vary from unit to unit. Morse and Meeker (1998) provide more information on the use of simulation methods to help plan accelerated degradation tests. Boulanger and Escobar (1994) also describe methods for planning accelerated degradation tests.

Epilogue

The material in this book is only part of a much larger picture. Our examples, like the particular applications from which they originated, focused on the reliability or durability of materials, components, and relatively simple systems. Each of these applications, however, was associated with larger reliability, economic, and business questions like:

- Is Bearing-A durable enough to use in a new automobile transmission or do the designers need to switch to the more durable (but somewhat more expensive) Bearing-B? With the prospect of selling millions of transmissions over several years, the decision has huge economic consequences.
- How much redundancy is required for a critical device in the repeater-transmitter subsystems for an undersea telecommunications system? Adding redundancy will add cost but also improve system reliability and affect life-cycle cost. Promised life-cycle cost is an extremely important part of contract negotiations. A critical input to the analysis is the life distribution of the device. An accelerated life test will be used to obtain needed information.

Numerous decisions like these are made in typical product design processes. The methods presented in this book are essential for providing some of the information needed for making such decisions in the face of uncertainty. We close with some commentary on the changing role of statistical methods in a Design for Reliability program and as part of an overall Reliability Assurance process.

THE CHALLENGES OF ACHIEVING HIGH RELIABILITY

In a conversation with one of us, an engineer commented that "reliability is much more difficult today than it used to be." He went on to explain that his company was facing real competition for the first time. They were having to reduce engineering safety factors and cut other corners to reduce cost. As a result, they were beginning to see many more field failures. New technology, new materials, higher customer expectations, more competition, and the need to get competitively priced products to the market quickly complicate decision making. Indeed, reliability engineering *is* more difficult today due to stringent cost constraints and enormous time pressures caused by increased competition.

The engineer's comment might suggest that there is a cost/reliability trade-off curve for product design and all one needs to do is to find the right place on the design curve. What is really needed, however, is to move off the old curve, find a better curve, and then optimize it. The "best curve" in this case depends on product design, manufacturing process design, the collection and use of appropriate reliability data, and other factors. There will always be cost/reliability trade-off issues. To be successful, products in competitive markets must provide high reliability at low cost. The best curves (i.e., designs or products) will be determined by good engineers who can properly use modern statistical and other analytical engineering methods.

Achieving high reliability is complicated by the fact that the impact of many reliability improvements will appear some time in the future (often some years). Thus it is extremely difficult to quantify the effect on the current year's bottom line. For most companies, this will require a change in metrics and mind-sets.

In spite of the difficult challenges involved, the leading companies in a number of industries have continued to improve performance and reliability while keeping cost low enough to maintain competitive prices. Examples include computer hardware, telecommunications systems, and automobiles.

HIGH RELIABILITY AT COMPETITIVE COST

The necessary constant pressure on product and process designs to reduce costs and/or improve performance has the potential to cause reliability problems. For example, reducing traditional safety factors mandates more careful engineering and statistical practice. Having large safety factors to protect against one failure mode may well have prevented other un-thought-of failure modes. A proposed design change for cost reduction may be analyzed for the known risks (or failure modes), but the unknown risks may not surface until product is in the field. There are also numerous instances where all available energy was devoted to addressing one failure mode; no resources or energy remained to address other failure modes that turned out to be the most troublesome in the field.

How can design engineers deal with failure modes that might otherwise be in the un-thought-of category? Three suggested ways are:

- Use careful, informed engineering (including knowledge of the product's use environment). Identify and prevent most potential failure modes before they have a chance to occur. See the discussion of FMEA, FMECA, design reviews, and related reliability management tools described in Section 15.5.4.
- Increase up-front experimentation and life testing to discover and eliminate other potential failure modes before they get into a final design. It might be necessary to do such testing at the component level, the subsystem level, or the system level.

- Collect and carefully scrutinize early information from the field. Compare with previous test results. Detect and fix product weaknesses before they cause serious problems.

As with the product design itself, the key is to find efficient implementation of these activities and to balance costs against potential risks. The biggest payoff is in building high reliability into the design before product introduction. The cost of detecting and eliminating failure modes increases as the product moves from conceptualization through design, development, testing, and production, and into the field. It is inevitable that there will be some product reliability problems. Reliability assurance processes should reduce risk by reducing the probability and severity of field-failure problems.

ADVANCING ENGINEERING PRACTICE TO ACHIEVE HIGH RELIABILITY

Today's engineers need to rely more heavily on modern tools like probabilistic design and risk analysis and less on the traditional easy-to-apply rules of thumb. This modern approach to engineering leads to a new set of concerns and costs. Issues of model adequacy and uncertainty in model inputs can be critical. More measurement, experimentation, systems analysis, and sensitivity analysis will be needed. Protection must still be provided where unacceptable uncertainty exists. To address these issues, some or many engineers need a command of basic experimental design and statistical concepts.

Although reliability practice is primarily an engineering discipline, statistics and other scientific disciplines (material science, physical chemistry, etc.) play crucial supporting roles. Engineers need specialized training to analyze reliability data (dealing with censored data, pitfalls of accelerated testing, difficulties of interpreting warranty data, multiple failure modes, physics of failure, etc.).

USEFUL TOOLS AND SOME SPECIFIC SUGGESTIONS

There is no simple solution or magic for the challenge of achieving high reliability at low cost. The process involves hard work. Some suggestions, related to the technical material in this book, include the following.

1. Electronic design engineers use databases containing up-to-date reliability-related information (component reliability, derating functions, materials properties, etc.) linked with reliability modeling software, embedded within CAD systems. More widespread development of such systems would lead to better design practices.
2. Computationally based models of physical phenomena and increased use of computer simulation have the potential to save time and money by reducing

reliance on expensive physical experimentation. Electronics and fracture mechanics are the important success stories here. Up-front investment in research and development is needed, but there are important potential payoffs.

3. Up-front testing of materials and components, as well as subsystems and systems, is needed to reduce uncertainty about product reliability. Such tests are an important part of any reliability assurance program. Deciding what to test and how much testing to do (number of units, test duration, and at what level of system integration) requires careful consideration to balance risks with costs. It is important to take into consideration the way in which the product is used by customers.

4. Efficiency in up-front testing (e.g., use of existing information, properly designed experiments) is important. Experimental effort should be focused where engineering uncertainty implies reliability uncertainty. For materials and components, the goal should be to assess the failure-time distribution and/or to determine allowable levels of stress. For subsystem and system tests in the product development stage, the goal should be to apply appropriate amounts of stresses in the right combinations to discover (and then fix) potential failure modes. Being able to identify failure modes that would not be expected to occur in actual operation is important and requires good engineering knowledge.

5. Experiments, in general, and robust-design experiments, in particular, have potential for leading to important improvements in product and process design and resulting better reliability. The robust-design methods of Taguchi (e.g., Phadke, 1989) are important here. Multifactor robust-design experiments (RDE) provide methods for systematic and efficient reliability improvement. These are often conducted on prototype units and subsystems and focus on failure modes involving interfaces and interactions among components and subsystems. Among many possible product-design factors that may affect a system's reliability, RDEs empirically identify the important ones and find levels of the product-design factors that yield consistent high quality and reliability. Graves and Menten (1996) provide an excellent description of experimental strategies that can be used to help design products with higher reliability. Other important references relating to RDEs are Condra (1993) and Hamada (1993, 1995a,b). Byrne and Quinlan (1993) present an interesting example to illustrate the concepts.

6. Product engineers, scientists, and statisticians can work more effectively together to develop experimental strategies for robust-design experiments for improving product performance and reliability. Those who follow the Taguchi approach seem to advocate running a larger number of different simple experiments (and the necessary confirmatory experiments) to obtain first-order improvements. Improvement is the goal. Traditional statistical/engineering approaches might recommend a more extensive sequential program of experimentation (perhaps requiring higher cost and more time) to gain more fundamental scientific understanding. The best approach depends on the potential for improvement, available resources, and long-term goals.

7. Field data are a vital resource. There will always be risk of failures with any product design. Field tracking is expensive and not always used. Warranty data usually have serious deficiencies and often come too late. Nevertheless, it is necessary to develop processes for the collection and appropriate use of field feedback to quickly discover and correct potential problems before they become widespread, thereby reducing overall risk. Field-data feedback should also be used to improve future designs. Important references describing the analysis of warranty and other field data include Amster, Brush, and Saperstein (1982), Suzuki (1985), Kalbfleisch and Lawless (1988), Robinson and McDonald (1991), Lawless and Kalbfleisch (1992), Lawless and Nadeau (1995), Lawless, Hu, and Cao (1995), Blischke and Murthy (1996), and Lawless (1998).

8. For field data, one general idea is to carefully monitor some number of units conveniently located in the field. To be most effective, the units should be operated in a reasonably use-intensive manner and failures should be reported promptly. All early failures should be analyzed carefully to determine cause, whether the same failure mode could be expected in the rest of the product population, and relevant actions that might be taken to eliminate the failure mode. Rather than just tracking failures, it is often useful, when practicable, to go out periodically and inspect and take measurements on field units (e.g., to assess degradation).

9. In some circumstances there is need to use limited reliability audit testing of ongoing production to catch the possible impact of changes in raw materials, supposedly innocuous design changes, and so on. This is especially important in today's manufacturing environment where many producers are often just assemblers, who rely heavily on components provided by vendors.

STATISTICS IS MUCH MORE THAN A COLLECTION OF FORMULAS

Those who have had one or two courses in statistics are often left with the impression that statistics is primarily a collection of analytical techniques and formulas. Choose the correct technique and formula and the problem is solved.

Viewed properly, statistics is the science of collecting and extracting useful information from raw data and of dealing with variability in quantitative information. Statistical tools provide the means for fitting and assessing the adequacy of models (physical or empirical). Statistical models are used to describe the relationships among variables as well as variability and uncertainty.

STATISTICS IS NOT MAGIC

The statistical methods described in this book are useful for planning reliability studies and extracting useful information from reliability data. Statistical methods also provide quantification of sampling uncertainty and allow planning statistical studies so that estimates and predictions can be obtained with a specified degree of statistical uncertainty.

There is, however, no magic in statistics. For example, we have heard more than once a question that asks something like this: "What kind of test can I run to demonstrate, with 95% confidence, that my system will have .99 reliability for its first year of operation when I only have, at most, three systems to test and the test has to be completed in two months?" The answer is that such a demonstration is impossible. Even if there were a given acceleration method providing an acceleration factor of six (which would be unlikely because for a complete system there are many failure modes with different acceleration factors and typically one cannot increase stress enough on a complete system to achieve an acceleration factor as large as six), one would need to test approximately 300 units for the two months with no failures to have a successful demonstration.

When demonstration of desired reliability is impossible, there is an inclination to take action on the basis of the best estimate. Then a minimal requirement would be that there be *no* failures. It is for this reason that it is easier to prove lack of reliability when you don't have it than it is to prove adequate reliability when you do have it.

Reliability demonstration for systems, once popular in the military and some other places, is difficult or impossible with today's higher reliability standards and cost sensitivity. In new industrial markets, "Reliability Assurance" processes are needed instead. The statistical methods in this book provide important tools for such Reliability Assurance processes. For more discussion of this topic, see Meeker and Hamada (1995).

APPENDIX A

Notation and Acronyms

In this book random variables are denoted by capital letters T, Y, and so on. A positive random variable is usually denoted by T and an unrestricted random variable is usually denoted by Y. Observed or realized values and dummy variables are denoted by lowercase letters t, y, and so on. Parameters are denoted by Greek letters such as β, θ, and γ. Estimates of these parameters are indicated by a hat as in $\widehat{\beta}$, $\widehat{\theta}$, and $\widehat{\gamma}$. We do not use different notation to distinguish between estimators, which are random variables, and estimates, which are the observed values of the estimators. Bootstrap-related quantities are denoted with an asterisk *; for example, $\widehat{\theta}^*$ is an estimate of θ obtained from a bootstrap sample. Posterior related quantities are denoted with a star $\star$; for example, $\widehat{\theta}^\star$ is an observation from the posterior density $f(\theta \mid \text{DATA})$. In an effort to use the same notation that is widely used and accepted in the statistical literature, there are a few symbols for which the use of a subscript or not indicates a difference in the meaning of the symbol (e.g., r versus r_i and β versus β_i). Also, notation defined and used locally, just in one section of the book, is not included in this appendix.

Some common symbols and their definitions follow.

- $\cap$ intersection of two sets [e.g., $A \cap B$ consists of elements that are in both A and B]
- $\cup$ union of two sets [e.g., $A \cup B$ consists of the elements that are in A or in B or both]
- $'$ matrix transpose operator
- $\square$ indicates a planning value used in choosing a test plan [e.g., $\theta^\square$ is a planning value for θ]
- $\widehat{}$ indicates an estimator [e.g., $\widehat{\theta}$ is an estimator of θ]
- $*$ indicates a bootstrap characteristic [e.g., $\widehat{\boldsymbol{\theta}}^*_j$ is the ML estimate of $\boldsymbol{\theta}$ computed from bootstrap sample j, DATA^*_j]
- $\star$ indicates a posterior characteristic [e.g., $\boldsymbol{\theta}^\star$ is an observation from the posterior $f(\boldsymbol{\theta} \mid \text{DATA})$]
- (j) indicates the jth order statistic [e.g., $t_{(1)}, \ldots, t_{(r)}$ are the first r failure times from a sample of n units]

[j]	indicates a vector with the single element j deleted [e.g., $\boldsymbol{\theta}_{[1]}$ is the vector $\boldsymbol{\theta}$ with the first element deleted]
$\sim$	indicates distributed as [e.g., $Z \sim \text{NOR}(0, 1)$]; but $[\mu, \widetilde{\mu}]$ is confidence interval for μ and $[\underset{\sim}{T}, \widetilde{T}]$ is prediction interval for T
$\dot{\sim}$	indicates *approximately* distributed [e.g., $Z_{\widehat{\theta}} \dot{\sim} \text{NOR}(0, 1)$]
$\approx$	indicates *approximately* equal to [e.g., $\Pr(Z_n \leq z_{(1-\alpha)}) \approx \Phi_{\text{nor}}(z_{(1-\alpha)})$]
α	significance level for a significance or hypothesis test; also $1 - \alpha$ is a confidence level for a statistical interval (e.g., a confidence interval)
β	Weibull distribution shape parameter (used on a limited basis)
$\beta_0, \beta_1, \ldots$	parameters in a linear or nonlinear regression model relating another parameter or response to explanatory variables
$\boldsymbol{\beta}$	vector of parameters; $\boldsymbol{\beta} = (\beta_1, \ldots, \beta_k)$
$\Gamma(\kappa)$	gamma function; $\Gamma(\kappa) = \int_0^\infty x^{\kappa-1} e^{-x} dx$
$\Gamma_I(z; \kappa)$	incomplete gamma probability function; $\Gamma_I(z; \kappa) = \int_0^z x^{\kappa-1} e^{-x} dx / \Gamma(\kappa)$
$\Gamma_I^{-1}(p; \kappa)$	inverse incomplete gamma probability function; $\Gamma_I[\Gamma_I^{-1}(p; \kappa); \kappa] = p$
γ	threshold parameter of the distribution of T [i.e., γ is the largest value such that $\Pr(T \leq \gamma) = 0$]
γ_2	coefficient of variation of a random variable; $\gamma_2 = \text{Var}(T)/\text{E}(T)$
γ_3	coefficient of skewness of a random variable; $\gamma_3 = \text{E}[(T - \text{E}(T))^2]/[\text{Var}(T)]^{3/2}$
$\gamma_0, \gamma_1, \ldots$	regression parameters in certain physical models
Δ	small positive quantity [e.g., $\pm \Delta_i$ represents roundoff error for observation i]
$\Delta_\mu(t)$	difference of mean cumulative functions at t; $\Delta_\mu(t) = \mu_1(t) - \mu_2(t)$
$\delta.(t_k)$	size of the risk set at time t_k in a sample of systems having recurrent events (such as repairs)
ϵ_{ij}	residual of observation j on unit i
ζ	standardized log time; $\zeta = [\log(t) - \mu]/\sigma$
ζ_c	standardized log censoring time; $\zeta_c = [\log(t_c) - \mu]/\sigma$
ζ_e	standardized log estimation time; $\zeta_e = [\log(t_e) - \mu]/\sigma$
η	Weibull scale parameter or characteristic life
Θ	parameter space [i.e., the set of all possible values of $\boldsymbol{\theta}$]
$\boldsymbol{\theta}$	vector parameter; $\boldsymbol{\theta} = (\theta_1, \ldots, \theta_r)$
$\boldsymbol{\theta}_1$	parameter vector subset [e.g., $\boldsymbol{\theta}_1$ is a subset of the elements of $\boldsymbol{\theta} = (\boldsymbol{\theta}_1, \boldsymbol{\theta}_2)$]
$\boldsymbol{\theta}_i$	parameters for the cdf of component i in a system
$\boldsymbol{\theta}_i^\star$	sample element i from the posterior, $f(\boldsymbol{\theta} \mid \text{DATA})$, for $\boldsymbol{\theta}$
θ	mean of the exponential distribution or a scale parameter for some other distributions

NOTATION AND ACRONYMS

κ	gamma distribution shape parameter
μ	location parameter of a location-scale distribution [e.g., mean of a normal distribution]
$\mu(t)$	mean cumulative function at time t; $\mu(t) = \mathrm{E}[N(t)]$
$\nu(t)$	system recurrence rate; $\nu(t) = d\mu(t)/dt$
$\boldsymbol{\pi}$	vector of unconditional failure probabilities; $\boldsymbol{\pi} = (\pi_1, \ldots, \pi_{m+1})$
π_i	multinomial probability (Chapters 2, 3); proportion allocated to experimental variable level or combination i (Chapter 20)
ρ	correlation between two random variables [e.g., $\rho_{\widehat{\theta}_1,\widehat{\theta}_2} = \mathrm{Cov}(\widehat{\theta}_1,\widehat{\theta}_2)/\sqrt{\mathrm{Var}(\widehat{\theta}_1)\mathrm{Var}(\widehat{\theta}_2)}$]
ρ_k	serial correlation (or autocorrelation) coefficient of lag-k
Σ	asymptotic covariance matrix [e.g., $\Sigma_{\widehat{\boldsymbol{\theta}}}$ is the asymptotic covariance matrix for $\widehat{\boldsymbol{\theta}}$]
σ	scale parameter for a location-scale distribution [e.g., standard deviation of a normal distribution]
τ_j	time elapsed between the $(j-1)$ and jth recurrence; $\tau_j = T_j - T_{j-1}$
Φ	cdf for a standardized location-scale distribution [i.e., $\mathrm{Pr}(Z \leq z) = \Phi(z)$, where $Z = (Y-\mu)/\sigma$]
$\Phi^{-1}(p)$	p quantile of random variable with cdf Φ
ϕ	pdf for a standardized location-scale distribution
$\chi^2_{(p;\nu)}$	p quantile for a chi-square random variable with ν degrees of freedom
$\mathcal{AF}$	acceleration factor [e.g., $\mathcal{AF}(\mathtt{temp},\mathtt{temp}_U,E_a)$ is the Arrhenius temperature acceleration factor and $\mathcal{AF}_{\mathrm{Ey}}(\mathtt{temp},\mathtt{temp}_U,E_a)$ is the Eyring temperature acceleration factor]
Acov	asymptotic covariance [e.g., $\mathrm{Acov}(\widehat{\theta}_1,\widehat{\theta}_2)$]
ADT	accelerated degradation test
ALT	accelerated life test
Ase	asymptotic standard error [e.g., $\mathrm{Ase}(\widehat{\theta}) = \sqrt{\mathrm{Avar}(\widehat{\theta})}$]
AT	accelerated test
Avar	asymptotic variance [e.g., $\mathrm{Avar}(\widehat{\theta})$]
B	number of bootstrap samples
BISA	Birnbaum–Saunders distribution
$\mathcal{C}$	parameter-free likelihood constant of proportionality
C	random censoring time
cdf	cumulative distribution function
Cov	covariance [e.g., $\mathrm{Cov}(\widehat{\theta}_1,\widehat{\theta}_2)$ is the covariance between $\widehat{\theta}_1$ and $\widehat{\theta}_2$]
$\mathcal{D}(t)$	level of degradation at time t
$\mathcal{D}_\mathrm{f}$	level of degradation defining failure [i.e., when $\mathcal{D}_\mathrm{f} > 0$, failure is defined to be the first time such that $\mathcal{D}(t) > \mathcal{D}_\mathrm{f}$; when $\mathcal{D}_\mathrm{f} < 0$, failure is the first time for which $\mathcal{D}(t) < \mathcal{D}_\mathrm{f}$]; $\mathcal{D}_\mathrm{f}$ could be either fixed or random

$\mathcal{D}_\infty$	asymptotic degradation level; $\mathcal{D}_\infty = \lim_{t\to\infty} \mathcal{D}(t)$
D	confidence interval half-width; $D = (\widetilde{g} - \underset{\sim}{g})/2$
D_T	specified target value for the confidence interval half-width
d_i	number of failures or deaths in observation interval i
$d_i(t_k)$	total number of recurrences for system i at time t_k
$d.(t_k)$	total number of recurrences for all systems at time t_k
$\bar{d}(t_k)$	average number of recurrences (across a population of systems) at time t_k
DATA	data for a complete sample of observational units
DATA$_j^*$	bootstrap sample j
data$_i$	data for the observational unit i in sample DATA
E	expected value or mean [e.g., E(T) is the expected value of T]
E_a	activation energy in units of electron volts (eV)
EGENG	extended generalized gamma distribution
$\mathcal{F}_{(p;\kappa,\nu)}$	p quantile of an F distribution with (κ, ν) degrees of freedom
$F(t)$	cdf of random variable [e.g., for T, $F(t) = \Pr(T \leq t)$]; when necessary, to distinguish among cdfs for different random variables we write, for example, $F_T(t) = \Pr(T \leq t)$
F_i	cdf of unit i in a system
F_T	cdf for a system
$f(t)$	pdf of random variable [e.g., for T, $f(t) = dF(t)/dt$]
$f(\boldsymbol{\theta} \mid \text{DATA})$	posterior pdf of $\boldsymbol{\theta}$ given the available data
$f(t \mid \text{DATA})$	posterior predictive pdf for T
g	real-valued function of $\boldsymbol{\theta}$ [e.g., $g(\boldsymbol{\theta})$]
$\boldsymbol{g}$	vector-valued function of $\boldsymbol{\theta}$ [e.g., $\boldsymbol{g}(\boldsymbol{\theta}) = [g_1(\boldsymbol{\theta}), g_2(\boldsymbol{\theta})]$]
GAM	gamma distribution
GENG	generalized gamma distribution
GENF	generalized F distribution
GETS	generalized threshold-scale distribution
GOMA	Gompertz–Makeham distribution
HPD	highest posterior density
HPP	homogeneous Poisson process
$h(t)$	hazard function of a random variable; $h(t) = f(t)/[1 - F(t)]$
$H(t)$	cumulative hazard function; $H(t) = \int_0^t h(x)\,dx$
hf	hazard function
I	Fisher information matrix [e.g., $I_{\boldsymbol{\theta}}$ is the Fisher information matrix for $\boldsymbol{\theta}$]
IC	integrated circuit
IGAU	inverse Gaussian distribution
iid	independent and identically distributed

NOTATION AND ACRONYMS 613

k_B	Boltzmann constant in units of electron volts per °C; $k_B = 8.6171 \times 10^{-5} \approx 1/11605$
k	number of explanatory variables [e.g., k is length of $\boldsymbol{\beta}$]
$\mathcal{L}$	log likelihood; $\mathcal{L} = \mathcal{L}(\boldsymbol{p}) = \mathcal{L}(\boldsymbol{p}; \text{DATA}) = \log[L(\boldsymbol{p})]$, or $\mathcal{L}(\boldsymbol{\theta}) = \mathcal{L}(\boldsymbol{\theta}; \text{DATA}) = \log[L(\boldsymbol{\theta})]$
$\mathcal{L}_i$	contribution of observation i to the log likelihood
L	likelihood [i.e., $L(\boldsymbol{p}) = L(\boldsymbol{p}; \text{DATA})$]
ℓ_i	number of observations left-censored at the upper endpoint of observation interval i
LEV	largest extreme value distribution
lev	indicates a standardized largest extreme value distribution [e.g., Φ_{lev} is the standardized largest extreme value cdf]
LFP	limited failure population
LOGIS	logistic distribution
logis	indicates a standardized logistic distribution [e.g., Φ_{logis} is the standardized logistic cdf]
logit	logistic transformation; $\text{logit}(p) = \log[p/(1-p)]$
LOGLOGIS	loglogistic distribution
LOGNOR	lognormal distribution
M	number of observations from prior $f(\boldsymbol{\theta})$
$M^\star$	number of observations from the posterior $f(\boldsymbol{\theta} \mid \text{DATA})$
MCF	mean cumulative function
ML	maximum likelihood
MTBF	mean time between failures in a repairable system
MTTF	mean time to failure for a replaceable unit
m	number of observation intervals; also, in Chapters 12 and 14 indicates the number of iid observations in a future sample
N	simulation size when using Monte Carlo to evaluate a complicated cdf
$N(s,t)$	cumulative number of recurrences in the interval (s,t)
$N(t)$	cumulative number of recurrences in the interval $(0,t)$ [i.e., $N(t) = N(0,t)$]
NHPP	nonhomogeneous Poisson process
NOR	normal distribution [e.g., $\text{NOR}(\mu, \sigma)$ indicates a normal distribution with mean μ and standard deviation σ]
nor	indicates a normal distribution [e.g., Φ_{nor} is the standardized normal cdf]
n	sample size
n_i	size of the risk set at the beginning of interval i
PH	proportional hazards

Pr	probability
p	conditional probability of failure in an interval (Chapters 2, 3) [e.g., $p_i = \Pr(t_{i-1} < T \leq t_i \mid T > t_{i-1})$]; more generally a probability or a proportion
$\boldsymbol{p}$	vector of conditional failure probabilities; $\boldsymbol{p} = (p_1, \ldots, p_m)$
pdf	probability density function [i.e., the derivative of the cdf]
Q	log-likelihood-ratio statistic; $Q = -2(\mathcal{L}_{\text{const}} - \mathcal{L}_{\text{unconst}})$
q	$q = 1 - p$ is conditional probability of survival in an interval
$\mathcal{R}_i$	rate of chemical reaction i
R	measure of precision for a confidence interval of a positive parameter; $R = \widetilde{g}/\widehat{g} = \widehat{g}/\underset{\sim}{g}$
R_T	target value for R
$R(\boldsymbol{\theta}_1)$	profile likelihood for parameter vector subset $\boldsymbol{\theta}_1$
r	number of failures out of n observations; also, number of parameters [e.g., r is the length of $\boldsymbol{\theta}$]
r_1	length of parameter vector subset $\boldsymbol{\theta}_1$
r_i	number of observations right-censored at the upper endpoint of the observation interval i
s	number of components in a multiple-component system
SAFT	scale-accelerated failure time
SD	standard deviation [e.g., $\text{SD}(T) = \sqrt{\text{Var}(T)}$]
se	standard error of an estimator [e.g., $\text{se}_{\widehat{\theta}}$ is the standard error of $\widehat{\theta}$]
$\widehat{\text{se}}$	estimated standard error
SEV	smallest extreme value distribution
sev	indicates a standardized smallest extreme value distribution [e.g., Φ_{sev} is the standardized smallest extreme value cdf]
$S(t)$	survival function; $S(t) = 1 - F(t)$
$S_i(t)$	survival function for component i in a system; $S_i(t) = 1 - F_i(t)$
T	positive random variable (usually time)
TTT	total time on test
t	a reported time (e.g., realization of T); also a dummy variable
t_c	specified censoring time
t_e	specified estimation time [e.g., time at which $F(t)$ or $h(t)$ is to be estimated]
t_i	upper endpoint of observation interval i or an exact observation
t_{i-1}	lower endpoint of observation interval i
$t_{(i)}$	ith order statistics [i.e., ith largest failure time in a sample]
t_{ij}	recurrence (e.g., failure, repair, or other event) time j for system i
t_p	p quantile of the random variable T; $F(t_p) = p$
$t_{.5}$	median of the random variable T; $F(t_{.5}) = .5$

$t_{(p;\nu)}$	p quantile for a random variable with Student's t cdf and ν degrees of freedom
`temp °C`	temperature in degrees Celsius
`temp K`	absolute temperature, Kelvin scale; `temp K` = `temp °C` + 273.15
`temp`$_U$ `°C`	temperature at use conditions in degrees Celsius
UNIF	uniform distribution [e.g., $Y \sim \text{UNIF}(a,b)$]
V	asymptotic variance or covariance factor [e.g., $V_{\widehat{\theta}_1, \widehat{\theta}_2} = n\text{Acov}(\widehat{\theta}_1, \widehat{\theta}_2)$ and $V_{\widehat{\theta}} = n\text{Avar}(\widehat{\theta})$]
Var	variance [e.g., $\text{Var}(T)$ is the variance of T]
WEIB	Weibull distribution
X^2_{MHB}	*Military Handbook* test statistic for trend in times between failures in a repairable system
x	vector of explanatory variables; $x = (x_1, \ldots, x_k)$
Y	unrestricted random variable, $-\infty < Y < \infty$ [e.g., $Y = \log(T)$]
y	a realization of Y; also a dummy variable
y_{ij}	observation j on unit i
y_p	p quantile of Y
Z	standard random variable [e.g., $Z = (Y - \mu)/\sigma$]
$Z_{\widehat{g}}$	studentized random variable for some parameter of function g of the parameters [e.g., $Z_{\widehat{g}} = (\widehat{g} - g)/\widehat{\text{se}}_{\widehat{g}}$]; used to derive approximate confidence intervals for $g = g(\boldsymbol{\theta})$
$Z_{\widehat{g}^*}$	studentized bootstrap statistic; $Z_{\widehat{g}^*} = (\widehat{g}^* - \widehat{g})/\widehat{\text{se}}_{\widehat{g}^*}$, where $\widehat{g}^* = g(\widehat{\boldsymbol{\theta}}^*)$; used to approximate the distribution of $Z_{\widehat{g}}$
Z_{LP}	Laplace test statistic for trend in times between failures in a repairable system
Z_{LR}	Lewis–Robinson test for trend in times between failures in a repairable system
$z_{(p)}$	p quantile of a standard normal; $\Phi_{\text{nor}}(z_{(p)}) = p$
$z_{\widehat{g}^*_{(p)}}$	p quantile of the bootstrap distribution of $Z_{\widehat{g}^*}$

APPENDIX B

Some Results from Statistical Theory

This appendix provides some useful tools and results from statistical theory. These tools facilitate the justification and extension of much of the methodology in the book. Section B.1 gives the basic theory on transformation of random variables that is used mainly in Chapters 4, 5, and 14. Section B.2 describes the "delta method," a useful method to obtain expressions for approximate variances of a function of random quantities as a function of the variances and covariances of the function arguments. Section B.3 gives a precise definition of expected and observed information matrices. Section B.4 lists general regularity conditions assumed in most of the book. Section B.5 provides the definition of convergence in distribution for random variables and gives examples of its use in this book. Section B.6 outlines general theory for ML estimation.

B.1 cdfs AND pdfs OF FUNCTIONS OF RANDOM VARIABLES

This section shows how to obtain expressions for the pdf and cdf of functions of random variables. Let U be a k-dimensional continuous random vector with pdf $f_U(u)$. We consider a k-dimensional transformation $V = g(U)$ with the following properties:

1. The function $v = g(u) = [g_1(u), \ldots, g_k(u)]$ is a one-to-one transformation.
2. The inverse function $u = g^{-1}(v) = [g_1^{-1}(v), \ldots, g_k^{-1}(v)]$ has continuous first partial derivatives with respect to v.
3. The Jacobian $J(v)$ of $g^{-1}(v)$ is nonzero, where

$$J(v) = \det \begin{vmatrix} \frac{\partial g_1^{-1}(v)}{\partial v_1} & \cdots & \frac{\partial g_k^{-1}(v)}{\partial v_1} \\ \vdots & \vdots & \vdots \\ \frac{\partial g_1^{-1}(v)}{\partial v_k} & \cdots & \frac{\partial g_k^{-1}(v)}{\partial v_k} \end{vmatrix}.$$

Then the pdf and cdf of V are

$$f_V(v) = f_U[\mathbf{g}^{-1}(v)]\,|J(v)|,$$

$$F_V(v) = \int_{\mathbf{x} \le v} f_U[\mathbf{g}^{-1}(\mathbf{x})]\,|J(\mathbf{x})|\,d\mathbf{x}.$$

For the scalar case (i.e., $k = 1$) the formulas simplify to

$$f_V(v) = f_U[g^{-1}(v)]\left|\frac{dg^{-1}(v)}{dv}\right|,$$

$$F_V(v) = \int_{-\infty}^{v} f_U[g^{-1}(x)]\left|\frac{dg^{-1}(x)}{dx}\right|dx$$

$$= \begin{cases} F_U\left[g^{-1}(v)\right] & \text{if } g \text{ is increasing} \\ 1 - F_U\left[g^{-1}(v)\right] & \text{if } g \text{ is decreasing.} \end{cases}$$

For illustration, consider the following special cases:

1. A one-dimensional transformation (i.e., $k = 1$). Let $U \sim \text{NOR}(\mu, \sigma)$ and consider the transformation $V = \exp(U)$. Then

$$f_U(u) = \frac{1}{\sigma}\phi_{\text{nor}}\left(\frac{u-\mu}{\sigma}\right), \quad -\infty < u < \infty,$$

and $g(u) = \exp(u)$, which implies that $g^{-1}(v) = \log(v)$. Consequently, $J(v) = 1/v$ and

$$f_V(v) = \frac{1}{v}f_U[\log(v)] = \frac{1}{v\sigma}\phi_{\text{nor}}\left[\frac{\log(v)-\mu}{\sigma}\right], \quad v > 0,$$

$$F_V(v) = \int_0^v \frac{1}{x\sigma}\phi_{\text{nor}}\left[\frac{\log(x)-\mu}{\sigma}\right]dx = \Phi_{\text{nor}}\left[\frac{\log(v)-\mu}{\sigma}\right], \quad v > 0.$$

Note that V has a LOGNOR(μ, σ) distribution and in the notation of Chapter 14, $f(v) = (1/v)f[\log(v)]$.

2. A bivariate transformation (i.e., $k = 2$). Let $\boldsymbol{U} = (U_1, U_2)$, where U_1 and U_2 are independent, $U_1 \sim \text{UNIF}[\log(a_1), \log(b_1)]$, and $U_2 \sim \text{NOR}(a_0, b_0)$. Consider finding the distribution of $\boldsymbol{V} = (V_1, V_2) = [U_1 - \Phi_{\text{sev}}^{-1}(p)\exp(U_2), \exp(U_2)]$.
   In view of the independence of U_1 and U_2,

$$f_U(u_1, u_2) = \frac{1}{\log(b_1/a_1)} \times \frac{1}{b_0}\phi_{\text{nor}}\left[\frac{u_2 - a_0}{b_0}\right],$$

where $\log(a_1) \le u_1 \le \log(b_1)$ and $-\infty < u_2 < \infty$. Using $\boldsymbol{v} = (v_1, v_2)$, direct computations give $g_1^{-1}(\boldsymbol{v}) = v_1 + \Phi_{\text{sev}}^{-1}(p)v_2$, $g_2^{-1}(\boldsymbol{v}) = \log(v_2)$, and $J(\boldsymbol{v}) = 1/v_2$. Thus

$$f_V(v_1, v_2) = \frac{1}{\log(b_1/a_1)} \times \frac{1}{v_2 b_0}\phi_{\text{nor}}\left[\frac{\log(v_2) - a_0}{b_0}\right],$$

where $\log(a_1) - \Phi_{\text{sev}}^{-1}(p)v_2 \le v_1 \le \log(b_1) - \Phi_{\text{sev}}^{-1}(p)v_2$, $v_2 > 0$. This is the same result as (14.2) with $U_1 = \log(t_p)$, $U_2 = \log(\sigma)$, $V_1 = \mu$, and $V_2 = \sigma$.

B.2 STATISTICAL ERROR PROPAGATION— THE DELTA METHOD

This section shows how to compute approximate expected values, variances, and covariances of functions of parameter estimators. Let $g(\boldsymbol{\theta})$ be a real-valued function of the parameters $\boldsymbol{\theta} = (\theta_1, \ldots, \theta_r)'$ and let $\widehat{\boldsymbol{\theta}} = (\widehat{\theta}_1, \ldots, \widehat{\theta}_r)'$ and $g(\widehat{\boldsymbol{\theta}})$ be estimates of $\boldsymbol{\theta}$ and $g(\boldsymbol{\theta})$, respectively. The objective is to obtain expressions or approximate expressions for $\mathrm{E}[g(\widehat{\boldsymbol{\theta}})]$ and $\mathrm{Var}[g(\widehat{\boldsymbol{\theta}})]$ as a function of $\mathrm{E}(\widehat{\theta}_i)$, $\mathrm{Var}(\widehat{\theta}_i)$, and $\mathrm{Cov}(\widehat{\theta}_i, \widehat{\theta}_j)$.

The simplest case is when $g(\widehat{\boldsymbol{\theta}})$ is a linear function of the $\widehat{\theta}_i$, say, $g(\widehat{\boldsymbol{\theta}}) = a_0 + \sum_{i=1}^{r} a_i \widehat{\theta}_i$, where the a_i are constants. To facilitate the development express $g(\widehat{\boldsymbol{\theta}})$ as

$$g(\widehat{\boldsymbol{\theta}}) = a_0 + \sum_{i=1}^{r} a_i \widehat{\theta}_i = b_0 + \sum_{i=1}^{r} b_i [\widehat{\theta}_i - \mathrm{E}(\widehat{\theta}_i)] \tag{B.1}$$

where $b_0 = a_0 + \sum_{i=1}^{r} a_i \mathrm{E}(\widehat{\theta}_i)$ and $b_i = a_i$, $i = 1, \ldots, r$. In this case, simple computations with expectations and variances give

$$\mathrm{E}[g(\widehat{\boldsymbol{\theta}})] = b_0,$$

$$\mathrm{Var}[g(\widehat{\boldsymbol{\theta}})] = \sum_{i=1}^{r} b_i^2 \, \mathrm{Var}(\widehat{\theta}_i) + \sum_{i=1}^{r} \sum_{\substack{j=1 \\ j \neq i}}^{r} b_i b_j \, \mathrm{Cov}(\widehat{\theta}_i, \widehat{\theta}_j).$$

When $g(\widehat{\boldsymbol{\theta}})$ is a smooth nonlinear function of the $\widehat{\theta}_i$ values and $g(\widehat{\boldsymbol{\theta}})$ can be approximated by a linear function of the $\widehat{\theta}_i$ values in the region with nonnegligible likelihood, it is still possible to apply the methodology above. The general procedure is known as the "delta method" or "statistical error propagation" and here we describe a simplified version of the methodology. For a more detailed account, see Hahn and Shapiro (1967, page 228) or Stuart and Ord (1994, page 350).

When $g(\boldsymbol{\theta})$ has continuous second partial derivatives with respect to $\boldsymbol{\theta}$, a first-order (i.e., keeping linear terms only) Taylor series expansion of $g(\widehat{\boldsymbol{\theta}})$ about $\boldsymbol{\mu} = [\mathrm{E}(\widehat{\theta}_1), \ldots, \mathrm{E}(\widehat{\theta}_r)]$ is given by

$$g(\widehat{\boldsymbol{\theta}}) \approx g(\boldsymbol{\mu}) + \sum_{i=1}^{r} \frac{\partial g(\boldsymbol{\theta})}{\partial \theta_i} [\widehat{\theta}_i - \mathrm{E}(\widehat{\theta}_i)], \tag{B.2}$$

where the partial derivatives of $g(\boldsymbol{\theta})$ with respect to the θ_i values are evaluated at $\boldsymbol{\mu}$.

Observe that equation (B.2) looks like equation (B.1) with

$$b_0 = g(\boldsymbol{\mu}) \quad \text{and} \quad b_i = \frac{\partial g(\boldsymbol{\theta})}{\partial \theta_i}, \quad i = 1, \ldots, r.$$

Consequently,

$$E[g(\widehat{\boldsymbol{\theta}})] \approx g(\boldsymbol{\mu}),$$

$$\text{Var}[g(\widehat{\boldsymbol{\theta}})] \approx \sum_{i=1}^{r} \left[\frac{\partial g(\boldsymbol{\theta})}{\partial \theta_i}\right]^2 \text{Var}(\widehat{\theta}_i)$$

$$+ \sum_{i=1}^{r} \sum_{\substack{j=1 \\ j \neq i}}^{r} \left[\frac{\partial g(\boldsymbol{\theta})}{\partial \theta_i}\right] \left[\frac{\partial g(\boldsymbol{\theta})}{\partial \theta_j}\right] \text{Cov}(\widehat{\theta}_i, \widehat{\theta}_j). \quad \text{(B.3)}$$

When the $\widehat{\theta}_i$ values are uncorrelated or when the covariances $\text{Cov}(\widehat{\theta}_i, \widehat{\theta}_j)$, $i \neq j$, are small when compared with the variances $\text{Var}(\widehat{\theta}_i)$, the last term on the right of equation (B.3) is usually omitted from the approximation.

The same ideas apply to vector-valued functions. For example, if $g_1(\boldsymbol{\theta})$ and $g_2(\boldsymbol{\theta})$ are two real-valued functions then

$$\text{Cov}[g_1(\widehat{\boldsymbol{\theta}}), g_2(\widehat{\boldsymbol{\theta}})] \approx \sum_{i=1}^{r} \left[\frac{\partial g_1(\boldsymbol{\theta})}{\partial \theta_i}\right] \left[\frac{\partial g_2(\boldsymbol{\theta})}{\partial \theta_i}\right] \text{Var}(\widehat{\theta}_i)$$

$$+ \sum_{i=1}^{r} \sum_{\substack{j=1 \\ j \neq i}}^{r} \left[\frac{\partial g_1(\boldsymbol{\theta})}{\partial \theta_i}\right] \left[\frac{\partial g_2(\boldsymbol{\theta})}{\partial \theta_j}\right] \text{Cov}(\widehat{\theta}_i, \widehat{\theta}_j). \quad \text{(B.4)}$$

In general, for a vector-valued function $\boldsymbol{g}(\boldsymbol{\theta})$ of the parameters such that all the second partial derivatives with respect to the elements of $\boldsymbol{\theta}$ are continuous

$$\text{Var}[\boldsymbol{g}(\widehat{\boldsymbol{\theta}})] \approx \left[\frac{\partial \boldsymbol{g}(\boldsymbol{\theta})}{\partial \boldsymbol{\theta}}\right]' \text{Var}(\widehat{\boldsymbol{\theta}}) \left[\frac{\partial \boldsymbol{g}(\boldsymbol{\theta})}{\partial \boldsymbol{\theta}}\right],$$

where $\partial \boldsymbol{g}(\boldsymbol{\theta})/\partial \boldsymbol{\theta} = [\partial g_1(\boldsymbol{\theta})/\partial \boldsymbol{\theta}, \partial g_2(\boldsymbol{\theta})/\partial \boldsymbol{\theta}, \ldots]$ is the matrix of gradient vectors of first partial derivatives of $\boldsymbol{g}(\boldsymbol{\theta})$ with respect to $\boldsymbol{\theta}$ and

$$\text{Var}(\widehat{\boldsymbol{\theta}}) = \begin{bmatrix} \text{Var}(\widehat{\theta}_1) & \text{Cov}(\widehat{\theta}_1, \widehat{\theta}_2) & \cdots & \text{Cov}(\widehat{\theta}_1, \widehat{\theta}_r) \\ & \text{Var}(\widehat{\theta}_2) & \cdots & \text{Cov}(\widehat{\theta}_2, \widehat{\theta}_r) \\ & & \ddots & \vdots \\ \text{symmetric} & & & \text{Var}(\widehat{\theta}_r) \end{bmatrix}$$

both evaluated at $\boldsymbol{\theta}$.

The delta method can provide good approximations for $E[g(\widehat{\boldsymbol{\theta}})]$ and $\text{Var}[g(\widehat{\boldsymbol{\theta}})]$. However, as indicated in more advanced textbooks, one needs to exercise caution in applying this method because the adequacy of the approximation depends on the validity of the Taylor approximation and the size of the remainder in the approximation. Simulation can be used to check the adequacy of the approximation.

REGULARITY CONDITIONS   621

B.3 LIKELIHOOD AND FISHER INFORMATION MATRICES

Let $\mathcal{L}(\boldsymbol{\theta}) = \sum_{i=1}^{n} \mathcal{L}_i(\boldsymbol{\theta})$ denote the total log likelihood for a specified model and data that will consist of n independent but not necessarily identically distributed observations. Here it is understood that $\mathcal{L}_i(\boldsymbol{\theta})$ is the contribution of the ith observation to the total log likelihood. Let $\widehat{\boldsymbol{\theta}}$ be the ML estimator of $\boldsymbol{\theta}$ with a sample of size n. This $\widehat{\boldsymbol{\theta}}$, when it exists, is the value of $\boldsymbol{\theta}$ that maximizes $\mathcal{L}(\boldsymbol{\theta})$. Let $\mathcal{I}(\boldsymbol{\theta})$ denote the large-sample (or limiting) average amount of information per observation. Then, in general,

$$\mathcal{I}(\boldsymbol{\theta}) = \lim_{n \to \infty} \left\{ \frac{1}{n} \mathrm{E}\left[-\frac{\partial^2 \mathcal{L}(\boldsymbol{\theta})}{\partial \boldsymbol{\theta} \, \partial \boldsymbol{\theta}'} \right] \right\} = \lim_{n \to \infty} \left\{ \frac{1}{n} \sum_{i=1}^{n} \mathrm{E}\left[-\frac{\partial^2 \mathcal{L}_i(\boldsymbol{\theta})}{\partial \boldsymbol{\theta} \, \partial \boldsymbol{\theta}'} \right], \right\} \quad \text{(B.5)}$$

where the expectation is with respect to the as of yet unobserved data. For large samples, the matrix $I_{\boldsymbol{\theta}} = n\mathcal{I}(\boldsymbol{\theta})$ approximately quantifies the amount of information that we "expect" to get from our future data. Intuitively, this can be seen because larger second derivatives of $\mathcal{L}(\boldsymbol{\theta})$ indicate more curvature in the likelihood, implying that the likelihood is more concentrated about its maximum. For a large class of model situations, including models with independent and identically distributed observations, $I_{\boldsymbol{\theta}}$ simplifies to the well-known Fisher information matrix for $\boldsymbol{\theta}$,

$$I_{\boldsymbol{\theta}} = \mathrm{E}\left[-\frac{\partial^2 \mathcal{L}(\boldsymbol{\theta})}{\partial \boldsymbol{\theta} \, \partial \boldsymbol{\theta}'} \right] = \sum_{i=1}^{n} \mathrm{E}\left[-\frac{\partial^2 \mathcal{L}_i(\boldsymbol{\theta})}{\partial \boldsymbol{\theta} \, \partial \boldsymbol{\theta}'} \right]. \quad \text{(B.6)}$$

$I_{\boldsymbol{\theta}}$ is often known as the Fisher information or "expected information" matrix for $\boldsymbol{\theta}$. When data are available, one can compute the "local" (or "observed information") matrix for $\boldsymbol{\theta}$ as

$$\widehat{I}_{\boldsymbol{\theta}} = -\frac{\partial^2 \mathcal{L}(\boldsymbol{\theta})}{\partial \boldsymbol{\theta} \, \partial \boldsymbol{\theta}'} = \sum_{i=1}^{n} \left[-\frac{\partial^2 \mathcal{L}_i(\boldsymbol{\theta})}{\partial \boldsymbol{\theta} \, \partial \boldsymbol{\theta}'} \right], \quad \text{(B.7)}$$

where the derivatives are evaluated at $\boldsymbol{\theta} = \widehat{\boldsymbol{\theta}}$.

In Section B.6.1, we explain that, under the standard regularity conditions, $n\Sigma_{\widehat{\boldsymbol{\theta}}} = n(I_{\boldsymbol{\theta}})^{-1}$ is the covariance matrix for the asymptotic distribution of $\sqrt{n}(\widehat{\boldsymbol{\theta}} - \boldsymbol{\theta})$ and an estimate of $I_{\boldsymbol{\theta}}$ can be used to estimate sampling variability in $\widehat{\boldsymbol{\theta}}$.

B.4 REGULARITY CONDITIONS

Each technical asymptotic result, such as the asymptotic distribution of an estimator, or a specific asymptotic property of an estimator, requires its own set of conditions on the model. For example, under a certain set of conditions it is possible to show that ML estimators are asymptotically normal. With additional conditions, it can be shown that ML estimators are also asymptotically efficient. The model, in this case, includes the underlying probability model for the process (e.g., a failure-time process) and for the observations process, such as inspections (when there is not continuous inspection) and characteristics of the censoring process. Lehmann (1983, Chapter 6), for example, gives precise regularity conditions in the context of "continuous inspection." Rao (1973, Section 5e) does the same assuming an underlying discrete multinomial observation scheme, like that outlined in Chapter 2. Although censoring is not explicitly treated in either of these references, the same asymptotic results hold under the standard kinds of noninformative censoring mechanisms as long as the average amount of information per sample [elements of $\mathcal{I}(\boldsymbol{\theta})$] does not decrease substantially as the sample size increases. For a modern and rigorous treatment of the asymptotic properties of ML estimators based on Type II censored data see Bhattacharyya (1985).

For a large set of cases known as "regular" cases, there are useful asymptotic results (see Appendix Section B.6) that apply when the pdf of T, $f(t; \boldsymbol{\theta})$ (or the pdf of a monotone transformation of T), satisfies certain conditions discussed below.

B.4.1 Regularity Conditions for Location-Scale Distributions

When Y [or a transformation of T such as $Y = \log(T)$] is location-scale with pdf, $f_Y(y; \boldsymbol{\theta}) = (1/\sigma)\phi[(y - \mu)/\sigma]$, $\boldsymbol{\theta} = (\mu, \sigma)$, $-\infty < y < \infty$, $-\infty < \mu < \infty$, $\sigma > 0$, the "regularity" conditions can be expressed as follows:

- $\phi(z) > 0$ for all $-\infty < z < \infty$.
- The following limits hold:

$$\lim_{z \to \pm\infty} \left[z^2 \times \frac{\partial \phi(z)}{\partial z} \right] = 0.$$

- The second derivative $\partial^2 \phi(z)/\partial z^2$ is continuous.
- The matrix

$$E\left\{ -\frac{\partial^2 \log[\phi(z)]}{\partial \boldsymbol{\theta} \, \partial \boldsymbol{\theta}'} \right\}$$

is positive definite and all its elements are finite.

These conditions are satisfied by the normal (lognormal), SEV (Weibull), and logistic (loglogistic) distributions. But they are not satisfied by distributions with a threshold parameter (see Section 5.10.1) because in these cases the points at which $f_Y(y; \boldsymbol{\theta}) > 0$ depend on the values of $\boldsymbol{\theta}$.

B.4.2 General Regularity Conditions

When T (or a monotone transformation of T) is not location-scale, an alternative set of regularity conditions are:

CONVERGENCE IN DISTRIBUTION 623

- The points t at which $f(t; \boldsymbol{\theta}) > 0$ do not depend on $\boldsymbol{\theta}$.
- The parameters are identifiable in the sense that $\boldsymbol{\theta}_1 \neq \boldsymbol{\theta}_2$ implies that the probability functions defined by $f(t; \boldsymbol{\theta}_1)$ and $f(t; \boldsymbol{\theta}_2)$ are not identically equal.
- The true parameter value $\boldsymbol{\theta}$ is in the interior of the parameter space Θ.
- The density $f(t; \boldsymbol{\theta})$ has third mixed partial derivatives with respect to the elements of $\boldsymbol{\theta}$ in a neighborhood of the true $\boldsymbol{\theta}$. Each one of these derivatives is bounded by a function that has finite expectations with respect to $f(t; \boldsymbol{\theta})$.
- For all $\boldsymbol{\theta}$ in a neighborhood of the true $\boldsymbol{\theta}$,

$$\mathrm{E}\left[\frac{\partial^2 \log f(t; \boldsymbol{\theta})}{\partial \boldsymbol{\theta}\, \partial \boldsymbol{\theta}'}\right] = \frac{\partial^2 \mathrm{E}[\log f(t; \boldsymbol{\theta})]}{\partial \boldsymbol{\theta}\, \partial \boldsymbol{\theta}'}$$

where the expectations are with respect to the data from $f(t; \boldsymbol{\theta})$.
- The elements of $\mathcal{I}(\boldsymbol{\theta})$ defined in (B.5) are finite and $\mathcal{I}(\boldsymbol{\theta})$ is positive definite.

These conditions are satisfied, for example, by the GAM, GENG, and BISA distributions of Chapter 5.

B.4.3 Asymptotic Theory for Nonregular Models

The standard regularity conditions hold for most of the models used in this book. One model for which the regularity conditions do not hold (such models are called "nonregular") is the threshold parameter distributions for which the range over which $f(t; \boldsymbol{\theta}) > 0$ depends on $\boldsymbol{\theta}$ (see Section 5.10.1). Having $\boldsymbol{\theta}$ on the boundary of Θ also leads to "nonregular" estimation. ML methods are still very useful for "nonregular" situations, but the statistical properties and asymptotic behavior in these cases are more complicated (e.g., limiting distributions may depend on $\boldsymbol{\theta}$). For such situations, it is still possible to find useful large-sample asymptotic results; see, for example, Smith (1985) and Woodroofe (1972, 1974).

B.5 CONVERGENCE IN DISTRIBUTION

In this section we use a subscript n to identify explicitly an estimator or quantity with properties that depend on the sample size n. Considering the sequence for increasing n facilitates the description of these properties when n gets large (i.e., when $n \to \infty$).

Convergence in distribution is an important concept for describing the behavior of estimators in large samples. For example, one is often interested in the statistical properties of the ML estimates $\widehat{\theta}_n$ of the scalar θ when the sample size n increases. In this case a common approach is to consider the studentized ratios

$$Z_n = Z_n(\theta) = \frac{\widehat{\theta}_n - \theta}{\widehat{\mathrm{se}}_{\widehat{\theta}_n}}, \quad n = 2, \ldots, \tag{B.8}$$

where $\widehat{\text{se}}_{\widehat{\theta}_n}$ is a consistent estimator of $\text{se}_{\widehat{\theta}_n}$. In general, the exact distribution of Z_n is complicated, depending on the model, actual parameter values, and sample size. But under the regularity conditions of Section B.4, if $Z_n(\theta)$ is evaluated at the true θ, then for all z

$$\lim_{n \to \infty} F_{Z_n}(z) = \Phi_{\text{nor}}(z).$$

Thus, for finite n, one can use the approximation

$$\Pr\left[z_{(\alpha/2)} < Z_n \leq z_{(1-\alpha/2)}\right] = F_{Z_n}[z_{(1-\alpha/2)}] - F_{Z_n}[z_{(\alpha/2)}]$$
$$\approx \Phi_{\text{nor}}[z_{(1-\alpha/2)}] - \Phi_{\text{nor}}[z_{(\alpha/2)}] = 1 - \alpha.$$

The adequacy of this approximation has to be studied (e.g., by simulation) for each individual problem but in general it works well for a large class of problems and moderate-to-large sample sizes.

More generally, we say that the sequence of scalars Z_n converges in distribution to the continuous random variable V if

$$\lim_{n \to \infty} F_{Z_n}(z) = F_V(z) \quad \text{for all } z,$$

where $F_V(z)$ is the cdf of V. Thus one can use the limiting distribution F_V to approximate the probabilities for finite n as follows:

$$\Pr(a < Z_n \leq b) = F_{Z_n}(b) - F_{Z_n}(a) \approx F_V(b) - F_V(a)$$

where a and b are specified constants. This approximation can be made as close as desired by taking large values of n. These ideas of convergence in distribution generalize to vector random variables; see, for example, Billingsley (1986, page 390).

For other examples, let $\widehat{\boldsymbol{\theta}}_n = (\widehat{\boldsymbol{\theta}}_{1n}, \widehat{\boldsymbol{\theta}}_{2n})$ be the ML estimate of a vector $\boldsymbol{\theta} = (\boldsymbol{\theta}_1, \boldsymbol{\theta}_2)$ with a sample of size n and suppose that the appropriate regularity conditions (Appendix Section B.4) hold.

- The profile likelihood of $\boldsymbol{\theta}_1$ is

$$R_n(\boldsymbol{\theta}_1) = \max_{\boldsymbol{\theta}_2} \left[\frac{L(\boldsymbol{\theta}_1, \boldsymbol{\theta}_2)}{L(\widehat{\boldsymbol{\theta}}_n)}\right].$$

The corresponding parameter subset log-likelihood-ratio statistic is $\text{LLR}_n(\boldsymbol{\theta}_1) = -2 \log[R_n(\boldsymbol{\theta}_1)]$. This statistic, when evaluated at the true $\boldsymbol{\theta}_1$, converges in distribution to a chi-square distribution with r_1 degrees of freedom, where r_1 is the number of parameters in $\boldsymbol{\theta}_1$.

- The parameter subset "Wald statistic" is

$$W_n(\boldsymbol{\theta}_1) = (\widehat{\boldsymbol{\theta}}_{1n} - \boldsymbol{\theta}_1)' \left(\widehat{\boldsymbol{\Sigma}}_{\widehat{\boldsymbol{\theta}}_{1n}}\right)^{-1} (\widehat{\boldsymbol{\theta}}_{1n} - \boldsymbol{\theta}_1).$$

$W_n(\boldsymbol{\theta}_1)$, evaluated at the true $\boldsymbol{\theta}_1$, converges in distribution to a chi-square random variable with r_1 degrees of freedom.

B.6 OUTLINE OF GENERAL ML THEORY

B.6.1 Asymptotic Distribution of ML Estimators

In this section, we assume that $\widehat{\boldsymbol{\theta}}$ is the ML estimate of $\boldsymbol{\theta}$ based on n observations and that the regularity conditions given in Section B.4 hold. Then it can be shown that $\sqrt{n}(\widehat{\boldsymbol{\theta}} - \boldsymbol{\theta})$ converges in distribution to a multivariate normal with mean zero and covariance matrix $\mathcal{I}^{-1}(\boldsymbol{\theta})$ where $\mathcal{I}(\boldsymbol{\theta})$ is defined in (B.5). In a convenient casual wording, we say that $\widehat{\boldsymbol{\theta}}$ is approximately normal with mean $\boldsymbol{\theta}$ and covariance matrix $\Sigma_{\widehat{\boldsymbol{\theta}}} = I_{\boldsymbol{\theta}}^{-1}$, where $I_{\boldsymbol{\theta}} = n\mathcal{I}(\boldsymbol{\theta})$. Asymptotic (large-sample) statistical theory shows that, under the standard regularity conditions, the elements of $\Sigma_{\widehat{\boldsymbol{\theta}}}$ are of the order of n^{-1}. This can be seen by noting that $n\Sigma_{\widehat{\boldsymbol{\theta}}}$ does not depend on n, following from the definition of $\mathcal{I}(\boldsymbol{\theta})$ in (B.5).

B.6.2 Asymptotic Covariance Matrix for Test Planning

For an assumed model if there is to be no censoring or truncation, and if the density approximation [equation (7.13)] is used for $L_i(\boldsymbol{\theta})$, then $\Sigma_{\widehat{\boldsymbol{\theta}}} = I_{\boldsymbol{\theta}}^{-1}$ is a function of the sample size n, the unknown parameters $\boldsymbol{\theta}$, and the levels of the explanatory variables (if any). Otherwise, $I_{\boldsymbol{\theta}}$ also depends on the type of censoring, truncation, rounding, and so on that will be encountered in the data. If any of these limitations on measurement or observation are random, then $I_{\boldsymbol{\theta}}$ depends on the distribution(s) of these limitations. Generally, the effect of roundoff or binning on the "correct likelihood" is not large (e.g., Meeker, 1986). The effect of censoring or truncation, however, can be substantial. The asymptotic covariance matrix $\Sigma_{\widehat{\boldsymbol{\theta}}}$ depends on the underlying model, including its parameters (but does not depend on data). Thus, for a specified model, if one has "planning values" for $\boldsymbol{\theta}$, it is generally straightforward to evaluate $\Sigma_{\widehat{\boldsymbol{\theta}}}$ numerically to compute the asymptotic variances of $\widehat{\boldsymbol{\theta}}$ and of smooth functions of $\widehat{\boldsymbol{\theta}}$ (see the details below) and these asymptotic variances are useful for planning experiments; see, for example, Escobar and Meeker (1994, 1995, 1998d) and Nelson (1990a, Chapter 6).

B.6.3 Asymptotic Distribution of Functions of ML Estimators

In general, one is interested in inferences on functions of $\boldsymbol{\theta}$. For example, consider a vector function $\boldsymbol{g}(\boldsymbol{\theta})$ of the parameters such that all the second derivatives with respect to the elements of $\boldsymbol{\theta}$ are continuous. The ML estimator of $\boldsymbol{g}(\boldsymbol{\theta})$ is $\widehat{\boldsymbol{g}} = \boldsymbol{g}(\widehat{\boldsymbol{\theta}})$. In large samples, $\boldsymbol{g}(\widehat{\boldsymbol{\theta}})$ is approximately normally distributed with mean $\boldsymbol{g}(\boldsymbol{\theta})$ and covariance matrix

$$\Sigma_{\widehat{\boldsymbol{g}}} = \left[\frac{\partial \boldsymbol{g}(\boldsymbol{\theta})}{\partial \boldsymbol{\theta}}\right]' \Sigma_{\widehat{\boldsymbol{\theta}}} \left[\frac{\partial \boldsymbol{g}(\boldsymbol{\theta})}{\partial \boldsymbol{\theta}}\right]. \qquad (B.9)$$

The approximation is based on the assumption that $\boldsymbol{g}(\widehat{\boldsymbol{\theta}})$ is approximately linear in $\widehat{\boldsymbol{\theta}}$ in the region near to $\boldsymbol{\theta}$. The approximation is better in large samples because then the variation in $\widehat{\boldsymbol{\theta}}$ is smaller and thus the region over which $\widehat{\boldsymbol{\theta}}$ varies is correspondingly smaller. If this region is small enough, the linear approximation will be adequate. See Section B.2 for more details.

For scalar g and θ the formula simplifies to

$$\operatorname{Avar}[g(\widehat{\theta})] = \left[\frac{\partial g(\theta)}{\partial \theta}\right]^2 \operatorname{Avar}(\widehat{\theta}),$$

where Avar is the asymptotic variance function. For example, if θ is positive and $g(\theta)$ is the logarithmic function, the asymptotic variance of $\log(\widehat{\theta})$ is $\operatorname{Avar}[\log(\widehat{\theta})] = \operatorname{Avar}(\widehat{\theta})/\theta^2$.

B.6.4 Estimating the Variance–Covariance Matrix of ML Estimates

Under mild regularity conditions (see Section B.4), $\widehat{\Sigma}_{\widehat{\boldsymbol{\theta}}} = (\widehat{I}_{\boldsymbol{\theta}})^{-1}$ is a consistent estimator of $\Sigma_{\widehat{\boldsymbol{\theta}}}$, where $\widehat{I}_{\boldsymbol{\theta}}$ is defined in (B.7). This "local" estimate of $\Sigma_{\widehat{\boldsymbol{\theta}}}$ is obtained by estimating the "expected" curvature in (B.6) by the "observed" curvature in (B.7). It is possible to estimate $\widehat{\Sigma}_{\widehat{\boldsymbol{\theta}}}$ directly by evaluating (B.6) at $\boldsymbol{\theta} = \widehat{\boldsymbol{\theta}}$, but this approach is rarely used because it is more complicated and has no clear advantage.

The "local" estimate of the covariance matrix of $\widehat{\boldsymbol{g}} = \boldsymbol{g}(\widehat{\boldsymbol{\theta}})$ can be obtained by substituting $\widehat{\Sigma}_{\widehat{\boldsymbol{\theta}}}$ for $\Sigma_{\widehat{\boldsymbol{\theta}}}$ in (B.9) giving

$$\widehat{\Sigma}_{\widehat{\boldsymbol{g}}} = \left[\frac{\partial \boldsymbol{g}(\boldsymbol{\theta})}{\partial \boldsymbol{\theta}}\right]' \widehat{\Sigma}_{\widehat{\boldsymbol{\theta}}} \left[\frac{\partial \boldsymbol{g}(\boldsymbol{\theta})}{\partial \boldsymbol{\theta}}\right], \qquad (B.10)$$

where the derivatives are again evaluated at $\boldsymbol{\theta} = \widehat{\boldsymbol{\theta}}$. For scalar g and θ the formula simplifies to

$$\widehat{\operatorname{Var}}[g(\widehat{\theta})] = \left[\frac{\partial g(\theta)}{\partial \theta}\right]^2 \widehat{\Sigma}_{\widehat{\theta}} = \left[\frac{\partial g(\theta)}{\partial \theta}\right]^2 \widehat{\operatorname{Var}}(\widehat{\theta}).$$

For example, if θ is positive and $g(\theta)$ is the logarithmic function, the local estimate of the variance of $\log(\widehat{\theta})$ is $\widehat{\operatorname{Var}}[\log(\widehat{\theta})] = \widehat{\operatorname{Var}}(\widehat{\theta})/\widehat{\theta}^2$ and $\widehat{\operatorname{se}}[\log(\widehat{\theta})] = \widehat{\operatorname{se}}(\widehat{\theta})/\widehat{\theta}$.

B.6.5 Likelihood Ratios and Profile Likelihoods

Assume that we want to estimate $\boldsymbol{\theta}_1$, from the partition $\boldsymbol{\theta} = (\boldsymbol{\theta}_1, \boldsymbol{\theta}_2)$. Let r_1 denote the length of $\boldsymbol{\theta}_1$. The profile likelihood for $\boldsymbol{\theta}_1$ is

$$R(\boldsymbol{\theta}_1) = \max_{\boldsymbol{\theta}_2}\left[\frac{L(\boldsymbol{\theta}_1, \boldsymbol{\theta}_2)}{L(\widehat{\boldsymbol{\theta}})}\right]. \qquad (B.11)$$

When the length of $\boldsymbol{\theta}_2$ is 0 (as in the exponential distribution in Chapter 7 or in Example 8.3), (B.11) is a relative likelihood for $\boldsymbol{\theta} = \boldsymbol{\theta}_1$. Otherwise we have a "maximized relative likelihood" for $\boldsymbol{\theta}_1$. In either case, $R(\boldsymbol{\theta}_1)$ is commonly known as a "profile likelihood" because it provides a view of the profile of $L(\boldsymbol{\theta})$ as viewed along a line that is perpendicular to the axes of $\boldsymbol{\theta}_1$.

OUTLINE OF GENERAL ML THEORY

- When $\boldsymbol{\theta}_1$ is of length 1, $R(\boldsymbol{\theta}_1)$ is a curve projected onto a plane.
- When $\boldsymbol{\theta}_1$ is of length 2 or more, $R(\boldsymbol{\theta}_1)$ is a surface projected onto a three-dimensional hyperplane.

In either case the projection is in a direction perpendicular to the coordinate axes for $\boldsymbol{\theta}_1$. When $\boldsymbol{\theta}_1$ is of length 1 or 2, it is useful to display $R(\boldsymbol{\theta}_1)$ graphically.

Asymptotically, $\text{LLR}_n(\boldsymbol{\theta}_1) = -2\log[R(\boldsymbol{\theta}_1)]$ when evaluated at the true $\boldsymbol{\theta}_1$, has a chi-square distribution with r_1 degrees of freedom. To do a likelihood-ratio significance test, we would reject the null hypothesis that $\boldsymbol{\theta} = \boldsymbol{\theta}_0$, at the α level of significance, if

$$\text{LLR}_n(\boldsymbol{\theta}_1) = -2\log[R(\boldsymbol{\theta}_0)] > \chi^2_{(1-\alpha;r_1)}.$$

B.6.6 Approximate Likelihood-Ratio-Based Confidence Regions or Confidence Intervals for the Model Parameters

An approximate $100(1 - \alpha)\%$ likelihood-ratio-based confidence region for $\boldsymbol{\theta}_1$ is the set of all values of $\boldsymbol{\theta}_1$ such that $\text{LLR}_n(\boldsymbol{\theta}_1) = -2\log[R(\boldsymbol{\theta}_1)] < \chi^2_{(1-\alpha;r_1)}$ or, equivalently, $R(\boldsymbol{\theta}_1) > \exp[-\chi^2_{(1-\alpha;r_1)}/2]$. Here $\boldsymbol{\theta}_1$ could be the full parameter vector, a single element of $\boldsymbol{\theta}$, or some other subset of $\boldsymbol{\theta}$. If one is interested in a scalar function $g(\boldsymbol{\theta})$, these same ideas can be applied after a reparameterization such that $g(\boldsymbol{\theta})$ is one of the parameters. Simulation studies for different applications and models (e.g., Ostrouchov and Meeker, 1988; Meeker, 1987; Vander Wiel and Meeker, 1990; Jeng and Meeker, 1998) have shown that in terms of closeness to the nominal confidence level, the likelihood-based intervals have important advantages over the standard normal-approximation intervals (discussed in Section B.6.7), especially when there is only a small number of failures in the data. Specifically, in repeated sampling, normal-approximation intervals tend to have actual confidence levels that are smaller than the nominal levels. Likelihood-ratio-based intervals tend to have confidence levels that are much closer to the nominal. Also see Meeker and Escobar (1995).

B.6.7 Approximate Confidence Regions and Intervals Based on Simple Asymptotic Approximations

The large-sample normal approximation for the distribution of ML estimators can be used to compute approximate confidence intervals (regions) for scalar (vector) functions of $\boldsymbol{\theta}$. In particular, an approximate $100(1 - \alpha)\%$ confidence region for $\boldsymbol{\theta}$ is the set of all values of $\boldsymbol{\theta}$ in the ellipsoid

$$(\widehat{\boldsymbol{\theta}} - \boldsymbol{\theta})' \left(\widehat{\Sigma}_{\widehat{\boldsymbol{\theta}}}\right)^{-1} (\widehat{\boldsymbol{\theta}} - \boldsymbol{\theta}) \leq \chi^2_{(1-\alpha;r)}, \tag{B.12}$$

where r is the length of $\boldsymbol{\theta}$. This is sometimes known as "Wald's method," but we will refer to it as the "normal-approximation" method. This confidence region (or interval) is based on the distributional result that, asymptotically, when evaluated at the true $\boldsymbol{\theta}$, the "Wald statistic"

$$W(\boldsymbol{\theta}) = (\widehat{\boldsymbol{\theta}} - \boldsymbol{\theta})' \left(\widehat{\Sigma}_{\widehat{\boldsymbol{\theta}}}\right)^{-1} (\widehat{\boldsymbol{\theta}} - \boldsymbol{\theta})$$

has a chi-square distribution with r degrees of freedom.

More generally, let $\boldsymbol{g}(\boldsymbol{\theta})$ be a vector function of $\boldsymbol{\theta}$. An approximate $100(1 - \alpha)\%$ normal-approximation confidence region for a r_1-dimensional subset $\boldsymbol{g}_1 = \boldsymbol{g}_1(\boldsymbol{\theta})$, from the partition $\boldsymbol{g}(\boldsymbol{\theta}) = [\boldsymbol{g}_1(\boldsymbol{\theta}), \boldsymbol{g}_2(\boldsymbol{\theta})]$, is the set of all the $\boldsymbol{g}_1$ values in the ellipsoid

$$(\widehat{\boldsymbol{g}}_1 - \boldsymbol{g}_1)' \left(\widehat{\Sigma}_{\widehat{\boldsymbol{g}}_1}\right)^{-1} (\widehat{\boldsymbol{g}}_1 - \boldsymbol{g}_1) \leq \chi^2_{(1-\alpha;r_1)},$$

where $\widehat{\boldsymbol{g}}_1 = \boldsymbol{g}_1(\widehat{\boldsymbol{\theta}})$ is the ML estimator of $\boldsymbol{g}_1(\boldsymbol{\theta})$ and $\widehat{\Sigma}_{\widehat{\boldsymbol{g}}_1}$ is the local estimate of the covariance matrix of $\widehat{\boldsymbol{g}}_1$. The estimate $\widehat{\Sigma}_{\widehat{\boldsymbol{g}}_1}$ can be obtained from the local estimate of $\Sigma_{\widehat{\boldsymbol{g}}}$ in equation (B.10). This confidence region (or interval) is based on the distributional result that the "Wald subset statistic," when evaluated at the true $\boldsymbol{g}_1$,

$$W(\boldsymbol{g}_1) = (\widehat{\boldsymbol{g}}_1 - \boldsymbol{g}_1)' \left(\widehat{\Sigma}_{\widehat{\boldsymbol{g}}_1}\right)^{-1} (\widehat{\boldsymbol{g}}_1 - \boldsymbol{g}_1),$$

has, asymptotically, a chi-square distribution with r_1 degrees of freedom. As shown in Meeker and Escobar (1995), this normal-approximation confidence region (or interval) can be viewed as a quadratic approximation for the log profile likelihood of $\boldsymbol{g}_1(\boldsymbol{\theta})$ at $\widehat{\boldsymbol{g}}_1$.

When $r_1 = 1$, $g_1 = g_1(\boldsymbol{\theta})$ is a scalar function of $\boldsymbol{\theta}$, an approximate $100(1 - \alpha)\%$ normal-approximation confidence interval is obtained from the familiar formula

$$[\underset{\sim}{g}_1, \widetilde{g}_1] = \widehat{g}_1 \pm z_{(1-\alpha/2)} \widehat{\text{se}}_{\widehat{g}_1}, \tag{B.13}$$

where $\widehat{\text{se}}_{\widehat{g}_1} = \sqrt{\widehat{\text{Var}}[g_1(\widehat{\boldsymbol{\theta}})]}$ is the local estimate for the standard error of $\widehat{g}_1$ and $z_{(1-\alpha/2)}$ is the $1 - \alpha/2$ quantile of the standard normal distribution.

APPENDIX C

Tables

Table C.1. Failure and censoring times of diesel generator fans
Table C.2. Distance to failure for 38 vehicle shock absorbers
Table C.3. Percent increase in resistance over time of carbon-film resistors
Table C.4. Life test comparison of two different snubber designs
Table C.5. Bearing-cage fracture data
Table C.6. Battery life test data
Table C.7. Bleed system failure data
Table C.8. Diesel engine age at times of replacement of valve seats
Table C.9. Locomotive age at time of replacement of braking grids
Table C.10. Temperature-accelerated life test data for Device-A
Table C.11. Computer program execution time versus system load
Table C.12. Low-cycle fatigue life of nickel-base superalloy specimens
Table C.13. Minutes to failure of mylar–polyurethane laminated DC HV insulating structure
Table C.14. Fatigue crack size as a function of number of cycles
Table C.15. Accelerated life test data on a new-technology integrated circuit device
Table C.16. Temperature- and voltage-accelerated life test data for tantalum capacitors
Table C.17. Percent increase in operating current for GaAs lasers tested at 80°C
Table C.18. Block error rates for magneto-optical data storage disks tested at 80°C and 85% relative humidity
Table C.19. Scar width caused by sliding metal wear for different applied weights
Table C.20. Normal distribution Fisher information, large-sample approximate variance–covariance matrix entries, and other factors for planning normal/lognormal distribution life tests with censored data

Table C.1. Failure and Censoring Times of Diesel Generator Fans

Hours	Status	Number of Fans	Hours	Status	Number of Fans
450	Failed	1	4850	Censored	4
460	Censored	1	5000	Censored	3
1150	Failed	2	6100	Censored	3
1560	Censored	1	6100	Failed	1
1600	Failed	1	6300	Censored	1
1660	Censored	1	6450	Censored	2
1850	Censored	5	6700	Censored	1
2030	Censored	3	7450	Censored	1
2070	Failed	2	7800	Censored	2
2080	Failed	1	8100	Censored	2
2200	Censored	1	8200	Censored	1
3000	Censored	4	8500	Censored	3
3100	Failed	1	8750	Censored	2
3200	Censored	1	8750	Failed	1
3450	Failed	1	9400	Censored	1
3750	Censored	2	9900	Censored	1
4150	Censored	4	10100	Censored	3
4300	Censored	4	11500	Censored	1
4600	Failed	1			

Data from Nelson (1982), page 133.

Table C.2. Distance to Failure for 38 Vehicle Shock Absorbers

Distance (km)	Failure Mode	Distance (km)	Failure Mode
6700	M1	17520	M1
6950	None	17540	None
7820	None	17890	None
8790	None	18450	None
9120	M2	18960	None
9660	None	18980	None
9820	None	19410	None
11310	None	20100	M2
11690	None	20100	None
11850	None	20150	None
11880	None	20320	None
12140	None	20900	M2
12200	M1	22700	M1
12870	None	23490	None
13150	M2	26510	M1
13330	None	27410	None
13470	None	27490	M1
14040	None	27890	None
14300	M1	28100	None

Data from O'Connor (1985), page 85.

Table C.3. Percent Increase in Resistance Over Time of Carbon-Film Resistors

Unit Number	Temperature (°C)	Initial Resistance	Hours			
			452	1030	4341	8084
1	83	217.97	.28	.32	.38	.62
2		217.88	.22	.24	.26	.38
3		224.67	.41	.46	.54	.81
4		215.92	.25	.29	.32	.48
5		219.88	.25	.26	.42	.57
6		219.63	.32	.36	.45	.58
7		218.27	.36	.41	.52	.70
8		217.27	.24	.28	.34	.55
9		219.98	.33	.40	.44	.85
11	133	218.05	.40	.47	.72	1.05
12		219.38	.88	1.19	2.06	3.15
13		218.35	.53	.64	.99	1.60
14		217.78	.47	.62	1.00	1.50
15		218.28	.57	.75	1.26	2.03
16		216.38	.55	.67	1.09	1.79
17		217.65	.78	.96	1.48	2.27
18		221.91	.83	1.12	1.96	3.29
19		218.47	.64	.80	1.23	1.84
20		217.59	.55	.74	1.29	2.03
21	173	216.31	.87	1.29	2.62	4.44
22		216.62	1.25	1.88	3.54	5.23
23		221.98	2.64	3.78	7.01	11.12
24		217.83	.98	1.36	2.66	4.42
25		217.30	1.62	2.34	3.82	6.14
26		216.75	1.59	2.41	3.46	6.75
27		220.39	2.29	2.24	6.30	8.34
28		216.26	.98	1.37	2.47	3.74
29		217.86	1.04	1.54	2.77	4.16
30		217.49	1.19	1.59	3.03	4.52

Data from Shiomi and Yanagisawa (1979).

Table C.4. Life Test Comparison of Two Different Snubber Designs

Old Design			New Design		
Hours	Status	Units	Hours	Status	Units
90	Failed	2	45	Censored	1
90	Censored	1	47	Failed	1
190	Censored	1	73	Failed	1
218	Censored	2	136	Censored	5
241	Censored	1	145	Failed	1
268	Failed	1	190	Censored	2
349	Censored	1	281	Censored	1
378	Censored	2	311	Failed	1
410	Failed	2	417	Censored	1
410	Censored	1	485	Censored	2
485	Failed	1	490	Failed	1
508	Failed	1	569	Censored	1
600	Censored	4	571	Failed	1
631	Failed	3	571	Censored	1
635	Failed	1	575	Failed	1
658	Failed	1	608	Failed	2
658	Censored	1	608	Censored	12
731	Failed	1	630	Failed	1
739	Failed	1	670	Failed	2
739	Censored	4	731	Censored	1
790	Failed	1	838	Failed	1
790	Censored	11	964	Failed	2
855	Failed	1	1164	Censored	7
980	Failed	2	1198	Failed	1
980	Censored	5	1198	Censored	1
			1300	Censored	3

Data from Nelson (1982), page 529.

Table C.5. Bearing-Cage Fracture Data

Hours	Status	Number of Units	Hours	Status	Number of Units
50	Censored	288	990	Failed	1
150	Censored	148	1009	Failed	1
230	Failed	1	1050	Censored	123
250	Censored	124	1150	Censored	93
334	Failed	1	1250	Censored	47
350	Censored	111	1350	Censored	41
423	Failed	1	1450	Censored	27
450	Censored	106	1510	Failed	1
550	Censored	99	1550	Censored	11
650	Censored	110	1650	Censored	6
750	Censored	114	1850	Censored	1
850	Censored	119	2050	Censored	2
950	Censored	127			

Data from Abernethy, Breneman, Medlin, and Reinman (1983), pages 43 and 47.

Table C.6. Battery Life Test Data

Ampere-Hours		Failure Mode				Number Failing	Number Censored
Lower	Upper	1	2	3	6		
0	50	0	0	0	1	1	5
50	100	0	0	0	0	0	6
100	150	1	0	0	0	1	1
150	200	0	3	0	1	4	6
200	250	0	0	1	0	1	2
250	300	1	0	0	0	1	1
300	350	0	0	0	1	1	2
350	400	1	2	0	1	4	2
450	500	0	3	1	0	4	3
500	550	0	1	0	1	2	1
550	600	1	0	0	1	2	0
600	650	0	0	1	0	1	0
650	700	1	1	0	0	2	1
700	750	0	0	1	0	1	0
800	850	2	0	1	0	3	0
850	900	0	0	0	0	0	1
950	1000	0	0	0	0	0	1
1000	1050	0	0	1	0	1	0
1050	1100	0	0	0	0	0	1
1100	1150	0	0	0	0	0	2
1150	1200	0	0	1	0	1	0
1300	1350	0	0	1	0	1	0
1500	1550	0	0	1	0	1	0
1650	1700	1	0	0	0	1	0
Total		8	10	9	5	33	35

Data from Morgan (1980).

Table C.7. Bleed System Failure Data

Hours	Status	Base D	Other Bases	All Bases
12	Censored	0	39	39
20	Censored	0	52	52
30	Censored	0	46	46
32	Failed	0	1	1
50	Censored	0	31	31
64	Failed	0	1	1
85	Censored	0	48	48
150	Censored	0	102	102
153	Failed	0	1	1
212	Failed	0	1	1
250	Censored	2	158	160
400	Censored	0	312	312
550	Censored	2	101	103
650	Censored	2	101	103
708	Failed	1	0	1
750	Censored	9	100	109
808	Failed	0	1	1
828	Failed	1	0	1
850	Censored	23	100	123
872	Failed	0	1	1
884	Failed	2	0	2
950	Censored	27	56	83
1013	Failed	1	0	1
1050	Censored	20	55	75
1082	Failed	1	0	1
1105	Failed	1	0	1
1150	Censored	22	56	78
1198	Failed	1	0	1
1249	Failed	1	0	1
1250	Censored	22	55	77
1251	Failed	1	0	1
1350	Censored	11	56	67
1405	Failed	0	1	1
1428	Failed	0	1	1
1450	Censored	11	53	64
1550	Censored	20	55	75
1568	Failed	0	1	1
1650	Censored	8	55	63
1750	Censored	4	55	59
1850	Censored	2	55	57
1950	Censored	3	152	155
2050	Censored	3	152	155
2150	Censored	1	0	1

Data adapted from a histogram and description in Abernethy, Breneman, Medlin, and Reinman (1983), pages 29–51.

Table C.8. Diesel Engine Age at Time of Replacement of Valve Seats

System ID	Days Observed	Engine Age at Replacement Time (Days)					System ID	Days Observed	Engine Age at Replacement Time (Days)		
251	761						403	593			
252	759						404	589	573		
327	667	98					405	606	165	408	604
328	667	326	653	653			406	594	249		
329	665						407	613	344	497	
330	667	84					408	595	265	586	
331	663	87					409	389	166	206	348
389	653	646					410	601			
390	653	92					411	601	410	581	
391	651						412	611			
392	650	258	328	377	621		413	608			
393	648	61	539				414	587			
394	644	254	276	298	640		415	603	367		
395	642	76	538				416	585	202	563	570
396	641	635					417	587			
397	649	349	404	561			418	578			
398	631						419	578			
399	596						420	586			
400	614	120	479				421	585			
401	582	323	449				422	582			
402	589	139	139								

Data from Nelson and Doganaksoy (1989).

Table C.9. Locomotive Age at Time of Replacement of Braking Grids

Batch 1					Batch 2					
Locomotive ID	Days Observed	Locomotive Age at Replacement (Days)			Locomotive ID	Days Observed	Locomotive Age at Replacement (Days)			
9100	730	462			9176	511	203	211	277	373
9102	724	364	391	548	9182	503	293			
9103	730	302	444	500	9190	470	173			
9106	730	250			9197	464	242			
9108	724	500			9199	464	39			
9110	724	88			9200	462	91			
9117	719	272	421	552 625	9201	461	119	148	306	
9124	710	481			9203	460	382			
9125	710	431			9207	434	250			
9126	710	367			9209	448	192			
9128	708	635	650		9212	448	369			
9134	700	402			9213	447	22			
9136	687	33			9216	441	54			
9138	687	287			9226	432	194			
9156	657	317	498		9235	419	61			
					9236	419	19	185		
					9238	416	187			
					9239	415	93	205	264	

Data from Doganaksoy and Nelson (1991).

Table C.10. Temperature-Accelerated Life Test Data for Device-A

Hours	Status	Number of Devices	Temperature (°C)
5000	Censored	30	10
1298	Failed	1	40
1390	Failed	1	40
3187	Failed	1	40
3241	Failed	1	40
3261	Failed	1	40
3313	Failed	1	40
4501	Failed	1	40
4568	Failed	1	40
4841	Failed	1	40
4982	Failed	1	40
5000	Censored	90	40
581	Failed	1	60
925	Failed	1	60
1432	Failed	1	60
1586	Failed	1	60
2452	Failed	1	60
2734	Failed	1	60
2772	Failed	1	60
4106	Failed	1	60
4674	Failed	1	60
5000	Censored	11	60
283	Failed	1	80
361	Failed	1	80
515	Failed	1	80
638	Failed	1	80
854	Failed	1	80
1024	Failed	1	80
1030	Failed	1	80
1045	Failed	1	80
1767	Failed	1	80
1777	Failed	1	80
1856	Failed	1	80
1951	Failed	1	80
1964	Failed	1	80
2884	Failed	1	80
5000	Censored	1	80

Data from Hooper and Amster (1990). Reprinted with permission. Copyright McGraw-Hill.

Table C.11. Computer Program Execution Time Versus System Load

Seconds	Load	Seconds	Load
123	2.74	110	.60
704	5.47	213	2.10
184	2.13	284	3.10
113	1.00	317	5.86
94	.32	142	1.18
76	.31	127	.57
78	.51	96	1.10
98	.29	111	1.89
240	.96		

Table C.12. Low-Cycle Fatigue Life of Nickel-Base Superalloy Specimens (in units of thousands of cycles to failure)

Pseudostress	k-Cycles	Status	Pseudostress	k-Cycles	Status
80.3	211.629	F	99.8	43.331	F
80.6	200.027	F	100.1	12.076	F
80.8	57.923	C	100.5	13.181	F
84.3	155.000	F	113.0	18.067	F
85.2	13.949	F	114.8	21.300	F
85.6	112.968	C	116.4	15.616	F
85.8	152.680	F	118.0	13.030	F
86.4	156.725	F	118.4	8.489	F
86.7	138.114	C	118.6	12.434	F
87.2	56.723	F	120.4	9.750	F
87.3	121.075	F	142.5	11.865	F
89.7	122.372	C	144.5	6.705	F
91.3	112.002	F	145.9	5.733	F

Cases marked with "F" are failures and cases marked with "C" are censored (unfailed). Data from Nelson (1990), page 272.

Table C.13. Minutes to Failure of Mylar–Polyurethane Laminated DC HV Insulating Structure

		Voltage Stress		
361.4 kV/mm	219.0 kV/mm	157.1 kV/mm	122.4 kV/mm	100.3 kV/mm
.10	15	49	188	606
.33	16	99	297	1012
.50	36	154.5	405	2520
.50	50	180	744	2610
.90	55	291	1218	3988
1.00	95	447	1340	4100
1.55	122	510	1715	5025
1.65	129	600	3382	6842
2.10	625	1656		
4.00	700	1721		

Data from Kalkanis and Rosso (1989). Reprinted with permission. Copyright Elsevier Science Ltd.

Table C.14. Fatigue Crack Size as a Function of Number of Cycles

	Millions of Cycles											
Unit	.00	.01	.02	.03	.04	.05	.06	.07	.08	.09	.10	.11	.12
1	.90	.95	1.00	1.05	1.12	1.19	1.27	1.35	1.48	1.64			
2	.90	.94	.98	1.03	1.08	1.14	1.21	1.28	1.37	1.47	1.60		
3	.90	.94	.98	1.03	1.08	1.13	1.19	1.26	1.35	1.46	1.58	1.77	
4	.90	.94	.98	1.03	1.07	1.12	1.19	1.25	1.34	1.43	1.55	1.73	
5	.90	.94	.98	1.03	1.07	1.12	1.19	1.24	1.34	1.43	1.55	1.71	
6	.90	.94	.98	1.03	1.07	1.12	1.18	1.23	1.33	1.41	1.51	1.68	
7	.90	.94	.98	1.02	1.07	1.11	1.17	1.23	1.32	1.41	1.52	1.66	
8	.90	.93	.97	1.00	1.06	1.11	1.17	1.23	1.30	1.39	1.49	1.62	
9	.90	.92	.97	1.01	1.05	1.09	1.15	1.21	1.28	1.36	1.44	1.55	1.72
10	.90	.92	.96	1.00	1.04	1.08	1.13	1.19	1.26	1.34	1.42	1.52	1.67
11	.90	.93	.96	1.00	1.04	1.08	1.13	1.18	1.24	1.31	1.39	1.49	1.65
12	.90	.93	.97	1.00	1.03	1.07	1.10	1.16	1.22	1.29	1.37	1.48	1.64
13	.90	.92	.97	.99	1.03	1.06	1.10	1.14	1.20	1.26	1.31	1.40	1.52
14	.90	.93	.96	1.00	1.03	1.07	1.12	1.16	1.20	1.26	1.30	1.37	1.45
15	.90	.92	.96	.99	1.03	1.06	1.10	1.16	1.21	1.27	1.33	1.40	1.49
16	.90	.92	.95	.97	1.00	1.03	1.07	1.11	1.16	1.22	1.26	1.33	1.40
17	.90	.93	.96	.97	1.00	1.05	1.08	1.11	1.16	1.20	1.24	1.32	1.38
18	.90	.92	.94	.97	1.01	1.04	1.07	1.09	1.14	1.19	1.23	1.28	1.35
19	.90	.92	.94	.97	.99	1.02	1.05	1.08	1.12	1.16	1.20	1.25	1.31
20	.90	.92	.94	.97	.99	1.02	1.05	1.08	1.12	1.16	1.19	1.24	1.29
21	.90	.92	.94	.97	.99	1.02	1.04	1.07	1.11	1.14	1.18	1.22	1.27

Data reported in Lu and Meeker (1993), read from Figure 4.52 in Bogdanoff and Kozin (1985), page 242.

Table C.15. Accelerated Life Test Data on a New-Technology Integrated Circuit Device

Hours		Status	Number of Devices	Temperature (°C)
Lower	Upper			
	1536	Right Censored	50	150
	1536	Right Censored	50	175
	96	Right Censored	50	200
384	788	Failed	1	250
788	1536	Failed	3	250
1536	2304	Failed	5	250
	2304	Right Censored	41	250
192	384	Failed	4	300
384	788	Failed	27	300
788	1536	Failed	16	300
	1536	Right Censored	3	300

Table C.16. Temperature- and Voltage-Accelerated Life Test Data for Tantalum Electrolytic Capacitors

Hours	Status	Number of Devices	Volts	Temperature (°C)
20	Failure	1	35.0	85
90	Failure	1	35.0	85
700	Failure	1	35.0	85
37000	Failure	1	35.0	85
37000	Censored	996	35.0	85
20	Failure	1	40.6	85
3600	Failure	1	40.6	85
9500	Failure	1	40.6	85
27000	Failure	1	40.6	85
27000	Censored	196	40.6	85
800	Failure	1	46.5	85
2800	Failure	1	46.5	85
2800	Censored	48	46.5	85
500	Failure	1	51.5	85
800	Failure	1	51.5	85
2400	Failure	1	51.5	85
10700	Failure	1	51.5	85
10700	Censored	49	51.5	85
100	Failure	1	46.5	45
1200	Failure	1	46.5	45
7500	Failure	1	46.5	45
20000	Failure	1	46.5	45
26000	Failure	1	46.5	45
27300	Failure	1	46.5	45
27300	Censored	496	46.5	45
1000	Failure	1	46.5	5
1000	Censored	174	46.5	5
25	Failure	1	62.5	5
50	Failure	1	62.5	5
165	Failure	1	62.5	5
500	Failure	1	62.5	5
620	Failure	1	62.5	5
720	Failure	1	62.5	5
820	Failure	1	62.5	5
910	Failure	1	62.5	5
980	Failure	1	62.5	5
1270	Failure	1	62.5	5
1600	Failure	1	62.5	5
2270	Failure	1	62.5	5
2370	Failure	1	62.5	5
4590	Failure	1	62.5	5
4880	Failure	1	62.5	5
7560	Failure	1	62.5	5
8730	Failure	1	62.5	5
12500	Failure	1	62.5	5
12500	Censored	156	62.5	5
8900	Failure	1	57.0	45
8900	Censored	49	57.0	45

Data from Singpurwalla, Castellino, and Goldschen (1975).

Table C.17. Percent Increase in Operating Current for GaAs Lasers Tested at 80°C

Time (hours)	Unit Number														
	1	2	3	4	5	6	7	8	9	10	11	12	13	14	15
250	.47	.71	.71	.36	.27	.36	.36	.46	.51	.41	.44	.39	.30	.44	.51
500	.93	1.22	1.17	.62	.61	1.39	.92	1.07	.93	1.49	1.00	.80	.74	.70	.83
750	2.11	1.90	1.73	1.36	1.11	1.95	1.21	1.42	1.57	2.38	1.57	1.35	1.52	1.05	1.29
1000	2.72	2.30	1.99	1.95	1.77	2.86	1.46	1.77	1.96	3.00	1.96	1.74	1.85	1.35	1.52
1250	3.51	2.87	2.53	2.30	2.06	3.46	1.93	2.11	2.59	3.84	2.51	2.98	2.39	1.80	1.91
1500	4.34	3.75	2.97	2.95	2.58	3.81	2.39	2.40	3.29	4.50	2.84	3.59	2.95	2.55	2.27
1750	4.91	4.42	3.30	3.39	2.99	4.53	2.68	2.78	3.61	5.25	3.47	4.03	3.51	2.83	2.78
2000	5.48	4.99	3.94	3.79	3.38	5.35	2.94	3.02	4.11	6.26	4.01	4.44	3.92	3.39	3.42
2250	5.99	5.51	4.16	4.11	4.05	5.92	3.42	3.29	4.60	7.05	4.51	4.79	5.03	3.72	3.78
2500	6.72	6.07	4.45	4.50	4.63	6.71	4.09	3.75	4.91	7.80	4.80	5.22	5.47	4.09	4.11
2750	7.13	6.64	4.89	4.72	5.24	7.70	4.58	4.16	5.34	8.32	5.20	5.48	5.84	4.83	4.38
3000	8.00	7.16	5.27	4.98	5.62	8.61	4.84	4.76	5.84	8.93	5.66	5.96	6.50	5.41	4.63
3250	8.92	7.78	5.69	5.28	6.04	9.15	5.11	5.16	6.40	9.55	6.20	6.23	6.94	5.76	5.38
3500	9.49	8.42	6.02	5.61	6.32	9.95	5.57	5.46	6.84	10.45	6.54	6.99	7.39	6.14	5.84
3750	9.87	8.91	6.45	5.95	7.10	10.49	6.11	5.81	7.20	11.28	6.96	7.37	7.85	6.51	6.16
4000	10.94	9.28	6.88	6.14	7.59	11.01	7.17	6.24	7.88	12.21	7.42	7.88	8.09	6.88	6.62

All percent increase values at time 0 are equal to 0.

Table C.18. Block Error Rates for Magneto-Optical Data Storage Disks Tested at 80°C and 85% Relative Humidity

Disk	Hours				
	0	500	1000	1500	2000
1	.621	.663	1.200	1.260	1.210
2	.624	.660	.733	1.010	1.840
3	.526	.562	.630	.841	.862
4	.444	.542	.573	.815	.903
5	1.330	1.430	1.430	1.590	1.750
6	.414	.456	.446	.606	.759
7	.435	.483	.541	.525	.615
8	.313	.382	.451	.515	.695
9	.824	.637	.806	1.220	1.450
10	.499	.642	.669	1.220	1.080
11	.467	.568	.690	.716	.844
12	.536	.626	.658	.759	.870
13	.865	.934	1.050	1.130	1.250
14	.398	.462	.557	.615	.737
15	.430	.499	.546	.610	.669
16	.308	.324	.371	.493	.658

Rates given are bytes with errors divided by the total number of bytes, times 10^5. Data from Murray (1993). Reprinted with permission. Copyright, Magnetics Society of Japan.

Table C.19. Scar Width (in microns) Caused by Sliding Metal Wear for Different Applied Weights

Weight (grams)	Unit	Cycles (hundreds)							
		2	5	10	20	50	100	200	500
10	1	3.2	4.1	4.5	4.7	5.8	6.8	7.7	9.6
	2	2.7	3.4	3.8	3.9	5.4	5.7	6.3	8.4
	3	2.1	2.7	3.1	3.3	4.0	4.6	5.7	6.6
	4	2.6	3.5	4.0	4.0	5.2	6.1	6.7	8.5
50	5	7.5	7.8	8.2	10.6	12.6	13.3	12.9	14.8
	6	7.5	8.1	9.8	10.9	14.8	16.1	17.3	20.2
	7	7.0	8.9	9.4	11.1	12.4	13.5	16.7	17.3
	8	7.8	8.9	10.0	11.5	13.7	16.2	16.2	21.0
100	9	12.5	15.4	17.2	20.5	24.1	27.0	29.4	37.9
	10	11.0	13.9	16.1	18.6	22.2	27.8	31.0	36.6
	11	13.0	15.1	18.6	20.2	23.9	29.7	31.5	39.6
	12	11.7	13.7	16.7	17.5	22.3	25.3	32.0	38.2

Table C.20. Normal Distribution Fisher Information, Large-Sample Approximate Variance–Covariance Matrix Entries, and Other Factors for Planning Normal/Lognormal Distribution Life Tests with Censored Data

ζ_c	$100\Phi_{nor}(\zeta_c)$	f_{11}	f_{22}	f_{12}	$\frac{1}{\sigma^2}V_{\hat{\mu}}$	$\frac{1}{\sigma^2}V_{\hat{\sigma}}$	$\frac{1}{\sigma^2}V_{(\hat{\mu},\hat{\sigma})}$	$\rho_{(\hat{\mu},\hat{\sigma})}$	$\frac{1}{\sigma^2}V_{\hat{\mu}|\sigma}$	$\frac{1}{\sigma^2}V_{\hat{\sigma}|\mu}$
−3.0	.13	.01467	.13583	−.04438	6001.31	647.931	1960.68	.99430	68.1891	7.36202
−2.8	.26	.02478	.20153	−.07015	2751.23	338.313	957.667	.99264	40.3532	4.96214
−2.6	.47	.04016	.28463	−.10589	1297.27	183.052	482.607	.99036	24.8990	3.51339
−2.4	.82	.06245	.38264	−.15260	628.580	102.590	250.686	.98718	16.0128	2.61345
−2.2	1.39	.09322	.48976	−.20998	312.728	59.5263	134.078	.98270	10.7269	2.04181
−2.0	2.28	.13371	.59734	−.27592	159.661	35.7402	73.7498	.97630	7.47860	1.67408
−1.8	3.59	.18451	.69536	−.34639	83.6383	22.1926	41.6638	.96706	5.41988	1.43811
−1.6	5.48	.24529	.77473	−.41570	44.9858	14.2432	24.1386	.95361	4.07682	1.29078
−1.4	8.08	.31476	.82978	−.47734	24.8920	9.44231	14.3192	.93401	3.17699	1.20513
−1.2	11.51	.39070	.86008	−.52495	14.2242	6.46160	8.68176	.90557	2.55948	1.16269
−1.0	15.87	.47022	.87084	−.55353	8.44766	4.56136	5.36957	.86502	2.12668	1.14831
−.8	21.19	.55009	.87193	−.56028	5.26120	3.31921	3.38069	.80899	1.81789	1.14688
−.6	27.43	.62719	.87550	−.54498	3.47293	2.48793	2.16185	.73546	1.59442	1.14221
−.4	34.46	.69881	.89314	−.50996	2.45318	1.91942	1.40071	.64550	1.43100	1.11964
−.2	42.07	.76293	.93338	−.45948	1.86310	1.52288	.91716	.54450	1.31073	1.07138
.0	50.00	.81831	1.00000	−.39894	1.51709	1.24145	.60523	.44101	1.22203	1.00000
.2	57.93	.86449	1.09172	−.33400	1.31180	1.03877	.40133	.34380	1.15675	.91599
.4	65.54	.90170	1.20294	−.26976	1.18876	.89108	.26658	.25901	1.10901	.83130
.6	72.57	.93069	1.32534	−.21026	1.11442	.78257	.17680	.18932	1.07447	.75452
.8	78.81	.95252	1.44973	−.15819	1.06923	.70251	.11667	.13462	1.04985	.68978
1.0	84.13	.96841	1.56779	−.11490	1.04168	.64344	.07634	.09325	1.03262	.63784
1.2	88.49	.97961	1.67317	−.08058	1.02488	.60004	.04936	.06294	1.02082	.59767
1.4	91.92	.98723	1.76212	−.05455	1.01467	.56847	.03141	.04136	1.01294	.56750
1.6	94.52	.99225	1.83336	−.03565	1.00852	.54583	.01961	.02643	1.00782	.54545
1.8	96.41	.99544	1.88766	−.02249	1.00485	.52990	.01197	.01641	1.00458	.52976
2.0	97.72	.99740	1.92712	−.01369	1.00270	.51896	.00712	.00987	1.00261	.51891
2.2	98.61	.99857	1.95450	−.00804	1.00147	.51166	.00412	.00576	1.00144	.51164
2.4	99.18	.99923	1.97267	−.00456	1.00078	.50693	.00231	.00325	1.00077	.50693
2.6	99.53	.99960	1.98420	−.00249	1.00040	.50398	.00126	.00177	1.00040	.50398
2.8	99.74	.99980	1.99121	−.00131	1.00020	.50221	.00066	.00093	1.00020	.50221
3.0	99.87	.99990	1.99530	−.00067	1.00010	.50118	.00033	.00047	1.00010	.50118
∞	100.00	1.00000	2.00000	.00000	1.00000	.50000	−.00000	.00000	1.00000	.50000

References

Aalen, O. (1976), Nonparametric inference in connection with multiple decrement models, *Scandinavian Journal of Statistics*, **3**, 15–27.

Aalen, O., and Husebye, E. (1991), Statistical analysis of repeated events forming renewal processes, *Statistics in Medicine*, **10**, 1227–1240.

Abernethy, R. B. (1996), *The New Weibull Handbook* (Second Edition), published by Robert B. Abernethy, 536 Oyster Road, North Palm Beach, FL 33408-4328.

Abernethy, R. B., Breneman, J. E., Medlin, C. H., and Reinman, G. L. (1983), *Weibull Analysis Handbook*, Air Force Wright Aeronautical Laboratories Technical Report AFWAL-TR-83-2079. Available from the National Technical Information Service, Washington, DC.

Akritas, M. G. (1986), Bootstrapping the Kaplan–Meier estimator, *Journal of the American Statistical Association,* **81**, 1032–1038.

Amster, S. J., Brush, G. G., and Saperstein, B. (1982), Planning and conducting field-tracking studies, *The Bell System Technical Journal*, **61**, 2333–2364.

Amster, S. J., and Hooper, J. H. (1983), Accelerated life tests with measured degradation data and growth curve models, paper presented at the American Statistical Association Annual Meetings, Toronto.

Anderson, P. K., Borgan, Ø., Gill, R. D., and Keiding, N. (1993), *Statistical Models Based on Counting Processes*, New York: Springer-Verlag.

Anscombe, F. J. (1964), Normal likelihood functions, *Annals of the Institute of Statistical Mathematics*, **16**, 1–19.

Ansell, J. I., and Phillips, M. J. (1989), Practical problems in the statistical analysis of reliability data, *Applied Statistics*, **38**, 205–231.

Ansell J. I., and Phillips, M. J. (1994), *Practical Methods for Reliability Data Analysis*, Oxford: Clarendon Press.

Ascher, H., and Feingold, H. (1984), *Repairable Systems Reliability*, New York: Marcel Dekker.

Atwood, C. L. (1984), Approximate tolerance intervals, based on maximum likelihood estimates, *Journal of the American Statistical Association,* **79,** 459–465.

Azem, A. (1995), *Software Reliability Determination for Conventional and Logic Programming*, New York: Walter de Gruyter.

Bagdonavičius, V., and Nikulin, M. (1995), Semiparametric models in accelerated life testing, *Queen's Papers in Pure and Applied Mathematics*, **98**, Kingston, Ontario: Queen's University.

Bagdonavičius, V., and Nikulin, M. (1998), Additive and multiplicative semiparametric models in accelerated life testing and survival analysis, *Queen's Papers in Pure and Applied Mathematics*, **108**, Kingston, Ontario: Queen's University.

Bailey, R. A., and Gilbert, R. A. (1981), STRIFE testing for reliability improvement, *Proceedings of the Institute of Environmental Sciences*, 119–121.

Bain, L. J., and Engelhardt, M. (1991), *Statistical Analysis of Reliability and Life Testing Models, Theory and Methods* (Second Edition), New York: Marcel Dekker.

Balakrishnan, N., Editor (1991), *The Logistic Distribution*, New York: Marcel Dekker.

Barlow, R. E., and Proschan, F. (1975), *Statistical Theory of Reliability and Life Testing*, New York: Holt, Rinehart, and Winston.

Barnett, V. (1976), Convenient plotting positions for the normal distribution, *Applied Statistics*, **25**, 47–50.

Baxter, L. A., and Tortorella, M. (1994), Dealing with real field reliability data: circumventing incompleteness by modeling & iteration. *1994 Proceedings Annual Reliability and Maintainability Symposium*, 225–262, New York: Institute of Electrical and Electronics Engineers.

Bayer, R. G. (1994), *Mechanical Wear Prediction and Prevention*, New York: Marcel Dekker.

Beckwith, J. P. (1979), Estimation of the strength remaining of a material that decays with time, Department of Mathematics and Computer Science, Michigan Technological University, Houghton, MI.

Beckwith, J. P. (1980), An estimator and design technique for the estimation of a rate parameter in accelerated testing, Department of Mathematics and Computer Science, Michigan Technological University, Houghton, MI.

Bendell, A. (1985), Proportional hazards modeling in reliability assessment, *Reliability Engineering*, **11**, 1975–1983.

Beran R. (1990), Calibrating prediction regions, *Journal of the American Statistical Association*, **85**, 715–723.

Berkson, J. (1966), Examination of randomness of α-particle emissions, in *Festschrift for J. Neyman, Research Papers in Statistics,* F. N. David, Editor. New York: John Wiley & Sons.

Bhattacharyya, G. K. (1985), The asymptotics of maximum likelihood and related estimators based on type II censored data, *Journal of the American Statistical Association*, **80**, 398–404.

Billingsley, P. (1986), *Probability and Measure* (Second Edition), New York: John Wiley & Sons.

Birnbaum, Z. W. (1979), *On the Mathematics of Competing Risks*, Department of Health, Education and Welfare Publication (PHS)79–1351. Available from the Superintendent of Documents, U.S. Government Printing Office, Washington, DC 20402.

Birnbaum, Z. W., and Saunders, S. C. (1969), A new family of life distributions, *Journal of Applied Probability*, **6**, 319–327.

Black, J. R. (1969), Electromigration—a brief survey and some recent results, *IEEE Transactions on Electronic Devices*, **ED-16**, 338–347.

Blischke, W. R., and Murthy, D. N. P., Editors (1996), *Product Warranty Handbook*, New York: Marcel Dekker.

Blom, G. (1958), *Statistical Estimates and Transformed Beta-Variables*, New York: John Wiley & Sons.

Blumenthal, S., Greenwood, J. A., and Herbach, L. (1973), The transient reliability behavior of series systems or superimposed renewal processes, *Technometrics*, **15**, 255–269.

Blumenthal, S., Greenwood, J. A., and Herbach, L. (1976), A comparison of the bad as old and superimposed renewal models, *Management Science*, **23**, 280–285.

Boccaletti, G., Borri, F. R., D'Esponosa, F., and Ghio, E. (1989), Accelerated Tests, Chapter 11 in *Microelectronic Reliability, Volume II, Reliability, Integrity Assessment and Assurance*, E. Pollino, Editor. Norwood, MA: Artech House.

Bogdanoff, J. L., and Kozin, F. (1985), *Probabilistic Models of Cumulative Damage*, New York: John Wiley & Sons.

Boulanger, M., and Escobar, L. A. (1994), Experimental design for a class of accelerated degradation tests, *Technometrics*, **36**, 260–272.

Box, G. E. P., and Draper, N. R. (1987), *Empirical Model-Building and Response Surfaces*, New York: John Wiley & Sons.

Box, G. E. P., and Tiao, G. C. (1973), *Bayesian Inference in Statistical Analysis,* Reading, MA: Addison-Wesley.

Boyko, K. C., and Gerlach, D. L. (1989), Time dependent dielectric breakdown of 210 Å oxides, *Proceedings of the IEEE International Reliability Physics Symposium*, **27**, 1–8.

Bro, P., and Levy, S. C. (1990), *Quality and Reliability Methods for Primary Batteries*, New York: John Wiley & Sons.

Brown, H. M., and Mains D. E (1979), *Accelerated Test Program for Sealed Nickel–Cadmium Spacecraft Batteries/Cells*, Technical Report WQEC/C 79-145. Available from the Department of the Navy, Naval Weapons Support Center, Weapons Quality Engineering Center, Crane, IN 47522.

Brownlee, K. A. (1960), *Statistical Theory and Methodology in Science and Engineering*, New York: John Wiley & Sons.

Buckland, S. T. (1985), Calculation of Monte Carlo confidence intervals, *Applied Statistics*, **34**, 296–301.

Byrne, D., and Quinlan, J. (1993), Robust function for attaining high reliability at low cost, *1993 Proceedings Annual Reliability and Maintainability Symposium*, 183–191, New York: Institute of Electrical and Electronics Engineers.

Carey, M. B. (1989), Challenges of reliability assessment based on degradation data: an example, paper presented at the American Statistical Association 150th Annual Meeting, Washington, DC.

Carey, M. B., and Koenig, R. H. (1991), Reliability assessment based on accelerated degradation: a case study, *IEEE Transactions on Reliability,* **40**, 499–506.

Carey, M. B., and Tortorella, M. (1988), Analysis of degradation data applied to MOS devices, paper presented at the 6th International Conference on Reliability and Maintainability, Strasbourg, France.

Casella, G., and Berger, R. L. (1990), *Statistical Inference*, Belmont, CA: Wadsworth.

Castillo, E. (1988), *Extreme Value Theory in Engineering*, New York: Academic Press.

Castillo, E., and Galambos, J. (1987), Lifetime regression models based on a functional equation of physical nature, *Journal of Applied Probability*, **24**, 160–169.

Chaloner, K., and Larntz, K. (1992), Bayesian design for accelerated life testing, *Journal of Statistical Planning and Inference*, **33**, 245–259.

Chan, C. K., Boulanger, M., and Tortorella, M. (1994), Analysis of parameter-degradation data using life-data analysis programs, *1994 Proceedings Annual Reliability and Maintainability Symposium*, 228–291, New York: Institute of Electrical and Electronics Engineers.

Chan, V., and Meeker, W. Q. (1998), A competing-risk limited failure population model for product failure times, Department of Statistics, Iowa State University, Ames, IA.

Chang, D. S. (1992), Analysis of accelerated degradation data in a two-way design, *Reliability Engineering and System Safety*, **39**, 65–69.

Cheng, R. C. H., and Iles, T. C. (1983), Confidence bands for cumulative distribution functions of continuous random variables, *Technometrics*, **25**, 77–86.

Cheng, R. C. H., and Iles, T. C. (1987), Corrected maximum likelihood in non-regular problems, *Journal of the Royal Statistical Society B*, **49**, 95–101.

Cheng, R. C. H., and Iles, T. C. (1988), One-sided confidence bands for cumulative distribution functions, *Technometrics*, **30**, 155–159.

Cheng, R. C. H., and Iles, T. C. (1990), Embedded models in three-parameter distributions and their estimation, *Journal of the Royal Statistical Society B*, **52**, 135–149.

Chernoff, H., and Lieberman, G. J. (1954), Use of normal probability paper, *Journal of the American Statistical Association*, **49**, 778–785.

Chhikara, R. S., and Folks, J. L. (1989), *The Inverse Gaussian Distribution*, New York: Marcel Dekker.

Chow, S. C., and Liu, J. P. (1995), *Statistical Design and Analysis in Pharmaceutical Science: Validation, Process Controls, and Stability*, New York: Marcel Dekker.

Christou, A. (1992), *Reliability of Gallium Arsenide MMICs*, New York: John Wiley & Sons.

Christou, A. (1994a), *Integrating Reliability into Microelectronics Manufacturing*, New York: John Wiley & Sons.

Christou, A. (1994b), *Electromigration and Related Electronic Device Degradation*, New York: John Wiley & Sons.

Cohen, A. C. (1991), *Truncated and Censored Samples. Theory and Applications*, New York: Marcel Dekker.

Cohen, A. C., and Whitten, B. J. (1988), *Parameter Estimation in Reliability and Life Span Models*, New York: Marcel Dekker.

Collett, D. (1994), *Modelling Survival Data in Medical Research*, New York: Chapman & Hall.

Condra, L. W. (1993), *Reliability Improvement with Design of Experiments*, New York: Marcel Dekker.

Confer, R., Canner, J., and Trostle, T. (1991), Use of highly accelerated life test (HALT) to determine reliability of multilayer ceramic capacitors, *Proceedings of the Electronic Components & Technology Conference*, **41**, 320–322.

Costa, J. M., and Mercer, A. D. (1993), *Progress in the Understanding and Prevention of Corrosion*, London: The Institute of Materials.

Cox, D. R. (1962), *Renewal Theory*, London: Methuen.

Cox, D. R. (1975), Prediction intervals and empirical Bayes confidence intervals, in *Perspectives in Probability and Statistics*, J. Gani, Editor. London: Academic Press.

Cox, D. R., and Hinkley, D. V. (1974), *Theoretical Statistics,* London: Chapman & Hall.

Cox, D. R., and Isham, V. (1980), *Point Processes,* London: Chapman & Hall.

Cox, D. R., and Lewis, P. A. W. (1966), *The Statistical Analysis of a Series of Events,* London: Chapman & Hall.

Cox, D. R., and Oakes, D. (1984), *Analysis of Survival Data*, London: Chapman & Hall.

Cox, D. R., and Snell, E. J., (1968), A general definition of residuals (with discussion), *Journal of the Royal Statistical Society B*, **30**, 248–275.

Cragnolino, G., and Sridhar, N., Editors (1994), *Application of Accelerated Corrosion Tests to Service Life Prediction of Materials*, ASTM STP 1194, Philadelphia: American Society for Testing and Materials.

Crow, E. L., and Shimizu, K. (1988), *Lognormal Distributions: Theory and Applications*, New York: Marcel Dekker.

Crow, L. (1982), Confidence interval procedures for the Weibull process with applications to reliability growth, *Technometrics*, **24**, 67–72.

Crowder, M. J., Kimber, A. C., Smith, R. L., and Sweeting, T. J. (1991), *Statistical Analysis of Reliability Data*, New York: Chapman & Hall.

Crowder, S. V. (1997), How to determine component-based preventive maintenance plans, SEMATECH Technology Transfer Document #92051125A-GEN, SEMATECH, Austin, TX.

Dale, C. J. (1985), Application of the proportional hazards model in the reliability field, *Reliability Engineering*, **10**, 5–25.

David, H. A. (1981), *Order Statistics* (Second Edition), New York: John Wiley & Sons.

David, H. A., and Moeschberger, M. L. (1978), *The Theory of Competing Risks*, London: Griffin.

Davidian, M., and Giltinan, D. M. (1995), *Nonlinear Models for Repeated Measurement Data,* London: Chapman & Hall.

Davis, D. J. (1952), An analysis of some failure data, *Journal of the American Statistical Association*, **47**, 113–150.

Day, N. E. (1969), Estimating the components of a mixture of normal distribution, *Biometrika,* **56**, 463–474.

Deming, W. E. (1975), On probability as a basis for action, *The American Statistician*, **29**, 146–152.

Derringer, G. C. (1982), Considerations in single and multiple stress accelerated life testing, *Journal of Quality Technology*, **14**, 130–134.

Desmond, A. F. (1986), On the relationship between two fatigue-life models, *IEEE Transactions on Reliability*, **R-35**, 167–169.

Doganaksoy, N. (1995), Likelihood ratio confidence intervals in life-data analysis, Chapter 20 in *Recent Advances in Life-Testing and Reliability*, N. Balakrishnan, Editor. Boca Raton, FL: CRC Press.

Doganaksoy, N., and Nelson, W. (1991), A method and computer program MCFDIFF to compare two samples of repair data, TIS report 91CRD172, General Electric Company Research and Development, Schenectady, NY.

Doganaksoy, N., and Schmee, J. (1993), Comparisons of approximate confidence intervals for distributions used in life-data analysis, *Technometrics*, **35**, 175–184.

Dowling, N. E. (1993), *Mechanical Behavior of Materials*, Englewood Cliffs, NJ: Prentice Hall.

Drapella, A. (1992), An extended mathematical model for failure kinetics, *Quality and Reliability Engineering International*, **8**, 371–373.

Drenick, R. F. (1960), The failure law of complex equipment, *Journal of the Society of Industrial and Applied Mathematics*, **8**, 680–690.

Efron, B. (1981), Censored data and the bootstrap, *Journal of the American Statistical Association*, **76**, 312–319.

Efron, B. (1982), *The Jackknife, the Bootstrap, and Other Resampling Plans*, Philadelphia: Society for Industrial and Applied Mathematics.

Efron, B. (1985), Bootstrap confidence intervals for a class of parametric problems, *Biometrika*, **72**, 45–58.

Efron, B., and Tibshirani, R. J. (1993), *An Introduction to the Bootstrap*, New York: Chapman & Hall.

Elandt-Johnson, R. C., and Johnson, N. L. (1980), *Survival Models and Data Analysis*, New York: John Wiley & Sons.

Engelhardt, M., and Bain, L. J. (1979), Prediction limits and two-sample problems with complete or censored Weibull data, *Technometrics*, **21**, 233–237.

Engelhardt, M., Bain, L. J., and Wright, F. T. (1981), Inferences on the parameters of the Birnbaum–Saunders fatigue life distribution based on maximum likelihood estimation, *Technometrics*, **23**, 251–256.

Epstein, B., and Sobel, M. (1953), Life testing, *Journal of the American Statistical Association*, **48**, 486–502.

Escobar, L. A., and Meeker, W. Q. (1992), Assessing influence in regression analysis with censored data, *Biometrics*, **48,** 507–528.

Escobar, L. A., and Meeker, W. Q. (1994), Fisher information matrix for the extreme value, normal, and logistic distributions and censored data, *Applied Statistics,* **43**, 533–540.

Escobar, L. A., and Meeker, W. Q. (1995), Planning accelerated life tests with two or more experimental factors, *Technometrics*, **37**, 411–427.

Escobar, L. A., and Meeker, W. Q. (1998a), Statistical prediction based on censored life data, Department of Statistics, Iowa State University, Ames, IA.

Escobar, L. A., and Meeker, W. Q. (1998b), Simultaneous confidence bands for models based on location-scale distributions and censored data, Department of Statistics, Iowa State University, Ames, IA.

Escobar, L. A., and Meeker, W. Q. (1998c), Stable parameterizations for maximum likelihood methods with censored data, Department of Statistics, Iowa State University, Ames, IA.

Escobar, L. A., and Meeker, W. Q. (1998d), Fisher information matrices with censoring, truncation, and explanatory variables, *Statistica Sinica*, **8**, 221–237.

Evans, M., Hastings, N., and Peacock, B. (1993), *Statistical Distributions* (Second Edition), New York: Wiley-Interscience.

Evans, R. A. (1977), Accelerated testing: the only game in town, *IEEE Transactions on Reliability,* **R-26**, 241.

Evans, R. A. (1989), Bayes is for the birds, *IEEE Transactions on Reliability,* **R-38**, 401.

REFERENCES

Evans, R. A. (1991), Accelerated testing, *IEEE Transactions on Reliability,* **R-40**, 497.

Everitt, B. S., and Hand, D. J. (1981), *Finite Mixture Distributions*, London: Chapman & Hall.

Eyring, H. (1980), *Basic Chemical Kinetics*, New York: John Wiley & Sons.

Eyring, H., Gladstones, S., and Laidler, K. J. (1941), *The Theory of Rate Processes*, New York: McGraw Hill.

Falls, L. W. (1970), Estimation of parameters in compound Weibull distributions, *Technometrics*, **12**, 399–407.

Farewell, V. T., and Cox, D. R. (1979), A note on multiple time scales in life testing, *Applied Statistics*, **28**, 73–75.

Farewell, V. T., and Prentice, R. L. (1977), A study of distributional shape in life testing, *Technometrics*, **19**, 69–75.

Faulkenberry, G. D. (1973), A method of obtaining prediction intervals, *Journal of the American Statistical Association*, **68**, 433–435.

Feinberg, A. A., and Windom, A. (1995), The reliability physics of thermodynamic aging, Chapter 13 in *Recent Advances in Life-Testing and Reliability*, N. Balakrishnan, Editor. Boca Raton, FL: CRC Press.

Fisher, R. A. (1925), Theory of statistical estimation, *Proceedings of the Cambridge Philosophical Society*, **22**, 700–725.

Fleming, T. R., and Harrington, D. P. (1992), *Counting Processes and Survival Analysis*, New York: John Wiley & Sons.

Friedman, L. B., and Gertsbakh, I. (1980), Maximum likelihood estimation in a minimum-type model with exponential and Weibull failure modes, *Journal of the American Statistical Association*, **75**, 460–465.

Fukuda, M. (1991), *Reliability and Degradation of Semiconductor Lasers and LED's*, Boston: Artech House.

Galambos, J. (1978), *The Asymptotic Theory of Extreme Order Statistics,* New York: John Wiley & Sons.

Geisser, S. (1993), *Predictive Inference: An Introduction*, New York: Chapman & Hall.

Gelfand, A. E., and Smith, A. F. M. (1990), Sampling-based approach to calculating marginal densities, *Journal of the American Statistical Association*, **85**, 398–409.

Gelfand, A. E., and Smith, A. F. M. (1992), Bayesian analysis of constrained parameter and truncated data problems using Gibbs sampling, *Journal of the American Statistical Association*, **87**, 523–532.

Gelman, A., Carlin, J. B., Stern, H. S., and Rubin, D. B. (1995), *Bayesian Data Analysis*, New York: Chapman & Hall.

Gentleman, R., and Geyer, C. J. (1994), Maximum likelihood for interval censored data: consistency and computation, *Biometrika*, **81**, 618–623.

Gertsbakh, I. B. (1989), *Statistical Reliability Theory*, New York: Marcel Dekker.

Gertsbakh, I. B., and Kordonsky, Kh. B. (1969), *Models of Failure*, English translation from the Russian version, New York: Springer-Verlag.

Giesbrecht, F., and Kempthorne, O. (1976), Maximum likelihood estimation in the three-parameter lognormal distribution, *Journal of the Royal Statistical Society B*, **38**, 257–264.

Gillen, K. T., and Clough, R. L. (1985), A kinetic model for predicting oxidative degradation rates in combined radiation-thermal environments, *Journal of Polymer Science, Polymer Chemistry Edition*, **23**, 2683–2707.

Gillen, K. T., and Mead, K. E. (1980), Predicting life expectancy and simulating age of complex equipment using accelerated aging techniques. Available from the National Technical Information Service, U. S. Department of Commerce, 5285 Port Royal Road, Springfield, VA 22151.

Gnedenko, B., and Ushakov, I. (1995), *Probabilistic Reliability Engineering*, New York: John Wiley & Sons.

Graves, S., and Menten, T. (1996), Designing experiments to measure and improve reliability, Chapter 11 in *Handbook of Reliability Engineering and Management* (Second Edition), W. G. Ireson, C. F Coombs, and R. Y. Moss, Editors. New York: McGraw Hill.

Griffiths, D. A. (1980), Interval estimation for the three-parameter lognormal distribution via the likelihood function, *Applied Statistics*, **29**, 58–68.

Guess, F. M., Usher, J. S., and Hodgson, T. J. (1991), Estimating system and component reliabilities under partial information on cause of failure, *Journal of Statistical Planning and Inference*, **29**, 75–85.

Hahn, G. J., and Meeker, W. Q. (1982a), Pitfalls and practical considerations in product life analysis, part 1: basic concepts and dangers of extrapolation, *Journal of Quality Technology*, **14**, 144–152.

Hahn, G. J., and Meeker, W. Q. (1982b), Pitfalls and practical considerations in product life analysis, part 2: mixtures of product populations and more general models, *Journal of Quality Technology*, **14**, 177–185.

Hahn, G. J., and Meeker, W. Q. (1991), *Statistical Intervals: A Guide for Practitioners*, New York: John Wiley & Sons.

Hahn, G. J., and Nelson, W. (1973), A survey of prediction intervals and their applications, *Journal of Quality Technology,* **5**, 178–188.

Hahn, G. J., and Shapiro, S. S. (1967), *Statistical Models in Engineering,* New York: John Wiley & Sons.

Hakim, E. B. (1989), *Microelectronic Reliability, Volume I, Reliability, Test and Diagnostics*, Norwood, MA: Artech House.

Hall, P. (1992), *The Bootstrap and Edgeworth Expansion*, New York: Springer-Verlag.

Hamada, M. (1993), Reliability improvement via Taguchi's robust design, *Quality and Reliability Engineering International*, **9**, 7–13.

Hamada, M. (1995a), Using statistically designed experiments to improve reliability and to achieve robust reliability, *IEEE Transactions on Reliability*, **R-44**, 206–215.

Hamada, M. (1995b), Analysis of experiments for reliability improvement and robust reliability, Chapter 9 in *Recent Advances in Life-Testing and Reliability*, N. Balakrishnan, Editor. Boca Raton, FL: CRC Press.

Hamada, M., and Wu, C. F. J. (1995), Analysis of censored data from fractionated experiments: a Bayesian approach, *Journal of the American Statistical Association*, **90**, 467–477.

Harter, H. L. (1977), A survey of the literature on the size effect of material strength, Report No. AFFDL-TR-77-11, Air Force Flight Dynamics Lab AFSC, Wright-Patterson AFB, OH 45433.

Harter, H. L. (1984), Another look at plotting positions, *Communications in Statistics, Part B—Simulation and Computation*, **13**, 1613–1633.

Hirose, H., and Lai, T. L. (1997), Inference from grouped data in three-parameter Weibull models with applications to breakdown voltage experiments, *Technometrics*, **39**, 199–210.

Hooper, J. H., and Amster, S. J. (1990), Analysis and presentation of reliability data, in *Handbook of Statistical Methods for Engineers and Scientists*, Harrison M. Wadsworth, Editor. New York: McGraw Hill.

Hosmer, D. W. Jr. (1973), A comparison of iterative maximum likelihood estimates of the parameters of a mixture of two normal distributions under three different types of sample, *Biometrics*, **29**, 761–770.

Howes, M. J., and Morgan, D. V. (1981), *Reliability and Degradation*, New York: John Wiley & Sons.

Høyland, A., and Rausand, M. (1994), *System Reliability Theory: Models and Statistics Methods*, New York: John Wiley & Sons.

Hudak, S. J. Jr., Saxena, A., Bucci, R. J., and Malcolm, R. C. (1978), Development of standard methods of testing and analyzing fatigue crack growth rate data, Technical Report AFML-TR-78-40, Westinghouse R & D Center, Westinghouse Electric Corporation, Pittsburgh, PA 15235.

Ireson, W. G. (1996), Reliability information collection and analysis, Chapter 10 in *Handbook of Reliability Engineering and Management* (Second Edition), W. G. Ireson, C. F. Coombs, and R. Y. Moss, Editors. New York: McGraw Hill.

Jeng, S. L., and Meeker, W. Q. (1998), Comparisons of Weibull distribution confidence intervals for type I censored data, Department of Statistics, Iowa State University, Ames, IA.

Jensen, K. L., and Meeker, W. Q. (1990), ALTPLAN: microcomputer software for developing and evaluating accelerated life test plans; paper presented at the Annual ASA Meetings, Anaheim, CA.

Jensen, F. (1995), *Electronic Component Reliability: Fundamentals, Modelling, Evaluation, and Assurance*, New York: John Wiley & Sons.

Jensen, F., and Petersen, N. E. (1982), *Burn-in: An Engineering Approach to Design and Analysis of Burn-in Procedures*, New York: John Wiley & Sons.

Johnson, N. L., Kotz, S., and Balakrishnan, N. (1994), *Continuous Univariate Distributions Volume 1*, New York: John Wiley & Sons.

Johnson, N. L., Kotz, S., and Balakrishnan, N. (1995), *Continuous Univariate Distributions Volume 2*, New York: John Wiley & Sons.

Joyce, W. B., Liou, K-Y, Nash, F. R., Bossard, P. R., and Hartman, R. L. (1985), Methodology of accelerated aging, *AT&T Technical Journal*, **64**, 717–764.

Kalbfleisch, J. D. (1971), Likelihood methods of prediction, in *Foundations of Statistical Inference; Proceedings of the Symposium on the Foundations of Statistical Inference*, V. P. Godambe and D. A. Sprott, Editors. Toronto: Holt, Rinehart and Winston of Canada.

Kalbfleisch, J. D., and Lawless, J. F. (1988), Estimation of reliability in field-performance studies, *Technometrics*, **30**, 365–388 (with discussion).

Kalbfleisch, J. D., and Lawless, J. F. (1992), Some useful statistical methods for truncated data, *Journal of Quality Technology*, **24**, 145–152.

Kalbfleisch, J. D., and Prentice, R. L. (1980), *The Statistical Analysis of Failure Time Data*, New York: John Wiley & Sons.

Kalkanis, G., and Rosso, E. (1989), The inverse power law model for the lifetime of a mylar–polyurethane laminated DC HV insulating structure, *Nuclear Instruments and Methods in Physics Research*, **A281**, 489–496.

Kaplan, E. L., and Meier, P. (1958), Nonparametric estimation from incomplete observations, *Journal of the American Statistical Association*, **53**, 457–481.

Kececioglu, D., and Sun, F. (1995), *Environmental Stress Screening: Its Quantification, Optimization and Management*, Englewood Cliffs, NJ: Prentice Hall.

Kempthorne, O., and Folks, L. (1971), *Probability, Statistics, and Data Analysis*, Ames, IA: Iowa State University Press.

Kennedy, W. J., and Gentle, J. E. (1980), *Statistical Computing*, New York: Marcel Dekker.

Klinger, D. J. (1991a), On the notion of activation energy in reliability: Arrhenius, Eyring, and thermodynamics, *1991 Proceedings of the Annual Reliability and Maintainability Symposium*, 295–300, New York: Institute of Electrical and Electronics Engineers.

Klinger, D. J. (1991b), Humidity acceleration factor for plastic packaged electronic devices, *Quality and Reliability Engineering International*, **7**, 365–370.

Klinger, D. J. (1992), Failure time and rate constant of degradation: an argument for the inverse relationship, *Microelectronics and Reliability*, **32**, 987–994.

Klinger, D. J., Nakada, Y., and Menendez, M. A. (1990), *AT&T Reliability Manual*, New York: Van Nostrand Reinhold.

Klion, J. (1992), *Practical Electronic Reliability Engineering*, New York: Van Nostrand Reinhold.

Knezevic, J. (1993), *Reliability, Maintainability and Supportability*, London: McGraw Hill.

Kordonsky, Kh. B., and Gertsbakh, I. B. (1993), Choice of the best time scale for system reliability analysis, *European Journal of Operational Research*, **65**, 235–246.

Kordonsky, Kh. B., and Gertsbakh, I. B. (1995a), System state monitoring and lifetime scales— I, *Reliability Engineering and System Safety*, **47**, 1–14.

Kordonsky, Kh. B., and Gertsbakh, I. B. (1995b), System state monitoring and lifetime scales— II, *Reliability Engineering and System Safety*, **49**, 145–154.

Kozlov, B. A., and Ushakov, I. A. (1970), *Reliability Handbook*, New York: Holt, Rinehart, and Winston.

Kulldorff, G. (1961), *Contributions to the Theory of Estimation from Grouped and Partially Grouped Samples*, Stockholm: Almqvist and Wiksell.

Kuo, W. (1984), Reliability enhancement through optimal burn-in, *IEEE Transactions on Reliability*, **R-33**, 145–156.

Kuo, W., Chien, W. T. K., and Kim, T. (1998), *Reliability and Stress Burn-in*, Netherlands: Kluwer Academic Publishers.

Kuo, L., and Yang, T. L. (1996), Bayesian computation for nonhomogeneous Poisson processes in software reliability, *Journal of the American Statistical Association*, **91**, 763–773.

Lagakos, S. W. (1979), General right censoring and its impact on the analysis of survival data, *Biometrics*, **35**, 139–156.

Landers, T. L., and Kolarik, W. J. (1987), Proportional hazards analysis of field warranty data, *Reliability Engineering*, **18**, 131–139.

Lawless, J. F. (1973), On estimation of safe life when the underlying distribution is Weibull, *Technometrics*, **15**, 857–865.
Lawless, J. F. (1982), *Statistical Models and Methods for Lifetime Data*, New York: John Wiley & Sons.
Lawless, J. F. (1986), A note of lifetime regression models, *Biometrika*, **73**, 509–512.
Lawless, J. F. (1998), Statistical analysis of product warranty data, *International Statistical Review* **66**, 41–60.
Lawless, J. F., Hu, J., and Cao, J. (1995), Methods for the estimation of failure distributions and rates from automobile warranty data, *Lifetime Data Analysis*, **1**, 227–240.
Lawless, J. F., and Kalbfleisch, J. D. (1992), Some issues in the collection and analysis of field reliability data, in *Survival Analysis: State of the Art*, 141–152, J. P. Klein and P. K. Goel, Editors. Netherlands: Kluwer Academic Publishers.
Lawless, J. F., and Nadeau, C. (1995), Some simple robust methods for the analysis of recurrent events, *Technometrics*, **37**, 158–168.
Lawless, J. F., and Thiagarajah, K. (1996), A point-process model incorporating renewals and time trends, with application to repairable systems, *Technometrics*, **38**, 131–138.
Le Cam, L. (1990), Maximum likelihood: an introduction, *International Statistical Review*, **58**, 153–171.
Lee, L. (1980), Testing adequacy of the Weibull and loglinear rate models for a Poisson process, *Technometrics*, **22**, 195–199.
Lehmann, E. L. (1983), *Theory of Point Estimation*, New York: John Wiley & Sons.
Lerch, C. F., and Meeker, W. Q. (1998), A Bayes approach to analyzing accelerated life test data, Department of Statistics, Iowa State University, Ames, IA.
Lewis, E. E. (1996), *Introduction to Reliability Engineering*, New York: John Wiley & Sons.
Li, G. (1995a), On nonparametric likelihood ratio estimation of survival probabilities for censored data, *Statistics & Probability Letters*, **25**, 95–104.
Li, G. (1995b), Nonparametric likelihood ratio estimation of probabilities for truncated data, *Journal of the American Statistical Association*, **90**, 997–1003.
Lieblein J., and Zelen, M. (1956), Statistical investigation of the fatigue life of deep-groove ball bearings, *Journal of Research, National Bureau of Standards*, **57**, 273–316.
Lindley, D. V. (1972), *Bayesian Statistics, A Review*, Philadelphia: Society for Industrial and Applied Mathematics.
Lindsey, J. K. (1996), *Parametric Statistical Inference*, Oxford: Clarendon Press.
Lindstrom, M. J., and Bates, D. M. (1990), Nonlinear mixed effects models for repeated measures data, *Biometrics*, **46**, 673–687.
Liu, S., Meeker, W. Q., and Escobar, L. A. (1998), A stable parameterization for fitting the generalized gamma distribution to censored data, Department of Statistics, Iowa State University, Ames, IA.
Lu, J. C., and Pantula, S. G. (1989), A repeated-measurements model for over-stressed degradation data, Department of Statistics, North Carolina State University, Raleigh, NC 27695-8203.
Lu, C. J., and Meeker, W. Q. (1993), Using degradation measures to estimate a time-to-failure distribution, *Technometrics*, **34**, 161–174.

Lu, C. J., Meeker, W. Q., and Escobar, L. A. (1996), A comparison of degradation and failure-time analysis methods of estimating a time-to-failure distribution, *Statistica Sinica*, **6**, 531–546.

LuValle, M. J. (1990), A note on experiment design for accelerated life tests, *Microelectronics and Reliability*, **30**, 591–603.

LuValle, M. J. (1993), Experimental design and graphical analysis for checking acceleration models, *Microelectronics and Reliability*, **33**, 741–763.

LuValle, M. J., and Hines, L. L. (1992), Using step stress to explore the kinetics of failure, *Quality and Reliability Engineering International*, **8**, 361–369.

LuValle, M. J., Welsher, T. L., and Mitchell, J. P. (1986), A new approach to the extrapolation of accelerated life test data, *The Proceedings of the Fifth International Conference on Reliability and Maintainability*, 620–635, Biarritz, France.

LuValle, M. J., Welsher, T. L., and Svoboda, K. (1988), Acceleration transforms and statistical kinetic models, *Journal of Statistical Physics*, **52**, 311–320.

Mann, N. R., Schafer, R. E., and Singpurwalla, N. D. (1974), *Methods for Statistical Analysis of Reliability and Life Data,* New York: John Wiley & Sons.

Martin, J. W. (1982), Time transformation functions commonly used in life testing analysis, *Durability of Building Materials,* **1**, 175–194.

Martz, H. F., and Waller, R. A. (1982), *Bayesian Reliability Analysis,* New York: John Wiley & Sons.

McCool, J. I. (1980), Confidence limits for Weibull regression with censored data, *IEEE Transactions on Reliability*, **R-29**, 145–150.

McPherson, J. W., and Baglee, D. A. (1985), Acceleration factors for thin gate oxide stressing, *Proceedings of the IEEE International Reliability Physics Symposium*, **23**, 1–5.

Mee, R., and Kushary, D. (1994), Prediction limits for the Weibull distribution utilizing simulation, *Computational Statistics and Data Analysis,* **17**, 327–336.

Meeker, W. Q. (1984), A comparison of accelerated life test plans for Weibull and lognormal distributions and Type I censored data, *Technometrics*, **26**, 157–171.

Meeker, W. Q. (1986), Planning life tests in which units are inspected for failure, *IEEE Transactions on Reliability*, **R-35**, 571–578.

Meeker, W. Q. (1987), Limited failure population life tests: application to integrated circuit reliability, *Technometrics*, **29**, 51–65.

Meeker, W. Q., and Escobar, L. A. (1993), A review of recent research and current issues in accelerated testing, *International Statistical Review*, **61**, 147–168.

Meeker, W. Q., and Escobar, L. A. (1994), Maximum likelihood methods for fitting parametric statistical models to censored and truncated data, in *Probabilistic and Statistical Methods in the Physical Sciences*, J. Stanford and S. Vardeman, Editors. New York: Academic Press.

Meeker, W. Q., and Escobar, L. A. (1995), Teaching about approximate confidence regions based on maximum likelihood estimation, *The American Statistician*, **49**, 48–53.

Meeker, W. Q., Escobar, L. A., and Hill, D. A. (1992), Sample sizes for estimating the Weibull hazard function from censored samples, *IEEE Transactions on Reliability*, **41**, 133–138.

Meeker, W. Q., Escobar, L. A., and Lu, C. J. (1998), Accelerated degradation tests: modeling and analysis, *Technometrics*, **40**, 89–99.

Meeker, W. Q., and Hahn, G. J. (1985), *How to Plan Accelerated Life Tests: Some Practical Guidelines*, Volume 10 of the ASQC Basic References in Quality Control: Statistical Techniques. Available from the American Society for Quality Control, 310 W. Wisconsin Ave., Milwaukee, WI 53203.

Meeker, W. Q., and Hamada, M. (1995), Statistical tools for the rapid development & evaluation of high-reliability products, *IEEE Transactions on Reliability*, **R-44**, 187–198.

Meeker, W. Q., and LuValle, M. J. (1995), An accelerated life test model based on reliability kinetics, *Technometrics*, **37**, 133–146.

Meeker, W. Q., and Nelson, W. (1975), Optimum accelerated life tests for Weibull and extreme value distributions, *IEEE Transactions on Reliability*, **R-24**, 321–332.

Meeker, W. Q., and Nelson, W. (1976), Weibull percentile estimates and confidence limits from singly censored data by maximum likelihood, *IEEE Transactions on Reliability*, **R-25**, 20–24.

Meeker, W. Q., and Nelson, W. (1977), Weibull variances and confidence limits by maximum likelihood for singly censored data, *Technometrics*, **19**, 473–476.

Meeter, C. A., and Meeker, W. Q. (1994), Optimum accelerated life tests with a nonconstant scale parameter, *Technometrics*, **36**, 71–83.

MIL-HDBK-189 (1981), *Reliability Growth Management*. Available from Naval Publications and Forms Center, 5801 Tabor Ave, Philadelphia, PA 19120.

MIL-HDBK-217E (1986), *Reliability Prediction for Electronic Equipment*. Available from Naval Publications and Forms Center, 5801 Tabor Ave, Philadelphia, PA 19120.

MIL-STD-883 (1985), *Test Methods and Procedures for Microelectronics*. Available from Naval Publications and Forms Center, 5801 Tabor Ave, Philadelphia, PA 19120.

MIL-STD-1629A (1980), *Failure Modes and Effects Analysis*. Available from Naval Publications and Forms Center, 5801 Tabor Ave, Philadelphia, PA 19120.

MIL-STD-2164 (1985), *Environmental Stress Screening Process for Electronic Equipment*. Available from Naval Publications and Forms Center, 5801 Tabor Ave, Philadelphia, PA 19120.

MINITAB (1997), Minitab User's Guide 2: Data Analysis and Quality Tools, Release 12, Minitab, Inc., State College, PA.

Morgan, B. J. T. (1984), *Elements of Simulation*, London: Chapman & Hall.

Morgan, C. B. (1980), Analyzing competing failure modes using statpac and a simple actuarial technique, General Electric Corporate Research and Development technical report 80CRD110, Schenectady, NY.

Morse, P. N., and Meeker, W. Q. (1998), Simulation methods for planning an accelerated degradation test, Department of Statistics, Iowa State University, Ames, IA.

Murray, W. P. (1993), Archival life expectancy of 3M magneto-optic media, *Journal of the Magnetics Society of Japan*, **17**, Supplement S1, 309–314.

Murray, W. P. (1994), Accelerated service life prediction of compact disks, in *Accelerated and Outdoor Durability Testing of Organic Materials, ASTM STP 1202*, W. D. Ketola and D. Grossman, Editors. Philadelphia: American Society for Testing and Materials, 263–271.

Murray, W. P., and Maekawa, K. (1996), Reliability evaluation of 3M magneto-optic media, *Journal of the Magnetics Society of Japan*, **20**, Supplement S1, 309–314.

Musa, J. D., Iannino, A., and Okumoto, K. (1987), *Software Reliability: Measurement, Prediction, Application*, New York: McGraw Hill.

Nair, V. N. (1981), Plots and tests for goodness of fit with randomly censored data, *Biometrika*, **68**, 99–103.

Nair, V. N. (1984), Confidence bands for survival functions with censored data: a comparative study, *Technometrics*, **26**, 265–275.

Nakamura, T. (1991), Existence of maximum likelihood estimates for interval-censored data from some three-parameter models with a shifted origin, *Journal of the Royal Statistical Society B*, **53**, 211–220.

Nash, J. C., and Quon, T. K. (1996), Software for modeling kinetic phenomena, *The American Statistician*, **50**, 368–378.

Nelson, W. (1969), Hazard plotting for incomplete failure data, *Journal of Quality Technology*, **1**, 27–52.

Nelson, W. (1972), Theory and application of hazard plotting for censored survival data, *Technometrics*, **14**, 945–966.

Nelson, W. (1973), Analysis of residuals from censored data, *Technometrics*, **15**, 697–715.

Nelson, W. (1975a), Analysis of accelerated life data with a mix of failure modes by maximum likelihood, *IEEE Transactions on Reliability*, **R-24**, 230–237.

Nelson, W. (1975b), Graphical analysis of accelerated life data with a mix of failure modes, *IEEE Transactions on Reliability*, **R-24**, 230–237.

Nelson, W. (1981), Analysis of performance degradation data from accelerated tests, *IEEE Transactions on Reliability*, **R-30**, 3, 149–155.

Nelson, W. (1982), *Applied Life Data Analysis,* New York: John Wiley & Sons.

Nelson, W. (1984), Fitting of fatigue curves with nonconstant standard deviation to data with runouts, *Journal of Testing and Evaluation*, **12**, 69–77.

Nelson, W. (1985), Weibull analysis of reliability data with few or no failures, *Journal of Quality Technology*, **17**, 140–146.

Nelson, W. (1988), Graphical analysis of system repair data, *Journal of Quality Technology*, **20**, 24–35.

Nelson, W. (1990a), *Accelerated Testing: Statistical Models, Test Plans, and Data Analyses*, New York: John Wiley & Sons.

Nelson, W. (1990b), Hazard plotting of left truncated life data, *Journal of Quality Technology*, **22**, 230–238.

Nelson, W. (1995a), Confidence limits for recurrence data—applied to cost or number of product repairs, *Technometrics*, **37**, 147–157.

Nelson, W. (1995b), Defect initiation and growth—a general statistical model & data analysis. Paper presented at the 2nd annual Spring Research Conference, sponsored by the Institute of Mathematical Statistics and the Physical and Engineering Section of the American Statistical Association, Waterloo, Ontario, Canada, June 1995.

Nelson, W. (1995c), Weibull prediction of a future number of failures, paper presented at the 1995 Joint Statistical Meetings, Orlando, FL, August 1995.

Nelson, W. (1998), Bibliography on accelerated test plans. Available from the author, 739 Huntingdon Dr., Schenectady, NY 12309-2917.

Nelson, W., and Doganaksoy, N. (1989), A computer program for an estimate and confidence limits for the mean cumulative function for cost or number of repairs of repairable products, TIS report 89CRD239, General Electric Company Research and Development, Schenectady, NY.

Nelson, W., and Doganaksoy, N. (1995), Statistical analysis of life or strength data from specimens of various size using the power-(log)normal model, Chapter 21 in *Recent Advances in Life-Testing and Reliability*, N. Balakrishnan, Editor. Boca Raton, FL: CRC Press.

Nelson, W., and Kielpinski, T. (1976), Theory for optimum censored accelerated life tests for normal and lognormal distributions, *Technometrics*, **18**, 105–114.

Nelson, W., and Meeker, W. (1978), Theory for optimum accelerated censored life tests for Weibull and extreme value distributions, *Technometrics*, **20**, 171–177.

Nelson, W., and Schmee, J. (1981), Prediction limits for the last failure of a (log) normal sample from early failures, *IEEE Transactions on Reliability,* **R-30**, 461–463.

Nelson, W., and Thompson, V. C. (1971), Weibull probability plots, *Journal of Quality Technology*, **3**, 45–50.

Neufelder, A. (1993), *Ensuring Software Reliability*, New York: Marcel Dekker.

Neter, J., Kutner, M. H., Nachtsheim, C. J., and Wasserman, W. (1996), *Applied Linear Statistical Models* (Fourth Edition), Homewood, IL: Richard D. Irwin.

O'Connor, P. D. T. (1985), *Practical Reliability Engineering* (Second Edition), New York: John Wiley & Sons.

Odeh, R. E., and Owen, D. B. (1980), *Tables for Normal Tolerance Limits, Sampling Plans, and Screening*, New York: Marcel Dekker.

Ostrouchov, G., and Meeker, W. Q. (1988), Accuracy of approximate confidence bounds computed from interval censored Weibull and lognormal data, *Journal of Statistical Computation and Simulation*, **29**, 43–76.

Owen, A. (1990), Empirical likelihood confidence regions, *The Annals of Statistics,* **18**, 90–120.

Parida, N. (1991), Reliability and life estimation from component fatigue failures below the go–no-go fatigue limit, *Journal of Testing and Evaluation*, **19**, 450–453.

Pascual, F. G., and Meeker, W. Q. (1997), Regression analysis of fatigue data with runouts based on a model with nonconstant standard deviation and a fatigue limit parameter, *Journal of Testing and Evaluation*, **25**, 292–301.

Pascual, F. G., and Meeker, W. Q. (1998a), Estimating fatigue curves with the random fatigue-limit model. Preprint, Department of Statistics, Iowa State University, Ames, IA.

Pascual, F. G., and Meeker, W. Q. (1998b), The modified sudden death test: planning life tests with a limited number of test positions. Preprint, Department of Statistics, Iowa State University, Ames, IA.

Patel, J. K. (1989), Prediction intervals—a review, *Communications in Statistics—Theory and Methods*, **18**, 2393–2465.

Pecht, M., Editor (1995), *Product Reliability, Maintainability, and Supportability Handbook*, Boca Raton, FL: CRC Press.

Peck, D. S. (1986), Comprehensive model for humidity testing correlation, *Proceedings of the IEEE International Reliability Physics Symposium*, **24**, 44–50.

Peck, D. S., and Zierdt, C. H. Jr. (1974), The reliability of semiconductor devices in the Bell System, *Proceedings of the IEEE*, **62**, 185–211.

Peto, R. (1973), Experimental survival curves for interval-censored data, *Applied Statistics*, **22**, 86–91.

Phadke, M. S. (1989), *Quality Engineering Using Robust Design*, Englewood Cliffs, NJ: Prentice Hall.

Pinheiro, J. C., and Bates, D. M. (1995a), Approximations to the loglikelihood function in the nonlinear mixed effects model, *Journal of Computational and Graphical Statistics*, **4**, 12–35.

Pinheiro, J. C., and Bates, D. M. (1995b), Mixed effects models, methods, and classes for S and S-PLUS. Department of Statistics, University of Wisconsin. Available from Statlib.

Pollino, E. (1989), *Microelectronic Reliability, Volume II, Reliability, Integrity Assessment and Assurance*, Norwood, MA: Artech House.

Prentice, R. L. (1974), A log gamma model and its maximum likelihood estimation, *Biometrika*, **61**, 539–544.

Prentice, R. L. (1975), Discrimination among some parametric models, *Biometrika*, **62**, 607–614.

Proschan, F. (1963), Theoretical explanation of observed decreasing failure rate, *Technometrics*, **5**, 375–383.

Rao, C. R. (1973), *Linear Statistical Inference and Its Applications*, New York: John Wiley & Sons.

Ripley, B. D. (1987), *Stochastic Simulation*, New York: John Wiley & Sons.

Robinson, J. A. (1983), Bootstrap confidence intervals in location-scale models with progressive censoring, *Technometrics*, **25**, 179–187.

Robinson, J. A. (1995), Standard errors for the mean number of repairs on systems from a finite population, Chapter 11 in *Recent Advances in Life-Testing and Reliability*, N. Balakrishnan, Editor. Boca Raton, FL: CRC Press.

Robinson, J. A., and McDonald, G. C. (1991), Issues related to field reliability and warranty data, in *Data Quality Control: Theory and Pragmatics*, G. E. Liepins and V. R. R. Uppuluri, Editors. New York: Marcel Dekker.

Ross, G. J. S. (1990), *Nonlinear Estimation*, New York: Springer-Verlag.

SAS Institute Inc. (1995), *JMP User's Guide*, Version 3.1, Cary, NC: SAS Institute.

SAS Institute Inc. (1997), SAS/QC Software: Changes and Enhancements for Release 6.12, Cary, NC: SAS Institute.

Schinner, C. (1996), Accelerated Testing, Chapter 12 in *Handbook of Reliability Engineering and Management* (Second Edition), W. G. Ireson, C. F. Coombs, and R. Y. Moss, Editors. New York: McGraw Hill.

Schmee, J., and Hahn, G. J. (1979), A simple method for regression analysis with censored data, *Technometrics*, **21**, 417–432.

Schneider, H. (1986), *Truncated and Censored Samples for Normal Populations*, New York: Marcel Dekker.

Schrödinger, E. (1915), Zur theorie der fall-und steigversuche an teilchen mit Brownscher bewegung, *Physikalische Zeitschrift*, **16**, 289–295.

Seber, G. A. F. (1977), *Linear Regression Analysis*, New York: John Wiley & Sons.

Seber, G. A. F., and Wild, C. J. (1989), *Nonlinear Regression*, New York: John Wiley & Sons.

Severini, T. A. (1991), On the relationship between Bayesian and non-Bayesian interval estimates, *Journal of the Royal Statistical Society B*, **53**, 611–618.

Shao, J., and Tu, D. (1995), *The Jackknife and Bootstrap*, New York: Springer-Verlag.

Shiomi, H., and Yanagisawa, T. (1979), On distribution parameter during accelerated life test for a carbon film resistor, *Bulletin of the Electrotechnical Laboratory*, **43**, 330–345 (in Japanese).

Shooman, M. L. (1983), *Software Engineering: Design, Reliability, and Management*, New York: McGraw Hill.

Singpurwalla, N. D. (1988a), An interactive PC-based procedure for reliability assessment incorporating expert opinion and survival data, *Journal of the American Statistical Association*, **83**, 43–51.

Singpurwalla, N. D. (1988b), Foundational issues in reliability and risk analysis, *SIAM Review*, **30**, 264–282.

Singpurwalla, N. D., Castellino, V. C., and Goldschen, D. Y. (1975), Inference from accelerated life tests using Eyring type re-parameterizations, *Naval Research Logistics Quarterly*, **22**, 289–296.

Smith, A. F. M., and Gelfand, A. E. (1992), Bayesian statistics without tears: a sampling–resampling perspective, *The American Statistician*, **46**, 84–88.

Smith, J. S. (1996), Physics of failure, Chapter 14 in *Handbook of Reliability Engineering and Management* (Second Edition), W. G. Ireson, C. F. Coombs, and R. Y. Moss, Editors. New York: McGraw Hill.

Smith, R. L. (1985), Maximum likelihood estimation in a class of nonregular cases, *Biometrika*, **72**, 67–90.

Smith, R. L., and Naylor, J. C. (1987), A comparison of maximum likelihood and Bayesian estimators for the three-parameter Weibull distribution, *Applied Statistics*, **36**, 358–369.

Snyder, D. L. (1975), *Random Point Processes*, New York: John Wiley & Sons.

S-PLUS, Statistical Sciences, Inc. (1996), S-PLUS *User's Manual, Volumes 1–2*, Version 3.4, Seattle: Statistical Sciences.

Sprott, D. A. (1973), Normal likelihoods and relation to a large sample theory of estimation, *Biometrika*, **60**, 457–465.

Starke, E. A., et al. (1996), *Accelerated Aging of Materials and Structures*, Publication NMAB-479, Washington, DC: National Academy Press.

Stuart A., and Ord, K. J. (1994), *Kendall's Advanced Theory of Statistics,* New York: Edward Arnold.

Sundararajan, C. (1991), *Guide to Reliability Engineering*, New York: Van Nostrand Reinhold.

Suzuki, K. (1985), Estimation method of lifetime based on the record of failures during the warranty period, *Journal of the American Statistical Association,* **80**, 66–72.

Suzuki, K., Maki, K., and Yokogawa, S. (1993), An analysis of degradation data of a carbon film and properties of the estimators, in *Statistical Sciences and Data Analysis*, K. Matusita, M. Puri, and T. Hayakawa, Editors. Utrecht, Netherlands: VSP, 501–511.

Thatcher, A. R. (1964), Relationships between Bayesian and confidence limits for prediction (with discussion), *Journal of the Royal Statistical Society B,* **26**, 176–210.

Thomas, R. (1964), When is a life test truly accelerated? *Electronic Design*, January 6, 64–71.

Thomas, D. R., and Grunkemeier, G. L. (1975), Confidence interval estimation of survival probabilities for censored data, *Journal of the American Statistical Association,* **70**, 865–871.

Thompson, W. A. (1981), On the foundations of reliability, *Technometrics*, **23**, 1–14.

Thompson, W. A. (1988), *Point Process Models with Applications to Safety and Reliability*, London: Chapman & Hall.

Titterington, D. M., Smith, A. F. M., and Makov, U. E. (1985), *Statistical Analysis of Finite Mixture Distributions*, New York: John Wiley & Sons.

Tobias, P. A., and Trindade, D. C. (1995), *Applied Reliability* (Second Edition), New York: Van Nostrand Reinhold.

Tomsky, J. (1982), Regression models for detecting reliability degradation, *1982 Proceedings of the Annual Reliability and Maintainability Conference*, 238–244, New York: Institute of Electrical and Electronics Engineers.

Trindade, D. C. (1991), Can burn-in screen wearout mechanisms? Reliability models of defective subpopulations—a case study, *Proceedings of the IEEE International Reliability Physics Symposium*, **29**, 260–263.

Tseng, S. T., Hamada, M., and Chiao, C. H. (1995), Using degradation data from a factorial experiment to improve fluorescent lamp reliability, *Journal of Quality Technology*, **27**, 363–369.

Tseng, S. T., and Wen, Z. C. (1997), Step-stress accelerated degradation analysis for highly reliable products, Institute of Statistics, National Tsing-Hua University.

Tseng, S. T., and Yu, H. F. (1997), A termination rule for degradation experiment, *IEEE Transactions on Reliability*, **46**, 130–133.

Turnbull, B. W. (1976), The empirical distribution function with arbitrary grouped, censored, and truncated data, *Journal of the Royal Statistical Society*, **38**, 290–295.

Tustin, W. (1990), Shake and bake the bugs out, *Quality Progress,* September, 61–64.

Tweedie, M. C. K. (1945), Inverse statistical variates, *Nature,* **155**, 453.

Ushakov, I., Editor (1994), *Handbook of Reliability Engineering*, New York: John Wiley & Sons.

Vander Wiel, S. A., and Meeker, W. Q. (1990), Accuracy of approximate confidence bounds using censored Weibull regression data from accelerated life tests, *IEEE Transactions on Reliability*, **R-39**, 346–351.

Viertl, R. (1988), *Statistical Methods for Accelerated Life Testing*, Göttingen: Vandenhoeck and Ruprecht.

Wald, A. (1947), *Sequential Analysis*, New York: John Wiley & Sons.

Wang, C. J. (1991), Sample size determination of bogey tests without failures, *Quality and Reliability Engineering International*, **7**, 35–38.

Weis, E. A., Caldararu, D., Snyder, M. M., and Croitoru, N. (1986), Investigating reliability attributes of silicon photodetectors, *Microelectronics and Reliability*, **26**, 1099–1110.

Weston, R., and Schwarz, H. A. (1972), *Chemical Kinetics*, Englewood Cliffs, NJ: Prentice Hall.

Weston, S. A., and Meeker, W. Q. (1990), Coverage probabilities of nonparametric simultaneous confidence bands for a survival function, *Journal of Statistical Simulation and Computation*, **32**, 83–97.

Wilk, M. B., Gnanadesikan, R., and Huyett, M. J. (1962a), Probability plots for the gamma distribution, *Technometrics*, **4**, 1–20.

Wilk, M. B., Gnanadesikan, R., and Huyett, M. J. (1962b), Estimation of the parameters of the gamma distribution using order statistics, *Biometrika*, **49**, 525–545.

Woodroofe, M. (1972), Maximum likelihood estimate of a translation parameter of a truncated distribution, *Annals of Mathematical Statistics*, **43**, 113–122.

Woodroofe, M. (1974), Maximum likelihood estimate of a translation parameter of a truncated distribution II, *Annals of Statistics*, **2**, 474–488.

Yanagisawa, T. (1997), Estimation of the degradation of amorphous silicon solar cells, *Microelectronics and Reliability*, **37**, 549–554.

Yokobori, T. (1951), Fatigue fracture in steel, *Journal of the Physical Society of Japan*, **6**, 81–86.

Yu, H. F., and Tseng, S. T. (1998), On-line procedure for terminating an accelerated degradation test, *Statistica Sinica*, **8**, 207–220.

Zelen, M. (1959), Factorial experiments in life testing, *Technometrics*, **1**, 269–288.

Author Index

Aalen, O., 42, 67
Abernethy, R. B., xviii, 144, 193, 250, 310, 633, 634
Akritas, M. G., 228
Amster, S. J., 383, 493, 578, 606, 637
Anderson, P. K., 67
Anscombe, F. J., 169
Ansell, J. I., 22, 420
Ascher, H., 415, 420
Atwood, C. L., 312
Azem, A., 420

Bagdonavičius, V., 460
Baglee, D. A., 485
Bailey, R. A., 519
Bain, L. J., 167, 284, 312
Balakrishnan, N., 93, 117, 118
Barlow, R. E., 118, 374, 389
Barnett, V., 149
Bates, D. M., 326, 567
Baxter, L. A., 22
Bayer, R. G., 489
Beckwith, J. P., 578
Bendell, A., 458
Beran, R., 312
Berger, R. L., 168
Berkson, J., 154
Bhattacharyya, G. K., 622
Billingsley, P., 624
Birnbaum, Z. W., 105, 389
Black, J. R., 487
Blischke, W. R., 606
Blom, G., 149
Blumenthal, S., 408
Boccaletti, G., 473, 487
Bogdanoff, J. L., 15, 260, 639
Borgan, O., 67
Borri, F. R. (*see* Boccaletti, G.)
Bossard, P. R. (*see* Joyce, W. B.)

Boulanger, M., 578, 601
Box, G. E. P., 15, 363
Boyko, K. C., 485
Breneman, J. E., 144, 193, 310, 633, 634
Bro, P., 489
Brown, H. M., 14
Brownlee, K. A., 50
Brush, G. G., 383, 606
Bucci, R. J., 15
Buckland, S. T., 228
Byrne, D., 605

Caldararu, D., 71
Canner, J., 519
Cao, J., 606
Carey, M. B., 320, 578
Carlin, J. B., 284, 363
Casella, G., 168
Castellino, V. C., 512, 641
Castillo, E., 93, 489
Chaloner, K., 559
Chan, C. K., 578
Chan, V., 590, 591, 592, 601
Chang, D. S., 578
Cheng, R. C. H., 118, 197, 284
Chernoff, H., 149
Chhikara, R. S., 103, 118
Chiao, C. H., 340
Chien, W. T. K., 521
Chow, S. C., 578
Christou, A., 489
Clough, R. L., 489
Cohen, A. C., 284
Collett, D., 445
Condra, L. W., 2, 605
Confer, R., 519
Costa, J. M., 489
Cox, D. R., 22, 110, 168, 300, 312, 420, 443, 460

665

Cragnolino, G., 489
Croitoru, N., 71
Crow, E. L., 93
Crow, L., 420
Crowder, M. J., 420, 444
Crowder, S. V., 340

D'Esponosa, F. (*see* Boccaletti, G.)
Dale, C. J., 458
David, H. A., 149, 389
Davidian, M., 340
Davis, D. J., 11, 12
Day, N. E., 284
Deming, W. E., 16, 517
Derringer, G. C., 529
Desmond, A. F., 108
Doganaksoy, N., 118, 197, 395, 400, 404, 420, 635, 636
Dowling, N. E., 319, 471
Drapella, A., 489
Draper, N. R., 15
Drenick, R. F., 408

Efron, B., 226, 227, 228, 312, 332, 333, 568
Elandt-Johnson, R. C., 42, 67
Engelhardt, M., 167, 284, 312
Epstein, B., 168
Escobar, L. A., 118, 169, 197, 241, 244, 250, 251, 284, 302, 306, 309, 312, 340, 442, 460, 529, 552, 559, 561, 578, 601, 625, 627, 628
Evans, M., 93, 118
Evans, R. A., 363, 489
Everitt, B. S., 118, 284
Eyring, H., 473

Falls, L. W., 284
Farewell, V. T., 22, 100, 118, 284
Faulkenberry, G. D., 312
Feinberg, A. A., 489
Feingold, H., 415, 420
Fisher, R. A., 168
Fleming, T. R., 67
Folks, J. L., 103, 118, 168, 284
Friedman, L. B., 169, 386
Fukuda, M., 489

Galambos, J., 93, 489
Geisser, S., 312, 364
Gelfand, A. E., 363, 364, 586
Gelman, A., 284, 363
Gentle, J. E., 93
Gentleman, R., 68
Gerlach, D. L., 485

Gertsbakh, I. B., 22, 169, 339, 386, 389
Geyer, C. J., 68
Ghio, E. (*see* Boccaletti, G.)
Giesbrecht, F., 284
Gilbert, R. A., 519
Gill, R. D., 67
Gillen, K. T., 487, 489
Giltinan, D. M., 340
Gladstones, S., 473
Gnanadesikan, R., 149, 285
Gnedenko, B., 389
Goldschen, D. Y., 512, 641
Graves, S., 605
Greenwood, J. A., 408
Griffiths, D. A., 284
Grunkemeier, G. L., 68
Guess, F. M., 386

Hahn, G. J., 16, 22, 49, 116, 130, 149, 161, 192, 250, 299, 312, 460, 517, 519, 535, 559, 584, 601, 619
Hakim, E. B., 489
Hall, P., 228
Hamada, M., 18, 340, 364, 605, 607
Hand, D. J., 118, 284
Harrington, D. P., 67
Harter, H. L., 149, 489
Hartman, R. L., (*see* Joyce, W. B.)
Hastings, N., 93, 118
Herbach, L., 408
Hill, D. A., 251
Hines, L. L., 489, 522
Hinkley, D. V., 168
Hirose, H., 284
Hodgson, T. J., 386
Hooper, J. H., 493, 578, 637
Hosmer, D. W., Jr., 284
Howes, M. J., 489
Høyland, A. 388, 389, 420
Hu, J., 606
Hudak, S. J., Jr., 15
Husebye, E., 42
Huyett, M. J., 149, 285

Iannino, A., 420
Iles, T. C., 118, 197, 284
Ireson, W. G., 380
Isham, V., 420

Jeng, S. L., 228, 627
Jensen, F., 488, 521
Jensen, K. L., 559
Johnson, N. L., 42, 93, 67, 117, 118
Joyce, W. B., 487

Kalbfleisch, J. D., 22, 42, 267, 271, 284, 312, 460, 606
Kalkanis, G., 479, 639
Kaplan, E. L., 67
Kececioglu, D., 521
Keiding, N., 67
Kempthorne, O., 168, 284
Kennedy, W. J., 93
Kielpinski, T., 558
Kim, T., 521
Kimber, A. C., 420
Klinger, D. J., 380, 473, 477, 486, 487, 488
Klion, J., 389
Knezevic, J., 339
Koenig, R. H., 320, 578
Kolarik, W. J., 458
Kordonsky, Kh. B., 22, 339
Kotz, S., 93, 117, 118
Kozin, F., 15, 260, 639
Kozlov, B. A., 389
Kulldorff, G., 42, 168
Kuo, L., 420, 423
Kuo, W., 521
Kushary, D., 302, 312
Kutner, M. H., 460

Lagakos, S. W., 42
Lai, T. L., 284
Laidler, K. J., 473
Landers, T. L., 458
Larntz, K., 559
Lawless, J. F., 4, 22, 42, 67, 149, 167, 168, 267, 268, 271, 284, 294, 312, 399, 410, 420, 445, 460, 606
Le Cam, L., 169
Lee, L., 395, 420
Lehmann, E. L., 168, 622
Lerch, C. F., 586, 601
Levy, S. C., 489
Lewis, E. E., 389
Lewis, P. A. W., 420
Li, G., 68
Lieberman, G. J., 149
Lieblein, J., 4
Lindley, D. V., 363
Lindsey, J. K., 168
Lindstrom, M. J., 326
Liou, K-Y (*see* Joyce, W. B.)
Liu, J. P., 578
Liu, S., 118, 284
Lu, C. J., 340, 578, 639
Lu, J. C., 578
LuValle, M. J., 13, 14, 320, 460, 477, 487, 489, 522, 529

Maekawa, K., 340, 578
Mains D. E., 14
Maki, K., 340, 471
Makov, U. E., 118, 284
Malcolm, R. C., 15
Mann, N. R., 473, 529
Martin, J. W., 460
Martz, H. F., 363
McCool, J. I., 461
McDonald, G. C., 380, 606
McPherson, J. W., 485
Medlin, C. H., 144, 193, 310, 633, 634
Mead, K. E., 487
Mee, R., 302, 312
Meeker, W. Q., 4, 6, 13, 14, 16, 22, 49, 67, 116, 118, 161, 169, 192, 197, 228, 241, 244, 250, 251, 284, 299, 302, 306, 309, 312, 320, 340, 442, 460, 517, 519, 529, 535, 550, 552, 559, 561, 578, 584, 586, 590, 591, 592, 594, 597, 601, 607, 625, 627, 628, 639
Meeter, C. A., 559
Meier, P., 67
Menendez, M. A., 380, 488
Menten, T., 605
Mercer, A. D., 489
Mitchell, J. P., 487
Moeschberger, M. L., 389
Morgan, B. J. T., 93
Morgan, C. B., 41, 633
Morgan, D. V., 489
Morse, P. N., 597, 601
Murray, W. P., 340, 578, 643
Murthy, D. N. P., 606
Musa, J. D., 420

Nachtsheim, C. J., 460
Nadeau, C., 399, 420, 606
Nair, V. N., 61, 67, 149
Nakada, Y., 380, 488
Nakamura, T., 118
Nash, F. R. (*see* Joyce, W. B.)
Nash, J. C., 489
Naylor, J. C., 284, 364
Nelson, W., xviii, 8, 12, 22, 67, 118, 149, 168, 171, 193, 195, 200, 241, 244, 251, 284, 312, 323, 339, 340, 389, 395, 397, 398, 399, 400, 404, 420, 429, 442, 443, 452, 460, 472, 473, 487, 488, 519, 521, 522, 523, 529, 547, 548, 549, 550, 558, 559, 578, 593, 596, 601, 625, 630, 632, 635, 636, 638
Neter, J., 460
Neufelder, A., 420
Nikulin, M., 460

O'Connor, P. D. T., 59, 389, 630
Oakes, D., 110, 460
Odeh, R. E., 192
Okumoto, K., 420
Ord, K. J., 619
Ostrouchov, G., 197, 627
Owen, A., 68
Owen, D. B., 192

Pantula, S. G., 578
Parida, N., 68
Pascual, F. G., 250, 594, 601
Patel, J. K., 312
Peacock, B., 93, 118
Pecht, M., 420
Peck, D. S., 487
Petersen, N. E., 521
Peto, R., 68
Phadke, M. S., 605
Phillips, M. J., 22, 420
Pinheiro, J. C., 326, 567
Pollino, E., 489
Prentice, R. L., 42, 100, 118, 284, 460
Proschan, F., 118, 374, 389, 585

Quinlan, J., 605
Quon, T. K., 489

Rao, C. R., 42, 168, 622
Rausand, M., 388, 389, 420
Reinman, G. L., 144, 193, 310, 633, 634
Ripley, B. D., 93
Robinson, J. A., 192, 228, 380, 420, 606
Ross, G. J. S., 442
Rosso, E., 479, 639
Rubin, D. B., 284, 363

Saperstein, B., 383, 606
Saunders, S. C., 105
Saxena, A., 15
Schafer, R. E., 473, 529
Schinner, C., 520
Schmee, J., 197, 312, 460
Schneider, H., 284
Schrödinger, E., 104
Schwarz, H. A., 473
Seber, G. A. F., 460
Severini, T. A., 364
Shao, J., 228
Shapiro, S. S., 130, 149, 619
Shimizu, K., 93
Shiomi, H., 471, 631
Shooman, M. L., 420

Singpurwalla, N. D., 363, 473, 512, 529, 641
Smith, A. F. M., 118, 284, 363, 364, 586
Smith, J. S., 488
Smith, R. L., 284, 364, 420, 623
Snell, E. J., 443
Snyder, D. L., 420
Snyder, M. M., 71
Sobel, M., 168
Sprott, D. A., 169
Sridhar, N., 489
Starke, E. A., 489
Stern, H. S., 284, 363
Stuart A., 619
Sun, F., 521
Sundararajan, C., 389
Suzuki, K., 340, 471, 606
Svoboda, K., 477
Sweeting, T. J., 420

Thatcher, A. R., 312
Thiagarajah, K., 410, 420
Thomas, D. R., 68
Thomas, R., 489
Thompson, V. C., 149
Thompson, W. A., 80, 420
Tiao, G. C., 363
Tibshirani, R. J., 226, 227, 228, 312, 332, 333, 568
Titterington, D. M., 118, 284
Tobias, P. A., 340, 488, 578
Tomsky, J., 339
Tortorella, M., 22, 578
Trindade, D. C., 284, 340, 488, 578
Trostle, T., 519
Tseng, S. T., 340, 578
Tu, D., 228
Turnbull, B. W., 68, 268, 284
Tustin, W., 521
Tweedie, M. C. K., 104

Ushakov, I. A., 389, 460
Usher, J. S., 386

Vander Wiel, S. A., 627
Viertl, R., 529

Wald, A., 104
Waller, R. A., 363
Wang, C. J., 250
Wasserman, W., 460
Weis, E. A., 71
Welsher, T. L., 477, 487
Wen, Z. C., 578
Weston, R., 473

AUTHOR INDEX

Weston, S. A., 67
Whitten, B. J., 284
Wild, C. J., 460
Wilk, M. B., 149, 285
Windom, A., 489
Woodroofe, M., 284, 623
Wright, F. T., 284
Wu, C. F. J., 364

Yanagisawa, T., 340, 471, 631
Yang, T. L., 420, 423
Yokobori, T., 260
Yokogawa, S., 340, 471
Yu, H. F., 340

Zelen, M., 4, 447, 448, 485
Zierdt, C. H., Jr., 487

Subject Index

See the Author Index for a listing of all referenced authors. See Appendix A for a listing of notation. Page numbers for some of the terms listed in Appendix A can be found in this Subject Index. Particular data sets and applications are listed under Examples. A listing of distributions and page of first introduction is given under Distribution, name. For pages giving technical information and applications of the more important distributions, look under the distribution name (e.g., Weibull, applications). When there are more than a few page references for a subject and if the first listed page is not the most appropriate for general first purposes, the suggested page number is set in bold.

Accelerated degradation test, *see* ADT
Accelerated life test, *see* ALT
Accelerated tests (AT), *see also* ADT; ALT
   models, 466, **469**
   other kinds of, 517
      burn-in, 520. *See also* Burn-in
      continuous operation, 518
      elephant tests, 519
      ESS, 519
      STRIFE, 519
   planning, Bayesian methods, 529
   practical considerations, 489, 529
   research issues, 529
   screening, 519
Acceleration models, 469
   current density, 487
   empirical, 469, 576
   generalized Eyring, 485
   guidelines, 487
   humidity, 487
   multiple stress variables, 484
   physical, 469
   temperature, 471–479, 485, 487. *See also* Arrhenius
   use rate, 470
   voltage-stress, 479–485
Actuarial estimate, 64
ADT models and analysis
   applications, 471, 578

approximate analysis, 574–577
chemical reaction, single-step, 472, 566
confidence intervals, 568
empirical models, 576
estimation, 566–569
function of parameters, estimation of, 566
Markov process for, 578
measurement error, 565
models, 565
one observation per unit, 578
random parameters, 565
repeated measures models, 578
starting values for ML iterations, 579
step-stress data, 578
versus ALT analysis, 569
ADT planning, 578, 597–601
ALT models and analysis, 493–517
   checking assumptions, 501
   data analysis strategy, 495
   functions of parameters, estimation of, 503–504
   given activation energy, 511
   interval data, 508
   ML fit
      comparison, individual versus model, 500
      individual groups, 496, 505, 509, 513
      overall model, 499, 506, 514
   multiple probability plot, 496–497, 505–506, 509, 513–514, 573–574
   potential pitfalls, 522–528

ALT models and analysis (*continued*)
   practical considerations, 487–489, 515
   residual analysis, 501, 507
   single accelerating variable, 495–512
   two accelerating variables, 512–515
   use conditions, estimation at, 503
ALT planning, *see also* Test planning
   allocation of units, 541
   Bayesian methods, 529, 559
   comparison of test plans, 554, 557
   compromise test plans, 545–547, 554–556
   constraint
      experimental region, 540, 555
      time, 535, 541, 555
   criteria, 541
   degenerate plans, 550
      splitting, 554–556
   evaluation of test plan properties, 538–540
   experimental region, 540
   levels, accelerating variables, 540
   model assumptions, 536
   multiple-variable plans, 547, 558
   nonconstant spread, 559
   one-variable plans, 536–547
   optimum plans
      lack of robustness, 544
      one-variable, 542–544
      splitting, 552–554
      theory, 559
      two-variable, 551–554
   planning information, 535
   practical aspects, 559
   prior information, using, 559
   software, 559
   theory, 529, 559, 625
   traditional test plans, 536–538
   two-variable plans, 547–559
   use conditions, test at, 541
   voltage-stress/temperature, 549
   voltage-stress/thickness, 547
Analytic studies, 16, 517
Arrhenius
   plot, 477, 491, 500, 511
   relationship, 472–479, 485–487, 498, 508, 514, 535, 549, 565, 570, 586
Asymptotic
   covariance matrix, 625
   ML theory
      nonregular models, 623
      regular models, 625
   standard error, 55, 162, 626
Asymptotic degradation level, 320, 475, 565
Audit, reliability testing, 606

Availability, 370
Average life, *see* Mean

B10, 193. *See also* Quantile
Bathtub curve, 29
Bayesian methods, 49, **343**, 344–368
   accelerated testing, 586–589
   applications in reliability, 363
   basic ideas, 344, 363
   Bayes's rule, 344
   compared with frequentist methods, 362, 364
   confidence intervals, 356–357
   highest posterior density, 356
   NHPP, 420
   point estimation, 356
   posterior pdf, 344, 350–355
      function of parameters, 357
      joint, 353
      marginal, 353
      numerical integration, 351
      using simulation, 351–353, 359–362
   practical considerations, 362–363
   prediction, 358–362
   prior information, 344
      cautions on the use, 363
      combining with likelihood, 350
      noninformative, 346, 587
      proper, 346
      sources, 347, 380
   Weibull three-parameter, 364
Birnbaum–Saunders distribution, 105
   comparison with lognormal, 260
   ML fitting, 260, 284
Boltzmann constant, 472
Bootstrap, 204–227
   discreteness of estimates, 221
   distribution of statistics, 220
   nonparametric, 207, 217–225
      few observations, 207
      limitation, 218
   parametric, 206–217
   potential problems, 220
   sampling methods, 206
Bootstrap confidence intervals, 210–227
   bootstrap-t, 209
   exact, 205
   function of parameters, 211, 215
   log-location, 212
   logit transformation, 216
   mean, 210
   methods for censored data, 228
   nonparametric, 217–225
   parametric, 206–217
   percentile method, 226–227, 332–333, 568–569

SUBJECT INDEX

software for, 228
system reliability, 381
transformations, importance of, 216
Type II censoring, 228
versus likelihood-based, 205, 212, 216, 228
versus normal-approximation, 205, 228
Burn-in, 6–8, **520**

Case studies, 582
cdf, 28
Censoring, 3, **34**
arbitrary, 65, 68, 135
assumptions, 35
biased inferences from, 42, 583
dangers with mixed population, 583
failure (Type II), 34
interval, 10–12, **34**, 35, 37–38, 132, 155, 508
left, 12, **38**
mechanisms, 34
multiply, 35
noninformative, 35
overlapping intervals, 68
random, 35
right, 12, **34**, 35, 38–39
time (Type I), 34
unbalanced, 585
uncertain censoring times, 64
Chemical reaction, rate of, 472
Coefficient of skewness, 78
Coefficient of variation, 78, 110
Comparison, two samples
ALT data, 525
life data, 8, 450, 583
recurrence data, 404, 636
Competing risks, *see* Multiple causes of failure
Conditional failure probabilities, 32, 268, 271
Confidence bounds, one-sided, 161, 187
comparison, time-censored data, 161, 187, 228
no failures, 167–168, 195
Confidence intervals
approximate, 49
bootstrap, *see* Bootstrap confidence intervals
conservative, 50
coverage probability, 49
exact, 49
failure probabilities, 164, 182–184, 331, 503–504. *See also* nonparametric
function of parameters, 50, 181–182, 504
half-width, 238
measures of precision, 238, 239
non log-location-scale distributions, 255
nonparametric, 49–52, 54–57
one-sided, *see* Confidence bounds

sampling error, relationship to, 49
significance tests, relationship to, 183
simultaneous, 60–63, 164, 197
system reliability, 381
Confidence intervals, likelihood based, 177
advantages, 197
function of parameters, 182
Monte-Carlo, using, 228, 601
parameters, 160, 179
theory for, 627
versus bootstrap-based, 212, 216, 218, 228
Confidence intervals, normal-approximation
failure probabilities, 55
function of parameters, 163, 189–191
improved, 165, 192, 402
relation to likelihood intervals, 169, 228
shortcomings, 186
theory for, 628
transformation, using, 163, 188, 190
Wald method, 628
Confidence level, 49
Confidence region
χ^2 calibration, 160, 179
parameters, 177, 179
Convergence in distribution, 623
Covariance matrix, 186
estimate, 187
large-sample approximation, 270
local estimate, 626
Coverage probability, 49, 228, 293
Cox–Snell residuals, *see* Regression
Cumulative hazard, 29, 45
nonparametric estimation, 67
Customer expectations, 602

Data, *see* Examples; Reliability data
Data analysis strategy, 21
Degradation, 317. *See also* ADT
approximate analysis, 336–340
comparison with traditional failure time analyses, 333–339
confidence intervals, 332–333
data, 15, 317, 642–643
destructive measurements, 339
failure time cdf
estimation of, 331
evaluation of, 328–330
hard failures, 328
level defining failure, 327, 565
likelihood, 326
limitations, 323
linear, 319, 329, 340, 478
measurement error, 323

Degradation (*continued*)
  models, 317, 326, 489
    deterministic, 318
    differential equations, 319–320
    relationship to failure-time, 318
    shape, 319–320
    soft failures, 327
    software, 326
    sudden failures, 324
    variability, sources of, 321–322
Delta method, 54, 237, 242, 381, **619**
Demonstration test plans, 247–250
Density function, 28
Design
  change for cost-reduction, 603
  experimental, *see* Test planning
  for reliability, 602
Distribution
  Bernstein, 339
  Birnbaum-Saunders (BISA), 105
  Burr type XII, 103, 110
  exponential (EXP), 79
  extended generalized gamma (EGENG), 101
  extreme value, 93
  Fréchet, 102, 115
  gamma, 98, 256
  Gaussian, *see* Normal
  generalized F (GENF), 102, 118
  generalized gamma (GENG), 99, 118
  generalized threshold-scale (GETS), 113
  Gompertz-Makeham (GOMA), 108
  inverse Gaussian (IGAU), 103
  largest extreme value (LEV), 86
  location-scale, 78
  logistic, 88
  log-location-scale, 78
  loglogistic, 89
  lognormal (LOGNOR), 82, 93
    three-parameter, 111
    two-parameter, 82
  normal (NOR), 80
  Pareto, 117
  power, 117
    lognormal, 118
    maximum-type, 87, **117**, 374–376
    minimum-type, 86, **117**, 370, 374
  smallest extreme value (SEV), 83
  Weibull (WEIB), 85
    three-parameter, 111
    two-parameter, 85
Distribution function, *see* cdf
Distributions
  continuous, 93, 118
  embedded, 112, 118

  mixture, compound, 116
  mixture, finite (discrete), 115, 118
  monotone hazard (IHR, DHR), 118
  threshold, *see* Threshold distributions
Drenick's theorem, 408
Dummy variable, *see* Regression,
    indicator-variables

Empirical models, caution in extrapolation, 442, 469, 489, 576
Enumerative studies, 16
Environmental
  effects in reliability, 17, 322
  stress screening, *see* ESS
Error propagation, statistical, *see* Delta method
Examples
  α-particle, 154, 209
  adhesive bond, 535
  alloy-A, *see* fatigue crack-size
  alloy-C, 276
  alloy T7987, 130, 279, 638
  ball bearing, 4, 139, 256, 294
  battery, 41, 64, 633
  battery cells, 14
  bearing, 203, 248
  bearing-cage, 193, 348, 633
  bleed system, 144, 634
  braking grids, locomotive, 404, 636
  break pad life, 267
  capacitor
    glass, 447, 449, 482
    tantalum, 512, 641
  ceramic ball bearings, 461
  chain links, 68
  chain reliability, 373
  circuit pack, 7, 267, 590
  component-A, 170
  component-B, 196
  computer laboratory, 421
  cylinder replacements, 400
  device-A, 493, 637
  device-B power output, 564, 598
  device-C, 529
  device-G, 383
  disk drives, 171
  earth-moving machine, 402
  electromigration, 198
  electronic system, 70
  error rates, storage disks, 340–341, 580, 643
  fan, diesel generator, 8, 166, 258, 277, 630
  fatigue crack-size, 15, 317, 639
  fatigue-fracture, 260
  heat exchanger tube, 8, 50, 53, 62, 134, 218
  horn, 202

SUBJECT INDEX

IC device, new-technology, 508, 587
insulating material, 167, 171
insulating structure, mylar-polyurethane, 479, 504, 639
insulation life, 239, 241, 244
integrated circuit, 4, 52, 198, 262, 640
jet engine turbine disk, 374
laser degradation, 324, 642
laser life data, 338
life-limiting component, 247
light bulb life, 238
low-cycle fatigue tests, 470
metallization failure mode, 472
metal wear, 574, 580, 643
modem reliability, 287, 372
printed circuit board, 13, 487
program execution time, 428, 638
resistors, carbon-film, 471, 578, 631
shock absorber, 59, 175, 183, 214, 630
silicon photodiode, 71
snubber design comparison, 452, 632
spacecraft power system, 380
superalloy nickel-based, 428, 593
titanium alloy, 151
transistors, 285
transmitter tube, 11, 147, 199
turbine wheel, 12, 65, 135
U.S.S. *Grampus* diesel engine, 394, 414
U.S.S. *Halfbeak* diesel engine, 415
ultrasonic inspection, 266
valve seats, diesel engine, 395, 635
Expected value, 77. *See also* Mean
Exponential (EXP) distribution, 79
    comparison, confidence interval procedures, 164
    inappropriate use, 80
    likelihood, 157
        correct versus density approximation, 166
    mixture of, 118, 583
    ML estimate for θ, 165
    no failures, analysis with, 167
    Poisson processes relationship, 80, 154, 407
    standard error for $\hat{\theta}$, 162
    times between system failures, 80, 407
Extended generalized gamma (EGENG) distribution, 101, 284
    compromise between lognormal and Weibull, 257
    ML fitting, 257
Extrapolation, 3, 76, 442, 576
Eyring relationship, 473
    compared with Arrhenius, 475

Failure modes, 389, *see* Multiple causes of failure
    effect analysis (FMEA), 387
    effect criticality analysis (FMECA), 387
Failure rate, 29
Fatigue failure models, 489
Fatigue-limit model, 593
    nonconstant standard deviation, 595
    random coefficients, 601
    S-N curves, 595
Fault trees, 387, 389
Field reliability data, 22
Field tracking, 310, 383, 606
Finite population correction, 402
Fisher information matrix, 237, 240, **621**, 644
    computation, 250–251
    definition, 162, 237, 621
    expected, 621
    exponential distribution, 253
    function of roundoff, binning, censoring, and truncation, 625
    LFP model, 251
    location-scale and log-location-scale distributions, 241, 250
    observed, 621, 626
    regression models, 251, 436
    test planning, 625
    truncated distributions, 251
FITs, 30
FMEA, 389
FMECA, 389
Fréchet distribution, 102, 115
Functions of random variables, distribution of, 617

Gamma distribution, 98
    comparison between lognormal and Weibull, 256
    ML fitting, 256
Gamma function, 85
    incomplete, 99
    inverse incomplete, 99
Generalized gamma (GENG) distribution, *see* Extended generalized gamma distribution
Generalized threshold-scale (GETS) distribution, 113, 118
Gompertz–Makeham (GOMA) distribution, 85, 108
Goodness of fit, 127. *See also* Simultaneous confidence bands for $F(t)$
Graphical estimation, 147
Greenwood's formula, 55
Guarantee parameter, *see* Threshold

SUBJECT INDEX

HALT, 519
Hazard
    average, 30
    constant, 79
    cumulative, 29, 45, 67
    estimation of, 191
    function, 7, **28**, 77
        decreasing, 80, 118
        increasing, 80, 85, 118
    plot, 149
    rate in FITs, 30
HPP, 407
    relationship to exponential, 80, 154, 407
    test for NHPP alternatives, 409–411, 420

Independent
    increments, 406
    times between recurrences, 408
Inference, basic ideas, 48
Influence analysis, 460. *See also* Sensitivity analysis
Inspection data, *see* Censoring, interval
Inspection interval, choosing, 10, 169
Interaction, 449, 485–486, 514, 547
Inverse-Gaussian (IGAU) distribution, 103
    ML fitting, 260
Inverse power relationship, 480–484

Jacobian, 619

Kaplan–Meier estimator, 67

Laplace trend test, 409
Largest extreme value (LEV) distribution, 86
Lewis–Robinson trend test, 409
Life table estimate, 64
Likelihood
    constant of proportionality, 37, 39
    contribution
        general data, 40
        interval-censored, 37
        left-censored, 38
        randomly censored, 41
        right-censored, 38, 174
        total, 39
    correct, 165, 169, 275, 277
    curvature, 162, 237, 621
    density approximation, inadequacy of, 169, 277, 284
    empirical, 68
    graphical display, 158, 255
    grouped data, 42
    inference, 36, 42

location-scale distribution, 174
non log-location-scale distributions, 255
nonregular models, 36, 623
poorly behaved, 255, 265–266, 280
profile
    function of parameters, 182–184
    parameters, 180–182, 265, 597
ratios, 626
ratio test, 185, 627
regularity conditions, 621–623
regular models, 36, 622
relationship to sample size, 158
specification, 36–41
Limited failure population (LFP), 263, 271, 284, 591
Logarithms base-10, 82, 127
Logistic
    distribution, 88
    likelihood, 174
    transformation, 56
Log likelihood
    contribution, 157. *See also* Likelihood
    large-sample approximation, 186
    partial derivatives, 197
    quadratic approximation, 186, 628
Loglogistic distribution, 88
Lognormal (LOGNOR) distribution, 82
    comparison with Weibull, 176, 257, 259, 269, 282
    estimation with given σ, 192
    induced failure time in ADT analysis, 573
    likelihood function, 174
    quantiles ML variance factors, 243
    special case of EGENG, 258
    three-parameter, 111, 273–274, 277, 279–283
LSINF, 251

Maintainability, 370
Markov model, 388–389, 578
Maximum likelihood, *see* ML estimation
MCF, 395–405
    comparison, 404–406
    confidence intervals, 398
        adequacy of, 402
    NHPP, 406
    nonparametric estimate, 397
    parametric estimate, 414–415
Mean cumulative function, *see* MCF
Mean time between failures (MTBF), 394, 408
Mean time to failure (MTTF), 77
Military Handbook test statistic, 409
Mixture distributions, 115–119, 284, 583
ML estimation
    asymptotic normality, 237, 625

SUBJECT INDEX

basic concepts, 153–159
covariance matrix
    functions of location-scale parameters, 242
    log-location-scale distributions, 240
    relationship to Fisher information, 237
    relationship to log likelihood curvature, 237, 621
given parameter, 192–197
large-sample approximations, 237
mixture distributions, 284
nonparametric, 53
standard error, approximate, 237
standard error, estimate of, 50, **55**, 166, 187, 189, 193, 436–438
theory, 168, **621**, 622–628
variance, approximate, 236
variance, factors for
    complicated censoring, 251
    functions of location-scale parameters, 242
    hazard, 246
    multiple censoring, 250
    quantiles, 243
    zero failures, 167, 195
Model adequacy, 501, 604
Multinomial
    cdf, 33
    failure-time model, 32
Multiple causes of failure, 35, 59, **382**, 523, 630, 634
    ALT, 523, 531
    incomplete information, 385, **525**, 591
    series model, 385

Nelson–Aalen estimator
    cumulative hazard, 67
    relation to hazard plot, 149
    theory, 67
NHPP, 407
    Bayesian methods, 420
    confidence intervals, 416
    generating pseudorandom realizations, 417
    loglinear model, 407
    ML estimation, 412
    power-model, 407
    prediction, 416
Nonhomogeneous Poisson process, *see* NHPP
Nonparametric bootstrap, *see* Bootstrap, nonparametric
Nonparametric estimation
    confidence interval, *see also* Confidence intervals
    cumulative hazard, 67
    failure probabilities
        left truncated, 268

multiply censored, 52, 57
right truncated, 270
singly censored, 47
Nonrepairable units, 19
Normal (NOR) distribution, 80, 452
    applications, 81
    likelihood function, 174
Numerical methods, 442

Observational data, 267, 606

P-P plot, 444
Parallel system
    cdf, 374
    component dependency, 375
    ML estimation, 381
    reliability as function of component reliability, 375
Parameter
    function of explanatory variables, 429. *See also* Regression
    guarantee, *see* Threshold
    location, 78
    scale, 78
    shape, 82, 85, 89, 110, 136
    threshold, 111
Parameterization, 3, 77, 90, 181, 285
Paris-rule, 319
Part count, 372
Performance degradation, 323, 325. *See also* Degradation
Peto-Turnbull estimator, 65, 68
PH model, 455–458, 460
    applications in reliability, 458
    not Weibull baseline, 457
    relationship to SAFT, 457
    semiparametric, 458
    time transformation, 456
    Weibull baseline, 457
Pitfalls
    ALT data analysis, 522–528
    life data analysis, 22, 601
Planning accelerated life test, *see* ALT planning
Planning life tests, *see* Test planning
Planning values, 232, 535, 548
    on probability paper, 233
Point process model, 394, 420
Poisson distribution, 195, 407
Poisson process, 406
    homogeneous, *see* HPP
    nonhomogeneous, *see* NHPP
Posterior, *see* Bayesian methods
Prediction intervals
    approximate, analytical, 312

Prediction intervals (*continued*)
  calibration, 300, 312
  complete lognormal data, 299
  coverage probability, 293
  exact simulation methods, 312
  exponential Type II censored data, 300
  information needed for, 291
  likelihood based, 312
  motivation, 290
  multiple samples, 304
  naive intervals, 293
  new sample, 290
  number of recurrences (NHPP), 416
  one-sided, 290
  pivotal method, 296
  probability prediction, 292
  simple cases, 298
  statistical prediction, 293
  two-sided, 290
  Type I censoring, 297
  Type II censoring, 296
  within sample, 290
Prior, *see* Bayesian methods
Probabilistic design, 604
Probability paper, *see* Probability plots
Probability plots
  applications, 141
  bend or curvature, 141, 144, 385, 585
  compare distributions, 132
  complete data, 149
  display planning values, 232
  estimates, graphical, 126, 144
  exponential, 124
  gamma, 137
  generalized gamma, 138
  given shape parameter, 136, 274
  given threshold parameter, 274
  goodness of fit, graphical, 127
  grid lines, 144
  linearizing a cdf, 123–127, 137–139
  location-scale-based distributions, 123
  lognormal, 125
  non log-location-scale distributions, 136–141
  normal, 125
  plotting positions, 128–129
    censored data, 132, 135
  reading parameter values from, 126
  simulation to assess variability, 141, 149
  simultaneous confidence bands, 127
  summary of plot scales, 142
  three-parameter Weibull, 137, 274
  unknown shape parameter, 136
  Weibull, 127

Product comparison, 450, *see also* Comparison
  combined analysis, 454
  separate analysis, 452
Product design processes, 602
Product limit estimator, *see* Nonparametric estimation
Profile likelihood, *see* Likelihood
Propagation of error, *see* Delta method
Proportional hazards model, *see* PH
Pseudo failure times, 337, 574
Pseudorandom samples
  continuous distribution, 91
  discrete distribution, 93
  efficient generation, 91
  exponential, 71
  failure-censored, 92
  generation, 93
  NHPP, 417
  of order statistics, 93
  time-censored, 92
  uniform, 91

Q-Q plots, 445
Quantile, 31, 77. *See also* Distribution

Random sample generation, *see* Pseudorandom samples
Recurrence, *see also* MCF
  comparison of two samples, 404
  data, 394, 400, 402, 421, 631
  nonparametric model, 395
  parametric models, 406–408
  rate, 395
  trend tests
    Laplace, 409
    Lewis–Robinson, 409
    MIL-HDBK-189, 409
Redundancy, 602
Regression, *see also* ADT; ALT
  checking assumptions, 443–447
  confidence intervals, 436
  Cox–Snell residuals, 443
  diagnostics, 445
  empirical models, 442
  examples, 13–15, 428–429
  indicator-variables, 450
  likelihood, 433
  models, failure-time, 429–435, 447–450, 455–460, 469
  multiple, 447
  nonconstant spread, 439
  product comparison, 450
  quadratic, 439

residual analysis, 443, 460, 501–502, 507
simple linear, 432
standard errors, 436
Regularity conditions, 621–623
Relative likelihood, 158, 175. *See also* Likelihood, profile
Reliability, 2, 28
assurance, 602
data
components, 19
distinguishing features, 3
examples, 4
nonrepairable units, 19
reasons for collecting, 2
sources, 380
synonyms, 3
environmental effects, 17
function, 28
growth, 420
improvement, 388, **601**, 602–606
quality, relationship to, 2
quantities of interest, 76
study planning, 20, 231
test, *see also* ALT
audit, 606
demonstration, 247
prototype, 467, 519
qualification, 467
screening, 467, **519**. *See also* Screening
Reliability data, sources of, 22
Reliability practice
modern approach, 604
useful tools, 604
Renewal processes, 408
Repairability, 370
Repairable system, 19. *See also* Recurrence data, 20, 394
models and analysis, 420
Residual analysis, *see* Regression, residual analysis
Risk set, 53
Robust-design, 605

S-N curves, 595
Safety factors, engineering, 602
SAFT, 430–431, 476–479
Sample size
effect on
inferences, 234
interval size, 158
likelihood shape, 158
needed to estimate
functions, positive, 239
functions, unrestricted, 238
hazard, 245–247, 251
log-location-scale parameters, 241
mean, 238–239
quantile, 242, 244, 251
shape parameter, 241
σ, 241
Sampling distribution, 49
Sampling error, 49
Screening, 270, 467, **519**, 520–521
Sensitivity analysis, 445, 489, 604
Serial correlation, test for, 411
Series system structure
cdf, 371, 385
component dependency, 371
hazard function, 371
reliability, 371
Weibull components, 372
Series-parallel system structure
component-level redundancy, 377
system-level redundancy, 377
Signal-to-noise ratio, 78
Significance level, 161
Simultaneous confidence bands for $F(t)$, 60, 67, 127, 197
goodness of fit, relation to, 127, 149
logit transformation based, 67
Skewed distribution, 111
ML fitting, 283
Smallest extreme value (SEV) distribution, 83–84, 279
Software
package capabilities, 22
to use with this book, xviii, 3
Software reliability, 419
Spread and skewness, parameter comparison, 110
Staggered entry, 8, 35, 193, 310
Standard deviation, 77
Standardized
log censoring time, 240
log estimation time, 245
Stationary increments, 407
Stress corrosion, 7
STRIFE, 519
Sudden death tests, 250
Superimposed renewal processes, 408
Survival function, 28
System
basic concepts, 370
cdf, 370

System (*continued*)
  structure
    bridge-system, 378
    k out of *s*, 379
    other, 386
    parallel system, 374–376
    series system, 370–374
    series-parallel, 376–378
System reliability, 369, 389
  component dependency, 386
  component importance, 388
  confidence intervals, 381
  estimation from component data, 380–386
  Markov models, 388
  state-space models, 388
  systems with repair, 386
System repair data, *see* Recurrence data

Target population, 15
Target process, 15
Taylor series, *see* Delta method
Temperature acceleration, *see* Acceleration, Arrhenius
  differential factor (TDF), 472
Test planning, 17. *See also* ADT planning; ALT planning
  approximate properties, 236–238
  demonstrate conformance, 247
  failure (Type II) censoring, 250
  non log-location-scale distributions, 251
  planning values, 232, 535, 548
    sources for, 236
    uncertainty in, 236
  simulation for, 233
Three parameter distributions, *see* Threshold distributions
Threshold distributions, 111, 118, 273. *See also* Generalized threshold-scale
  correct likelihood, 275
  density approximation inadequacy for ML fitting, 284
  embedded models, 112–113, 277, 284
  ML fitting, 276, 284
  probability plotting, 284
Time acceleration, *see* Acceleration

Time scales, 18, 22, 523
Time transformations, 460. *See also* Acceleration
  general, 459
  PH, 456
  SAFT, 430
Total time on test, 166, 240
Trend tests, 409
Truncated data, 41, 68, 266, 284
  distributions, 266
  examples, 266, 270
  Fisher information matrix, 284
  left, 266
  likelihood, 268, 271
  ML fitting, 269
  nonparametric estimation, 268, 270, 284
  right, 270
Two-sample comparison, *see* Comparison, two samples

Use-rate acceleration, 17, 468, **470**

Variance, 77
  factors for ML estimates, 240, 243, 246
    algorithm to compute, 241
Variance-covariance matrix, *see* Covariance matrix
Voltage acceleration, 479–485
  mechanism, 480

Wald statistic, *see* Confidence intervals, normal-approximation
Warranty data, 270, 287, 380, 606
Wear, 489. *See also* Examples, wear
Weibull distribution, 85
  alternative parameterization, 86
  applications, 86
  comparison with lognormal, 177, 257, 259, 269, 282
  likelihood function, 174
  maximum likelihood equations, 201
  special case of EGENG, 257
  three parameter, 111, 276–277, 282

Zero-failure
  confidence bounds, 147, 195
  demonstration plans, 247

WILEY SERIES IN PROBABILITY AND STATISTICS
ESTABLISHED BY WALTER A. SHEWHART AND SAMUEL S. WILKS

Editors
Vic Barnett, Ralph A. Bradley, Noel A. C. Cressie, Nicholas I. Fisher, Iain M. Johnstone, J. B. Kadane, David G. Kendall, David W. Scott, Bernard W. Silverman, Adrian F. M. Smith, Jozef L. Teugels; J. Stuart Hunter, Emeritus

Probability and Statistics Section

*ANDERSON · The Statistical Analysis of Time Series
ARNOLD, BALAKRISHNAN, and NAGARAJA · A First Course in Order Statistics
ARNOLD, BALAKRISHNAN, and NAGARAJA · Records
BACCELLI, COHEN, OLSDER, and QUADRAT · Synchronization and Linearity: An Algebra for Discrete Event Systems
BASILEVSKY · Statistical Factor Analysis and Related Methods: Theory and Applications
BERNARDO and SMITH · Bayesian Statistical Concepts and Theory
BILLINGSLEY · Convergence of Probability Measures
BOROVKOV · Asymptotic Methods in Queuing Theory
BRANDT, FRANKEN, and LISEK · Stationary Stochastic Models
CAINES · Linear Stochastic Systems
CAIROLI and DALANG · Sequential Stochastic Optimization
CONSTANTINE · Combinatorial Theory and Statistical Design
COVER and THOMAS · Elements of Information Theory
CSÖRGŐ and HORVÁTH · Weighted Approximations in Probability Statistics
CSÖRGŐ and HORVÁTH · Limit Theorems in Change Point Analysis
DETTE and STUDDEN · The Theory of Canonical Moments with Applications in Statistics, Probability, and Analysis
*DOOB · Stochastic Processes
DRYDEN and MARDIA · Statistical Analysis of Shape
DUPUIS and ELLIS · A Weak Convergence Approach to the Theory of Large Deviations
ETHIER and KURTZ · Markov Processes: Characterization and Convergence
FELLER · An Introduction to Probability Theory and Its Applications, Volume 1, *Third Edition,* Revised; Volume II, *Second Edition*
FULLER · Introduction to Statistical Time Series, *Second Edition*
FULLER · Measurement Error Models
GELFAND and SMITH · Bayesian Computation
GHOSH, MUKHOPADHYAY, and SEN · Sequential Estimation
GIFI · Nonlinear Multivariate Analysis
GUTTORP · Statistical Inference for Branching Processes
HALL · Introduction to the Theory of Coverage Processes
HAMPEL · Robust Statistics: The Approach Based on Influence Functions
HANNAN and DEISTLER · The Statistical Theory of Linear Systems
HUBER · Robust Statistics
IMAN and CONOVER · A Modern Approach to Statistics
JUREK and MASON · Operator-Limit Distributions in Probability Theory
KASS and VOS · Geometrical Foundations of Asymptotic Inference

*Now available in a lower priced paperback edition in the Wiley Classics Library.

Probability and Statistics (Continued)

    KAUFMAN and ROUSSEEUW · Finding Groups in Data: An Introduction to Cluster Analysis
    KELLY · Probability, Statistics, and Optimization
    LINDVALL · Lectures on the Coupling Method
    McFADDEN · Management of Data in Clinical Trials
    MANTON, WOODBURY, and TOLLEY · Statistical Applications Using Fuzzy Sets
    MARDIA and JUPP · Statistics of Directional Data, *Second Edition*
    MORGENTHALER and TUKEY · Configural Polysampling: A Route to Practical Robustness
    MUIRHEAD · Aspects of Multivariate Statistical Theory
    OLIVER and SMITH · Influence Diagrams, Belief Nets and Decision Analysis
    *PARZEN · Modern Probability Theory and Its Applications
    PRESS · Bayesian Statistics: Principles, Models, and Applications
    PUKELSHEIM · Optimal Experimental Design
    RAO · Asymptotic Theory of Statistical Inference
    RAO · Linear Statistical Inference and Its Applications, *Second Edition*
    *RAO and SHANBHAG · Choquet-Deny Type Functional Equations with Applications to Stochastic Models
    ROBERTSON, WRIGHT, and DYKSTRA · Order Restricted Statistical Inference
    ROGERS and WILLIAMS · Diffusions, Markov Processes, and Martingales, Volume I: Foundations, *Second Edition;* Volume II: Îto Calculus
    RUBINSTEIN and SHAPIRO · Discrete Event Systems: Sensitivity Analysis and Stochastic Optimization by the Score Function Method
    RUZSA and SZEKELY · Algebraic Probability Theory
    SCHEFFE · The Analysis of Variance
    SEBER · Linear Regression Analysis
    SEBER · Multivariate Observations
    SEBER and WILD · Nonlinear Regression
    SERFLING · Approximation Theorems of Mathematical Statistics
    SHORACK and WELLNER · Empirical Processes with Applications to Statistics
    SMALL and McLEISH · Hilbert Space Methods in Probability and Statistical Inference
    STAPLETON · Linear Statistical Models
    STAUDTE and SHEATHER · Robust Estimation and Testing
    STOYANOV · Counterexamples in Probability
    TANAKA · Time Series Analysis: Nonstationary and Noninvertible Distribution Theory
    THOMPSON and SEBER · Adaptive Sampling
    WELSH · Aspects of Statistical Inference
    WHITTAKER · Graphical Models in Applied Multivariate Statistics
    YANG · The Construction Theory of Denumerable Markov Processes

Applied Probability and Statistics Section

    ABRAHAM and LEDOLTER · Statistical Methods for Forecasting
    AGRESTI · Analysis of Ordinal Categorical Data
    AGRESTI · Categorical Data Analysis
    ANDERSON, AUQUIER, HAUCK, OAKES, VANDAELE, and WEISBERG · Statistical Methods for Comparative Studies
    ARMITAGE and DAVID (editors) · Advances in Biometry
    *ARTHANARI and DODGE · Mathematical Programming in Statistics
    ASMUSSEN · Applied Probability and Queues
    *BAILEY · The Elements of Stochastic Processes with Applications to the Natural Sciences

*Now available in a lower priced paperback edition in the Wiley Classics Library.

Applied Probability and Statistics (Continued)
BARNETT and LEWIS · Outliers in Statistical Data, *Third Edition*
BARTHOLOMEW, FORBES, and McLEAN · Statistical Techniques for Manpower Planning, *Second Edition*
BATES and WATTS · Nonlinear Regression Analysis and Its Applications
BECHHOFER, SANTNER, and GOLDSMAN · Design and Analysis of Experiments for Statistical Selection, Screening, and Multiple Comparisons
BELSLEY · Conditioning Diagnostics: Collinearity and Weak Data in Regression
BELSLEY, KUH, and WELSCH · Regression Diagnostics: Identifying Influential Data and Sources of Collinearity
BHAT · Elements of Applied Stochastic Processes, *Second Edition*
BHATTACHARYA and WAYMIRE · Stochastic Processes with Applications
BIRKES and DODGE · Alternative Methods of Regression
BLOOMFIELD · Fourier Analysis of Time Series: An Introduction
BOLLEN · Structural Equations with Latent Variables
BOULEAU · Numerical Methods for Stochastic Processes
BOX · Bayesian Inference in Statistical Analysis
BOX and DRAPER · Empirical Model-Building and Response Surfaces
BOX and DRAPER · Evolutionary Operation: A Statistical Method for Process Improvement
BUCKLEW · Large Deviation Techniques in Decision, Simulation, and Estimation
BUNKE and BUNKE · Nonlinear Regression, Functional Relations and Robust Methods: Statistical Methods of Model Building
CHATTERJEE and HADI · Sensitivity Analysis in Linear Regression
CHOW and LIU · Design and Analysis of Clinical Trials: Concepts and Methodologies
CLARKE and DISNEY · Probability and Random Processes: A First Course with Applications, *Second Edition*
*COCHRAN and COX · Experimental Designs, *Second Edition*
CONOVER · Practical Nonparametric Statistics, *Second Edition*
CORNELL · Experiments with Mixtures, Designs, Models, and the Analysis of Mixture Data, *Second Edition*
*COX · Planning of Experiments
CRESSIE · Statistics for Spatial Data, *Revised Edition*
DANIEL · Applications of Statistics to Industrial Experimentation
DANIEL · Biostatistics: A Foundation for Analysis in the Health Sciences, *Sixth Edition*
DAVID · Order Statistics, *Second Edition*
*DEGROOT, FIENBERG, and KADANE · Statistics and the Law
DODGE · Alternative Methods of Regression
DOWDY and WEARDEN · Statistics for Research, *Second Edition*
DUNN and CLARK · Applied Statistics: Analysis of Variance and Regression, *Second Edition*
ELANDT-JOHNSON and JOHNSON · Survival Models and Data Analysis
EVANS, PEACOCK, and HASTINGS · Statistical Distributions, *Second Edition*
FLEISS · The Design and Analysis of Clinical Experiments
FLEISS · Statistical Methods for Rates and Proportions, *Second Edition*
FLEMING and HARRINGTON · Counting Processes and Survival Analysis
GALLANT · Nonlinear Statistical Models
GLASSERMAN and YAO · Monotone Structure in Discrete-Event Systems
GNANADESIKAN · Methods for Statistical Data Analysis of Multivariate Observations, *Second Edition*
GOLDSTEIN and LEWIS · Assessment: Problems, Development, and Statistical Issues
GREENWOOD and NIKULIN · A Guide to Chi-Squared Testing
*HAHN · Statistical Models in Engineering
HAHN and MEEKER · Statistical Intervals: A Guide for Practitioners

*Now available in a lower priced paperback edition in the Wiley Classics Library.

Applied Probability and Statistics (Continued)

HAND · Construction and Assessment of Classification Rules
HAND · Discrimination and Classification
HEIBERGER · Computation for the Analysis of Designed Experiments
HINKELMAN and KEMPTHORNE: · Design and Analysis of Experiments, Volume 1: Introduction to Experimental Design
HOAGLIN, MOSTELLER, and TUKEY · Exploratory Approach to Analysis of Variance
HOAGLIN, MOSTELLER, and TUKEY · Exploring Data Tables, Trends and Shapes
HOAGLIN, MOSTELLER, and TUKEY · Understanding Robust and Exploratory Data Analysis
HOCHBERG and TAMHANE · Multiple Comparison Procedures
HOCKING · Methods and Applications of Linear Models: Regression and the Analysis of Variables
HOGG and KLUGMAN · Loss Distributions
HOLLANDER and WOLFE · Nonparametric Statistical Methods
HOSMER and LEMESHOW · Applied Logistic Regression
HØYLAND and RAUSAND · System Reliability Theory: Models and Statistical Methods
HUBERTY · Applied Discriminant Analysis
JACKSON · A User's Guide to Principle Components
JOHN · Statistical Methods in Engineering and Quality Assurance
JOHNSON · Multivariate Statistical Simulation
JOHNSON and KOTZ · Distributions in Statistics
Continuous Multivariate Distributions
JOHNSON, KOTZ, and BALAKRISHNAN · Continuous Univariate Distributions, Volume 1, *Second Edition*
JOHNSON, KOTZ, and BALAKRISHNAN · Continuous Univariate Distributions, Volume 2, *Second Edition*
JOHNSON, KOTZ, and BALAKRISHNAN · Discrete Multivariate Distributions
JOHNSON, KOTZ, and KEMP · Univariate Discrete Distributions, *Second Edition*
JUREČKOVÁ and SEN · Robust Statistical Procedures: Aymptotics and Interrelations
KADANE · Bayesian Methods and Ethics in a Clinical Trial Design
KADANE AND SCHUM · A Probabilistic Analysis of the Sacco and Vanzetti Evidence
KALBFLEISCH and PRENTICE · The Statistical Analysis of Failure Time Data
KELLY · Reversability and Stochastic Networks
KHURI, MATHEW, and SINHA · Statistical Tests for Mixed Linear Models
KLUGMAN, PANJER, and WILLMOT · Loss Models: From Data to Decisions
KLUGMAN, PANJER, and WILLMOT · Solutions Manual to Accompany Loss Models: From Data to Decisions
KOVALENKO, KUZNETZOV, and PEGG · Mathematical Theory of Reliability of Time-Dependent Systems with Practical Applications
LAD · Operational Subjective Statistical Methods: A Mathematical, Philosophical, and Historical Introduction
LANGE, RYAN, BILLARD, BRILLINGER, CONQUEST, and GREENHOUSE · Case Studies in Biometry
LAWLESS · Statistical Models and Methods for Lifetime Data
LEE · Statistical Methods for Survival Data Analysis, *Second Edition*
LePAGE and BILLARD · Exploring the Limits of Bootstrap
LINHART and ZUCCHINI · Model Selection
LITTLE and RUBIN · Statistical Analysis with Missing Data
MAGNUS and NEUDECKER · Matrix Differential Calculus with Applications in Statistics and Econometrics
MALLER and ZHOU · Survival Analysis with Long Term Survivors
MANN, SCHAFER, and SINGPURWALLA · Methods for Statistical Analysis of Reliability and Life Data

*Now available in a lower priced paperback edition in the Wiley Classics Library.

Applied Probability and Statistics (Continued)

McLACHLAN and KRISHNAN · The EM Algorithm and Extensions
McLACHLAN · Discriminant Analysis and Statistical Pattern Recognition
McNEIL · Epidemiological Research Methods
MEEKER and ESCOBAR · Statistical Methods for Reliability Data
MILLER · Survival Analysis
MONTGOMERY and PECK · Introduction to Linear Regression Analysis, *Second Edition*
MYERS and MONTGOMERY · Response Surface Methodology: Process and Product in Optimization Using Designed Experiments
NELSON · Accelerated Testing, Statistical Models, Test Plans, and Data Analyses
NELSON · Applied Life Data Analysis
OCHI · Applied Probability and Stochastic Processes in Engineering and Physical Sciences
OKABE, BOOTS, and SUGIHARA · Spatial Tesselations: Concepts and Applications of Voronoi Diagrams
PANKRATZ · Forecasting with Dynamic Regression Models
PANKRATZ · Forecasting with Univariate Box-Jenkins Models: Concepts and Cases
PIANTADOSI · Clinical Trials: A Methodologic Perspective
PORT · Theoretical Probability for Applications
PUTERMAN · Markov Decision Processes: Discrete Stochastic Dynamic Programming
RACHEV · Probability Metrics and the Stability of Stochastic Models
RÉNYI · A Diary on Information Theory
RIPLEY · Spatial Statistics
RIPLEY · Stochastic Simulation
ROUSSEEUW and LEROY · Robust Regression and Outlier Detection
RUBIN · Multiple Imputation for Nonresponse in Surveys
RUBINSTEIN · Simulation and the Monte Carlo Method
RUBINSTEIN and MELAMED · Modern Simulation and Modeling
RYAN · Statistical Methods for Quality Improvement
SCHUSS · Theory and Applications of Stochastic Differential Equations
SCOTT · Multivariate Density Estimation: Theory, Practice, and Visualization
*SEARLE · Linear Models
SEARLE · Linear Models for Unbalanced Data
SEARLE, CASELLA, and McCULLOCH · Variance Components
STOYAN, KENDALL, and MECKE · Stochastic Geometry and Its Applications, *Second Edition*
STOYAN and STOYAN · Fractals, Random Shapes and Point Fields: Methods of Geometrical Statistics
THOMPSON · Empirical Model Building
THOMPSON · Sampling
TIJMS · Stochastic Modeling and Analysis: A Computational Approach
TIJMS · Stochastic Models: An Algorithmic Approach
TITTERINGTON, SMITH, and MAKOV · Statistical Analysis of Finite Mixture Distributions
UPTON and FINGLETON · Spatial Data Analysis by Example, Volume 1: Point Pattern and Quantitative Data
UPTON and FINGLETON · Spatial Data Analysis by Example, Volume II: Categorical and Directional Data
VAN RIJCKEVORSEL and DE LEEUW · Component and Correspondence Analysis
WEISBERG · Applied Linear Regression, *Second Edition*
WESTFALL and YOUNG · Resampling-Based Multiple Testing: Examples and Methods for p-Value Adjustment
WHITTLE · Systems in Stochastic Equilibrium
WOODING · Planning Pharmaceutical Clinical Trials: Basic Statistical Principles

*Now available in a lower priced paperback edition in the Wiley Classics Library.

WOOLSON · Statistical Methods for the Analysis of Biomedical Data
*ZELLNER · An Introduction to Bayesian Inference in Econometrics

Texts and References Section

AGRESTI · An Introduction to Categorical Data Analysis
ANDERSON · An Introduction to Multivariate Statistical Analysis, *Second Edition*
ANDERSON and LOYNES · The Teaching of Practical Statistics
ARMITAGE and COLTON · Encyclopedia of Biostatistics: Volumes 1 to 6 with Index
BARTOSZYNSKI and NIEWIADOMSKA-BUGAJ · Probability and Statistical Inference
BERRY, CHALONER, and GEWEKE · Bayesian Analysis in Statistics and
  Econometrics: Essays in Honor of Arnold Zellner
BHATTACHARYA and JOHNSON · Statistical Concepts and Methods
BILLINGSLEY · Probability and Measure, *Second Edition*
BOX · R. A. Fisher, the Life of a Scientist
BOX, HUNTER, and HUNTER · Statistics for Experimenters: An Introduction to
  Design, Data Analysis, and Model Building
BOX and LUCEÑO · Statistical Control by Monitoring and Feedback Adjustment
BROWN and HOLLANDER · Statistics: A Biomedical Introduction
CHATTERJEE and PRICE · Regression Analysis by Example, *Second Edition*
COOK and WEISBERG · An Introduction to Regression Graphics
COX · A Handbook of Introductory Statistical Methods
DILLON and GOLDSTEIN · Multivariate Analysis: Methods and Applications
DODGE and ROMIG · Sampling Inspection Tables, *Second Edition*
DRAPER and SMITH · Applied Regression Analysis, *Third Edition*
DUDEWICZ and MISHRA · Modern Mathematical Statistics
DUNN · Basic Statistics: A Primer for the Biomedical Sciences, *Second Edition*
FISHER and VAN BELLE · Biostatistics: A Methodology for the Health Sciences
FREEMAN and SMITH · Aspects of Uncertainty: A Tribute to D. V. Lindley
GROSS and HARRIS · Fundamentals of Queueing Theory, *Third Edition*
HALD · A History of Probability and Statistics and their Applications Before 1750
HALD · A History of Mathematical Statistics from 1750 to 1930
HELLER · MACSYMA for Statisticians
HOEL · Introduction to Mathematical Statistics, *Fifth Edition*
JOHNSON and BALAKRISHNAN · Advances in the Theory and Practice of Statistics: A
  Volume in Honor of Samuel Kotz
JOHNSON and KOTZ (editors) · Leading Personalities in Statistical Sciences: From the
  Seventeenth Century to the Present
JUDGE, GRIFFITHS, HILL, LÜTKEPOHL, and LEE · The Theory and Practice of
  Econometrics, *Second Edition*
KHURI · Advanced Calculus with Applications in Statistics
KOTZ and JOHNSON (editors) · Encyclopedia of Statistical Sciences: Volumes 1 to 9
  wtih Index
KOTZ and JOHNSON (editors) · Encyclopedia of Statistical Sciences: Supplement
  Volume
KOTZ, REED, and BANKS (editors) · Encyclopedia of Statistical Sciences: Update
  Volume 1
KOTZ, REED, and BANKS (editors) · Encyclopedia of Statistical Sciences: Update
  Volume 2
LAMPERTI · Probability: A Survey of the Mathematical Theory, *Second Edition*
LARSON · Introduction to Probability Theory and Statistical Inference, *Third Edition*
LE · Applied Survival Analysis
MALLOWS · Design, Data, and Analysis by Some Friends of Cuthbert Daniel
MARDIA · The Art of Statistical Science: A Tribute to G. S. Watson

*Now available in a lower priced paperback edition in the Wiley Classics Library.

Texts and References (Continued)
    MASON, GUNST, and HESS · Statistical Design and Analysis of Experiments with
        Applications to Engineering and Science
    MURRAY · X-STAT 2.0 Statistical Experimentation, Design Data Analysis, and
        Nonlinear Optimization
    PURI, VILAPLANA, and WERTZ · New Perspectives in Theoretical and Applied
        Statistics
    RENCHER · Methods of Multivariate Analysis
    RENCHER · Multivariate Statistical Inference with Applications
    ROSS · Introduction to Probability and Statistics for Engineers and Scientists
    ROHATGI · An Introduction to Probability Theory and Mathematical Statistics
    RYAN · Modern Regression Methods
    SCHOTT · Matrix Analysis for Statistics
    SEARLE · Matrix Algebra Useful for Statistics
    STYAN · The Collected Papers of T. W. Anderson: 1943–1985
    TIERNEY · LISP-STAT: An Object-Oriented Environment for Statistical Computing
        and Dynamic Graphics
    WONNACOTT and WONNACOTT · Econometrics, *Second Edition*

WILEY SERIES IN PROBABILITY AND STATISTICS
ESTABLISHED BY WALTER A. SHEWHART AND SAMUEL S. WILKS

Editors
*Robert M. Groves, Graham Kalton, J. N. K. Rao, Norbert Schwarz,
Christopher Skinner*

Survey Methodology Section

    BIEMER, GROVES, LYBERG, MATHIOWETZ, and SUDMAN · Measurement
        Errors in Surveys
    COCHRAN · Sampling Techniques, *Third Edition*
    COX, BINDER, CHINNAPPA, CHRISTIANSON, COLLEDGE, and KOTT (editors) ·
        Business Survey Methods
    *DEMING · Sample Design in Business Research
    DILLMAN · Mail and Telephone Surveys: The Total Design Method
    GROVES and COUPER · Nonresponse in Household Interview Surveys
    GROVES · Survey Errors and Survey Costs
    GROVES, BIEMER, LYBERG, MASSEY, NICHOLLS, and WAKSBERG ·
        Telephone Survey Methodology
    *HANSEN, HURWITZ, and MADOW · Sample Survey Methods and Theory,
        Volume 1: Methods and Applications
    *HANSEN, HURWITZ, and MADOW · Sample Survey Methods and Theory,
        Volume II: Theory
    KASPRZYK, DUNCAN, KALTON, and SINGH · Panel Surveys
    KISH · Statistical Design for Research
    *KISH · Survey Sampling
    LESSLER and KALSBEEK · Nonsampling Error in Surveys
    LEVY and LEMESHOW · Sampling of Populations: Methods and Applications
    LYBERG, BIEMER, COLLINS, de LEEUW, DIPPO, SCHWARZ, TREWIN (editors) ·
        Survey Measurement and Process Quality
    SKINNER, HOLT, and SMITH · Analysis of Complex Surveys

*Now available in a lower priced paperback edition in the Wiley Classics Library.